Basic metric unit	
meter	m
liter	L
gram	g
Celsius degree	°C

Prefix	Symbol	
kilo	k	1000
hecto	h	100
deka	da	10
deci	d	0.1
centi	c	0.01
milli	m	0.001

English/Metric Conversions

Since metric units of measure will soon replace English units of measure, it is helpful to know which metric units to use in place of the English units.

Unit of measure	In place of (English)	
Length	inch foot yard mile	millimeter centimeter meter kilometer
Capacity	fluid ounce pint quart gallon	milliliter liter
Weight	ounce pound ton	gram kilogram metric ton
Temperature	degree Fahrenheit	degree Celsius

NOTE: Metric conversion factors are given in the tables found inside the back cover.

FUNDAMENTALS OF MATHEMATICS

FUNDAMENTALS OF MATHEMATICS

FIFTH EDITION

WILLIAM M. SETEK, JR.

Monroe Community College

MACMILLAN PUBLISHING COMPANY
NEW YORK

Photo credits: frontispiece, E. R. Degginger; Ch. 1, Steuben Glass; Chs. 2 and 3, Alan Carey/The Image Works; Ch. 4, © Jerry Wachter/Focus on Sports; Ch. 5, Ken Lax for Macmillan; Ch. 6, Howard Miller Clock Company; Ch. 7, Copyright, The Trustees of the British Museum; Ch. 8, © Mark Antman/The Image Works; Ch. 9, Andy Levin/Black Star; Ch. 10, E. R. Degginger; Ch. 11, © Brent Jones; Ch. 12, Alan Carey/The Image Works.

Printed in the United States of America

Macmillan Publishing Company
866 Third Avenue, New York, New York 10022

Library of Congress Cataloging-in-Publication Data

Setek, William M.
Fundamentals of mathematics/William M. Setek, Jr.—5th ed.
p. cm.
Includes index.
ISBN 0-02-409201-0
1. Mathematics—1961- I. Title.
QA39.2.S48 1989
510—dc 19 88-25230
CIP

Printing: 2 3 4 5 6 7 8 Year: 9 0 1 2 3 4 5 6 7 8

To my wife, **Addie,** for her encouragement,
understanding, assistance, and patience throughout
this project, and to my sons, **Scott** and **Joe,**
who helped in their own special way

PREFACE

In writing this fifth edition of *Fundamentals of Mathematics,* I have tried to reflect the changing approach to teaching liberal arts mathematics courses. The student population of colleges has changed; there is a greater diversity of students enrolled in the typical liberal arts mathematics course. Today, such a course may enroll students ranging from recent high school graduates to mature students with a wide variety of mathematical backgrounds. Motivation and interest vary greatly among these students, and many of them suffer from "math anxiety." Consequently, the course content has become more diversified.

The only prerequisite for this text is a working knowledge of arithmetic. The approach is intuitive. The text contains an abundance of completely worked-out examples with systematic step-by-step solutions; there are no gaps or "magic" solutions. I have found that this type of experience provides the student with confidence and competence when doing homework or test problems.

Organization of the Text

The text is divided into twelve chapters. Chapters 1 and 2 develop the basic ideas of sets and logic from an intuitive standpoint. Generous use is made of Venn diagrams and truth tables. Chapters 3 and 4 introduce the student to probability and statistics. Chapter 5 gives a thorough treatment of the metric system, emphasizing both metric–metric and metric–English conversions. Chapters 6, 7, and 8 are designed to broaden students' ideas about mathematics by exposing them to various mathematical systems and systems of numeration and the structure of the real number system. Chapter 9 gives the student an introduction to algebra, including experience in solving elementary linear equations, graphing equations and inequalities, solving word problems, and solving quadratic equations by means of the quadratic formula. Chapter 10 provides an introduction to many different topics in geometry including perimeter, area, volume, and surface area.

Chapter 11 covers a number of mathematical topics of use to students in their role as consumers. Chapter 12 introduces the student to the computer, including practice in writing programs using the BASIC language.

Features of the Fifth Edition

The fifth edition of *Fundamentals of Mathematics* reflects many improvements suggested by instructors and students who used the fourth edition. Chapter 1 (Sets) has been reworked and updated. Expanded explanations, additional exercises, and updated exercises have been added to Chapter 2 (Logic). Chapter 3 (Probability) contains additional worked-out examples and additional homework exercises. Chapter 4 (Statistics) contains a revised normal curve section. Chapters 5 (Metrics), 6 (Mathematical Systems), and 7 (Systems of Numeration) have been revised, updated, and contain additional exercises. Chapter 8 (Sets of Numbers and Their Structure) contains additional exercises, expanded explanations, and a new section on scientific notation. Chapter 9 (An Introduction to Algebra) contains a new section on quadratic equations. Chapter 10 (An Introduction to Geometry) has been reworked and includes those topics required on competency examinations given in some states. Chapter 11 (Consumer Mathematics) has been updated, expanded, and reworked to meet the greater interest in applications of mathematics to problems related to finances. Chapter 12 (An Introduction to Computers) has also been expanded and updated.

As with the fourth edition, the emphasis in this text is on encouraging the student to participate actively—to do mathematics by working examples and problems, as described in *To the Student.* To this end, the more than 600 worked-out examples in this text have been thoroughly reviewed for clarity and effectiveness, the many exercises have been reevaluated, and new examples and exercises have been added. The chapter review exercises are carefully designed to test the learning objectives given at the beginning of each chapter, and the 25 questions in each *Chapter Quiz* further test the learning objectives. The lists of selected words serve as a vocabulary check for each chapter, and a glossary of all these terms is found at the back of the book.

Typically the review exercises and chapter quizzes should be done in preparation for the unit or chapter exams. These have been found to serve as excellent preparation for the examinations.

To further encourage students to become active participants, each exercise set concludes with a Just for Fun problem. These problems range from serious extensions of mathematical ideas in the text to light-hearted puzzles and "brain teasers." They have been chosen primarily for their ability to capture student interest, and users of previous editions have been pleased with the results. Student interest also is enhanced by Historical Notes and Notes of Interest in every chapter.

Course Outlines

Since liberal arts mathematics is not a well-defined course and its content varies from one school to the next, several suggested course outlines follow. Essentially, the chapters of this text are designed to be independent of one another so that the topics can be covered in any order. A student who fails to master the material in one chapter will not necessarily be at a disadvantage when a new topic is begun. Many users of earlier editions began their course with Chapter 1 (Sets), others began with Chapter 3 or 4 (Probability or Statistics), and still others began with other chapters that suited the needs of their classes.

SUGGESTED COURSE OUTLINES

Chapters	Possible Omissions	Time Allotment
1–6, 12	Secs. 2.8–2.11, 3.7–3.9	one semester
2–6, 11, 12	Secs. 2.8–2.11, 3.7–3.9	one semester
3–6, 9–12	Secs. 3.7–3.9, 9.11	one semester
1, 2, 5–8, 12	Secs. 2.8–2.11, 8.7, 8.8	one semester
1, 2, 4, 9–12	none	one semester
1–12	Chapters 1, 6, 10	two semesters

Pedagogical Aids

Each chapter begins with learning objectives and a list of the symbols that will be introduced in the chapter. A generous set of exercises, graded in level of difficulty, follows each section. In addition, each chapter concludes with a summary, a vocabulary check, a set of review exercises, and a chapter quiz. The review exercises are organized so that they test each learning objective in order; more challenging exercises that require the student to master several objectives are placed at the end of the review exercise set. Almost all exercises may be done with pencil and paper, but some are more readily done with a calculator. We have marked these problems with the symbol ▯. Starred (⋆) exercises are considered optional as they are more challenging.

The answers to all odd-numbered exercises (including their multiple parts) are given for each section, along with all answers to the chapter review exercises and chapter quizzes. Therefore, assignments can be made with confidence in either fashion: with or without answers available.

Supplementary Materials

A number of supplementary aides are available. Two of the most significant are a comprehensive *Student Study Guide* and a computer disk that can assist both students and instructors. The Study Guide has been developed to be used in conjunction with this text. Each section of the Study Guide directly corresponds to each section of the textbook. The material contained in the Study Guide represents a streamlined version of the material from the textbook. Each section has been carefully written so that it can provide added depth and insight to important topics and concepts. All examples presented in the Study Guide correspond to selected odd-numbered problems from the textbook. The Study Guide also assumes the role of a solutions manual. In addition, the appendix of the Study Guide contains a list of notation and symbols used in the text, tables of information, a summary of important geometric facts and formulas, various topics that are not formally addressed in the textbook, and useful information regarding the College Level Academic Skills Test (CLAST).

The computer disk has been prepared using the learning objectives at the beginning of each chapter. It can be used by the instructor to produce examinations with a multitude of test questions that are different but equal. It can also be used to produce multiple questions for a specific objective, thus serving as a learning tool for the student.

The Instructor's Manual contains answers to the even-numbered exercises for each section, teaching suggestions, sets of more challenging exercises for each chapter, suggested projects and student activities, and a list of films and readings related to the topics in the text. A test package containing three examinations for each chapter is also contained in the Instructor's Manual.

Acknowledgments

I am grateful and indebted to those users of the fourth edition who provided me with many valuable suggestions and constructive criticisms—in particular my students and colleagues at Monroe Community College—and those who reviewed the manuscript for this edition: Warren Burch, Brevard Community College; Sharon R. Butler-Davis, Pikes Peak Community College; Carl Carlson, Moorhead State University; Marte Carter, Central Missouri State University; Jane Edgar, Brevard Community College; Judy D. Halchin, Eastern Illinois University; George Kelley, Indian Hills Community College; Marjorie Midyett, East Tennessee State University; June Shanholtzer, Eastern Illinois University; Melvin C. Somers, Norwich University; Linda Sons, Northern Illinois University.

I would like to thank those at Macmillan Publishing Company, particularly Elisabeth Belfer, for their enthusiastic interest and support throughout the project.

A special note of appreciation goes to my wife, Addie, and to my sons, Scott and Joe. To Pam Dretto, thank you for excellent work. Special thanks to David Rogachefsky for his support, perseverance, and eagle eye. A special recognition for Florence Whittaker who helped in so many ways, particularly verifying *all* the solutions.

W. M. S.

CONTENTS

CHAPTER 5

AN INTRODUCTION TO THE METRIC SYSTEM 243

CHAPTER 6

MATHEMATICAL SYSTEMS 283

CHAPTER 7

SYSTEMS OF NUMERATION 327

CHAPTER 8

SETS OF NUMBERS AND THEIR STRUCTURE 379

CHAPTER 9

AN INTRODUCTION TO ALGEBRA 441

CHAPTER 10

AN INTRODUCTION TO GEOMETRY 511

TO THE STUDENT

This book is designed to help you learn some mathematics, regardless of your mathematical background. It is written so that you can understand, appreciate, and even enjoy areas of mathematics to which you may or may not have been exposed. But, in order for this to occur, you must use this book. Someone once said:

I hear and I forget
I see and I remember
I do and I understand

Mathematics is not a spectator subject; it is a participation sport—you must actively use the text. Read it with pencil in hand. Work the illustrative examples. There are more examples in this text than any other of this nature. Their purpose is to help you understand the material and learn by doing. Make use of the wide margins—they are designed for scratch work.

The Objectives, Chapter Summaries, Chapter Review Exercises, Chapter Quizzes, and Vocabulary Checks are designed to highlight the contents of each chapter, and to help you check your progress. [Starred (⋆) exercises are considered optional as they are more challenging. Exercises marked ▯ are more readily done with a calculator.] The Historical Notes and Notes of Interest are designed to provide you with some insight into the development of mathematics. Finally, the Just for Fun problems are just that. They are provided as a change of pace. Some are relevant, some are not.

A Student Study Guide is also available. It contains additional problems, explanations, and worked-out solutions.

I hope you will find reading and using this book a worthwhile and enjoyable endeavor. Good luck!

I welcome any and all comments. Feel free to write and let me know your thoughts and reactions to this text.

William M. Setek, Jr.
Monroe Community College
Rochester, New York 14623

CHAPTER 1

SETS

AFTER STUDYING THIS CHAPTER YOU WILL BE ABLE TO DO THE FOLLOWING:

1. Describe the meaning of the word **set**, and write a given set in two ways.
2. Identify **well-defined sets**, **finite sets**, and **infinite sets.**
3. Identify **equal sets**, **equivalent sets**, and **disjoint sets.**
4. Find the **subsets** and **proper subsets** of a given set.
5. Identify a **universal set** and find the **complement** of any set contained in some universal set.
6. Find the **intersection** and **union** of two or more sets.
7. Draw **Venn diagrams** to show the relationship between sets.
8. Show a **one-to-one correspondence** between any two equivalent sets and find the **cardinality** of sets.
9. Use Venn diagrams to solve survey problems.
10. Determine the **Cartesian product** ($A \times B$) of two sets A and B.

SYMBOLS FREQUENTLY USED IN THIS CHAPTER

{ }	braces, used to enclose members of a set
$\in$	"is an element of"
$\notin$	"is not an element of"
...	proceed in the indicated pattern
$\varnothing$	the empty set, also denoted by { }
$\subset$	"is a proper subset of"
$\subseteq$	"is a subset of"
$\not\subseteq$	"is not a subset of"
U	the universal set
A'	the complement of A
$\cap$	intersection
$\cup$	union
$n(A)$	the cardinal number of set A
(a, b)	the ordered pair a and b
$=$	"is equal to"
$A \times B$	the Cartesian product of sets A and B
$a \mid b$	a such that b

1.1 INTRODUCTION

The concept and properties of a *set* are often used in mathematics. Students study some form of set theory at all levels of mathematics, from grade school through college and graduate school. It has been said that the use of sets is the one unifying idea that unites all of the different branches of mathematics.

We do not define *set,* as it is an intuitive concept. This is not unusual; you may recall that in geometry we do not attempt to define what we mean by a *point* or a *line.* A set may be thought of as a collection of objects. These objects are called **elements,** or **members,** of the set.

It is not uncommon for us to use the concept of set in everyday experiences. You have probably examined some of the following.

A set of dishes
A set of tires for a car
A set of silverware
A set of encyclopedias
A set of golf clubs

In mathematics you may have discussed some of the following.

The set of counting numbers
The set of integers
A set of points
A set of solutions for an equation
A set of ordered pairs

It is important for you to remember that a set is an intuitive concept. But you should also be aware that there are specific properties of sets. We are going to examine some basic properties, rules, and operations that pertain to sets.

1.2 NOTATION AND DESCRIPTION

We do not have to describe a set by a name as we did in the introduction; we can also describe it by listing or naming its elements. Braces, { }, are used to enclose the members of a set when we list them. Note that the only correct fence or enclosure is braces, not parentheses, ( ), or brackets, [ ].

If we are talking about the set of vowels in the English alphabet, we may denote this set of vowels as $\{a, e, i, o, u\}$. The listing of the elements in a different order does not change the set: we could also write the set as $\{i, o, u, e, a\}$, and it would still be the set of vowels in the English alphabet. When we list the elements in a set, as we have just done, it is called the **roster form** of the set.

Sets are usually denoted by capital letters such as A, B, or C. Therefore, we could write

$$A = \{a, e, i, o, u\}$$

to indicate that set A contains the elements a, e, i, o, u.

We shall use $\in$ to indicate that elements are members of a set. The symbol $\in$ is read "is a member of" or "is an element of," and the notation $\notin$ is read "is *not* an element of." Using our previous example $A = \{a, e, i, o, u\}$, we may say that $a \in A$, $e \in A$, $u \in A$, $2 \notin A$, and $z \notin A$.

If a set contains many elements, we often use three dots, . . . , called an **ellipsis,** to indicate that there are elements in the set that have not been written down. The following are some examples of sets where we list some elements and then use an ellipsis to indicate that the pattern is to be continued indefinitely.

$$N = \{1, 2, 3, 4, \ldots\},\ A = \{5, 10, 15, \ldots\},\ W = \{0, 1, 2, 3, \ldots\}$$

The set N is called the set of **counting numbers** (or natural numbers), whereas the set W is called the set of **whole numbers** (note the addition of the element zero). Using the sets N, W, and A, we can say that $2 \in N$, $2 \in W$, $2 \notin A$, and $0 \notin N$, $0 \in W$, $0 \notin A$.

We can also use an ellipsis in another manner when listing elements in a set, to indicate that some elements are missing in the listing. Consider the set K consisting of those counting numbers from 1 through 100. We can make use of the ellipsis to list the set K as

$$K = \{1, 2, 3, \ldots, 98, 99, 100\}$$

This notation tells us the first element in the set, some of the succeeding elements, and the last element in the set.

When we use three dots in the roster form of listing elements, we must list some of the elements so that the pattern can be

HISTORICAL NOTE

Georg Cantor (1845–1918) is considered the father of set theory because he first developed this branch of mathematics. His ideas on set theory met with public ridicule, particularly his idea of a set with an infinite number of elements. This was considered revolutionary by mathematicians of his time. Cantor died in a mental institution in Germany at the age of 73. Many believe that his mental breakdowns were caused by the attacks on his work made by other mathematicians.

determined. Remember that the ellipsis means that the listing of the elements will continue in the indicated pattern.

If we use the three dots to indicate that the pattern continues indefinitely, as in $N = \{1, 2, 3, 4, \ldots\}$, then we have what is known as an **infinite set** because the set has an unlimited number of elements. The pattern is unending, and the list of elements goes on and on. If we have a set like $A = \{1, 2, 3, \ldots, 10\}$—where the ellipsis shows an indicated pattern, and we know the last element of the pattern and how many elements are in the set—then we have a **finite** set. An infinite set has an unlimited number of elements; a finite set has a last element, *and* we can count the number of elements in the set.

EXAMPLE 1 Is set $A = \{1, 2, 3, 4, \ldots\}$ an infinite set or a finite set?

SOLUTION Set $A = \{1, 2, 3, 4, \ldots\}$ is an infinite set. It has an unlimited number of elements.

EXAMPLE 2 Is set $B = \{a, b, c, d, \ldots, x, y, z\}$ an infinite set or a finite set?

SOLUTION Set $B = \{a, b, c, d, \ldots, x, y, z\}$ is a finite set. From the pattern indicated, we might assume that set B is the set of letters in the alphabet, and that there are 26 elements in the set.

USDA photo

Is the set of contented cows a well-defined set?

We should be able to determine whether or not any given element is a member of any given set; that is, any set that we consider should be **well defined.** There should not be any ambiguity as to whether an element belongs to a set. The following are some sets that are not well defined. Why?

The set of interesting courses you can take
The set of nice people in your class
The set of good instructors in your school

None of these sets is well defined because there is no common agreement as to what is meant by "interesting courses," "nice people," or "good instructors."

EXAMPLE 3 Is the set of big people a well-defined set?

SOLUTION No, it is an ill-defined set. What is meant by big people—height, power, money? The word *big* is ambiguous.

EXAMPLE 4 Given $A = \{2, 4, 6, 8, \ldots\}$, is set A a well-defined set?

SOLUTION Yes, set A is a well-defined set. Set A is the set of even counting numbers beginning with 2 and proceeding on. From the given description we can ascertain that $100 \in A$. Note that set A is a well-defined infinite set.

It is sometimes cumbersome to write a word description of a set, and it also is sometimes awkward to describe a set by listing all of its elements in roster form. There is another method that we can use to describe a set, called **set-builder** notation.

Consider set $A = \{2, 4, 6, 8, \ldots\}$. We have described set A as the set of even counting numbers. We could say that A is the set of all even counting numbers. We can also say A = the set of all x's (or any other letter) such that x is an even counting number. We can refine this to

$$A = \{x\text{'s such that } x \text{ is an even counting number}\}$$

In set notation we use a vertical line (|) to stand for "such that." Hence, we now have $A = \{x \mid x \text{ is an even counting number}\}$. This is read as "$A$ is the set of all x such that x is an even counting number." The set-builder notation is commonly used in mathematics when discussing sets.

EXAMPLE 5 What does $A = \{x \mid x \text{ is a Great Lake}\}$ mean?

SOLUTION This set is described in set-builder notation. Set A is the set of all x such that x is a Great Lake. The vertical line after the x stands for "such that." The x after the first brace tells us that we are considering all x's, that is, all of the Great Lakes.

EXAMPLE 6 List the elements of $\{x \mid x$ is a vowel in the word *Westhampton*$\}$.

SOLUTION We have the set of all x such that x is a vowel in the word *Westhampton.* The vowels are *a, e, i, o, u,* and those that appear in *Westhampton* are *e, a,* and *o*.

A set does not have to contain elements that are related. It may be that the only thing that the elements in a set have in common is that they are in the same set. For example, consider the sets $A = \{\triangle, \square, a, 2\}$ and $B = \{\text{red, blue, } 1, 1{,}000, \text{XII}\}$.

Sets may contain a definite number of elements or an unlimited number of elements. It is also true that a set may contain *no* elements. If a set does not contain any elements, it still contains a definite number of elements, namely, zero elements. Consider the set of lobsters that live in Lake Ontario. This set has no elements: there are no lobsters living in Lake Ontario. If we were to list the elements for this set, we would have to put nothing between the braces. We would have { }. This is an **empty set,** also called the **null set.** The empty set is usually denoted by { }, but another common symbol is ∅.

When you denote the empty set you may use either symbol, but do not use both together.

{ } or ∅ correct
{∅} WRONG!

The notation {∅} is incorrect because the set is not empty: the symbol ∅ is inside the braces. It is false that there is nothing inside the braces; hence the set cannot be empty.

EXAMPLE 7 List the elements in the following set.

The set of months containing 33 days

SOLUTION There are no months that have 33 days; therefore, there are no elements in the set. Hence the set of months containing 33 days is the empty set, or null set, denoted by { } or ∅.

Two sets that contain exactly the same elements are said to be **equal** sets. If we are given $A = \{a, e, i, o, u\}$ and $B = \{i, o, u, a, e\}$, then we can say that $A = B$. These two sets contain exactly the same elements and therefore they are equal.

Two sets that contain exactly the same number of elements are **equivalent** sets. If we are given $A = \{a, e, i, o, u\}$ and $B = \{1, 2, 3, 4, 5\}$, we say that A is equivalent to B. Both sets contain five elements and hence they are equivalent—but they are not equal.

EXAMPLE 8 Are the following sets equal?

a. $A = \{d, a, b\}$ $B = \{b, a, d\}$
b. $C = \{1, 2, 3, 4, \ldots\}$ $D = \{5, 10, 15, \ldots\}$

SOLUTION a. Yes, sets A and B are equal sets because they contain exactly the same elements. The order of the listing of the elements does not change the set.

b. No, sets C and D are not equal sets. They do not contain exactly the same elements. For example, from the set descriptions we know that $6 \in C$, but $6 \notin D$, because set D contains only those counting numbers that are multiples of 5.

EXAMPLE 9 Are the following sets equivalent?

a. $A = \{d, a, b\}$ $B = \{b, a, d\}$
b. $C = \{1, 2, 3, \ldots\}$ $D = \{l, o, v, e\}$

SOLUTION a. Sets A and B are equivalent because they each contain three elements.
b. Set C contains an unlimited number of elements while D contains four elements. Hence sets C and D are not equivalent.

EXERCISES FOR SECTION 1.2

1. Are the following statements true or false?
   a. $2 \in \{1, 2, 3, 4\}$
   b. $8 \in \{2, 4, 6, \ldots\}$
   c. $\{m, o, r, e\} = \{r, o, m, e\}$
   d. $\{1, 2, 3\}$ is equivalent to $\{4, 5, 6\}$.
   e. $12 \in \{x \mid x \text{ is a counting number}\}$
   f. $0 \in \{\ \}$

2. Are the following statements true or false?
   a. $4 \notin \{1, 2, 3, 4, 5\}$
   b. $M \in \{a, b, c, d, \ldots, z\}$
   c. $\{\ \} = \varnothing$
   d. $\{1, 2, 3\}$ is both equivalent to and equal to $\{3, 1, 2\}$.
   e. March $\in \{x \mid x \text{ is a month of the year}\}$
   f. $\{1, 2| = (1, 2)$

3. Are the following statements true or false?
   a. $\{\varnothing\}$ is a finite set.
   b. $\{\varnothing\}$ is an empty set.
   c. $\{O\}$ is an empty set.
   d. The set of students enrolled in this course is a finite set.
   e. The set of big cars manufactured in the United States is a well-defined set.
   f. Equivalent sets are equal sets.

4. Are the following statements true or false?
   a. $\{t, a, p\} = \{p, a, t\}$
   b. $\{a, b, c\}$ is equivalent to $\{x, y, z\}$.
   c. $\{a, b, c, \ldots, z\}$ is a finite set.
   d. Equal sets are equivalent sets.
   e. $6 \in \{x \mid x \text{ is a whole number}\}$
   f. $95 \in \{5, 10, 15, \ldots\}$

5. List the elements of each set in roster form.
   a. The set of Great Lakes.
   b. The set of states whose name begins with the word *New*.
   c. The set of states whose names begin with the letter A.
   d. The set of states whose names begin with the letter B.
   e. The set of states whose names end with the letter o.

6. List the elements of each set in roster form.
   a. The set of days of the week.
   b. The set of days with names containing the letter s.
   c. The set of days with names containing the letter x.
   d. The set of months containing 31 days.
   e. The set of months containing 32 days.

7. List the elements of each set in roster form.
   a. $\{x \mid x$ is a letter in the English alphabet$\}$
   b. $\{x \mid x$ is a Great Lake$\}$
   c. $\{x \mid x$ is the capital of Canada$\}$
   d. $\}x \mid x + 3 = 5\}$
   e. $\{x \mid x + 1 = x\}$

8. List the elements of each set in roster form.
   a. $\{x \mid x$ is a counting number less than 5$\}$
   b. $\{x \mid x$ is a counting number greater than 5$\}$
   c. $|x \mid x$ is a whole number less than 5$\}$
   d. $\{x \mid x$ is a month whose name contains an $r\}$
   e. $\{x \mid x$ is a month whose name does not contain an $r\}$

9. Write each set in set-builder notation.
   a. {Monday, Tuesday, Wednesday, Thursday, Friday, Saturday, Sunday}
   b. $\{a, e, i, o, u\}$
   c. $\{1, 3, 5, 7, \ldots\}$
   d. $\{2, 3, 5, 7, 11, 13, 17, \ldots\}$

10. Write the following in set-builder notation.
    a. $\{0, 1, 2, 3, \ldots\}$
    b. $\{1, 2, 3, \ldots\}$
    c. $\{2, 4, 6, \ldots\}$
    d. $\{a, b, c, \ldots, z\}$

11. Is the empty set a well-defined set? Explain your answer.

12. If $A = \{1, 2, 3, 4, \ldots\}$ and $B = \{5, 10, 15, 20, \ldots\}$, is A equivalent to B?

JUST FOR FUN

We have considered the set {a, e, i, o, u} in our discussion. Can you find a word that contains all of these vowels in the order they are listed? (The vowels may be separated by other letters.)

1.3 SUBSETS

Many times two or more sets contain some, but not all, of the same elements. Consider the set of positive even whole numbers, $A = \{2, 4, 6, 8, \ldots\}$, and the set of positive whole numbers, $B = \{1, 2, 3, 4, \ldots\}$. We can see that $4 \in A$ and $4 \in B$; similarly, we note that $10 \in A$ and $10 \in B$. In fact, every element that is in set A is also contained in set B. Therefore, we can say that set A is **contained in** set B, or, symbolically, we can write

$$A \subseteq B$$

When a set A is contained in another set B, we say that A is a *subset* of B.

Given any two sets A and B, if every element in A is also an element in B, then A is a subset of B.

Mark Antman/The Image Works, Inc.

How many subsets does the set of rowers in this boat have?

Since every set is a subset of itself, we have to be careful in our notation. A subset of a given set that is *not* the set itself is called a *proper subset*. If set A is a proper subset of set B, then two conditions must be satisfied: first, A must be a subset of set B; second, set B must contain at least one element that is not found in set A. If A is a proper subset of B, then we say that A is **properly contained in** B, and we write

$$A \subset B$$

If A is a subset of B, and there is at least one element in B not contained in A, then A is a proper subset of B.

In other words, A is a proper subset of B, if and only if A is a subset of B and A does not equal B.

If A is a subset of B, but not necessarily a proper subset of B, then we denote this by

$$A \subseteq B$$

A proper subset contains at least one less element than the parent set. That is, if $A \subset B$ then $A \subseteq B$ and the number of elements in A is less than the number of elements in B. The notation $A \not\subset B$ means that A is not a proper subset of B.

Consider the sets $A = \{1, 2, 3\}$ and $B = \{1, 2, 3, 4\}$. We can say that $A \subset B$ since each element in A is also an element in B, and there is at least one element in B not contained in A. But we cannot say that $B \subset A$. Why? Set B is not a subset of A because $4 \in B$, but $4 \notin A$. Hence $B \not\subset A$, but $A \subset B$. In order to show that B was not a subset of A, we found an element in B that was not in A.

Consider the empty set, $\{\ \}$. The empty set has no elements. This means that it is impossible to find an element in the empty set that is not in set A. Since the empty set has no elements there are none that can fail to be elements of A. Hence the empty set is a subset of A. By the same reasoning the empty set is a subset of the set B. In fact, the empty set is a subset of every set.

EXAMPLE 1 Determine all the possible subsets of the set $\{a, b\}$.

SOLUTION From our discussion on subsets, we know that every set is a subset of itself: thus $\{a, b\}$ is a subset. We also know that the empty set is a subset of all sets, so we have $\{\ \}$. Are there any others? Yes; namely, $\{a\}$ and $\{b\}$. It appears that a complete list of subsets of $\{a, b\}$ is

$$\{a, b\}, \{\ \}, \{a\}, \{b\}$$

EXAMPLE 2 Determine all the possible subsets of $\{a, b, c\}$.

SOLUTION We know that we can list the set itself, $\{a, b, c\}$, and the empty set, { }. But what about the others? If we are to proceed in a manner that has some order, then we should probably first consider the subsets that would be obtained by taking the elements one at a time, then two at a time, and so on. Note that taking zero elements at a time gives us the empty set.

Since the set $\{a, b, c\}$ contains three elements, we would have the following.

Zero at a Time	One at a Time	Two at a Time	Three at a Time
{ }	$\{a\}$	$\{a, b\}$	$\{a, b, c\}$
	$\{b\}$	$\{a, c\}$	
	$\{c\}$	$\{b, c\}$	

There are eight subsets for the given set. Remember that, since $\{a, b\}$ is the same set as $\{b, a\}$, we do not list both of these, because we would not wish to list the same subset twice.

The sets in the two examples we just considered contained two and three elements, respectively, and it was not too difficult to determine all of the subsets for each set. In order to determine the number of subsets for any given set with n elements we may use the following rule.

If a set contains n elements, then it has 2^n subsets.

The set $\{a, b\}$ contains two elements, so it has 2^2, or $2 \times 2 = 4$, possible subsets. The set $\{a, b, c\}$ contains three elements, so it has 2^3, or $2 \times 2 \times 2 = 8$, possible subsets. Note that this rule also works for the empty set { }. The empty set contains zero elements, so we have 2^0, which is equal to one. The subsets of the empty set are the set itself, and the empty set, which is again the set itself; hence we only list it once, and have only one subset.

When we discuss sets, we often refer to some general set that contains all the other sets under consideration. If we are discussing a set of numbers, this set could be generated from many other sets—the set of counting numbers, the set of whole numbers, the set of integers, and so on. To avoid confusion, it is necessary to know what general set the elements are taken from. This general set is called the **universal set,** and it contains all of the elements being considered in the given discussion or problem. The universal set can change from problem to problem, depending on the nature of the set being discussed. We usually denote the universal set in any problem by the capital letter U.

For example, the universal set $U = \{0, 1, 2, 3, 4, 5, 6, 7, 8, 9\}$ contains the digits 0 through 9. In a discussion using this universal set, we would only consider those sets whose elements are members of U. For example, $A = \{0, 2, 4\}$ might be discussed, but $C = \{2, b, c\}$ would not be, because not all the elements of C are elements of U.

The **complement** of a set A is the set of all the elements in the given universal set, U, that are not in the set A. The notation for the complement of A is A'. Some texts use the notation $\bar{A}$ for the complement of A, but we shall use the "prime" notation. In order to find the complement of a set, we must be given a universal set U.

EXAMPLE 3 Given $U = \{a, e, i, o, u\}$ and $A = \{i, o, u\}$, find A'.

SOLUTION The set A', the complement of A, is the set of elements that are in U, but not in A. These elements are a and e. Hence we have $A' = \{a, e\}$.

EXAMPLE 4 Given $U = \{a, e, i, o, u\}$, $B = \{a, i\}$, and $C = \varnothing$, find

a. B'
b. C'

SOLUTION

a. The complement of B, B', is the set of elements that are in U, but not in B. These elements are e, o, and u. Therefore $B' = \{e, o, u\}$.
b. Set C is the empty set and therefore contains no elements, and its complement is the set of elements in U that are not in C. In this case, the complement of C is all the elements in the universal set. Therefore, $C' = \{a, e, i, o, u\} = U$.

EXERCISES FOR SECTION 1.3

1. Tell whether each statement is true or false.
   a. $\{1, 2\} \subset \{1, 2, 3, 4, 5\}$
   b. $0 \subset \varnothing$
   c. $\{2, 4, 6\} \subset \{6, 4, 2\}$
   d. $\{a, b\} \subseteq \{b, a\}$
   e. $\{0\} \subset \varnothing$
   f. $\{\varnothing\} = \{\ \}$

2. Tell whether each statement is true or false.
   a. $\{a, b, c\} \subset \{a, b, c\}$
   b. $\{5, 10\} \subset \{1, 2, 3, \ldots\}$
   c. $\{1, 3, 5, \ldots\} \subset \{1, 2, 3, \ldots\}$
   d. $\{5, 10\} \subset \{2, 4, 6, \ldots\}$
   e. $\{\varnothing\} \subset \{\ \}$
   f. $\{2, 4, 6\} \subseteq \{6, 2, 4\}$

3. Tell whether each statement is true or false.
   a. $3 \subset \{1, 3, 5, \ldots\}$
   b. $\{b, a, t\} \subset \{t, a, b\}$
   c. $\{2\} \in \{2, 4, 6\}$
   d. $\{d\} \subset \{a, b, c, \ldots, z\}$
   e. $\{2\} \subseteq \{2, 4, 6\}$
   f. $\{B\} \subset \{a, b, c\}$

4. List all possible subsets of each set.
   a. $\{5, 10\}$ b. $\{a, b, c, d\}$
   c. $\{x, y, z\}$ d. $\{a, e, i, o, u\}$

5. List all possible subsets of each set.
   a. $\{10, 4\}$ b. $\{m, a, t, h\}$
   c. $\{i, o, u\}$ d. $\varnothing$

6. If $U = \{0, 1, 2, 3, 4, 5, 6, 7, 8, 9\}$, find
   a. $\{0, 1, 2\}'$ b. $\{0, 2, 4, 6, 8\}'$
   c. $\{0, 5\}'$ d. $\{\ \}'$
   e. $\{1, 3, 5, 7, 9\}'$ f. U'

7. If $U = \{m, e, t, r, i, c\}$, find
   a. $\{m, e, t\}'$ b. $\{r, i, c\}'$
   c. $\{m, e, i, c\}'$ d. $\{e, r, i, c\}'$
   e. $\{\ \}'$ f. $\{m, e, t, r, i, c\}'$

8. If any two sets A and B are equal, is it true that A is a proper subset of B? Explain your answer.

9. Mr. Reed has a nickel, dime, quarter, and half-dollar in his pocket. How many different sums of money may he select as a tip for the paperboy? What is the largest tip he can give?

10. If you have three bills, a one-dollar bill, a five-dollar bill, and a ten-dollar bill, and you are allowed to select any number of these bills, how many different sums of money can you select? What is the greatest amount of money you can select? What is the least amount of money you can select?

JUST FOR FUN Can you find two words in the English language that rhyme with the word *orange*?

1.4 SET OPERATIONS

In the preceding sections, we discussed sets to some extent and considered various properties of sets and subsets. Now we are ready to examine set operations that will enable us to combine sets.

In arithmetic, we have operations such as addition and subtraction that enable us to combine numbers. In this section, we shall consider the intersection and union of sets, and we shall also do some more work with the complement of a set.

If we have two sets A and B, then the *intersection* of A and B is a set of elements that are members of both A and B. The notation for A intersection B is $A \cap B$.

In other words, the *intersection* of sets A and B results in another set, denoted by $A \cap B$. This resulting set is composed of elements that are common to both A and B.

EXAMPLE 1 Given $A = \{1, 2, 3, 4, 5\}$ and $B = \{2, 4, 6, 8\}$, find $A \cap B$.

SOLUTION The elements that are in A and also in B are 2 and 4. Hence $A \cap B = \{2, 4\}$.

EXAMPLE 2 Given $A = \{1, 2, 3, \ldots, 10\}$ and $B = \{2, 4, 6, \ldots\}$, find $A \cap B$.

SOLUTION We see that set A contains the numbers 1 through 10, whereas set B contains the positive, even whole numbers. The elements common to both sets are 2, 4, 6, 8, and 10, therefore $A \cap B = \{2, 4, 6, 8, 10\}$.

EXAMPLE 3 Given $A = \{2, 4, 6, 8\}$ and $B = \{a, e, i, o, u\}$, find $A \cap B$.

SOLUTION Examining sets A and B, we see that there are no elements common to both. Therefore, the intersection of these two sets is the empty set. Hence $A \cap B = \{\ \}$, or $A \cap B = \emptyset$.

Two sets whose intersection is the empty set are said to be **disjoint.** Disjoint sets have no elements in common.

EXAMPLE 4 Given $U = \{a, e, i, o, u\}$, $A = \{a, e, o\}$, and $B = \{e, i, o\}$, find $(A \cap B)'$.

SOLUTION We are asked to find $(A \cap B)'$. As in arithmetic and algebra, when we have parentheses, we first do the work inside the parentheses, and then perform the operation(s) outside the parentheses. So we must find $A \cap B$, and then find the complement of our answer. $A \cap B = \{e, o\}$. The complement of $\{e, o\}$ is all the elements that are in U, but not in $A \cap B$. Therefore the complement of $\{e, o\}$ is $\{a, i, u\}$, and

$$(A \cap B)' = \{e, o\}' = \{a, i, u\}$$

EXAMPLE 5 Given $A = \{1, 2, 4, 5, 6, 7\}$, $B = \{1, 3, 5, 7, 9\}$, and $C = \{2, 4, 6, 7, 8\}$, find $(A \cap B) \cap C$.

SOLUTION First we find $A \cap B$, then find the intersection of that with C. Since $A \cap B = \{1, 5, 7\}$,

$$(A \cap B) \cap C = \{1, 5, 7\} \cap C = \{1, 5, 7\} \cap \{2, 4, 6, 7, 8\} = \{7\}.$$

The *union* of sets A and B is the set of all the elements that are members of either set A or set B, or both. When we list the elements in the union of two sets, we list all of the elements in set A and all of the elements in set B, but if an element is in both sets, we list it only once. Therefore, the union of sets A and B is the set of elements that are elements of at least one of the two sets. The notation for A union B is $A \cup B$.

If we have two sets A and B, then the *union* of A and B, $A \cup B$, is the set of elements that are members of A, or members of B, or members of both A and B.

EXAMPLE 6 Given $A = \{1, 2, 3, 4, 5\}$ and $B = \{2, 4, 6, 8\}$, find $A \cup B$.

SOLUTION The elements of A are 1, 2, 3, 4, 5. The elements of B are 2, 4, 6, 8. The union of A and B is the set of all these elements, because all of these elements belong either to set A or to set B, or to both. Therefore, $A \cup B = \{1, 2, 3, 4, 5, 6, 8\}$.

Note that 2 and 4 are members of both sets, but we list each only once in our solution. Also, it is not necessary to list the elements in order, but it does provide a means of checking that all elements are included.

EXAMPLE 7 Given $A = \{1, 3, 5, \ldots\}$ and $B = \{2, 4, 6, \ldots\}$, find $A \cup B$.

SOLUTION We have the set of odd counting numbers and the set of even counting numbers. Hence the union of these two sets is the set of counting numbers: $A \cup B = \{1, 2, 3, 4, 5, 6, \ldots\}$.

EXAMPLE 8 Given $U = \{a, e, i, o, u\}$, $A = \{a, e, o\}$, and $B = \{e, i, o\}$, find $A' \cup B'$.

SOLUTION We want to find the union of two sets, A' and B'. First we must find the complement of each set, and then the union of the two complements. $A' = \{i, u\}$ and $B' = \{a, u\}$. Therefore $A' \cup B' = \{i, u, a\}$.

EXAMPLE 9 Given $U = \{a, e, i, o, u\}$, $A = \{a, e, o\}$, and $B = \{e, i, o\}$, find $(A \cup B)'$.

SOLUTION This problem is a little different from the previous example. This time we want $(A \cup B)'$, which is the complement of $A \cup B$. Recall that we must first perform the operation inside the parentheses, and then perform the operation outside the parentheses. First we find $A \cup B = \{a, e, i, o\}$; then we find the complement of $\{a, e, i, o\}$, which is $\{u\}$. Hence $(A \cup B)' = \{a, e, i, o\}' = \{u\}$.

Note that Examples 8 and 9 show that $A' \cup B' \neq (A \cup B)'$. It is important to remember this, and to exercise care in reading problems and computing solutions. It is also true that $A' \cap B' \neq (A \cap B)'$. As we shall see later, $(A \cap B)' = A' \cup B'$, and $(A \cup B)' = A' \cap B'$.

There are many different combinations of operations in set theory. However, if you do each operation as it is indicated, you will be able to do any problem involving set operations.

It should be noted that the words *and, or,* and *not* correspond to operations in set theory. The intersection of two sets A and B is the set of elements in set A *and* set B. The union of two sets A and B is the set of elements in set A *or* set B *or* both. The complement of set A is the set of elements in the universal set U, but *not* in set A.

$$A \cap B = \{x \mid x \in A \text{ and } x \in B\}$$
$$A \cup B = \{x \mid x \in A \text{ or } x \in B\}$$
$$A' = \{x \mid x \in U \text{ and } x \notin A\}$$

EXERCISES FOR SECTION 1.4

1. For each pair of sets, find $A \cap B$ and $A \cup B$.
   a. $A = \{2, 4, 6, 8\}$, $B = \{1, 3, 4, 6, 7\}$
   b. $A = \{a, b, c\}$, $B = \{d, e, f\}$
   c. $A = \{a, e, i, o, u\}$, $B = \{a, e, i\}$
   d. $A = \{g, i, a, n, t, s\}$, $B = \{j, e, t, s\}$
   e. $A = \{5, 10, 15, \ldots\}$, $B = \{10, 20, 30, \ldots\}$
   f. $A = \{1, 3, 5, 7, \ldots\}$, $B = \{2, 4, 6, 8, \ldots\}$

2. Find $A \cap B$ and $A \cup B$ for each pair of sets.
   a. $A = \{\text{dog, cat, pig}\}$, $B = \{\text{fox, dog, cow}\}$
   b. $A = \{1, 2, 3, \ldots\}$, $B = \{2, 4, 6, 8, \ldots\}$
   c. $A = \{\$, ?, !\}$, $B = \{\$, ¢, ?\}$
   d. $A = \{a, b, c\}$, $B = \{x, y, z\}$
   e. $A = \{1, 3, 5\}$, $B = \varnothing$
   f. $A = \{m, a, t, h\}$, $B = \{e, a, s, y\}$

3. Given the sets $U = \{1, 2, 3, 4, 5, 6, 7\}$, $A = \{2, 4, 6, 7\}$, and $B = \{1, 3, 5, 6, 7\}$, find
   a. A'
   b. B'
   c. $A' \cap B'$
   d. $A' \cup B'$
   e. $(A \cap B)'$
   f. $(A \cup B)'$

4. Given the sets $U = \{a, e, i, o, u\}$, $A = \{a, e, u\}$, and $B = \{i, o, u\}$, find
   a. A' b. B'
   c. $A' \cap B'$ d. $A' \cup B'$
   e. $(A \cap B)'$ f. $(A \cup B)'$
5. If $U = \{1, 2, 3, \ldots, 10\}$, $A = \{2, 3, 4, 5\}$, $B = \{4, 5, 6, 7, 8\}$, and $C = \{4, 6, 7, 8, 9\}$, find
   a. $A' \cap B'$ b. $A' \cup B'$
   c. $(A \cap B)'$ d. $(A \cup B)'$
   e. $(A \cap B) \cap C$ f. $A \cup (B \cap C)$
   g. $(A \cap B)' \cup C$ h. $A' \cap (B' \cap C')$
6. If $U = \{0, 1, 2, 3, 4, 5, 6, 7, 8, 9\}$, $A = \{1, 3, 4, 5, 7\}$, $B = \{2, 3, 4, 5, 6\}$, and $C = \{0, 2, 4, 6, 8, 9\}$, find
   a. $(A \cap B)'$ b. $A' \cap B$
   c. $(A \cup B)'$ d. $A' \cup B'$
   e. $(A \cap B) \cup C$ f. $(A \cup B) \cap C$
   g. $A' \cap (B \cup C)$ h. $(A \cup B)' \cap C'$
7. If $U = \{m, e, t, r, i, c\}$, $A = \{m, e, r\}$, $B = \{e, i, c\}$, and $C = \{m, t, c\}$, is each of the following statements true or false?
   a. $B \cup C = U$ b. $(A \cup B) \subset U$
   c. $A' \cap B' = \varnothing$ d. $(B \cap C) \subset A$
   e. $(A \cap B)' = A' \cap B'$
   f. $(A \cup B)' = A' \cup B'$
8. If $U = \{a, e, i, o, u\}$, $A = \{i, o, u\}$, $B = \{e, i, o\}$, and $C = \{a, i, o\}$, is each of the following statements true or false?
   a. $A \cap B = C$ b. $A \cup B = U$
   c. $B \cup C = U$ d. $A' \cap B' = \varnothing$
   e. $(A \cap B) \subset C$ f. $(B \cap C) \subseteq A$

JUST FOR FUN

A snake is stuck at the bottom of a 30-foot well. It can climb up 3 feet every hour, but at the end of each hour it stops to rest and slips back 2 feet. At this rate, how long will it take for the snake to get out of the well?

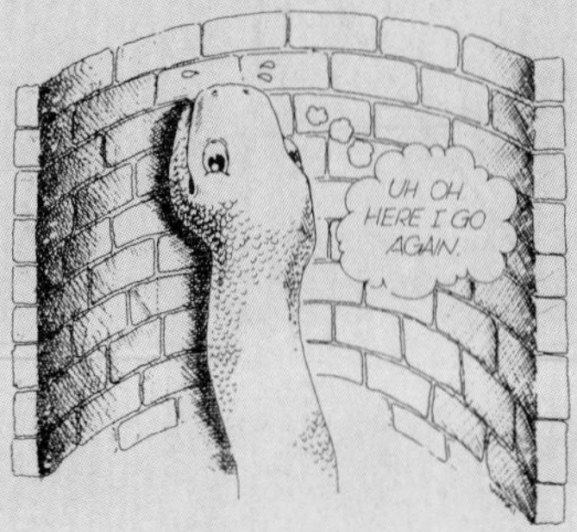

1.5 PICTURES OF SETS (VENN DIAGRAMS)

It is sometimes useful to represent sets and relationships between sets by means of a picture or diagram. This is almost always done by using a rectangle to represent the universal set and a circle or circles inside the rectangle to represent the set or sets being considered in the discussion. It is understood that the elements in the set are inside the circle that represents the set.

EXAMPLE 1 Let $U = \{a, e, i, o, u\}$ and $A = \{a, e, i\}$. Make a picture to show the relationship between A and U.

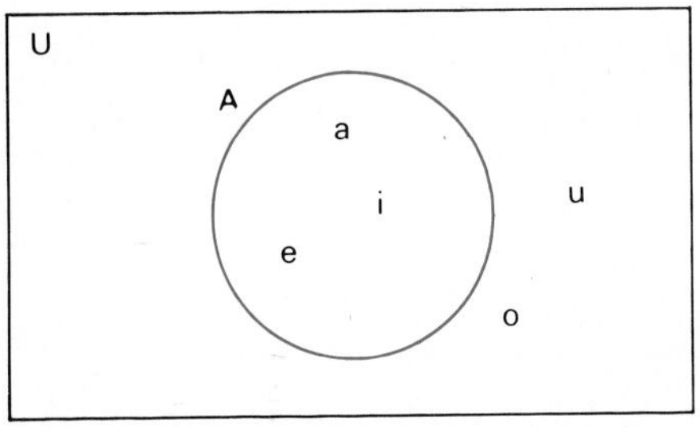

FIGURE 1
Note that the elements of the universal set that are in A are inside the circle, while those not in A are inside the rectangle, but outside the circle.

SOLUTION The diagram that represents the given information is Figure 1.

These picture representations of sets are called **Venn diagrams.** They were developed by John Venn (1834–1923), who made great contributions to modern mathematics. His diagrams are used to picture relationships in set theory and logic.

EXAMPLE 2 Use a Venn diagram to show the relationship $B \subset A$.

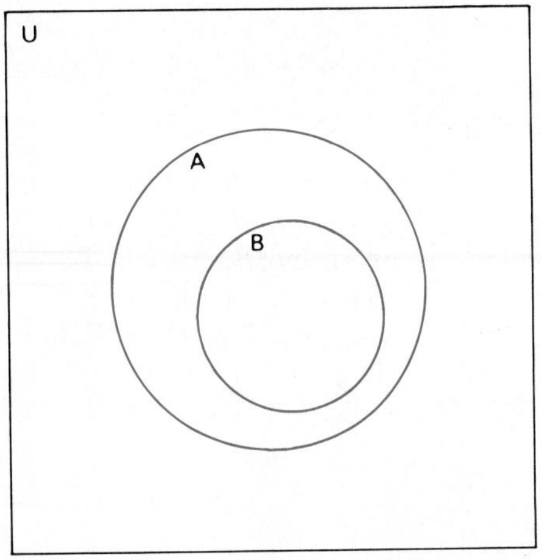

FIGURE 2
$B \subset A$ (*Note:* This is more correctly called an *Euler diagram*, as Leonhard Euler (1707–1783) first used circles to represent sets. Venn later added refinements to produce what we shall use here.)

SOLUTION First, the sets under consideration are understood to be in a universal set U. More importantly, $B \subset A$. This means that all of the elements contained in B are also in A; hence our solution is the diagram in Figure 2.

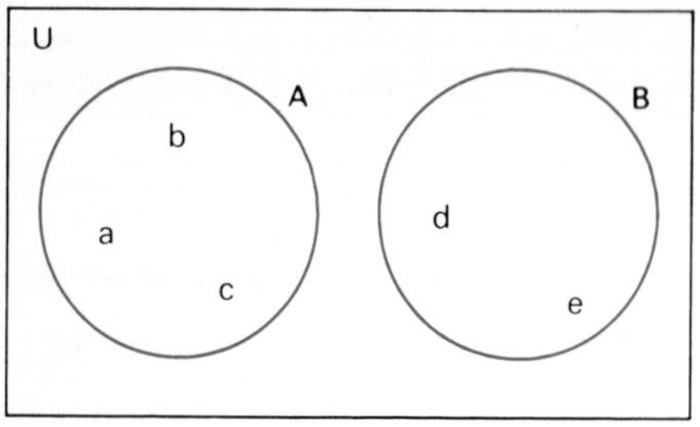

FIGURE 3
Sets A and B are disjoint.

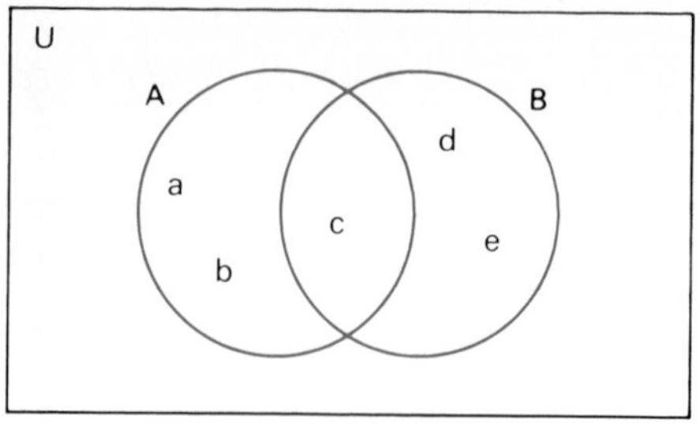

FIGURE 4
Sets A and B have the element c in common.

If we are considering two sets A and B, there are two other distinct pictures we could use to show a relationship between A and B. Suppose that $U = \{a, b, c, d, e\}$, $A = \{a, b, c\}$, and $B = \{d, e\}$. We see that A and B are disjoint sets. We illustrate this in Figure 3.

But suppose that $A = \{a, b, c\}$ and $B = \{c, d, e\}$; now B and A have an element, c, in common. This is shown in Figure 4.

Note that the two sets overlap, which allows us to illustrate that c is an element common to both set A and set B.

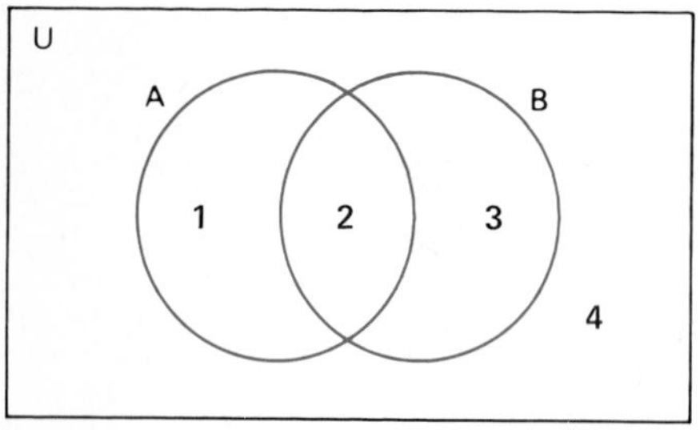

FIGURE 5

A Venn diagram that we will use quite often is shown in Figure 5. This type of illustration allows us to show all relations that might exist between the two sets A and B. The overlapping technique is used because it is the most efficient; it allows us to consider all of the possibilities that might exist. Although we have shown sets A and B overlapping, it does not mean that the two sets intersect.

We have assigned numbers to each region in the diagram in Figure 5. This enables us to discuss easily the diagram and its various parts. Region 2, for example, represents the intersection of A and B, $A \cap B$. In using the overlapping diagram, we are not concerned with whether or not sets A and B are disjoint; we just want to identify the region that represents $A \cap B$. The two most

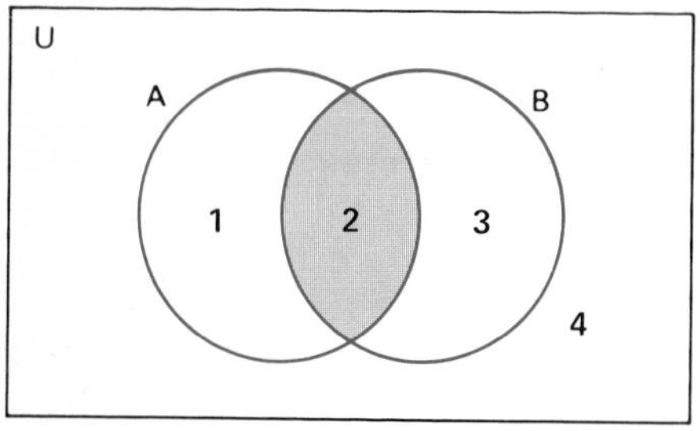

FIGURE 6
Region 2 represents $A \cap B$.

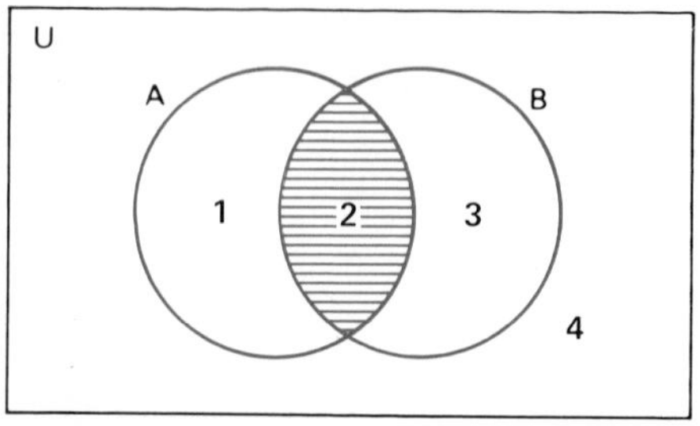

FIGURE 7
Region 2 represents $A \cap B$.

common methods of doing this are shading the region and using stripes (a type of shading) to distinguish the region. See Figures 6 and 7.

We know that region 2 represents $A \cap B$, but what region or regions represent $A \cup B$? Since $A \cup B$ is the set of elements that are elements or A or B or both, we must have all the elements in both sets. Therefore the Venn diagram showing $A \cup B$ would have regions 1, 2, and 3 shaded, as in Figure 8.

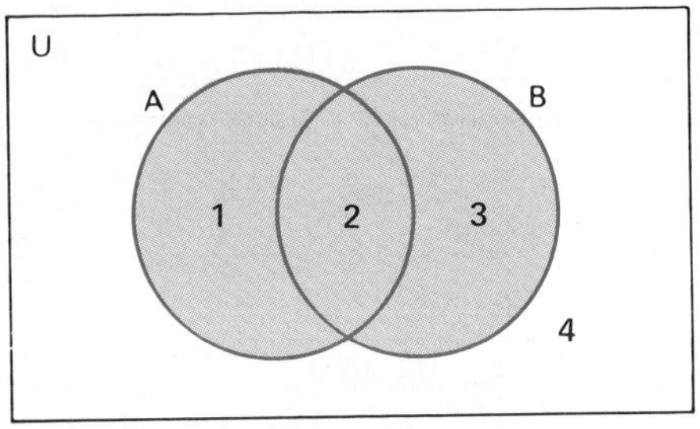

FIGURE 8
Regions 1, 2, 3, represent $A \cup B$.

EXAMPLE 3 Use a Venn diagram to show $(A \cup B)'$.

SOLUTION We must determine what region or regions represent $(A \cup B)'$. In the preceding discussion, we determined that regions 1, 2, and 3 represent $A \cup B$; we now want to determine $(A \cup B)'$, the complement of $A \cup B$. This region must be in the universal set, but not in $A \cup B$; hence it must be region 4. See Figure 9.

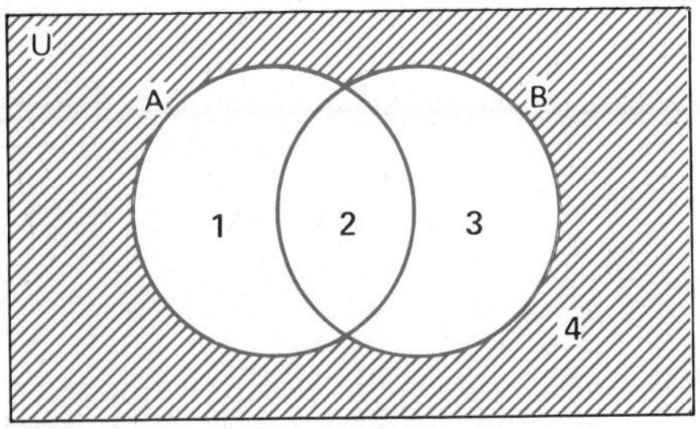

FIGURE 9
Region 4 represents $(A \cup B)'$.

Note that in Figure 9 we used stripes to identify a particular region in our Venn diagram. We are not concerned about what elements are in the region, or even whether there are any elements in it; we just want to distinguish it from the other regions in some way. If we have to identify more than one region in a diagram, we should use a different type of shading for each part of the problem. Each set of stripes should run in a different direction (such as horizontal or vertical), as shown in Example 4.

EXAMPLE 4 Use a Venn diagram to show $A \cap B$.

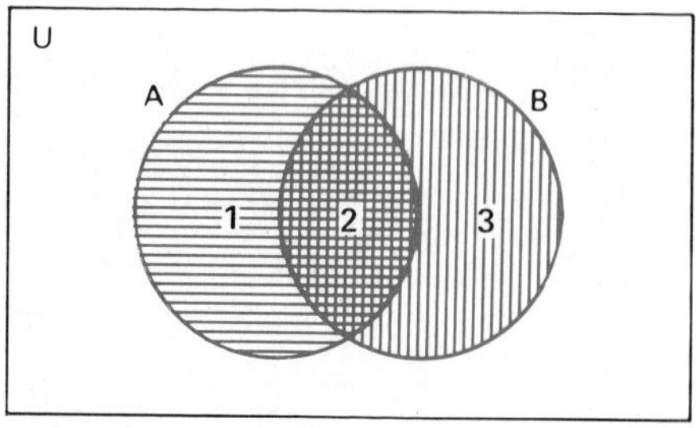

FIGURE 10
Region 2 represents $A \cap B$.

SOLUTION In Figure 10, regions 1 and 2 are in A; we shade these with horizontal stripes. Set B is made up of regions of 2 and 3; we shade set B with vertical stripes. Since we are looking for the intersection of A and B, we want the region with stripes that are common to both A and B. Note that this occurs in region 2, which is shaded with both horizontal and vertical stripes. Therefore, region 2 represents $A \cap B$.

In Example 4, the regions of A and B were shaded separately, and the intersection of the two sets was the shading common to both sets. For the union of A and B, $A \cup B$, we would have done the problem in the same manner, but we would have taken all of the shaded areas for our answer.

EXAMPLE 5 Use a Venn diagram to show that $(A \cap B)' = A' \cup B'$.

SOLUTION In order to illustrate that the statement is true, we must use two Venn diagrams. We will let one diagram represent the left side of the equation and the second diagram represent the right side, and then show that the final results for each diagram have the same regions shaded.

The set $(A \cap B)'$ is the complement of $A \cap B$. The region that would satisfy the stated problem must be in U, but not in $A \cap B$. The region satisfying this consists of regions 1, 3, and 4 in Figure 11.

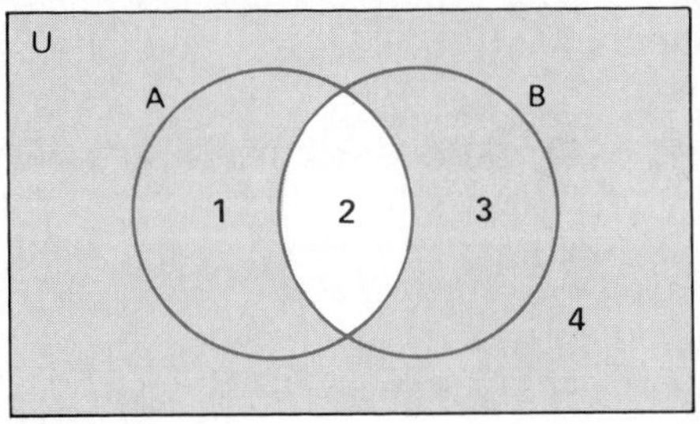

FIGURE 11
Regions 1, 3, 4 represent $(A \cap B)'$.

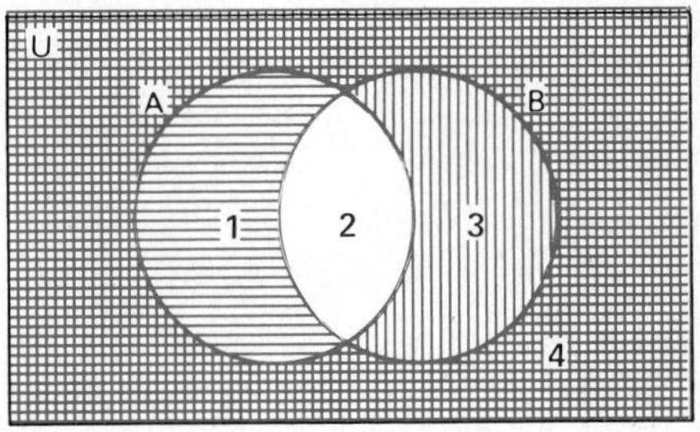

FIGURE 12
Regions 1, 3, 4 represent $A' \cup B'$.

The set A' is represented by regions 3 and 4 in Figure 12; we shade this with a vertical set of lines. The set B' is represented by regions 1 and 4, and we shade this with a horizontal set of lines.

The set $A' \cup B'$ is determined by shading the region for A' vertically, shading the region for B' horizontally, and then taking all of the shaded areas. See Figure 12.

Now note that the shaded region for $(A \cap B)'$ in Figure 11 is the same as the shaded region for $A' \cup B'$ in Figure 12. Hence we have shown that $(A \cap B)' = A' \cup B'$.

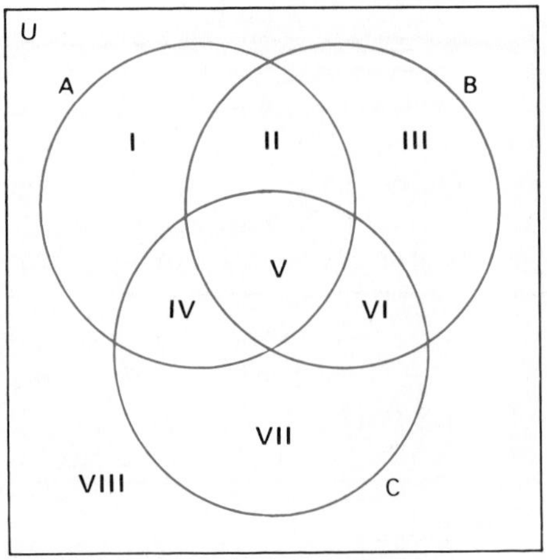

FIGURE 13

Venn diagrams may also be used when working with three sets. Figure 13 is a typical Venn diagram involving three sets A, B, and C. We show the three sets overlapping as this allows us to see what relationships might exist between A and B, B and C, A and C, and so on. Because there are many possible combinations for such a problem, we have to use care in shading the proper region. For example, consider the following sets: $A' \cap (B \cup C)$ and $(A \cap B) \cup C'$. Before we attempt to illustrate such sets, let us agree that we shall number the separate regions as shown in Figure 13. Note that we used Roman numerals in numbering the regions. We could have used Arabic numerals (1, 2, 3, . . .), but, as you will see later, the use of Roman numerals enables us to avoid some confusion in discussing Venn diagrams. We shall use Roman numerals from now on.

Examine the Venn diagram shown in Figure 13. What regions represent $A \cup B$? We see that the regions comprising $A \cup B$ are I, II, III, IV, V, and VI. If we are asked to shade $A \cup B$, then these are the regions we should shade.

Regions II and V represent $A \cap B$; regions IV and V represent $A \cap C$; and regions V and VI represent $B \cap C$. Now what region represents $(A \cap B) \cap C$? That is, what region is common to all three sets? Examining Figure 13, we see that it is region V.

What does region VIII represent? It is the region inside U, but outside of all three sets. Hence, it must be the complement of the union of all three sets. Therefore region VIII is the complement of $A \cup B \cup C$; we write this as $(A \cup B \cup C)'$.

Now that we are somewhat familiar with the different regions, let us consider some examples using shading.

EXAMPLE 6 Use a Venn diagram to show $A \cap (B \cup C)$.

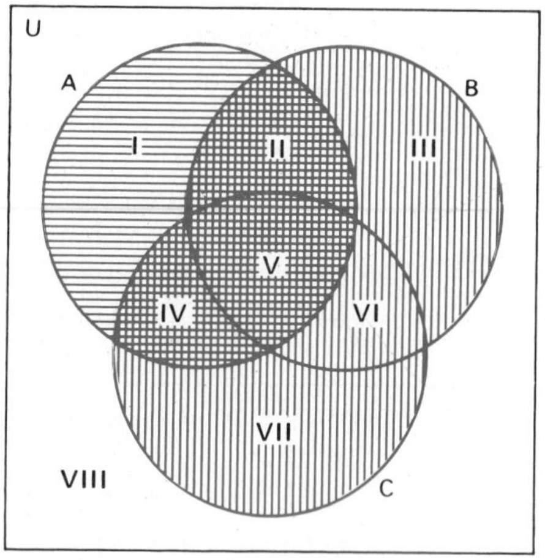

FIGURE 14
Regions II, IV, V represent $A \cap (B \cup C)$.

SOLUTION We shall first shade $(B \cup C)$ with vertical lines. In Figure 14, $B \cup C$ is regions II, III, V, VI, IV, and VII. We then shade A with horizontal lines. This problem asks for $A \cap (B \cup C)$, the intersection of A with the union of B and C. Therefore our answer appears in the regions with double lines, namely, regions II, IV, and V.

EXAMPLE 7 Use a Venn diagram to show $A \cup (B \cap C)$.

SOLUTION We shall shade $(B \cap C)$ with horizontal lines. In Figure 15, the regions shaded for $B \cap C$ are V and VI. Set A is shaded with vertical lines. Since we want the union of A with $(B \cap C)$, our answer is regions I, II, IV, V, and VI.

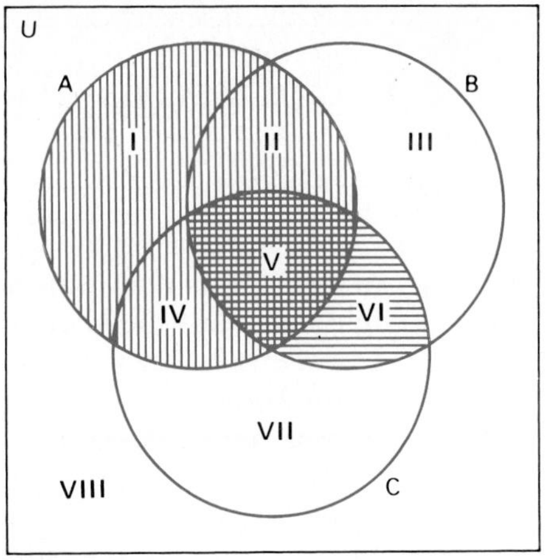

FIGURE 15
Regions I, II, IV, V, VI represent
$A \cup (B \cap C)$.

EXAMPLE 8 Show that $A \cup (B \cap C) = (A \cup B) \cap (A \cup C)$.

SOLUTION We shall show that $A \cup (B \cap C) = (A \cup B) \cap (A \cup C)$ by comparing the corresponding Venn diagrams. Recall that if they both contain the same regions for a final answer, then the sets that the regions represent are equal. We have already determined the regions for $A \cup (B \cap C)$ (see Figure 15); they are I, II, IV, V, and VI.

Now we must represent $(A \cup B) \cap (A \cup C)$ by means of a Venn diagram. We first shade $(A \cup B)$ with horizontal lines and then shade $(A \cup C)$ with vertical lines, as shown in Figure 16. We

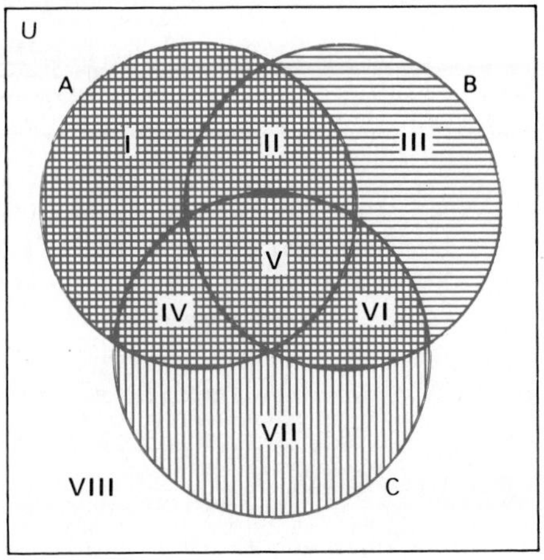

FIGURE 16
Regions I, II, IV, V, VI represent
$(A \cup B) \cap (A \cup C)$.

want the intersection of these two, so our answer is regions I, II, IV, V and VI. These are precisely the same regions that we got for $A \cup (B \cap C)$. Therefore we have shown that $A \cup (B \cap C) = (A \cup B) \cap (A \cup C)$.

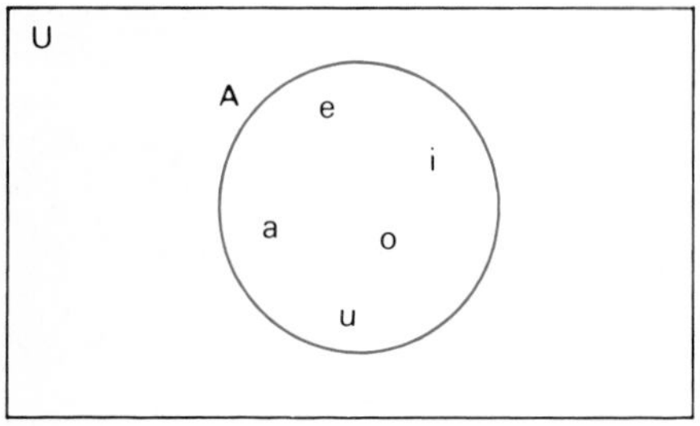

FIGURE 17

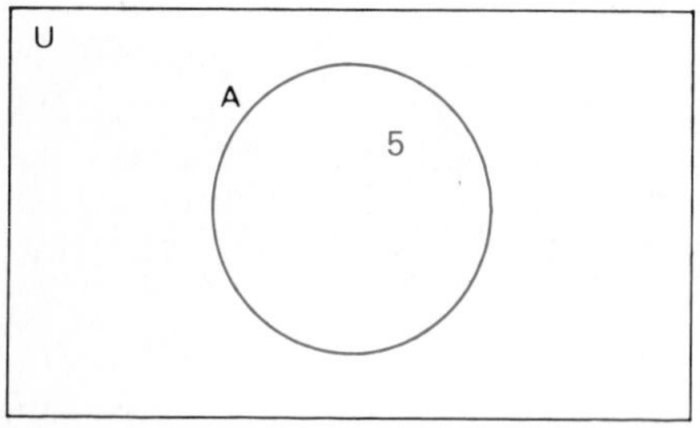

FIGURE 18

The reason we used Roman numerals to designate the regions of a set is that many times we may want to use Arabic numerals to tell how many members are in a set. This is called the **cardinality** of a set. A **cardinal number** tells us "how many," as opposed to an **ordinal number,** which tells us "what position." (Examples of ordinal numbers are first, second, third, fourth, fifth, and so on.)

A cardinal number gives the number of elements in a set. The empty set contains no elements, so its cardinal number is zero. We may say that the *cardinality* of the empty set is zero. The set $A = \{a, e, i, o, u\}$ has five elements; hence its cardinality is 5.

We shall use the notation $n(A)$ to stand for the cardinality of A. Therefore, we may say $n(A) = 5$. But suppose set A is in a Venn diagram—how could we tell how many members it has? One solution is to list the elements in the circle. Another solution is to write the Arabic numeral 5 to indicate that set A contains five elements. See Figures 17 and 18.

Although you may feel that neither technique is better than the other, suppose set A contained all the letters of the alphabet. Would you list all 26 letters separately in the circle? Carrying this idea a bit further, suppose the set contains all the students in your school. If we are concerned only with the cardinality of this set, then it is much more efficient to write the Arabic numeral for that number than to list all of the elements.

EXAMPLE 9 Use Figure 19 to find the cardinality of set A, that is, $n(A)$.

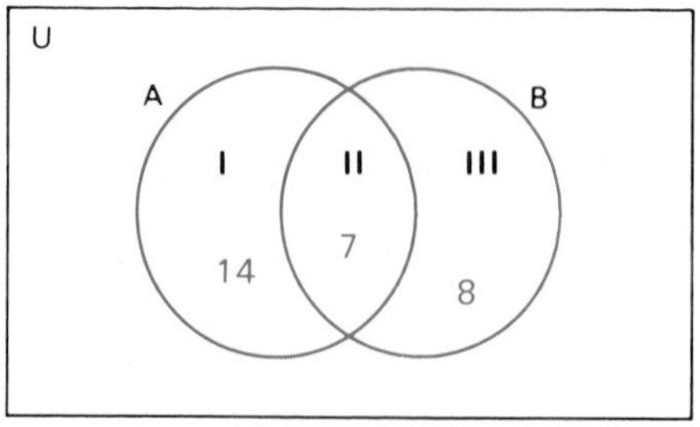

FIGURE 19

SOLUTION In Figure 19, set A is composed of regions I and II. The number of elements in region I is 14, and the number of elements in region II is 7. The total number of elements in set A is 21. Therefore,

$$n(A) = 21$$

EXAMPLE 10 Use Figure 20 to find $n(A \cap B)$, $n(B \cup C)$, and $n(U)$.

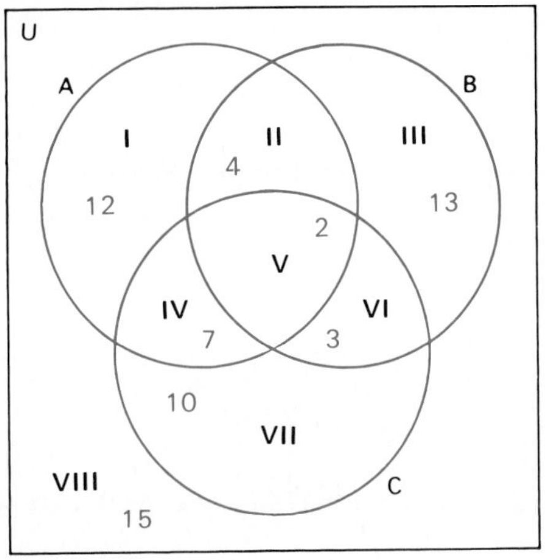

FIGURE 20

SOLUTION In Figure 20, we see that $A \cap B$ is composed of regions II and V. The total number of elements in these two regions is 6. Therefore,

$$n(A \cap B) = 6$$

For $B \cup C$, we use regions II, III, V, VI, IV, and VII. The total number of elements in all of these regions is 39. Recall that in doing set union problems, we only list an element once even if it appears in two sets. Similarly, we need only count the elements once in a region even if the region is in both sets. Hence,

$$n(B \cup C) = 39$$

To find the cardinality of the universal set, we must find the total number of elements in all eight regions. Adding these numbers, we find that the cardinality of U is 66, or $n(U) = 66$.

Recall that two sets that contain the same number of elements are said to be equivalent sets. If two sets are equivalent, then they have the same cardinality, and the elements of one set can be paired with the elements of the other set. Consider the following two sets, A and B.

$$A = \{a, b, c\}$$
$$B = \{\text{bat, cat, rat}\}$$

The two sets have the same cardinality, so we can match the elements of the two sets with each other. We could pair *a* with *bat*, *b* with *cat*, and *c* with *rat*. We can show this by

$$\begin{matrix} \{a, & b, & c\} \\ \updownarrow & \updownarrow & \updownarrow \\ \{\text{bat}, & \text{cat}, & \text{rat}\} \end{matrix}$$

This is not the only way that we could have paired the elements of the two sets; there are others. For example: *a* with *rat*, *b* with *bat*, and *c* with *cat*. A pairing of the elements of two sets is called a **one-to-one correspondence** between the sets. When two sets are in a one-to-one correspondence, it means that each element of the first set is paired with just one element of the second set, and each element of the second set is paired with just one element of the first set. Note that two sets that are in one-to-one correspondence must be equivalent sets and have the same cardinality.

NOTE OF INTEREST

Consider the sets $A = \{1, 2, 3, 4, \ldots\}$ and $B = \{2, 4, 6, 8, \ldots\}$. We can describe both A and B as infinite sets; each set has an unlimited number of elements. What is the cardinality of each set? At first glance, it may appear that A contains a greater number of elements than B, but does it? The cardinal number of the set of counting numbers is called "aleph-null" or "aleph-zero," and is designated by $\aleph_0$. It is the first letter of the Hebrew alphabet. It can be shown that the set $B = \{2, 4, 6, \ldots\}$ can be placed in a one-to-one correspondence with the set of counting numbers (set A), which implies that they contain the same number of elements. Thus, the cardinally of B is also $\aleph_0$. A and B are called countably infinite sets. The cardinal number of all countably infinite sets is $\aleph_0$.

EXAMPLE 11 In how many ways can a one-to-one correspondence be established between $A = \{a, b, c\}$ and $B = \{\text{bat, cat, rat}\}$?

SOLUTION We have already discussed two different one-to-one correspondences for these two sets, but we shall start from the beginning. The letter *a* can be paired with any of the three words *bat, cat,* or *rat* in set B. Now, if *a* is paired, then *b* can be paired with any of the *two* remaining words. With *a* and *b* paired, *c* has to be paired with the remaining word. Therefore, the total number of one-to-one correspondences is

$$3 \times 2 \times 1 = 6$$

Here are the six different one-to-one correspondences.

$$\begin{matrix} \{a, & b, & c\} \\ \updownarrow & \updownarrow & \updownarrow \\ \{\text{bat}, & \text{cat}, & \text{rat}\} \end{matrix} \quad \begin{matrix} \{a, & b, & c\} \\ \updownarrow & \updownarrow & \updownarrow \\ \{\text{cat}, & \text{bat}, & \text{rat}\} \end{matrix} \quad \begin{matrix} \{a, & b, & c\} \\ \updownarrow & \updownarrow & \updownarrow \\ \{\text{rat}, & \text{bat}, & \text{cat}\} \end{matrix}$$

$$\begin{matrix} \{a, & b, & c\} \\ \updownarrow & \updownarrow & \updownarrow \\ \{\text{bat}, & \text{rat}, & \text{cat}\} \end{matrix} \quad \begin{matrix} \{a, & b, & c\} \\ \updownarrow & \updownarrow & \updownarrow \\ \{\text{cat}, & \text{rat}, & \text{bat}\} \end{matrix} \quad \begin{matrix} \{a, & b, & c\} \\ \updownarrow & \updownarrow & \updownarrow \\ \{\text{rat}, & \text{cat}, & \text{bat}\} \end{matrix}$$

EXAMPLE 12 Can the two sets $A = \{a, e, i, o, u\}$ and $B = \{2, 4, 6, 8\}$ be placed in a one-to-one correspondence?

SOLUTION No, the two sets do not have the same cardinality, because the cardinality of set A is 5, while the cardinality of set B is 4. If we tried pairing the elements of the two sets, we would have an element left over, and it would have to be paired with an element that had already been matched.

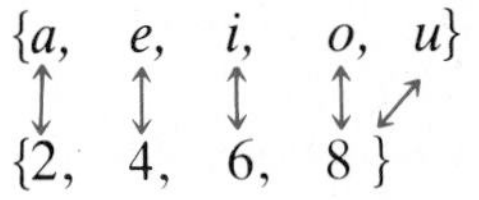

not one-to-one

EXERCISES FOR SECTION 1.5

1. Illustrate each of the following with a Venn diagram. Number the regions in your diagram as shown in Examples 3–5 and list the regions that make up your answer.
   a. $A \cap B$
   b. $A' \cap B'$
   c. $A \cup B$
   d. $A' \cup B'$
   e. $(A' \cap B')'$
   f. $(A' \cup B')'$

2. Illustrate each of the following with a Venn diagram. Number the regions in your diagram as shown in Examples 3–5 and list the regions that make up your answer.
   a. $A \cap B'$
   b. $A' \cup B$
   c. $A' \cap B'$
   d. $(A \cup B)'$
   e. $(A \cap B)'$
   f. $A' \cup B'$

3. Use a Venn diagram to illustrate each of the following. Number the regions in your diagram as shown in Examples 6–10 and list the regions that make up your answer.
   a. $A \cap (B \cap C)$
   b. $A \cup (B \cup C)$
   c. $(A \cap B) \cup C$
   d. $(A \cap B) \cup (A \cap C)$
   e. $(A \cup B) \cap (A \cup C)$
   f. $(A' \cap B') \cup C$

4. Use a Venn diagram to illustrate each of the following. Number the regions in your diagram as shown in Examples 6–10 and list the regions that make up your answer.
   a. $(A \cap B) \cap C$
   b. $(A \cup B) \cup C$
   c. $(A \cup B) \cap C$
   d. $(A \cap B) \cap (B \cap C)$
   e. $(B \cup C) \cap (A \cup C)$
   f. $(C' \cup A') \cap B$

5. Use Venn diagrams to show that each of the following statements is true.
   a. $A \cap B = (A' \cup B')'$
   b. $A \cup B = (A' \cap B')'$
   c. $(A \cup B)' = A' \cap B'$
   d. $(A \cap B)' = A' \cup B'$
   e. $A \cap (B \cup C) = (A \cap B) \cup (A \cap C)$
   f. $A \cup (B \cap C) = (A \cup B) \cap (A \cup C)$

6. Use Figure 21 to find the following cardinalities.
   a. $n(A)$
   b. $n(B)$
   c. $n(A \cap B)$
   d. $n(A \cup B)$
   e. $n(A')$
   f. $n(B')$
   g. $n(A' \cap B')$
   h. $n(A' \cup B')$

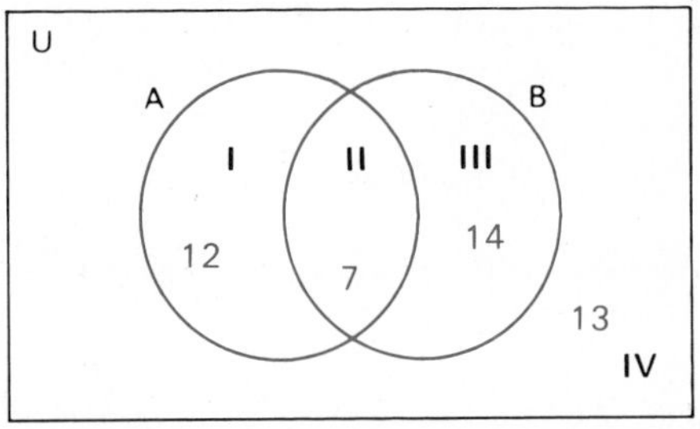

FIGURE 21

7. Use Figure 22 to find the following cardinalities.
   a. $n(A)$ b. $n(B)$
   c. $n(C)$ d. $n(U)$
   e. $n(B \cup C)$ f. $n(B \cap C)$
   g. $n(A \cap B \cap C)$ h. $n(A \cup B \cup C)$

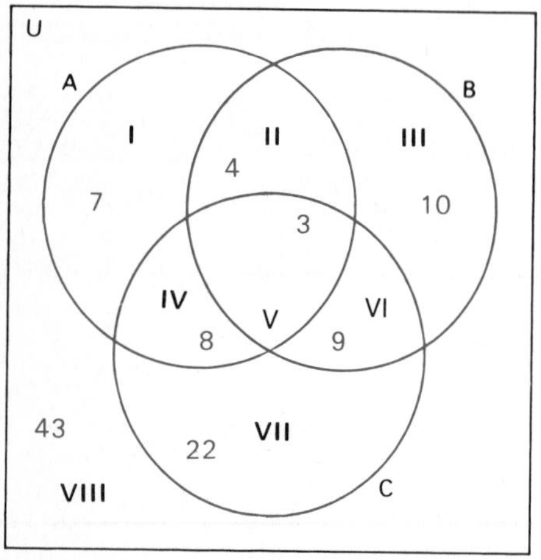

FIGURE 22

8. Show a one-to-one correspondence between $A = \{5, 10\}$ and $B = \{15, 20\}$. In how many ways can you do this?

9. In how many ways can one show a one-to-one correspondence between $A = \{i, o, u\}$ and $B = \{x, y, z\}$? Show one of these correspondences.

10. Show a one-to-one correspondence between $A = \{\text{Bob, Joe, Cy, Ted}\}$ and $B = \{5, 10, 15, 20\}$. In how many ways can you do this?

11. In how many ways can a one-to-one correspondence be established between $A = \{m, a, t, h\}$ and $B = \{f, u, n\}$? Explain your answer.

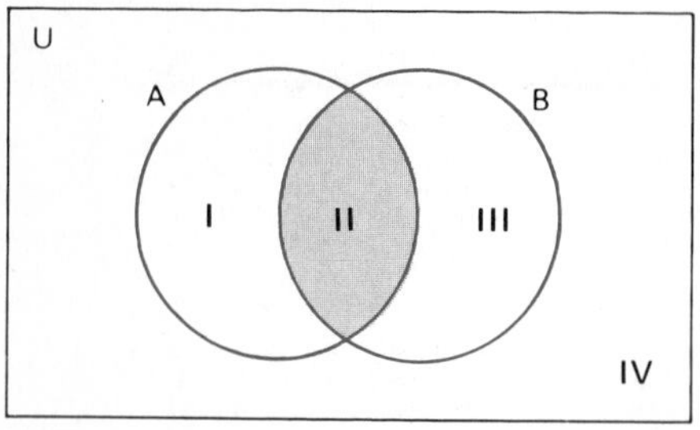

FIGURE 23

12. For each of the following diagrams, use set notation to describe the situation shown. For example, Figure 23 shows $A \cap B$.

a.

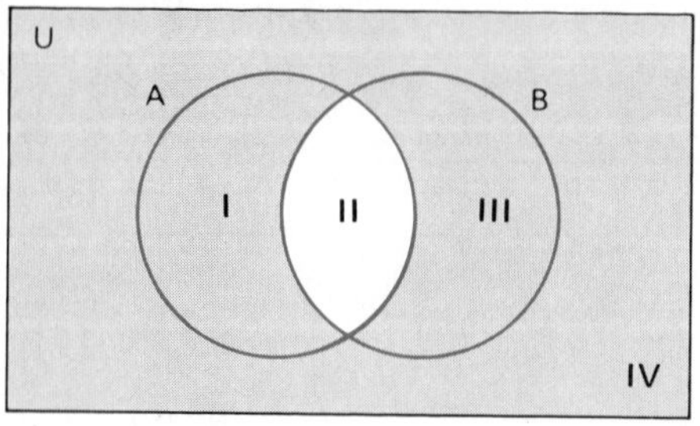

b.

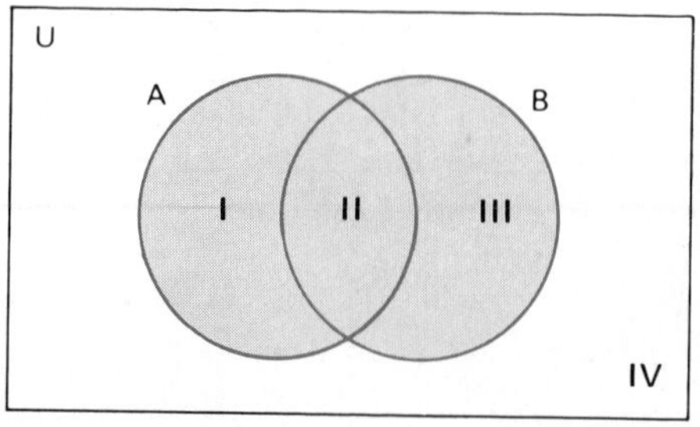

c.

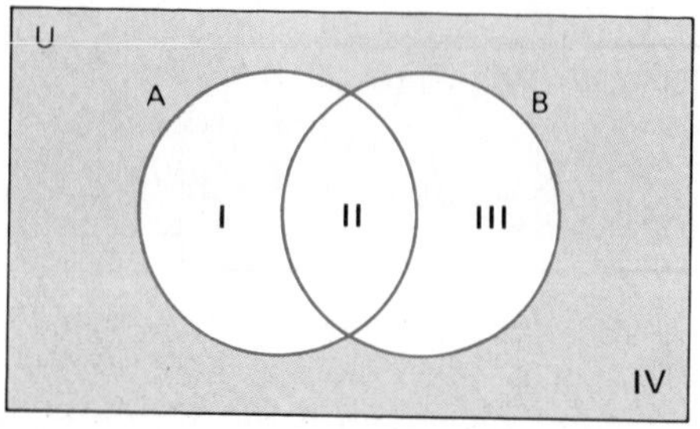

d.

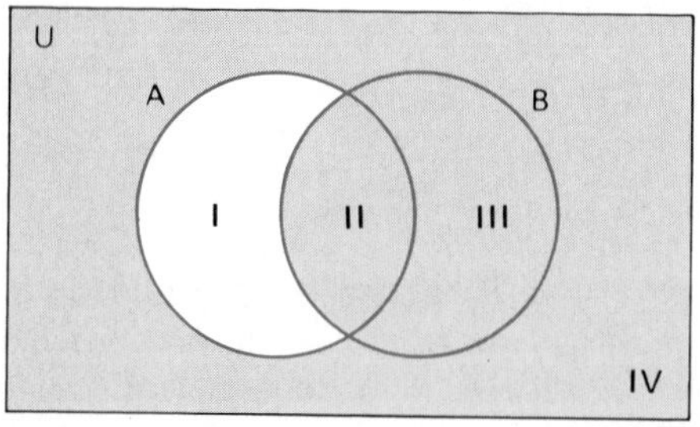

e.

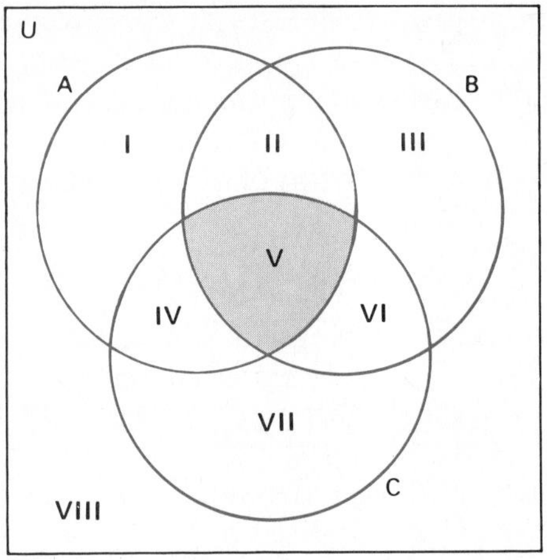

f.

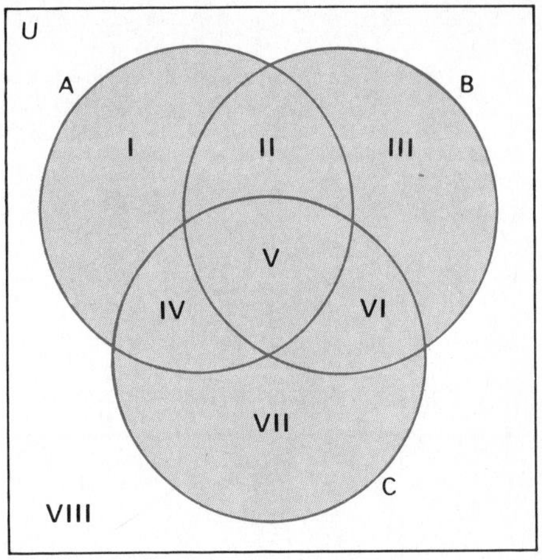

Reprinted by permission of UFS, Inc.

JUST FOR FUN

A Venn diagram with one set has two regions; a diagram with two sets has four regions; and one with three sets has eight regions. How many regions would a Venn diagram with four sets have? Can you draw such a diagram?

1.6 AN APPLICATION OF SETS AND VENN DIAGRAMS

Our knowledge of sets and Venn diagrams is particularly useful in solving problems involving overlapping sets of data on individuals. Consider Example 1.

EXAMPLE 1 In a certain group of 100 customers at Phil's Pizza Palace, 60 customers ordered cheese and pepperoni on their pizza. Altogether 80 customers ordered a pizza with cheese on it, and 72 customers ordered pizza with pepperoni on it.

a. How many customers ordered cheese on their pizza, but no pepperoni?
b. How many customers ordered pepperoni on their pizza, but no cheese?
c. How many customers in the group of 100 customers did not order either type of pizza?

SOLUTION Because we have two pizza toppings to consider, cheese and pepperoni, we can draw a Venn diagram illustrating these two sets. See Figure 24a. We label the sets C (cheese) and P (pepperoni) and also number the regions.

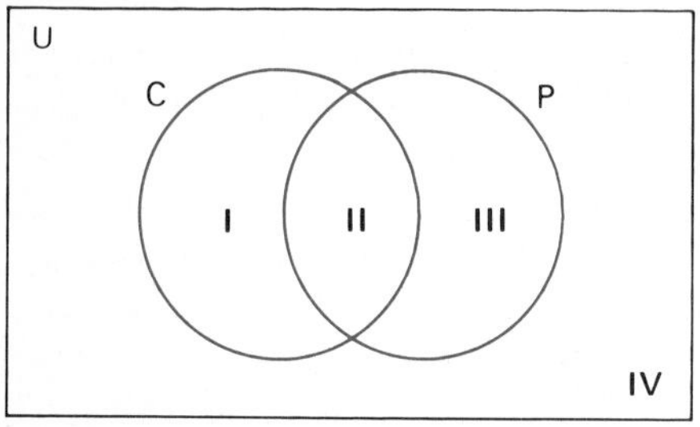

FIGURE 24a

There were 60 customers who ordered pizza with cheese *and* pepperoni. The *and* is our clue that these customers belong in the intersection of C and P, namely, region II. Altogether 80 customers ordered a pizza with cheese on it, but we cannot place 80 in region I. If we put 80 in region I, since we already have 60 in region II, that would give us 140 customers in set C—and there are only 100 customers total. The 60 customers who ordered cheese and pepperoni on their pizza are also counted as part of the 80 customers who ordered cheese on their pizza. If there are 80 customers in set C, and we already have 60 customers in region II (part of C), then we have 20 in region I.

Seventy-two customers are in set P, and we know 60 of them are in region II; hence we write 12 in region III. We now have the diagram in Figure 24b.

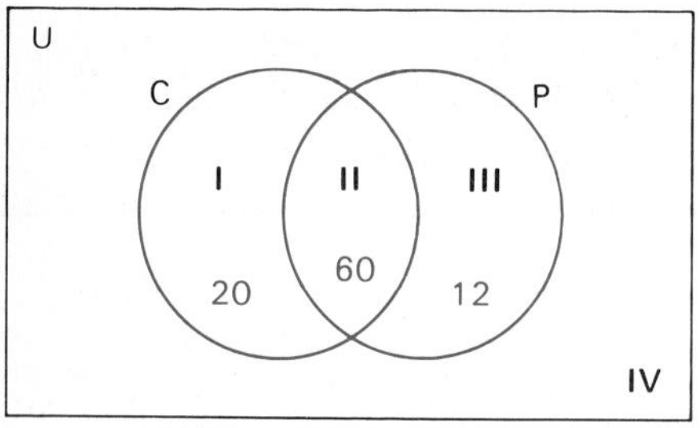

FIGURE 24b

Adding up the total number of people in regions I, II, and III, we see that there are only 92 customers in our diagram. We had a total of 100 customers in our universal set. Where are the other 8? As they are not contained in set P or set C, but are in the universal set, they must be in region IV. Therefore, we place 8 customers in region IV. The completed diagram is shown in Figure 24c.

Now that we have our completed diagram, we can answer the original questions.

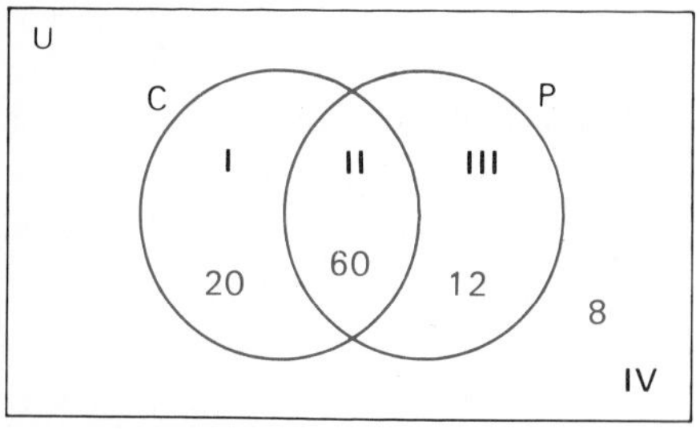

FIGURE 24c

a. How many of these customers ordered cheese on their pizza, but no pepperoni? These are the customers in region I and therefore the answer is 20.
b. How many of these customers ordered pepperoni on their pizza, but no cheese? These customers are in region III and the answer is 12.
c. How many customers in this group of 100 customers did not order either type of pizza? These customers are in region IV and we see that the answer is 8.

Note that in determining the numbers that went in the different regions of the diagram in Example 1, we started with the most specific piece of information. We first placed the 60 customers that were in the intersection of the two sets, and then proceeded with the completion of the diagram. Once a diagram is completed, we can examine it for information and answer the questions that are asked. But we must use care in entering the data in our Venn diagram. If the data are entered correctly, then we can answer the questions by reading directly from the Venn diagram.

EXAMPLE 2 In a certain group of 75 students, 16 students are taking psychology, geology, and English; 24 students are taking psychology and geology; 30 students are taking psychology and English; and 22 students are taking geology and English. However, 7 students are taking only psychology, 10 students are taking only geology, and 5 students are taking only English.

a. How many of these students are taking psychology?
b. How many of these students are taking psychology and English, but not geology?
c. How many students in this group are not taking any of the three subjects?

SOLUTION This problem involves overlapping sets of data, so we may represent the data by means of a Venn diagram. There are three subjects, so we will have three circles (see Fig. 25a). We label the sets *P* (psychology), *G* (geology), and *E* (English) and number the regions.

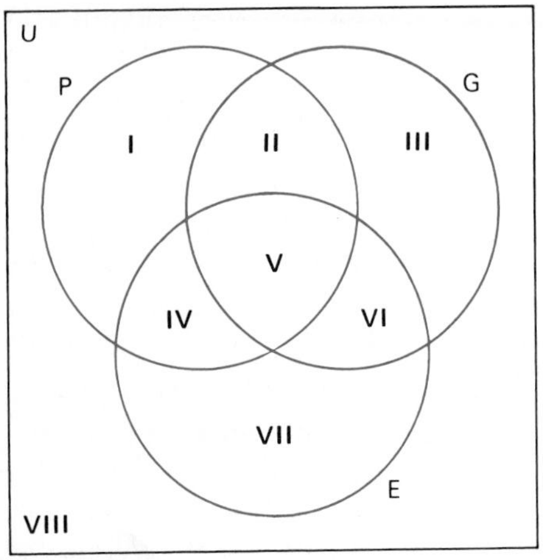

FIGURE 25a

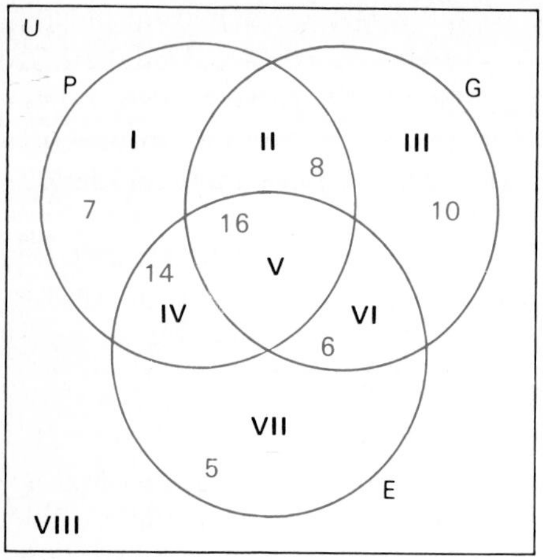

FIGURE 25b

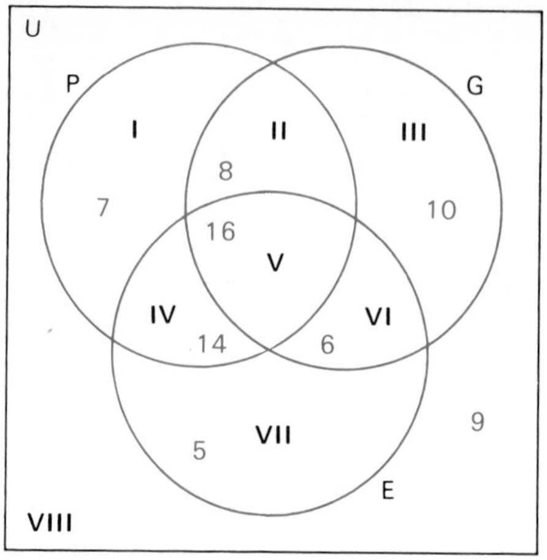

FIGURE 25c

It is best to start with the most specific piece of information and work backward from it. The most specific information we have is that 16 students are taking all three subjects (the intersection of all three sets.) These students would have to appear in region V. If 24 students are taking psychology and geology (the intersection of P and G), and we already know that 16 are in region V, then we need to place 8 of them in region II. Why not put 24 there? If we did that, then we would have 24 in region II and 16 in region V, which would give us a total of 40 students taking psychology and geology. That is not the case!

Since 30 students are in English and psychology, we place 14 in region IV. Similarly, we place 6 in region VI. Since 7 students are taking only psychology, they belong in region I; similarly, we place 10 students in region III and 5 students in region VII. We now have the diagram in Figure 25b.

Counting the total number of students in the circles, we get 66. But there are 75 students in our group and therefore we place 9 students in region VIII to complete our diagram. See Figure 25c.

Now we can readily answer the questions by reading directly from the Venn diagram.

a. How many of these students are taking psychology? These students are in set P, so we add the numbers of students in regions I, II, IV, and V. The answer is 45.
b. How many of these students are taking psychology and English, but not geology? These are the students in region IV, and the answer is 14.
c. How many students in this group are not taking any of the three subjects? These students are in region VIII, and the answer is 9.

EXAMPLE 3 At a meeting of 50 car dealers, the following information was obtained: 12 dealers sold Buicks, 15 dealers sold Oldsmobiles, 16 dealers sold Pontiacs, 4 dealers sold both Buicks and Oldsmobiles, 6 dealers sold both Oldsmobiles and Pontiacs, 5 dealers sold both Buicks and Pontiacs, and 1 dealer sold all three brands of cars.

a. How many dealers sold Buicks and neither of the other two brands?
b. How many of the dealers at the meeting did not sell any of these cars?

SOLUTION We may obtain answers to these questions by completing a Venn diagram and reading our answers directly from the figure (see Fig. 26). It is important to remember to start with the most specific piece of information, that which belongs in the intersection of all three sets (in this case, the one dealer who sold all three brands of cars). After using this information, we proceed to the data that belong in the intersection of two sets, and continue working backward.

In our diagram, we labeled the circles Buick, Oldsmobile, and Pontiac in that order, but we could have changed the order if we had so chosen. The resulting information would still be the same, and we would still be able to answer the questions.

Reading the information from the Venn diagram, we can answer the questions.

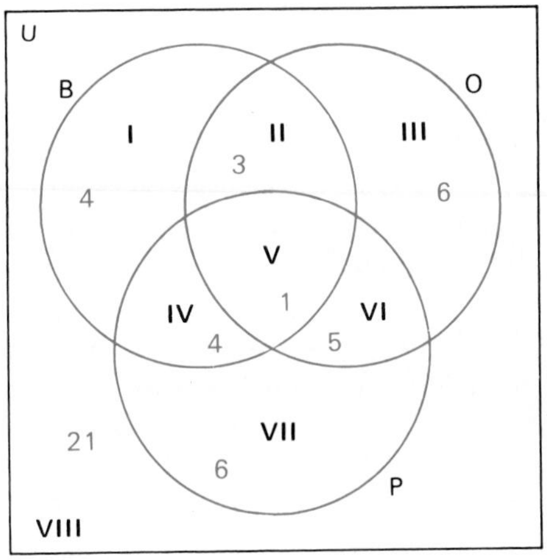

FIGURE 26
B = the set of Buicks, O = the set of Oldsmobiles, and P = the set of Pontiacs.

a. How many dealers sold Buicks and none of the other brands? These are the dealers in region I, and the answer is 4.
b. How many of the dealers at the meeting did not sell any of these cars? These are the dealers in region VIII; they are not members of any of the three sets, and the answer is 21.

EXERCISES FOR SECTION 1.6

1. Seventy-five students participated in a survey at a local college and the following data were collected: there were 27 students taking accounting, 26 taking psychology, and 41 taking statistics. Twelve students were taking accounting and psychology, 13 students were taking accounting and statistics, and 17 were taking psychology and statistics. Four students were taking all three courses.

   a. How many students were taking only psychology?
   b. How many students in the group were not taking any of the three subjects?
   c. How many students were taking accounting and statistics, but not psychology?
   d. How many students were taking just one of these subjects?

2. In a certain Las Vegas casino, a survey of 125 gamblers was taken and the following data were collected: 71 gamblers played roulette, 72 gamblers played poker, and 80 gamblers played blackjack, while 33 gamblers played roulette and poker, 42 gamblers played roulette and blackjack, and 47 gamblers played poker and blackjack. Eleven gamblers played all three games.
   a. How many of these gamblers played only blackjack?
   b. How many of these gamblers played poker and blackjack, but not roulette?
   c. How many gamblers in this group did not play any of these three games?
   d. How many gamblers did not play poker?

3. In a survey of 78 residents in a certain dormitory, the following data were collected: 35 residents had radios in their rooms, 33 residents had televisions in their rooms, and 31 residents had tape players in their rooms. Sixteen residents had both tape players and televisions in their rooms, 14 residents had televisions and radios in their rooms, and 14 residents had radios and tape players in their rooms. Five residents had radios, televisions, and tape players in their rooms.
   a. How many residents had none of these items in their rooms?
   b. How many residents did not have televisions in their room?
   c. How many residents had only radios in their rooms?
   d. How many residents did not have radios in their rooms?

4. In analyzing the scoring for college football teams in a particular conference, the following facts were gathered: 70 players had scored touchdowns, 44 players had scored points after touchdowns (PATs), and 32 players had scored field goals, while 19 players had scored both touchdowns and PATs, 16 players had scored touchdowns and field goals, and 21 players had scored both PATs and field goals. Six players had scored in all three ways.
   a. How many players scored only touchdowns?
   b. How many players scored field goals and PATs but not touchdowns?
   c. How many players scored touchdowns or field goals?
   d. How many players did not score a PAT?

5. At a subway stop in New York City, 125 people were asked what newspaper they read. Forty-six people read the *Times,* 43 people read the *Post,* and 65 people read the *News.* Nineteen people read the *Times* and the *Post,* 18 people read the *Post* and the *News,* and 11 people read the *Times* and the *News.* Seven people read all three papers.
   a. How many people read only the *News?*
   b. How many people did not read any of the papers?
   c. How many people read the *Times* or the *News?*
   d. How many people read the *Post* and the *News,* but not the *Times?*

6. At a local bank, 100 customers were asked what bank services they used. Fifty-two people in the group had savings accounts, 52 people had checking accounts, and 57 people had bank charge cards, while 23 people had savings and checking accounts, 25 people had checking accounts and bank charge cards, and 24 people had savings accounts and bank charge cards. Eleven people used all three types of services—checking accounts, savings accounts, and charge cards.
   a. How many of these people had only checking accounts?
   b. How many of these people did not have savings accounts?
   c. How many of these people did not have bank charge cards?
   d. How many of these people had savings accounts or checking accounts?

7. In the college union, 130 students were surveyed as to what board games they played. The following data were collected: 57 played backgammon, 89 played checkers, and 76 played chess, while 35 played checkers and backgammon, 50 played chess and checkers, and 30 played backgammon and chess. Twenty students played all three games.
   a. How many of these students played only chess?

b. How many of these students did not play chess?
c. How many students in this group played at least one of these games?
d. How many students in this group played backgammon or checkers?

8. At a clambake, a survey of 100 people was taken to determine their preferences in shellfish. Sixty-eight people in the group liked lobsters, 60 people liked clams, and 56 people liked scallops, while 36 people liked lobsters and clams, 38 people liked clams and scallops, and 37 people liked lobsters and scallops. Twenty-four people liked all three kinds of shellfish.
a. How many of the people liked only scallops?
b. How many of the people liked lobsters and clams, but not scallops?
c. How many of the people liked lobsters or clams?
d. How many of the people did not like clams?

9. Many fish markets also sell clams. Usually these stores sell three different sizes of clams: cherrystones, littlenecks, and chowders. During a certain week, the following data were collected regarding the sale of these size clams: 40 customers purchased cherrystone clams, 47 customers purchased littleneck clams, and 32 customers purchased chowder clams, while 18 customers purchased cherrystones and littlenecks, 14 customers purchased littlenecks and chowders, and 9 customers purchased cherrystones and chowders. Four customers purchased all three types of clams.
a. How many customers purchased clams during the week?
b. How many customers purchased only littlenecks?
c. How many customers purchased cherrystones or littlenecks?
d. How many customers purchased only one type of clam?

10. An advertising agency conducted a survey of 50 retail outlets and the following data were collected: 24 merchants advertised on radio, 20 merchants advertised on television, and 27 merchants advertised in the newspapers. Eleven merchants advertised on radio and television, 10 merchants advertised on television and in the newspapers, and 9 merchants advertised on radio and in the newspapers. Four merchants advertised through all three media.
a. How many merchants advertised only on television?
b. How many merchants did not use any of these media?
c. How many merchants advertised on radio or television?
d. How many merchants advertised on radio and television, but not in the newspapers?

11. A used-car dealer must complete an inventory of the cars on his lot. He notes that he has 22 compact, two-door, standard-transmission cars. Of the 50 standard-transmission cars on the lot, 28 are classified as compact, while 30 are two-door. Also, of the 47 two-door cars on the lot, 31 are classified as compact. The dealer also notes that he has 44 compact cars on his lot and 15 large, four-door, automatic-transmission cars.
a. How many cars are there on the lot altogether?
b. How many compact cars have standard transmission, but are not the two-door type?
c. How many of the two-door cars with standard transmission are not compact?
d. How many of the standard-transmission cars are not compact?

12. In a recent survey of 300 people regarding television programming, the following information was gathered: 160 people watched ABC, 150 people watched CBS, and 150 people watched NBC, while 90 people watched both ABC and CBS, 70 people watched CBS and NBC, and 100 people watched ABC and NBC. Forty people watched all three networks.
a. How many people watched ABC or NBC?
b. How many people watched only one of the networks?
c. How many people did not watch any of the networks?
d. How many people did not watch NBC?

13. A statistician reported to his employer that he had gathered the following information: in a survey of 40 households in a certain tract, 36 households had a video cassette recorder (VCR), 36 had two cars, and 21 owned a camper, while 22 households had both a VCR and two cars, 19 had a VCR and a camper, and 17 had a camper and two cars. Six households had a VCR, two cars, and a camper. The statistician was promptly fired by his employer. Why?

14. An independent survey agency was hired by the Metropolitan Transit Authority (MTA) to find out how people commuted to their jobs. The agency interviewed 1,000 commuters and submitted the following report: 631 people came to work by car, 554 people came to work by bus, and 759 came to work by subway. Also, 373 people came to work by a combination of car and bus, 301 people came to work by bus and subway, and 268 people came to work by car and subway, while 231 people used all three means of transportation to get to work. The MTA refused to accept the report, stating that it was inaccurate. Why?

JUST FOR FUN Here is a problem about a set of people. See if you can solve it. An accountant, attorney, architect, and author all belong to the same club. Their names, although not necessarily in order, are Jack, Joe, Sue, and Sharon. The following is known about these people: Jack and the attorney are not on speaking terms with Sue. Joe and the author are good friends. Sue and the architect live in the same apartment complex. The accountant is good friends with Sharon and the author. Given this much information, determine the profession of each person.

1.7 CARTESIAN PRODUCTS

As the last set operation in this chapter, we shall consider the *Cartesian product* of two sets (named after René Descartes, a seventeenth-century French mathematician and philosopher). The Cartesian product is unique among set operations in that it produces new elements that are not members of the universal set. These new elements are called *ordered pairs*.

An **ordered pair** is a pair of objects where one element is considered first and the other element is considered second. Which element is first and which element is second is important. If we have a pair of socks, it doesn't matter which sock we put on first. A pair of socks is *not* an ordered pair. But if we have a sock and a shoe, it does matter which we put on first: try putting a sock on over a shoe!

In mathematics, if we wish to discuss an ordered pair of elements consisting of a and b, we use the notation (a, b). This notation tells us that we want to consider a first and b second. Suppose a stands for "walk a blocks north" and b stands for "walk b blocks west." Now consider the ordered pair (3, 2). This would mean "walk 3 blocks north and then walk 2 blocks west." This is certainly different from (2, 3), which would mean "walk 2 blocks north and then walk 3 blocks west." We would not arrive at the same place with these two ordered pairs.

Given two sets A and B, the Cartesian product of A and B, denoted by $A \times B$ (read "A cross B"), is the set of all possible ordered pairs such that the first element of the ordered pair is an element of A and the second element of the ordered pair is an element of B.

EXAMPLE 1 Given $A = \{\text{Joe, Scott}\}$ and $B = \{p, q\}$, find $A \times B$, the Cartesian product of A and B.

SOLUTION The Cartesian product $A \times B$ is the set of all possible ordered pairs such that the first element is an element of A and the second element is an element of B. We must pair the elements in A, Joe and Scott, with the elements in B, p and q. Joe may be paired with p or q; hence, the possible ordered pairs with Joe are (Joe, p), (Joe, q). Now we do the same thing for Scott; the possible ordered pairs are (Scott, p) and (Scott, q). Therefore we have

$$A \times B = \{(\text{Joe}, p), (\text{Joe}, q), (\text{Scott}, p), (\text{Scott}, q)\}$$

Note that $A \times B$ gives us a set where the elements are ordered pairs. These are not members of the universal set, since the universal set consists of single elements such as p, q, Joe, and Scott.

EXAMPLE 2 Given $A = \{4, 8\}$ and $B = \{a, b, c\}$, find $A \times B$.

SOLUTION The first element in our ordered pairs must come from A. Therefore we pair 4 with a, then 4 with b, etc. Then we do the same for 8. Hence, we have

$$A \times B = \{(4, a), (4, b), (4, c), (8, a), (8, b), (8, c)\}$$

EXAMPLE 3 Given $A = \{4, 8\}$ and $B = \{a, b, c\}$, find $B \times A$.

SOLUTION This is similar to the last example, but here we want to find $B \times A$. Remember $B \times A$ is not the same as $A \times B$. In $B \times A$, we want the set of all possible ordered pairs such that the first element is an element of B and the second element is an element of A. Here we take the elements in B, a, b, and c, and pair them with the elements in A, 4 and 8. Therefore,

$$B \times A = \{(a, 4), (a, 8), (b, 4), (b, 8), (c, 4), (c, 8)\}$$

By comparing the results of Examples 2 and 3, we see that $A \times B \neq B \times A$. The two sets contain different ordered pairs, as $(4, a)$ is not the same as $(a, 4)$.

Note that in Example 1 the number of elements in $A \times B$ was 4—that is, $n(A \times B) = 4$—and in Example 2, $n(A \times B) = 6$. If we want to determine the number of elements in a Cartesian product $A \times B$ (the cardinality of the set), we take the number of elements in A and multiply it by the number of elements in B. If set A has m elements in it and set B has n elements in it, then the number of elements in $A \times B$ is $m \times n$. In other words,

$$n(A \times B) = n(A) \times n(B)$$

This provides us with a handy check in computing $A \times B$ because we can use the cardinality to check to see if we have all of the possible ordered pairs.

EXAMPLE 4 Find $A \times A$ if $A = \{1, 2, 3\}$.

SOLUTION To find $A \times A$, we pair each element in A with every element in A. Therefore we have

$$A \times A = \{(1, 1), (1, 2), (1, 3), (2, 1), (2, 2), (2, 3), (3, 1), (3, 2), (3, 3)\}$$

Since there are 3 elements in A, we should have $3 \times 3 = 9$ elements in $A \times A$. Checking our answer, we see that we do have 9 different ordered pairs in $A \times A$.

The Cartesian product of two sets, $A \times B$, gives us a set of ordered pairs. This set of ordered pairs may be pictured by means of an **array** or **lattice.** Consider the following example.

EXAMPLE 5 Find $A \times B$ if $A = \{a, b, c\}$ and $B = \{d, e\}$.

SOLUTION There are 3 elements in A and 2 elements in B; hence there are $3 \times 2 = 6$ ordered pairs in the Cartesian product $A \times B$. Pairing each element of A with every element in B, we have

$$A \times B = \{(a, d), (a, e), (b, d), (b, e), (c, d), (c, e)\}$$

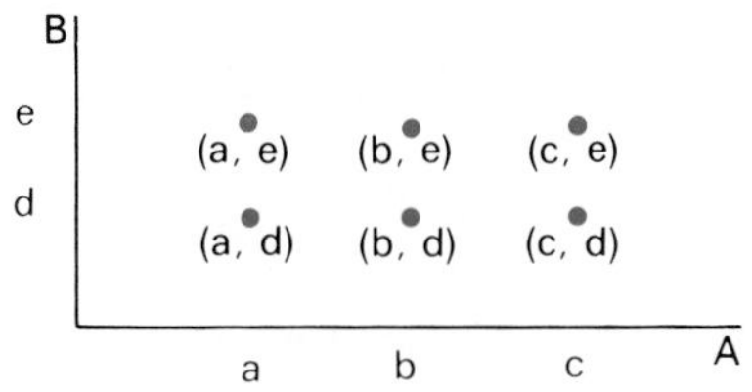

FIGURE 27

Figure 27 shows the array of ordered pairs for this example. In the lattice, a dot represents an ordered pair.

Note that the vertical axis represents set B with the elements d and e. The horizontal axis represents set A with elements a, b, and c. It is traditional to use the horizontal axis to represent the first element in an ordered pair.

One other way of picturing the formation of a Cartesian product is a *tree diagram*. A **tree diagram** consists of a number of branches that illustrate the possible pairings in $A \times B$, as in Figure 28.

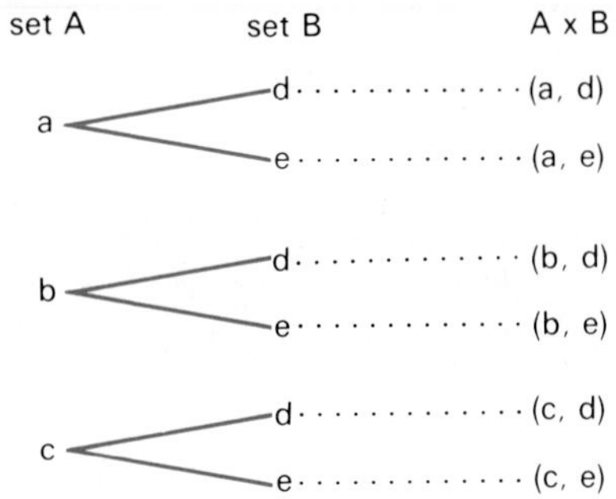

FIGURE 28

EXERCISES FOR SECTION 1.7

1. If $A = \{a, b, c\}$ and $B = \{10, 20\}$, find $A \times B$, $B \times A$, and $n(A \times B)$.

2. If $A = \{?, \&, !\}$ and $B = \{a, b, c\}$, find $A \times B$, $B \times A$, and $n(A \times B)$.

3. If $C = \{2, 4, 6\}$ and $D = \{1, 3, 5\}$, find $C \times D$, $D \times C$, and $n(C \times D)$.

4. If $V = \{a, e, i, o, u\}$ and $Z = \{x, y, z\}$, find $V \times Z$, $Z \times V$, and $n(V \times Z)$.

5. Let $T = \{t, f\}$. Find $T \times T$ and $n(T \times T)$.

6. Let $A = \{c, d, e\}$ and $B = \{1, 2, 3\}$. Make a lattice showing $A \times B$.

7. Let $A = \{4, 5, 6\}$ and $B = \{x, y\}$. Make a lattice showing $A \times B$.

8. Let $A = \{a, b, c, d\}$ and $B = \{x, y, z\}$. Make a tree diagram showing $B \times A$.

9. If $A = \{x, y\}$ and $B = \{a, e, i, o, u\}$, make a tree diagram showing $A \times B$.

10. If $U = \{1, 2, 3, 4\}$, find $U \times U$ and $n(U \times U)$.

11. Find each of the following, given $A = \{a, b\}$, $B = \{b, c, d\}$, and $C = \{c, d, e\}$.
 a. $A \times B$ b. $A \times C$
 c. $n(B \times C)$ d. $(A \cap B) \times C$
 e. $C \times (A \cup B)$ f. $A \times (B \cap C)$

12. If $A = \{1, 2, 3\}$, $B = \{3, 4, 5, 6\}$, and $C = \{4, 5, 6, 7\}$, find each of the following:
 a. $A \times B$ b. $n(B \times C)$
 c. $(A \cap C) \times B$ d. $(B \cup C) \times A$
 e. $(B \cap C) \times A$ f. $(A \cap B) \times (B \cap C)$

13. Given that $A = \{1, 2\}$ and $B = \{3, 4\}$, does $A \times B = B \times A$? Why or why not?

JUST FOR FUN Given: $142857 \times 2 = 285714$, $142857 \times 3 = 428571$, $142857 \times 4 = 571428$, $142857 \times 5 = 714285$, $142857 \times 6 = 857142$, then $142857 \times 7 = ?$

Summary

Given any two sets A and B, A is a *subset* of B if every element in A is also an element in B. If A is a subset of B and there is at least one element in B that is not contained in A, then A is a *proper subset* of B. The *intersection* of A and B, $A \cap B$, is a set of elements that are elements of both A and B. The *union* of set A and B, $A \cup B$, is a set of all elements that are elements of A or B or both. The *complement* of set A, A', is the set of all the elements in the given universal set, U, that are not in set A.

Equal sets are sets that contain exactly the same elements, whereas *equivalent sets* are sets that have the same number of elements, that is, the same cardinality. If two sets are equivalent, then their elements may be placed in a *one-to-one correspondence*.

Venn diagrams are picture representations of sets. The universal set is represented by a rectangle, and circles inside the rectangle represent the sets being considered. In a Venn diagram, we can show the sets overlapping one another, so that all possible combinations may be considered. Set operations, such as $A \cap B$, are represented by the intersection of the circles in a Venn diagram. We can determine if different set expressions, such as $(A \cap B)'$ and $A' \cup B'$, are equivalent by examining the corresponding Venn diagrams to see if both expressions determine the same regions.

Venn diagrams may also be used to answer questions when we are given certain data and wish to determine more information from the data. A typical example is the survey type of equation discussed in Section 1.6.

The Cartesian product of sets A and B, denoted by $A \times B$, is the set of all possible ordered pairs such that the first element of the ordered pair is an element of A and the second element of the ordered pair is an element of B.

Vocabulary Check

set	union	equal sets	whole number	proper subset	complement
element	cardinality	equivalent sets	disjoint sets	Venn diagram	intersection
finite	infinite	Cartesian product	subset	universal set	natural number

Review Exercises for Chapter 1

1. Describe in your own words what a *set* is.

2. Write each set in two ways.
   a. The set of states whose names begin with the letter T.
   b. The set of Great Lakes.
   c. All even whole numbers.
   d. All positive whole numbers.
   e. All counting numbers less than 8.

3. State whether each sentence is true or false.
   a. The set of good instructors is a well-defined set.
   b. $\{1, 2, 3, \ldots\}$ is a well-defined set.
   c. $\{2, 4, 6, 8, \ldots\}$ is not a well-defined set.
   d. The set of books in a bookstore is a finite set.
   e. $\{a, b, c, \ldots, z\}$ is an infinite set.
   f. $\varnothing$ is a finite set.

4. State whether each sentence is true or false.
   a. $\{t, e, a, m\} = \{m, e, a, t\}$
   b. $\varnothing = \{\ \}$
   c. $\{l, o, v, e\}$ is equivalent to $\{h, a, t, e\}$.
   d. $\{1, 2, 3, \ldots\}$ is equivalent to $\{a, b, c, \ldots, z\}$.
   e. If $A = \{1, 3, 5, \ldots\}$ and $B = \{2, 4, 6, \ldots\}$, then sets A and B are disjoint sets.
   f. $\{m, a, t, h, i, s, f, u, n\} \subset \{a, b, c, \ldots, z\}$

5. State whether each of the following is true or false.
   a. $\{1, 7\} \subset \{1, 2, 3, \ldots\}$
   b. $\{a, b\} \subseteq \{a, b, c\}$
   c. $\{1, 3, 5, \ldots\} \subset \{1, 2, 3, \ldots\}$
   d. $\varnothing \subseteq \{a, b\}$
   e. Disjoint sets have at least one element in common.
   f. The set $\{m, a, t, h\}$ has 32 possible subsets.

6. List all the possible subsets of $\{i, o, u\}$.

7. Let $U = \{0, 1, 2, 3, 4, 5\}$, $A = \{0, 1, 3, 5\}$, $B = \{1, 3, 4, 5\}$, and $C = \{0, 2, 4, 5\}$. Find each of the following.
   a. $A \cap B$
   b. $B \cup C$
   c. $A' \cap B'$
   d. $B' \cup C'$
   e. $(A \cap B) \cup C$
   f. $A \cup (B \cap C)$
   g. $(A' \cap B')' \cup C$
   h. $(A \cap B)' \cup C'$

JUST FOR FUN

Mr. Apple and Mr. Jack have an 8-gallon container full of cider. They also have two empty containers whose capacities are 5 gallons and 3 gallons, respectively. How can cider be poured, using all three containers, so that Mr. Apple and Mr. Jack each have 4 gallons of cider?

8. Use a Venn diagram to illustrate each of the following sets. List the regions that make up your answer.
 a. $A \cup B$ b. $(A \cap B)'$
 c. $A' \cap B'$ d. $A \cap (B \cup C)$
 e. $A \cup (B \cap C)$ f. $(A \cap C) \cup B'$

9. Show a one-to-one correspondence between $A = \{1, 2, 3\}$ and $B = \{5, 10, 15\}$. In how many ways can you do this?

10. Use Figure 29 to find each cardinality.
 a. $n(A)$ b. $n(B)$
 c. $n(C')$ d. $n(A \cap B)$
 e. $n(A \cup C)$ f. $n(B \cup C)$
 g. $n(U)$ h. $n(A' \cap B)$

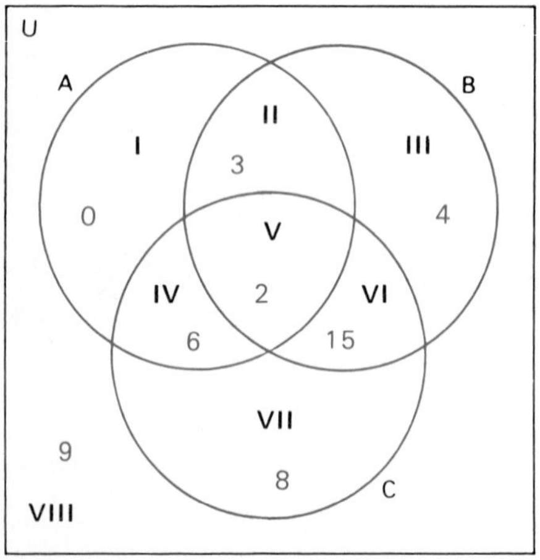

FIGURE 29

11. In a survey of 77 students at a local college, the following information was gathered: 45 students had a class at eight o'clock in the morning, 43 students had a class at nine o'clock in the morning, 47 students had a class at 10 o'clock in the morning. Twenty-six students had an 8:00 A.M. and a 9:00 A.M. class, 24 students had an 8:00 A.M. and a 10:00 A.M. class, and 25 students had a 9:00 A.M. and a 10:00 A.M. class. Eleven students had classes at 8:00, 9:00, and 10:00 A.M.
 a. How many of the students did not have a class at any of the three times?
 b. How many students had a class at 9:00 A.M. and 10:00 A.M., but not at 8:00 A.M.?
 c. How many students did not have a class at 8:00 A.M.?
 d. How many students had a class at 8:00 A.M. or 10:00 A.M.?

12. Let $A = \{m, a, t, h\}$ and $B = \{e, a, s, y\}$.
 a. Show a one-to-one correspondence between A and B.
 b. In how many ways can you do this?
 c. List all possible subsets of A.
 d. Find $A \times B$.
 e. Find $n(A \times B)$.

13. When are the following statements true?
 a. $A \cap B = A$ b. $A \cup B = A \cap B$
 c. $(A \cap B)' = A' \cup B'$ d. $B \subseteq \varnothing$
 e. $(A \cup B)' = A' \cap B'$

Chapter Quiz

State whether each of the following is true or false.

1. $\{m, o, n, e, y\} = \{r, i, c, h\}$
2. $\{a, b, c\}$ is equivalent to $\{x, y, z\}$.
3. $\{a, b, c\} \subseteq \{c, a, b\}$
4. $\{0\} = \varnothing$
5. $44 \in \{1, 3, 5, \ldots\}$
6. $44 \notin \{x \mid x \text{ is a natural number}\}$
7. $44 \in \{x \mid x \text{ is a whole number}\}$
8. $A \times B = B \times A$
9. If $A = \{a, b, c, d\}$, then A has at most eight subsets.
10. If A and B are disjoint sets, then A and B have only one element in common.
11. If A and B are any two disjoint sets, then it is impossible to show a one-to-one correspondence between them.

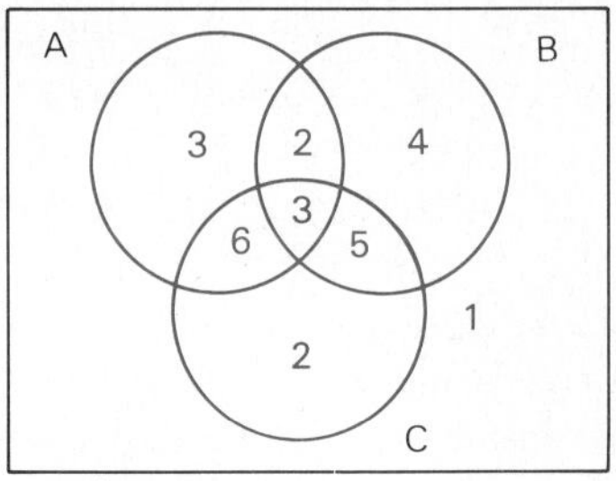

FIGURE 30

Use Figure 30 to answer questions 12–19.

12. $n(A) = n(B)$
13. $n(B \cup C) = 22$
14. $n(A' \cap B') = 2$
15. $n(B \cap C)' = 14$
16. $n(A \cup C) = 21$
17. $n(C') = 11$
18. $n(A \cap B)' = 21$
19. $n(A' \cup C') = 15$

For questions 20–25 use the following information to answer the questions.

$U = \{u, l, i, k, e, m, a, t, h\}$, $A = \{l, i, k, m, t\}$
$B = \{u, i, e, a, h\}$, $C = \{l, i, m, a\}$

20. $B \cap C = \{i, m, a\}$
21. $A \cup B \cup C = U$
22. $A' \cap B' = \{i\}$
23. $(A \cap B)' = A' \cup B'$
24. $C' \cup (A \cap B) = \{u, i, k, e, m, t, h\}$
25. $A' \cup B' = \{u, e, a, h, l, k, m, t\}$

HISTORICAL NOTE

René Descartes (1596–1650) was a French mathematician-philosopher who revolutionized mathematical concepts and initiated modern mathematics with his creation of coordinate (plane analytic) geometry. Descartes also developed views of philosophy that were revolutionary in their break with the past and caused him to be known as the father of modern philosophy.

As a young boy he was sent to a Jesuit school. It was there that he developed (because of delicate health) his lifelong habit of lying in bed as late as he pleased. As an adult he slept ten hours each night and never let anyone disturb him before noon.

Descartes lived in various countries in Europe before settling in Holland, where he pursued his mathematical studies and philosophical contemplations. In 1637 he published a treatise outlining his views on philosophy. It was in the appendix of this book that he introduced analytic geometry, and the Cartesian plane, for which he is famous today.

According to legend, Descartes was led to the contemplation of analytic geometry while watching a fly crawl about the ceiling near a corner of his room. He noted that the path of the fly could be described, if he knew a relation connecting the fly's distance from the two adjacent walls. The technical terms *coordinates*, *abscissa*, and *ordinate* used in analytic geometry today were contributed by Liebniz in 1692.

CHAPTER 2

LOGIC

AFTER STUDYING THIS CHAPTER YOU WILL BE ABLE TO DO THE FOLLOWING:

1. Distinguish between **simple statements** and **compound statements,** and identify a compound statement as a **negation, conjunction, disjunction, conditional,** or **biconditional.**
2. Write compound verbal statements in **symbolic form** by using letters for the simple statements and the proper logical connectives, and write symbolic statements as English sentences.
3. Construct a **truth table** for a given compound statement containing up to three different variables.
4. Determine whether two statements are **logically equivalent** by means of truth tables or De Morgan's law.
5. Determine whether an argument is **valid** or **invalid** by means of truth tables using up to three variables.
6. Use Venn diagrams to determine whether two statements are **consistent** or **inconsistent,** and also use them to determine whether an argument is valid or invalid.
7. Identify the various forms of the **conditional statement—converse, inverse,** and **contrapositive**—and write a conditional statement in any of these forms.

★ 8. Symbolize verbal **conditional** and **biconditional statements** stated in the only if, necessary, sufficient, necessary and sufficient, and if and only if forms.

★ 9. Identify **series** and **parallel connections** in a switching network.

★10. Write a symbolic statement for a given network and construct a **switching network** for a given symbolic statement.

(*Note:* ★ indicates optional material.)

NOTATION FREQUENTLY USED IN THIS CHAPTER

$P \wedge Q$	P and Q	$\sim P$	*not* P; it is false that P; it is not the case that P
$P \vee Q$	P or Q	iff	if and only if
$P \veebar Q$	P or Q, but not both	$\equiv$	is the same as
$P \rightarrow Q$	if P, then Q		
$P \leftrightarrow Q$	P if and only if Q		

2.1 INTRODUCTION

What is logic? It is often defined as the science of thinking and reasoning correctly. Many times false assumptions are made about things or people because some of us misinterpret statements. The reader or listener understands something other than what the writer or speaker has written or said. An understanding of logic and its uses will help us to avoid these pitfalls and increase our skills in analytical thinking.

We should always phrase our statements so that they express our meaning exactly. An old proverb states, "Say what you mean and mean what you say." All too often statements are misconstrued, sometimes deliberately. The following are some examples of statements that have hidden implications or may be misinterpreted:

It's good to see you here on time!
I am glad to see you sober today!
Four dentists out of five said they used "Smiley" toothpaste.
One thousand heroin addicts admitted that they had used marijuana.

Symbolic logic enables us to make some complex problems manageable. The symbols in logic, like those in other areas of mathematics (algebra, for example), are important aids to our thinking. Many large corporations, such as insurance companies, have begun to use symbolic analysis to check contracts and policies for loopholes and inconsistencies. But the major use of logic is still in mathematics, where it has had a powerful influence on the development of ideas.

Logic was first studied extensively by Aristotle in approximately 400 B.C., which means traditional logic is over 2,000 years old. Some other famous contributors to the development of logic were Leonhard Euler (1707–1783), George Boole (1815–1864), John Venn (1834–1923), and Bertrand Russell (1872–1970). These dates show that modern logic is relatively young; it is only a little over 200 years old.

HISTORICAL NOTE

It was not until the early twentieth century that logic came to be considered part of the study of mathematics.

George Boole (1815–1864), an English mathematician, was primarily responsible for this acceptance. His book *Laws of Thought* developed logic as an abstract mathematical system. It consisted of undefined terms, binary operations, and rules for using these operations on the terms. Today, we know these terms as **propositions**; the operations are the **conjunction, disjunction,** and **negation.**

Using this idea, he was able to use logic as a special kind of algebra. The system follows many of the rules of ordinary arithmetic. Today Boolean algebra is used extensively in the design of computers.

The advantage of Boole's symbolic notation is that many errors in reasoning are greatly reduced. The ambiguities of language are avoided by the use of symbols because once a problem is translated into symbolic notation, the solution becomes mostly mechanical.

2.2 STATEMENTS AND SYMBOLS

In preparing for any task, we must first equip ourselves with the proper tools and the "know-how" required to do the job. The first thing that we must be able to do is identify and symbolize sentences. In logic we concern ourselves only with those sentences that are either true or false, but not both. A **statement** is a declarative sentence that is either true or false (but not both true and false).* We shall not concern ourselves with sentences that cannot be assigned a true or false value. (Sentences of this nature are usually questions or commands.)

Note that it is not possible to assign a true or false value to the following.

Did you do the assignment?
Hand in your paper.
Is it raining?
Close the door when you leave.
Stop the car!

The following statements are either true or false.

February has 30 days.
$4 + 2 = 3 \times 2$
Ronald Reagan was President of the United States.
Ottawa is the capital of Canada.
Tomorrow is Saturday.

There are other types of sentences that cannot be assigned a true or false value. The sentence "I am lying to you" is one example. Suppose it is true that I am lying to you; then—if I am lying—the sentence is false. On the other hand, assume that the sentence is false. If that is the case, then I am not lying, so the sentence is true. This is known as a **paradox.**

Another example of a paradox is "All rules have exceptions." This rule negates itself. It says that the rule itself must have exceptions and therefore cannot be true. Many people like the paradox about the little boy who is concerned about God. He has been told that God can do anything. The boy then asks, "If that is the case, then can God make a stone so big that he can't move it?"

Remember that in logic we concern ourselves with statements, that is, sentences that are true or false, but not both. The basic type of statement in logic is called a *simple* statement. A **simple statement** is a complete sentence that conveys one thought with no connecting words. The following are examples of simple statements.

Five is a counting number.
The Raiders have won the Super Bowl.
Sally was late for class.
Today is Monday.
Ronald Reagan won the 1984 presidential election.

* Many texts distinguish between a *proposition* and a *statement.* A **proposition** is defined as a statement that is either true or false, but not both. We shall use both terms interchangeably, as this is the only type of statement we shall consider.

NOTE OF INTEREST

Consider the following

> In a certain village there is a man who is a barber; the barber shaves all and only those men in the village who do not shave themselves.

The question is, does the barber shave himself?

Note that any man in the village is shaved by the barber if and only if he does not shave himself. Can you see the problem? The barber shaves himself if and only if he does not. Hence, we have another example of a paradox.

Now, if we take simple statements and put them together using connecting words, we form sentences that are known as **compound** or **complex** statements. The basic connectives are *and, or, if . . . then . . ., if and only if,* and the negation *not.*

The word *not* does not connect two simple statements, but it is still thought of as a connective. It negates a simple statement. Some logicians do not like to call a negated simple statement a compound statement, but if it is no longer simple, then it must be compound. Therefore we shall think of a compound statement as a sentence that is formed by connecting one or more simple statements with a connective.

A simple statement such as "Today is Monday" is no longer simple if we say "Today is not Monday" or "It is false that today is Monday." The original simple statement has been negated, so we call the newly formed compound statement a **negation.**

When we connect two simple statements using the word *and,* we have a compound statement that is called a **conjunction.** The sentence "Today is Monday *and* tomorrow is Wednesday" is a conjunction. Remember that we are not concerned about the meaning of the sentence, only what type of statement it is. Consider

All looms are booms and all booms are zooms.

This statement is a conjunction, even though we cannot make too much sense out of it.

Sometimes the word *but* will be used in place of *and* in a sentence.

Bonnie was early and Clyde was late.

could be written as

Bonnie was early, but Clyde was late.

The connective *or* is used in forming a compound statement called a **disjunction.** The following are some examples of disjunctions.

Either he took my coat or someone stole it.
I will pass history or I will be sad.
Today will be sunny or the weather forecast is wrong.

The connective *if . . . then . . .* is used in compound statements referred to as **conditionals.** An example of a conditional is

If you do your homework, then you will pass the exam.

The statement between the *if* and *then* ("you do your homework") is called the **antecedent** of the conditional. The part of the sentence that follows *then* ("you will pass the exam") is called the

consequent. As with other connectives, there are variations in writing conditional statements. Two of the more common variations are illustrated by the following examples.

If someone was late, it was Benny.
We will win the game if Jackson doesn't play for them.

In the first sentence, *then* was omitted, but it is understood to be there. In the second sentence, we switched the two parts around and also omitted *then*. Nonetheless, both of these statements are conditional.

If two sides of a triangle are equal, then two angles of the triangle are equal, and if two angles of a triangle are equal, then two sides of the triangle are equal.

You may remember the above statement from the study of geometry. It is the conjunction of two conditional statements where the antecedent and consequent of the first statement have been switched in the second. This type of sentence is usually stated as

Two sides of a triangle are equal if and only if two angles of the triangle are equal.

This type of statement is called a **biconditional.** It has the advantage of shortening the original statement. An abbreviation for *if and only if* is *iff;* we shall sometimes use this abbreviation.

Remember that a biconditional statement is the conjunction of two conditional statements where the antecedent and consequent of the first statement have been switched in the second.

You should be able to identify a simple statement or a compound statement. If the statement is compound, then it must be one of the following: negation, conjunction, disjunction, conditional, or biconditional.

In mathematics and the English language, many statements are lengthy and cumbersome. We would all tire quickly if we had to copy this page word for word. In logic, this problem is overcome by using symbols to represent simple statements.

It is traditional in algebra to use x, y, and z as symbols for variables and a, b, and c as symbols for constants. In logic, we normally use the letters P, Q, R, and S, and sometimes A and B, to represent statements. Other letters may also be used if needed.

Consider the statement "Today is Friday." We shall let the letter P represent this simple statement. We shall also let Q stand for the statement "I have a test." Hence we have

$$P = \textit{Today is Friday}$$

$$Q = \textit{I have a test}$$

If we combine the two statements to form a conjunction, "Today is Friday and I have a test," we could symbolize this as

$$P \text{ and } Q$$

It certainly seems odd to have the simple statements symbolized, but not the connective. However, we will find that there are symbols for each of the connectives.

Ampersand, &, is the typewriter symbol for *and*. In logic, it is more common to use an inverted $\vee$, that is, $\wedge$, to represent *and*. Some of you may be familiar with this symbol as a *caret*. Using this symbol, the statement "Today is Friday and I have a test" can be completely symbolized as

$$P \wedge Q \quad \text{(conjunction)}$$

If we have a statement *not P,* and P stands for "Today is Friday," it is awkward to say

Not today is Friday.

We would be more comfortable if we said

It is not the case that today is Friday.

We probably would be even more comfortable if we said

Today is not Friday.

When we use the word *not* in a sentence, we are negating the original statement. The logical symbol most commonly used to show negation is a *tilde*, which looks like this: $\sim$. The tilde is a diacritical mark used in some languages. If we let P stand for "Today is Friday," then the statement "Today is not Friday" would be symbolized as

$$\sim P \quad \text{(negation)}$$

Remember that $\sim P$ may also be interpreted as

Not P
It is false that P
It is not the case that P

EXAMPLE 1 Let P = *Today is Monday* and Q = *I am tired.* Write each of the following statements in symbolic form.

a. Today is not Monday.
b. Today is Monday and I am tired.

c. Today is Monday and I am not tired.
d. Today is not Monday and I am tired.
e. Today is not Monday and I am not tired.

SOLUTION

a. This statement is the negation of P, and hence it would be symbolized as $\sim P$.
b. The statement is a conjunction of P and Q and therefore would be symbolized as $P \wedge Q$.
c. This is also a conjunction, but here we have Q negated. The proper symbolization is $P \wedge \sim Q$.
d. This is similar to statement c, but now P is negated, and we have $\sim P \wedge Q$.
e. This time each part of the statement is negated. We would symbolize this as $\sim P \wedge \sim Q$.

We shall see later that statements such as $\sim P \wedge Q$ and $\sim(P \wedge Q)$ are not the same and must be interpreted differently.

A disjunction is a compound statement consisting of two statements connected by the word *or*. The symbol for this connective is $\vee$. If we let

$$P = \textit{Today is Monday}$$

$$Q = \textit{Tomorrow is Wednesday}$$

Then the statement

Today is Monday, or tomorrow is Wednesday.

is symbolized as

$$P \vee Q \quad \text{(disjunction)}$$

Consider the statement

Either two is not even, or three is not odd.

In this case we would let

$$P = \textit{Two is even}$$

$$Q = \textit{Three is odd}$$

and the compound statement would be symbolized as

$$\sim P \vee \sim Q$$

EXAMPLE 2 Let P = *Today is Monday* and Q = *Yesterday was Sunday*. Write each of the following statements in symbolic form.

a. Either today is Monday, or yesterday was Sunday.
b. Yesterday was not Sunday, or today is Monday.
c. Either today is not Monday, or yesterday was not Sunday.

SOLUTION a. This statement is a disjunction since we have the connective *or;* the word *either* also tells us that we have a disjunction. The correct symbolization is $P \vee Q$.
b. This is also a disjunction, but here we have the statements interchanged and Q is negated. Therefore the statement should be symbolized as $\sim Q \vee P$.
c. Here each part of the compound statement is negated; we would symbolize this as $\sim P \vee \sim Q$.

It should be noted that the word *or* can be used in two different ways in a sentence. For example, consider the following statements.

The weather forecast calls for rain or snow.
I will get an A or B for this course.

The first statement illustrates the **inclusive** use of *or,* since it might rain, it might snow, or it might do both. The second statement illustrates the **exclusive** use of *or,* since it is not possible for both things to occur. That is, the grade for the course is an *A* or a *B,* but not both. The symbol commonly used for the exclusive *or* is $\veebar$. Hence *P or Q but not both* is symbolized as $P \veebar Q$. Unless otherwise noted, we shall assume that *or* is used in the inclusive sense.

A conditional is a statement that implies something. The symbol used in mathematics for implication is $\rightarrow$. The statement $P \rightarrow Q$ is usually interpreted as

If P, then Q.

or, equivalently,

P implies Q.

Consider the statement

If the Browns win the championship, then I'll eat my hat.

If we let

P = *The Browns win the championship*

Q = *I'll eat my hat*

the compound statement is symbolized as

$$P \rightarrow Q \quad \text{(conditional)}$$

Let us examine a conditional statement where the antecedent and consequent are negated. If we let

P = *x is negative*

Q = *x is less than zero*

then $\sim P \rightarrow \sim Q$ is interpreted as

If x is not negative, then x is not less than zero.

A biconditional is the conjunction of two conditional statements where the antecedent and consequent of the first statement have been switched in the second. The symbol for the connective in a biconditional is $\leftrightarrow$.

Consider the statement

Skating is permitted if and only if the ice is 6 inches thick.

Let

P = *Skating is permitted*

Q = *The ice is 6 inches thick*

the compound statement is symbolized as

$$P \leftrightarrow Q \quad \text{(biconditional)}$$

We know from the preceding discussion that this statement is the same as the conjunction of two conditionals, and therefore we are aware that if skating is permitted, then the ice is 6 inches thick, and if the ice is 6 inches thick, then skating is permitted; that is,

$$P \leftrightarrow Q \equiv (P \rightarrow Q) \wedge (Q \rightarrow P)$$

(The symbol $\equiv$ means *is the same as*.)

PAR/NYC photo by John Schultz

EXAMPLE 3 Let P = *I was late* and Q = *My car broke down.* Write each of the following statements in symbolic form.

a. If I was late, then my car broke down.
b. If my car broke down, then I was late.
c. I was late if and only if my car broke down.

SOLUTION

a. This statement is a conditional; the key is the connective *if ... then. ...* We would symbolize this as $P \rightarrow Q$.
b. This is similar to statement *a,* but here we have "my car broke down" as the antecedent, and therefore we symbolize the statement as $Q \rightarrow P$.
c. Statement *c* is a biconditional, as indicated by the phrase "if and only if," and the correct symbolization is $P \leftrightarrow Q$.

EXAMPLE 4 Let P = *Today is Monday,* Q = *Yesterday was Sunday.* Write each of the following statements in words.

a. $P \wedge Q$
b. $\sim P \vee Q$
c. $P \rightarrow \sim Q$
d. $\sim P \leftrightarrow \sim Q$
e. $\sim P \wedge \sim Q$
f. $P \rightarrow P \wedge \sim Q$

SOLUTION

a. This is a conjunction and we may write this directly from the symbols: "Today is Monday and yesterday was Sunday."
b. This is a disjunction with the first part negated: "Today is not Monday, or yesterday was Sunday."
c. This is a conditional statement with the consequent negated: "If today is Monday, then yesterday was not Sunday."
d. This statement is a biconditional with both parts negated: "Today is not Monday if and only if yesterday was not Sunday."
e. This conjunction could be written as "Today is not Monday and yesterday was not Sunday." Another correct interpretation is "Neither is today Monday, nor was yesterday Sunday." A *neither-nor* statement is a conjunction where both parts are negated. That is, a sentence of the form "neither P nor Q" is symbolized as $\sim P \wedge \sim Q$.
f. This statement is a conditional whose consequent is a conjunction: "If today is Monday, then today is Monday and yesterday was not Sunday."

At this point, you should be familiar with the following types of statements and their connective symbols.

Type of Statement	Connective	Symbol
Negation	not	$\sim$
Conjunction	and	$\wedge$
Disjunction	or	$\vee$
Conditional	if ... then ...	$\rightarrow$
Biconditional	if and only if iff	$\leftrightarrow$

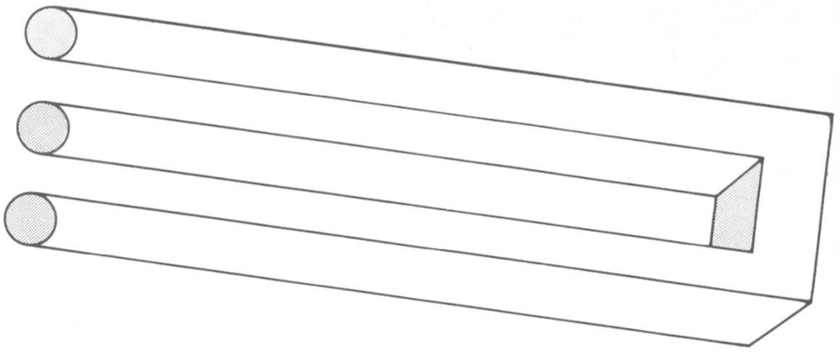

FIGURE 1
A visual paradox.

EXERCISES FOR SECTION 2.2

1. Identify each of the following sentences as a simple statement, compound statement, or neither. Classify each compound statement as a negation, conjunction, disjunction, conditional, or biconditional.
   a. Today is Friday.
   b. It is false that Scott is in class.
   c. You may play tennis here iff you are a member of the club.
   d. Joey is not going to Canada during vacation.
   e. If Addie went swimming, then Julia went sailing.
   f. Close the door when you leave.
   g. Both Ruth and Horace are members of the council.
   h. Either Bill is here, or he did not come to school.

2. Identify each sentence as a simple statement, compound statement, or neither. Classify each compound statement as a negation, conjunction, disjunction, conditional, or biconditional.
   a. If $3 + 4 = 8$, then $9 - 2 = 6$.
   b. When in doubt, punt!
   c. A student may take Math 176 iff he has successfully completed Math 175.
   d. Neither Hugh nor Bill is here.
   e. Addie is not late, and Julia is not early.
   f. Mary went swimming or cycling.
   g. Yesterday was Sunday.
   h. Norma was not in class.

3. Let P = *Polly is good* and Q = *Johnny is good*, and let us agree that *bad* = *not good*. Write each of the following statements in symbolic form.
   a. Both Polly and Johnny are good.
   b. Either Johnny or Polly is good.
   c. Polly is not good, or Johnny is bad.
   d. If Polly is good, then Johnny is bad.
   e. It is false that Polly is bad.
   f. Johnny is good iff Polly is good.

4. If "Polly is good" and "Johnny is good" are true statements, which of the statements in Exercise 3 do you think are true?

5. Let P = *Algebra is difficult* and Q = *Logic is easy*. Write each of the following statements in symbolic form.
   a. Algebra is difficult, or logic is easy.
   b. Logic is not easy, and algebra is difficult.
   c. It is false that logic is not easy.
   d. Logic is easy iff algebra is difficult.
   e. If algebra is difficult, then logic is easy.
   f. Neither is algebra difficult nor is logic easy.

6. If "Algebra is difficult" and Logic is easy" are true statements, which of the statements in Exercise 5 do you think are true?

7. Write the following statements in symbolic form using the letters in the parentheses.
   a. Either I sink this putt, or I lose the match. (P, M)
   b. Five is greater than zero, and 5 is positive. (G, P)
   c. If you do not attend class, then you will be dropped from the course. (A, D)
   d. Either the bus is late, or my watch is not working correctly. (B, W)
   e. Two equals 1 iff 3 is greater than 4. (T, F)
   f. Smith will raise taxes if he is elected. (R, E)

8. Write the following statements in symbolic form using the letters in parentheses.
   a. It is false that Sally did not pass math. (S)
   b. You cannot go to the concert, but I have a ticket. (C, T)
   c. Neither Don nor Bob likes football. (D, B)
   d. If you like ice cream, then you'll love sherbet. (I, S)
   e. I can go to the show iff I finish my homework. (S, H)
   f. I can go to the show if I finish my homework. (H, S)

9. Let P = *I like algebra* and Q = *I like geometry*. Write each of the following statements in words.
   a. $P \wedge Q$
   b. $P \rightarrow \sim Q$
   c. $P \vee Q$
   d. $P \vee \sim Q$
   e. $\sim P \wedge \sim Q$
   f. $P \leftrightarrow Q$

10. Let P = *Tom is tense* and Q = *Freddy is ready*. Write each of the following statements in words.
   a. $P \vee \sim P$
   b. $P \leftrightarrow \sim Q$
   c. $\sim P \vee Q$
   d. $\sim P \wedge \sim Q$
   e. $\sim Q \rightarrow P$
   f. $P \rightarrow \sim Q$

JUST FOR FUN *What month has 28 days?*

2.3 DOMINANCE OF CONNECTIVES

Up to this point in our discussion, except for some of the propositions in the exercises, we have encountered few compound sentences with multiple connectives. We have examined a few statements that contained a negation in addition to another connective, but that is all.

Jack Spratt/The Image Works

Suppose that we have a statement such as

A. I will go swimming or I will go cycling, and I will go to the movies.

This statement is a conjunction because the comma separates the sentence into two parts with the connective *and.* But if the comma were omitted, we would not know whether the statement was a disjunction or a conjunction. Some of us would interpret it as

B. I will go swimming, or I will go cycling and I will go to the movies.

whereas others would interpret it as it is stated in A.

We need punctuation marks in writing statements in order to make sense out of them. The following example points out this need. Try to make sense out of this statement.

Tom is here or Jim left and Bob came late.

Unless this statement is correctly punctuated, we do not know if it is a disjunction or a conjunction.

Mathematics uses parentheses as punctuation marks and so does logic. Some statements do not need parentheses, whereas others do; some even need more than one set.

If we let

P = *I will go swimming*

Q = *I will go cycling*

R = *I will go to the movies*

then statement A would be symbolized as $(P \vee Q) \wedge R$, and statement B would be symbolized as $P \vee (Q \wedge R)$.

It is superfluous to use parentheses in statements like the following.

$$\sim P \qquad P \wedge Q \qquad P \vee Q \qquad P \rightarrow Q \qquad P \leftrightarrow Q$$

When we interpret a statement, we shall interpret it exactly as written. That will enable us to determine the symbolic form of the statement.

EXAMPLE 1 Identify each of the following statements as a negation, conjunction, disjunction, conditional, or biconditional.

a. Either the Jets won, or the Bills won and the Browns lost.
b. If John goes to college and Jack goes to art school, then their father will have to take a loan.
c. José is here or Larry is here, and it is hot.
d. It is false that if I took French II, then I had taken French I before.

SOLUTION

a. This statement is a disjunction, as the comma indicates.
b. Statement b is a conditional statement because of the word *if*.
c. Statement c is a conjunction, as the comma indicates.
d. The phrase *it is false that* identifies statement d as a negation, even though it contains a conditional. *It is false that* negates everything that follows.

When we want to translate symbolic statements into words, we need parentheses. We will also adopt the convention of the dominance of connectives. In logic, some connectives are considered more dominant than others. The following is a list of the connectives in their dominant order, as found in most logic books; the most dominant connective is listed first.

1. Biconditional $\leftrightarrow$
2. Conditional $\rightarrow$
3. Conjunction $\wedge$, disjunction $\vee$
4. Negation $\sim$

Note that conjunction and disjunction are listed together. They are of equal value. If a compound statement contains both of these connectives and no others, we must use parentheses to designate it as a conjunction or a disjunction.

Also, a symbol outside parentheses dominates or outranks any symbol inside the parentheses. For example, $\sim(P \rightarrow Q)$ is a negation and not a conditional.

EXAMPLE 2 Identify each symbolic statement.

a. $P \wedge Q \leftrightarrow R$
b. $(P \vee Q) \wedge R$
c. $\sim P \vee Q \rightarrow R \wedge S$
d. $\sim(P \wedge Q)$
e. $\sim(P \rightarrow Q \vee R)$
f. $P \vee (Q \rightarrow R)$

SOLUTION

a. The statement $P \vee Q \leftrightarrow R$ is a biconditional because the double arrow is the dominant connective and there are no parentheses.
b. The parentheses in $(P \vee Q) \wedge R$ separate the statement at the connective; hence it is a conjunction.
c. In $\sim P \vee Q \rightarrow R \wedge S$ there are no parentheses and none are needed, as the conditional arrow is stronger than any of the other connectives in the statement. The statement is a conditional. Note that $\sim P \vee Q$ is the antecedent and $R \wedge S$ is the consequent.
d. At first glance it might appear that $\sim(P \wedge Q)$ is a conjunction, but it isn't because the negation sign is outside the parentheses. If the statement were $\sim P \wedge Q$, it would be a conjunction, but with the parentheses it is a negation.
e. The negation sign in $\sim(P \rightarrow Q \vee R)$ takes precedence because it is outside the parentheses and the arrow is inside. Therefore it is a negation.
f. The parentheses divide the statement into two major parts and the connective outside the parentheses is dominant; hence it is a disjunction.

EXAMPLE 3 Let P = *I can go*, Q = *You can go*, and R = *Lew can go*. Write each of the following symbolic statements in words.

a. $P \wedge (Q \vee R)$
b. $P \vee Q \rightarrow R$
c. $\sim(P \wedge Q)$
d. $\sim P \vee \sim Q$

SOLUTION a. Because of the parentheses, this statement is a conjunction, so we write "I can go, and you can go or Lew can go."
b. The arrow is the dominant connective, so statement *b* is a conditional: "If I can go or you can go, then Lew can go."
c. Statement *c* is a negation; hence we have "It is false that I can go and you can go."
d. This statement is a disjunction with each part negated. Therefore we have "I cannot go or you cannot go."

Parentheses are used in logic to tell us what type of statement we are considering. If there are no parentheses, then we follow the convention of the dominance of connectives. The biconditional ($\leftrightarrow$) is the strongest connective, followed by the conditional ($\rightarrow$), and then conjunction ($\wedge$) and disjunction ($\vee$). The conjunction and disjunction are of equal value. The negation ($\sim$) is the weakest connective.

EXERCISES FOR SECTION 2.3

1. Add parentheses in each statement to form the type of compound statement indicated. If none are needed, indicate that fact.
   a. Negation: $\sim P \wedge Q \rightarrow R$
   b. Conditional: $\sim P \wedge Q \rightarrow R$
   c. Conjunction: $\sim P \wedge Q \rightarrow R$
   d. Biconditional: $P \wedge Q \leftrightarrow R$
   e. Disjunction: $\sim P \vee Q \wedge R$
   f. Conjunction: $P \wedge Q \leftrightarrow R$

2. Add parentheses in each statement to form the type of compound statement indicated. If none are needed, indicate that fact.
   a. Negation: $\sim P \rightarrow \sim Q$
   b. Biconditional: $\sim P \leftrightarrow Q \vee R$
   c. Disjunction: $P \vee Q \rightarrow R \wedge S$
   d. Conditional: $P \vee Q \rightarrow R \wedge S$
   e. Conjunction: $P \vee Q \rightarrow R \wedge S$
   f. Negation: $\sim P \wedge Q \rightarrow \sim R$

3. Let P = *Algebra is difficult*, Q = *Logic is easy*, and R = *Latin is interesting*. Use appropriate connectives and parentheses to symbolize each statement.
   a. If logic is easy and algebra is difficult, then Latin is interesting.
   b. Latin is interesting and algebra is difficult, or logic is easy.
   c. It is false that logic is easy and algebra is difficult.
   d. Either logic is easy and Latin is interesting, or algebra is difficult.
   e. Algebra is difficult iff Latin is interesting and logic is easy.
   f. Neither is algebra difficult, nor is Latin interesting.

4. Let P = Tom is tense, Q = *Sam is sulky*, and R = *Freddy is ready*. Use appropriate connectives and parentheses to symbolize each statement
   a. If Tom is tense and Sam is sulky, then Freddy is ready.
   b. Either Tom is tense and Sam is sulky, or Freddy is ready.
   c. It is false that Tom is tense and Sam is sulky.
   d. Tom is tense iff Sam is sulky and Freddy is ready.
   e. Tom is tense or Freddy is ready, and Sam is sulky.
   f. If Freddy is ready, then it is false that Tom is tense and Sam is sulky.

5. Using the suggested notation, symbolize each statement completely.
   a. If I pass the exam, then I will pass the course and I will graduate. (E, C, G)
   b. Either Scott took the book, or Joe took it and I am upset. (S, J, I)
   c. If you cut class, then you will miss the exam and receive a zero. (C, E, Z)
   d. It is not the case that today is not Friday. (F)
   e. Neither Sue nor Lou is a good swimmer. (S, L)

6. Using the suggested notation, symbolize each statement completely.
   a. Today is not Friday iff tomorrow is Sunday. (F, S)
   b. Paul likes strawberry ice cream and Juanita likes strawberry ice cream, or Connie made a mistake. (P, J, C)
   c. We will be happy if Barbara is elected chairperson. (B, H)
   d. It is false that I was late and missed the test. (L, T)
   e. Today is Friday, and if today is Friday then I get paid. (F, P)

7. Let P = *Algebra is difficult*, Q = *Logic is easy*, and R = *Latin is interesting*. Write each symbolic statement in words.
   a. $P \wedge (Q \vee R)$
   b. $P \wedge Q \rightarrow R$
   c. $P \vee (Q \wedge R)$
   d. $P \wedge (Q \rightarrow R)$
   e. $\sim(P \wedge Q)$
   f. $\sim P \leftrightarrow Q \wedge \sim R$

8. Let P = *Harry studies*, Q = *Harry will pass*, and R = *Harry will succeed*. Write each symbolic statement in words.
   a. $P \rightarrow Q \vee R$
   b. $\sim(P \wedge Q)$
   c. $(P \rightarrow Q) \vee R$
   d. $Q \wedge R \leftrightarrow P$
   e. $\sim P \wedge \sim Q$
   f. $P \wedge (Q \vee \sim R)$

Reprinted by permission: Tribune Media Services

JUST FOR FUN Even the least-interested sports fan knows that a football field is 100 yards long (120 if we count the end zones), but do you know how wide it is?

2.4 TRUTH TABLES

Because a statement in logic is one which may be true or false, we must be able to determine the truth or falsity of a given statement under given conditions.

If we are given a simple statement *P*, we know that *P* must be either true or false, but not both. If *P* is a true statement, then we say that the **truth value** of *P* is *true;* is *P* is false, then its truth value is *false*. Now, what happens if we negate *P*? If *P* is true, then $\sim P$ must be false, and if *P* is false, then $\sim P$ must be true. This type of analysis is shown in Table 1. This type of table is called a **truth table.** A truth table gives us the truth value of a compound statement for each possible combination of the truth or falsity of the simple statements within the compound statement.

TABLE 1

P	$\sim P$
T	F
F	T

The conjunction $P \wedge Q$ contains two simple statements. *P* has two possible truth values, namely, true or false, and *Q* has two possible values. How many possible values does a compound statement such as $P \wedge Q$ have? We know that *P* and *Q* could both be true, or they could both be false. That gives us two possibilities. There is also the case where one could be true and the other false, say, *P* true and *Q* false. But remember that *Q* could be true and *P* false. Therefore we have four possibilities. We will not go through this type of reasoning every time we want to construct a truth table. If there are n simple statements in a compound statement, then there are 2^n possible true–false combinations. The statement $P \wedge Q$ has two simple statements and, therefore, we have $2^2 = 2 \times 2 = 4$ possible true–false combinations. The statement $P \wedge Q \rightarrow R$ has $2^3 = 2 \times 2 \times 2 = 8$ possible true–false combinations.

So far, a truth table for $P \wedge Q$ would look like Table 2a.

TABLE 2a

P	Q	$P \wedge Q$
T	T	
T	F	
F	T	
F	F	

Now let us examine a statement that will enable us to complete the truth table for a conjunction $P \wedge Q$. Consider the following statement.

Today is Tuesday and I have a math class.

Let

P = *Today is Tuesday*

Q = *I have a math class*

The statement is a conjunction and is symbolized as $P \wedge Q$. When is this compound statement true? In order for the whole statement to be true, we must have both parts true. If the person speaking said the statement on any day other than Tuesday, he would not be telling the truth. This would take care of the possibilities, F–T, F–F. We have already agreed that the T–T combination (both parts true) yields a true. Now, if the person making the statement said it on Tuesday, but didn't have math class, then he would be making a false statement. Hence, the T–F possibility yields a false. The completed truth table appears in

TABLE 2b

P	Q	$P \wedge Q$
T	T	T
T	F	F
F	T	F
F	F	F

Table 2b. This table shows that a conjunction is only true when both parts are true; otherwise it is false.

When is a disjunction true? If someone says that he or she has been swimming or cycling, then the speaker is telling the truth if he or she has been swimming and not cycling. Likewise, the speaker is telling the truth when he or she has been cycling and not swimming.

Since we are using the inclusive *or* in logic, the person is also telling the truth when he or she has been swimming and also cycling. Table 3 shows the truth table for a disjunction. It shows that a disjunction is true unless both parts are false.

TABLE 3

P	Q	$P \vee Q$
T	T	T
T	F	T
F	T	T
F	F	F

Recall from Section 2.2 that the exclusive *or,* that is, one event *or* the other, but not both, is symbolized by $P \veebar Q$. This means we want P or Q, but not both at the same time. The truth table for the exclusive *or* is shown in Table 4. The statement $P \veebar Q$ is true only when exactly one part is true; otherwise, it is false.

TABLE 4

P	Q	$P \veebar Q$
T	T	F
T	F	T
F	T	T
F	F	F

EXAMPLE 1 Construct a truth table for $\sim(P \vee Q)$.

SOLUTION Note that the parentheses tell us that the statement is a negation. The whole statement is to be negated; therefore, the negation is the final column completed in the table. The first thing we must do is list the truth values for the variables. We have numbered the columns so that you can follow, in order, the step-by-step process.

P	Q	$\sim$	$(P$	$\vee$	$Q)$
T	T		T		T
T	F		T		F
F	T		F		T
F	F		F		F
1	2		1		2

Since $\vee$ is the connective inside the parentheses, we complete the column for this connective, column 3.

P	Q	$\sim$	$(P$	$\vee$	$Q)$
T	T		T	T	T
T	F		T	T	F
F	T		F	T	T
F	F		F	F	F
1	2		1	3	2

Now, to complete the truth table, we must negate the statement inside the parentheses (the disjunction). To negate the statement, we negate column 3. This gives us column 4 and completes the table.

P	Q	$\sim$	$(P$	$\vee$	$Q)$
T	T	F	T	T	T
T	F	F	T	T	F
F	T	F	F	T	T
F	F	T	F	F	F
1	2	4	1	3	2

Note that we fill in the columns headed by variables before filling in those columns headed by connectives. We fill in the column of the least dominant connective before filling in the columns of the more dominant connectives.

EXAMPLE 2 Construct a truth table for $\sim(P \wedge \sim Q)$.

SOLUTION This problem is similar to Example 1, but it is a little more involved. The statement is a negation, so that is the last column completed in our truth table. Listing the truth values for the variables, we have the following table.

P	Q	$\sim$	$(P$	$\wedge$	$\sim$	$Q)$
T	T		T			T
T	F		T			F
F	T		F			T
F	F		F			F
1	2		1			2

The statement inside the parenthesis is a conjunction, but before we can complete the column for the connective we must negate Q, because the conjunction is of P with *not* Q.

P	Q	$\sim$	$(P$	$\wedge$	$\sim$	$Q)$
T	T		T		F	T
T	F		T		T	F
F	T		F		F	T
F	F		F		T	F
1	2		1		3	2

Now, to figure out the value of the conjunction, we compare columns 1 and 3. This gives us column 4, which is the column that we negate to get column 5 and complete the table.

P	Q	$\sim$	$(P$	$\wedge$	$\sim$	$Q)$
T	T	T	T	F	F	T
T	F	F	T	T	T	F
F	T	T	F	F	F	T
F	F	T	F	F	T	F
1	2	5	1	4	3	2

The completed truth table (column 5) tells us that the statement $\sim(P \wedge \sim Q)$ is true in all cases except when P is true and Q is false.

Remember, a negation ($\sim$) changes the value of the statement to its opposite. That is, negate T, you get F, negate F, you get T. A conjunction ($\wedge$) is only true when both parts are true, and a disjunction ($\vee$) is true unless both parts are false.

EXERCISES FOR SECTION 2.4

In Exercises 1–16, construct a truth table for each statement.

1. $\sim(P \wedge Q)$
2. $\sim P \wedge Q$
3. $\sim P \wedge \sim Q$
4. $P \vee \sim P$
5. $P \wedge \sim P$
6. $\sim P \vee \sim Q$
7. $P \vee \sim Q$
8. $\sim P \vee Q$
9. $\sim(P \vee \sim Q)$
10. $\sim(\sim P \vee \sim Q)$
11. $\sim P \vee (P \wedge \sim Q)$
12. $Q \wedge (\sim P \vee Q)$
13. $\sim P \veebar Q$
14. $\sim(P \veebar \sim Q)$
15. $\sim(\sim P \veebar \sim Q)$
16. $P \veebar \sim Q$
17. Show that each statement has the same truth value as the *exclusive or,* $P \veebar Q$ (that is, truth values $F\ T\ T\ F$).
    a. $(P \vee Q) \wedge (\sim P \vee \sim Q)$
    b. $(P \wedge \sim Q) \vee (\sim P \wedge Q)$

JUST FOR FUN Can you number the rest of the squares in the figure so that each row, column, and diagonal totals 15? No number may be used more than once.

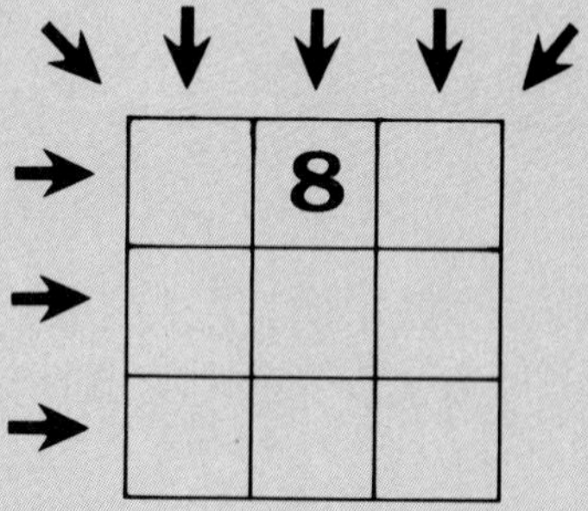

2.5 MORE TRUTH TABLES—CONDITIONAL AND BICONDITIONAL STATEMENTS

We are often confronted with the *if . . . then . . .* type of statement. When is the statement $P \rightarrow Q$ true, and when is it false? We shall examine a completed truth table shortly. But let us first try to justify the entries by means of a discussion. Try to follow the discussion closely so that the resulting table appears natural.

Suppose you are told the following by a counselor.

If you pass biology, then you will graduate.

Now, the statement says that if you pass biology, then you will graduate, regardless of what you may have done previously, or will do later on. Let

Dan Chidester/The Image Works

P = *You pass biology*

Q = *You will graduate*

Since we have two variables in the compound statement $P \rightarrow Q$, there are four possible true–false combinations. First let's examine the case where P is true and Q is true. That is, you do pass biology and you do graduate. The counselor's statement must be a true one; it is certainly not a lie. (If the counselor had lied, then the statement would be false.)

Suppose you do pass biology, but you do not graduate. This is the case where P is true and Q is false. Has the counselor told the truth? No, the original statement is not true. Therefore, the statement $P \rightarrow Q$ is false when the antecedent is true and the consequent is false.

Now, let's consider the case where P is false and Q is true. In this instance, we have the situation where you did not pass biology, but you still graduated. Examine the original statement: "If you pass biology, then you will graduate." Did the counselor lie to you? No, the counselor said what would happen if you did pass biology, not what would happen if you did not pass. There may be other ways of meeting the graduation requirements, such as taking another course. In this case, the counselor's original statement does not apply. A lie would be false statement, but since the counselor did not lie to you, the original statement must be true! Therefore, the statement $P \rightarrow Q$ is true when P is false and Q is true.

The last case to consider is when the antecedent and consequent are both false, that is, P is false and Q is false. You did not pass biology and you did not graduate. The original statement said, "If you pass biology, then you will graduate." In this instance, you did not pass biology and you did not graduate. Did the counselor lie to you? No, the counselor said what would happen if you *passed* biology. As we stated earlier, a lie would be a false statement; if the counselor did not lie, then the original statement must be true.

We are now aware that a conditional is true in all cases except one: a conditional is false when the antecedent is true and the consequent is false; otherwise it is true. The truth table for the conditional statement is shown in Table 5.

TABLE 5

P	Q	$P \rightarrow Q$
T	T	T
T	F	F
F	T	T
F	F	T

Consider the statement

If the moon is made of green cheese, then this is a mathematics book.

You may balk at the sentence as being silly or absurd. But the point here is that we want to use logic to determine whether it is true or false—without being influenced by our own emotions or prejudices. Is the statement true or false? It is true! Why? The reason that the statement is true in this particular case is that the antecedent is false and the consequent is true. We see from the truth table for a conditional that *F–T* yields a *T*. Remember, the only time a conditional statement is false is when the antecedent is true and the consequent is false.

Another conditional statement to consider is

If 2 = 5, then 6 = 9.

Is this statement true or false? It is true. We can determine this from our truth table: Since the antecedent is false, the statement has to be true.

EXAMPLE 1 Construct a truth table for $P \rightarrow \sim Q$.

SOLUTION The statement is a conditional; the arrow is the most dominant connective. Therefore, the last column completed in the truth table will be under the arrow. We first list the truth values for the variables.

P	Q	P	$\rightarrow$	$\sim$	Q
T	T	T			T
T	F	T			F
F	T	F			T
F	F	F			F
1	2	1			2

Next, we fill in the column for the least dominant connective, the negation.

P	Q	P	$\rightarrow$	$\sim$	Q
T	T	T		F	T
T	F	T		T	F
F	T	F		F	T
F	F	F		T	F
1	2	1		3	2

Column 3 is derived from column 2. In order to figure out the truth values for the arrow, we must compare columns 1 and 3 (the truth values for the antecedent and consequent). Thus we obtain column 4, which completes the table.

P	*Q*	*P*	→	~	*Q*
T	*T*	*T*	*F*	*F*	*T*
T	*F*	*T*	*T*	*T*	*F*
F	*T*	*F*	*T*	*F*	*T*
F	*F*	*F*	*T*	*T*	*F*
1	2	1	4	3	2

EXAMPLE 2 Construct a truth table for $P \wedge Q \rightarrow \sim P$.

SOLUTION This statement is a conditional statement since there are no parentheses and the arrow is the dominant connective. The antecedent is $P \wedge Q$ and the consequent is $\sim P$. The computed truth table is

P	*Q*	*P*	∧	*Q*	→	~	*P*
T	*T*	*T*	*T*	*T*	*F*	*F*	*T*
T	*F*	*T*	*F*	*F*	*T*	*F*	*T*
F	*T*	*F*	*F*	*T*	*T*	*T*	*F*
F	*F*	*F*	*F*	*F*	*T*	*T*	*F*
1	2	1	3	2	5	4	1

Column 3 is the truth value for the antecedent and column 4 is the truth value for the consequent. We compare columns 3 and 4 (using the rules for the conditional) to obtain column 5. This tells us that the statement $P \wedge Q \rightarrow \sim P$ is true in all cases except when P and Q are both true.

Recall from Section 2.2 that a biconditional is the conjunction of two conditional statements where the antecedent and consequent of the first conditional have been switched in the second. Therefore, a biconditional statement $P \leftrightarrow Q$ is the same as $(P \rightarrow Q) \wedge (Q \rightarrow P)$. Using this information, let us construct the truth table for the biconditional $P \leftrightarrow Q$ by constructing a table for the statement $(P \rightarrow Q) \wedge (Q \rightarrow P)$. If we figure out the truth table for the conjunction, we will know the truth values for

$P \leftrightarrow Q$, since they are equivalent. The first steps in our truth table are shown in Table 6a.

TABLE 6a

P	Q	$(P$	$\rightarrow$	$Q)$	$\wedge$	$(Q$	$\rightarrow$	$P)$
T	T	T	T	T		T	T	T
T	F	T	F	F		F	T	T
F	T	F	T	T		T	F	F
F	F	F	T	F		F	T	F
1	2	1	3	2		2	4	1

So far we have worked out both sides of the conjunction. Note that each side is a conditional, but the parentheses tell us that $\wedge$ is the dominant connective. Our final step is to compare columns 3 and 4 using the rules for a conjunction. This leads to column 5, which completes the truth table (see Table 6b).

TABLE 6b

P	Q	$(P$	$\rightarrow$	$Q)$	$\wedge$	$(Q$	$\rightarrow$	$P)$
T	T	T	T	T	T	T	T	T
T	F	T	F	F	F	F	T	T
F	T	F	T	T	F	T	F	F
F	F	F	T	F	T	F	T	F
1	2	1	3	2	5	2	4	1

Table 6b shows that the conjunction of $(P \rightarrow Q)$ and $(Q \rightarrow P)$ is true when P and Q are both true, and when P and Q are both false. Because this statement is equivalent to $P \leftrightarrow Q$, we also know the truth value of $P \leftrightarrow Q$. We may say that a biconditional is true when P and Q are both true and when P and Q are both false. That is, a biconditional is true when both parts have the same truth value; otherwise it is false. The truth table for the biconditional appears in Table 7.

TABLE 7

P	Q	$P \leftrightarrow Q$
T	T	T
T	F	F
F	T	F
F	F	T

EXAMPLE 3 Construct a truth table for $(\sim P \vee Q) \leftrightarrow (P \rightarrow Q)$.

SOLUTION

P	Q	$(\sim$	P	$\vee$	$Q)$	$\leftrightarrow$	$(P$	$\rightarrow$	$Q)$
T	T	F	T	T	T	T	T	T	T
T	F	F	T	F	F	T	T	F	F
F	T	T	F	T	T	T	F	T	T
F	F	T	F	T	F	T	F	T	F
1	2	3	1	4	2	6	1	5	2

As with the other truth tables, we listed the possibilities for P and Q first (columns 1 and 2). Our next step was column 3. Then we compared columns 3 and 2 to get column 4 (the disjunction). On the right side, we compared 1 and 2 to get column 5 (the conditional). Our final step was to compare columns 4 and 5 to obtain column 6. Here we used the rule for the biconditional.

Note that column 6 in Example 3 contains only T's. This tells us that the compound statement $(\sim P \vee Q) \leftrightarrow (P \rightarrow Q)$ is true for all cases, regardless of the truth values of the variables P and Q. A compound statement which is true for any combination of truth values of the variables in the statement is called a **tautology.**

A statement that yields all F's, that is, a statement that is always false, is called a **contradiction.** A common example of a contradiction is $P \wedge \sim P$.

A biconditional that is a tautology tells us something else, too. In Example 3, the truth values for columns 4 and 5 are exactly the same. Columns 4 and 5 represent the truth values for the left and right members of the biconditional. Two statements are logically equivalent if they have identical truth tables. Therefore, a biconditional that is a tautology tells us that the left and right members are logically equivalent. In some logic texts, the biconditional is referred to as the **equivalence,** and therefore in Example 3 we have a tautological equivalence.

EXAMPLE 4 How many cases have to be considered in order to construct a truth table for the statement $(P \wedge Q) \rightarrow R$?

SOLUTION Note that this compound statement contains three simple statements. How many true–false combinations are there? Recall that if there are n simple statements in a compound statement, then there are 2^n possible true–false combinations. So we have $2^3 = 2 \times 2 \times 2 = 8$ possible cases to consider.

EXAMPLE 5 Construct a truth table for $(P \wedge Q) \rightarrow R$.

SOLUTION There are eight true–false possibilities to consider in our truth table. We should therefore have eight lines in our table so we can consider all possibilities for the three variables, P, Q, and R. We have to consider "all true," "two true and one false," "one true and two false," and "all false." Here is a listing of all such possibilities for P, Q, and R.

P	Q	R
T	T	T
T	T	F
T	F	T
T	F	F
F	T	T
F	T	F
F	F	T
F	F	F

After noting that a pattern occurs here, we can easily remember how to construct the possibilities. The first column has four T's followed by four F's. The second column then has two T's, two F's, two T's, two F's. The third column is T, F, T, F, and so on.

Next we complete the truth table.

P	Q	R	$(P$	$\wedge$	$Q)$	$\rightarrow$	R
T	T	T	T	T	T	T	T
T	T	F	T	T	T	F	F
T	F	T	T	F	F	T	T
T	F	F	T	F	F	T	F
F	T	T	F	F	T	T	T
F	T	F	F	F	T	T	F
F	F	T	F	F	F	T	T
F	F	F	F	F	F	T	F
1	2	3	1	4	2	5	3

Remember, a conditional ($\rightarrow$) is true in all cases except when the antecedent is true and the consequent is false. A biconditional ($\leftrightarrow$) is true when both parts have the same truth value; otherwise it is false.

EXERCISES FOR SECTION 2.5

In Exercises 1–20, construct a truth table for each symbolic statement.

1. $P \rightarrow Q$
2. $\sim(P \rightarrow Q)$
3. $\sim P \rightarrow \sim Q$
4. $P \rightarrow \sim Q$
5. $\sim P \rightarrow Q$
6. $\sim P \leftrightarrow Q$
7. $\sim P \leftrightarrow \sim Q$
8. $P \wedge Q \rightarrow P$
9. $P \vee Q \rightarrow \sim Q$
10. $P \rightarrow Q \vee \sim P$
11. $(P \rightarrow Q) \vee P \rightarrow Q$
12. $P \wedge (\sim P \vee Q) \rightarrow Q$
13. $P \wedge Q \leftrightarrow P \vee Q$
14. $\sim P \wedge \sim Q \leftrightarrow \sim(P \vee Q)$
15. $(P \vee Q) \wedge R$
16. $P \vee (Q \wedge R)$
17. $(P \wedge Q) \vee (P \wedge R)$
18. $P \wedge (Q \vee \sim R)$
19. $P \leftrightarrow Q \vee R$
20. $P \wedge \sim(Q \vee R) \rightarrow P \vee Q$
21. Determine whether $\sim(P \vee Q)$ is logically equivalent to $\sim P \wedge \sim Q$.
22. Determine whether $\sim(P \wedge Q)$ is logically equivalent to $\sim P \vee \sim Q$.
23. Are $P \wedge \sim Q$ and $\sim(\sim P \vee Q)$ logically equivalent?
24. Are $\sim P \vee Q$ and $\sim(P \wedge \sim Q)$ logically equivalent?
25. In Example 3 we found that $\sim P \vee Q$ is logically equivalent to $P \rightarrow Q$. Using this equivalence, rewrite each of the following statements. For example,

    If today is Monday, then tomorrow is Tuesday.

    can be rewritten as

    Today is not Monday or tomorrow is Tuesday.

    a. If the tide is out, then we can go clamming.
    b. If Bill drove his van, then he brought the packages.
    c. If today is Wednesday, then tomorrow is not Friday.
    d. Either 2 does not equal 3, or 4 equals 6.
    e. Bob didn't pass the test, or he is unhappy about something else.

JUST FOR FUN If 3 cats kill 3 rats in 3 minutes, then how many cats will it take to kill 30 rats in 30 minutes?

2.6 DE MORGAN'S LAW AND EQUIVALENT STATEMENTS

In the preceding section, we were introduced to logically equivalent statements. Two statements that have exactly the same truth values were said to be *logically equivalent.* In Example 3 of Section 2.5, we showed that $P \rightarrow Q$ is logically equivalent to $\sim P \vee Q$. We can write this as

$$P \rightarrow Q \equiv \sim P \vee Q$$

We determined this by constructing the truth table for the statement $(\sim P \vee Q) \leftrightarrow (P \rightarrow Q)$, and obtaining all T's for the biconditional (a tautology). By means of this technique, we can always determine if one statement is logically equivalent to another.

Sometimes when we are given a statement we can create another statement that is logically equivalent to it. We can change a disjunction, a conjunction, or the negation of one of these to an equivalent statement by means of a rule known as **De Morgan's law.** To illustrate how this works, consider the statement

1. *Neither the Giants nor the Browns won.*

We know from previous discussions that this is the same as

2. *The Giants did not win and the Browns did not win.*

Can we restate this in still another way? Consider

3. *It is not the case that either the Giants or Browns won.*

Let P = *The Giants won* and Q = *The Browns won;* then statement 2 would be symbolized as $\sim P \wedge \sim Q$, and statement 3 would be symbolized as $\sim(P \vee Q)$. Are these two statements logically equivalent? We can determine if they are by means of a truth table (Table 8). Let's examine the truth table for $(\sim P \wedge \sim Q) \leftrightarrow \sim(P \vee Q)$. If it is a tautology, then we will know that the two statements are equivalent.

TABLE 8

P	Q	$(\sim$	P	$\wedge$	$\sim$	$Q)$	$\leftrightarrow$	$\sim$	$(P$	$\vee$	$Q)$
T	T	F		F	F		T	F	T	T	T
T	F	F		F	T		T	F	T	T	F
F	T	T		F	F		T	F	F	T	T
F	F	T		T	T		T	T	F	F	F
1	2	3		5	4		8	7	1	6	2

HISTORICAL NOTE

Augustus De Morgan (1806–1871) was born of British parents in India, but spent most of his life in and around London, England.

As a youngster De Morgan had to endure intense ridicule from his classmates as he had only one eye. Consequently, he developed an indifferent attitude toward social activities and a dislike for the out of doors, athletics, and animals. De Morgan, instead, turned to books. At an early age he was literally a bookworm, he spoke five languages fluently, played the flute masterfully, and was well on his way to becoming a mathematician.

De Morgan became known and respected as an energetic worker, a prolific writer, and an excellent instructor. His total dedication to his work won him an appointment as the first professor of mathematics at University College, London.

His attitudes on religious freedom were reflected on a number of occasions throughout his life. While in college, De Morgan was denied a candidacy for a fellowship and was also refused the right to continue on to the M.A. degree because he would not take and sign a theological test. Later in life he resigned his professorship in protest against the preference given to members of the church of England in the selection of books for the college and the religious philosophy of the college itself. He was a man of integrity and unafraid of making personal sacrifices for a cause or a principle in which he believed.

Column 8 of Table 8 shows us that the given statement is a tautology, thereby verifying the fact that statements 2 and 3 are logically equivalent.

De Morgan's law enables us to create an equivalent statement when we are given a certain type of statement. To begin with, we must have some form of a disjunction or conjunction, or the negation of one of these. In order to create an equivalent statement using De Morgan's law, we must perform the following three steps.

1. Negate the whole statement.
2. Negate each statement that makes up the disjunction or conjunction.
3. Change the conjunction to a disjunction or the disjunction to a conjunction.

EXAMPLE 1 Use De Morgan's law to create a statement equivalent to $P \wedge Q$.

SOLUTION We first negate the whole statement, which gives us $\sim(P \wedge Q)$. Next we negate each part, and that yields $\sim(\sim P \wedge \sim Q)$. We then make the third and final change, changing $\wedge$ to $\vee$, which gives us $\sim(\sim P \vee \sim Q)$. Hence $P \wedge Q \equiv \sim(\sim P \vee \sim Q)$.

EXAMPLE 2 Use De Morgan's law to create a statement equivalent to $\sim(\sim P \vee Q)$.

SOLUTION We are given $\sim(\sim P \vee Q)$. First we negate the whole statement, which makes it $\sim\sim(\sim P \vee Q)$ or $(\sim P \vee Q)$. Now we negate each part, that is, $\sim\sim P \vee \sim Q$, which leaves us with $P \vee \sim Q$. (Note that $\sim\sim P \equiv P$; $\sim\sim P$ is called a **double negation.**) We apply the third step and change $\vee$ to $\wedge$, which yields $P \wedge \sim Q$. Finally, we have $\sim(\sim P \vee Q) \equiv P \wedge \sim Q$.

EXAMPLE 3 Use De Morgan's law to write a statement equivalent to the statement "It is false that Allan abhors anatomy and Lucy loves Latin."

SOLUTION Let A = *Allan abhors anatomy* and L = *Lucy loves Latin;* then the given sentence may be symbolized as $\sim(A \wedge L)$. Now we create an equivalent statement using De Morgan's law: $\sim(A \wedge L) \equiv \sim A \vee \sim L$. Translating the new statement, we have "Either Allan does not abhor anatomy or Lucy does not love Latin."

EXAMPLE 4 Use De Morgan's law to write a statement equivalent to the statement "Sally did not stay and Quincy quit."

SOLUTION Let S = *Sally stayed* and Q = *Quincy quit*. The statement may be symbolized as $\sim S \wedge Q$. Using De Morgan's law, we have $\sim S \wedge Q \equiv \sim(S \vee \sim Q)$. Translating the equivalent statement, we have "It is not the case that Sally stayed or that Quincy did not quit."

EXAMPLE 5 Use De Morgan's law to create a statement equivalent to $\sim(P \rightarrow Q)$.

SOLUTION At first glance, it may seem that we cannot do this problem, because we do not have a disjunction or conjunction. However, recall that in Example 3 of Section 2.5 we discovered that $P \rightarrow Q \equiv \sim P \vee Q$. Therefore, we can take the given statement and rewrite it as $\sim(\sim P \vee Q)$. Now we are ready to use De Morgan's law: $\sim(\sim P \vee Q) \equiv P \wedge \sim Q$. Therefore $\sim(P \rightarrow Q) \equiv P \wedge \sim Q$.

Note that Example 5 verifies that a statement such as

It is false that if today is Friday, then I get paid

is equivalent to

Today is Friday and I do not get paid.

It is interesting to note that the logical operation *and* corresponds to the set operation *intersection*. It is also the case that *or* corresponds to set *union*, and that *not* corresponds to set *complementation*. Two of the most common examples of De Morgan's law for equivalent statements are

$$\sim(P \wedge Q) \equiv \sim P \vee \sim Q$$

$$\sim(P \vee Q) \equiv \sim P \wedge \sim Q$$

The corresponding expressions in set theory are

$$(A \cap B)' = A' \cup B'$$

$$(A \cup B)' = A' \cap B'$$

These two statements can be shown to be true by means of Venn diagrams (see Example 5 and Exercise 5c in Sec. 1.5). Therefore, you can see that set operations are similar to some of those in logic. The basic difference is in the notation.

Remember that in logic we can only use De Morgan's law on some form of a disjunction, conjunction, or negation of one of these. Table 9 gives a list of logically equivalent statements. Their equivalence can be verified by means of truth tables.

TABLE 9
EQUIVALENT STATEMENTS

De Morgan's law	$\sim(P \wedge Q) \equiv \sim P \vee \sim Q$
	$\sim(P \vee Q) \equiv \sim P \wedge \sim Q$
Implication	$P \rightarrow Q \equiv \sim P \vee Q$
Contraposition	$P \rightarrow Q \equiv \sim Q \rightarrow \sim P$
Biconditional	$P \leftrightarrow Q \equiv (P \rightarrow Q) \wedge (Q \rightarrow P)$
Association	$(P \wedge Q) \wedge R \equiv P \wedge (Q \wedge R)$
	$(P \vee Q) \vee R \equiv P \vee (Q \vee R)$
Distribution	$P \wedge (Q \vee R) \equiv (P \wedge Q) \vee (P \wedge R)$
	$P \vee (Q \wedge R) \equiv (P \vee Q) \wedge (P \vee R)$
Idempotent	$P \wedge P \equiv P$
	$P \vee P \equiv P$

EXERCISES FOR SECTION 2.6

In Exercises 1–10, use De Morgan's law to create a statement equivalent to each given statement.

1. $\sim(\sim P \wedge Q)$
2. $\sim(P \vee \sim Q)$
3. $P \wedge Q$
4. $P \vee Q$
5. $P \wedge \sim Q$
6. $\sim P \vee \sim Q$
7. $\sim(\sim P \vee \sim Q)$
8. $\sim(\sim P \wedge \sim Q)$
9. $\sim(P \rightarrow \sim Q)$
10. $\sim[P \vee (Q \wedge R)]$

In Exercises 11–22, use De Morgan's law to rewrite each statement.

11. It is not the case that John went to the party or Janie went to the movies.
12. Neither Bill nor Mary is tall.
13. I did not pass the test, or I studied too much.
14. Either Sandra or Julia cut class.
15. It is false that logic is dull and not interesting.
16. Neither David nor Neal is a good swimmer.
17. Either the bus is late, or my watch is not working correctly.
18. You cannot go to the game and I cannot get a ticket.
19. It is not the case that x is greater than zero and x is negative.
20. It is false that today is Monday or tomorrow is Wednesday.

★21. If the wind doesn't come up, then we can't sail.

★22. If you do not attend class, then you will be dropped from the course.

Use De Morgan's law on the set expressions in Exercises 23–30 to create equivalent expressions.

23. $(A \cap B)'$
24. $(A \cup B)'$
25. $A' \cap B'$
26. $A' \cup B$
27. $A \cap B$
28. $(A' \cap B)'$
29. $(A \cup B')'$
30. $(A' \cap B')'$

JUST FOR FUN Which straight-line segment is longer?

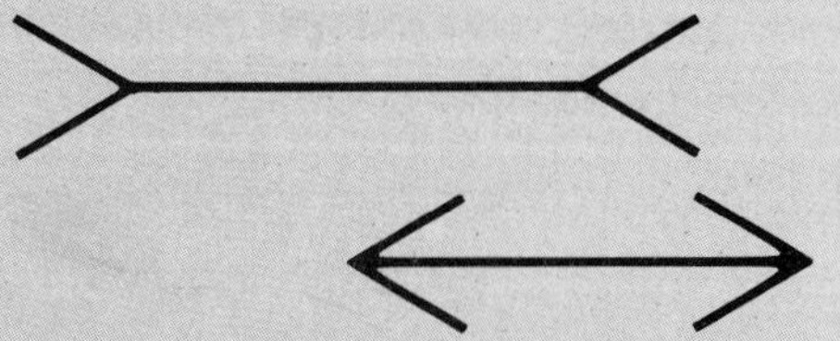

2.7 VALID ARGUMENTS

We shall now review another application of the truth table, for by means of a truth table we can determine if an argument is valid or not.

An **argument,** or **proof,** consists basically of two parts: the given statements, which are called premises, and the conclusion. You may recall proofs from geometry. A proof or argument is said to be **valid** if the conclusion follows logically from the premises. What does this mean? This is really another way of saying that an argument is valid if the premises imply the conclusion. We can take the premises of an argument and connect them using conjunction, and then use this compound statement as the antecedent of a conditional statement of which the conclusion is the consequent. If this conditional, or implication, is a tautology, then the argument is valid. A tautological implication tells us that the premises do indeed imply the conclusion.

Consider the following argument.

If the given figure is a square, then it is a rectangle.
(Major Premise)
The given figure is a square.
(Minor Premise)
Therefore, it is a rectangle.
(Conclusion)

Let

$$P = \textit{The given figure is a square}$$

$$Q = \textit{It is a rectangle}$$

Then the argument would be symbolized as

$$\begin{array}{l} P \rightarrow Q \\ P \\ \hline Q \end{array}$$

If we connect the premises using conjunction, and then imply the conclusion, the corresponding conditional statement is as follows.

$$(P \to Q) \wedge P \to Q$$

premises *conclusion*

Is this argument valid? Does the conclusion follow logically from the premises? That is, is $(P \to Q) \wedge P \to Q$ true for all possible values of P and Q? Table 10 shows the truth table for $(P \to Q) \wedge P \to Q$.

Note: If a symbolized statement has more than one letter in it, then we enclose it in parentheses when we construct the corresponding conditional statement. We do this to avoid confusion as to which is the dominant connective.

TABLE 10

P	Q	$(P \to Q)$	$\wedge$	P	$\to$	Q
T	T	T	T	T	T	T
T	F	F	F	T	T	F
F	T	T	F	F	T	T
F	F	T	F	F	T	F
1	2	3	4	1	5	2

Note that we skipped a couple of steps in Table 10. Our major concern is column 5, where we see that the implication is true in all cases, therefore verifying the fact that the premises imply the conclusion in all cases. Hence, the argument is valid.

Maybe you had already guessed that. However, we are trying to convey what a valid argument is, and how to determine validity. It should be noted from Table 10 that a valid argument can have either a true or a false conclusion. The truth or falsehood of the conclusion does not determine the validity of the argument. Also, the validity of an argument does not guarantee the truth of its conclusion. But if the premises are true, a valid argument has to have a true conclusion. If you have an argument in which all the premises are true, and the conclusion is false, then the argument is **invalid.**

EXAMPLE 1 Determine whether the following argument is valid or invalid.

If I study, then I will pass math.
I didn't pass math.
Therefore, I didn't study.

SOLUTION Let P = *I study* and Q = *I will pass math.* The argument in symbolic form would appear as

$$\begin{array}{l} P \rightarrow Q \\ \underline{\sim Q} \\ \sim P \end{array}$$

If we connect the premises using conjunction and imply the conclusion, the corresponding conditional statement is

$$(P \rightarrow Q) \wedge \sim Q \rightarrow \sim P$$

premises *conclusion*

(Note that the first premise is in parentheses because it has more than one letter in it.) The truth table for this conditional statement is

P	Q	$(P \rightarrow Q)$	$\wedge$	$\sim$	Q	$\rightarrow$	$\sim$	P
T	T	T	F	F		T	F	
T	F	F	F	T		T	F	
F	T	T	F	F		T	T	
F	F	T	T	T		T	T	
1	2	3	6	4		7	5	

Column 7 tells us that we have a tautology, and therefore the argument is valid. The premises do imply the conclusion.

The argument used in Example 1 illustrates one of the basic rules of inference in logic: the **law of contraposition,** or **modus tollens.** If the statement $P \rightarrow Q$ is true, and $\sim Q$ is known to be true, then $\sim P$ must be true.

EXAMPLE 2 Determine whether the following argument is valid or invalid.

If you are healthy, then you are wealthy.
You are wealthy.
Therefore, you are healthy.

SOLUTION Let P = *You are healthy* and Q = *You are wealthy.* The argument in symbolic form would be

$$\begin{array}{l} P \to Q \\ \underline{Q \quad\quad} \\ P \end{array}$$

If we connect the premises using conjunction and imply the conclusion, the corresponding conditional statement is

$$(P \to Q) \land Q \to P$$

premises *conclusion*

The corresponding truth table is

P	Q	$(P \to Q)$	$\land$	Q	$\to$	P
T	T	T	T	T	T	T
T	F	F	F	F	T	T
F	T	T	T	T	F	F
F	F	T	F	F	T	F
1	2	3	4	2	5	1

The resulting conditional statement is not a tautology (see column 5). The premises do not imply the conclusion in all cases, and therefore the argument is invalid. Note that in the third horizontal line of the truth table we have a case where the premises are true, but the conclusion is false. A valid argument does not allow us to go from true premises to a false conclusion.

An invalid argument is one in which the conclusion does not logically follow from the premises. That is, the form of the argument does not permit only true conclusions to be logically derived from true premises. The form of the invalid argument in Example 2 is

$$\begin{array}{l} P \to Q \\ \underline{Q \quad\quad} \\ P \end{array}$$

This form of incorrect reasoning is commonly called the **fallacy of affirming the consequent.** If the consequent Q of a conditional statement $P \to Q$ is true, it does *not* follow that the antecedent P must be true. The statement $P \to Q$ can still be true, even if the consequent Q is false.

Can an argument be valid if one or more of the premises is false and the conclusion is false? Consider the argument in Example 3.

EXAMPLE 3 Determine whether the following argument is valid or invalid.

If the moon is made of green cheese, then two equals one.
The moon is made of green cheese.
Therefore, two equals one.

SOLUTION Let P = *The moon is made of green cheese* and Q = *Two equals one.* The argument in symbolic form would appear as

$$\begin{array}{l} P \rightarrow Q \\ P \\ \hline Q \end{array}$$

If we connect the premises using conjunction and imply the conclusion, the corresponding conditional statement is

$$\underbrace{(P \rightarrow Q) \wedge P}_{\textit{premises}} \rightarrow \underbrace{Q}_{\textit{conclusion}}$$

The corresponding truth table is

P	Q	$(P \rightarrow Q)$	$\wedge$	P	$\rightarrow$	Q
T	T	T	T	T	T	T
T	F	F	F	T	T	F
F	T	T	F	F	T	T
F	F	T	F	F	T	F
1	2	3	4	1	5	2

Column 5 consists of all T's, so the conditional resulting from the argument is a tautology. Hence, the argument is valid. The symbolic argument used in this example illustrates another of the basic rules in logic. It is called the **law of detachment,** or **modus ponens.**

We have just examined an example of a valid argument, parts of which are false (refer back to the truth table in Example 3). There is a difference between the validity of an argument and its truth: an argument is valid because of its form—that is, the manner in which the conclusion is derived from the premises—not because of the meaning of the statements in it.

EXAMPLE 4 Determine whether the following argument is valid or invalid.

Either the bank is closed, or it is not after three o'clock.
It is not after three o'clock.
Therefore, the bank is not closed.

SOLUTION Let P = *The bank is closed* and Q = *It is after three o'clock*. The argument in symbolic form is

$$\begin{array}{l} P \vee \sim Q \\ \underline{\sim Q} \\ \sim P \end{array}$$

If we connect the premises using conjunction and imply the conclusion, the corresponding conditional statement is

$$(P \vee \sim Q) \wedge \sim Q \rightarrow \sim P$$

premises *conclusion*

(Note that the first premise is in parentheses because it has more than one letter in it.) The corresponding truth table is

P	Q	$(P \vee \sim Q)$	$\wedge$	$\sim Q$	$\rightarrow$	$\sim P$
T	T	T	F	F	T	F
T	F	T	T	T	F	F
F	T	F	F	F	T	T
F	F	T	T	T	T	T
1	2	5	6	3	7	4

The resulting conditional statement is not a tautology (see column 7). The premises do not imply the conclusion in all cases. Therefore, the argument is invalid.

In order to test whether an argument is valid or invalid, we need only consider the one conditional statement whose antecedent is the conjunction of all the premises of the argument and whose consequent is the conclusion of the argument. If this conditional statement is a tautology, then the argument is valid. If it is not a tautology, then the argument is invalid.

EXERCISES FOR SECTION 2.7

Symbolize each of the following arguments (using the suggested notation) and, by means of a truth table, determine whether the argument is valid or invalid. State your answer.

1. If I pass the test, then I will quit coming to class. (P, Q)
   I quit coming to class.
   Therefore, I passed the test.

2. If Harry was late, then he missed the exam. (H, M)
   Harry was late.
   Therefore, Harry missed the exam.

3. If I pass math, then I will graduate. (P, G)
   I graduated.
   Therefore, I passed math.

4. If I pass math, then I will graduate. (P, G)
   I did not pass math.
   Therefore, I did not graduate.

5. I will graduate iff I pass math. (G, P)
   I graduated.
   Therefore, I passed math.

6. It will be sunny or cloudy today. (S, C)
   It isn't sunny.
   Therefore, it will be cloudy.

7. Addie and Bill will be at the party. (A, B)
   Bill was at the party.
   Therefore, Addie was at the party.

8. If two divides eight, then three divides seven. (E, S)
   Three does not divide seven.
   Therefore, two does not divide eight.

9. You may attend the concert iff you purchased a ticket. (A, T)
   You purchased a ticket.
   Hence, you may attend the concert.

10. If it snows, Benny will go skiing. (S, B)
    It did not snow.
    Therefore, Benny did not go skiing.

11. Pat or Sandy will bring the doughnuts. (P, S)
    Pat did not bring the doughnuts.
    Therefore, Sandy did not bring the doughnuts.

12. You can play tennis here if you are a member. (T, M)
    You play tennis here.
    Therefore, you are a member.

13. If it rains, Bobby takes his umbrella to school. (R, B)
    It did not rain.
    Therefore, Bobby did not take his umbrella to school.

14. If you finish the exam early, you may leave early. (E, L)
    You did not finish the exam early.
    Therefore, you may not leave early.

★15. Paul did not study, or he is bluffing. (P, B)
    If he is bluffing, then he will cut class. (B, C)
    Therefore, Paul did not study, or he will cut class.

★16. If I get a job, then I will not be able to study. (J, S)
    It is false that I will fail history and that I will not get a job. (F, J)
    Therefore, if I study, then I will not fail the course.

★17. If a is greater than b, then b is greater than c. (A, B)
    If b is greater than c, then c is greater than d. (B, C)
    Therefore, if a is greater than b, then c is greater than d.

★18. If the heavenly body is Mars, then it is near Venus. (M, V)
    If it is not near Venus, then it is not close to Saturn. (V, S)
    Therefore, either it is not close to Saturn, or the heavenly body is Mars.

JUST FOR FUN You are given six line segments of equal length. You are to construct four triangles by using these six line segments. Can you do it?

2.8 PICTURING STATEMENTS WITH VENN DIAGRAMS (OPTIONAL)

Venn diagrams can also be used to determine whether certain kinds of arguments are valid or invalid. The arguments that we shall consider here are called *syllogisms*. A **syllogism** is an argument that contains three statements: the **major premise,** the **minor premise,** and the **conclusion.**

Recall that a valid argument is one in which the conclusion follows logically from the premises. The conclusion is derived from the premises according to the laws of logic. The following is an example of a syllogism that is a valid argument. The conclusion follows from the premises.

All mathematics students are ambitious.	(Major Premise)
No ambitious people are lazy.	(Minor Premise)
Therefore, no mathematics students are lazy.	(Conclusion)

Note that each of the statements in the example contains a quantifier such as *all* or *no*.

The syllogisms that we shall consider here contain quantified statements. There are four such types of statements.

1. The **universal affirmative** statement states that "All *A*'s are *B*'s." For example, "All students are scholars."
2. The **universal negative** statement states that "No *A*'s are *B*'s." For example, "No students are scholars."
3. The **particular affirmative** statement states that "Some *A*'s are *B*'s." For example, "Some students are scholars."
4. The **particular negative** statement states that "Some *A*'s are not *B*'s." For example, "Some students are not scholars."

Do you think that the following syllogism is a valid argument? Does the conclusion follow from the premises?

All golfers are swingers.	(Major Premise)
No hackers are swingers.	(Minor Premise)
Therefore, no hackers are golfers.	(Conclusion)

The preceding example syllogism is a valid argument. You will see why shortly, but first we must be able to symbolize (picture) a statement using Venn diagrams.

HISTORICAL NOTE

John Venn (1834–1923) was an Englishman who devoted his life to the study and teaching of both mathematics and religion. He was a mathematical scholar and ordained a deacon in his church. Two of his better known works are *The Logic of Chance* and *Symbolic Logic*. At the age of 69, Venn was appointed President of Gonville and Cains College, near London.

John Venn developed his diagrams to give sensible illustrations of the relation of terms and properties to each other. In his own words, his purpose was "to save time, avoid unpleasant drudgery and to be sure to avoid mistakes and oversights." Venn devised his own diagrams because he found Euler diagrams inadequate and ineffective, even for the propositions of formal logic.

George Boole significantly influenced the development of Venn's diagrams, for they are founded on Boole's systems of logic. However, Boole did not use such diagrams, nor did he indicate or suggest their introduction.

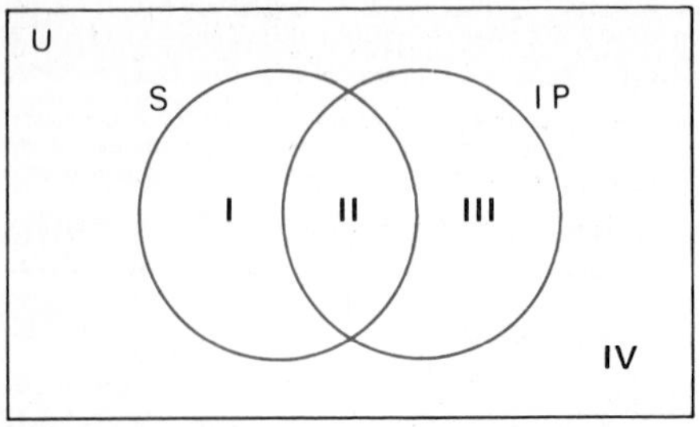

FIGURE 2
S = the set of students, and IP = The set of industrious people.

Consider the statement "All students are industrious." We have a set of students and a set of industrious people. Hence, we may draw two circles and label them (see Fig. 2). It doesn't matter which circle is on the left, but it is convenient to take them in the order given.

Note that we overlap the circles so that all possibilities may be considered. What do we know from the statement "All students are industrious"? We know that if a person is a student, then he or she is industrious; that is, the set of students is a subset of the set of industrious people. It does not follow that if a person is industrious, then he or she is a student.

How are we going to picture our statement? The statement tells us that all the students are in the "industrious" circle; that is, all the elements in the set containing students are in the set containing industrious people. So, they are in region II, and therefore region I is empty. Probably your first inclination is to shade region II, that is, the set $S \cap IP$, as shown in Figure 3. This is because all the elements in set S are also in set IP.

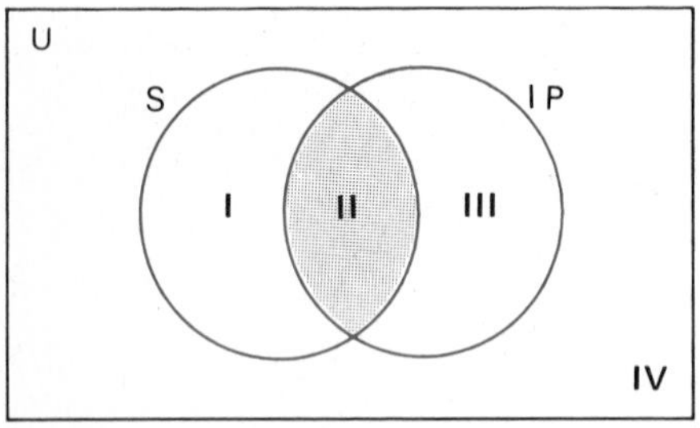

FIGURE 3

However, suppose one or more of the sets we are picturing do not contain any elements. For example, consider the statement "All purple people-eaters are one-eyed." This is a universal affirmative statement; it is of the form "All A's are B's." To picture this statement, we must consider the set of purple people-eaters, which is an empty set. The shading technique used in Figure 3 would not be appropriate for the statement "All purple people-eaters are one-eyed," because shading region II would indicate that the set of purple people-eaters has members, which is not true. Therefore, we must use another way of picturing statements in Venn diagrams, since we must have a method that works for all cases.

A technique that does work for all cases is to distinguish the sets (regions) that *cannot* have elements in them, according to the given statement. For example, in constructing a picture of the statement "All purple people-eaters are one-eyed," we note that if purple people-eaters exist, then they would also all be in the set of one-eyed objects (region II in Fig. 4). But whether they exist or not, region I is known to be empty. According to the given statement, there are no purple people-eaters that are not one-eyed. Therefore we can eliminate region I, that is, we can cross it out. In Figure 4, we have indicated that region I is empty by shading it.

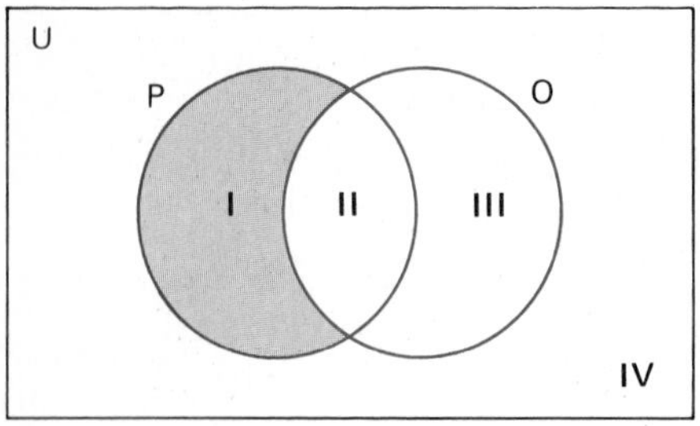

FIGURE 4
P = the set of purple people-eaters, and O = the set of one-eyed objects.

Now, back to the original statement under consideration, "All students are industrious." From this statement we know that there is nothing in the set of students that is not also in the set of industrious people. That is, there are no students in set S who are not also members of set IP. Hence, region I is empty and we can

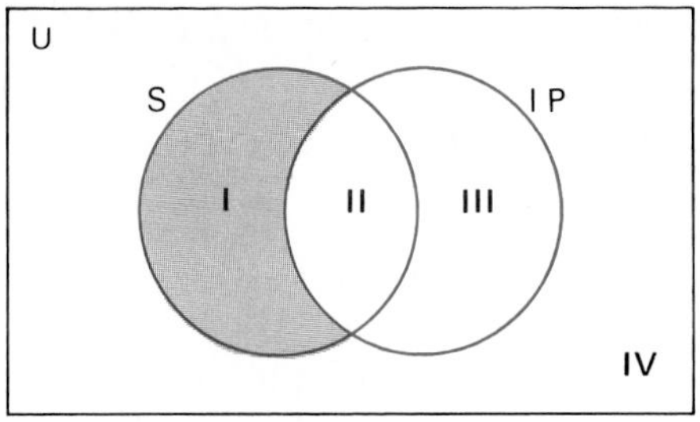

FIGURE 5
S = the set of students, and IP = the set of industrious people.

eliminate it, or cross it out. The statement "All students are industrious" may be pictured as in Figure 5.

Because the set of all students is contained in the set of industrious people, the set of students is a subset of the set of industrious people. You may have considered picturing the given statement as in Figure 6. Here we have one set whose elements are contained in another. This picture is also correct and it may seem more reasonable to you than Figure 5, but remember that we shall be picturing more than one statement in a diagram. It will be more efficient to handle them by the elimination technique than as a circle within a circle. The overlapping circles in a Venn diagram and the elimination technique will allow us to immediately determine whether an argument is valid or not.

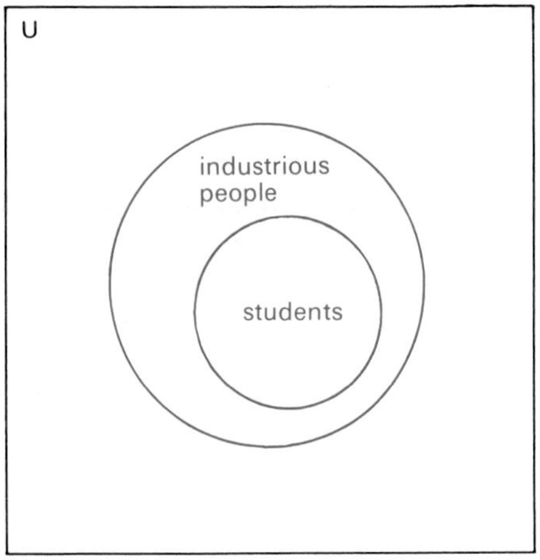

FIGURE 6

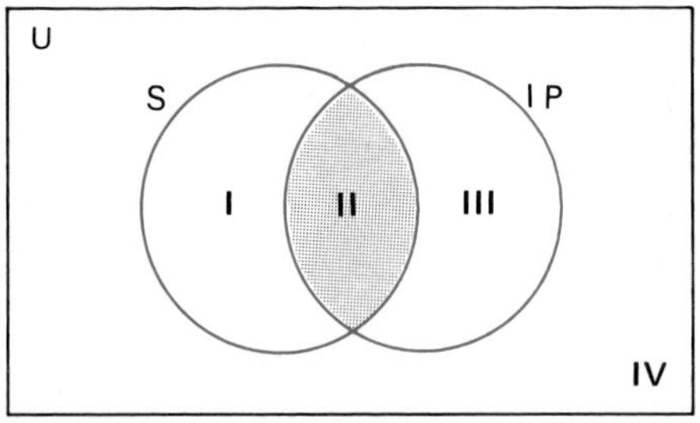

FIGURE 7
S = the set of students, and IP = the set of industrious people.

We have discussed the diagram of the universal affirmative statement. Now let's examine the next one, the universal negative statement. Consider the statement "No students are industrious." Again we have two sets, *students* and *industrious people,* but what region do we eliminate? There are no elements in the set of students that are also in the set of industrious people. Therefore region II in Figure 7 should be crossed out because it contains no elements. Hence the statement "No students are industrious" would be pictured as in Figure 7.

If two statements have the same diagram, the two statements are logically equivalent. The statements have the same meaning because they have the same picture in the diagram and the same region eliminated. The statement "No *A*'s are *B*'s" is logically equivalent to "No *B*'s are *A*'s," but how do "All *A*'s are *B*'s" and "All *B*'s are *A*'s" compare? Are they logically equivalent? By

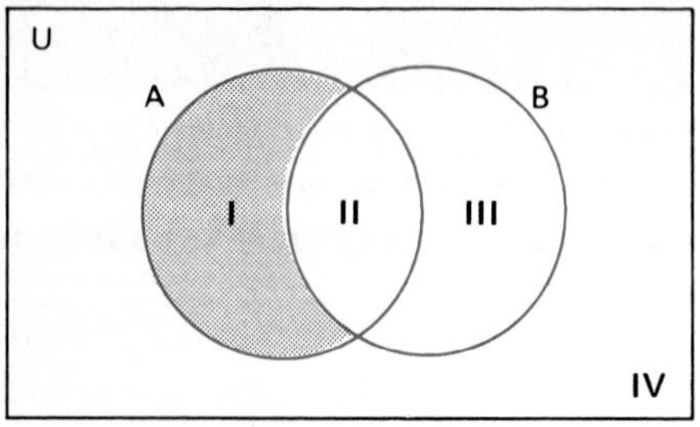

FIGURE 8a
All *A*'s are *B*'s.

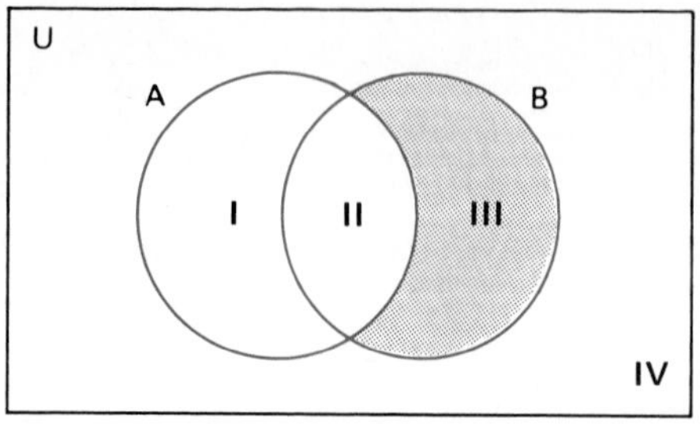

FIGURE 8b
All *B*'s are *A*'s.

checking diagrams of these two statements you will see that they are not! (See Figs. 8a and 8b.)

So far we have pictured the universal affirmative and universal negative statements. Consider the statement "Some students are industrious." We have the set of students and the set of industrious people, and from the given statement we know that there are some students that are industrious. We are discussing "some" students. In logic, the word *some* is interpreted to mean "There is at least one," but there could be more. The word *some* does not specify a certain number, so we just maintain that there is at least one. Since there are some students who are industrious, we know that there is at least one student who is industrious. This one student is common to both sets. We diagram this by showing that the intersection of the two sets is not empty. We place something in the intersection (region II) to show that it is not empty, namely, an *X*. Therefore the statement "Some students are industrious" would be pictured as in Figure 9. Note that when we cross out a region in a Venn diagram, then there are no elements in that region; but when we place an *X* in a region, then there is at least one element in that region.

Consider the particular negative statement "Some students are not industrious." This statement tells us that there is at least

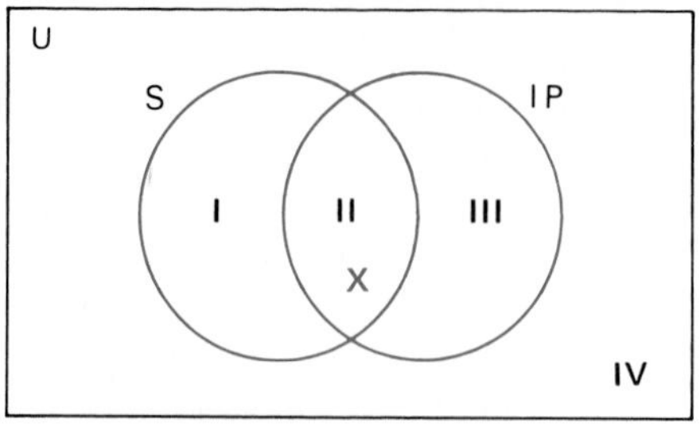

FIGURE 9
S = the set of students, and *IP* = the set of industrious people.

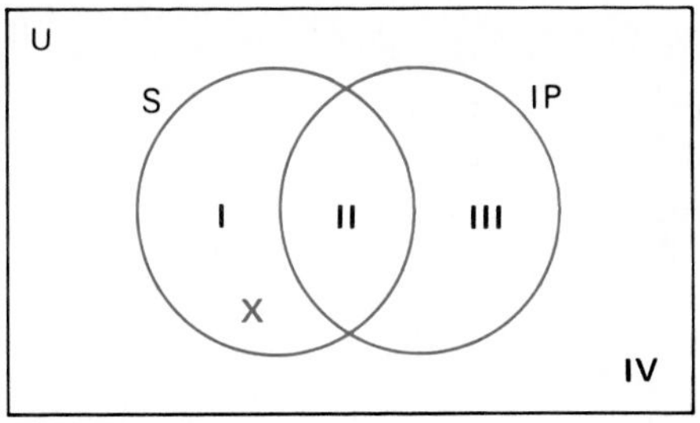

FIGURE 10
S = the set of students, and *IP* = the set of industrious people.

one element in the set of students that is not in the set of industrious people. Therefore the intersection of the set of students and the set of nonindustrious people is not empty. Hence we would place an X in region I because it contains at least one element that is a student and is not industrious. The proper diagram for "Some students are not industrious" is shown in Figure 10.

Using examples, we have diagrammed the four types of statements that may appear in a syllogism: universal affirmative, universal negative, particular affirmative, and particular negative (see Figs. 11a, 11b, 11c, and 11d).

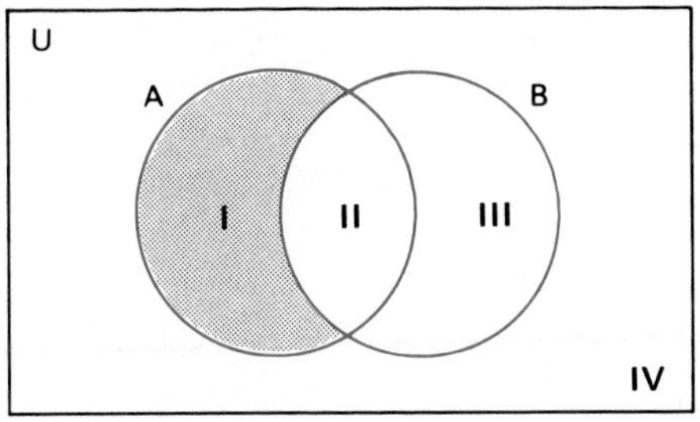

FIGURE 11a
Universal affirmative: All *A*'s are *B*'s.

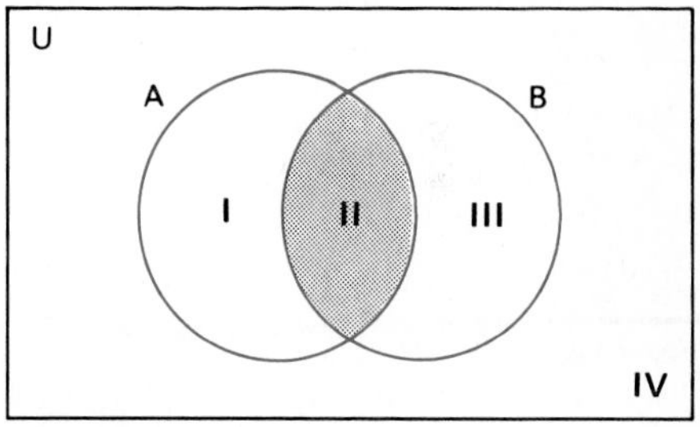

FIGURE 11b
Universal negative: No *A*'s are *B*'s.

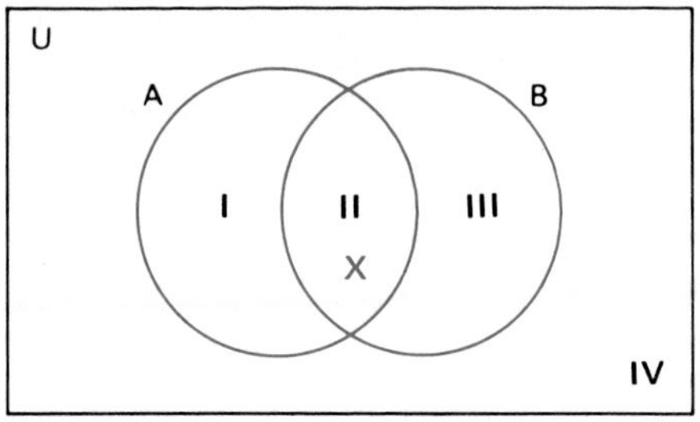

FIGURE 11c
Particular affirmative: Some *A*'s are *B*'s.

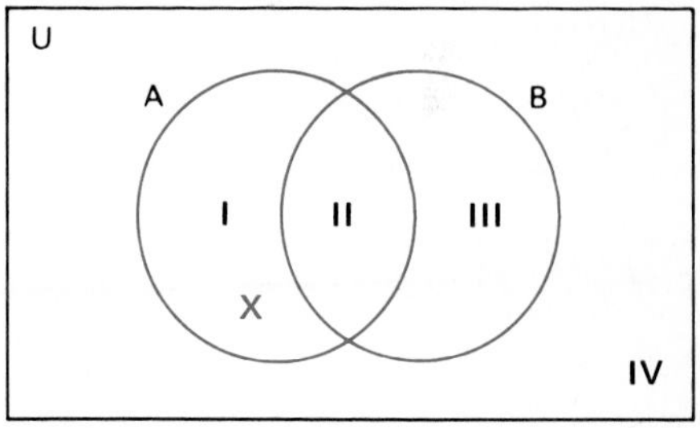

FIGURE 11d
Particular negative: Some *A*'s are not *B*'s.

EXAMPLE 1 Identify and diagram the following.

a. All bugs are pesky.
b. Some bugs are pesky.
c. No bugs are pesky.
d. Some pesky things are not bugs.

SOLUTION Let B = The set of bugs, P = The set of pesky things.

a. Universal affirmative; see Figure 12a.
b. Particular affirmative; see Figure 12b.
c. Universal negative; see Figure 12c.
d. Particular negative; see Figure 12d.

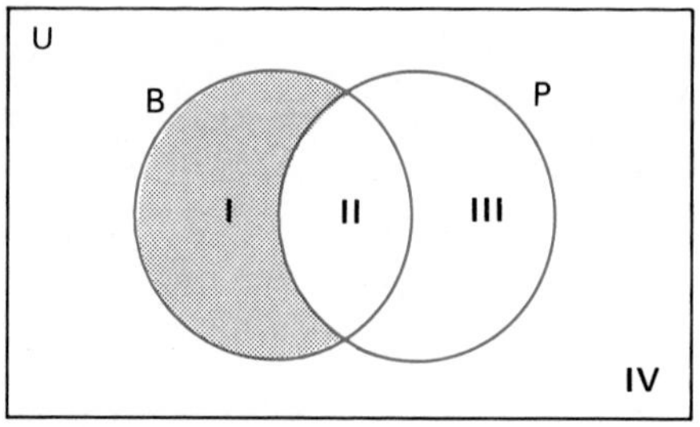

FIGURE 12a
All bugs are pesky.

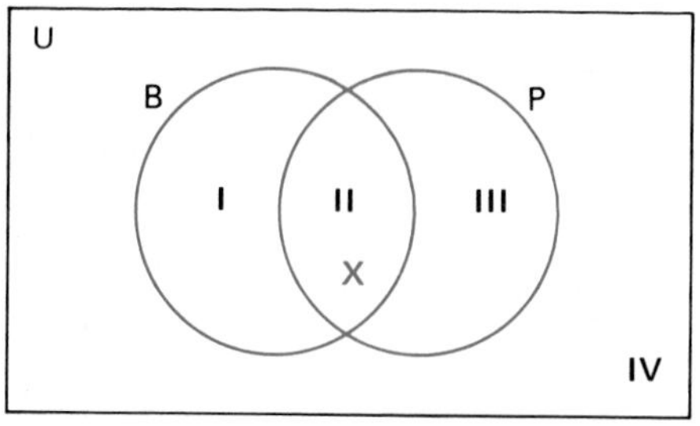

FIGURE 12b
Some bugs are pesky.

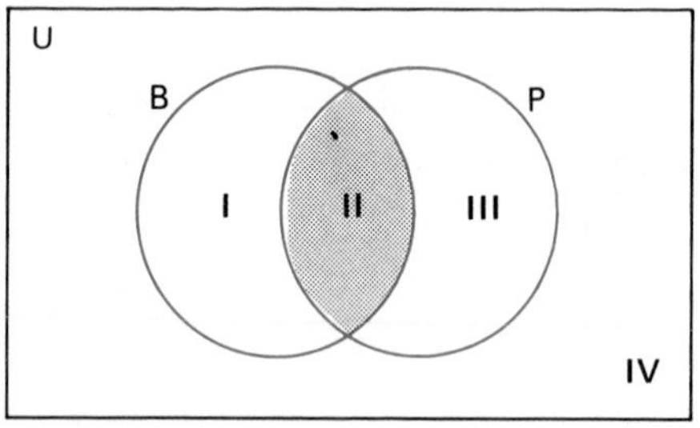

FIGURE 12c
No bugs are pesky.

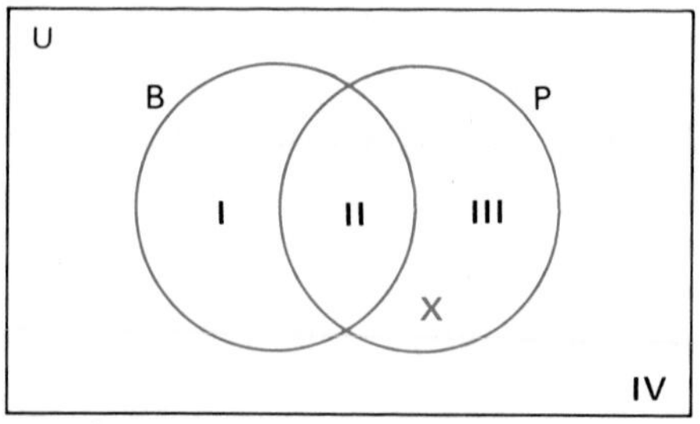

FIGURE 12d
Some pesky things are not bugs.

Statements that cannot be true together are said to be **inconsistent.** Given two statements, "Today is Monday" and "Today is not Monday," we see that these two statements contradict each other and cannot be true at the same time. Statements that ***can*** be true together are **consistent** statements; they do not contradict each other. Many times it is difficult to determine whether two statements are consistent or not. Let us see how we may use Venn diagrams to do this.

EXAMPLE 2 Determine whether the following pair of statements is consistent or inconsistent.

1. *Some students are lazy.*
2. *No students are lazy.*

SOLUTION We construct a Venn diagram and diagram both statements in the same picture (see Fig. 13). The shading in region II represents statement 2 *(No students are lazy),* and the X in region II represents statement 1 *(Some students are lazy).* We have both crossed out a region and placed an X in the region. When we cross out a region to picture a statement, it means that there are no elements in that particular region. If we place an X in a region, then there is at least one element in that region. This is a contradiction; the two statements cannot be pictured in the same diagram. They are inconsistent. Two statements that are not consistent cannot be true at the same time.

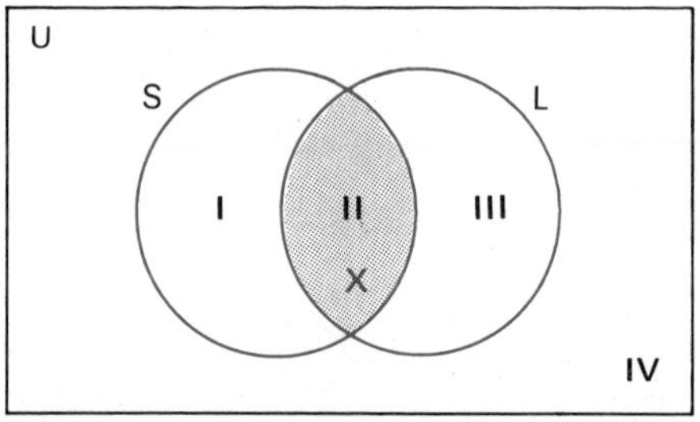

FIGURE 13
S = the set of students, and L the set of lazy people.

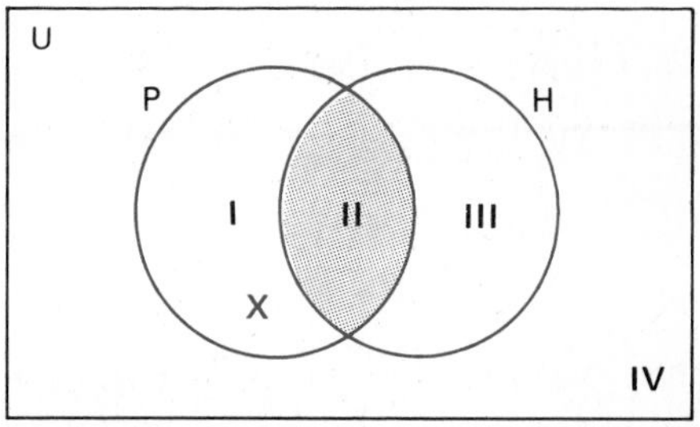

FIGURE 14
P = the set of politicians, and H the set of honest people.

EXAMPLE 3 Determine whether the following pair of statements is consistent or inconsistent.

1. *No politicians are honest.*
2. *Some politicians are not honest.*

SOLUTION We construct a Venn diagram and diagram statements 1 and 2 in the same picture (see Fig. 14). The shading in region II represents statement 1 *(No politicians are honest),* and the X in region I represents statement 2 *(Some politicians are not honest).* Because we are able to picture both statements in the same diagram without any contradiction, we may conclude that the statements are consistent.

EXERCISES FOR SECTION 2.8

Identify each statement as *universal affirmative, universal negative, particular affirmative,* or *particular negative,* and then diagram it by means of a Venn diagram.

1. No heroes are imps.
2. All girls are heroes.
3. Some girls are not imps.
4. No cheaters are happy.
5. Some taxpayers are happy.
6. All numbers are interesting.
7. Some letters are not interesting.
8. All math courses are interesting.
9. Some math courses are challenging.
10. All politicians are sly.
11. Some politicians are not honest.
12. All Republicans are politicians.
13. No math teachers are compassionate.
14. All math teachers are compassionate.
15. Some cars are not gas eaters.
16. Some gamblers are losers.
17. Some losers are not gamblers.
18. All math teachers are dull.
19. No math teachers are boring.
20. Some tests are difficult.

Use a Venn diagram to detemine whether each pair of statements is consistent or inconsistent.

21. Some kind people are clever.
    No clever people are kind.
22. Some math courses are interesting.
    No math courses are interesting.
23. No math teachers are compassionate.
    All math teachers are compassionate.
24. All dogs are barkers.
    Some barkers are not dogs.
25. Some gamblers are losers.
    Some losers are not gamblers.
26. Some tests are not easy.
    All tests are easy.
27. No math courses are boring.
    Some math courses are boring.
28. All students are procrastinators.
    No procrastinators are students.
29. All politicians are sly.
    No sly people are politicians.
30. Some real numbers are rational.
    Some rational numbers are real.
31. All logic students are gullible.
    No logic students are gullible.
32. No Democrats are politicians.
    Some politicians are not Democrats.
33. Some tests are long.
    Some tests are not long.
34. Some girls are not imps.
    No imps are girls.
35. Some houses are homes.
    Some homes are houses.

JUST FOR FUN

If it takes 5 seconds for a clock to strike 5, how long does it take to strike 10?

2.9 VALID ARGUMENTS AND VENN DIAGRAMS (OPTIONAL)

Smithsonian Institution

This machine, on display at the Smithsonian Institution in Washington, D.C., can be used to evaluate the validity of syllogisms.

An argument consists of premises and a conclusion. An argument is said to be valid if the conclusion follows from the premises. Arguments may contain more than two premises, but here we are only going to consider syllogisms. Recall that a syllogism contains a major premise, a minor premise, and a conclusion. An argument is valid if the conclusion follows from the premises.

You should be aware of the difference between truth and validity. An argument may be valid even though your own knowledge tells you that the conclusion is false. But if the conclusion follows from the premises, then the argument is valid. On the other hand, a conclusion may be true and the argument invalid. You may know that a certain conclusion is true, but if it does not follow from the premises, the argument is not valid.

Venn diagrams are useful to determine the validity of syllogisms. An argument is valid if the conclusion follows from the premises. How do we determine if a conclusion follows from the premises? We diagram the premises in a Venn diagram, and if the conclusion is shown in the diagram of the premises without any ambiguity, then the argument is valid. If it is possible to diagram the premises without at the same time showing the conclusion, then the argument is invalid. Let's examine some examples to see how this technique is used.

EXAMPLE 1 Determine whether the following argument is valid.

All students are industrious.
No dropouts are industrious.
Therefore, no dropouts are students.

SOLUTION First we construct a Venn diagram with three overlapping circles, one for the set of students, one for the set of industrious people, and one for the set of dropouts (see Fig. 15). Next we diagram each of the premises in the argument. "All students are industrious" is pictured by crossing out regions I and IV. "No dropouts are industrious" is pictured by crossing out regions V and VI. You will note that the conclusion, "No dropouts are students," is already pictured in the diagram as a result of diagramming the premises. This tells us that the argument is valid. (If the conclusion is shown in the diagram of the premises without any ambiguity, the argument is valid.) See Figure 16 for the completed diagram.

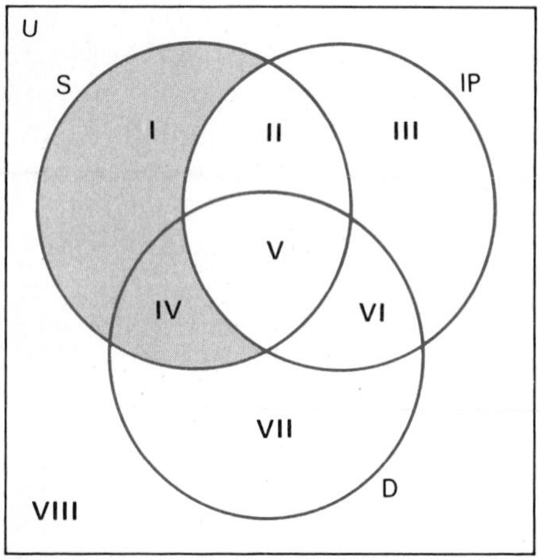

FIGURE 15
S = the set of students, IP = the set of industrious people, and D = the set of dropouts.

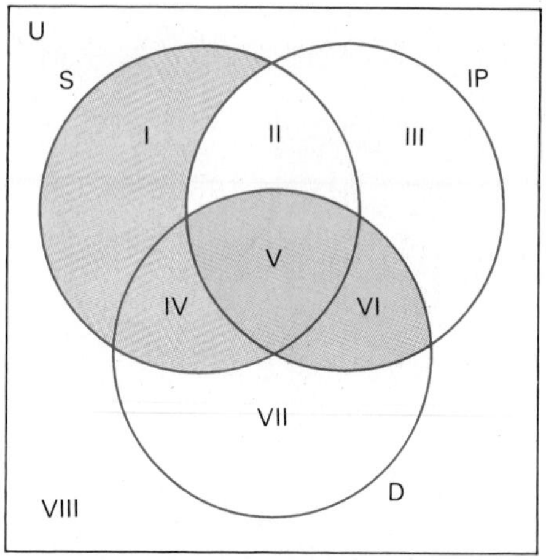

FIGURE 16

The premises of a valid argument *must* show the conclusion when they are diagrammed. We do not diagram the conclusion; its picture must be a result of diagramming the premises. If we are able to diagram the premises without showing the conclusion, then the argument is invalid.

EXAMPLE 2 Determine whether the argument is valid.

No generals are pacifists.
No pacifists are soldiers.
Hence, no generals are soldiers.

SOLUTION We diagram the premises in Figure 17 to see if the conclusion follows from the premises. We have three circles: generals, pacifists, and soldiers. "No generals are pacifists" is pictured by eliminating region II and V. "No pacifists are soldiers" is pictured by eliminating regions V and VI. We already have crossed out region V, so we need only do region VI. The conclusion, "No generals are soldiers," is not pictured in the diagram. If it were, then regions IV and V would be crossed out—and region IV is not. The conclusion does not follow from the premises; therefore the argument is invalid.

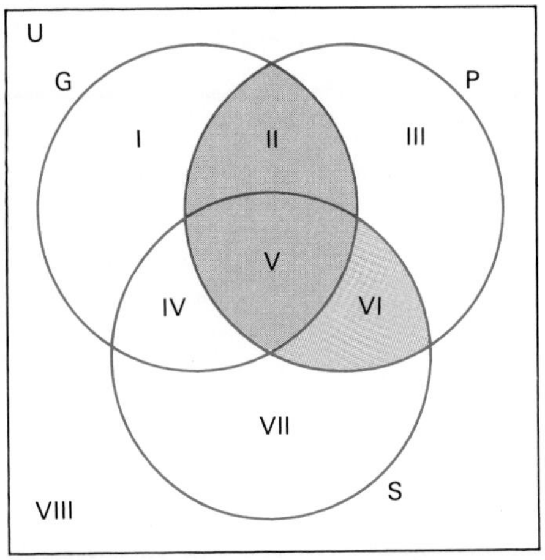

FIGURE 17
G = the set of generals, P = the set of pacifists, and S = the set of soldiers

EXAMPLE 3 Determine whether the argument is valid.

All generals are soldiers.
Some generals are fighters.
Therefore, some soldiers are fighters.

SOLUTION We diagram the premises to see if we also obtain a picture of the conclusion (see Fig. 18). The statement "All generals are soldiers" is diagrammed by eliminating regions I and IV. The second premise, "Some generals are fighters," tells us that there are some elements in set G that are also in set F. We have a choice of

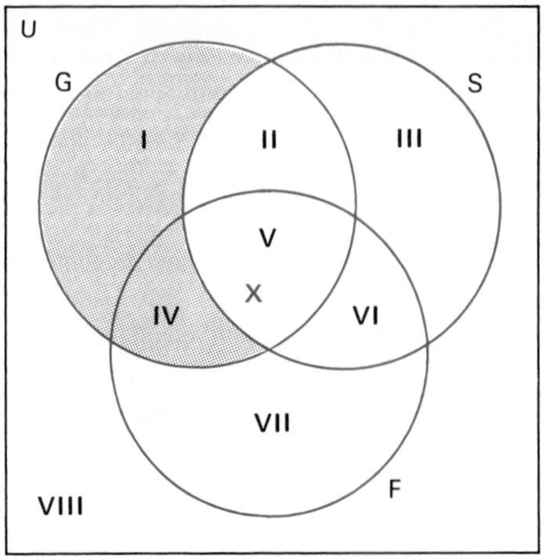

FIGURE 18
G = the set of generals,
S = the set of soldiers, and
F = the set of fighters.

placing an X in region IV or V. We must place the X in region V, since we have already eliminated region IV in diagramming the first premise. Now we see that the conclusion, "Some soldiers are fighters," is already diagrammed in Figure 18. The conclusion does follow from the given premises, so the argument is valid.

EXAMPLE 4 Determine whether the argument is valid.

All fishermen are patient.
Some fishermen are not liars.
Hence, some liars are not fishermen.

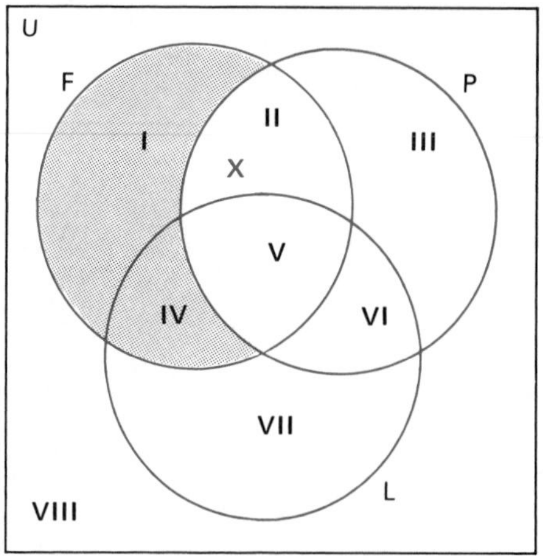

FIGURE 19
F = the set of fishermen,
P = the set of patient people, and
L = the set of liars.

HISTORICAL NOTE

Charles Lutwidge Dodgson (1832–1893), or Lewis Carroll, as we know him, was born in England. As a child he was fascinated with math and solving puzzles. His mathematical career began with teaching at Oxford. In later years, most of his time was spent writing, solving puzzles, and lecturing at Christ Church. *Alice in Wonderland* and *Through the Looking Glass* are two of his most famous works.

Carroll posed many problems in his work. His poetry, puns, and novels have been read by many mathematicians and philosophers. They have studied the logic in his work extensively, and many of them have been inspired by Carroll.

Following is a puzzle typical of those found in his writings. See if you can solve it without looking at the answer, which is given below the rhyme.

Dreaming of apples on a wall,
and dreaming often dear,
I dreamed that, if I counted all,
—how many would appear?

The answer is 10 (dreaming of-*ten*).

SOLUTION The major premise, "All fishermen are patient," is diagrammed in Figure 19 by eliminating regions I and IV. "Some fishermen are not liars" tells us that there are some elements in set F that are not in set L. We must place the X in region II, since region I has already been eliminated. The conclusion, "Some liars are not fishermen," is not pictured in the diagram. This conclusion does not follow from the premises. Hence, the argument is invalid.

Note that by placing an X in region II we could have shown the statement "Some fishermen are not liars," but that is not the desired conclusion.

If it is possible to diagram the premises without showing the conclusion at the same time, then the argument is invalid. This may occur when we diagram one particular statement and have an option of placing an X in more than one region (only after the first premise has been already diagrammed).

To summarize, *an argument is valid if the conclusion is shown without any ambiguity as soon as the premises are diagrammed.* We do not diagram the conclusion; its picture must be a result of diagramming the premises. If it is possible to diagram the premises without at the same time showing the conclusion; then the argument is invalid.

EXERCISES FOR SECTION 2.9

Use Venn diagrams to determine whether the following arguments are valid or invalid. State your answer.

1. No logic students are gullible.
   Some gullible people are superstitious.
   Hence, no logic students are superstitious.
2. All weightlifters are strong.
   Some football players are strong.
   So, some football players are weightlifters.
3. All coins are valuable.
   Some coins are old.
   Therefore, some old things are valuable.
4. All cars are gas eaters.
   Some gas eaters are four-wheeled.
   So, some cars are four-wheeled.
5. All cars have four wheels.
   No bikes have four wheels.
   Therefore, no cars are bikes.
6. Some jets are gigantic.
   Some gigantic things are heavy.
   Hence, some jets are heavy.
7. All books are readable.
   Some readable things are not interesting.
   So, some books are not interesting.
8. All mathematics students are clever.
   Some history students are clever.
   Therefore, some mathematics students are history students.
9. All logic students are adroit.
   Some logic students are ambitious.
   So, some ambitious people are adroit.
10. All logic students are adroit.
    Some logic students are ambitious.
    Hence, some adroit people are ambitious.

11. All logic students are adroit.
All adroit people are ambitious.
Therefore, some ambitious people are logic students.

12. All gamblers are losers.
Some quarterbacks are not losers.
So, some quarterbacks are not gamblers.

13. All gamblers are losers.
No losers are lucky.
Hence, no gamblers are lucky.

14. No winners are losers.
Some gamblers are losers.
Therefore, some gamblers are not winners.

15. All winners are gamblers.
Some gamblers are losers.
Hence, some gamblers are winners.

16. No skyjackers are sane.
Some writers are sane.
Hence, some skyjackers are not writers.

17. Some gamblers are not kibitzers.
Some kibitzers are not winners.
Therefore, some gamblers are not winners.

18. All politicians are leaders.
Some politicians are lawyers.
Hence, some lawyers are leaders.

19. Some teachers are dull.
All dull people are boring.
Therefore, some teachers are boring.

20. Some instructors are understanding.
All understanding people are tolerant.
So, some instructors are tolerant.

21. All warriors are brave.
No cowards are brave.
Hence, no cowards are warriors.

22. No circles are squares.
No squares are triangles.
Therefore, no circles are triangles.

23. All squares are rectangles.
Some rectangles are parallelograms.
So, some squares are parallelograms.

24. All handball players are agile.
No weightlifters are agile.
Hence, no handball players are weightlifters.

25. All actors are vain.
Some vain people are not greedy.
So, some actors are not greedy.

26. All vain people are greedy.
Some actors are greedy.
Therefore, some actors are not vain.

JUST FOR FUN Do you dial O for operator? No, you dial zero. What letters are missing on the standard telephone dial?

2.10 THE CONDITIONAL (OPTIONAL)

All math students have, at one time or another, encountered rules or theorems stated as conditionals. One of the first theorems proved in high school geometry is the conditional. "If two sides of a triangle are equal, then the angles opposite these sides are equal."

In this section, we shall examine the conditional statement in detail. We shall concern ourselves with statements that are equivalent to *If P, then Q,* and the various forms that these statements may take.

Suppose we are given two statements, P and Q, and we are asked to construct a conditional statement using P and Q, or $\sim P$ and $\sim Q$. What are the possible variations? After some thought, you would probably come up with the following list (their names are included).

1.	$P \rightarrow Q$	Statement
2.	$Q \rightarrow P$	*Converse* of statement 1
3.	$\sim P \rightarrow \sim Q$	*Inverse* of statement 1
4.	$\sim Q \rightarrow \sim P$	*Contrapositive* of statement 1
5.	$\sim(P \rightarrow Q)$	*Negation* of statement 1

Note, however, that statement 5, $\sim(P \rightarrow Q)$, is not a conditional statement, but the *negation* of a conditional statement. Therefore, we will only concern ourselves with the first four statements. By comparing their truth tables, we can discover which statements are logically equivalent.

TABLE 11

Statement			*Converse*			*Inverse*			*Contrapositive*		
P	$\rightarrow$	Q	Q	$\rightarrow$	P	$\sim P$	$\rightarrow$	$\sim Q$	$\sim Q$	$\rightarrow$	$\sim P$
T	*T*	*T*	*T*	*T*	*T*	*F*	*T*	*F*	*F*	*T*	*F*
T	*F*	*F*	*F*	*T*	*T*	*F*	*T*	*T*	*T*	*F*	*F*
F	*T*	*T*	*T*	*F*	*F*	*T*	*F*	*F*	*F*	*T*	*T*
F	*T*	*F*	*F*	*T*	*F*	*T*	*T*	*T*	*T*	*T*	*T*

From Table 11 we see that a conditional statement is logically equivalent to its contrapositive. Hence, when we have a statement such as "If I study, then I shall pass math," we know that it is equivalent to "If I did not pass math, then I did not study."

The conditional statement is dangerous in the hands of the uninformed. A common error made by some people in reasoning about mathematics or everyday occurrences is that they assume the converse or inverse of a given statement has the same truth value as the given statement. This is particularly the case when the given conditional statement seems to be obviously true. Checking the truth tables in Table 11, we see that the inverse and converse of a statement are not always true when the given statement is true.

In some of the advertising that we see, hear, or read, the

advertiser wants us to believe that the inverse or converse of a premise is true. Consider the statement

For a cleaner wash, insist on Sudsy Soap.

This translates to "If you want a cleaner wash, then use Sudsy Soap." The advertiser wants you to use Sudsy Soap, but he or she also wants you to believe that if you do not use Sudsy Soap, then your wash will not be clean; hence, you had better buy some Sudsy Soap.

Consider another conditional statement,

If x is greater than zero, then x is not negative.

This statement is true. Its converse is

If x is not negative, then x is greater than zero.

This statement is not true because x could be zero (zero is neither positive nor negative), and therefore x would not be greater than zero. Thus we see that an inverse or converse does not follow just because the given statement is true.

Recall that *then* is omitted in a conditional statement when it is thought to be obvious. Consider the following.

If anybody was late, it had to be Dave.

EXAMPLE 1 Write the converse, inverse, and contrapositive of the statement

If you use Sudsy Soap, then your clothes are clean.

SOLUTION Let P = *You use Sudsy Soap* and Q = *Your clothes are clean.* We can symbolize the given statement as $P \to Q$.

The converse of $P \to Q$ is $Q \to P$, or "If your clothes are clean, then you use Sudsy Soap." Note that even if the given statement is true, the converse is not necessarily true.

The inverse of $P \to Q$ is $\sim P \to \sim Q$, which may be stated as "If you do not use Sudsy Soap, then your clothes are not clean." This statement is also not necessarily true.

The contrapositive of $P \to Q$ is $\sim Q \to \sim P$, "If your clothes are not clean, then you do not use Sudsy Soap." If the given statement is true, then the contrapositive is also true because it is logically equivalent to the given statement.

EXAMPLE 2 Write the converse, inverse, and contrapositive of the statement

If today is Sunday, then yesterday was not Friday.

SOLUTION Let P = *Today is Sunday* and Q = *Yesterday was Friday.* The given statement is symbolized as $P \to \sim Q$.

The converse is $\sim Q \rightarrow P$, "If yesterday was not Friday, then today is Sunday."

The inverse of $P \rightarrow \sim Q$ is $\sim P \rightarrow \sim\sim Q$. Since we have a double negation, $\sim\sim Q$, we may interpret it as Q; that is, $\sim\sim Q \equiv Q$. Hence the inverse $\sim P \rightarrow \sim\sim Q$ is the same as $\sim P \rightarrow Q$: "If today is not Sunday, then yesterday was Friday."

The contrapositive of $P \rightarrow \sim Q$ is $\sim\sim Q \rightarrow \sim P$, which is the same as $Q \rightarrow \sim P$: "If yesterday was Friday, then today is not Sunday."

The conditional causes a great deal of confusion and trouble for many people because they do not understand it. We shall use an example to show the various ways that a conditional may be stated. You should already be aware that $P \rightarrow Q$ can be translated as

If P, then Q.
If P, Q.
Q if P.

In the following discussion we shall see that $P \rightarrow Q$ may also be translated in other ways. Consider the statement

If you are a citizen of Buffalo, then you are a citizen of New York State.

P = *You are a citizen of Buffalo*

Q = *You are a citizen of New York State*

Then the statement would be symbolized as $P \rightarrow Q$.

Now, suppose someone says "You are a citizen of Buffalo only if you are a citizen of New York State." What do we know if you live in New York State? We know that you live in the state of New York, but you may live in Buffalo, Rochester, Syracuse, New York City, or a thousand other places. But what do we know if you live in Buffalo? We know that you are a citizen of New York State because you are a citizen of Buffalo *only if* you are a citizen of New York State. *If* you live in Buffalo, *then* you must live in New York State.

Another statement that may help you understand *P only if Q* is

It snows only if it is cold.

It the thermometer tells us it's cold, do we know anything? We know it's cold, but we do not know if it will snow. But suppose we look out the window and see that it is snowing. Then we can deduce that it is cold. Why? Because we know that it snows only if it is cold. That is, *if* it snows, *then* it is cold. We would conclude that *P only if Q* may be symbolized as $P \rightarrow Q$.

Consider a variation of the citizenship statement, "Being a citizen of Buffalo is sufficient for being a citizen of New York State." This statement means that in order to be a citizen of New York State, it is enough to be a citizen of Buffalo, since every citizen of the city of Buffalo is also a resident of the state of New York. Do you have to do anything else? No, because *if* you are a citizen of Buffalo, *then* you are a citizen of New York State.

Suppose an instructor tells you that doing your homework every day is sufficient for passing the course. What do you know? Do you have to do your homework every day? The instructor did not say that you had to do your homework every day; but he did say that if you did, then you would pass the course. You could pass the course some other way, perhaps by passing all of the exams. But you know that if you do your homework every day, then you will pass the course. Suppose you did not pass the course? We may conclude that you did not do your homework every day. That is, if you did not pass the course, then you did not do your homework every day. But this is the contrapositive of "If you do your homework every day, then you will pass the course." We may conclude from these discussions that *P is sufficient for Q* may be symbolized as $P \rightarrow Q$.

Another variation of "If you are a citizen of Buffalo, then you are a citizen of New York State" is the following sentence: "Being a citizen of New York State is necessary for being a citizen of Buffalo." This means that being a citizen of New York State is a necessary condition for being a citizen of Buffalo. Do you have to be a citizen of Buffalo in order to be a citizen of New York? No, but you have to be a citizen of New York State in order to be a citizen of Buffalo. Since you have to be a citizen of New York State, we might conclude that

If you are not a citizen of New York State, then you are not a citizen of Buffalo.

But that is the contrapositive of

If you are a citizen of Buffalo, then you are a citizen of New York State.

Therefore, we may conclude that *Q is necessary for P* may be symbolized as $P \rightarrow Q$.

Another example that may aid you in understanding *Q is necessary for P* is the following.

Oxygen is necessary for fire.

Given this statement, we know that we cannot have fire unless we have oxygen. That is, if we do not have oxygen, then we can't have a fire. Again, we can apply the contrapositive to this, and we have the statement.

If we have fire, then we have oxygen.

In summary, we have the fact that *Q is necessary for P* is symbolized as $P \rightarrow Q$.

Consider the statement

If you studied Algebra II, then you studied Algebra I.

Three other sentences that have the same meaning are

You studied Algebra II only if you studied Algebra I.
Studying Algebra II is sufficient for studying Algebra I.
Studying Algebra I is necessary for studying Algebra II.

When we have the conditional statement $P \rightarrow Q$, it is usually interpreted as *If P, then Q,* but three other common equivalent wordings are

P only if Q.
P is sufficient for Q.
Q is necessary for P.

The following is a list of equivalent wordings of the conditional statement $P \rightarrow Q$. Bear in mind that the list is a sampling and not intended to be a list of all possible variations, but all statements listed are interpretations of the original conditional statement.

1. *If P, then Q.*
2. *If P, Q.*
3. *Q if P.*
4. *P implies Q.*
5. *Q is implied by P.*
6. *Q whenever P.*
7. *P only if Q.*
8. *P is sufficient for Q.*
9. *Q is necessary for P.*

(All nine: $P \rightarrow Q$)

The following examples are conditional statements using some of the variations in the previous list.

1. If the Mets won, then the Dodgers lost.
2. If the Mets won, the Dodgers lost.
7. The Mets won only if the Dodgers lost.
8. The Mets' winning is sufficient for the Dodgers' losing.
9. The Dodgers' losing is necessary for the Mets' winning.

EXAMPLE 3 Let P = *Today is Saturday* and Q = *Tomorrow is Sunday*. Write each of the following statements in symbolic form.

a. If today is Saturday, then tomorrow is Sunday.
b. If tomorrow is not Sunday, then today is not Saturday.
c. Tomorrow is Sunday if today is Saturday.
d. Today's being Saturday is sufficient for tomorrow's being Sunday.
e. Tomorrow's being Sunday is necessary for today's being Saturday.
f. Today is not Saturday only if tomorrow is not Sunday.
g. It is false that today's being Saturday is sufficient for tomorrow's being Sunday.
h. Tomorrow's not being Sunday is sufficient for today's not being Saturday.

SOLUTION a. $P \rightarrow Q$ b. $\sim Q \rightarrow \sim P$ c. $P \rightarrow Q$
d. $P \rightarrow Q$ e. $P \rightarrow Q$ f. $\sim P \rightarrow \sim Q$
g. $\sim(P \rightarrow Q)$ h. $\sim Q \rightarrow \sim P$

Because a conditional statement may be written in various ways, the same is true for a biconditional statement. Consider the statement.

A number is positive if and only if that number is greater than zero.

From the discussion on truth tables in Section 2.5, we know that this statement may be rewritten as a conjunction.

If a number is positive, then the number is greater than zero and if a number is greater than zero, then the number is positive.

Let P = *A number is positive* and Q = *A number is greater than zero.* Then the above statement would be symbolized as

$$(P \rightarrow Q) \wedge (Q \rightarrow P)$$

For the first part of the conjunction, we can say *P is sufficient for Q,* and for the second part, *P is necessary for Q.* Or we could switch them around: *P is necessary for Q* and *P is sufficient for Q,* or simply P is necessary and sufficient for Q.

$$P \leftrightarrow Q \quad \begin{cases} P \text{ if and only if } Q \\ P \text{ is necessary and sufficient for } Q \end{cases}$$

EXAMPLE 4 Using the suggested notation, write the following in symbolic form:

a. You may play on this course if and only if you are a member of Oak Hill. (C, O).
b. Citizens become irate whenever, and only whenever, their taxes are raised. (C, T)
c. Being a student here is a necessary and sufficient condition for obtaining a yearbook. (S, Y)
d. Quickness and coordination are necessary and sufficient for being a pole vaulter. (Q, C, P)

SOLUTION a. $C \leftrightarrow O$ b. $C \leftrightarrow T$ c. $S \leftrightarrow Y$ d. $(Q \wedge C) \leftrightarrow P$

EXERCISES FOR SECTION 2.10

1. Use the suggested notation to write the converse, inverse, and contrapositive for each of the following statements in symbolic form.
   a. If Jim studies, then he will pass. (S, P)
   b. We will go skating if the pond freezes over. (F, S)
   c. If the Sabres do not make the playoffs, then the fans will be sad. (S, F)
   d. If logic is easy, then geometry is not difficult. (L, G)
   e. If today isn't Thursday, then yesterday wasn't Wednesday. (T, W)

2. "If the storm is a hurricane, then its winds are traveling at least 74 miles per hour." Given that this is a true statement, which of the following must also be true? (Use your knowledge of the conditional.)
   a. If the storm is not a hurricane, then its winds are not traveling at least 74 miles per hour.
   b. If the storm's winds are traveling at least 74 miles per hour, then the storm is a hurricane.
   c. It is false that if the storm's winds are not traveling at least 74 miles per hour, then the storm is not a hurricane.
   d. If the storm's winds are not traveling at least 74 miles per hour, then the storm is not a hurricane.
   e. The storm is not a hurricane, or the storm's winds are traveling at least 74 miles per hour.
   f. The storm's winds are not traveling at least 74 miles per hour, or the storm is not a hurricane.

3. Let P = *It is a logic course* and Q = *It is interesting*. Write each of the following statements in symbolic form.
   a. It is a logic course only if it is interesting.
   b. If it isn't interesting, then it isn't a logic course.
   c. The fact that it is a logic course is sufficient for it to be interesting.
   d. Only if it is interesting is it a logic course.
   e. Being a logic course is necessary and sufficient for it to be interesting.

4. Use the suggested notation to write the following in symbolic form.
   a. A student can be successful in mathematics only if he does his homework and pays attention in class. (S, H, P)
   b. A sufficient condition for a student to flunk is that he not attend class. (S, A)
   c. A necessary condition for me to go swimming is that the water must be warm or the air temperature must be 80 °F. (S, W, T)

d. If a person is contributing to society, he is contributing to society only if he is helping his fellow man. (C, H)
e. Only exams are boring. (E, B)

5. Let P = *He works hard* and Q = *He is happy*. Write each of the following statements in symbolic form.
   a. He works hard only when he is happy.
   b. If he isn't working hard, then he isn't happy.
   c. His working hard is necessary for his being happy.
   d. He works hard iff he is happy.
   ★e. He is not happy unless he works hard.

6. Sneaky Sam made the following promise to Gullible Gary: "I will help you only if you give me five dollars." Gary gave Sam five dollars and Sam walked away. Did Sam break his promise? Why or why not?

7. An instructor told a student "I will pass you only if you come to class every day." The student came to class every day and still failed the course. Does the student have a legitimate gripe? Why or why not?

2.11 SWITCHING NETWORKS (OPTIONAL)

A practical application of logic is the design of switching networks. Switching networks may also be thought of as electrical circuits. Electrical circuits, such as those in light switches and computers, are familiar objects of everyday experience. You do not need any previous knowledge of electricity to understand the material presented in this section.

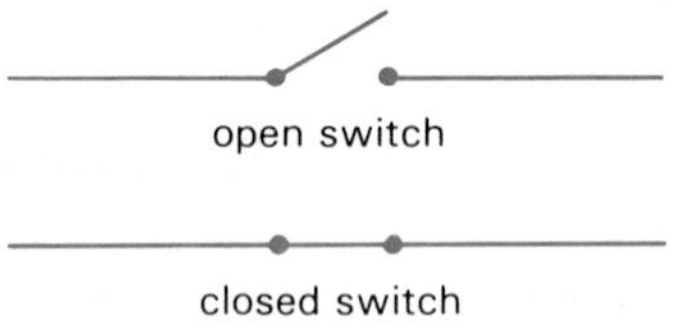

FIGURE 20

A *switching network* is an arrangement of wires and switches connecting two terminals. A *switch* can exist in two possible positions. A switch is "on" if electricity can flow through it (the switch is closed), and a switch is "off" if electricity cannot flow through it (the switch is open). See Figure 20.

An example of a switching network is the circuit attached to a doorbell. If the button is depressed, the switch becomes closed, electricity flows through it, and the doorbell rings. Normally the button is not depressed, the switch is open, and electricity cannot flow through it; hence the doorbell does not ring. A simple circuit for a doorbell appears in Figure 21.

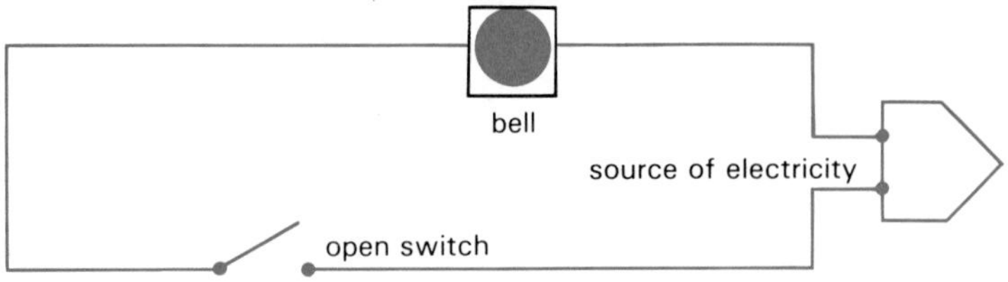

FIGURE 21

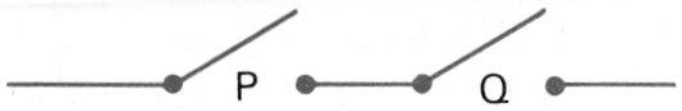

FIGURE 22
Series connection

It should be noted that most switching networks encountered in everyday experience usually consist of more than one switch. Two switches can be arranged in a switching network in two different ways. The two arrangements are *series* connection and *parallel* connection. Figure 22 shows two switches P and Q connected in series. In a series connection, electricity will flow through the network only when both switches P and Q are closed. Electricity will not flow through the network if one or both of the switches are open. Does this remind you of anything in logic? Note that a series connection resembles the conjunction $P \wedge Q$. A conjunction is true only when both parts are true; otherwise, it is false. The series connection in Figure 22 will work (allow electricity to flow through) only when both switches are closed; otherwise, current will not pass through the network.

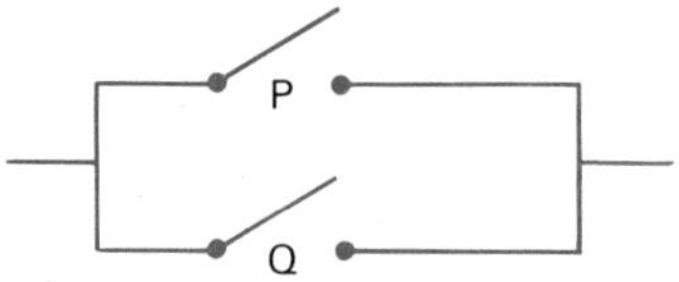

FIGURE 23
Parallel connection

Figure 23 shows two switches in a parallel connection. In a parallel connection, electricity will flow through the network if P is closed, or Q is closed, or if both P and Q are closed. The only time that current will not pass through this network is when both switches are open. A parallel connection corresponds to the disjunction $P \vee Q$. A disjunction is true when P is true, or Q is true, or when both P and Q are true. The only time a disjunction is false is when both parts are false. The only time that current will fail to pass through a parallel connection is when both switches are open.

In some networks there are switches whose positions (open–closed) are determined by another switch. That is, when one switch is open, the other switch is in the opposite position (closed). Similarly, when one switch is closed, the other switch is open. Two switches that always have opposite positions are said to be *complementary switches.* If we have two complementary switches, we can name one of them P and its complement P'. Figures 24 and 25 illustrate complementary switches in series and parallel connections, respectively.

FIGURE 24

Note that in Figure 24 we have complementary switches in a series connection. If P is open, then P' is closed; and if P is closed, then P' is open. One of the two switches will always be open. Therefore, it is impossible for electricity to flow through this network. In Figure 25 we have complementary switches in a parallel connection. Electricity will always flow through this network. One of the switches will always be closed. Again, notice the similarity of complementary switches with a statement and its negation in logic. If P is true, then $\sim P$ is false, and if P is false, then $\sim P$ is true. A statement and its negation always have opposite truth values.

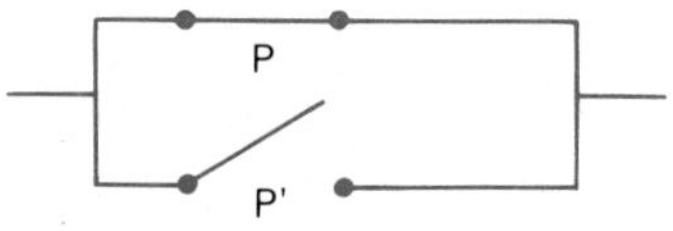

FIGURE 25

So far we have discussed series, parallel, and complementary switches. Let's examine these switches and their corresponding

FIGURE 26

truth tables. But, first let us agree that the notation in Figure 26 will represent a switch whose position is not known. That is, we do not know if switch P is open or closed.

Table 12 shows the logic truth table corresponding to the switching network in Figure 27. Current will flow only when P and Q are closed.

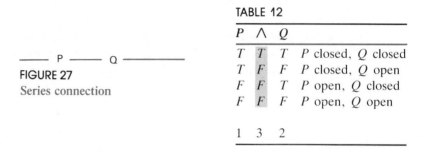

FIGURE 27
Series connection

TABLE 12

P	$\wedge$	Q	
T	T	T	P closed, Q closed
T	F	F	P closed, Q open
F	F	T	P open, Q closed
F	F	F	P open, Q open
1	3	2	

Table 13 shows the truth table corresponding to the switching network in Figure 28. Current will flow when P is closed, or Q is closed, or both P and Q are closed.

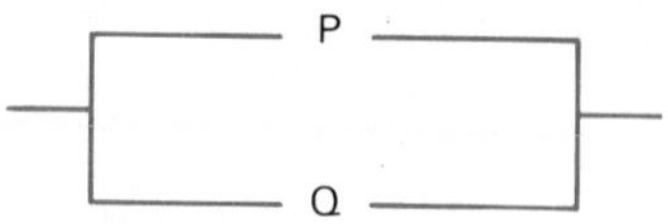

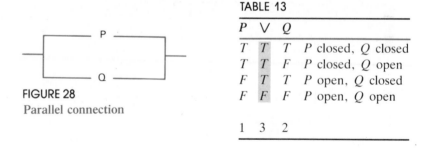

FIGURE 28
Parallel connection

TABLE 13

P	$\vee$	Q	
T	T	T	P closed, Q closed
T	T	F	P closed, Q open
F	T	T	P open, Q closed
F	F	F	P open, Q open
1	3	2	

Figures 29 and 30 show series and parallel connection of complementary switches. In Figure 29, current cannot flow through: when P is closed, P' is open, and vice versa. In Figure 30, current will always flow: when P is closed, P' is open, and vice versa. Note, the logic symbol for P' is $\sim P$.

FIGURE 29
Series connection of complementary switches

TABLE 14

P	$\wedge$	$\sim$	P
T	F	F	T
F	F	T	F
1	3	2	1

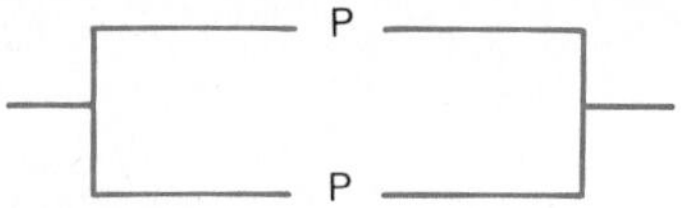

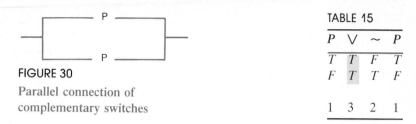

FIGURE 30
Parallel connection of complementary switches

TABLE 15

P	$\vee$	$\sim$	P
T	T	F	T
F	T	T	F
1	3	2	1

We have already discussed complementary switches, P and P'. These two switches will always have opposite positions to each other. When P is open, P' is closed, and when P is closed, P' is open. There also exist switches that are set up to operate together: when one is open the other is open; and when one is closed, the other is closed; they are always in the same position. When switches are set up to operate together in a network, we use the same letter to designate each such switch. Figure 31 shows a network that has two such switches. We have labeled each of these with the letter P.

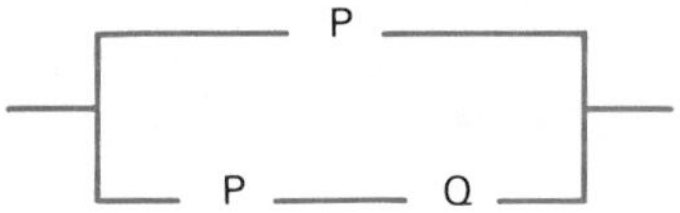

FIGURE 31

You will note in Figure 31 that we have a network that is more involved than the previous examples we have considered. Earlier in this chapter we took simple statements and worked out truth tables for them; then we combined the simple statements into compound statements. We are about to do the same thing with networks. Figure 31 is a parallel connection, but within the parallel network is a series connection. The P, Q connection in the bottom wire is a series connection. From our previous discussion, we know that this can be represented as $P \wedge Q$. Since the top wire contains switch P and the whole network is parallel (current will flow through either the top wire or bottom wire), the entire network may be represented by $P \vee (P \wedge Q)$.

EXAMPLE 1 Write a symbolic statement for the network shown in Figure 32.

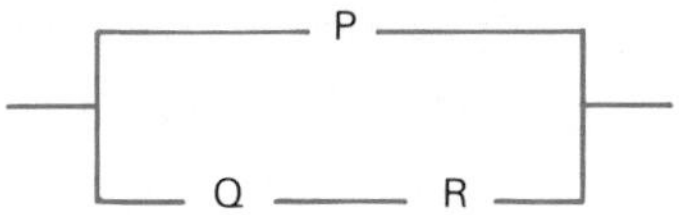

FIGURE 32

SOLUTION The network shown is parallel. The bottom wire contains switches Q and R, and they are connected by a series connection, so we can write $Q \wedge R$. Switch P is in parallel with the rest of the network; therefore, the symbolic statement is $P \vee (Q \wedge R)$.

EXAMPLE 2 Write a symbolic statement for the network shown in Figure 33.

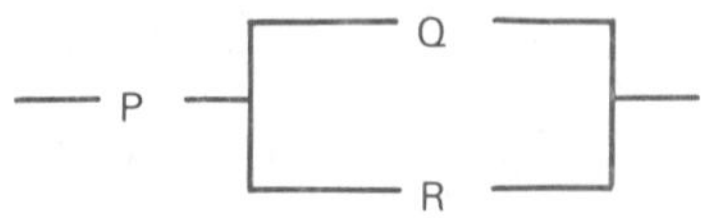

FIGURE 33

SOLUTION Switches Q and R are connected in parallel so we write $Q \vee R$. Note that switch P is in a series connection with the rest of the network. Therefore, the symbolic statement is $P \wedge (Q \vee R)$.

EXAMPLE 3 Write a symbolic statement for the network shown in Figure 34.

FIGURE 34

SOLUTION We have two sets of parallel switches; P and Q are in parallel, as are P' and R. But, note that the sets of parallel switches are in series connection with each other. Therefore, we can symbolize this network as $(P \vee Q) \wedge (\sim P \vee R)$. Note that we used the logic symbol $\sim P$ for the switch P'.

EXAMPLE 4 Write a symbolic statement for the network shown in Figure 35.

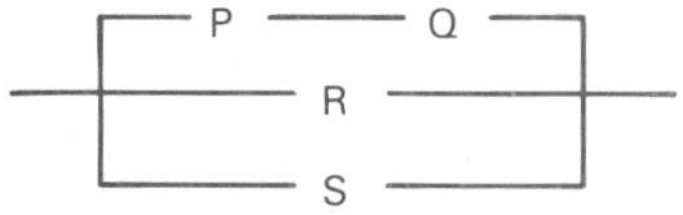

FIGURE 35

SOLUTION Note that the given circuit is a parallel one. There are three wires that are parallel to each other. The top wire contains two switches P and Q which are set in a series connection, so we have $P \wedge Q$ for the top wire. Since each wire is in parallel with the others, we can symbolize this network as $(P \wedge Q) \vee R \vee S$.

So far we have been considering given networks and then writing symbolic statements for them. We can also construct (draw) networks that correspond to given symbolic statements. When we interpret such statements, we must remember that the connective $\vee$ means a parallel connection, the connective $\wedge$ means a series connection, and the negation symbol $\sim P$ means we have a switch that is complementary to switch P. In a circuit we denote this complementary switch by P'.

EXAMPLE 5 Draw a network representing the statement $(P \wedge Q) \vee (R \wedge S)$.

SOLUTION The statement $(P \wedge Q) \vee (R \wedge S)$ is a disjunction. The connective $\vee$ is the dominant connective. Therefore, the circuit will contain two parallel wires. Each wire contains two switches set in a series connection, $P \wedge Q$ and $R \wedge S$. Hence, the network is shown in Figure 36.

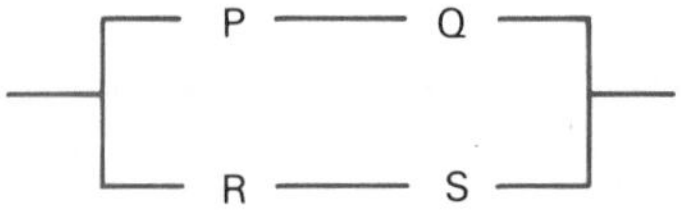

FIGURE 36

EXAMPLE 6 Draw a network to represent the following statement: $(P \vee Q) \wedge [P \vee (R \wedge \sim Q)]$.

SOLUTION The parentheses and brackets indicate that the statement $(P \vee Q) \wedge [P \vee (R \wedge \sim Q)]$ is a conjunction. Therefore, the entire network will be in series. The first part, $(P \vee Q)$, is a parallel connection, as is $[P \vee (R \wedge \sim Q)]$. Inside of $[P \vee (R \wedge \sim Q)]$ we have a series connection, $R \wedge \sim Q$. The resulting network is shown in Figure 37. Note that $\sim Q$ appears in the network as Q'.

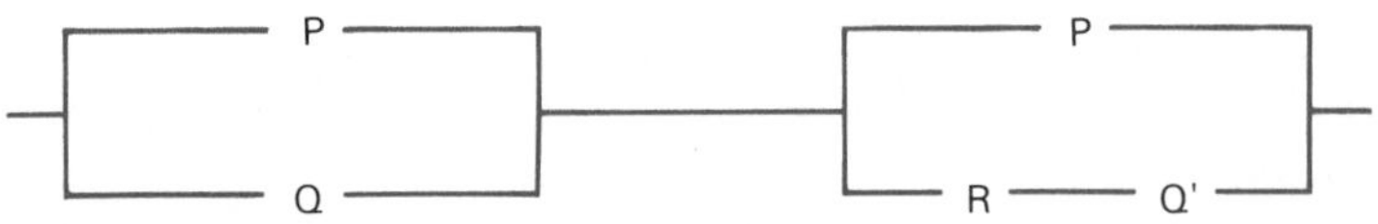

FIGURE 37

We have used logic to develop switching networks, and we may use logic to simplify a network. When we simplify a network, we reduce the number of switches in it. In practical terms, a person or company can save money if the number of switches

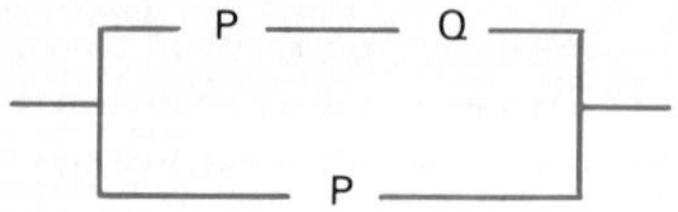

FIGURE 38

required in a network is reduced. Two networks are considered to be equivalent when both networks function in the same manner. That is, if current flows through the first network, then it flows through the second, and if current does not flow through the first, then it does not flow through the second. Consider the network shown in Figure 38. The symbolic statement for this network is $(P \wedge Q) \vee P$. Constructing a truth table for the statement $(P \wedge Q) \vee P$, we have Table 16.

TABLE 16

P	Q	$(P$	$\wedge$	$Q)$	$\vee$	P
T	T	T	T	T	T	T
T	F	T	F	F	T	T
F	T	F	F	T	F	F
F	F	F	F	F	F	F
1	2	1	3	2	4	1

The answer, or final result, is column 4. A switch is closed when it has a T value and open when it has an F value. Current will flow through this network whenever switch P is closed. In Table 16 horizontal rows 1 and 2 indicate that this is the case. Similarly, current will not flow through the network when P is open. Therefore, a simpler network need only contain switch P. Thus, the network $(P \wedge Q) \vee P$ in Figure 38 is equivalent to the network in Figure 39, namely, P.

FIGURE 39

EXAMPLE 7 Find and draw a simplified network equivalent to the one shown in Figure 40.

FIGURE 40

SOLUTION We must first write a symbolic statement for the network shown in Figure 41. It is a parallel network with series connections in the top and bottom wire. Therefore, the symbolic statement is $(P \wedge Q) \vee (P \wedge R)$.

P	Q	R	$(P$	$\wedge$	$Q)$	$\vee$	$(P$	$\wedge$	$R)$
T	T	T	T	T	T	T	T	T	T
T	T	F	T	T	T	T	T	F	F
T	F	T	T	F	F	T	T	T	T
T	F	F	T	F	F	F	T	F	F
F	T	T	F	F	T	F	F	F	T
F	T	F	F	F	T	F	F	F	F
F	F	T	F	F	F	F	F	F	T
F	F	F	F	F	F	F	F	F	F
1	2	3	1	4	2	6	1	5	3

Next, we construct a truth table for this statement. Column 6 is our answer, and we see that the statement is true in the first three horizontal lines in our table. Current will flow through the network when switch P is closed and Q *or* R is closed. Hence, a symbolic statement for an equivalent network is $P \wedge (Q \vee R)$. This network is shown in Figure 41. We have reduced the number of switches by one. This could save a manufacturing company a lot of money.

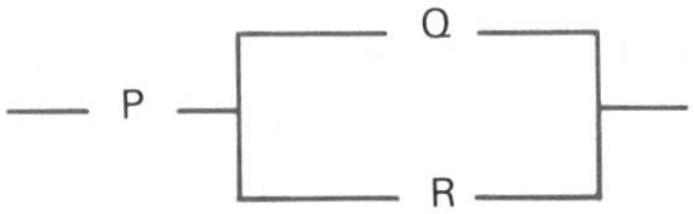

FIGURE 41

ALTERNATE SOLUTION Recall that in Section 2.6 we discussed logically equivalent statements. Two logical equivalences that we mentioned were the distributive properties:

$$P \wedge (Q \vee R) \equiv (P \wedge Q) \vee (P \wedge R)$$
$$P \vee (Q \wedge R) \equiv (P \vee Q) \wedge (P \vee R)$$

Note that the symbolic statement for the network given in Figure 41 is $(P \wedge Q) \vee (P \wedge R)$. This is part of the first distributive property above. Therefore, $(P \wedge Q) \vee (P \wedge R)$ may be simplified to the equivalent statement $P \wedge (Q \vee R)$ by means of the distributive property. This is the desired result for the simplified, equivalent network shown in Figure 41.

EXAMPLE 8 Given the network shown in Figure 42 find and draw a simplified, equivalent network using the distributive property.

FIGURE 42

SOLUTION We must first write a symbolic statement for the given network. It is a series network with parallel switches in the series. The symbolic statement is $(P \lor Q) \land (P \lor R)$. By means of the distributive property of Section 2.6, we have $(P \lor Q) \land (P \lor R) \equiv P \lor (Q \land R)$. The network corresponding to the simplified, equivalent statement is shown in Figure 43.

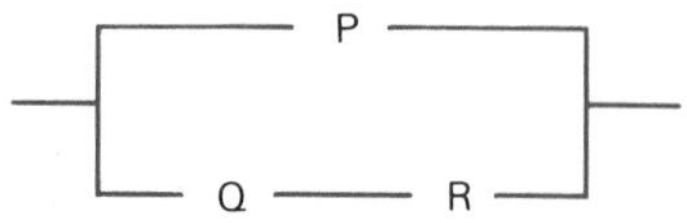

FIGURE 43

The next problem is a traditional one in switching networks. It is usually called the "hall light problem." We have a hall light with a switch at either end of the hall. We want to control the light from either switch. That is, we want to be able to turn the light on (or off) by means of either switch. We want current to pass through the network when we operate either of the switches; we also want the current to cease when we operate either of the switches. That is, we want the value of the network to change whenever we change the position of one of the switches. Let's call the switches P and Q. Listing the possibilities for P and Q, we have Table 17.

TABLE 17

	P	Q
1.	T	T
2.	T	F
3.	F	T
4.	F	F

In the case represented by line 1 in Table 17, P and Q are both closed, and current could definitely flow. It would cease if we changed the position of P or Q, as in the cases corresponding to lines 2 and 3. Now, the fourth case (line 4 in Table 17) could be the result of changing the position of either P or Q in lines 2 and 3, and in this case we would want the current to flow through the network.

In other words, the hall light will be on when P and Q are both T, or when P and Q are both F. We need a parallel network that will function in this manner. (A series network will not work when a switch is open, that is, F.) If P and Q are both true, then $(P \land Q)$ is true, and if P and Q are both false, then $(\sim P \land \sim Q)$ is true. Thus, the symbolic statement that behaves in the desired manner is $(P \wedge Q) \vee (\sim P \wedge \sim Q)$, as shown in Table 18. The corresponding network is shown in Figure 44.

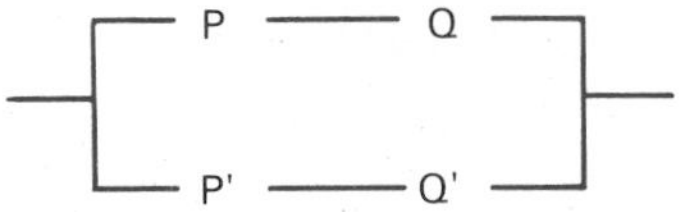

FIGURE 44

TABLE 18

P	Q	$(P$	$\wedge$	$Q)$	$\vee$	$(\sim P$	$\wedge$	$\sim Q)$
T	T	T	T	T	T	F	F	F
T	F	T	F	F	F	F	F	T
F	T	F	F	T	F	T	F	F
F	F	F	F	F	T	T	T	T
1	2	1	3	2	7	4	6	5

EXERCISES FOR SECTION 2.11

For exercises 1–10, write a symbolic statement for the networks shown.

1. 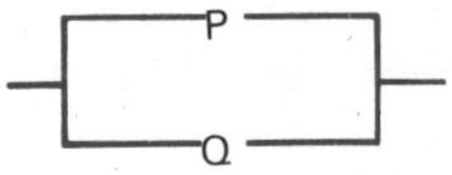

2. 

3. 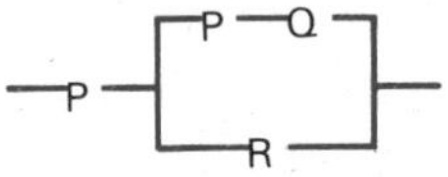

4. 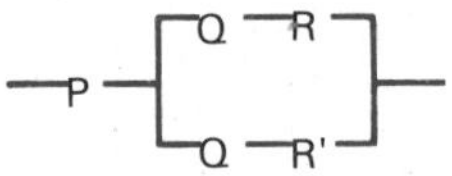

5. 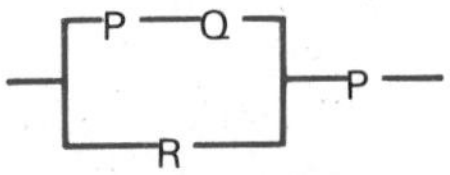

6. 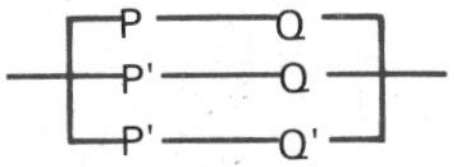

7. 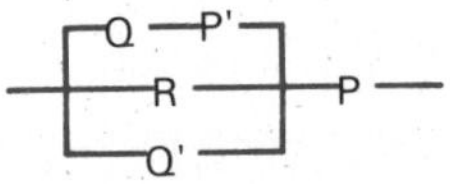

8. 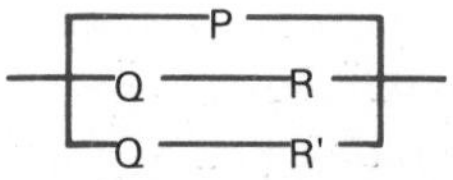

9. 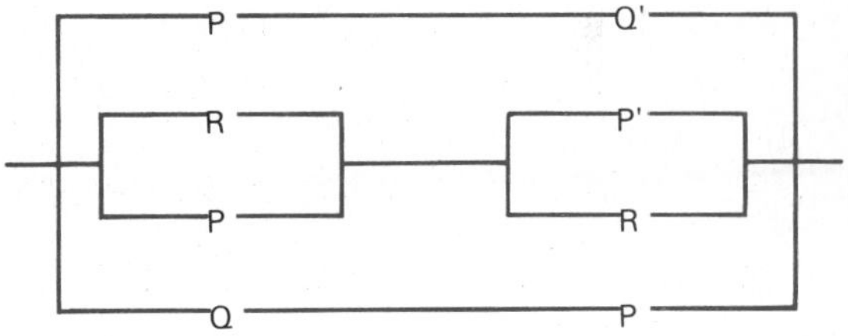

10. 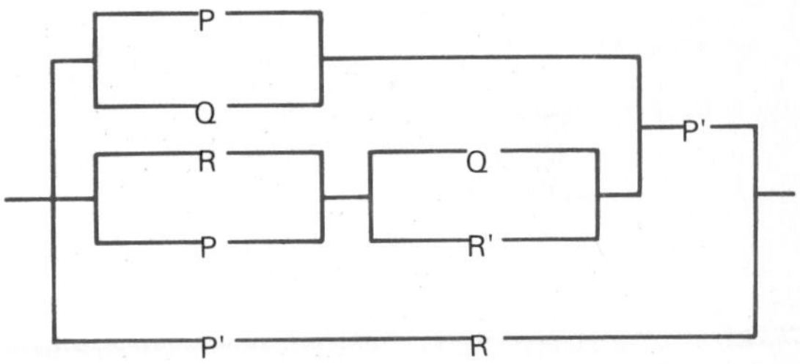

For exercises 11–20, construct the switching networks that correspond to the given statements.

11. $P \wedge (Q \vee R)$
12. $P \vee (Q \wedge R)$
13. $(P \vee Q) \wedge (P \vee R)$
14. $(P \wedge Q) \vee (P \wedge R)$
15. $(P \vee Q) \wedge (\sim P \vee R)$
16. $(P \wedge Q) \wedge [(P \wedge R) \vee (\sim Q \wedge P)]$
17. $P \vee [Q \wedge (P \vee \sim Q)]$
18. $(P \wedge \sim Q) \vee [(P \wedge Q) \vee (\sim P \wedge Q)]$

★19. $P \to Q$

★20. $P \to (Q \wedge R)$

21. Construct switching networks that are equivalent to and also simpler than the given networks.

a.
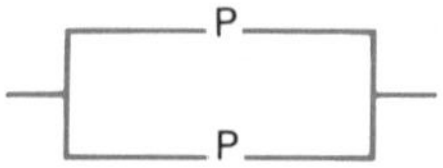

b.
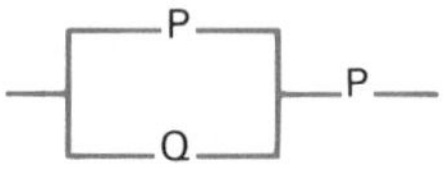

c.

d.
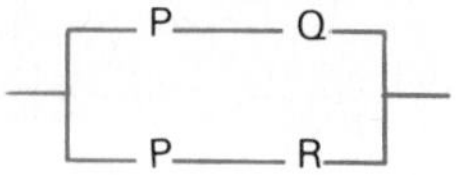

22. A committee of three (Bob, Joe, and Fred) must vote yes or no on an important proposal. They want to do this secretly; design a switching network for indicating a majority vote. Write the symbolic statement and draw the network.

JUST FOR FUN Which is bigger?

3 7

Summary

In this chapter, we were introduced to modern logic. We considered only *statements* (*propositions*) that are either true or false, but not both.

Besides examining simple statements, we also studied compound statements of the following types: *negation*, *conjunction*, *disjunction*, *conditional*, and *biconditional*.

We learned that letters such as P, Q, R, and S, and sometimes A, B, and C, are traditionally used to symbolize simple statements in logic. The connectives $\wedge$ (*and*), $\vee$ (*or*), $\to$ (*if . . . then . . .*), $\leftrightarrow$ (*if and only if*), and $\sim$ (*not*) are used to connect the simple statements in a compound statement. Parentheses are used to tell us what type of compound statement we are considering. When parentheses are omitted, we follow the convention of the dominance of connectives. The biconditional ($\leftrightarrow$) is the most dominant connective, followed by the conditional ($\to$), and then the equally dominant conjunction ($\wedge$) and disjunction ($\vee$). The negation ($\sim$) is the least dominant connective.

Truth tables show us all possible truth values for given compound statements. A statement that is always true is called a *tautology* and a statement that is always false is called a *contradiction*. Truth tables enable us to determine *logical equivalence* between two statements. We may also use *De Morgan's law* to create logically equivalent statements from conjunctions and disjunctions.

Truth tables are also used to determine whether an argument is *valid* or *invalid*. An argument is valid if the conclusion follows from the premises. To test whether an argument is valid, we first connect the premises by means of conjunction and then connect the resulting statement to the conclusion to form a conditional. Next we complete the truth table for this conditional, and if the truth table shows that the conditional is a tautology, then the argument is valid. If the conditional is not a tautology, then the argument is invalid.

Venn diagrams can be used to picture certain kinds of quantified statements. These statements are of four types: the *universal affirmative, universal negative, particular affirmative,* and *particular negative.* We can use Venn diagrams to determine whether two statements are *consistent.* Statements that cannot be true at the same time are *inconsistent.*

We can also use Venn diagrams to determine whether a syllogism (one type of argument) is valid. To determine if a syllogism is valid, we picture the premises in a Venn diagram, and if the conclusion is shown without any ambiguity as soon as the premises are diagrammed, then the argument is valid. If it is possible to diagram the premises without showing the conclusion at the same time, then the argument is invalid.

A conditional statement may be worded in various ways. Statements such as *P only if Q, P is sufficient for Q,* and *Q is necessary for P* are all logically equivalent to *If P, then Q,* and may be symbolized as $P \to Q$. Logic enables us to take statements of this type and symbolize them in some form of $P \to Q$.

A conditional statement is logically equivalent to its *contrapositive.* When a conditional statement is true, its *converse* does not have to be true; and when a conditional statement is true, its *inverse* does not have to be false.

An application of the principles of logic is the development of switching networks. A *switching network* is an arrangement of wires and switches connecting two terminals. A switch is on (closed) if current can flow through it, and off (open) if current cannot flow through it. Switches can be arranged in a *series* connection (*P and Q*), or a *parallel* connection (*P or Q*), or a combination of series and parallel connections. *Complementary switches* (P and P') are pairs of switches that are always in opposite positions. A series connection corresponds to a conjunction in logic, a parallel connection corresponds to the disjunction, and complementary switches correspond to a statement and its negation.

JUST FOR FUN

Let us agree that all special agents always tell the truth and all spies always lie. Once three people appeared in a court of law before a judge. The first accused person spoke, but the judge was unable to hear what the accused had said. The judge asked the second accused person what the first had said. Number 2 replied, "She said she is a spy." The judge then asked the third accused person whether number 2 was a special agent or a spy. Number 3 replied, "Number 2 is a special agent." Determine whether each accused person was a special agent or a spy.

Vocabulary Check

statement	iff	consistent	inverse
negation	contradiction	syllogism	contrapositive
conjunction	biconditional	premise	implication
disjunction	logically equivalent	tautology	conclusion
condition	valid	inconsistent	inclusive "or"
De Morgan	invalid	converse	paradox

Review Exercises for Chapter 2

1. Identify each sentence as a simple statement, compound statement, or neither. If the statement is compound, then classify it as a negation, conjunction, disjunction, conditional, or biconditional.
   a. Either Hugh sold his car, or he traded it in.
   b. Is it raining?
   c. Yesterday was Friday.
   d. Scott stayed home, but Joe went to the show.
   e. Carlos did not bring his notes to class.
   f. If Mary went swimming, then she did not study.

2. By means of the appropriate connectives and parentheses, symbolize each statement, using the given symbols for the simple statements.

   P = *Paula is polite,* Q = *Sally is quiet,* and R = *Ralph is right.*

   a. Paula is polite, but Sally is quiet.
   b. If Sally is not quiet, then Ralph is not right or Paula is polite.
   c. Neither is Paula polite nor is Sally quiet.
   d. Paula is polite and Sally is quiet, if Ralph is right.
   e. Either it is false that both Paula is polite and Sally is quiet, or Ralph is right.
   f. Paula is polite if and only if Sally is quiet or Ralph is right.

3. Let P = *Sam is sulky,* Q = *Tom is tense,* and R = *Freddy is ready.* Write each statement in words.
   a. $P \wedge (Q \vee R)$ b. $(P \wedge Q) \vee R$
   c. $P \vee Q \rightarrow R$ d. $P \vee (Q \rightarrow R)$
   e. $R \wedge Q \rightarrow \sim P$ f. $\sim P \leftrightarrow Q \wedge \sim R$

4. Construct a truth table for each statement.
   a. $\sim P \rightarrow Q$
   b. $P \vee \sim Q$
   c. $P \vee Q \leftrightarrow P$
   d. $P \vee Q \rightarrow R$
   e. $P \wedge \sim Q \rightarrow Q \vee R$
   f. $\sim(P \wedge \sim Q) \rightarrow \sim(\sim P \vee Q)$

5. Use De Morgan's law to rewrite each statement.
   a. It is false that today is Monday and tomorrow is Sunday.
   b. Neal is not first, or David is second.
   c. Hugh is painting or cutting the grass.
   d. It is not the case that Norma went to the store or Laurie went swimming.
   e. If mathematics is difficult, then logic is easy.

6. Determine whether $\sim P \vee Q$ is logically equivalent to $\sim(P \wedge \sim Q)$.

7. Use a truth table to determine whether the following arguments are valid or invalid. State your answer.
   a. If Paula is not polite, then Sally is quiet. (P, S)
      It is false that Sally is quiet.
      Therefore, Paula is not polite.
   b. If Sam is sulky then Tom is tense. (S, T)
      If Tom is tense, then Freddy is ready. (T, F)
      Hence if Sam is sulky, then Freddy is ready.

*8. Use Venn diagrams to determine whether each pair of statements is consistent or inconsistent.

a. No liars are virtuous.
Some liars are virtuous.

b. No liars are virtuous.
All liars are virtuous.

c. Some cars are gas guzzlers.
Some gas guzzlers are not cars.

d. All mathematics students are industrious.
Some mathematics students are not industrious.

★9. Use Venn diagrams to determine whether each argument is valid or invalid. State your answer.

a. All logic students are mathematics students.
No mathematics students are gullible.
Hence no logic students are gullible.

b. All logic students are mathematics students.
Some logic students are not gullible.
Therefore some gullible people are not mathematics students.

c. No golfers are sane.
All students are sane.
Therefore no students are golfers.

d. Some television programs are stupid.
All stupid things are useless.
So some television programs are useless.

e. All commercials are deceiving.
Some ads are deceiving.
Hence some ads are commercials.

★10. Using the suggested notation, write in symbolic form the converse, inverse, and contrapositive of each conditional statement.

a. If I pass math, then I will graduate. (P, G)

b. If I pass the test, then I will not come to class. (P, C)

c. If you do not attend class, then you will not pass the course. (A, C)

★11. A young man promised his girl friend, "I will marry you only if I get a job." He eventually got a job and then married someone else. Is the young man a heel? Why or why not? Explain your answer by using logic.

12. Given that "If I study logic then I will not be gullible" is a true statement.

a. Write another statement (in words) that you know is *true* from the above.

b. Write another statement (in words) that you know is false from the above.

★13. Let P = *It is a gas enter* and Q = *It is a car*. Write each of the following statements in symbolic form.

a. If it is a gas eater, then it is a car.

b. Being a car is necessary for it to be a gas eater.

c. Not being a car is sufficient for its not being a gas eater.

d. It isn't a car only if it isn't a gas eater.

e. It isn't a gas eater unless it is a car.

★14. Write a symbolic statement for each network.

a.

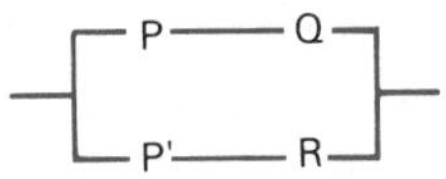

b.

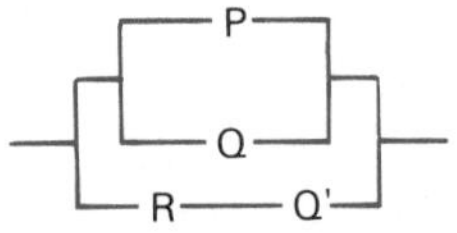

c.

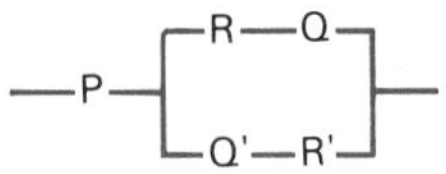

d.

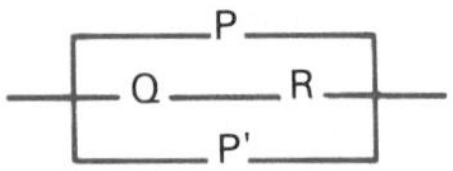

★15. Construct the switching networks that correspond to the given statements.

a. $P \wedge (Q \vee R)$

b. $P \vee (Q \wedge R)$

c. $(P \wedge Q) \vee (P \wedge R)$

d. $(P \wedge Q) \vee [(P \wedge Q) \vee R]$

Note: ★ indicates optional material.

Chapter Quiz

Indicate whether each statement is true or false.

1. The sentence "Is your homework finished?" is considered to be a statement in logic.
2. "Today is Monday or my watch is wrong." is an example of a disjunction.
3. "If someone was late, it was Benny." is an example of a conditional.
4. Given P = *Today is Monday,* and Q = *I am tired,* then $\sim P \wedge Q$ can be written as "Today is not Monday, but I am tired."
5. The answer column of the truth table for $P \rightarrow \sim Q$ is T, F, T, F.
6. $P \wedge (Q \vee \sim P)$ is a tautology.
7. "If you cut class, then you missed the exam" is logically equivalent to "Either you did not cut class or you missed the exam."
8. "It is false that you did not study and you passed the exam" is logically equivalent to "You did study or you did not pass the exam."
9. The following argument is valid.

   If I study then I will pass.
   I did not study.
   Therefore, I did not pass.
10. A valid argument can have either a true or a false conclusion.
11. Statements that cannot be true at the same time are inconsistent.
12. The following argument is valid.

    All ants are busy creatures.
    All busy creatures are creative.
    Therefore, some ants are creative.
13. "If I do not pass this course, then I cannot graduate" is the converse of "If I do pass this course, then I can graduate."
14. "If you work hard, then you will succeed" is the contrapositive of "If you don't succeed, then you didn't work hard."
15. "If you attend class, then you will pass" is the inverse of "If you did not pass, then you did not attend class."
16. "It snows only if it is cold" is logically equivalent to "If it snows, then it is cold."
17. The biconditional is the most dominant connective and the least dominant connective.
18. The disjunction is a more dominant connective than the conjunction.
19. Given $\sim(P \wedge Q \rightarrow \sim P)$, the answer column in the truth table for this statement would be found under the arrow.
20. Given P = *Today is Monday* and Q = *I am tired,* then $\sim P \rightarrow Q$ can be written as "Today is not Monday if I am tired."
21. The answer column of the truth table for $P \vee Q \rightarrow \sim R$ is F, T, F, T, F, T, T, T.
22. $P \wedge \sim Q \rightarrow Q \vee R$ is a contradiction.
23. The statement $\sim P \wedge Q \rightarrow \sim R$ is an example of a negation.

★24. Given the two premises

All students are industrious.
No politicians are industrious.

A conclusion that will make the argument valid is

Some students are politicians.

★25. Given the symbolic statement

$$(P \wedge Q) \vee (\sim P \wedge \sim Q)$$

the corresponding networks would be described as a series connection.

JUST FOR FUN

Bill, the bewildered builder, discovered that a large picture window had been broken in his big new beautiful building. Bill knew that three workers were on the premises when the window was broken—Bob, Bart, and Barry. The workers' professions were painter, mason, and carpenter. But Bill did not know which man did which job, although he did know that one had committed the foul deed. He also knew the painter always told the truth, the mason never told the truth, and the carpenter always told one true statement and one false statement.

Barry said "Bart didn't do it."
"Bob did it."
Bob said "I didn't do it."
"Bart did it."
Bart said "I didn't do it."
"Barry did it."

Using the true-false idea, help Bill discover the culprit's name and profession.

CHAPTER 3

PROBABILITY

AFTER STUDYING THIS CHAPTER YOU WILL BE ABLE TO DO THE FOLLOWING:

1. Compute the **probability** that an event A will occur.
2. Compute the probability of *not* A, given the probability of A.
3. Use the **fundamental counting principle** to determine how many ways two or more events can occur together.
4. Construct a **sample space** showing the possible outcomes for an experiment.
5. Construct a **tree diagram** for an experiment and use the tree to list the possible outcomes of the experiment.
6. Compute the **odds** in favor of or against an event.
7. Compute the **mathematical expectation** for an event.
8. Determine whether two events are **dependent** or **independent,** and determine whether two events are **mutually exclusive.**
9. Compute the probability of the event A *and* B and the probability of the event A *or* B, given the probability of A and of B.

★10. Determine the number of **permutations** that can occur for n things taken r at a time, and the number of **combinations** that can occur for n things taken r at a time.

★11. Compute the probabilities of events involving unordered arrangements and a large number of possible outcomes.

(*Note:* ★ indicates optional material)

NOTATION FREQUENTLY USED IN THIS CHAPTER

$P(A)$	the probability that event A will occur
$A : B$, A *to* B, $\frac{A}{B}$	ratio of A and B
$N!$	n factorial $\quad n! = n \times (n-1) \times \ldots \times 3 \times 2 \times 1$
${}_nP_r$	the number of permutations of n things taken r at a time
${}_nC_r$	the number of combinations of n things taken r at a time

3.1 INTRODUCTION

"I'll bet you!" This is a phrase that many of us have used at one time or another—most people have bet on something. It might have been on the results of an election or what the weather will be like tomorrow. Maybe you have bet with someone on the outcome of a World Series or Super Bowl game. Millions of people make some sort of a wager on major horse races such as the Kentucky Derby. You may have even bet with someone on how you would do on an exam.

Most people are introduced to the topic of probability through betting or games of chance. Even games that we played as small children—or play now—involve the use of a pair of dice. Even if you have never made a friendly bet or played any dice games (remember Monopoly?), you have heard the weather forecaster mention that there is an "80% chance" of showers for the weekend. Probability is with us every day.

Probability has been studied by mathematicians for a long time. The concept of probability was first formally studied in the sixteenth century. At that time, it was the outgrowth of a study on gambling and games of chance.

Today, probability is still used to help people understand games of chance such as blackjack, craps, and lotteries. But there are certainly other uses of probability. Insurance companies are concerned with the probable life expectancy of their policy holders. Surveyers of public opinion use probability to determine the results of their polls: the Harris and Gallup polls arrive at their results by means of probability, as do the various television polls. Biology (genetics), astronomy, and manufacturing are some other areas that make extensive use of probability theory.

The topics covered in this chapter will provide you with a basis for understanding probability and some everyday applications.

3.2 DEFINITION OF PROBABILITY

Suppose we are given one die from a pair of dice (Fig. 1). Upon examination, we see that it is a six-sided solid cube in which one side has one dot, another side has two dots, and so on, until the last side has six dots.

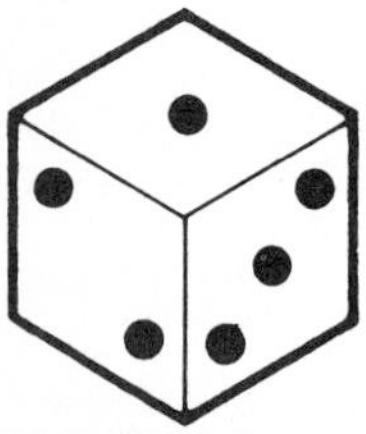

FIGURE 1

When we toss a die, we are performing an **experiment** to see which set of dots, or number, will turn face up. The number of dots that are on the top surface when we toss the die is called the **outcome** of the experiment. Each number has an equal chance of occurring, so we say that each outcome—1, 2, 3, 4, 5, or 6—is **equally likely** to occur.

The set of all possible outcomes of an experiment is called a **sample space.** For example, when the experiment is tossing a die, the sample space is {1, 2, 3, 4, 5, 6}. An **event** is any subset of the sample space: {3} and {1, 6} are both events.

If we toss a die, what is the probability of obtaining a 3? If we are interested in the probability of obtaining a 3, then we are actually interested in the probability that the event {3} will occur. Therefore, before we proceed any further, let us define what we mean by the probability that an event A will occur.

If an experiment has a total T of equally likely possible outcomes, and if exactly S of them are considered successful (or favorable)—that is, they are members of event A—then the probability that event A will occur, denoted by $P(A)$, is

$$P(A) = \frac{\text{number of successful outcomes}}{\text{total number of all possible outcomes}} = \frac{S}{T}$$

A helpful way to remember this probability formula, $P(A) = S/T$, is to remember that to find the probability in favor of an event A, we find the number of *successful* outcomes and divide it by the *total* number of possible outcomes.

Hence the probability of obtaining a 3 in a toss of one die is $\frac{1}{6}$. We see that there is only one possible successful outcome, a 3. There are a total of six possible outcomes, namely, 1, 2, 3, 4, 5, and 6. To compute the probability of obtaining a 3 for this experiment, we have

$$P(3) = \frac{1}{6} \qquad \frac{\text{number of successful outcomes}}{\text{total number of possible outcomes}}$$

EXAMPLE 1 Find the probability of getting a heads when you flip a quarter.

SOLUTION There are two different possible outcomes (heads, tails), and only one of these outcomes (heads) is successful. Hence the probability in favor of obtaining heads is $\frac{1}{2}$.

$$P(\text{heads}) = \frac{1}{2} \qquad \frac{\text{number of successful outcomes}}{\text{total number of possible outcomes}}$$

EXAMPLE 2 Find the probability of obtaining a number greater than 4 on a single toss of a die.

SOLUTION There are six different possible outcomes and two of these outcomes {5, 6} are successful. Hence the probability in favor of obtaining a number greater than 4 is $\frac{2}{6}$.

$$P(\text{number greater than 4}) = \frac{2}{6} = \frac{1}{3}$$

EXAMPLE 3 Find the probability of drawing a king (one pick) from a shuffled standard deck of 52 cards. (A standard deck of cards is the most common type of deck used in most card games. If you are not familiar with such a deck, see Fig. 2, which shows all 52 cards.)

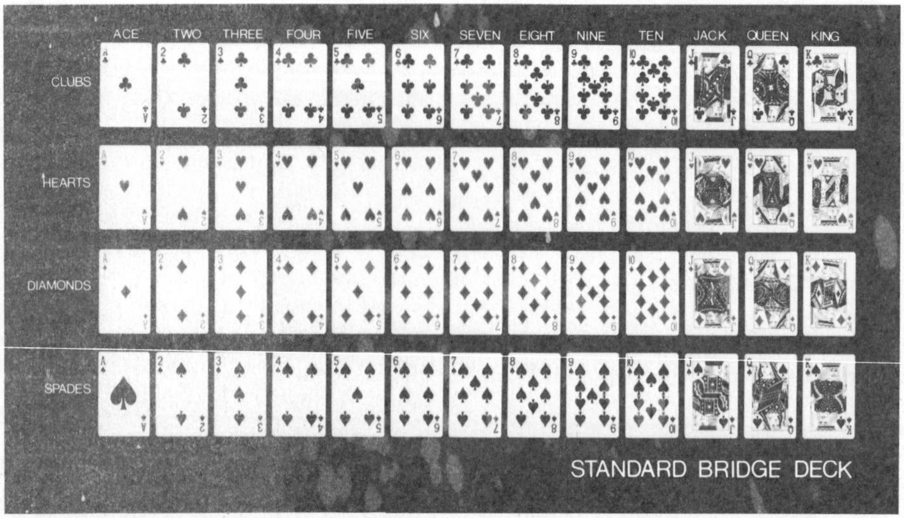

FIGURE 2
A standard deck of 52 cards.

SOLUTION There are 52 different possible outcomes. Four of these outcomes are successful: king of spades, king of hearts, king of clubs, king of diamonds. Therefore the probability in favor of obtaining a king is $\frac{4}{52}$.

$$P(K) = \frac{4}{52} = \frac{1}{13}$$

EXAMPLE 4 Find the probability of drawing a jack or a queen from a shuffled standard deck of 52 cards.

SOLUTION There are 52 different possible outcomes and eight of these are considered successful outcomes, because if a jack *or* a queen is picked, we are successful. Hence the probability in favor of obtaining a jack or a queen is $\frac{8}{52}$.

$$P(J \text{ or } Q) = \frac{8}{52} = \frac{2}{13}$$

EXAMPLE 5 Find the probability of obtaining a 7 on a single toss of one die.

SOLUTION There are six different possible outcomes and none of these outcomes would produce a 7. That is, zero of these outcomes would be successful. The probability in favor of obtaining a 7 on a single toss of one die is $\frac{0}{6}$, or 0.

$$P(7) = \frac{0}{6} = 0$$

When an event cannot possibly succeed, we say it is an **impossible event.** The probability of an impossible event is zero.

$$P(\text{impossible event}) = \frac{0}{T} = 0 \qquad (T \neq 0)$$

HISTORICAL NOTE

The laws of chance as we know them today began with interests in gambling. A French nobleman, the Chevalier de Méré, asked a mathematician friend of his, Blaise Pascal (1623–1662) "how to split the pot in a dice game that has to be discontinued."

Pascal pondered the problem for some time and then relayed it to another mathematician friend, Pierre de Fermat (1601–1665). From the correspondence and research of Pascal and Fermat regarding various gambling situations the theory of probability has evolved.

If the compartments on this roulette wheel are numbered 1–36, what is the probability that the winning number will be 7?

Ewing Galloway

EXAMPLE 6 Find the probability of getting heads or tails when you flip a quarter.

SOLUTION You can't lose! There are two different outcomes (heads, tails) and both of these outcomes are successful. Hence the probability in favor of obtaining heads or tails is $\frac{2}{2}$, or 1.

$$P(\text{heads or tails}) = \frac{2}{2} = \frac{1}{1} = 1$$

NOTE OF INTEREST

No one expects a coin to fall heads once in every two tosses; however, in a large number of tosses the results tend to even out. For a coin to fall heads 50 consecutive times it would take 1 million people tossing coins 10 times a minute for 40 hours per week. This is so unlikely to happen that it would only occur once every nine centuries.

When an event is sure to occur, then success is inevitable, and we say that the probability is 1. This is sometimes called certainty.

$$P(\text{certain event}) = \frac{T}{T} = 1 \qquad (T \neq 0)$$

For any event, $P(A) \geq 0$ and $P(A) \leq 1$, so we may write

$$0 \leq P(A) \leq 1$$

If an event can never occur, then its probability is 0.
If an event is certain to occur, then its probability is 1.

EXAMPLE 7 Find the probability of drawing an ace or a spade from a shuffled standard deck of 52 cards.

SOLUTION Your initial answer might be $\frac{17}{52}$, but this is not correct! There are 52 different possible outcomes, and there are 4 aces and 13 spades; hence, adding 4 and 13, we would have $\frac{17}{52}$. Why isn't this correct? If we examine the possible successful outcomes, we see there are indeed 13 spades, but one of them is an ace! We include the ace with our 13 spades, leaving 3 other aces to include in our successful number of outcomes. If you count the aces first, then you have included the ace of spades, and there are 12 spades left to include in the successful number of outcomes. Altogether, there are 16 successful outcomes, and the probability of drawing an ace or a spade is $\frac{16}{52}$, or $\frac{4}{13}$.

$$P(\text{ace or spade}) = \frac{16}{52} = \frac{4}{13}$$

EXAMPLE 8 One card is drawn at random from a shuffled standard deck of 52 cards. Find the probability that the card selected is *not* a king.

SOLUTION There are 52 different possible outcomes. There are 4 kings in a deck, so the other 48 cards are not kings, and these are the successful outcomes. Hence the probability that the card selected is not a king is $\frac{48}{52}$.

$$P(\textit{not } K) = \frac{48}{52} = \frac{12}{13}$$

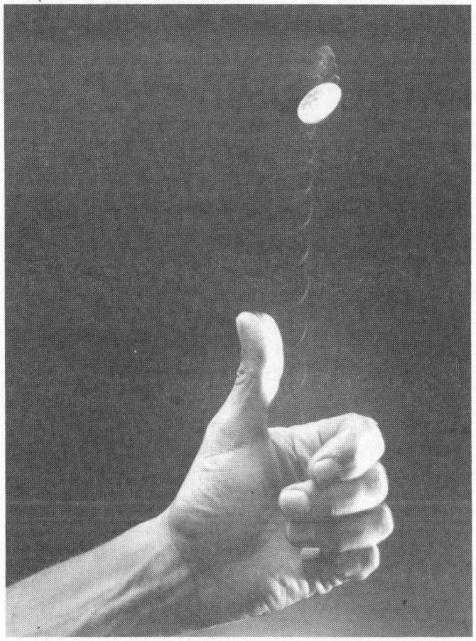

Richard Megna © Fundamental Photographs

The sum of the probability that an event will occur and the probability that it will not occur is 1. Therefore, we can solve Example 8 in another way.

Be sure to remember that if

$$P(A) = \frac{S}{T}$$

then

$$P(not\ A) = 1 - \frac{S}{T} \qquad (T \neq 0)$$

Alternate solution to Example 8

The probability of selecting a king is $\frac{4}{52}$ or $\frac{1}{13}$. Therefore the probability that the card selected is *not* a king is

$$P(not\ K) = 1 - \frac{1}{13} = \frac{13}{13} - \frac{1}{13} = \frac{12}{13}$$

EXERCISES FOR SECTION 3.2

1. On the single toss of one die, find the probability of obtaining
   a. a 4
   b. an odd number
   c. an even number
   d. a number less than 4
   e. a number greater than 4
   f. an odd or an even number
2. On the single toss of one die, find the probability of obtaining
   a. a number divisible by 3 (Example: 6 is divisible by 3 because 3 divides 6 evenly; that is, the remainder is zero)
   b. a number divisible by 5
   c. a number divisible by 2
   d. a number divisible by 1
   e. a number less than 1
   f. a number less than 7
3. On a single draw from a shuffled standard deck of 52 cards, find the probability of obtaining
   a. the ace of spades
   b. the two of hearts
   c. a deuce (2)
   d. a red card
   e. a diamond
   f. a red jack
4. On a single draw from a shuffled standard deck of 52 cards, find the probability of obtaining
   a. a spade
   b. a club
   c. a spade or a club
   d. a spade and a club
   e. a three or a heart
   f. a three and a heart (Be careful on this one!)
5. On a single draw from a shuffled standard deck of 52 cards, find the probability of obtaining
   a. a picture card (jack, queen, king)
   b. a picture card or a heart
   c. a jack or a heart
   d. a jack and a heart (Be careful on this one.)

e. a one-eyed jack (jack of spades or jack of hearts)
f. a king with an axe (king of diamonds)

6. Harry Hose keeps all of his socks in the top drawer of his bureau. In the drawer there are four blue socks, six black socks, seven brown socks (he lost one in the laundry), and four red socks. Harry reaches in and pulls a sock out at random. Find the probability that the sock chosen is
   a. brown
   b. blue
   c. red
   d. black
   e. not brown
   f. neither brown nor blue

7. Gloria Glove keeps all of her mittens on the top shelf of her hall closet. On the shelf are four blue mittens, six brown mittens, and four green mittens. Gloria reaches up and pulls a mitten out at random. Find the probability that the mitten chosen is
   a. blue or brown
   b. blue or green
   c. not red
   d. green or red
   e. neither blue nor green

8. If it is dark in the hall and the light doesn't work, what is the greatest number of mittens Gloria (see Exercise 7) will have to pull out to make sure that she has two mittens of the same color?

9. Veronica has separated the 12 pages from last year's calendar, put them in a paper bag, and shaken them around. She will draw a page at random. Find the probability of obtaining
   a. a page with a month beginning with *A*
   b. a page with a month beginning with *M*
   c. a page with a month beginning with *B*
   d. a page with a month ending with *y*
   e. a page with a month containing 31 days
   f. a page with a month containing 30 days

10. Charlie's wallet contains a one-dollar bill, a five-dollar bill, a ten-dollar bill, and a twenty-dollar bill. If Charlie chooses one bill at random, find the probability that the bill chosen is
   a. even
   b. odd
   c. greater than one dollar
   d. less than five dollars
   e. less than 20 dollars
   f. divisible by 3

JUST FOR FUN What is the answer to each problem?

a. $\frac{0}{1} = ?$ b. $\frac{1}{0} = ?$ c. $\frac{0}{0} = ?$

3.3 SAMPLE SPACES

Many times when we want to compute the probability of some event, the total number of possible outcomes of the experiment is not easy to determine. Consider the following examples.

EXAMPLE 1 A quarter is flipped and a die is tossed. What is the probability of obtaining heads or an odd number?

SOLUTION A person might reason that there are two outcomes with the quarter and six outcomes with the die, so altogether there are eight possible outcomes. Also, this same person might reason that there is one successful outcome with the quarter (heads) and three successful outcomes with the die $\{1, 3, 5\}$, so there are four successful outcomes altogether. Therefore the probability of getting heads or an odd number is $\frac{4}{8}$. This is WRONG!

In fact, there are 12 possible outcomes. The quarter may turn up two ways, heads or tails. If it lands heads up, then the heads can be matched with six different outcomes of the die. Namely, H—1, H—2, H—3, H—4, H—5, H—6. The same thing can happen with tails. If tails come up, then we could have T—1, T—2, T—3, T—4, T—5, T—6. Altogether, we have 12 *different* outcomes. Making a list of these outcomes we have

H—1	T—1
H—2	T—2
H—3	T—3
H—4	T—4
H—5	T—5
H—6	T—6

How many of these possible outcomes are successful? Remember, we wanted heads or an odd number. Six of the total outcomes contain a head, so they have to be considered successful. Three of the remaining possible outcomes are also successful: T—1, T—3, T—5. Hence, there are nine successful outcomes, and the probability of obtaining heads or an odd number is $\frac{9}{12}$, or $\frac{3}{4}$.

Recall from Section 3.2 that a list of all possible outcomes for an experiment is called a sample space. A method to determine the number of outcomes for two experiments together is to multiply the number of ways one can occur by the number of ways the other can occur. In Example 1, a quarter can come up two ways (assuming it can't land on its edge), and a die can turn up six ways. Therefore, we have $2 \times 6 = 12$ outcomes for both experiments together.

In general, if one experiment has m different outcomes and a second experiment has n different outcomes, then the first and second experiments performed together have $m \times n$ different outcomes. This idea may be extended if there are other experiments to follow: we would have $m \times n \times r \times \cdots \times t$ outcomes. This is often called the **fundamental counting principle.**

EXAMPLE 2 List a sample space showing all possible outcomes when a pair of dice is tossed.

SOLUTION First of all, how many outcomes will we have in our sample space? There are two dice and each die has six possible outcomes. Using the counting principle, we have $6 \times 6 = 36$ total possible outcomes.

Let one die be blue and the other white, so that we can distinguish the two dice. If a 1 comes up on the white die, it can be paired with each of the six numbers on the blue die; that is, (1, 1), (1, 2), (1, 3), (1, 4), (1, 5), and (1, 6). We can do this in turn for each of the numbers that comes up on the white die, pairing it with all the numbers on the blue die. This pairing is illustrated in Figure 3.

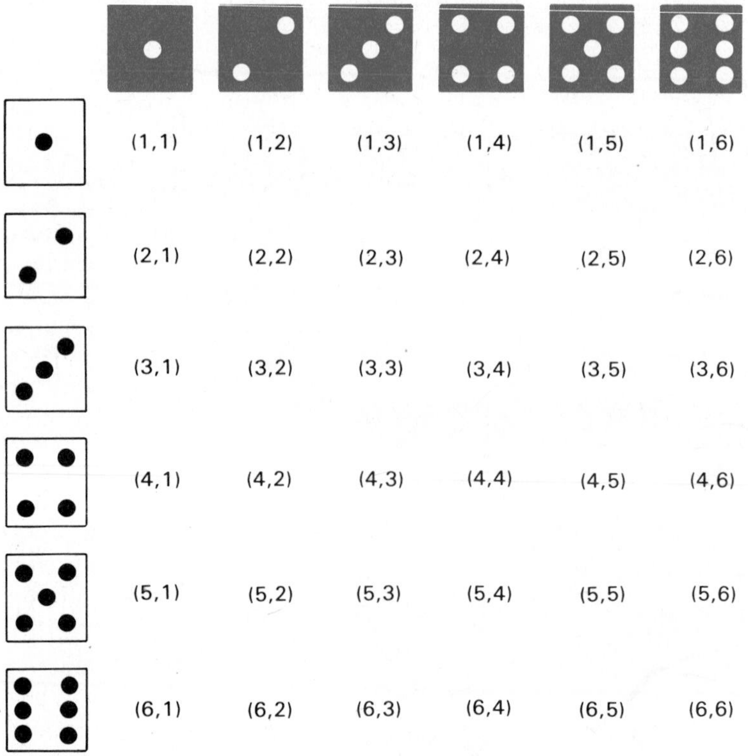

FIGURE 3

Harold M. Lambert

In games where a pair of dice is used, the sum of the two numbers is the primary concern. Note that there is only one outcome where the sum of the dice is 2 (the vernacular term is "snake eyes"); this is also the case for 12 ("box cars"). The sums that may be obtained by rolling a pair of dice are: 2, 3, 4, 5, 6, 7, 8, 9, 10, 11, 12. These are the only possible totals with two dice.

Examining the sample space illustrated in Figure 3, we see that we have the following number of ways to obtain each of the sums.

Sum:	2	3	4	5	6	7	8	9	10	11	12
	|	|	|	|	|	|	|	|	|	|	|
Number of ways:	1	2	3	4	5	6	5	4	3	2	1

Note that the sum 7 can occur in six ways and all of the other sums can occur in fewer ways. It is interesting (and helpful) to note that 6 and 8 can occur in five ways, the 5 and 9 can occur in four ways, and so on. This is a symmetric pattern, and it should help you to remember the total number of ways each sum can occur.

EXAMPLE 3 Use the sample space for the total number of possible outcomes when a pair of dice is tossed to find the probability of obtaining a sum of 7 on a single toss of a pair of dice.

SOLUTION There are 36 total possible outcomes and six of these outcomes are successful, because there are six ways of obtaining a 7. Hence the probability of obtaining a 7 is $\frac{6}{36}$ or $\frac{1}{6}$.

$$P(7) = \frac{6}{36} = \frac{1}{6}$$

Note that a 7 is the most likely outcome when a pair of dice is tossed. It is believed that this is one of the reasons why many people consider 7 their lucky number. It keeps coming up for them more often than other numbers.

EXAMPLE 4 A popular dice game in gambling establishments is "over and under." This is a game where the bettor may bet that the dice (when flipped, rolled, or tossed) will total over 7 or under 7. Larry decides to play this game, and he wants to bet on "under" because he feels that "under" should win. Is he right?

SOLUTION Since Larry bet on "under," he will win if any of the following sums comes up: 2, 3, 4, 5, 6. According to the sample space, there are five ways for a 6 to occur, four ways for a 5 to occur, three ways for a 4 to occur, two ways for a 3 to occur, and one way for a 2 to occur. Hence, there are 15 outcomes that can be considered successful out of a possible 36 outcomes. Therefore, the probability of the dice turning up under 7 is $\frac{15}{36}$ or $\frac{5}{12}$.

$$P(\text{under } 7) = \frac{15}{36} = \frac{5}{12}$$

Is this a good bet? Out of a total of 36 possible outcomes, only 15 are considered successful. The other 21 outcomes can turn Larry into a loser. The probability that the dice will *not* be under 7 is $\frac{21}{35}$; since the sum of the probability in favor of an event and the probability against an event is 1, if $P(\text{under } 7) = \frac{15}{36}$, then $P(\textit{not}$ under 7) is $1 - \frac{15}{36}$, or $\frac{36}{36} - \frac{15}{36} = \frac{21}{36}$. Hence, he is more likely to lose than win.

What happens if Larry changes his mind and decides to bet "over"? Has he got a better chance of winning? No! If we compute the probability of "over 7," we get $\frac{15}{36}$, and again Larry is more likely to lose because the probability of "not over 7" is $\frac{21}{36}$.

What happens if Larry decides to bet both "over" and "under"? The best he can do is get his own money back because if he wins on one side, he loses on the other side. Note also that he could be a double loser if 7 comes up because 7 is neither "over" nor "under"!

As we learned in Chapter 1, one of the uses of Venn diagrams is to show and separate numerical data that has been compiled. Venn diagrams can also be used to solve probability problems. Consider Examples 5 and 6.

EXAMPLE 5 At a recent college registration, 100 students were interviewed. Eighty of the students stated that they had registered for a mathematics course, 14 of the students stated that they had registered for a history course, and 5 of the students stated that they had registered for a mathematics course and a history course. What is the probability that a student in this survey registered only for history?

SOLUTION A Venn diagram using the given information is shown in Figure 4. With it, we can summarize the information that 11 of the students did not register for either a mathematics course or a history course, 9 students registered only for history, and 75 students registered only for mathematics.

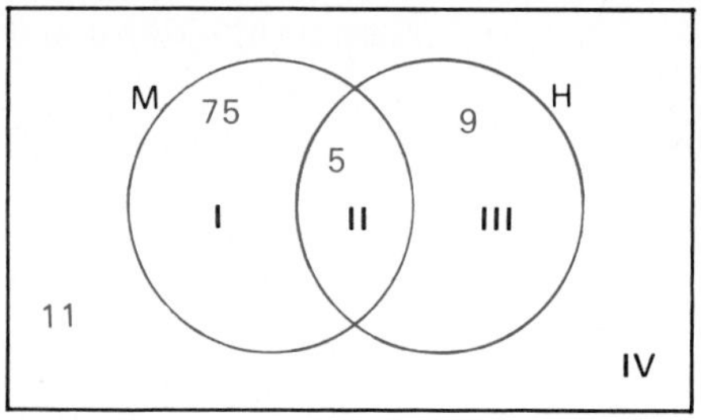

FIGURE 4
M = mathematics, and H = history.

This information can also be used to solve the probability problem. We note that the total number of students is 100, whereas 9 of them are registered only for history. Hence, the answer is $\frac{9}{100}$.

EXAMPLE 6 In a certain group of 75 students, it has been determined that 16 students are taking statistics, chemistry, and psychology; 24 students are taking statistics and chemistry; 30 students are taking statistics and psychology; 22 students are taking chemistry and psychology; 6 students are taking only statistics; 9 students are taking only chemistry; and 5 students are taking only psychology.

a. What is the probability that a student is not taking any of the three subjects?
b. What is the probability that a student is taking chemistry?

SOLUTION We first complete the necessary Venn diagram (see Fig. 5).

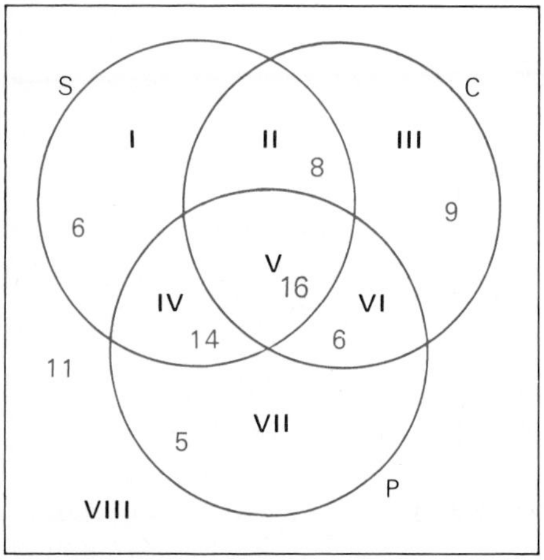

FIGURE 5
S = statistics, C = chemistry, and
P = psychology.

After completing the diagram, we can answer the questions.

a. The probability that a student is not taking any of the three subjects is $\frac{11}{75}$.
b. The probability that a student is taking chemistry is $\frac{39}{75}$.

The answer to question *b* is obtained by adding the number of students in each partition of the chemistry circle. Hence there are 39 students taking chemistry.

EXERCISES FOR SECTION 3.3

1. A quarter is flipped and a die is tossed (see Example 1, this section).
   a. How many outcomes are possible?
   b. Construct a sample space showing the possible outcomes.
   c. What is the probability of obtaining heads and a 6?
   d. What is the probability of obtaining heads and a 7?
   e. What is the probability of obtaining heads or a 7?
   f. What is the probability of obtaining tails and an even number?

2. A nickel is flipped and a dime is flipped.
   a. How many outcomes are possible?
   b. Construct a sample space showing the possible outcomes.
   c. What is the probability of obtaining two heads?
   d. What is the probability of obtaining two tails?
   e. What is the probability of obtaining exactly one head?
   f. What is the probability of obtaining no heads?

3. A box contains a one-dollar bill, a five-dollar bill, a ten-dollar bill, a twenty-dollar bill, and a fifty-dollar bill. Two bills are chosen at random in succession. The first bill is not replaced before the second is drawn.
   a. How many outcomes are possible?
   b. Construct a sample space for this experiment.
   c. What is the probability that the first bill is even?
   d. What is the probability that the second bill is even?
   e. What is the probability that both bills are even?
   f. What is the probability that neither bill is even?

4. A bag contains five balls numbered 0 through 4. Two balls are chosen at random in succession. The first ball is replaced before the second is chosen.
   a. How many outcomes are possible?
   b. Construct a sample space for this experiment.
   c. What is the probability that the first ball has an odd number?
   d. What is the probability that the second ball has an odd number?
   e. What is the probability that both balls have odd numbers?
   f. What is the probability that neither ball has an odd number?

5. A quarter is flipped and a die is tossed (see Example 1, this section). Find the probability of obtaining
   a. tails and a number less than 4
   b. tails and a number less than 1
   c. tails or a number less than 10
   d. heads and a number divisible by 2
   e. heads or a number divisible by 2
   f. tails or a number divisible by 7.

6. Using the sample space for the possible outcomes when a pair of dice is tossed (see Example 2, this section), find the probability that
   a. the same number appears on both dice

b. the number on one die is twice the number on the other die
c. the sum of the numbers is greater than 7
d. the sum of the numbers is at least 8
e. the sum of the numbers is greater than 1
f. the sum of the numbers is less than 1.

7. Using the sample space for the possible outcomes when a pair of dice is tossed (see Example 2, this section), find the probability that
a. the sum of the numbers is 7
b. the sum of the numbers is 11
c. the sum of the numbers is 1
d. the sum of the numbers is even
e. the sum of the numbers is odd or even
f. the sum of the numbers is divisible by 5.

8. In a certain mathematics class, there are 30 students; the table below shows the distribution of students: male, female, freshmen, and sophomores.

	Freshmen	Sophomores
Male	7	6
Female	8	9

The names of all class members are placed on tags and the tags are placed in a box. A tag is selected at random. Find the probability of choosing
a. a freshman
b. a sophomore
c. a female
d. a male
e. a female sophomore
f. a male freshman

9. The table below shows the number of students in a swimming class. To complete the course, each student must pass a skills test.

	Freshmen	Sophomores
Female	6	2
Male	5	7

The test is begun by selecting a student at random. Find the probability of choosing
a. a female
b. a male
c. a freshman
d. a sophomore
e. a male sophomore
f. a female freshman

10. In a survey of 50 contestants at a track meet, the following information was obtained: 12 contestants were entered in both a running event and a field event, 32 contestants were entered in a running event, and 30 contestants were entered in a field event.
a. Show the results of this survey in a Venn diagram.
b. Find the probability that a contestant is entered in a running event only.
c. Find the probability that a contestant is entered in a field event only.

11. At a meeting of 50 car salespeople, the following information was obtained: 12 salespeople sold Plymouths, 15 salespeople sold Fords, and 16 salespeople sold Chevrolets. Four salespeople sold Plymouths and Fords, six salespeople sold Fords and Chevrolets, and five salespeople sold both Plymouths and Chevrolets. Two salespeople sold all three kind of cars.

Using a Venn diagram, find the probability that a salesperson at this meeting sold
a. Plymouths only
b. Fords only
c. Chevrolets only
d. Plymouths and Fords, but not Chevrolets
e. Fords and Chevrolets, but not Plymouths
f. Fords or Plymouths

12. In a survey of 75 students who registered for courses, the following data were collected: 27 students were taking statistics, 26 were taking history, and 41 were taking English. Twelve students were taking statistics and history, 13 students were taking statistics and English, and 17 students were taking history and English. Four students were taking all three courses.

Use a Venn diagram to find the probability that a student participating in this survey is taking

a. only statistics
b. only English
c. statistics and English, but not history
d. statistics or English
e. none of the three subjects
f. exactly one of these three subjects

JUST FOR FUN Given the digits 1, 2, 3, 4, 5, 6, and 7, construct an addition problem whose sum is 100. You may use each digit only once.

3.4 TREE DIAGRAMS

A sample space consists of all possible outcomes for a particular experiment. A technique that shows us the sample space for two or more experiments that are performed together is a **tree diagram.**

A tree diagram consists of a number of "branches" that illustrate the possible outcomes for the experiments. We may read the possibilities directly from the branches. The following examples illustrate the use of a tree diagram to obtain a sample space.

EXAMPLE 1 When a coin is flipped, it may turn up heads or tails. How many different outcomes are possible when two coins are tossed, and what are the possible outcomes?

SOLUTION There are two experiments and each experiment has two possible outcomes (heads or tails). Using the counting principle, we have $2 \times 2 = 4$ possible outcomes.

In constructing the tree diagram, remember that the first experiment has two possible outcomes and each of these may be matched with the two possible outcomes of the second experiment. Hence we have

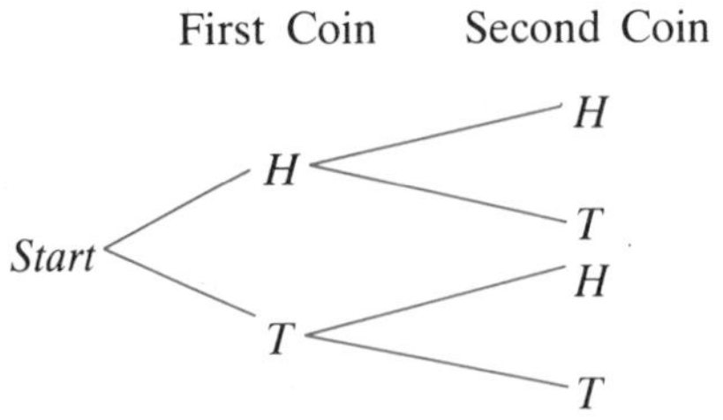

Possible Outcomes	
First Coin	Second Coin
H	H
H	T
T	H
T	T

From the tree diagram we may obtain our sample space. The first branch gives us *H—H,* the second branch *H—T,* and so on, until we have the complete sample space: *H—H, H—T, T—H,* and *T—T.*

EXAMPLE 2 Mr. Examination is preparing a quickie quiz for his mathematics class to see if the students did their assignment. The quiz is to consist of three true-false questions. How many different arrangements of the answers are possible? What are the possible outcomes?

SOLUTION We have three questions and each question has two possible outcomes (true or false). Using the counting principle, we compute $2 \times 2 \times 2 = 8$ total possible outcomes.

We can determine the various outcomes by means of a tree diagram. Remember that the quickie quiz consists of three questions and the answer to each question is either true or false.

NOTE OF INTEREST

1 1
1 2 1
1 3 3 1
1 4 6 4 1
1 5 10 10 5 1

This array of numbers is known as "Pascal's triangle." We can construct such an array by the following method. In each row, the first and last numbers are always 1. Any number in the array is the sum of the two numbers on the line above that it lies between. Five rows are shown (it may be extended).

The sum of the numbers in any row yields the total number of combinations possible within that group. For example, to find the probability of any boy-girl combination in a family of four children, we examine the fourth row. Note the sum of the numbers is 16, the number of different outcomes. The numbers at the ends of the row represent the chances of the least likely events, that is, all boys or all girls: 1 in 16. The next number from the end represents the next type of combination, three boys and one girl or vice versa: 4 in 16. Finally, the chances of having two boys and two girls are 6 in 16.

Pascal did not invent the triangle that bears his name; some Chinese works of the early 1300s contain such an array of numbers (see page 335). However, because he discovered and applied many of its properties, this triangular array has become known as Pascal's triangle.

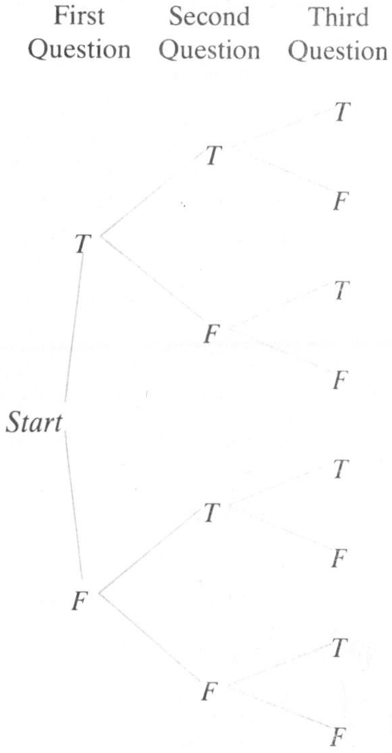

Possible Outcomes		
First Question	Second Question	Third Question
T	T	T
T	T	F
T	F	T
T	F	F
F	T	T
F	T	F
F	F	T
F	F	F

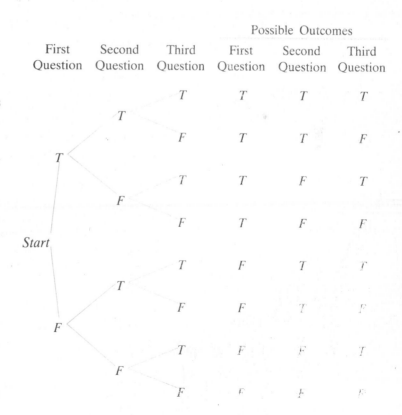

The sample space is listed beside the tree diagram.

EXAMPLE 3 David, a student in Mr. Examination's class, did not do the assignment. David took the quiz and guessed at the answers. David answered the questions $F—T—F$. What is the probability that he answered all three questions correctly?

SOLUTION The sample space in Example 2 shows each of the eight possible outcomes. Only one of these contains all three correct answers. Hence, the probability that $F—T—F$ is the correct combination is $\frac{1}{8}$.

EXERCISES FOR SECTION 3.4

1. Use a tree diagram to find the sample space showing the possible arrangements of boys and girls in a family with exactly two children.
   a. What is the probability that both children are boys?
   b. What is the probability that both children are girls?
   c. What is the probability that the first child is a boy and the second child is a girl?
   d. What is the probability that at least 1 of the children is a girl?
   e. What is the probability that none of the children are girls?

2. A box contains a one-dollar bill, a five-dollar bill, a ten-dollar bill, and a twenty-dollar bill. Two bills are chosen in succession without replacement. Use a tree diagram to list the sample space for this experiment.
   a. What is the probability that both bills are even?
   b. What is the probability that neither bill is even?
   c. What is the probability that exactly one of the bills is even?
   d. What is the probability that at least one of the bills is even?
   e. What is the probability that the total value of the bills chosen is \$30?

3. Use a tree diagram to list the sample space showing the possible arrangements of boys and girls in a family with exactly three children.
   a. What is the probability that all three children are boys?
   b. What is the probability that two children are boys and one is a girl?
   c. What is the probability that at least one of the children is a girl?
   d. What is the probability that none of the children are boys?
   e. What is the probability that all three children are of the same sex?
   f. If a family has three children, all boys, what is the probability that a fourth child would be a girl?

4. Use a diagram to list the sample space showing the possible arrangements of heads and tails when four coins are tossed. Then use the sample space to find the probability that
   a. all four coins come up heads
   b. exactly two coins come up heads
   c. exactly three coins come up tails
   d. at most two coins come up heads
   e. at least two coins come up heads
   f. no more than three coins come up tails

5. An urn contains one blue and three orange balls. An experiment consists of selecting two balls in succession, without replacement. Use a tree diagram to list the sample space for this experiment. (*Hint:* Balls of the same color are indistinguishable, so you may find it helpful to use the symbols O_1, O_2, and O_3 to distinguish the three orange

balls.) Then use the sample space to find the probability that
a. both balls are orange
b. at least one ball is orange
c. the first ball is blue
d. the second ball is blue
e. at least one ball is blue
f. one ball is blue and the other is orange

6. The Knicks won the first game against the Lakers in a basketball playoff. If the first team to win four games is the winner, construct a tree diagram for this playoff. (*Hint:* Some of the branches in the tree diagram will end sooner than others, as one of the teams will have won four games.)

7. A cafeteria menu has the following items. A student selects one item from each group.

Group A	Group B	Group C
Hot dog	Fruit cup	Apple
Hamburger	Salad	Jello
Peanut butter and jelly		Ice cream

a. Make a tree diagram to list the sample space for the possible luncheon choices.
b. What is the probability that a student selects a hot dog?
c. What is probability that a student selects the fruit cup?
d. What is the probability that a student's lunch consists of a hamburger, salad, and ice cream?
e. What is the probability that a student's lunch consists of a hot dog, fruit cup, and jello or ice cream?
f. What is the probability that a student's lunch consists of a hot dog or hamburger together with salad and jello?

8. Scott can travel from Rochester, N.Y., to New York City by bus, train, car, or airplane. He can travel from New York to London, England, by airplane or ship.
a. How many choices for the trip from Rochester to London are there?
b. Make a tree diagram to list the sample space for the possible outcomes.
c. What is the probability that Scott will select a bus?
d. What is the probability that Scott will select a ship?
e. What is the probability that Scott will select an airplane?
f. If Scott must return from London to New York by airplane, and from New York to Rochester by airplane, how many ways may he make the round trip?

★9. The Comets and Barons are two evenly matched teams. Each has an equally likely chance of winning a game. Suppose the two teams are in a playoff series that is won by winning two out of three games.
a. Make a tree diagram to list the sample space for the possible outcomes of the series.
b. What is the probability that the Barons will win the series?
c. What is the probability that the Barons will win the first two games?
d. What is the probability that the Comets will win two straight games?
e. What is the probability that the Comets will win at least one game?
f. What is the probability that the Comets will win the last two games?

★10. A quarter is flipped and the outcome is noted. If a heads results, then a die is rolled. If a tails results, then the quarter is flipped a second time.
a. How many different outcome are possible?
b. Make a tree diagram to list the sample space for the possible outcomes.
c. What is the probability of obtaining a heads?
d. What is the probability of obtaining an odd number?
e. What is the probability of obtaining no heads?
f. What is the probability of obtaining at least one tails?

B. C. by permission of Johnny Hart and Creators Syndicate, Inc.

JUST FOR FUN Long-playing records are approximately 12 inches in diameter and spin at $33\frac{1}{3}$ revolutions per minute. How many grooves are on such a record?

3.5 ODDS AND EXPECTATION

What odds will you give me?
The Browns are odds-on favorites to win the title.
The odds on Dead Last are 100 to 1.
The odds on getting a royal flush in poker are 649,739 to 1.

All of these expressions mention the word *odds*. We all have a nodding acquaintance with odds and probably have come in contact with them at some time or another. Most of us know that if the odds on a bet are 3 to 1 (3 : 1) and we bet a dollar and win, then we will win three dollars. But there is more to it than that.

Odds are usually expressed as a ratio, for example, 3:2. But they may also expressed as 3 to 2, or even as a fraction, $\frac{3}{2}$.

Odds can occur in two ways. We may discuss the *odds in favor of* an event or the *odds against* an event.

Odds are computed by finding a ratio. The **odds in favor of an event A** are found by taking the probability that an event A will occur and dividing it by the probability that an event A will not occur. We may state this as

$$\text{Odds in favor of } A = \frac{\text{probability that } A \text{ will occur}}{\text{probability that } A \text{ will not occur}}$$

EXAMPLE 1 What are the odds in favor of obtaining a sum of 7 when a pair of dice is tossed once?

SOLUTION Using our definition, we first find the probability of obtaining a sum of 7 when a pair of dice is tossed. It is $\frac{6}{36}$. Next, we find the probability of not getting a 7. Recall that if $P(A) = S/T$, then $P(not\ A) = 1 - S/T$. Hence, we have

$$P(7) = \frac{6}{36} \quad \text{and} \quad P(not\ 7) = 1 - \frac{6}{36} = \frac{36}{36} - \frac{6}{36} = \frac{30}{36}$$

Also, if there are 6 ways in 36 that we may obtain a 7, then the remaining 30 ways in 36 is the probability of not obtaining a 7.

Now we construct our ratio according to the definition.

$$\text{Odds in favor of } A = \frac{\text{probability that } A \text{ will occur}}{\text{probability that } A \text{ will not occur}}$$

$$\text{Odds in favor of a } 7 = \frac{\frac{6}{36}}{\frac{30}{36}} = \frac{6}{36} \times \frac{36}{30} = \frac{6}{30} = \frac{1}{5}$$

The odds in favor of obtaining a 7 are $\frac{1}{5}$. But remember that the odds may also be stated as 1 to 5 or 1 : 5.

What does this mean? If we roll a pair of dice, over the long run we should obtain a 7 once out of six times. The other five times we would lose.

EXAMPLE 2 Find the odds in favor of obtaining two heads in a single toss of two coins.

SOLUTION The probability in favor of obtaining two heads in a single toss of two coins is $\frac{1}{4}$. We can obtain this from the sample space $\{H—H, H—T, T—H, T—T\}$. The probability against two heads is

$1 - \frac{1}{4} = \frac{4}{4} - \frac{1}{4} = \frac{3}{4}$. Therefore we have

$$\frac{\frac{1}{4}}{\frac{3}{4}} = \frac{1}{4} \times \frac{4}{3} = \frac{1}{3}$$

The odds in favor of two heads are 1 to 3, or 1:3.

If we change the ratio in Example 2 around (3 to 1, or 3:1), we will have the odds *against* obtaining two heads in a single toss of two coins. We must always make certain that we find the odds in favor of an event, if that is what we are asked to find. If we are asked to find the **odds against an event A,** then we should use the following ratio.

$$\text{Odds against } A = \frac{\text{probability that } A \text{ will not occur}}{\text{probability that } A \text{ will occur}}$$

We can also find the odds in favor of A and then reverse the ratio.

EXAMPLE 3 What are the odds against obtaining an 11 when a pair of dice is tossed once?

SOLUTION Using the definition, we still find the probability in favor of getting an 11 and the probability against getting an 11 when a pair of dice is tossed. The probability of getting an 11 is $\frac{2}{36}$, and the probability of not getting an 11 is $1 - \frac{2}{36} = \frac{36}{36} - \frac{2}{36} = \frac{34}{36}$. Now we construct our ratio according to the definition.

$$\text{Odds against an } 11 = \frac{\frac{34}{36}}{\frac{2}{36}} = \frac{34}{36} \times \frac{36}{2} = \frac{34}{2} = \frac{17}{1}$$

The odds against obtaining an 11 are $\frac{17}{1}$, which we would usually write as 17 to 1, or 17:1. Therefore, from the odds (17:1), we can see that we probably are not going to get an 11 when we roll the dice. Over the long run, we should get an 11 once out of every 18 tries. This is not a very good percentage of successful outcomes.

Odds enable us to play and bet fairly on games. We have just computed the odds against obtaining an 11 on the toss of a pair of dice, and it was 17 to 1. Suppose that Nevada Nellie is rolling a pair of dice and she bets she will roll an 11. If she bets a dollar, then according to the odds, she should receive $17 if she does toss an 11. This would make the game a "fair" one. Nevada Nellie should get an 11 once every 18 tries, over the long run. Hence, in 18 tries she should lose $17, $1 on each of 17 failures, and she should win once, and that one time should pay $17.

Two things are usually against a shooter like Nevada Nellie. One, the owner of the game (the house) normally cuts down the odds a little when they pay a winner, to 15 to 1 or 14 to 1. Two, Nevada Nellie has to play over the long run in order to regain her losses, which means she has to play for a long time. People like Nevada Nellie usually lose all their money first, or just get tired and bored, and then quit. The house can keep its game going because the casinos are open 24 hours a day.

EXAMPLE 4 If the odds in favor of obtaining a 7 when a pair of dice is tossed are 1 : 5, what is the probability of obtaining a 7?

SOLUTION We already know that $P(7) = \frac{6}{36}$ or $\frac{1}{6}$, but how do we obtain the answer when we are given the odds?

If the odds in favor of event A are $B : C$, then the probability of A is B divided by $B + C$, that is

$$P(A) = \frac{B}{B + C}$$

Hence, for this particular problem, we have $B = 1$, $C = 5$, and

$$P(7) = \frac{1}{1 + 5} = \frac{1}{6}$$

EXAMPLE 5 If the odds against the Dodgers winning the pennant are 8 : 5, what is the probability that the Dodgers will win the pennant?

SOLUTION We can use the formula given in Example 4, but first, because the odds are given as 8 : 5 against, we restate them as 5 : 8 in favor. Now we can apply our formula.

If $B : C$ equals odds in favor of A, then $P(A) = B/(B + C)$, $B = 5$, $C = 8$, and the probability that the Dodgers will win the pennant is $\frac{5}{5+8} = \frac{5}{13}$.

HISTORICAL NOTE

Girolamo Cardano (1501–1576), also known as Jerome Cardan, is probably one of the most extraordinary people in the history of mathematics. He was born in Pavia, Italy, the illegitimate son of a local jurist.

He was a man of many contrasts. He began his professional life as a physician and during this time he also studied, taught, and wrote mathematics. His book, *Liber de Ludo Aleae* (Book on Games of Chance), is a gambler's handbook and is considered the first book on probability. In this book Cardan discussed the number of successes versus the number of outcomes, mathematical expectation, the different outcomes for two dice, and the different outcomes for three dice.

Later in his career Cardan held important chairs at the universities of Pavia and Bologna. Eventually he became a distinguished astrologer, and at one time he was imprisoned for publishing a horoscope of Christ's life. Once, in a fit of rage, he cut off the ears of his younger son.

Cardan's life came to a dramatic end. Years before, he had made an astrological prediction of the date of his death. When the day arrived and he was still alive, he committed suicide to make his prediction come true.

Mathematical probability and odds lead to another related topic: **mathematical expectation.** We can describe expectation (or expected value) using an example, as follows: Suppose that Lucky Louie is betting on a certain dice game, and his probability of winning is P, and if Louie wins then he receives M dollars. We would say that Lucky Louie's *mathematical expectation* is $P \times M$. You may not understand mathematical expectation yet, but at least we have a method for finding it. First, let's describe the way we find mathematical expectation a little more formally. Let

$$M = \textit{the amount that will be won if an event occurs}$$

$$P = \textit{the probability that the event will occur}$$

Then

$$\textit{Expectation} = P \times M$$

Expectation tells us the expected value or "fair" price to pay to play a game, if the game can be described by the probability of equally likely outcomes. It also gives us the *average* amount of winnings we can expect for each game if we play a great many games.

Suppose Lucky Louie is playing a dice game in an Atlantic City casino and he is betting on 7 coming up. If a 7 comes up, then Louie receives a payoff of \$18. According to the formula, Lucky Louie's expectation is

$$E = P(7) \times \$18$$

$$E = \frac{6}{36} \times \$18 = \$3$$

That is, Lucky Louie should be willing to bet \$3 for the privilege of betting on 7. But unfortunately, the smallest denomination chip used at the casino is \$5. Hence, Lucky Louie must bet at least \$5 each time he plays the game. We have already computed his expectation, and it is \$3. If he has to bet \$5 each time he plays, then Lucky Louie can expect to lose an average of \$2 on each \$5 bet over the long run.

EXAMPLE 6 The Brighton Fire Department is running a raffle in which the prize for the lucky ticket is \$1,000. If 5,000 tickets are sold at \$2 each, what is the expectation of Hugh, who buys one ticket?

SOLUTION The probability of having the winning ticket for this person is $\frac{1}{5000}$ and Hugh's expectation is

$$E = \frac{1}{5{,}000} \times \$1{,}000.00 = \frac{\$1.00}{5} = \$.20$$

Twenty cents represents a fair price to pay for a ticket. It appears that \$2.00 is too much to pay for a ticket, but this is how organizations make money.

EXAMPLE 7 Suppose Hugh changes his mind and decides to buy five tickets in the raffle for the \$1,000 prize. Does this change his expectation?

SOLUTION Hugh has increased only his chances of winning. Now the probability of winning is $\frac{5}{5000}$, and his expectation is

$$E = \frac{5}{5{,}000} \times \$1{,}000.00 = \frac{5{,}000}{5{,}000} = \$1.00$$

Hugh's mathematical expectation is \$1.00 for five tickets; it is still \$.20 per ticket.

EXAMPLE 8 A bag contains two red, five white, and three green balls. A prize of \$10 is given if a red ball is drawn and a prize of \$1 is given if a green ball is drawn. Nothing is won if a white ball is drawn. What is the expectation?

SOLUTION First we must compute the probability of winning. The probability of drawing a red ball is $\frac{2}{10}$, and the probability of drawing a green ball is $\frac{3}{10}$. The expectation for the red ball is

$$E(\text{red}) = \frac{2}{10} \times \$10.00 = \frac{20}{10} = \$2.00$$

The expectation for the green ball is

$$E(\text{green}) = \frac{3}{10} \times \$1.00 = \frac{3}{10} = \$0.30$$

The total expectation is the sum of the two previous expectations. Therefore, we would have

$$E = E(\text{red}) + E(\text{green}) = \$2.00 + \$0.30 = \$2.30$$

Remember that the \$2.30 represents a "fair" price to pay for the privilege of playing the game. It also represents the average amount you should expect to win when you play a great many games—that is, if play is over the long run—since these are equally likely events. Suppose you played the game 10 times. You should expect to win two times on red, which equals \$20.00 in prizes; you should win three times on green, which equals \$3.00 in prizes; and the other five times you should receive nothing. So, in 10 games you should receive \$23.00. In a fair game, you should also bet \$23.00, which is \$2.30 per try.

EXERCISES FOR SECTION 3.5

1. In a single toss of a pair of dice, find the odds (the number referred to is the sum of the numbers on the dice)
   a. in favor of obtaining a 7
   b. in favor of obtaining an 11
   c. in favor of obtaining a 12
   d. against obtaining a 12
   e. against obtaining a 6
   f. against obtaining a 10

2. In a single toss of a pair of dice, find the odds (the number referred to is the sum of the numbers on the dice)
   a. in favor of obtaining a 3
   b. in favor of obtaining a 5
   c. against obtaining a 5
   d. against obtaining a double
   e. in favor of obtaining a 7 or 11
   f. against obtaining a number other than 7

3. On a single draw from a shuffled standard deck of 52 bridge cards, find the odds
   a. in favor of drawing an ace
   b. in favor of drawing a club
   c. in favor of drawing a red card
   d. against drawing a deuce (2)
   e. in favor of drawing a picture card (jack, queen, or king)
   f. against drawing a picture card (jack, queen, or king) or a diamond

4. On a single draw from a shuffled standard deck of 52 bridge cards, find the odds
   a. in favor of drawing a queen
   b. against drawing a queen
   c. in favor of drawing a black card
   d. against drawing a red card
   e. in favor of drawing a heart
   f. against drawing a heart or a diamond

5. In a single toss of two coins, make a sample space and find the odds
   a. in favor of obtaining two tails
   b. against obtaining two heads
   c. in favor of obtaining exactly one tail
   d. against obtaining no tails
   e. in favor of obtaining at least one tail
   f. in favor of obtaining a head and a tail

6. Find the probability that event A will happen if the odds are
   a. 7:5 in favor of A
   b. 1:1 in favor of A
   c. 3:2 against A
   d. 5:9 against A

7. The odds against the Yankees winning the pennant are 7:2. What is the probability that the Yankees will win the pennant?

8. The odds against Laura winning a certain raffle are 99:1. What is the probability that Laura will win?

9. In order to win a game, Nelson must throw a 7 in a single toss of a pair of dice. Larry bets that Nelson will; Benny bets that Nelson won't. What are the

odds for and against this event? Do the odds favor Larry or Benny?

10. A player will win $18 if he or she throws a double on the first toss of a pair of dice. What are the odds in favor of this player winning? What is a fair price to pay to play this game?

11. A box contains three one-dollar bills, two five-dollar bills, one ten-dollar bill, and one twenty-dollar bill. If you reach in and select a bill at random, what is your mathematical expectation?

12. The Association for Conservation is awarding a $1,000 cash prize to the winner of a raffle. A total of 5,000 tickets are sold for $2 each. What is the mathematical expectation for a person who buys five tickets?

13. Many states conduct daily lotteries of one type or another. "Numbers" is a popular type. A player selects any three-digit number from 000 to 999 (any number that seems lucky to the player). In New York a player may select a number for $1. If the player wins, the payoff is $500. What is a player's chances of winning? What is a fair price to pay for a ticket? What can you conclude about the cost of playing this lottery?

14. The New York State Lottery offered a weekly prize of $50,000. A person bought one ticket which cost $.50. The winning ticket can be any six-digit number from 000,000 to 999,999. What were the person's chances of winning? What was a fair price to pay for it?

15. A special New York State Lottery offers a grand prize of $100,000, three second prizes of $10,000, and ten third prizes of $1,000. The winning ticket can be any six-digit number from 000,000 to 999,999. If a person buys one ticket, what is a fair price to pay for the ticket?

16. Four envelopes are placed in a bag. Each envelope contains a single bill: a one-dollar bill, a five-dollar bill, a ten-dollar bill, and a twenty-dollar bill, respectively. The envelopes are mixed, and if you pay a certain amount, you may reach in and select one envelope at random. Would you be willing to pay $7 to select an envelope? Why?

17. A magazine subscription company runs a contest where first prize is $100,000, second prize is $50,000, and third prize is $25,000. If the company receives 1,000,000 entries, what is the expectation for a person who enters the contest once? If the entrant must use a first class stamp to send in the entry, what can you conclude about the cost of entering the contest?

18. One thousand raffle tickets are sold for a first prize of $300, a second prize of $150, and a third prize of $50. What is a fair price to pay for a raffle ticket? If Mary bought a raffle ticket for $1, what can you conclude about the cost of entering this raffle?

JUST FOR FUN How many cubes are shown in the figure?

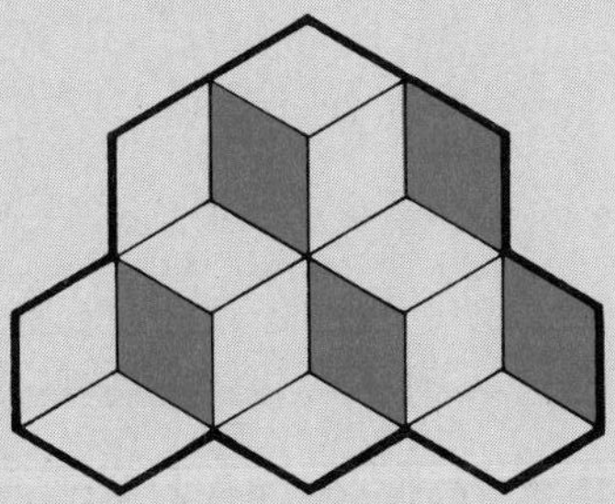

3.6 COMPOUND PROBABILITY

Many events are **compound events,** made up of two or more simpler events. If we draw a card from a shuffled deck of cards, look at it, replace it, shuffle the cards again, and then draw another card, we have made two single simple drawings; but together they constitute a compound event. The problem in this example could be stated as, "What is the probability of drawing two red cards from a shuffled deck of 52 cards, if the first card is replaced?" We can also think of this as "What is the probability of drawing a red card *and then another* red card from a shuffled deck of 52 cards, if the first card is replaced?"

This probability problem leads us to two important questions:

1. How do we compute compound probabilities?
2. How does replacement influence the probability?

First, let us figure out how to compute compound probabilities. Recall the *counting principle;* it said that if one experiment has m different outcomes and a second experiment has n different outcomes, then the first and second experiments performed together have $m \times n$ different outcomes. Thus, if the outcome of the first experiment does not influence the outcome of the second, we can use this idea to compute the probability of events occurring in succession: that is, we find the product of the probabilities of the two events. In general, we have

$$P(A \text{ and } B) = P(A) \times P(B)$$

Note that this rule may be extended for more than two events.

Second, let us decide how replacement influences the probability. In the above problem, the probability that the second card is red is the same as the probability that the first card is red, because we replaced the first card. The occurrence of the second event is **independent** of the first event; that is, the first event does not influence the probability of the second event.

Let us see why this is the case. The probability that the first card is red is $\frac{26}{52}$, or $\frac{1}{2}$, because there are 26 red cards (hearts and diamonds) out of a total of 52. The probability that the second card is red is also $\frac{26}{52}$, or $\frac{1}{2}$, because we replaced the first card. Therefore the probability of drawing two red cards in succession, with replacement, is $\frac{26}{52} \times \frac{26}{52}$, or $\frac{1}{2} \times \frac{1}{2} = \frac{1}{4}$ or 0.25.

Suppose the first card had not been replaced. The probability that the second card is red then becomes $\frac{25}{51}$. Why? Because we know that there is one less card in the deck; hence, the denominator of our fraction must be 51. But the 25 in the numerator confuses some students. Why not 26? How do we know that the first card is red? We

NOTE OF INTEREST

The term *probability* as applied to precipitation forecasting (for example, a 40% chance of precipitation) is the percentage chance that at least one-hundredth inch of precipitation (rain or the liquid equivalent of snow or other frozen precipitation) will fall at any selected point in the area and time period covered by the forecast.

A 60% probability indicates a 6 in 10 chance of precipitation, and a 4 in 10 chance of no precipitation, at any location in the forecast area. Forecasts usually cover a 12-hour period and a moderate-sized metropolitan area. Usually no differentiation is made for points within the forecast area.

If there is a 10% chance of precipitation, the qualifying forecast usually has no mention of precipitation. If there is a 20% chance of precipitation, the qualifying forecast calls for "slight or small chance of" precipitation. When there is a 30, 40, or 50% chance of precipitation, the qualifying forecast calls for a "chance of" precipitation. If there is a 60 or 70% chance of precipitation, the qualifying forecast states it is "likely that" precipitation will occur. If there is an 80 or 90% chance of precipitation, then there is no qualifying forecast term; precipitation is virtually assured.

don't necessarily know that is is red, but we must assume that it is, because the probability of success (that is, of drawing two red cards in succession) depends on the first card being red. The probability of the second card being red is **dependent** on the probability of the first card being red. The first event *did* influence the second event. Two events are **dependent** if the occurrence of one affects the occurrence of the other.

Therefore, the probability of drawing two red cards in succession, without replacement is

$$\frac{26}{52} \times \frac{25}{51} = \frac{1}{2} \times \frac{25}{51} = \frac{25}{102} \text{ or } 0.245$$

EXAMPLE 1 What is the probability of being dealt two hearts in succession, without replacement, from a shuffled bridge deck?

SOLUTION This implies that we should use the formula

$$P(A \text{ and } B) = P(A) \times P(B)$$

These are dependent events. The probability that the first card is a heart is $\frac{13}{52}$, but the probability that the second card is a heart is $\frac{12}{51}$. Hence the probability that both cards are hearts is

$$\frac{13}{52} \times \frac{12}{51} = \frac{156}{2652} = \frac{1}{17}$$

EXAMPLE 2 A quarter is flipped three times. What is the probability of obtaining three tails in succession?

SOLUTION The flipping of a coin is an independent event. What happens the first time does not affect what will happen the next time. The probability that a flip will yield a tail is $\frac{1}{2}$. To compute the probability of obtaining three tails in a row we have

$$P(T \text{ and } T \text{ and } T) = P(T) \times P(T) \times P(T)$$

$$P(3 \text{ tails}) = \frac{1}{2} \times \frac{1}{2} \times \frac{1}{2} = \frac{1}{8}$$

EXAMPLE 3 On a certain Sunday in October, the probability that the Rams will win their football game is 0.6 and the probability that the Browns will win their football game is 0.4. Assuming that they are not playing each other, what is the probability that both teams will win their games on this given Sunday?

SOLUTION The probability that the Rams will win does not affect the probability that the Browns will win. Hence, we may consider these events to be independent, and we may compute the probability that the Rams *and* Browns will both win by multiplying their respective probabilities. Therefore, we have

$$P(\text{Rams and Browns will win}) = 0.6 \times 0.4 = 0.24$$

The next question is how to compute the probability that *A or B* will occur.

For example, if a quarter is flipped and a die is tossed, what is the probability of obtaining heads *or* an odd number? To solve this problem previously, we generated a sample space (see Example 1, Sec. 3.3) and selected those outcomes that were considered to be successful. We could solve this problem by means of a sample space because we only had to consider 12 outcomes, but when the total possible number of outcomes becomes rather large we shall see that it is much more efficient to use a formula. Before computing the probability of *A or B* by means of a formula, we must first familiarize ourselves with a different kind of set of events.

These are *mutually exclusive events*. Events that are **mutually exclusive** are events that *cannot* happen at the same time: only one of the events can occur at any one time. For example, when we flip a coin, we can get heads or tails. Only one of these—not both—can occur for any one flip. If we roll a die, we can get an odd or an even number with one roll, but not both. These are mutually exclusive events. If two or more events cannot happen at the same time, they are mutually exclusive.

How do we compute the probability of *A or B?* Recall our discussion of the union of two sets *A* and *B* in Chapter 1: The *union* of sets *A* and *B* is the set of all elements that are elements of *A or B,* or elements of both. When we count the elements of the union of two sets, we found the elements of the two sets, but if an element is in both sets, we only count it once.

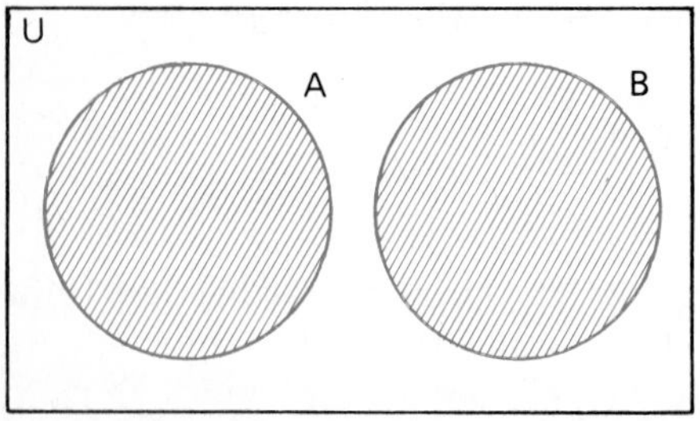

FIGURE 6

If two events are mutually exclusive, then they are disjoint (have no common elements). The Venn diagram for this situation is shown in Figure 6.

The probability of *A or B,* which we will denote by $P(A \text{ or } B)$, is the same as the probability of *A* union *B,* denoted by $P(A \cup B)$. From our diagram and the previous discussion, we have

$$P(A \cup B) = P(A) + P(B) = P(A \text{ or } B)$$

That is, in order to compute $P(A \text{ or } B)$, where A and B are mutually exclusive events, we add $P(A)$ and $P(B)$.

Suppose A and B are not mutually exclusive events; then what is $P(A \text{ or } B)$? In this case, A and B are not mutually exclusive events and therefore have a nonempty intersection. The Venn diagram for this situation is shown in Figure 7.

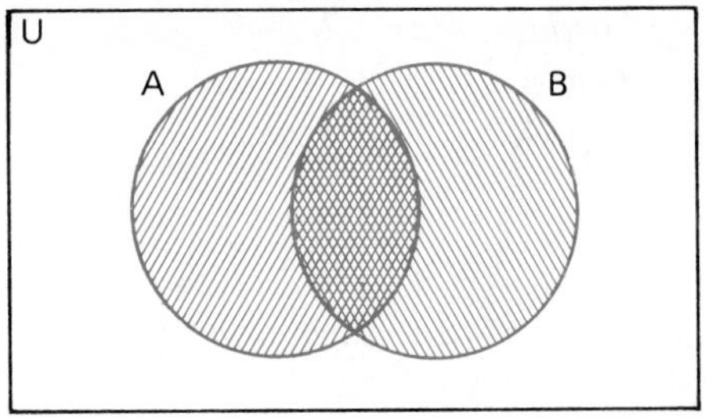

FIGURE 7

Here again, $P(A \text{ or } B)$ is the same as $P(A \cup B)$, but remember that this time the events are not mutually exclusive. Hence we would have

$$P(A \cup B) = P(A) + P(B) - P(A \cap B)$$

We subtract $P(A \cap \mathrm{B})$, the probability of the intersection of A and B, from $P(A) + P(B)$ because $P(A \cap B)$ is included twice in $P(A) + P(B)$. Recall that the *intersection* of a set A and a set B is the set of elements common to both A *and* B. So $P(A \cap B)$ is the same as $P(A \text{ and } B)$, which is the probability of the outcomes which are common to both events A and B. Therefore, if A and B are not mutually exclusive events,

$$P(A \text{ or } B) = P(A) + P(B) - P(A \text{ and } B)$$

It should be noted that this equation is true even when A and B are mutually exclusive events; then $P(A \text{ and } B) = 0$, because A and B are disjoint, and so the intersection is empty. Then we have $P(A \text{ or } B) = P(A) + P(B) - 0$.

EXAMPLE 4 In a single toss of a pair of dice, find the probability of obtaining a 7 or 11.

SOLUTION A 7 and an 11 are mutually exclusive events. Hence, we have

$$P(\{7\} \cup \{11\}) = P(7 \text{ or } 11) = P(7) + P(11)$$
$$= \frac{6}{36} + \frac{2}{36} = \frac{8}{36} = \frac{2}{9}$$

EXAMPLE 5 One card is drawn from a shuffled deck of 52 cards. Find the probability that the card selected is a heart or a jack.

SOLUTION We must first determine if the events are mutually exclusive or not. Since a card may be selected that is both a heart and a jack, namely, the jack of hearts, the events are not mutually exclusive.

Therefore we should use the formula

$$P(A \text{ or } B) = P(A) + P(B) - P(A \text{ and } B)$$

and we have

$$P(\text{heart or jack}) = P(H) + P(J) - P(H \text{ and } J)$$

$$= \frac{13}{52} + \frac{4}{52} - \frac{1}{52} = \frac{16}{52} = \frac{4}{13}$$

EXAMPLE 6 A quarter is flipped and a die is tossed. What is the probability of obtaining heads or a 6?

SOLUTION These events are not mutually exclusive. We can get heads on the quarter and a 6 on the die at the same time.

We can compute probability problems of this type in a variety of ways. One technique is to rely on a sample space and read our answer directly from it. For this problem we would have

(*H*—1)	(*H*—4)	*T*—1	*T*—4
(*H*—2)	(*H*—5)	*T*—2	*T*—5
(*H*—3)	(*H*—6)	*T*—3	(*T*—6)

The outcomes that are circled in the sample space are the successful outcomes, so $P(H \text{ or } 6) = \frac{7}{12}$.

Another way to solve this problem is to use the formula for events that are not mutually exclusive, namely, $P(A \text{ or } B) = P(A) + P(B) - P(A \text{ and } B)$. Then

$$P(H \text{ or } 6) = P(H) + P(6) - P(H \text{ and } 6)$$

$$P(H \text{ or } 6) = \frac{1}{2} + \frac{1}{6} - \frac{1}{12}$$

$$= \frac{6}{12} + \frac{2}{12} - \frac{1}{12} = \frac{7}{12}$$

EXERCISES FOR SECTION 3.6

1. A bag contains three red, two green, and one blue marble. A marble is drawn and then replaced, after which a second marble is drawn. Find the probability that
   a. the first is red and the second is blue
   b. the first is green and the second is red
   c. the first is blue and the second is green
   d. both are red
   e. both are green
   f. both are blue

2. Each of the numbers 1 through 5 is painted on a separate ball. The five balls are placed in a bag and it is shaken to mix up the balls. A ball is drawn and then replaced, after which a second ball is drawn. Find the probability that
   a. the first ball has an odd number and the second ball has an even number
   b. the first ball has a number less than 4 and the second ball has a number greater than 2
   c. the first ball has a number less than 4 and so does the second
   d. both balls contain even numbers
   e. both balls contain odd numbers
   f. the sum of the numbers on both balls will be 10

3. In a certain political science class there are 20 students. The table below shows the distribution of students: male, female, freshmen, and sophomores.

	Freshmen	Sophomores
Male	4	7
Female	3	6

Mr. Harris, the instructor, wants to have a student panel discussion regarding recent events. If the panel is to consist of three students randomly chosen from the class, find the probability that the panel will consist of
   a. all freshmen
   b. all sophomores
   c. all females
   d. all males
   e. a freshman and then two sophomores
   f. a sophomore and then two freshmen

4. The table that follows shows the number of students in a backgammon tournament. The tournament schedule is set by selecting each of the 32 players at random.

	Freshmen	Sophomores	Juniors	Seniors
Female	2	4	4	5
Male	3	4	7	3

Find the probability that
   a. the first two contestants are female
   b. the first two contestants are sophomores
   c. the first two contestants are juniors
   d. the first contestant selected is a female and the second contestant selected is a male
   e. the first four contestants selected are female seniors
   f. the first two contestants selected are freshmen and the next two contestants selected are seniors

5. One card is randomly selected from a shuffled deck of 52 cards and then a quarter is flipped. Find the probability of obtaining
   a. a red card and heads
   b. a red card or heads
   c. a picture card and tails
   d. a picture card or tails
   e. a red picture card and heads
   f. a red picture card or tails

6. One card is randomly selected from a shuffled deck of 52 cards and then a die is rolled. Find the probability of obtaining
   a. a red card and a 1
   b. a red card or a 1
   c. a picture card and a 1
   d. a picture card or a 1
   e. a red picture card and a 1
   f. a red picture card or a 1

7. Two cards are selected at random from a shuffled standard deck of 52 cards, without replacement. Find the probability that
   a. both cards are picture cards
   b. the first card is an ace and the second card is a picture card
   c. the first card is the ace of spades and the second card is a picture card
   d. both cards are kings
   e. both cards are diamonds
   f. both cards are of the same denomination

8. A bag contains 25 marbles, of which eight are red marbles, eight are green, four are blue, and five are clear (colorless).
   a. What is the probability that a marble drawn at random is green?
   b. What is the probability of drawing seven blue marbles in succession if there is no replacement after each draw?

c. What is the probability of drawing four blue marbles in succession if there is replacement after each draw?
d. If two marbles are drawn in succession (without replacement), what is the probability that both marbles are red?
e. If two marbles are drawn in succession (without replacement), what is the probability that the first marble is clear and the second marble is blue?

9. An ice chest contains five cans of ginger ale, three cans of orange soda, and four cans of root beer. If three cans are drawn at random from the chest without replacement, find the probability of getting
a. three cans of root beer
b. three cans of orange
c. three cans of ginger ale
d. a can of orange soda, then a can of root beer, then a can of ginger ale
e. a can of orange soda, then a can of ginger ale, then a can of root beer
f. no root beer

10. One die is rolled, and then another die is rolled. Find the probability of obtaining
a. a 6 on the first die and a 5 on the second die
b. a 6 on the first die or a 5 on the second die
c. the same number on both dice
d. an even number on the first die and an odd number on the second die
e. an even number on the first die or an odd number on the second die
f. an even number on the first die and 6 on the second die

11. A track coach must decide who is going to run on the school's relay team. He must choose four runners from a list of six equally fast runners: Nenno, Nelson, Gilligan, Neanderthal, Clar, and Connelly. He decides that the easiest way to make the decision is draw names from a hat in succession. (Note that there is no replacement since the coach needs four different runners.) Find the probability that
a. he chooses Neanderthal first
b. he chooses Neanderthal first and Nenno second
c. he chooses Nelson or Gilligan first
d. his third choice is a person whose name begins with *N*, given that Connelly and Clar were already chosen
e. Nenno is chosen first, Neanderthal is chosen second, Nelson is chosen third, and Gilligan is chosen to run the anchor leg (fourth).

12. Each of the numbers 0 through 9 is painted on a separate ball. The 10 balls are put in a can and the can is shaken to mix up the balls. Find the probability of
a. drawing an even-numbered ball
b. drawing a ball numbered 4 and then another numbered 4 on two successive draws with replacement
c. drawing a ball numbered 4 and then another numbered 4 on two successive draws without replacement
d. drawing a ball numbered 4 on the first draw, but not on the second draw, with replacement
e. drawing an even number on the first draw and an odd number on the second draw, without replacement.

13. A bingo caller has a machine that contains 75 balls. The balls are marked in the following manner.

$$\begin{array}{l} B1,\ B2,\ B3,\ \ldots,\ B15 \\ I16,\ I17,\ I18,\ \ldots,\ I30 \\ N31,\ N32,\ N33,\ \ldots,\ N45 \\ G46,\ G47,\ G48,\ \ldots,\ G60 \\ O61,\ O62,\ O63,\ \ldots,\ O75 \end{array}$$

The balls are mixed and each ball is drawn at random.
a. What is the probability of drawing a ball that has a *B* on it?
b. What is the probability of drawing three *B* balls in succession, if there is no replacement?
c. What is the probability of drawing a ball that has a double number (11, 22, . . .) on it?
d. What is the probability of drawing a ball that has a *B* or a *G* on it?

e. What is the probability of drawing two balls, without replacement, so that the first one has a B on it and the second one has an I or N on it?
f. What is the probability of drawing a ball that has a number on it that is not divisible by 5?

14. A package of flower seeds contains 10 seeds, six seeds for red flowers and four seeds for yellow flowers. A seed is randomly selected and planted and then another seed is randomly selected and planted. Find the probability that
a. both seeds will result in red flowers
b. both seeds will result in yellow flowers
c. one of each color is selected

15. Given that events A and B are independent, and $P(A) = 0.4$ and $P(B) = 0.5$, find
a. $P(A \text{ and } B)$ b. $P(A \text{ or } B)$
c. $P(\text{not } A)$ d. $P(\text{not } B)$
e. $P(A \text{ and not } B)$ f. $P(\text{not } A \text{ or } B)$

16. Given that events C and D are mutually exclusive, and $P(C) = 0.6$ and $P(D) = 0.3$, find
a. $P(C \text{ and } D)$ b. $P(C \text{ or } D)$
c. $P(\text{not } C)$ d. $P(\text{not } D)$
e. $P(C \text{ or not } D)$ f. $P(\text{not } C \text{ and not } D)$

JUST FOR FUN

Try the following multiple-choice test and answer the questions after the test.

1. In 1956, Mickey Mantle led the major leagues in home runs with a total of
a. 42 b. 48 c. 52 d. 49
2. The recipient of the 1975 Nobel Peace Prize was
a. Anwar Sadat b. Andrei Sakharov
c. Henry Kissinger d. Willy Brandt
3. Franklin D. Roosevelt was born in
a. 1884 b. 1890 c. 1887 d. 1882
4. In 1979, the company that spent the most money on advertising was
a. Sears b. Philip Morris
c. Procter and Gamble d. Coca Cola
5. Kansas, the Sunflower State, entered the union in
a. 1861 b. 1846 c. 1816 d. 1836

Questions pertaining to the test

1. What is the probability of guessing correctly on any one given question?
2. What is the probability of guessing correctly on all of the questions?
3. What is the probability of guessing incorrectly on any one given question?
4. What is the expected number of correct answers?

Answers to multiple choice test

1. c 2. b 3. d 4. c 5. a

3.7 COUNTING, ORDERED ARRANGEMENTS, AND PERMUTATIONS (OPTIONAL)

What is counting? When we count something, we want to know "how many." Often we have to count all of the possible outcomes for probability problems. The basic definition of the probability that an event A will occur is the *number* of successful outcomes of the experiment divided by the total *number* of possible outcomes. In Section 3.3, we introduced the *counting principle* to enable us to efficiently count the possible outcomes for an experiment.

Counting Principle: If one experiment has m different outcomes, and a second experiment has n different outcomes, then the first and second experiments performed together has $m \times n$ different outcomes. This may be extended, if there are other experiments to follow, to $m \times n \times r \times \cdots \times t$.

How many different telephone numbers are possible in your local calling area? Every phone number consists of seven digits (10, if you also count the area code). As a counting example, we shall examine just the last four digits. There are four numbers of outcomes to multiply, one for each digit.

$$m \times n \times r \times s$$

In how many ways can each digit be chosen? Each digit may be any one of the numerals 0, 1, 2, 3, 4, 5, 6, 7, 8, 9, and therefore, there are 10 possibilities for each digit. (We are assuming that each numeral may be repeated in more than one digit.) Hence each digit may be chosen in any one of 10 different ways, assuming no restrictions on the use of 0 or 1, so we have

$$10 \times 10 \times 10 \times 10 = 10{,}000$$

different telephone numbers in a local calling area or exchange. If we want to find out the number of telephone numbers in one general area which share the same area code, we can expand this problem to seven digits.

EXAMPLE 1 How many different license plates can be made if each license plate is to consist of three letters followed by two digits? (Assume replacement.)

SOLUTION Here order is important. There are five slots to fill. The first three slots are to be filled by letters of the alphabet; because we have replacement, there are 26 choices for each of the first three slots.

Each of the last two slots may be filled by any one of the 10 numerals, 0, 1, 2, 3, 4, 5, 6, 7, 8, 9. Therefore, the solution is

$$26 \times 26 \times 26 \times 10 \times 10 = 1{,}757{,}600$$

EXAMPLE 2 How many different license plates can be made if each license plate is to consist of three letters followed by two digits, and no letters or digits may be repeated?

SOLUTION This problem is similar to Example 1, but now we have a restriction that no repetition is allowed. We still have 26 choices for the first slot in the license plant, but there are only 25 choices for the second slot because we have already used a letter in the first slot. Similarly, there are 24 choices for the third slot, because we used one letter of the 25 in the second slot. The same thing happens when we fill in the slots for the digits. Hence, the solution is

$$26 \times 25 \times 24 \times 10 \times 9 = 1{,}404{,}000$$

EXAMPLE 3 Marie is planning her schedule for next semester. She must take the following five courses: English, history, geology, psychology, and mathematics.

a. In how many different ways can Marie arrange her schedule of courses?
b. How many of these schedules have mathematics listed first?

SOLUTION a. Since Marie has to take five different courses, there are five time slots to consider.

$$m \times n \times r \times s \times t$$

There are five choices for the first time slot, four for the second, and so on. The solution is

$$5 \times 4 \times 3 \times 2 \times 1 = 120$$

b. If mathematics is to be listed first, then this is a restriction, and there is only one choice for the first time slot. Mathematics is used to fill the first slot, so we have four choices for the second slot, three for the third slot, and so on. The solution is

$$1 \times 4 \times 3 \times 2 \times 1 = 24$$

EXAMPLE 4 Given the set of digits {1, 3, 4, 5, 6}.

a. How many three-digit numbers can be formed?
b. How many three-digit numbers can be formed if the number must be even?
c. How many three-digit numbers can be formed if the number must be even and no repetition of digits is allowed?

SOLUTION a. There are three places, or slots, to consider. Each slot can be filled by any one of the five digits. Therefore, the solution is

$$5 \times 5 \times 5 = 125$$

b. The three-digit number must be even, so it must end either in a 4 or 6, the even digits in our given set. This is a restriction, so we should attend to it first: we have two choices, 4 or 6, for the last place. The counting principle gives us

$$\_ \times \_ \times 2$$

three-digit numbers. We have five choices for the first place and five choices for the second. Hence, we have

$$5 \times 5 \times 2 = 50$$

three-digit numbers.

c. There are two restrictions here: The number must be even, and no digit may be repeated. We still fill in the last place with 2 choices.

$$\_ \times \_ \times 2$$

But now there are only four choices for the first place, because one of the digits has been used in the last place. There are then three choices for the second place, and we have

$$4 \times 3 \times 2 = 24$$

three-digit numbers.

If a woman has individual photos of each of her three children, Mary, Scott, and Joe, how many ways can she arrange these photos in a row on her desk? From our previous discussion, we know that there are $3 \times 2 \times 1 = 6$ different possible outcomes, and they are

Mary—Scott—Joe
Mary—Joe—Scott

Scott—Joe—Mary
Scott—Mary—Joe
Joe—Scott—Mary
Joe—Mary—Scott

Each of these arrangements is different. The pictures are in a definite order, and the order in which the pictures are arranged is important. When we have a group of things arranged in a definite order, we have a *permutation*. A **permutation** is a particular ordering of the elements of a given set.

Permutations may use all the elements from a given set (as we did in our example), or only a certain number of them. If we are given the digits 1, 2, 3, 4, and 5, how many two-digit numbers can we form? This is a permutation (arrangement) of five things taken two at a time. In this case, we would have $5 \times 4 = 20$ different arrangements. If we have n items and want to count the possible arrangements using r of them at a time, we say that we want the number of **permutations of n things taken r at a time.** The notation for this is

$_nP_r$ the number of permutations of n things taken r at a time

The notation $_nP_r$ says that we want to fill r places or slots from a total group of n things. Recall that $_5P_2$ is a permutation of five things taken two at a time, which is $5 \times 4 = 20$. Similarly, $_6P_4$ is a permutation of six things taken four at a time, and $_6P_4 = 6 \times 5 \times 4 \times 3 = 360$. If we have $_nP_r$, a permutation of n things taken r at a time, then we have r slots and must fill them from n things. The first slot may be filled in n different ways, the second slot may be filled in $n - 1$ different ways, the third slot may be filled in $n - 2$ different ways, and so on. The computation of the number of permutations of n things taken r at a time is indicated below, along with two examples that we discussed previously.

$$_5P_2 = 5 \times 4 = 20$$
$$_6P_4 = 6 \times 5 \times 4 \times 3 = 360$$

$$_nP_r = \underbrace{\underset{\text{1st slot}}{n} \times \underset{\text{2nd slot}}{(n-1)} \times \underset{\text{3rd slot}}{(n-2)} \times \underset{\text{4th slot}}{(n-3)} \times \cdots \times \underset{\text{rth (last) slot}}{(n-r+1)}}_{\text{number of ways each slot may be filled}}$$

This leads to the formula

$$_nP_r = n \times (n - 1) \times (n - 2) \times (n - 3) \times \cdots \times (n - r + 1)$$

or

$$ {}_nP_r = n(n-1)(n-2)(n-3)(n-4)\cdots(n-r+1) $$

From this formula we now know that

$$ \underbrace{{}_8P_3}_{n \text{ things},\ r \text{ slots}} = \underbrace{8}_{n} \times \underbrace{(8-1)}_{(n-1)} \times \underbrace{(8-3+1)}_{(n-r+1)} = 8 \times 7 \times 6 $$

What happens when $n = r$? If we examine ${}_nP_r$ when $n = r$, we have ${}_nP_r = n(n-1)(n-2)\cdots(n-r+1)$ and the factor $(n-r+1)$ becomes $(n-n+1)$, or just 1. Thus when $n = r$, we have ${}_nP_n = n(n-1)(n-2)(n-3)\cdots(3)(2)(1)$. The product of all the numbers from n down to and including 1 is called n **factorial.** The symbol for n factorial is $n!$ A table of factorials (Table 1) appears in the Appendix: Some examples are

(n factorial) $n! = n(n-1)(n-2)(n-3)\cdots(3)(2)(1)$
(3 factorial) $3! = 3 \times 2 \times 1$ or $(3)(2)(1)$ or $3 \cdot 2 \cdot 1 = 6$
(5 factorial) $5! = 5 \times 4 \times 3 \times 2 \times 1$ or $(5)(4)(3)(2)(1)$
or $5 \cdot 4 \cdot 3 \cdot 2 \cdot 1 = 120$

Zero factorial, 0!, cannot be defined by the rule used in the preceding examples. We agree to let 0! equal 1 as a matter of convenience. This special definition enables us to produce a simpler formula for ${}_nP_r$.

Examine the following examples.

$$ {}_6P_2 = 6 \cdot 5 = 30 \quad \text{or} \quad {}_6P_2 = \frac{6!}{4!} = \frac{6 \cdot 5 \cdot 4 \cdot 3 \cdot 2 \cdot 1}{4 \cdot 3 \cdot 2 \cdot 1} $$

$$ = \frac{720}{24} = 30 $$

$$ {}_5P_3 = 5 \cdot 4 \cdot 3 = 60 \quad \text{or} \quad {}_5P_3 = \frac{5!}{2!} = \frac{5 \cdot 4 \cdot 3 \cdot 2 \cdot 1}{2 \cdot 1} $$

$$ = \frac{120}{2} = 60 $$

In general, we have

$$ {}_nP_r = \frac{n!}{(n-r)!} \qquad (r \leq n) $$

For example,

$$_5P_3 = \frac{5!}{(5-3)!} = \frac{5!}{2!} = \frac{5 \cdot 4 \cdot 3 \cdot 2 \cdot 1}{2 \cdot 1} = 60$$

$$_6P_2 = \frac{6!}{(6-2)!} = \frac{6!}{4!} = \frac{6 \cdot 5 \cdot 4 \cdot 3 \cdot 2 \cdot 1}{4 \cdot 3 \cdot 2 \cdot 1} = 30$$

$$_4P_4 = \frac{4!}{(4-4)!} = \frac{4!}{0!} = \frac{4 \cdot 3 \cdot 2 \cdot 1}{1} = 24$$

Note that 0! appears in the denominator of the last example, $_4P_4$. In order for our formula to work for this case, we must define $0! = 1$.

EXAMPLE 5 Evaluate each of the following.

a. 4! b. 8! c. $_7P_4$ d. $_8P_3$ e. $_{10}P_{10}$ f. $_5P_0$

SOLUTION a. $4! = 4 \cdot 3 \cdot 2 \cdot 1 = 24$

b. $8! = 8 \cdot 7 \cdot 6 \cdot 5 \cdot 4 \cdot 3 \cdot 2 \cdot 1 = 40{,}320$

c. $_7P_4 = \frac{7!}{(7-4)!} = \frac{7!}{3!} = \frac{7 \cdot 6 \cdot 5 \cdot 4 \cdot \cancel{3} \cdot \cancel{2} \cdot 1}{\cancel{3} \cdot \cancel{2} \cdot 1} = 840$

d. $_8P_3 = \frac{8!}{(8-3)!} = \frac{8!}{5!} = \frac{8 \cdot 7 \cdot 6 \cdot \cancel{5} \cdot \cancel{4} \cdot \cancel{3} \cdot \cancel{2} \cdot 1}{\cancel{5} \cdot \cancel{4} \cdot \cancel{3} \cdot \cancel{2} \cdot 1} = 336$

e. $_{10}P_{10} = \frac{10!}{(10-10)!} = \frac{10 \cdot 9 \cdot 8 \cdot 7 \cdot 6 \cdot 5 \cdot 4 \cdot 3 \cdot 2 \cdot 1}{1}$

$= 3{,}628{,}800$

f. $_5P_0 = \frac{5!}{(5-0)!} = \frac{5!}{5!} = \frac{\cancel{5} \cdot \cancel{4} \cdot \cancel{3} \cdot \cancel{2} \cdot 1}{\cancel{5} \cdot \cancel{4} \cdot \cancel{3} \cdot \cancel{2} \cdot 1} = 1$

EXAMPLE 6 If seven people board an airplane, and there are nine aisle seats, in how many ways can they be seated if they all choose aisle seats?

SOLUTION This is a permutation of nine things taken seven at a time. So we have

$$_9P_7 = \frac{9!}{(9-7)!} = \frac{9!}{2!} = \frac{9 \cdot 8 \cdot 7 \cdot 6 \cdot 5 \cdot 4 \cdot 3 \cdot \cancel{2} \cdot 1}{\cancel{2} \cdot 1} = 181{,}440$$

EXAMPLE 7 A disc jockey can play eight records in a 30-minute segment of her show. For a particular 30-minute segment, she has 12 records to select from. In how many ways can she arrange her program for the particular segment?

SOLUTION This is a permutation of 12 things taken eight at a time. Therefore we have

$$_{12}P_8 = \frac{12!}{(12-8)!} = \frac{12!}{4!}$$

$$= \frac{12 \cdot 11 \cdot 10 \cdot 9 \cdot 8 \cdot 7 \cdot 6 \cdot 5 \cdot \not{4} \cdot \not{3} \cdot \not{2} \cdot 1}{\not{4} \cdot \not{3} \cdot \not{2} \cdot 1}$$

$$= 19{,}958{,}400$$

We now know that $_3P_3$ is equal to 6. This means that there are six distinct arrangements of three things taken three at a time. Let's now look at a slightly different problem. Consider the word *ALL* and the number of arrangements that can be made from the letters in it. There are three letters, and therefore the number of arrangements is 3! Let's call the first L in the word L_1 ("L sub one") and the second L in the word L_2 ("L sub two"). Now we can distinguish between the two L's. The arrangements of the three letters are

1. $A\ L_1\ L_2$	4. $L_2\ A\ L_1$
2. $A\ L_2\ L_1$	5. $L_1\ L_2\ A$
3. $L_1\ A\ L_2$	6. $L_2\ L_1\ A$

If we use the subscript notation as above, we can see the different arrangements. When we interchange the L's, we technically have a different arrangement. But without the subscripts, it is impossible to tell if the L's have been moved, and we have

1. $A\ L\ L$	4. $L\ A\ L$
2. $A\ L\ L$	5. $L\ L\ A$
3. $L\ A\ L$	6. $L\ L\ A$

Since we cannot tell if the L's have been interchanged, the arrangements are not all distinct. Arrangements 1 and 2 are the same, as are 3 and 4, and 5 and 6. Hence, there are only three distinct arrangements, namely, *ALL, LAL, and LLA*. The two L's can be arranged in 2! ways. In order to obtain the number of distinct arrangements, we must divide $_3P_3 = 3!$ by 2! That is, there are

$$\frac{3!}{2!} = \frac{3 \cdot 2 \cdot 1}{2 \cdot 1} = 3 \qquad \text{distinct arrangements from the letters in } ALL$$

If we want to find the number of **distinct arrangements of** n things when p of the things are alike, then we divide $n!$ by $p!$.

EXAMPLE 8 How many distinct arrangements can be formed from all the letters of *GENESEE?*

SOLUTION There are seven letters. If they were all different, we would have 7! arrangements, but there are 4 *E*'s, which can be arranged in 4! = 24 ways. So we divide 7! by 4! and the answer is

$$\frac{7!}{4!} = \frac{7 \cdot 6 \cdot 5 \cdot \cancel{4} \cdot \cancel{3} \cdot \cancel{2} \cdot 1}{\cancel{4} \cdot \cancel{3} \cdot \cancel{2} \cdot 1} = 210$$

Suppose, in addition to p things of the same kind, we also have q things alike, and r things alike, and so on. If this occurs, then we just extend our computation to

$$\frac{n!}{p!q!\cdots}$$ the number of distinct arrangements of n things when p things are alike, q things are alike, and so on.

EXAMPLE 9 In how many distinct ways can the letters of *OSMOSIS* be arranged?

SOLUTION There are seven letters in the word *OSMOSIS,* and if they were all different we would have 7! arrangements. Since there are 2 *O*'s, we must divide by 2!, and since there are 3 *S*'s, we must divide by 3!. Hence we have

$$\frac{7!}{2!3!} = \frac{7 \cdot 6 \cdot 5 \cdot 4 \cdot \cancel{3} \cdot \cancel{2} \cdot 1}{2 \cdot 1 \cdot \cancel{3} \cdot \cancel{2} \cdot 1} = 420$$

EXERCISES FOR SECTION 3.7

1. Given the set of digits {4, 5, 6, 7, 8, 9}, how many three-digit numbers can be formed if no digit can be repeated? How many three-digit numbers can be formed if repetition is allowed?
2. If there are 50 contestants in a beauty pageant, in how many ways can the judges award first and second prizes?
3. The Riverhead Sports Car Club has 30 members. A slate of officers consists of a president, a vice president, a secretary, and a treasurer. If a person can hold only one office, in how many ways can a set of officers be formed?
4. A baseball manager has eight pitchers and three catchers on his squad. In how many ways can the manager select a starting battery (pitcher and catcher) for a game?

5. A conference room has four doors. In how many ways can a person enter and leave the conference room by a different door?

6. Given the set of digits {5, 6, 7, 8, 9}, how many four-digit numbers can be formed if no digit can be repeated? How many of these will be odd? How many of these will be divisible by 5? How many of these will be over 6,000? How many will be over 5,000?

7. How many license plates can be made if each plate must consist of two letters followed by four digits, if we assume no repetition? How many are possible if the first digit cannot be zero? How many are possible if the first letter cannot be *O* and the first digit cannot be zero?

8. How many four-letter words can be formed from the set of letters {*m*, *o*, *n*, *e*, *y*}? Assume that any arrangement of letters is a word. How many four-letter words can be formed if the first letter must be *y* and the last letter must be *m*? (Assume no repetition.)

9. The Rochester Tennis Club is having a mixed-doubles tournament. If eight women and their husbands sign up for the tournament, how many mixed-doubles teams are possible? How many can be formed if no woman is paired with her husband?

10. At Finger Lakes Race Track, there are eight horses in each race. The daily double consists of picking the winning horses in the first and second races. If a bettor wanted to purchase all possible daily double tickets, how many would he have to purchase?

11. Evaluate the following.
a. $3!$ b. $5!$ c. $1!$
d. $\frac{7!}{4!}$ e. ${}_5P_2$ f. ${}_4P_4$

12. Evaluate the following.
a. $4!$ b. $6!$ c. $0!$
d. $\frac{100!}{99!}$ e. ${}_{50}P_2$ f. ${}_6P_0$

13. On a naval vessel, a signal can be formed by running up two flags on a flag pole, one above the other. If there are 10 different flags to choose from, how many different signals can be formed?

14. How many different ways can seven students be seated in seven seats on a subway car?

15. A baseball manager must hand in the lineup card before the game begins. The manager has decided which nine players will play, but not the batting order. In how many ways can a batting order be chosen?

16. Given the set of digits {3, 4, 5, 6, 7, 8, 9}, how many different numbers between 3,000 and 6,000 can be written using these digits if repetition of digits is allowed?

17. In how many ways can a basketball coach select a guard and then a center from a squad of 12 players?

18. A dictionary, an almanac, a catalog, and a diary are to be placed on a shelf. In how many ways can they be arranged?

19. In how many distinct ways can the letters of each word be arranged?
a. *ALGEBRA* b. *STATISTICS*
c. *CALCULUS* d. *SCIENCE*

20. In how many distinct ways can the letters of each word be arranged?
a. *OHIO* b. *ALABAMA*
c. *MISSISSIPPI* d. *CONNECTICUT*

21. Given the set of digits {1, 2, 3, 4, 5}, how many different numbers consisting of four digits can be formed when repetition is allowed?

22. A traveling book salesperson has five copies of a certain statistics book, four copies of a certain geometry book, and three copies of a certain calculus book. If these books are to be stored on a shelf in the salesperson's van, how many distinct arrangements are possible?

23. Almost all students have a social security number. In many schools, a student's social security number is also his or her ID number. How many possible social

security numbers are there? (Assume repetition of digits. A social security number contains nine digits.)

24. Telephone numbers consist of seven digits: three digits for the exchange, followed by four more digits. In order to call long distance, you must also use an are code, which consists of three more digits. How many long-distance telephone numbers are there if the first digit cannot be zero or 1, and the fourth digit cannot be zero or 1? (Assume repetition of digits.)

Reprinted by permission of LIFS, Inc.

JUST FOR FUN Name the southernmost state in the United States.

3.8 COMBINATIONS (OPTIONAL)

In Section 3.7, we were concerned with the counting of ordered arrangements, or permutations. The techniques developed can be applied to situations like the following: If a tennis squad has five players on it, the coach can select the first singles player in five ways and the second singles player in four ways. Hence, there are $5 \times 4 = 20$ different ways that the coach can select the first and second singles players. This type of choice is a permutation of five things taken two at a time.

Now, let's consider a slightly different type of choice. Suppose the five members of the squad are Stan, Bill, Mike, Alice, and Pam, and we want to know in how many ways the coach can select a doubles team from this group of five players. This is different because when the coach selects a doubles team, it does not matter which person is chosen first and which person is chosen second. That is, the order of choice is not important. The 20 different ordered arrangements are

Stan, Bill	Bill, Stan	Mike, Stan
Stan, Mike	Bill, Mike	Mike, Bill
Stan, Alice	Bill, Alice	Mike, Alice
Stan, Pam	Bill, Pam	Mike, Pam

Alice, Stan	Pam, Stan
Alice, Bill	Pam, Bill
Alice, Mike	Pam, Mike
Alice, Pam	Pam, Alice

Note that the doubles team of Stan and Bill is the same doubles team as that of Bill and Stan. This is also the case for Alice and Mike and Mike and Alice. If we eliminate the duplicate doubles teams, we have

Stan, Bill	Bill, Mike
Stan, Mike	Bill, Alice
Stan, Alice	Bill, Pam
Stan, Pam	Mike, Alice
Alice, Pam	Mike, Pam

There are 10 different doubles teams that the coach can select. The 20 distinct ordered arrangements of two people are reduced to 10 groups of two people when we disregard order.

The problem we have just discussed can be classified as a *combination* problem. A **combination** is a distinct group of objects without regard to their arrangement. Committees are good examples of combinations. We are only concerned with who is on a committee, not with who is first, who is second, and so on.

Consider the set of letters $\{x, y, z\}$. How many different three-letter arrangements can be formed from this set? The answer is 3!, or 6, and they are

x, y, z	y, z, x	z, x, y
x, z, y	y, x, z	z, y, x

How many combinations (distinct groups) can be formed from this set? We see that all six arrangements constitute the same group of elements. Therefore there is only one combination. We can say that the number of combinations of three things taken three at a time is 1. Symbolically, we can write

$$_3C_3 = 1$$

In the doubles problem, we had combinations of five things taken two at a time. The number of these combinations was equal to 10; hence

$$_5C_2 = 10$$

Note that

$$_5C_2 = \frac{_5P_2}{2!} = \frac{5!}{(5-2)!2!} = \frac{5!}{3!2!} = \frac{5 \cdot 4 \cdot \cancel{3} \cdot \cancel{2} \cdot 1}{\cancel{3} \cdot 2 \cdot 1 \cdot \cancel{2} \cdot 1} = 10$$

Focus on Sports

and

$$_3C_3 = \frac{_3P_3}{3!} = \frac{3!}{(3-3)!3!} = \frac{3!}{0!3!} = 1$$

In order to obtain the number of distinct combinations, we must eliminate the different ordered arrangements within each group, and we do this by division. From the set of five tennis players, we have five choices for the first selection and four for the second, but those two selections can order themselves in 2! ways, so we must divide by 2!

The general formula for the number of combinations of n things taken r at a time is

$$_nC_r = \frac{n!}{(n-r)!r!}$$

Recall the formula for the number of permutations of n things taken r at a time.

$$_nP_r = \frac{n!}{(n-r)!}$$

If you know this formula, then the formula for $_nC_r$ is easy to remember because it is similar to $_nP_r$. If

$$_nP_r = \frac{n!}{(n-r)!} \quad \text{and} \quad _nC_r = \frac{n!}{(n-r)!r!}$$

then

$$_nP_r = {_nC_r} \cdot r! \quad \text{or} \quad _nC_r = \frac{_nP_r}{r!} = \frac{n!}{(n-r)!r!}$$

You should be thoroughly familiar with both formulas.

$$_nP_r = \frac{n!}{(n-r)!}$$ the number of *permutations* of n things taken r at a time.

$$_nC_r = \frac{n!}{(n-r)!r!}$$ the number of *combinations* of n things taken r at a time.

Note: Some books use the notation $\binom{n}{r}$ to represent the number of combinations of n things taken r at a time. We shall not use this notation, but be aware that $\binom{n}{r} = \frac{n!}{(n-r)!r!}$

EXAMPLE 1 Two co-captains are to be selected from the starting five for a basketball team. In how many ways can this be done?

SOLUTION This is a combination problem since order is not important. We have a combination of five things taken two at a time. Therefore, we have

$$ {}_5C_2 = \frac{5!}{(5-2)!2!} = \frac{5!}{3!2!} = \frac{5 \cdot 4 \cdot \cancel{3} \cdot \cancel{2} \cdot 1}{\cancel{3} \cdot 2 \cdot 1 \cdot \cancel{2} \cdot 1} = 10 $$

EXAMPLE 2 The student association each year selects a council consisting of seven members. If there are 10 candidates for the seven-member council, how many different councils may be elected?

SOLUTION Since order is not important, we treat this as a combination of 10 things taken seven at a time. Therefore, we have

$$ \begin{aligned} {}_{10}C_7 &= \frac{10!}{(10-7)!7!} \\ &= \frac{10!}{3!7!} \\ &= \frac{10 \cdot 9 \cdot 8 \cdot \cancel{7} \cdot \cancel{6} \cdot \cancel{5} \cdot \cancel{4} \cdot \cancel{3} \cdot \cancel{2} \cdot 1}{3 \cdot 2 \cdot 1 \cdot \cancel{7} \cdot \cancel{6} \cdot \cancel{5} \cdot \cancel{4} \cdot \cancel{3} \cdot \cancel{2} \cdot 1} = 120 \end{aligned} $$

EXAMPLE 3 How many different poker hands can be dealt from a standard deck of 52 cards? (Here we assume a poker hand consists of five cards.)

SOLUTION Order is not important, since a poker hand consisting of king of hearts, king of clubs, queen of clubs, jack of hearts, and king of spades is the same as one consisting of king of hearts, queen of clubs, king of spades, jack of hearts, and king of clubs. We have a combination of 52 things taken five at a time; that is,

$$ \begin{aligned} {}_{52}C_5 &= \frac{52!}{(52-5)!5!} = \frac{52!}{47!5!} = \frac{52!}{5!47!} \\ &= \frac{52 \cdot 51 \cdot 50 \cdot 49 \cdot 48 \cdot \cancel{47!}}{5 \cdot 4 \cdot 3 \cdot 2 \cdot 1 \cdot \cancel{47!}} = 2{,}598{,}960 \end{aligned} $$

EXAMPLE 4 How many committees can be selected from four teachers and 100 students if each committee must have two teachers and three students?

SOLUTION This combination problem is somewhat different from the others we have considered. Here we must choose two teachers from four and three students from 100 to form our committee. Hence we must use the counting principle. If we can choose the teachers in t ways and the students in s ways, then together the choosing can be done in $t \times s$ ways. Therefore the number of committees consisting of two teachers and three students may be found by multiplying ${}_4C_2$ (four teachers taken two at a time) by ${}_{100}C_3$ (100 students taken three at a time).

$$ {}_4C_2 = \frac{4!}{(4-2)!2!} = \frac{4!}{2!2!} = \frac{4 \cdot 3 \cdot \not{2} \cdot 1}{2 \cdot 1 \cdot \not{2} \cdot 1} = 6 $$

$$ {}_{100}C_3 = \frac{100!}{(100-3)!3!} = \frac{100!}{97!3!} = \frac{100 \cdot 99 \cdot 98 \cdot \not{97}!}{3 \cdot 2 \cdot 1 \cdot \not{97}!} = 161{,}700 $$

Therefore

$$ {}_4C_2 \cdot {}_{100}C_3 = 6 \cdot 161{,}700 = 970{,}200 $$

In doing problems of this nature, we must carefully examine the problem to determine if we are doing a permutation problem or a combination problem. If we want distinct arrangements, then we are doing permutations. If the distinct arrangements are not to be counted, only the different groups, then we are doing combinations. Always be careful of special conditions in the problem. Try to take care of these first; then proceed with the rest of the problem.

EXERCISES FOR SECTION 3.8

1. Evaluate the following.
   a. ${}_5C_3$ b. ${}_5C_2$ c. ${}_7C_4$
   d. ${}_7C_3$ e. ${}_{10}C_{10}$ f. ${}_{10}C_0$

2. Evaluate the following.
   a. ${}_8C_5$ b. ${}_8C_3$ c. ${}_{52}C_2$
   d. ${}_{52}C_{50}$ e. ${}_{52}C_5$ f. ${}_{52}C_{47}$

3. If the Xerox Corporation has to transfer four of its 10 junior executives to a new location, in how many ways can the four executives be chosen?

4. A newspaper boy discovers in delivering his papers that he is three papers short. He has eight houses left to deliver to, but only five papers left. In how many ways can he deliver the remaining newspapers?

5. Alice has a penny, a nickel, a dime, a quarter, and a half dollar. She may spend any three coins. In how many ways can Alice do this? What is the most money she can spend using just three coins?

6. Joe has to take a math exam that consists of 10 questions. He must answer only seven of the 10 questions. In how many ways can Joe choose the seven questions? If he must answer the first and last questions and still only answer a total of seven, in how many ways can he do this?

7. In a mathematics class of 15 students, 10 students must do problems at the board on a given day. In how many ways can the 10 students be chosen?

8. A football coach has 40 candidates out for the squad. In how many ways can a starting 11 be selected without regard to the position that a candidate will play? (Indicate your answer; do not evaluate.)

9. At registration, a student needs two more courses to complete her schedule. If there are seven possible courses left to pick from, in how many ways can she choose the two courses?

10. A committee of 11 people, six women and five men, is forming a subcommittee that is to be made up of two women and three men. In how many ways can the subcommittee be formed?

11. A baseball squad consists of eight outfielders and seven infielders. If the baseball coach must choose three outfielders and four infielders, in how many ways can this be done?

12. An urn contains six blue balls and four orange balls.
    a. In how many ways can we select a group of three balls?
    b. In how many ways can we select two blue balls and one orange ball?
    c. In how many ways can we select two orange balls and one blue ball?

13. From a group of 12 sprinters and 10 distance runners, a medley relay team is to be formed. The relay team must consist of two sprinters and two distance runners. How many possible medley relay teams are there?

14. Don has to take a history exam that consists of 15 multiple-choice questions and five essay questions. If Don has to answer 10 multiple-choice questions and two essay questions, in how many ways can he choose them?

15. A student belongs to a record club. This month she has to purchase two records and three tapes. If there are 10 records and 10 tapes to choose from, in how many ways can she choose her purchases?

16. If six points are drawn on a plane, no three of which are on the same straight line, how many straight lines can be formed? (Two points determine a line.)

17. How many different committees, each composed of two Democrats, two Republicans, two Liberals, and one Conservative, can be formed from 12 Democrats, 11 Republicans, 5 Liberals, and 3 Conservatives?

18. The Speaker of the House wants to appoint a committee consisting of three representatives from New York, three representatives from California, two representatives from Ohio, and three representatives from Illinois. How many different committees can be formed if eight representatives from New York, ten representatives from California, five representatives from Ohio, and six representatives from Illinois are eligible?

JUST FOR FUN Is a "combination" lock really a combination lock? Why or why not?

3.9 MORE PROBABILITY (OPTIONAL)

We can utilize the counting principle, permutations, and combinations in solving many probability problems. Some of them are similar to those that we discussed previously, whereas others are a little more involved.

Consider the discussion on pages 150–51. We want to find the probability of being dealt two red cards in succession, without replacement, from a shuffled deck of 52 cards. We found that the probability was

$$\frac{26}{52} \cdot \frac{25}{51} = \frac{650}{2652} = \frac{25}{102}$$

There are 26 red cards out of 52 for the first card, and then 25 red cards out of 51 remaining cards for the second card. We multiplied the two probabilities together to find the probability of getting a red card *and* a red card.

Another way to solve this problem is by combinations. We are not concerned with the order in which the cards appear, just as long as they are red. In how many ways can two cards be chosen from a deck of 52? We have

$$_{52}C_2 = \frac{52!}{(52-2)!2!} = \frac{52!}{50!2!} = \frac{52 \cdot 51 \cdot \cancel{50!}}{\cancel{50!} \cdot 2 \cdot 1} = \frac{52 \cdot 51}{2 \cdot 1} = 1326$$

Camerique

A poker hand consisting of the ten, jack, queen, king, and ace of any one suit is called a *royal flush*. The probability of being dealt a royal flush is

$$\frac{4}{_{52}C_5} \quad \text{or} \quad \frac{4}{2,598,960}$$

In how many ways can two red cards be chosen? This is the number of successful outcomes. There are 26 red cards in the deck, and a successful outcome is any combination of two of these cards. Therefore, we compute

$$_{26}C_2 = \frac{26!}{(26-2)!2!} = \frac{26!}{24!2!} = \frac{26 \cdot 25 \cdot \cancel{24!}}{\cancel{24!} \cdot 2 \cdot 1} = \frac{26 \cdot 25}{2 \cdot 1} = 325$$

Hence the probability of being dealt two red cards in succession is

$$\frac{_{26}C_2}{_{52}C_2} = \frac{325}{1326} = \frac{25}{102}$$

Let us look at some other examples.

EXAMPLE 1 A student has to complete registration for the next semester by choosing three more courses. The courses left to choose from are five humanities courses and four science courses. If the three courses are chosen at random, what is the probability that they will all be humanities courses?

SOLUTION There are nine courses to choose from, and the student must choose three of them. There are $_9C_3$ ways of choosing three courses. The three humanities courses may be chosen from the five offered in $_5C_3$ ways. Hence we have

$$_5C_3 = \frac{5!}{(5-3)!3!} = \frac{5!}{2!3!} = \frac{5 \cdot 4 \cdot \cancel{3!}}{2 \cdot \cancel{3!}} = 10$$

$$_9C_3 = \frac{9!}{(9-3)!3!} = \frac{9!}{6!3!} = \frac{9 \cdot 8 \cdot 7 \cdot \cancel{6!}}{\cancel{6!} \cdot 3 \cdot 2 \cdot 1} = 84$$

$$P(3 \text{ humanities}) = \frac{_5C_3}{_9C_3} = \frac{10}{84}$$

EXAMPLE 2 Find the probability of being dealt a hand in five-card poker that is all spades. (A *flush* is a hand where all the cards are of the same suit: all hearts, all spades, all diamonds, or all clubs.)

SOLUTION There are 52 cards in a deck, and there are $_{52}C_5$ possible ways of being dealt five cards. There are 13 spades in a deck, and a successful outcome is being dealt any five of these spades. There are ${}_{13}C_5$ ways of being dealt five spades. Therefore

$$_{13}C_5 = \frac{13!}{(13-5)!5!} = \frac{13!}{8!5!} = \frac{13 \cdot 12 \cdot 11 \cdot 10 \cdot 9 \cdot \cancel{8!}}{\cancel{8!} \cdot 5 \cdot 4 \cdot 3 \cdot 2 \cdot 1}$$

$$= 1287$$

$$_{52}C_5 = \frac{52!}{(52-5)!5!} = \frac{52!}{47!5!}$$

$$= \frac{52 \cdot 51 \cdot 50 \cdot 49 \cdot 48 \cdot \cancel{47!}}{\cancel{47!} \cdot 5 \cdot 4 \cdot 3 \cdot 2 \cdot 1} = 2{,}598{,}960$$

$$P(5 \text{ spades}) = \frac{_{13}C_5}{_{52}C_5} = \frac{1287}{2{,}598{,}960} = \frac{33}{66{,}640}$$

EXAMPLE 3 Find the probability of being dealt a *full house* (three cards of one denomination and two of another) consisting of kings over deuces (three kings and two deuces).

SOLUTION There are 52 cards in a deck and there are ${}_{52}C_5$ possible ways of being dealt five cards. A successful outcome consists of getting three kings and two deuces. There are four of each kind of card in a deck and so the number of ways of getting three kings is ${}_4C_3$. The number of ways of getting two deuces is ${}_4C_2$. Therefore the total number of ways of obtaining three kings and two deuces is ${}_4C_3 \cdot {}_4C_2$. So we have

$$_4C_2 = \frac{4!}{(4-2)!2!} = \frac{4!}{2!2!} = \frac{4 \cdot 3 \cdot \cancel{2} \cdot 1}{2 \cdot 1 \cdot \cancel{2} \cdot 1} = 6$$

$$_4C_3 = \frac{4!}{(4-3)!3!} = \frac{4!}{1!3!} = \frac{4 \cdot \cancel{3} \cdot \cancel{2} \cdot 1}{1 \cdot \cancel{3} \cdot \cancel{2} \cdot 1} = 4$$

$$_{52}C_5 = 2{,}598{,}960 \qquad \text{(see Example 2)}$$

and the probability of obtaining a full house consisting of three kings and two deuces is

$$P(3 \text{ kings and 2 twos}) = \frac{_4C_3 \cdot {}_4C_2}{_{52}C_5} = \frac{24}{2{,}598{,}960} = \frac{1}{108{,}290}$$

Table 1 provides some interesting information about five-card poker. It shows the number of various kinds of hands possible when a 52-card deck is used and nothing is wild. Note that the total number of different hands is 2,598,960, which is ${}_{52}C_5$.

NOTE OF INTEREST

The odds against dealing a bridge hand consisting of 13 cards of one suit are 158,753,389,899 to 1. The odds against each of the four players receiving a complete suit (a perfect deal) are 2,235,197,406,895,366,368,301,559,999 to 1.

TABLE 1

Straight flush	40
Four of a kind	624
Full house	3,744
Flush	5,108
Straight	10,200
Three of a kind	54,912
Two pairs	123,552
One pair	1,098,240
No pair	1,302,540
Total	2,598,960

EXERCISES FOR SECTION 3.9

1. Find the probability of being dealt two queens when you are dealt two cards from a shuffled deck of 52 cards.
2. Find the probability of being dealt three kings when you are dealt three cards from a shuffled deck of 52 cards.
3. You are dealt three cards from a shuffled deck of 52 cards. Find the probability that all three cards are hearts.
4. You are dealt five cards from a shuffled deck of 52 cards. Find the probability that all five cards are picture cards (king, queen, or jack). What is the probability that none of the cards are picture cards? (Indicate your answers; do not evaluate.)
5. On the track team of the York Athletic Club, there are 8 sprinters and 10 distance runners. A relay team consisting of four people must be chosen at random (without regard to who runs first, second, etc.). Find the probability that the relay team will
   a. consist of sprinters only
   b. consist of distance runners only
   c. consist of two sprinters and two distance runners
   d. consist of three sprinters and one distance runner
   e. contain at least one sprinter
6. You are dealt five cards from a shuffled deck of 52 cards. Find the probability that
   a. all five cards are aces
   b. Three cards are aces and two cards are picture cards
   c. four cards are aces and one card is a picture card
7. A five-man committee is to be formed at random from seven Democrats and nine Republicans. Find the probability that the committee will consist of
   a. all Democrats
   b. all Republicans
   c. two Democrats and three Republicans
   d. two Republicans and three Democrats
   e. four Democrats and one Republican
   f. four Republicans and one Democrat
8. A football coach has five guards and seven tackles trying out for the squad. As a final cut, five of these players will be cut at random. Find the probability that the group of players cut will
   a. consist of guards only
   b. consist of tackles only
   c. consist of three guards and two tackles
   d. consist of four tackles and one guard
   e. consist of three tackles and two guards
   f. consist of four guards and one tackle
9. In how many ways can 13 cards (a bridge hand) be selected from a deck of 52 cards? Find the probability of being dealt a bridge hand that consists of

all spades. (Indicate your answer, but do not evaluate.)

10. A case of soda pop contains six bottles of root beer, six bottles of orange soda, seven bottles of cola, and five bottles of ginger ale. If you select three bottles at random, find the probability that
    a. all three bottles are cola
    b. all three bottles are root beer
    c. two bottles are ginger ale and one bottle is cola
    d. one bottle is orange soda and two bottles are root beer
    e. one bottle is cola, one bottle is orange soda, and one bottle is ginger ale
    f. one bottle is cola, one bottle is orange soda, and one bottle is root beer.

11. A committee consisting of four people is to be selected at random from a group of seven people that includes Bob and Carol and Ted and Alice. Find the probability that the committee will consist of Bob, Carol, Ted, and Alice.

12. A jury (12 people) is to be randomly selected from a group consisting of six men and eight women. Find the probability that the jury will
    a. consist of women only
    b. consist of men only
    c. consist of six women and six men
    d. consist of seven women and five men
    e. consist of seven men and five women
    f. contain at least one man

13. An urn contains six orange balls, four blue balls, and three red balls. If three balls are drawn at random, indicate (but do not evaluate) the probability of selecting
    a. three orange balls
    b. three blue balls
    c. three red balls
    d. two orange balls and one blue ball
    e. two blue balls and one red ball
    f. two red balls and one orange ball

14. Pocket billiards (pool) is played with 15 balls numbered 1 through 15. The balls with numbers greater than eight are striped, whereas the rest are solid colors. Fast Eddie, a pool hustler, sinks two balls with a single shot. Find the probability that the balls made by Fast Eddie are
    a. both striped
    b. both solids
    c. 1 striped and 1 solid
    d. the number 1 ball and the number 15 ball

JUST FOR FUN

Take a standard deck of 52 cards and shuffle them as much as you want. Start with the top card and turn each card over one at a time. Is it a good bet that you will get two cards in a row that are of the same denomination and the same color (for example, 3 of clubs and 3 of spades, 7 of hearts and 7 of diamonds)?

15. Find the probability of being dealt a flush (five cards of the same suit) in hearts in five-card poker.

16. What is the probability of being dealt a flush in hearts or spades in five-card poker?

17. Find the probability of being dealt two queens and three jacks in five-card poker.

18. Find the probability of being dealt two queens, two kings, and one ace in five-card poker.

19. Find the probability of being dealt a full house (three of a kind, together with two of a kind) in a five-card poker.

Summary

A *sample space* is the set of all possible outcomes to a given experiment. An *event* is any subset of a sample space. If an experiment has a total of T *equally likely* outcomes, and if exactly S of these outcomes are considered successful—that is, they are the members of the event A—then the probability that event A will occur is

$$P(A) = \frac{\text{number of successful outcomes}}{\text{total number of all possible outcomes}} = \frac{S}{T}$$

When two or more experiments are performed together, the *counting principle* can sometimes be used to determine the total number of possible outcomes in the sample space. If one experiment has m different outcomes, and a second experiment has n different outcomes, then the first and second experiments performed together have $m \times n$ different outcomes. A *tree diagram* may be used to determine a sample space for a particular problem, because it illustrates the possible outcomes for an experiment.

The *odds in favor of an event A* are found by taking the probability that the event A will occur and dividing it by the probability that the event A will not occur:

$$\text{Odds in favor of } A = \frac{\text{probability that } A \text{ will occur}}{\text{probability that } A \text{ will not occur}}$$

Odds are usually expressed as a ratio: if the odds in favor of A are $\frac{3}{2}$, then we may express the odds as 3 to 2, or 3:2. If the odds in favor of A are 3:2, then the *odds against A* are 2:3.

Mathematical expectation is found by multiplying P, the probability that an event will occur, times M, the amount that will be won if the event occurs.

$$E = P \times M$$

Mathematical expectation tells us what is a fair price to pay to play a game.

Independent events are events where the occurrence of one event does not affect the occurrence of a second event. *Mutually exclusive events* are events that cannot happen at the same time. To compute the probability of two or more events occurring together, we multiply the various probabilities together. That is,

$$P(A \text{ and } B) = P(A) \times P(B)$$

When we have mutually exclusive events and we want to compute the probability of A or B, we have

$$P(A \text{ or } B) = P(A) + P(B) \quad \text{(where } A \text{ and } B \text{ are mutually exclusive)}$$

If A and B are not mutually exclusive events and we want to compute the probability of A or B, we have

$$P(A \text{ or } B) = P(A) + P(B) - P(A \text{ and } B)$$
(where A and B are *not* mutually exclusive)

Permutations are ordered arrangements of things. The number of permutations of n things taken r at a time is denoted by ${}_nP_r$ and

$${}_nP_r = \frac{n!}{(n-r)!}$$

Combinations are distinct groups of things without regard to their arrangement. The number of combinations of n things taken r at a time is denoted by ${}_nC_r$ and

$$_nC_r = \frac{n!}{(n-r)!r!}$$

Permutations and combinations are useful in solving many probability problems, particularly those that involve a great many possible outcomes, such as the various probabilities for different poker hands. (For more examples, see Sec. 3.9.)

Vocabulary Check

probability	expectation	dependent events	★combinations
sample space	★permutations	equally likely events	★factorial
tree diagram	independent events	fundamental counting principle	mutually exclusive

Review Exercises for Chapter 3

1. On a single draw from a bag containing four red, six blue, and three green balls, find the probability of obtaining
   a. a red ball
   b. a blue ball
   c. a red or a green ball
   d. a ball that is not red
   e. a ball that is not green
   f. a ball that is neither red nor blue
2. A pair of dice is tossed. Find the probability that
   a. the sum of the numbers is 7
   b. the sum of the numbers is not 7
   c. the same number appears on both dice
   d. the sum of the numbers is 7 or 11
   e. the sum of the numbers is greater than 7
   f. the sum of the numbers is not even
3. Mary is choosing a pair of running shoes. In her size, the running shoes come in four different styles and six different colors. How many different pairs of running shoes can Mary choose from?
4. By means of a tree diagram, list the sample space showing the possible arrangements when three coins are tossed.
   a. Find the probability that all three coins are heads.
   b. Find the probability that at least one coin is heads.
   c. Find the probability that no coins are heads.
5. On a single draw from a shuffled deck of 52 cards, find the odds
   a. in favor of drawing a king
   b. against drawing a king
   c. in favor of drawing a picture card (jack, queen, king)
   d. in favor of drawing a club or jack
   e. in favor of drawing a club and a jack
   f. against drawing the ace of spades
6. If the odds are 8 to 1 against the Giants winning the Super Bowl, what is the probability that the Giants will win the Super Bowl?
7. The probability that the Bruins will win the Stanley Cup is $\frac{3}{11}$. Find the odds in favor of the Bruins winning the Stanley Cup.
8. Five thousands tickets are sold for a drawing on a yacht valued at $10,000. If a woman buys one ticket, what is her expectation?
9. A fraternity sold 500 raffle tickets at $2 each on a color television set valued at $400. If Joe Kool buys five tickets, what is his mathematical expectation?
10. A bag contains five balls numbered 1 through 5. Two balls are chosen in succession. The first ball is replaced before the second is drawn. Are these events *dependent* or *independent?* Why?
11. The object of a game is to obtain a 7 or 11 with a single toss of a pair of dice. Are these events *mutually exclusive?* Why?

12. The New York State Lottery Commission runs a daily lottery where a player chooses a three-digit number (any three digits from 000 through 999). A lottery ticket costs \$1 and if a player chooses the winning number, the payoff (prize) is \$500.

   a. How many outcomes are possible?
   b. What is a fair price to pay for a ticket?
   c. What can you conclude about the cost of a lottery ticket?

13. Two cards are randomly selected in succession from a shuffled deck of 52 cards, without replacement. Find the probability that
   a. both cards are red
   b. both cards are three's
   c. the first card is red and the second card is black
   d. the first card is a heart and the second card is a club
   e. the first card is a picture card and the second card is not a picture card
   f. the first card is an ace and the second card is a king

14. One card is randomly selected from a shuffled deck of 52 cards and then a die is rolled. Find the probability of obtaining
   a. a black card and a 1
   b. a black card or a 1
   c. a queen and a 1
   d. a queen or a 1
   e. a red ace and a 1
   f. a red ace or a 1

15. There is a game called *poker dice,* which is based on rolling five dice. The resulting outcome is then treated as a poker hand.
   a. How many different outcomes are possible?
   b. What is the probability of getting five of a kind?

(*Note:* Exercises 16–26 are based on the optional topics in this chapter, as indicated by ⋆.)

⋆16. How many license plates can be made if each plate must consist of two letters followed by three digits? (Assume no repetition.) How many are possible if the first letter cannot be *O* and the first digit cannot be zero? How many are possible if the letter *Q* cannot be used and zero cannot be used?

⋆17. How many different four-letter "words" (that is, arrangements of four letters) can be formed from the letters of the alphabet if each letter can only be used once and none of the vowels *a, e, i, o, u* may be used?

⋆18. In a certain collegiate basketball conference, there are 10 teams and each team plays every other team in the conference twice. How many league games are played in a season?

⋆19. How many distinct arrangements are possible in using all the letters of the word *ECOLOGY?*

⋆20. Evaluate the following.

a. $4!$	b. $0!$	c. $\frac{6!}{3!2!}$
d. ${}_5P_2$	e. ${}_6P_2$	f. ${}_nP_0$

⋆21. Evaluate the following.

a. ${}_5C_2$	b. ${}_6C_2$	c. ${}_7C_4$
d. ${}_7C_3$	e. ${}_nC_0$	f. ${}_nC_n$

⋆22. From five teachers and 50 students, how many committees can be selected if each committee is to have two teachers and three students?

⋆23. From a group of five freshmen, six sophomores, four juniors, and three seniors, a staff of three freshmen, three sophomores, two juniors, and two seniors is to be chosen for the school's radio station. In how many ways can this be done?

⋆24. Three balls are drawn simultaneously at random

from a bag containing four red, four blue, and two yellow balls. Find the probability that
a. all three balls are blue
b. all three balls are red
c. two balls are yellow and one is red
d. two balls are red and one is blue
e. two balls are red and one is yellow

★25. You are dealt five cards from a shuffled deck of 52 cards. Indicate (but do not evaluate) the probability that
a. all five cards are red
b. none of the five cards is red
c. all five cards are picture cards
d. none of the five cards is a picture card
e. three of the cards are picture cards and two are not

★26. In the game of five-card poker, find the probability of being dealt a hand that is
a. four of a kind
b. a full house
c. a flush

JUST FOR FUN

Do you think that two people in your class have the same birthday (month and day)? Try it; you might be surprised. If there are 25 people in your class, the probability that two people have the same birthday is greater than 0.5. If there are 50 people in your class, the probability that two people have the same birthday is very close to 1.0 (certainty).

Chapter Quiz

Indicate whether each statement is true or false.

1. If an event is certain to occur, then its probability is 1.
2. If an event can never occur, then its probability is 0.
3. If an event may occur or may not occur, then its probability is $\frac{1}{2}$.
4. The probability of drawing an ace or a heart from a shuffled standard deck of 52 cards is $\frac{4}{13}$.
5. The probability of obtaining 7 or 11 when a pair of dice is tossed is $\frac{2}{9}$.
6. The set of all possible outcomes of an experiment is called an *event.*
7. The odds in favor of obtaining a 7 when a pair of dice is tossed are 1:6.
8. The odds against obtaining an 11 when a pair of dice is tossed are 17:1.
9. The more tickets you purchase for a raffle, the more chances you have of winning.
10. The more tickets you purchase for a raffle increases the mathematical expectation of a ticket.
11. Two events, A and B, are *independent* events if the occurrence of one event does not affect the probability of the other.
12. Drawing an ace from a regular deck of playing cards and then drawing a king from the same deck without replacing the first card are considered independent events.
13. If two or more events can happen at the same time, they are *mutually exclusive.*
14. If a quarter is flipped and a die is tossed, then the probability of obtaining a heads or a 6 is $\frac{8}{12}$.
15. If two coins are tossed, then the probability that at least one heads appears is 0.75.
16. A die is rolled, and then one card is randomly

HISTORICAL NOTE

Blaise Pascal (1623–1662) was born in France. He displayed signs of mathematical genius at an early age. By the time he was 12, he had discovered many of the theorems of elementary geometry. At 14, he participated in weekly meetings of French mathematicians from which the French Academy ultimately arose in 1666. Pascal provided the first proofs of theorems in projective geometry.

Courtesy International Business Machines Corporation

By the time he was 20, Pascal had built the first mechanical adding machine, and furthermore sold over 50 of these machines to businessmen in the community. Some of these machines are still preserved in a museum in Paris. See page 649 for a photograph of this device.

Pascal, together with Pierre de Fermat, another French mathematician, developed the basic mathematical theory of probability by solving a problem given to him by a gambler friend, the Chevalier de Méré.

In approximately 1650 Pascal decided to abandon his work in mathematics and science and devote himself to religious contemplations. Also, most of his life was spent with physical pain as he suffered from acute dyspepsia and at times was paralyzed. It is interesting to note that Pascal has also been credited with the invention of the one-wheeled wheelbarrow as we know it today.

selected from a shuffled deck of 52 cards. The probability of obtaining a 1 and an ace is $\frac{1}{78}$.

17. A special lottery offers a prize of $1000. The winning ticket can be any four-digit number from 0000 to 9999. If a person buys a ticket, a fair price to pay for the ticket is $1.

18. The odds in favor of obtaining two heads in a single toss of two coins are 1:4.

19. The odds against Neal winning a certain raffle are 999:1, the probability that Neal will win is 1:1000.

20. A coin is flipped and a die is tossed. The probability of obtaining a tails or an even number is 8:12.

(*Note:* Questions 21–25 are based on the optional topics in this chapter, as indicated by ⋆).

⋆21. Given the set of digits $\{1, 3, 4, 5, 6\}$, 50 three-digit numbers can be formed that are even.

⋆22. $3! = 2! + 1!$

⋆23. The letters in *TOMATO* can be arranged in 90 distinct ways.

⋆24. Two cochairpersons are to be selected from a group of five eligible people. This can be done in 20 ways.

⋆25. ${}_7C_3 = {}_7C_4$

CHAPTER 4

STATISTICS

AFTER STUDYING THIS CHAPTER YOU WILL BE ABLE TO DO THE FOLLOWING:

1. State four measures of **central tendency** and distinguish among them.
2. Compute the **mean, median, mode,** and **midrange** for a given set of data.
3. State two measures of **dispersion** and distinguish between them.
4. Compute the **range** and **standard deviation** for a given set of data.
5. Find the **percentile** or **quartile** of a single datum in a given set of data.
6. Construct **circle graphs**.
7. Construct a **frequency distribution table** and **histogram** or **frequency polygon** from a given set of data.
8. Determine what percentage of normally distributed data is within a given number of **standard deviations** from the mean.

SYMBOLS FREQUENTLY USED IN THIS CHAPTER

$\bar{x}$ mean (read "x bar")
a^2 a squared (for example, $3^2 = 3 \times 3 = 9$)
$\sqrt{a}$ the positive square root of a (for example, $\sqrt{9} = 3$)
σ standard deviation (lowercase Greek letter sigma)

4.1 INTRODUCTION

Whenever we watch television, listen to the radio, or read newspapers, magazines, or books, we encounter statistics. We can find statistics in articles on business, the state of the economy, politics, science, education, sports, and many other subjects. In order to understand the information that is presented, we must possess some understanding of statistics.

Most people first encounter statistics in elementary school when they take standardized achievement tests: schools want to know what the "average" student has learned or how a student compares to other students who have taken the standardized tests.

A **statistic** is a number derived from a set of data that (in some way) is used to represent or describe the data. Unfortunately, statistics are often misused or abused. Some people maintain that a statistician can prove whatever he or she wants to. A more familiar quote is "Figures don't lie, but liars figure." Adults and children alike are inundated by statistics. All too often they have no real understanding of the facts presented.

Advertising is one area where statistics are often abused. Statistics may be distorted by the manner in which the facts are gathered or by the manner in which the facts are presented. Statistics can be distorted in diagrams by emphasizing the wrong facts, exaggerating comparisons, or simply not showing all of the data.

In this chapter, we shall discuss the basic concepts of statistics in order to prepare you better to understand and interpret statistics when you encounter them. It has often been said that if a person is going to be an intelligent member of today's society, he or she must possess some understanding of statistics. As stated by H. G. Wells, "Statistical thinking will one day be as necessary for efficient citizenship as the ability to read and write."

4.2 AVERAGE—THE MEASURE OF CENTRAL TENDENCY

One of the first concepts of statistics that people encounter, and one that is familiar to everyone, is the concept of "average." Whether a person can calculate an average or not, he or she has an intuitive idea of what it is. We are familiar with phrases such as "average miles per gallon," "average precipitation for the month," the "batting average" of a baseball player, and a "student's average" for a set of test scores.

The term *average* is used in different ways by different people. An average is a *measure of central tendency*. A **measure of central** tendency describes a set of data by locating the middle region of the set. The common measures of central tendency are the *arithmetic mean,* the *median,* the *mode,* and the *midrange*. The arithmetic mean is usually referred to simply as the *mean*. Each of the four measures of central tendency—mean, median, mode, or midrange—is an average because each describes the middle region of the data. Each one has its advantages and disadvantages; in a given situation, it may be more desirable to use one as opposed to the other three.

The mean (arithmetic mean) is the most familiar type of average that we encounter. You have probably found the mean for a set of data before, but instead of calling it a mean, you probably called it an average.

The *mean* for a set of data, scores, or facts is found by determining the *sum* of the data and dividing this sum by the total number of elements in the set.

The scores for five students on a quiz are 40, 20, 30, 25, and 15. To find the mean score for this group of students, we first find the sum of the scores.

$$40 + 20 + 30 + 25 + 15 = 130$$

We then divide the sum by 5, the number of scores.

$$\frac{130}{5} = 26$$

The mean for a set of data is often represented by the notation $\bar{x}$ (read "*x* bar"). Therefore, we can write $\bar{x} = 26$ for our given set of data. Even though the mean score is 26, note that none of the scores is 26. There are two scores, 40 and 30, above the mean and three scores, 25, 20, and 15, below the mean. This single measure of central tendency is used to represent the whole group of five students. It gives us information as to how the group of students performed, but it does not tell us anything about a particular student in the group.

NOTE OF INTEREST

A *sample* or a *census* is often used to provide data for a statistical project. A *sample* is a study of only a portion of the items or people to be used in the analysis. To find the average height of the students in your college, you would probably use a sample of the students. The sample should be representative of all the students attending your college. The basketball team would probably not be a representative sample for finding the heights of all students.

A *census* is a study or count of every item or person that is to be used in the analysis. For example, if you wanted to know the brand of calculator that is most popular with the students in your mathematics class, you would ask each student.

EXAMPLE 1 Find the mean for the set of test scores

71, 75, 60, 84, 71, 63, 66

SOLUTION The sum of the scores is $71 + 75 + 60 + 84 + 71 + 63 + 66 = 490$, and we have seven scores; hence we divide 490 by 7.

$$\bar{x} = \frac{490}{7} = 70$$

EXAMPLE 2 At the Surf and Sand Restaurant, a waitress earns \$10 a night in tips for five nights during the week. But on weekends (Saturday and Sunday) she earns \$20 a night in tips. What is her average daily earnings in tips?

SOLUTION We must find the total income and then divide this sum by the number of elements in our set of data. We have $5 \times \$10 = \$50$, $2 \times \$20 = \$40$, and total tips are $\$40 + \$50 = \$90$. Now we divide \$90 by 7.

$$\bar{x} = \frac{\$90}{7} \approx \$12.86 \qquad \text{(daily average earnings in tips)}$$

In Example 1, the mean score of 70 tells us that the average score for the set was 70, but it does not necessarily represent any particular score. Note that the mean of 70 is somewhat centrally located in the set of data. However, a mean can be misleading in describing data because it can be affected by extreme values in the data. Example 3 illustrates this.

EXAMPLE 3 Find the mean for the set of scores

82, 81, 80, 87, 20

SOLUTION The sum of the scores is $82 + 81 + 80 + 87 + 20 = 350$; hence the mean is

$$\bar{x} = \frac{350}{5} = 70$$

The mean is only 70, although most of the scores are in the 80's.

Although the average, or mean, of the scores in Example 3 was 70, it is obvious that this number does not give a very accurate

NOTE OF INTEREST

A *random sample* is one that you pick by chance. That is, an item is selected at *random* from a group of items if the selection process is such that each item in the group has an equal chance to be selected. For example, if you put the name of each person in your class on the same size card, put all the cards in a box, shake, shuffle, or toss them until they are all mixed up, and then select three cards without looking, you would have a random sample of three names from your class. If you wanted to know the opinions of students in your college, you would draw a random sample from the names of all the students registered at your college. You would not pick a sample of the students in your mathematics class, in the bookstore, or at a sporting event. Each of these places would give you a sample of a certain type of student, not students in general, or at random.

The method of selecting the sample might introduce a *bias*. This in turn can lead to the misuse of statistics found in the samples that are used as the basis for statistical research.

idea of the typical score. Suppose, however, that we arrange the scores in Example 3 in order, from lowest to highest.

20, 80, 81, 82, 87

The middle number in this arrangement is 81. This number called the *median* of the data, gives a more accurate idea of the typical score in cases where the data include extreme values. The **median** can be described as the middle value of a set of data when the data are listed in order. The median for a set of data is found by arranging the data (numbers) in sequential order and finding the middle number.

The scores for five students on a quiz are 40, 20, 30, 24, and 15. To find the median, we must first arrange the scores in sequential order, that is, 15, 20, 24, 30, 40. The median (middle number) for these five scores is 24.

When we have an even number of scores, it is customary to use the number halfway between the two middle scores (i.e., the mean of the two middle numbers) as the median.

For example, consider the following set of scores.

{40, 20, 30, 24, 28, 15}

What is the median? Arranging the scores in order gives us the following: 15, 20, 24, 28, 30, 40. Because there are six pieces of data (an even number), we must find the mean of the two middle numbers. The two middle numbers are 24 and 28. Therefore, the median is

$$\frac{24 + 28}{2} = \frac{52}{2} = 26$$

The median is the measure of central tendency that determines the middle of a given set of data, that is, the number, value, or score such that the number of scores below the median is the same as the number of scores above the median.

EXAMPLE 4 Find the median for the given set of data.

13, 16, 14, 12, 20, 19, 10, 18

SOLUTION Arranging the numbers in sequential order, we have

10, 12, 13, 14, 16, 18, 19, 20

Note that there are eight data (an even number), so we must find the mean of the two middle numbers, 14 and 16.

$$\frac{14 + 16}{2} = \frac{30}{2} = 15$$

Therefore the median is 15. Four of the pieces of data are less than 15, and four of the pieces of data are greater than 15.

EXAMPLE 5 O. J. Simpson played for the Buffalo Bills from 1969 to 1977. Following is a list of the number of touchdowns he scored each of these years.

2, 5, 5, 6, 12, 3, 16, 8, 0

Find the median number of touchdowns that O. J. scored during this period.

SOLUTION When we arrange the data in sequential order, we have

0, 2, 3, 5, 5, 6, 8, 12, 16

There are nine pieces of data, so the median is the fifth one, 5. There are four pieces of data before it, and four pieces of data after it.

EXAMPLE 6 A group of students, Tom, Carlos, Janie, Irene, and Frank, reported that they had earned the following amounts during summer vacation.

Tom	$800
Carlos	$900
Janie	$300
Irene	$900
Frank	$600

Find the mean and median summer income for this group of students.

SOLUTION a. To find the mean income, we must find the total income and then divide this sum by 5 (the number of students).

$$\$800 + \$900 + \$300 + \$900 + \$600 = \$3{,}500$$

$$\frac{\$3{,}500}{5} = \$700 \qquad \text{(mean income)}$$

b. To find the median income, we must arrange the amounts in sequential order and find the middle amount. (The number of students is odd.)

$300, $600, $800, $900, $900

The middle amount, $800, is the median income.

In Example 6 we calculated both the mean and the median for a given set of data. Notice, however, that although the mean was $700 and the median was $800, more people in the group earned $900 than earned any other amount. This figure, $900, is called the *mode* of the data in Example 6.

The *mode* for a given set of data is that number, item, or value that occurs *most frequently*.

The scores for five students on a quiz are 40, 24, 30, 24, and 15. The mode for this set of scores is 24 because it occurs twice, whereas each of the other scores occurs only once.

There are times when the mode is the most meaningful of the three measures of central tendency. Consider the case of Harry, who operates Harry's Hamburger Haven. During a typical 8-hour shift, Harry sells approximately 2,400 hamburgers. This is a mean of 300 hamburgers per hour. But Harry's customers do not come into his Hamburger Haven at the same rate each hour. He is much busier some hours than others. In planning how many hamburgers to have ready each hour, Harry is interested in what hour(s) he sells the most hamburgers.

You should be aware that a set of data always has a mean and a median, but does not necessarily have a mode. Consider the set of data 1, 2, 3, 4, 5. The mean is 3 and the median is 3, but there is no mode, because there is no number that occurs most frequently.

It is also the case that a set of data can have more than one mode. Consider the set of data 1, 2, 3, 3, 4, 5, 5, 6. We see that the number 3 occurs twice and the number 5 occurs twice. Hence, there are two modes, 3 and 5. Since this given set of data has two modes, we refer to it as **bimodal.**

EXAMPLE 7 Find the mode for the following data.

12, 10, 11, 13, 11, 14, 13, 11, 17

SOLUTION The mode is 11. It occurs three times in the set of data.

EXAMPLE 8 Find the mode for the following data.

18, 16, 13, 14, 12, 10, 11, 15, 17, 19

SOLUTION There is no mode for this set of data because each value occurs only once.

EXAMPLE 9 Following is a list of scores that Terrie received on a series of agility tests.

52, 77, 74, 82, 74, 104, 83

Find (a) the mean, (b) the median, and (c) the mode for this set of data.

SOLUTION a. The sum of the scores is 546, the mean is

$$\frac{546}{7} = 78$$

b. Arranging the scores in order, we have

52, 74, 74, 77, 82, 83, 104

The median is 77.

c. The mode is 74.

In Example 9, we calculated the mean, median, and mode for a given set of data. Another measure that may be used to give an approximation for a measure of central tendency is the *midrange*. We can think of the **midrange** as the value midway between the end points of the set of data. Note, this is not the same as the median, although it may sometimes have the same value.

The midrange is found by adding the least value in the given set of data to the greatest value and dividing the sum by 2.

$$\text{Midrange} = \frac{L + G}{2}$$

In Example 9, the least value that occurs is 52 and the greatest value is 104. Therefore, the midrange for Example 9 is

$$\frac{52 + 104}{2} = \frac{156}{2} = 78$$

This is also the value of the mean in Example 9, but this is a coincidence. The midrange is used to estimate the central tendency of the data. Weather forecasters use the midrange when computing the average temperature for a given day. That is, they add the lowest temperature reading to the highest temperature reading, divide the sum by two, and report the result as the average temperature.

EXAMPLE 10 Listed below are the maximum speeds achieved by various animals.

Animal	Mph	Animal	Mph
Cheetah	70	Quarter horse	47
Antelope	61	Elk	45
Wildebeest	50	Cape dog	45
Lion	50	Coyote	43
Gazelle	50	Gray fox	42

Find the mean, median, mode, and midrange for this set of animals.

SOLUTION

$$\text{Mean} = \frac{70 + 61 + 50 + 50 + 50 + 47 + 45 + 45 + 43 + 42}{10}$$

$$= \frac{503}{10} = 50.3$$

There are 10 pieces of data (an even number), so we must find the mean of the two middle numbers, 50 and 47.

$$\frac{50 + 47}{2} = \frac{97}{2} = 48.5 \qquad \text{(median)}$$

Hence, the median is 48.5. Five pieces of data are less than 48.5 and five pieces of data are greater than 48.5. The *mode* is 50. It occurs three times in the set of data.

$$\text{Midrange} = \frac{42 + 70}{2} = \frac{112}{2} = 56$$

You should be aware that each of these measures is a useful measuring device. There are instances when one of them (mean, median, mode, midrange) is more representative than the others. The mean is not a true indication of the "average" if there are values in the given set of data that are extreme values at one end or the other. A typical example that illustrates this phenomenon is one that involves wages. Consider the annual salaries of the employees of the Custom Moving Company. The manager earns \$28,000 per year, one driver earns \$11,000 per year, and another driver earns \$10,000 per year. One helper earns \$6,000 per year, while three other helpers each earn \$5,000 per year. The mean salary for these seven people is computed as follows.

$$\begin{aligned}\$28{,}000 + \$11{,}000 + \$10{,}000 + \$6{,}000 \\ + \$5{,}000 + \$5{,}000 + \$5{,}000 = \$70{,}000\end{aligned}$$

$$\frac{\$70{,}000}{7} = \$10{,}000$$

Carrying the discussion a little further, suppose the owner earns \$50,000 a year from the Custom Moving Company. The mean salary for all eight people would be

$$\begin{aligned}\$50{,}000 + \$28{,}000 + \$11{,}000 + \$10{,}000 + \$6{,}000 \\ + \$5{,}000 + \$5{,}000 + \$5{,}000 = \$120{,}000\end{aligned}$$

$$\frac{\$120{,}000}{8} = \$15{,}000$$

The mean salary is \$15,000. This certainly is not a sensible representation of the average salary. This is a case where the mean should not be used to describe the situation for the given set of data. In contract bargaining, the management would probably like to use the mean (\$15,000) as a basis for negotiations, whereas the union would probably like to use the mode (\$5,000). Remember that the mean is not a true indication of the average if the given set of data contains extreme values at one end.

The median is the middle value (number). The number of data below the median is the same as the number of data above the median. The median is not affected by extreme values in the set of data. But it may not be a true representation of the average if the data occur in distinct, separate groups.

The mode is that item that occurs most frequently in a set of data. There can be times when a set of data has no mode, that is, when each value occurs an equal number of times. Other times a set of data can have more than one mode, that is, when two or more different values occur the same number of times. The mode can be misleading at times: it does not take into account the other numbers in the set of data, as do the mean and median.

The midrange is the value midway between the end points of the set of data. It too can be misleading at times; it does not take into account the values or how many pieces of data are contained in a set of data.

Many times information is presented in some form of a table. Consider the following table, which depicts the salary distribution for a certain company, Ketes Enterprises.

Annual Income	Number Receiving This Income
$100,000	1
$ 50,000	2
$ 20,000	8
$ 15,000	10
$ 10,000	4

The mean, median, and mode income may be determined from the given information. But we must be careful how we compute them. To find the mean income, we divide the total salary by the total number of employees. What is the total salary? To find the total salary, we multiply the amount of income by the number of people receiving that income and then sum these totals. That is,

$$
\begin{aligned}
\$100{,}000 \times 1 &= \$100{,}000 \\
\$50{,}000 \times 2 &= \$100{,}000 \\
\$20{,}000 \times 8 &= \$160{,}000 \\
\$15{,}000 \times 10 &= \$150{,}000 \\
\$10{,}000 \times 4 &= \$\ 40{,}000 \\
\hline
&\ \$550{,}000 \quad \text{Total annual income}
\end{aligned}
$$

Note: Total number of employees = 25

Therefore, the mean income is

$$\frac{\$550{,}000}{25} = \$22{,}000$$

The median income is determined by the number of pieces of data: 25. Since we have an odd number, the middle piece of data, the 13th, is the median. Note that the salaries are already ranked in the table and the 13th employee's salary is the median income.

We may begin from the top or bottom. Regardless of how we begin, we see that the 13th ranking income is \$15,000. Hence, the median income is \$15,000.

The mode is the item that occurs most frequently in a set of data. The table indicates that 10 people earned \$15,000, which is a greater number than any of the other salaries. Therefore, the mode income is \$15,000.

$$\text{The midrange} = \frac{\$10{,}000 + \$100{,}000}{2} = \frac{\$110{,}000}{2} = \$55{,}000$$

Care must be exercised when deriving statistical data from tables. At first glance, it would appear that the median income is \$20,000 and the mean income is (\$100,000 + \$50,000 + \$20,000 + \$15,000 + \$10,000) = \$195,000 ÷ 5 = \$39,000. This is not correct!

EXERCISES FOR SECTION 4.2

In Exercises 1–10, find the mean, median, mode, and midrange for each set of data. (Round off any decimal answer to the nearest tenth.)

1. 1, 2, 3, 4, 4, 5, 9, 12
2. 10, 16, 14, 10, 17
3. 1, 2, 3, 4, 5, 6, 7, 8, 9, 10
4. 1, 1, 5, 5, 7, 10, 12, 11, 10
5. 2, 4, 8, 10, 6, 12
6. 5, 7, 13, 1, 3, 9, 11, 15
7. 11, 99, 77, 88, 66, 44, 55, 22, 33
8. 99, 19, 89, 29, 79, 39, 69, 49, 59
9. 1492, 1776, 1941, 1812
10. 1984, 2004, 1974, 1954, 2004

11. Listed below are the home-run leaders of the American League for the years 1977–1987

Year	Player	Number
1977	Jim Rice	39
1978	Jim Rice	46
1979	Gorman Thomas	45
1980	Reggie Jackson	41
1981	Eddie Murray	22
1982	Reggie Jackson	39
1983	Jim Rice	39
1984	Tony Armas	43
1985	Darrell Evans	40
1986	Jesse Barfield	40
1987	Mark McGwire	49

Find the (a) mean (b) median, (c) mode, and (d) midrange for the number of home runs.

12. Here are the numbers of tornadoes that occurred in the United States for the years 1970–1979.

Year	Number
1970	653
1971	888
1972	741
1973	1,102
1974	947
1975	920
1976	835
1977	852
1978	788
1979	852

Find the (a) mean, (b) median, (c) mode, and (d) midrange for the number of tornadoes.

13. Following is a table of the 10 tallest dams in the United States.

Name	Location	Height (feet)
Oroville	California	770
Hoover	Nevada	726
Sallisaw Creek	Oklahoma	721
Dworshak	Idaho	717
Glen Canyon	Arizona	710
New Bullards	California	635
New Melons	California	625
Swift	Washington	610
Mossyrock	Washington	606
Shasta	California	602

Find the (a) mean, (b) median, (c) mode, and (d) midrange for the various heights.

14. An art teacher grades all student projects 1, 2, or 3, as follows.

1—excellent
2—acceptable
3—unacceptable (must be done over)

Given a set of data consisting of all the grades assigned to student projects during the term, which measure of central tendency would you use to determine the grade received by the greatest number of projects?

© *Robert Kalman/The Image Works*

Based on this photograph, do you think the median of the heights of the members of this class would be equal to the mean of their heights?

15. Women employees of a certain firm have complained of sex discrimination in the company's pay scale. If the women win their case, the firm will have to raise the salaries of all underpaid employees. The employees and their salaries are listed below.

Ms. O'Brien	$25,000	Mr. Ponti	$15,000
Ms. Jones	$10,000	Mr. Hansen	$15,000
Ms. Chung	$10,000	Mr. Steinberg	$15,000

a. What is the mean salary for all employees?
b. What is the mean salary for women? For men?
c. What is the median salary for women? For men?
d. What is the midrange salary for women? For men?
e. If you were a lawyer acting for the firm, which measure of central tendency would you use to

describe the average salaries of male and female employees? Which measure of central tendency would you use if you were a lawyer arguing on behalf of Ms. Jones and Ms. Chung?

16. In Sam's physics course, a mean score of 80 on 10 tests is necessary for a grade of B. Sam's mean score for the 10 tests was 79 and his instructor, Ms. Molecule, gave him a grade of C. Sam protested that because he was so close, his instructor should give him the "one lousy point" and hence a B. Did Sam need only one point for a B? Explain your answer.

17. In a certain week, the New York Mets played eight baseball games. The number of runs they scored in the respective games were 3, 2, 0, 6, 9, 4, 5, and 3. Find the mean, median, and mode for this set of data. Did the Mets win all eight games?

18. The mean score on a set of 10 scores is 71. What is the sum of the 10 test scores?

19. The mean score on a set of 13 scores is 77. What is the sum of the 13 test scores?

20. The mean score on four of a set of five scores is 75. The fifth score is 90. What is the sum of the five scores? What is the mean of the five scores?

21. Two sets of data are given: the first set of data has 10 scores with a mean of 70, and the second set of data has 20 scores with a mean of 80. What is the mean for both sets of data combined?

22. Janet and Larry took the same courses last semester: calculus (4 credits), geology (4 credits), English (3 credits), history (3 credits), and physical education (1 credit). Janet received A, A, B, C, C, respectively, and Larry received C, C, B, A, A, respectively. Janet and Larry bet a dinner as to who would have the higher average. Janet maintains she won, whereas Larry maintains that they are even because they both got two A's, one B, and two C's. Who is right? (*Hint:* A grade-point average is found as follows: Allow 4 points for an A, 3 points for a B, and 2 points for a C. Multiply the number of points equivalent to the letter grade received in each course by the number of credits for the course to arrive at the total points earned in each course. Divide the sum of the points by the total number of credit hours. The answer is the grade-point average.)

23. The table below gives the annual salary distribution for the Ronolog Corporation. Using the information provided in the table, find the following.
a. the mean annual income
b. the median annual income
c. the mode annual income
d. the midrange annual income

Annual Income	Number Receiving This Income
$110,000	1
60,000	1
25,000	7
22,000	10
19,000	8
16,000	13

24. The table below indicates the ages of a group of 30 people that entered a record store on a Saturday morning. Using the information provided in the table, find the following.
a. the mean age
b. the median age
c. the mode age
d. the midrange age

Age	Number of People
17	4
18	5
19	7
20	5
21	4
22	2
23	3

25. On the first day of class last semester, 30 students were asked for the one-way distance from home to

college (to the nearest mile). The table below provides the resulting data. Using the information provided in the table, find the following.

a. the mean distance
b. the median distance
c. the mode distance
d. the midrange distance

Distance (miles)	Number of People
30	2
26	1
22	2
18	5
14	6
10	7
6	5
2	2

26. The fastest scheduled train runs for different passenger trains in the world are listed at the right. Using the information provided in the table, find the following.

a. the mean speed
b. the median speed
c. the mode speed
d. the midrange speed

Country	From	To	Speed (mph)
France	Paris	Macon	134.0
Japan	Nagoya	Yokohama	112.3
Great Britain	Peterborough	Stevenage	101.0
USA	Wilmington	Baltimore	100.1
West Germany	Hamm	Bielefeld	96.2
USA	Rensselaer	Hudson	88.4
USA	Newark	Philadelphia	87.8
Italy	Rome	Chiusi	83.5

JUST FOR FUN If you are given 10 pennies and three coffee cups, can you place all of the pennies in the three coffee cups so that there is an odd number of coins in each coffee cup? Each cup has to contain at least one penny.

4.3 MEASURES OF DISPERSION

In the previous section, we discussed measures of central tendency: mean, median, mode, and midrange. They are called measures of central tendency because each of them tells us something about the average of the data, that is, where the data tend to center or cluster. Measures of dispersion tell us how much the data tend to disperse or scatter, that is, the spread of the data. One measure of dispersion is the *range*.

The *range* for a set of data is found by subtracting the smallest value from the largest value in the given set of data.

Consider the respective quiz scores in a history class for Cathy and Juanita.

Juanita: 72, 74, 74, 77, 80, 83, 86
Cathy: 58, 74, 74, 77, 81, 82, 100

Computing the mean score for Juanita, we have

$$\frac{72 + 74 + 74 + 77 + 80 + 83 + 86}{7} = \frac{546}{7} = 78$$

The mean score for Cathy is

$$\frac{58 + 74 + 74 + 77 + 81 + 82 + 100}{7} = \frac{546}{7} = 78$$

The mean score for each is 78. Comparing further, we see that the median for each is 77, the mode for each is 74, and the midrange for each is 79. Cathy and Juanita have the same measures of central tendency for their quiz scores. The one significantly different thing about their scores is the range: the range for Juanita's scores is $86 - 72 = 14$, whereas the range for Cathy's scores is $100 - 58 = 42$.

Cathy's scores had the greater range. Because Cathy and Juanita cannot be compared by means of the measures of central tendency, we might use the range. But we must be careful. We might argue that Juanita is the more consistent of the two because her score range is only 14, but we might also argue that Cathy showed the most improvement because her score range is 42.

EXAMPLE 1 Find the mean and the range for the two sets of test scores.

Test A: 75, 75, 70, 70, 70, 65, 65, 65, 55, 40
Test B: 70, 70, 67, 66, 65, 65, 65, 62, 60, 60

SOLUTION Test A: Mean $= \dfrac{75+75+70+70+70+65+65+65+55+40}{10}$

$$= \frac{650}{10} = 65$$

Range $= 75 - 40 = 35$

$$\text{Test B: Mean} = \frac{70+70+67+66+65+65+65+62+60+60}{10}$$

$$= \frac{650}{10} = 65$$

$$\text{Range} = 70 - 60 = 10$$

Note that both test A and test B had a mean score of 65, but the range of test A was 35, whereas the range for test B was only 10. There were both higher and lower grades on test A. There is less dispersion for the scores on test B.

The range, though easily computed, is not considered to be the best measure of dispersion because it only involves the use of two extreme values. It does not tell us anything about the remaining values in our set of data.

One measure of dispersion (variation) that considers all scores in a given set of data is the **standard deviation.**

In order to find the standard deviation for a given set of data, we must perform a number of tasks. First we must find the **deviation from the mean** for each value in the given set of data. As an example of this process, consider the set of scores {60, 65, 70, 70, 70}. The mean for this set of data is 67. The deviation from the mean is found by subtracting the mean from each value in the set of data.

The deviation for 70 is $70 - 67 = 3$.
The deviation for 65 is $65 - 67 = -2$.
The deviation for 60 is $60 - 67 = -7$.

This is summarized in Table 1.

After we have found the deviation from the mean for each score, we square each of the deviations (see Table 2). We need to do this in order to find a value that is useful in measuring dispersion or deviation about the mean. Note that the sum of the

TABLE 1

	Value	Deviation from the Mean
	70	70 − 67 = 3
	70	70 − 67 = 3
	70	70 − 67 = 3
	65	65 − 67 = −2
	60	60 − 67 = −7
Sum	335	0

$$\text{Mean} = \frac{335}{5} = 67$$

TABLE 2

	Value	Deviation	$(\text{Deviation})^2$
	70	3	9
	70	3	9
	70	3	9
	65	−2	4
	60	−7	49
Sum	335	0	80

deviations is zero. This is true for any set of data. Therefore you can use this information to check the accuracy of your work, particularly the accuracy of your mean.

Now that we have squared each deviation, we have a set of positive values. The sum of these values in Table 2 is 80. We next find the mean of these squared deviations, called the **variance.** The mean of the squared values is $80 \div 5 = 16$. The variance also indicates dispersion, but we only mention it in passing; we need the variance in order to find the standard deviation. We find the standard deviation by finding the principal square root of the variance. Hence for our example,

$$\text{Standard deviation} = \sqrt{16} = 4$$

The standard deviation for a set of data is usually represented by the lower case Greek letter σ, called *sigma.* Therefore we can write $\sigma = 4$ for our given set of data.

Admittedly, there is a lot of work involved in finding the standard deviation, but, if it is done in an orderly and precise manner, it is not difficult.

To find the *standard deviation* for a given set of data, we must follow these steps.

1. **Find the *mean* for the given set of data.**
2. **Find the *deviation from the mean* for each value in the set of data.**
3. **Square each deviation.**
4. **Find the mean of the squared deviations.**
5. **Find the principal square root of this number.**

EXAMPLE 2 Find the standard deviation for the following data.

$$\{5, 7, 9, 13, 16\}$$

SOLUTION We follow the steps outlined previously.

1. The mean is

$$\frac{5 + 7 + 9 + 13 + 16}{5} = \frac{50}{5} = 10$$

2. The deviation from the mean for each value is

$$5 - 10 = -5, \quad 7 - 10 = -3, \quad 9 - 10 = -1,$$

$$13 - 10 = 3, \quad 16 - 10 = 6$$

3. Now we square each deviation.

$$(-5)^2 = 25, (-3)^2 = 9, (-1)^2 = 1, (3)^2 = 9, (6)^2 = 36$$

4. To find the mean of the squared deviations, we sum the squares and divide by 5 (the number of values).

$$\frac{25 + 9 + 1 + 9 + 36}{5} = \frac{80}{5} = 16$$

5. We now find the principal square root of this number: $\sqrt{16} = 4$
6. Hence

$$\text{Standard deviation} = \sigma = 4$$

The fact that we obtained a standard deviation of 4, as we did in the discussion prior to this example, is pure coincidence.

EXAMPLE 3 Find the standard deviation for the given set of data.

$$\{20, 22, 26, 26, 28, 34\}$$

SOLUTION We combine the necessary operations in the table below.

	Value	Deviation from the Mean	$(\text{Deviation})^2$
	34	$34 - 26 = 8$	64
	28	$28 - 26 = 2$	4
	26	$26 - 26 = 0$	0
	26	$26 - 26 = 0$	0
	22	$22 - 26 = -4$	16
	20	$20 - 26 = -6$	36
Sum	156	0	120

$$\text{Mean} = \frac{156}{6} = 26$$

$$\text{Mean of squared deviations} = \frac{120}{6} = 20 \qquad \text{(variance)}$$

$$\text{Standard deviation} = \sigma = \sqrt{20} \approx 4.5$$

The standard deviation for the set of data given in Example 3 is approximately 4.5. We approximated $\sqrt{20}$ by means of Table 2 in the Appendix.

EXAMPLE 4 Nancy took seven tests in her math class. Following is her set of test scores.

$$\{72, 74, 74, 77, 82, 83, 84\}$$

Find the standard deviation for her set of test scores.

SOLUTION We combine the necessary operations in the table below.

	Score	Deviation	$(\text{Deviation})^2$
	84	6	36
	83	5	25
	82	4	16
	77	−1	1
	74	−4	16
	74	−4	16
	72	−6	36
Sum	546	0	146

$$\text{Mean} = \frac{546}{7} = 78$$

$$\text{Mean of squared deviations} = \frac{146}{7} = 20.86 \qquad \text{(variance)}$$

$$\text{Standard deviation} = \sigma = \sqrt{20.86} \approx 4.6$$

Note that in Example 4 the mean of the squared deviations did not come out evenly; that is, $\frac{146}{7}$ was rounded off to 20.86 and then we approximated the square root of 20.86 by 4.6, using Table 2 in the Appendix.

A large standard deviation indicates that the data are scattered widely about the mean, whereas a small standard deviation indicates that the data are closely grouped about the mean. As an example, consider the following two sets of data.

$A = \{4, 5, 6, 7, 8\}$	$B = \{2, 4, 6, 8, 10\}$
The mean for set A is 6 and the standard deviation is $\sqrt{2} \approx 1.41$	The mean for set B is 6 and the standard deviation is $\sqrt{8} \approx 2.82$

Now that we are able to calculate the standard deviation, what can we do with it? Suppose that Tony scored 78 on a mathematics exam. The mean score for this exam was 73 and the standard deviation was 5. What does Tony know? First, he scored above the mean, that is, he obtained a better than average score on his exam. Secondly, he scored one standard deviation above the mean. On the next exam in his mathematics class, Tony scored 47. Did he do better or worse on the second test? In order to answer this, we must know the mean and the standard deviation. Suppose the second exam had a mean of 41 and a standard deviation of 3. Therefore, Tony again scored above the mean and, in fact, he scored 2 standard deviations above the mean $[41 + (2 \times 3) = 47]$, which indicates that he performed better on the second exam than he did on the first.

EXAMPLE 5 Sue and Jim took their midterm exams in their respective statistics courses. Sue is in Mr. Data's class, where she scored 79. In Mr. Data's class the mean was 75 and the standard deviation was 4. Jim is in Ms. Mode's class, where he also scored 79. But in Ms. Mode's class the mean was 73 and the standard deviation was 3. Who performed better in their respective class?

SOLUTION Sue's score was one standard deviation above the mean $(75 + 4 = 79)$, while Jim's score was two standard deviations above the mean for his class $[73 + (2 \times 3) = 79]$. Jim performed better compared to his classmates in Ms. Mode's class than did Sue compared to her classmates in Mr. Data's class.

EXAMPLE 6 Frank, Louie, Janie, and Irene are all in the same history class. Their scores for the first two exams in their history class are listed below.

	Exam 1	**Exam 2**
Frank	78	66
Louie	66	66
Janie	90	70
Irene	72	70

The first exam had a mean of 78 and a standard deviation of 6, whereas the second exam had a mean of 66 and a standard deviation of 4. Which of the four people improved on the second exam, which did worse, and which performed the same, in relation to the rest of the class?

SOLUTION On each of the exams, Frank scored the mean score (78 on exam 1 and 66 on exam 2), so he performed the same on both exams. Louie scored 66 on the first exam, two standard deviations below the mean [78 − (2 × 6) = 66]. On the second exam, he also scored 66, but that was the mean for exam 2; hence Louie improved. Janie scored two standard deviations above the mean on the first exam [78 + (2 × 6) = 90], whereas on the second exam she only scored one standard deviation above the mean [66 + (1 × 4) = 70]. Hence Janie did worse on the second exam compared to the first. On the first exam, Irene scored one standard deviation below the mean [78 − (1 × 6) = 72], whereas on the second exam, she scored one standard deviation above the mean [66 + (1 × 4) = 70]. Therefore Irene's performance improved on the second exam.

EXERCISES FOR SECTION 4.3

In Exercises 1–10, find the standard deviation for each set of data. (Round off any decimal answer to the nearest tenth.)

1. {8, 10, 10 16}
2. {2, 4, 6, 8}
3. {16, 13, 13, 12, 9, 9}
4. {13, 11, 11, 9, 9, 7}
5. {38, 44, 46, 48, 48, 48, 50}
6. {20, 22, 23, 23, 24, 24, 25}
7. {1, 2, 7, 10, 15}
8. {11, 13, 17, 19}
9. {68, 68, 70, 72, 66, 73, 78, 65, 80, 70}
10. {82, 80, 81, 87, 86, 86, 88, 84, 82, 84}
11. A sample of 10 bowlers was taken in a tournament. The number of strikes recorded for each bowler in his or her first game was as follows: {2, 3, 4, 5, 5, 6, 3, 2, 7, 3}. For this set of data, find
    a. mean
    b. median
    c. mode
    d. range
    e. midrange
    f. standard deviation
12. On a physics lab quiz, the following scores were made in a class of 10 students: 80, 75, 88, 95, 90, 72, 67, 73, 78, 82. For this set of scores, find
    a. mean
    b. median
    c. mode
    d. range
    e. midrange
    f. standard deviation
13. The maximum recorded life spans (to the nearest year) of 10 different animals are listed below.

Baboon	36	Beaver	21
Polar bear	35	Dog (domestic)	20
Cat (domestic)	28	Horse	46
Elephant (African)	60	Lion	25
Gorilla	39	Cow	30

For this set of life spans, find the following.
    a. mean
    b. median

c. mode d. range
e. midrange f. standard deviation

□★14. The table below lists the estimated annual earnings of selected athletes for the year 1987.

Athlete	Sport	Earnings
Moses Malone	Basketball	$2,145,000
Jim Rice	Baseball	$2,412,500
Wayne Gretzky	Hockey	$ 717,250
Jim Kelly	Football	$1,400,000
Martina Navratilova	Tennis	$1,905,841
Don Mattingly	Baseball	$1,975,000
Michael Spinks	Boxing	$4,000,000
Patrick Ewing	Basketball	$1,500,000
Mike Tyson	Boxing	$2,333,332

For this set of estimated earnings, find the following. (Round off any decimal answer to the nearest dollar.)
a. mean b. median
c. mode d. range
e. midrange f. standard deviation

□★15. The back nine holes of the Southampton Golf Club have the following lengths in yards: 150, 370, 310, 340, 200, 450, 490, 420, 375. For this set of measurements, find the following.
a. mean b. median
c. mode d. range
e. midrange f. standard deviation

16. Hugh, Norma, Ruth, Joe, and Doris are all in the same statistics class. Their scores for the first two exams in their statistics class are listed. The first exam had a mean of 84 and a standard deviation of 6, whereas the second exam had a mean of 78 and a standard deviation of 4.

	Exam 1	Exam 2
Hugh	84	78
Norma	90	74
Ruth	66	78
Joe	78	70
Doris	84	78

a. Who improved on the second exam?
b. Who did the poorest on the second exam?
c. Who performed the same on both exams?

17. Two history exams were given. On the first exam, the mean was 74 and the standard deviation was 8. On the second exam, the mean was 48 and the standard deviation was 10. The scores of six students who took the exams are listed below.

	Exam 1	Exam 2
Eric	82	48
Maria	78	53
Rudy	70	58
Maureen	58	58
Jeff	74	53
Mark	90	78

a. Who improved on the second exam?
b. Who improved the most on the second exam?
c. Who did not improve on the second exam?
d. Considering both exams, which student did the poorest?
e. Who performed the same on both exams?
f. Considering both exams, which student has the best grade so far?

JUST FOR FUN How many triangles are there in this figure?

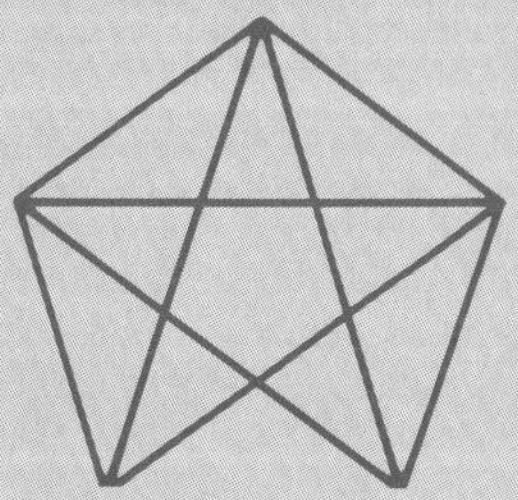

4.4 MEASURES OF POSITION (PERCENTILES)

HISTORICAL NOTE

Carl Friedrich Gauss (1777–1855) was one of the greatest mathematicians of the nineteenth century. Along with Archimedes and Isaac Newton, he is considered one of the three greatest mathematicians of all time.

Gauss had many interests, including astronomy, physics, and mathematics. He discovered a technique for computing the orbits of the asteroids and helped develop electromagnetic theory. In mathematics, his major contributions were in the areas of number theory, theory of functions, probability, and statistics.

The "prince of mathematicians," as he was called by his contemporaries, loved to work long, difficult problems. As a result of this penchant, he was able to predict, after weeks of calculations, the correct orbit of the asteroid known as Ceres.

Consider the case of Sue and Dale, who are applying for a job: both people have to fill out an application form. On the form, there is a question which asks for the applicant's high school class rank. Sue's rank in her class was 30th, whereas Dale's class rank was 10th. That is, Sue was 30th in her class, whereas Dale was 10th in his class. At first glance it seems that Dale was the better student and probably should get the job. But this is not necessarily the case, because Sue and Dale went to different high schools. We do not have enough information for a fair comparison.

Suppose Sue's rank was 30th in a class of 300, whereas Dale's rank was 10th in a class of 50. This points out that rank is useless as a measure of position unless we know how many are in the total group. It would appear that Sue's rank is similar to Dale's. In order to describe their relative positions better, we can find the **percentile rank,** or **percentile,** for Sue and Dale.

In Sue's class, there were 300 − 30 = 270 people ranked below her. Therefore

$$\frac{270}{300} = \frac{27}{30} = \frac{9}{10} = 0.90 = 90\%$$

of Sue's classmates were ranked lower than she was. We can say that Sue was in the 90th percentile. Again, this means that 90% of her classmates were ranked below her.

Computing Dale's percentile rank, we find that 50 − 10 = 40 people rank below him. Hence

$$\frac{40}{50} = \frac{4}{5} = 0.80 = 80\%$$

of Dale's classmates were ranked lower than he was. Dale was in the 80th percentile in his class.

Percentiles divide sets of data into 100 equal parts; hence 100% is the basis of measure. Practically every student has encountered a percentile rank at one time or another. Every graduating senior has a percentile rank in his or her graduating class. Standardized achievement tests, intelligence tests, perception tests, etc., give results in terms of percentiles. When a student is told that he scored at the 77th percentile on an exam, he then knows that he scored better than 77% of those that took the exam or that 77% of those taking the exam scored lower than he did.

Note: In each of these examples, it is assumed that each measure is distinct. That is, no two of the measures are the same.

EXAMPLE 1 In a class of 100 students, Bob has the rank of 12th. What is Bob's percentile rank in the class?

SOLUTION There are $100 - 12 = 88$ students ranked below Bob. Hence, we compare 88 to 100.

$$\frac{88}{100} = 0.88 = 88\%$$

Bob's percentile rank is 88; he is in the 88th percentile.

EXAMPLE 2 In a class of 120 students, Sam has the rank of 32nd. What is Sam's percentile rank?

SOLUTION $120 - 32 = 88$

$$\frac{88}{120} = \frac{22}{30} = \frac{11}{15} = 0.73 = 73\%$$

Sam's percentile rank is 73.

EXAMPLE 3 In a statistics class of 40 students, an exam was given and Pam scored at the 75th percentile. How many students scored lower than Pam?

SOLUTION Since Pam scored at the 75th percentile, 75% of the students scored less than she did.

$$75\% = 0.75, \qquad 0.75 \times 40 = 30$$

Therefore 30 students scored lower than Pam on the exam.

EXAMPLE 4 In a class of 30 students, Larry has a percentile rank of 70. What is Larry's rank in the class?

SOLUTION Larry's percentile rank indicates that 70% of the students have scores lower than Larry's on the tests given to the class.

$$70\% \text{ of } 30 = 0.70 \times 30 = 21$$

Twenty-one students have scored lower than Larry. Hence, Larry's rank in the class is ninth, because $30 - 21 = 9$.

To determine at what percentile a student scored on an exam, we need to find the number of students that scored below the individual and divide that number by the total number of students in the class.

EXAMPLE 5 The scores of 10 students on a math quiz were

4, 10, 12, 14, 16, 18, 20, 20, 22, 22

Tracy's score on the quiz was 16. What is her percentile rank?

SOLUTION Since Tracy's score was 16, there were four scores lower than hers. We divide 4 by 10, the total number of students that took the quiz: $4 \div 10 = 0.40 = 40\%$. Therefore, Tracy's percentile rank is 40; she is in the 40th percentile.

EXAMPLE 6 The scores of 10 students on a history quiz were

8, 11, 13, 15, 17, 19, 20, 21, 24, 25

Mike's score on the quiz was 21.

a. What is the rank (from the top) for Mike's score?
b. What is Mike's percentile rank?

SOLUTION
a. There were two scores higher than Mike's, 24 and 25. His score of 21 was third.
b. Since Mike's score was third, there were seven scores lower than his score of 21. We divide 7 by 10: $7 \div 10 = 0.70 = 70\%$. Hence Mike is in the 70th percentile.

The 25th percentile, 50th percentile, and 75th percentile are probably the most commonly used percentiles in educational testing. They are unique in that they divide sets of data into fourths, or quarters; hence they are referred to as **quartiles.**

Each set of data has three quartiles. The procedure for determining the value of the quartiles is the same as for percentiles. The **first quartile** for a set of data is that value which has 25%, or one-fourth, of the data (scores) below it. The **second quartile** is that value which has 50%, or one-half, of the data (scores) below it. The **third quartile** is that value which has 75%, or three-fourths, of the data (scores) below it.

EXAMPLE 7 Given the following set of scores: {60, 70, 72, 73, 73, 80, 82, 84, 84, 85, 87, 88}. Find

a. the first quartile
b. the median
c. the third quartile

SOLUTION a. The first quartile is that value which has 25% of the scores below it. There are 12 scores; 25% (one-fourth) of 12 is 3. So the value for the first quartile is the fourth score (three scores will lie below it), which is 73.

b. The median is that measure that determines the middle of a given set of data. There are 12 scores (an even number), so we must find the mean of the sixth and seventh scores.

$$\frac{80 + 82}{2} = 81$$

The median is 81.

c. The third quartile is that value which has 75% of the scores below it; 75% (three-fourths) of 12 is 9. The value for the third quartile is the tenth score (nine scores lie below it), which is 85.

EXAMPLE 8 Given the following set of data: {30, 35, 36, 37, 38, 39, 40, 41}. Find

a. the first quartile
b. the third quartile

SOLUTION a. There are eight data; $\frac{1}{4}$ of 8 = 2. Hence, the third piece of data from the bottom, 36, is the first quartile, or 25th percentile.

b. Three-fourths of 8 is 6. Therefore, the seventh piece of data from the bottom, 40, is the third quartile, or 75th percentile.

B.C. by permission of Johnny Hart and Creators Syndicate, Inc.

You should be aware that a datum or score has no significance or meaning by itself. If Helen reported that she scored 92 on her last exam in history, many people would jump to the conclusion that she did well on the exam. But that might not be the case. If Helen got 92 points out of a possible 200, then the score has more meaning, and we would conclude that Helen did not do well on the exam. On the other hand, if Helen got 92 points out of a possible 100, then we could conclude that she did do well on the exam. We could also better judge Helen's performance on the exam if we knew her score's location relative to the other students' scores on the exam. If, for instance, Helen reported that her score was at the 90th percentile, then we know that 90% of the scores are below Helen's.

EXERCISES FOR SECTION 4.4

1. In a class of 200 students, Julia has a rank of 12th. What is Julia's percentile rank in the class?
2. In a senior class of 300 students, Erin has a rank of 30th. What is Erin's percentile rank in the class?
3. In a ten-kilometer "fun-run" Dave finished fourteenth. If 200 contestants entered the race, what was Dave's percentile rank?
4. Sharon scored at the 90th percentile on a certain skills test in her gymnastics class of 20 students. What is her rank in the class?
5. In a statistics class of 30 students, an exam was given and Eddie scored at the 80th percentile. How many students scored lower than Eddie?
6. In Hugh's mathematics class, there are 32 students including Hugh. On the last exam, Hugh's score was at the third quartile. How many students scored lower than Hugh?
7. In a class of 40 students, Don has a percentile rank of 80. What is Don's rank in the class?
8. Doris is ranked at the first quartile in her senior class of 300 students. What is her rank in the class?
9. Jessie is ranked at the third quartile in her economics class of 60 students. What is her rank in the class?
10. The following data represent the heights in inches

of the starting five for the Dunkem basketball team: 73, 78, 80, 82, 85.

a. What is the rank (from the top) of the height of 82 inches?

b. What is the percentile rank of the height of 82 inches?

11. Given the 12 scores 62, 72, 74, 75, 75, 82, 85, 86, 86, 87, 89, and 90.

a. What is the rank (from the top) of a score of 87?

b. What is the percentile rank of a score of 87?

c. What is the percentile rank of a score of 89?

d. What score is at the first quartile?

12. Given the 20 scores 62, 64, 66, 69, 70, 72, 74, 75, 80, 82, 83, 86, 87, 88, 90, 91, 92, 94, 97, and 98.

a. What is the rank (from the top) of 87?

b. What is the percentile rank of a score of 87?

c. What score is at the first quartile?

d. What score is at the third quartile?

e. What score is at the 90th percentile?

f. What score is at the 80th percentile?

13. Roberto is ranked 75th in his senior class of 200 students. Larry is in the same senior class and Larry has a percentile rank of 75. Of the two seniors, who has the higher standing in the class?

14. Sarah is ranked eighth in her mathematics class of 35 students. Helen is in the same mathematics class, and Helen has a percentile rank of 85. Who ranks higher, Sarah or Helen?

15. When asked how he did on a mid-term exam, Ricci replied that he scored at the "100th percentile." What is your reaction to this statement?

16. Daniel is ranked at the third quartile in his English class of 60 students. Peter is in the same English class, and he is ranked 13th. Who ranks higher, Daniel or Peter?

17. In a class of 50 students, Jay has a rank of 50th. What is his percentile rank?

JUST FOR FUN Almost every word that begins with a *q* has a *u* as the second letter. Can you name two *q* words that do not have *u* as the second letter? Can you name one?

4.5 PICTURES OF DATA

It has been said that a picture is worth a thousand words. It is true that most of us can gain information more easily from pictures than from written material. Architects and contractors use drawings (blueprints) to exchange information. Large corporations use various types of graphs in their reports to stockholders to show how the company has performed. Newspapers and magazines use graphs to show the reader some types of information, usually the presentation of data.

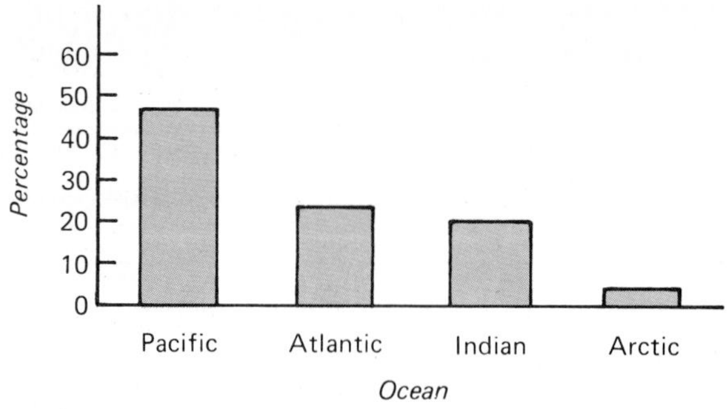

FIGURE 1
Percentage of world's water

Graphs are used in mathematics to show relationships between sets of numbers. Graphs are useful in the field of statistics because they can show the relationships in a set of data. In this section we shall examine the most common types of graphs used in statistics: the *bar graph,* the *circle* or *pie graph,* the *histogram,* and the *frequency polygon* or *line graph.*

A **bar graph** consists of a series of bars of uniform width, with some form of measure on a vertical or horizontal axis. Figure 1 is an example of a typical bar graph.

EXAMPLE 1 Using Figure 1, answer the following.

a. What percentage of the world's water does the Indian Ocean contain?
b. Approximately what percentage of the world's water does the Pacific Ocean contain?

SOLUTION a. We locate the top of the bar representing the Indian Ocean and then read across to the corresponding percentage. The Indian Ocean contains 20% of the world's water.
b. Approximately 47%.

The **circle** or **pie graph** is another type of graph that is used quite often. It is particularly useful in illustrating how a whole quantity is divided into parts. Budgets are often illustrated in this manner. Circle graphs used by government agencies often show how the tax dollar is spent or how the tax dollar is collected. Figure 2 is an example of a typical circle graph.

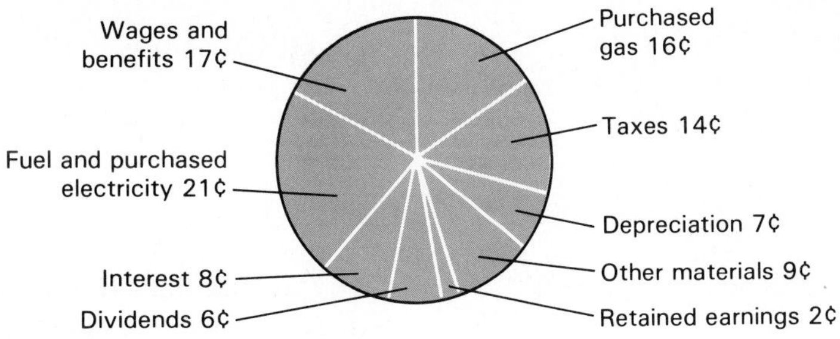

FIGURE 2
Where your utility dollar goes

EXAMPLE 2 Using Figure 2, answer the following.

a. How much of every utility dollar is used for wages and benefits?
b. How much of every utility dollar is used for taxes?

SOLUTION a. 17¢ b. 14¢

When we want to show how a whole quantity is divided into parts, we can use a circle graph. Consider the case of Sam Student. On any given weekday, Sam spends 8 hours sleeping, 6 hours in school, 4 hours doing homework, and 3 hours eating and doing odd jobs. He uses the remaining 3 hours for recreation, leisure, and miscellaneous. We can illustrate how Sam spends his time for a 24-hour period by means of a circle graph.

First recall that a circle contains 360 degrees. In order to illustrate the given data, we must find the percentage of a 24-hour day spent in each activity and then multiply the percentages by 360 degrees, so that we can divide the circle into proportional parts.

Sleeping: 8 hours $\frac{8}{24} = \frac{1}{3}$ $\frac{1}{3}$ of $360° = 120°$

School: 6 hours $\frac{6}{24} = \frac{1}{4}$ $\frac{1}{4}$ of $360° = 90°$

Homework: 4 hours $\frac{4}{24} = \frac{1}{6}$ $\frac{1}{6}$ of $360° = 60°$

Meals, chores:	3 hours	$\frac{3}{24} = \frac{1}{8}$	$\frac{1}{8}$ of $360° = 45°$
Miscellaneous:	3 hours	$\frac{3}{24} = \frac{1}{8}$	$\frac{1}{8}$ of $360° = 45°$
	24 hours		360°

We then use a protractor to construct the corresponding angles in a circle. Figure 3 shows the completed circle graph. It does not matter how large or how small the circle is, since every circle represents 100% of the data and can be divided into proportional parts.

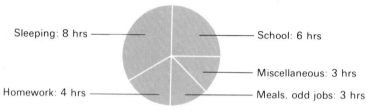

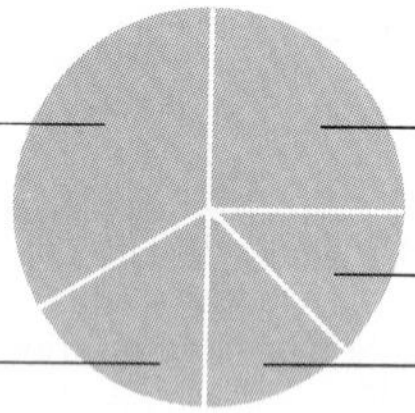

FIGURE 3

Circle graphs are used to show how a whole quantity is divided into parts; however, they are not as useful when we wish to picture other types of data. Consider the following set of 35 scores for an exam in a statistics class.

42, 48, 43, 47, 46, 45, 42
49, 50, 42, 44, 48, 47, 42
42, 48, 50, 45, 46, 42, 49
45, 44, 48, 49, 45, 47, 46
46, 46, 44, 46, 49, 50, 45

Examining these scores does not give us much information because the scores have been presented in the order of their occurrence. We probably could find the highest and lowest score without too much trouble, but we can get a better understanding of the scores if we do the following: first, list the scores in order from highest to lowest, and then count how many of each score we have.

The number of times a score occurs is the **frequency** of the score, and Table 3 is called a **frequency distribution.** In this particular case, we have ungrouped frequency distribution because each score value in the distribution is listed separately.

Note that Table 3 contains nine categories or classes. We can reduce this number by creating classes that contain more than one

TABLE 3

Score	Tally	Frequency
50	///	3
49	////	4
48	////	4
47	///	3
46	~~////~~ /	6
45	~~////~~	5
44	///	3
43	/	1
42	~~////~~ /	6

TABLE 4

Score	Tally	Frequency
50–51	///	3
48–49	~~////~~ ///	8
46–47	~~////~~ ////	9
44–45	~~////~~ ///	8
42–43	~~////~~ //	7

score. For example, Table 4 has five classes for the same data we used in Table 3. Note that each class interval is the same width, and they do not overlap. Also, each piece of data belongs in exactly one class. Table 4 is called a **grouped frequency distribution.** Grouped frequency distributions representing the same set of data may be different. That is, the classes or intervals constructed by one person may be different from those done by someone else. But regardless of the preference of the individual, three guidelines should be followed.

1. A range of 5–12 classes is desirable to avoid difficulty in interpretation of the data.
2. Classes should not overlap; each piece of data belongs to only one class.
3. Each class should be the same width.

The frequency distribution or frequency table in Table 3 lists all the different scores and the frequency with which each score occurs. This frequency distribution can be illustrated by means of a **histogram.** A histogram consists of series of bars that are drawn all with the same width on the horizontal axis, and uniform units on the vertical axis. The frequencies are shown on the vertical axis (see Fig. 4).

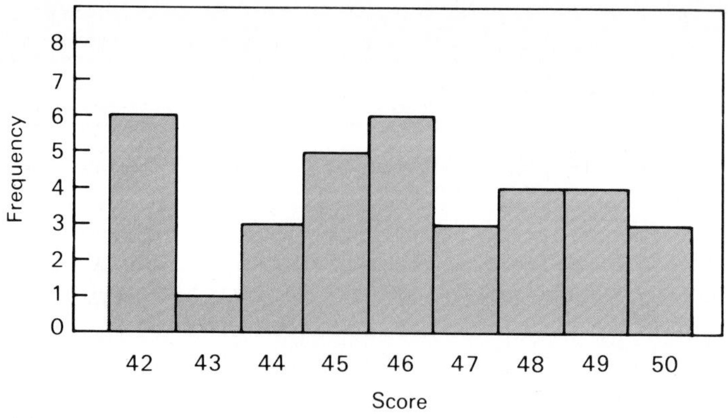

FIGURE 4

It is important that each bar in a histogram should be drawn in proportion to the frequency of values that occur. Note that both the horizontal and vertical scales have been labeled completely.

Figure 5 is a histogram for the grouped frequency distribution in Table 4. Note that the bars in a histogram meet one another. The reason for this is that these bars represent classes that are consecutive. That is, one class is 42–43, whereas the next is 44–45, and so on. The frequency is shown by the height of the vertical bar—the higher the bar, the greater the frequency.

A **frequency polygon,** sometimes called a **line graph,** can also be used to graph a frequency distribution. It is constructed in much the same manner as a histogram, using the same kind of vertical and horizontal scales. If we connect the midpoints of the

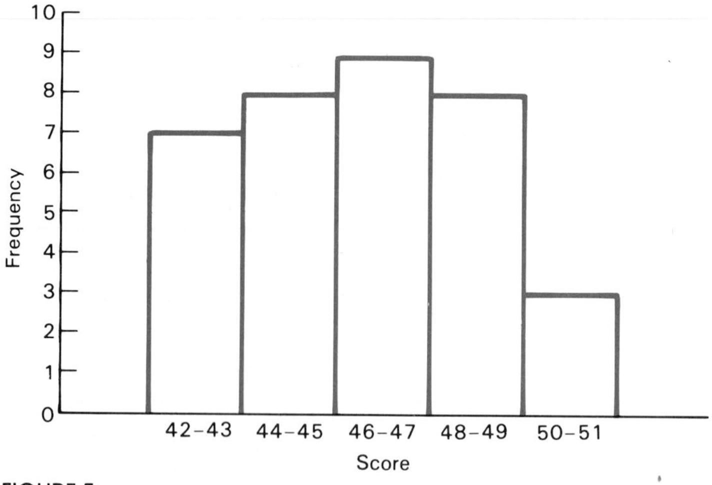

FIGURE 5

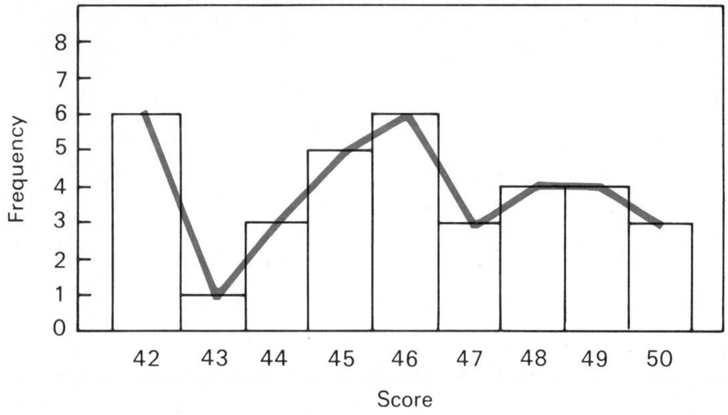

FIGURE 6

top of each bar in a histogram, the resulting line graph, shown in Figure 6, is a frequency polygon. Figure 6 uses the same data as Figure 4.

EXAMPLE 3 A survey of 30 customers was made at Dan's Donut Shop. Following is the set of 30 breakfast checks, in cents, for each customer. Construct a histogram for these data.

92, 93, 90, 91, 84, 85, 86, 87, 89, 88
93, 93, 84, 85, 85, 92, 92, 91, 88, 87
86, 92, 92, 86, 90, 86, 88, 88, 87, 86

SOLUTION First we construct a frequency distribution for the given data.

Amount of Check	Tally	Frequency
93	///	3
92	~~////~~	5
91	//	2
90	//	2
89	/	1
88	////	4
87	///	3
86	~~////~~	5
85	///	3
84	//	2

Next, we construct a histogram, making sure that the frequencies are shown on the vertical axis in uniform intervals, and that all of the bars have the same width. The result is shown in Figure 7.

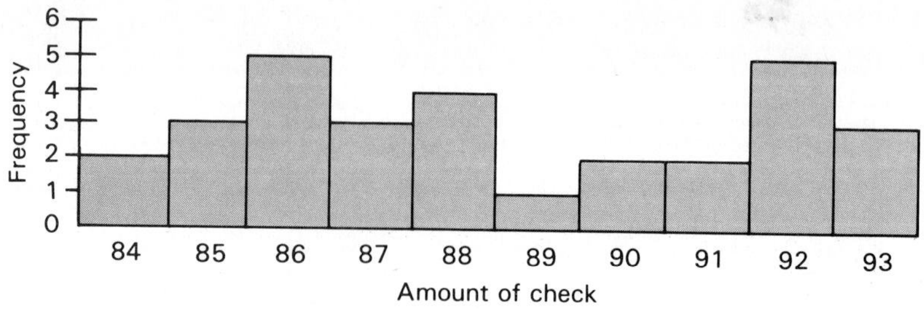

FIGURE 7

EXAMPLE 4 A student rolled a die 30 times and obtained the results shown below. Construct a frequency polygon for these data.

6, 5, 4, 4, 5, 6, 1, 2, 1, 6, 4, 3, 3, 3, 4,
2, 2, 5, 6, 4, 1, 2, 4, 3, 5, 5, 3, 3, 4, 2

SOLUTION First we construct a frequency distribution for the given data.

Number	Tally	Frequency
6	////	4
5	~~////~~	5
4	~~////~~ //	7
3	~~////~~ /	6
2	~~////~~	5
1	///	3

To aid us in drawing the frequency polygon, we construct a histogram for the data and then connect the midpoints of the tops of each bar in the histogram. The resulting line graph is the frequency polygon shown in Figure 8.

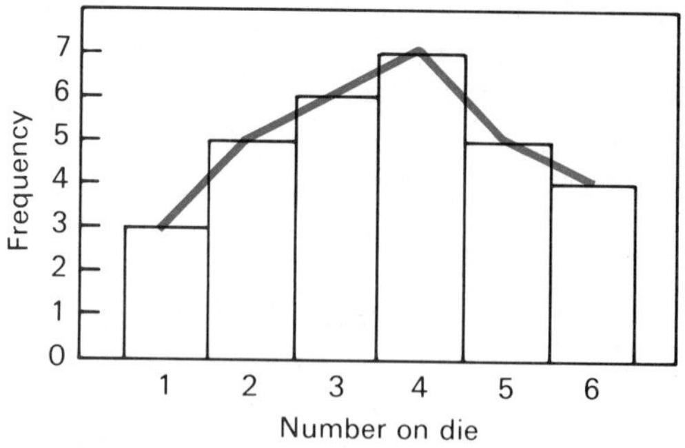

FIGURE 8

EXERCISES FOR SECTION 4.5

1. In a survey of 180 students, it was determined that 90 students came to school by bus, 60 students came by car, and 30 students walked. Construct a circle graph representing this information.

2. The Garcia family has a budget. Each month they use their income in the following manner: 30% for food, 25% for household expenses, 20% for transportation, 10% for savings, 5% for entertainment, and 10% for unexpected expenses. Construct a circle graph representing this information.

3. The Land of Taxes obtains its money from taxes. Each dollar that the Land of Taxes collects is obtained in the following manner: 25¢ comes from personal income taxes, 25¢ comes from corporate income taxes, 15¢ comes from excise taxes, 20¢ comes from sales taxes, 10¢ comes from highway taxes, and 5¢ comes from miscellaneous taxes. Construct a circle graph representing this information.

4. The following table indicates the approximate land area of the seven continents. Construct a circle graph (to the nearest degree) representing this information.

Continent	Area in Square Miles	Percentage of World's Land
Asia	16,988,000	29.5
Africa	11,506,000	20.0
North America	9,390,000	16.3
South America	6,795,000	11.8
Europe	3,745,000	6.5
Australia	2,975,000	5.2
Antarctica	5,500,000	9.6

5. For the fiscal year 1988, the government of Gotham City received the indicated amount of revenue from the following sources.

Federal aid	$70,000,000	Licenses	$10,000,000
State aid	$30,000,000	Sales tax	$20,000,000
Property tax	$60,000,000	Other	$10,000,000

Construct a circle graph representing this information.

6. A student rolled a die 40 times; the results are shown below. Construct a frequency distribution and a histogram to represent these data.

6, 4, 3, 5, 6, 2, 4, 4, 5, 2, 1, 5, 2, 4, 1, 6, 6, 6, 1, 5
2, 6, 3, 1, 6, 1, 6, 2, 1, 2, 2, 4, 4, 6, 4, 1, 3, 4, 5, 5

7. The heights of 28 students in a mathematics class were recorded (in inches), as shown below. Construct a frequency distribution and a frequency polygon to represent these data.

65, 69, 64, 65, 70, 71, 75, 60, 60, 61, 68, 70, 67, 69
67, 70, 67, 66, 67, 71, 70, 72, 70, 66, 68, 70, 72, 64

8. On final-exam day in a statistics class last semester, 40 students were asked for the number of hours of study they spent preparing for the exam. The resulting data were

3, 6, 6, 6, 14, 8, 10, 9, 4, 1
5, 6, 2, 7, 12, 8, 10, 12, 5, 6
5, 3, 7, 9, 12, 9, 10, 11, 4, 6
9, 4, 8, 8, 11, 13, 10, 11, 6, 4

a. Construct a grouped frequency distribution of the data by using 1–3 as the first class.
b. Using the frequency distribution from part *a*, construct a histogram that represents the data.

9. A police radar unit measured the speed of 25 cars on a certain street. The resulting data were

23, 38, 24, 26, 18
23, 52, 30, 45, 27
28, 25, 28, 37, 29
33, 27, 34, 36, 32
23, 18, 23, 38, 21

a. Construct a grouped frequency distribution of the data by using 15–19 as the first class.
b. Using the frequency distribution from part *a*, construct a frequency polygon that represents the data.

10. The following test scores were received by 33 students in a statistics class.

56, 91, 85, 66, 72, 81, 60, 90, 70, 71, 77
84, 75, 58, 89, 67, 98, 96, 70, 87, 74, 64
64, 59, 87, 73, 91, 63, 86, 81, 72, 72, 73

a. Construct a grouped frequency distribution for these scores using the intervals 95–99, 90–94, 85–89, and so on.
b. Use the frequency distribution from part *a* to construct a histogram that represents the data.

11. A student tossed three coins together 32 times and after each toss recorded the number of heads. The table represents the results. Construct a frequency polygon that represents these data.

Number of Heads	Frequency
0	4
1	12
2	12
3	4

12. A survey of 40 students was made in the cafeteria to determine the cost (in cents) of each student's lunch, as shown below. Construct a frequency distribution and a frequency polygon for these data.

92, 93, 86, 93, 93, 92, 90, 84, 92, 91
85, 86, 84, 85, 90, 86, 92, 86, 87, 92
87, 92, 88, 89, 91, 88, 87, 88, 88, 86
93, 84, 86, 84, 85, 86, 85, 92, 92, 88

13. The weights (in pounds) of 50 elementary school students are given as follows.

62, 43, 62, 90, 84, 78, 46, 53, 44, 92
65, 73, 61, 66, 76, 53, 58, 87, 83, 71
94, 87, 83, 71, 96, 64, 58, 77, 76, 58
85, 74, 68, 63, 47, 68, 86, 75, 77, 71
90, 42, 84, 84, 53, 58, 84, 62, 68, 74

a. Construct a grouped frequency distribution for these weights using the intervals 95–99, 90–94, 85–89, and so on.
b. Using the frequency distribution from part *a*, construct a frequency polygon that represents the data.

14. Given the following distribution.

Scores	Frequency
56–60	2
61–65	4
66–70	6
71–75	7
76–80	6
81–85	10
86–90	6
91–95	5
96–100	4

a. Construct a histogram and a frequency polygon for these data.
b. What do you think a "good guess" would be for the mode?
c. What do you think a "good guess" would be for the median?

15. The circle graph in Figure 9 is titled "Where Your Tax Dollar Goes." Does the graph give enough information? Why or why not?

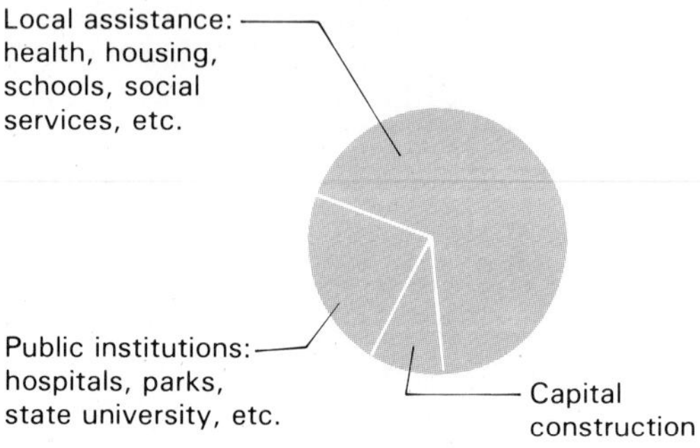

FIGURE 9

16. What can you deduce from the statement "Statistics show that more men than women are involved in auto accidents"?

17. Toss a die 30 times and tabulate the results.
a. Construct a frequency distribution showing the results.

b. Construct a histogram that represents the results.
c. Construct a frequency polygon that represents the results.

18. Toss three coins together 32 times and tabulate the number of heads appearing on each toss, that is, 0, 1, 2, or 3 heads.
a. Construct a frequency distribution showing the results.
b. Construct a frequency polygon that represents the results.
c. Construct a percentage distribution for part *a*.
d. Compare the percentages in *c* with the theoretical percentages: 0 heads—12.5%, 1 head—37.5%, 2 heads—37.5%, 3 heads—12.5%.
e. Toss the three coins 32 more times and tabulate the results with those of part *a*. Construct a frequency polygon for the 64 tosses.

JUST FOR FUN Can you make 10 by using just nine matches?

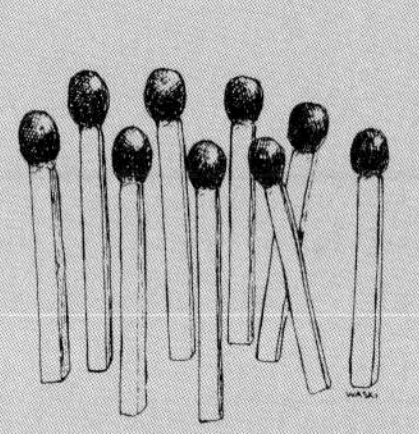

4.6 THE NORMAL CURVE

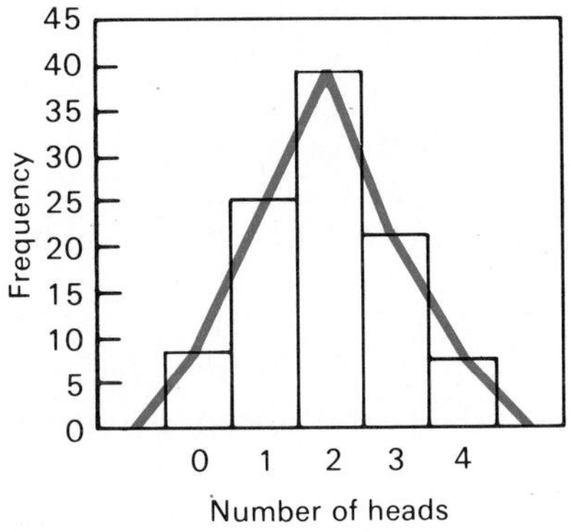

FIGURE 10

Two students, Scott and Joe, decided to toss four coins 100 times, record the number of heads appearing for each toss, and then construct a frequency polygon representing the outcomes. Their results were as follows.

1, 2, 2, 4, 2, 2, 1, 1, 0, 3, 1, 0, 2, 2, 3, 1, 4, 2, 2, 3
2, 1, 1, 4, 2, 2, 2, 1, 2, 4, 2, 3, 2, 2, 2, 1, 2, 2, 3, 2
2, 0, 3, 3, 2, 1, 1, 3, 2, 1, 4, 0, 2, 3, 2, 1, 2, 1, 3, 1
1, 3, 3, 3, 3, 3, 2, 1, 2, 2, 1, 1, 2, 1, 0, 3, 2, 3, 2, 3
3, 3, 2, 2, 1, 0, 2, 1, 2, 1, 4, 1, 4, 0, 0, 2, 1, 3, 2, 2

After recording the number of heads appearing on each toss, Scott and Joe next constructed the frequency distribution in Table 5. After constructing the frequency distribution, they constructed the frequency polygon in Figure 10.

TABLE 5

Number of Heads	Tally	Frequency
0	~~||||~~ |||	8
1	~~||||~~ ~~||||~~ ~~||||~~ ~~||||~~ ~~||||~~	25
2	~~||||~~ ~~||||~~ ~~||||~~ ~~||||~~ ~~||||~~ ~~||||~~ ~~||||~~ ||||	39
3	~~||||~~ ~~||||~~ ~~||||~~ ~~||||~~ |	21
4	~~||||~~ ||	7

Computing the mean for this experiment, we find it is 1.94. The mode is 2, as is the median. If Scott and Joe had increased the number of tosses, their outcomes would have more closely approached the outcomes for tossing four coins according to the laws of probability. For example, they tabulated 39 outcomes out of 100 where two heads appeared. Recall from Exercise 4b, Section 3.4 that if four coins are tossed, the probability of obtaining two heads is $\frac{6}{16}$, or 0.375, which is the same as 375 out of 1,000. This ratio is quite close to the results that Scott and Joe got ($\frac{39}{100}$). Note that the frequency polygon in Figure 10 is similar in shape to the bell-shaped curve in Figure 11.

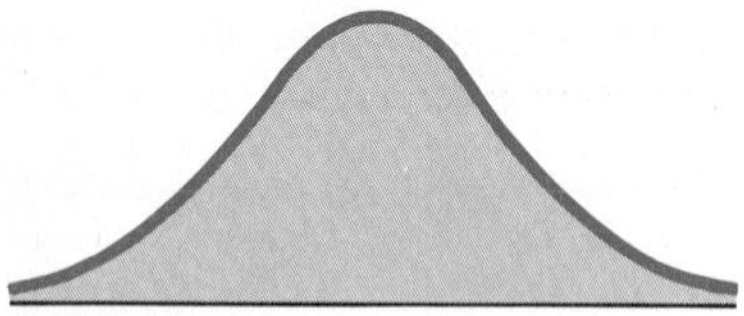

FIGURE 11

There is one particular bell-shaped curve that has been studied extensively by statisticians and named the **normal curve.** The normal curve has some unique properties. If we have a **normal distribution,** then the mean, median, and mode all have the same value, and all occur exactly at the center of the distribution. For normally distributed data, it can also be shown that *approximately* 68% of the data will be included within an interval of one standard deviation about the mean—that is, from 1 standard deviation above the mean to one standard deviation below the mean. For two standard deviations about the mean, that is, from two standard deviations above the mean to two standard deviations below the mean, approximately 95% of the data will be included. Practically all of the data, *about* 99.7%, will be included in the interval from three standard deviations above the mean to three standard deviations below the mean. The given percentages for a normal curve are illustrated in Figure 12. The area under the curve represents all of the frequencies for some normal distribution.

Note that the tails of the curve do not touch the horizontal axis and they will not, no matter how far they may be extended. Data that lie more than three standard deviations from the mean are rare.

An interesting occurrence in natural phenomena is that the data in large samples are often distributed so that the frequency

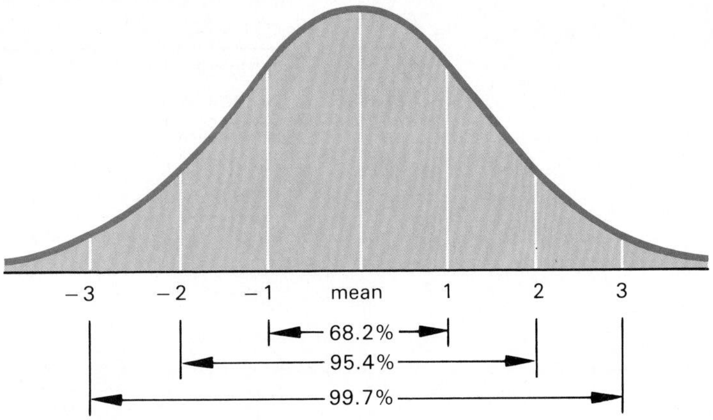

FIGURE 12

polygon approximates the normal curve. A classic example is the distribution of intelligence quotient (IQ) scores for the entire population of a city, state, or country.

EXAMPLE 1 The weights of the first-graders in an elementary school are found to be approximately normally distributed with a mean of 60 pounds and a standard deviation of 5 pounds.

a. What percentage of the students in this group weigh between 55 and 65 pounds?
b. What percentage of these students weigh between 55 and 70 pounds?

SOLUTION Given that the weights are approximately normally distributed with a mean of 60 and a standard deviation of 5, we can draw a normal curve for this group of first-graders, as shown in Figure 13.

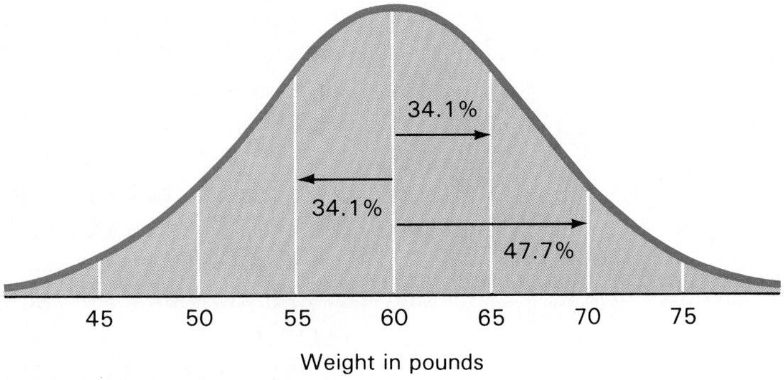

FIGURE 13

TABLE 6

Amount of Standard Deviation from Mean (in one direction only)	Percentage of Data	Amount of Standard Deviation from Mean (in one direction only)	Percentage of Data
0.1	4.0	1.6	44.5
0.2	7.9	1.7	45.5
0.3	11.8	1.8	46.4
0.4	15.5	1.9	47.1
0.5	19.2	2.0	47.7
0.6	22.6	2.1	48.2
0.7	25.8	2.2	48.6
0.8	28.8	2.3	48.9
0.9	31.6	2.4	49.2
1.0	34.1	2.5	49.4
1.1	36.4	2.6	49.5
1.2	38.5	2.7	49.65
1.3	40.3	2.8	49.7
1.4	41.9	2.9	49.8
1.5	43.3	3.0	49.87

Table 6 indicates what percentage of the data is contained between various portions of a standard deviation from the mean.

a. Using Figure 13, we see that 34.1 + 34.1 = 68.2% of the students weigh between 55 and 65 pounds.
b. 34.1 + 47.7 = 81.8% of the students weigh between 55 and 70 pounds.

EXAMPLE 2 The results of an exam in a statistics class are approximately normally distributed, with a mean of 70 and a standard deviation of 10. The instructor decides that a student will receive an A on the exam if the student scores more than two standard deviations above the mean. What is the lowest grade a student can get and still receive an A?

SOLUTION Because the scores are approximately normally distributed with a mean of 70 and a standard deviation of 10, we can draw a normal curve for this class (see Fig. 14).

A score of 90 is exactly two standard deviations above the mean of 70. But the instructor said that only those students scoring *more* than two standard deviations above the mean would receive an A. Hence a student must have a score greater than 90 in order to receive an A.

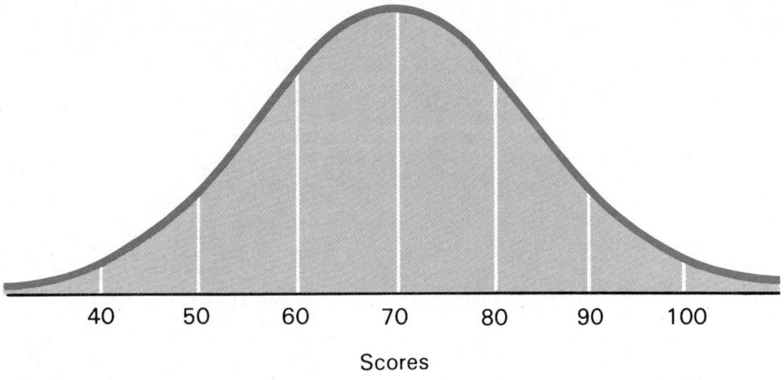

FIGURE 14

EXAMPLE 3 A statistician took a survey of the mileage on 1,000 cars in a parking lot. The mileage on each car was recorded. A frequency polygon of the mileages *approximated* the normal curve in Figure 15.

© *Brent Jones*

a. What is the mean mileage for this set of cars?
b. What is the standard deviation?
c. What percentage of the cars had less than 20,000 miles on the odometer?
d. How many cars had more than 50,000 miles on the odometer?

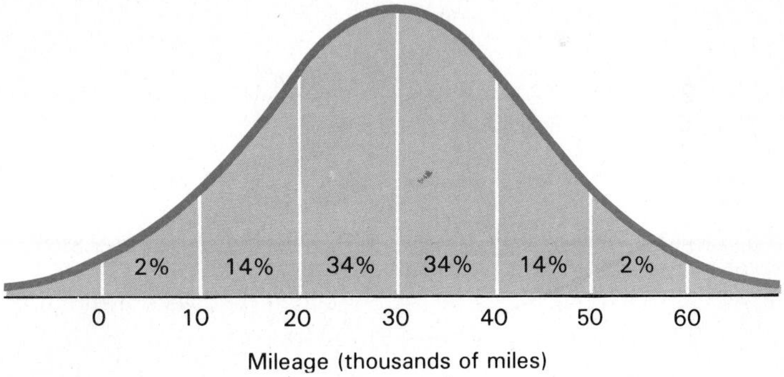

FIGURE 15

SOLUTION

a. Since the data are approximated by the normal curve in Figure 15, we see that the mean is 30,000 miles.
b. Vertical lines have been drawn on the graph to indicate the standard deviations from the mean. The standard deviation is 10,000 miles.
c. Since the mean is 30,000 miles and the standard deviation is 10,000 miles, those cars with less than 20,000 miles on the odometer must be more than one standard deviation below the mean. The data approximate the normal curve, so 50%

of the data lie to the left of the mean. Figure 15 shows that 34% of the data are between 20,000 and 30,000. Hence, 50% − 34% = 16% of the cars have been driven less than 20,000 miles.

d. Cars with more than 50,000 miles on the odometer are more than two standard deviations above the mean. Note that 50% of the data lie to the right of the mean, and 48% of the data lie between 30,000 and 50,000. Hence, 50% − 48% = 2% of the cars have more than 50,000 miles. Now, since there are 1000 cars, we have 2% of 1000 = 0.02 × 1000 = 20 cars.

EXAMPLE 4 Testing indicates that the lifetimes of 1,000 Dimlite light bulbs are approximately distributed with a mean of 100 hours and a standard deviation of 8. A frequency polygon of the lifetimes approximated the normal curve in Figure 16. How many of these 1,000 Dimlite bulbs will last

a. more than 116 hours? b. less than 92 hours?

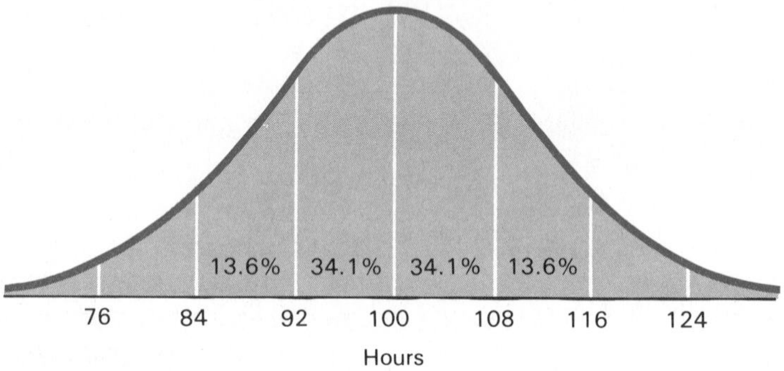

FIGURE 16

SOLUTION Since the lives of the light bulbs are approximately normally distributed with a mean of 100 and a standard deviation of 8, the normal curve shown in Figure 16 reflects the lifetimes of this set of bulbs.

a. The light bulbs lasting more than 116 hours are those that lie more than two standard deviations above the mean. Figure 16 shows that 34.1% + 13.6% = 47.7% of the data are between 100 and 116 hours. Since the data are approximately normally distributed, 50% of the data are greater (or above) 100 hours. Hence, 50% − 47.7% = 2.3% of the 1000 Dimlite bulbs will last more than 116 hours; 2.3% of 1000 = 0.023 × 1000 = 23 bulbs.

b. The light bulbs lasting less than 92 hours are those that lie more than one standard deviation below the mean. That is,

$50\% - 34.1\% = 15.9\%$ of the 1000 bulbs will last less than 92 hours; 15.9% of $1000 = 0.159 \times 1000 = 159$ bulbs.

In an earlier discussion, it was pointed out that approximately 68% of normally distributed data are included within an interval of plus or minus one standard deviation about the mean, or 34.1% of the data are between the mean and one standard deviation above the mean. Similarly, approximately 95% of normally distributed data are included within an interval of plus or minus two standard deviations about the mean, and 99.7% of the data are included within three deviations about the mean. Sometimes we may wish to consider a portion of the data that is a fractional part of a standard deviation from the mean. Table 6 indicates what percentages of the data are contained between various tenths of a standard deviation from the mean.

Many problems require fractional parts of standard deviations. Consider the following example. In a statistics class, the scores on the last exam were approximately normally distributed. The instructor decided to assign a grade of C for any score within 1.2 standard deviations of the mean. What percentage of the class will receive a grade of C? Table 6 indicates that 38.5% of the data are within 1.2 standard deviations of the mean on only one side of the mean. Since the normal curve is symmetric, we have the same percentage on the other side, and we therefore double the percentage: $2 \times 38.5 = 77\%$ of the class will receive a grade of C.

EXAMPLE 5 Testing indicates that the lifetimes of a new shipment of 1,000 Dimlite light bulbs are approximately normally distributed with a mean of 100 hours and a standard deviation of 10. How many of the light bulbs will last between 83 and 117 hours?

SOLUTION This problem is similar to Example 4, but the boundaries involve fractional parts of the standard deviations from the mean. The mean is 100, and the standard deviation is 10.

$$100 - 83 = 17, \qquad \frac{17}{10} = 1.7$$

and

$$117 - 100 = 17, \qquad \frac{17}{10} = 1.7$$

Therefore, both boundaries are 1.7 standard deviations from the mean. Table 6 indicates that 45.5% of the data lie within 1.7

standard deviations of the mean. We have 1.7 standard deviations on either side of the mean; hence 2 × 45.5 = 91%, or 910 light bulbs, will last between 83 and 117 hours.

HISTORICAL NOTE

Many of the symbols that we now use in mathematics were adopted through repeated use by different people, so their exact origins are not known. For example, the sign for plus (+) is believed to have been adapted from the symbol for "and," the ampersand, &. After repeated writing, the symbol for "and" began to appear as + instead of &. The plus sign did not originate with mathematicians—it was first used by merchants in the 1400s to indicate excess. For example, if a bolt of cloth was supposed to contain 10 yards of cloth but it actually contained 13 yards, then a merchant had an excess of 3 yards and would mark "10 + 3" on the package.

The exact origin of the symbol for equality (=) is known. It first appeared in an English algebra book *Whetstone of Witte* (1557) written by Robert Recorde. The author stated that he chose a pair of horizontal line segments to represent equality because "noe 2 thynges can be moare equable." Recorde's symbol for equality was greeted cautiously and not universally accepted until the early 1700s.

It should be noted at this point that the position of a piece of data in terms of the number of standard deviations it is located from the mean is called the *Z-score* or *standard score*. The **Z-score** is found by the formula

$$Z = \frac{\text{piece of data} - \text{mean}}{\text{standard deviation}}$$

Therefore, for this example we have

$$Z = \frac{117 - 100}{10} = \frac{17}{10} = 1.7$$

and

$$Z = \frac{83 - 100}{10} = \frac{-17}{10} = -1.7$$

Data that lie below the mean will always have negative Z-scores. This is an indication of the position of the data with respect to the mean. Similarly, data that lie above the mean will always have positive Z-scores. We will still use Table 6 to find the percentages of data contained between various tenths of a standard deviation from the mean, whether we call them Z-scores or the amounts of standard deviation from the mean.

EXAMPLE 6 The final-exam results in a statistics class were approximately normally distributed. Grades were assigned in the following manner.

A for a score more than 1.6 standard deviations above the mean
B for a score between 1.1 and 1.6 standard deviations above the mean
C for a score between 1.1 standard deviations above the mean and 1.1 standard deviations below the mean
D for a score between 1.1 and 1.6 standard deviations below the mean
F for a score more than 1.6 standard deviations below the mean

What percentage of the class received each grade?

SOLUTION

a. Table 6 indicates that 44.5% of the population lies within 1.6 standard deviations above the mean. Hence, 50 − 44.5 = 5.5% of the class was above 1.6 standard deviations from the mean and received an A.
b. Since 44.5% of the population lies within 1.6 standard deviations and 36.4% lies within 1.1 standard deviations, we subtract 44.5 − 36.4 to determine that 8.1% of the class received a grade of B.
c. Since 36.4% of the class lies within 1.1 standard deviations from the mean, and we want those both above and below the mean, we multiply 36.4 by 2 to determine that 72.8% of the class received a grade of C.
d. This is similar to part *b*, but we are concerned with standard deviations below the mean. Hence, we have 44.5 − 36.4 = 8.1% of the class received a grade of D.
e. Calculations like those in part *a* show that 5.5% of the class received a grade of F.

NOTE OF INTEREST

Table 7 gives a percentile-rank interpretation for a normal distribution. For example, consider a national competency exam in mathematics. If the scores are approximately normally distributed with a mean of 100 and a standard deviation of 10, then a score of 110 on the exam would yield a percentile rank of 84. That is, it is likely or probable that 84% of the people taking the exam would rank below a person having a score of 110.

TABLE 7

Number of Standard Deviations from the Mean	Percentile Rank in a Normal Distribution
+2.0	98
+1.5	93
+1.0	84
+0.5	69
0	50
−0.5	31
−1.0	16
−1.5	7
−2.0	2

EXERCISES FOR SECTION 4.6

1. The IQ scores for a certain group of elementary school students are approximately normally distributed with a mean of 100 and a standard deviation of 10.
   a. What percentage of the students have IQ scores between 90 and 110?
   b. What percentage of the students have IQ scores between 80 and 120?
   c. What percentage of the students have IQ scores between 70 and 130?
2. An examination is given to all entering students at a certain college. The scores are approximately normally distributed with a mean of 100 and a standard deviation of 15.
   a. What percentage of the students have scores between 85 and 115?

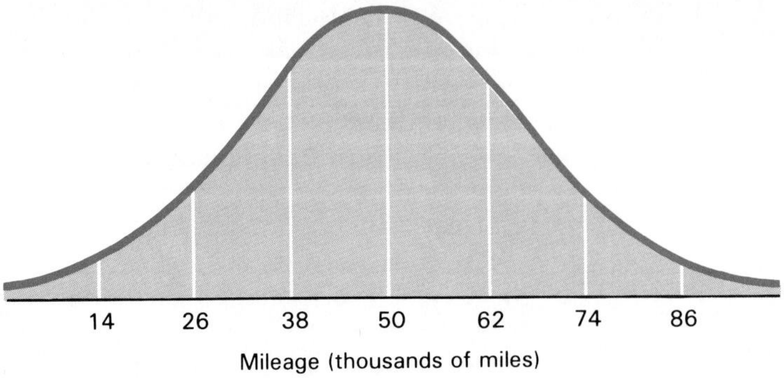

FIGURE 17

b. What percentage of the students have scores between 70 and 130?
c. What percentage of the students have scores between 55 and 145?
d. What percentage of the students have scores above 115?

3. A group of students in a statistics class recorded the mileages of 2,000 cars in the student parking lot at their college. A frequency polygon of the mileages approximated the normal curve in Figure 17.
   a. What is the mean mileage for this set of cars?
   b. What is the standard deviation?
   c. What percentage of cars had less than 62,000 miles on the odometer?
   d. How many had more than 62,000 miles on the odometer?

4. A statistician recorded the heights of 1,000 students at a particular college. A frequency polygon of the heights approximated the normal curve in Figure 18.
   a. What is the mean height for this set of students?
   b. What is the standard deviation?
   c. What percentage of the students were more than 5 feet tall and less than 6 feet tall?
   d. How many students were less than 5 feet tall?

5. Testing indicates that the lifetimes of a shipment of disposable butane lighters are approximately normally distributed with a mean of 1,000 lights and a standard deviation of 100. A shipment contains 5,000 of these lighters.
   a. Approximately how many of the lighters will light more than 1,100 times?

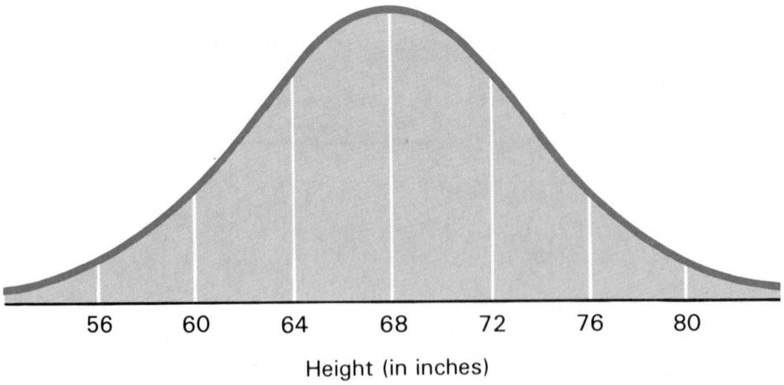

FIGURE 18

b. Approximately how many of the lighters will light more than 1,200 times?
c. Approximately how many of the lighters will light less than 700 times?
d. Approximately how many of the lighters will light between 800 and 1,100 times?

6. The scores on a statistics exam for a class of 50 students were approximately normally distributed with a mean of 30 and a standard deviation of 5.
   a. Approximately how many of the students scored above 35?
   b. Approximately how many of the students scored above 40?
   c. Approximately how many of the students scored below 25?
   d. Approximately how many of the students scored between 20 and 35?

7. It has been shown that the lifetimes for a certain type of belted radial tire are approximately normally distributed with a mean of 40,000 miles and a standard deviation of 5,000 miles.
   a. What percentage of these tires will last more than 50,000 miles?
   b. What percentage of these tires will last less than 35,000 miles?
   c. What percentage of these tires will last between 35,000 and 50,000 miles?
   d. What percentage of these tires will last between 30,000 and 55,000 miles?

8. The manufacturer of ZAP television sets guarantees its picture tube for 2 years. Testing indicates that the lifetimes of ZAP picture tubes are approximately normally distributed with a mean of 3 years and a standard deviation of 6 months ($\frac{1}{2}$ year). What percentage of ZAP television sets will have to be replaced under the guarantee?

9. The scores on a mathematics exam were approximately normally distributed, and the instructor assigned grades in the following manner.

   A for a score more than 1.7 standard deviations above the mean
   B for a score between 1.1 and 1.7 standard deviations above the mean
   C for a score between 1.1 standard deviations above the mean and 1.2 standard deviations below the mean
   D for a score between 1.2 and 1.8 standard deviations below the mean
   F for a score more than 1.8 standard deviations below the mean

   What percentage of the class received each grade?

10. Given that the scores on a certain exam are approximately normally distributed with a mean of 50 and a standard deviation of 10, use Table 6 (page 226) to find
    a. the percentage of scores greater than 55
    b. the percentage of scores greater than 65
    c. the percentage of scores greater than 71
    d. the percentage of scores between 45 and 55
    e. the percentage of scores between 33 and 44
    f. the percentage of scores less than 44.

11. The heights of the students at a local school are approximately normally distributed with a mean height of 64 inches and a standard deviation of 4 inches. Use Table 6 (page 226) to find
    a. the percentage of students between 62 and 66 inches tall
    b. the percentage of students between 58 and 66 inches tall
    c. the percentage of students over 6 feet tall
    d. the percentage of students shorter than 5 feet 6 inches
    e. the percentage of students between 5 feet 6 inches and 6 feet tall.

12. IQs for a class of mathematics students are approximately normally distributed with a mean of 100 and a standard deviation of 10. Use Table 6 (page 226) to find
    a. the percentage of students with an IQ between 95 and 115
    b. the percentage of students with an IQ between 93 and 119
    c. the percentage of students with an IQ between 115 and 119
    d. the percentage of students with an IQ greater than 122
    e. the percentage of students with an IQ less than 117.

13. A survey was taken among students in the college cafeteria to find out how many hours a week they spent studying for courses for which they were currently enrolled. The data were approximately normally distributed with a mean of 24 hours and a standard deviation of 5 hours. Use Table 6 in this section to find
   a. the percentage of students that studied more than 28 hours per week
   b. the percentage of students that studied more than 33 hours per week
   c. the percentage of students that studied less than 16 hours per week
   d. the percentage of students that studied between 22 and 30 hours per week
   e. the percentage of students that studied between 28 and 33 hours per week.

14. The coffee machines in a certain snack bar are supposed to dispense 5 ounces of coffee in each cup. The amount of coffee in each cup is approximately normally distributed with a mean of 5.0 ounces and a standard deviation of 0.5 ounce.
   a. What percentage of cups contain between 4 and 5.5 ounces of coffee?
   b. What percentage of cups contain between 5 and 5.8 ounces of coffee?
   c. What percentage of cups contain between 4.8 and 5.8 ounces of coffee?
   d. What percentage of cups will overflow if 6-ounce cups are used in the machine?

15. A survey indicates that a postmarking machine in a certain post office cancels an average (mean) of 65 letters per minute, with a standard deviation of 10 letters per minute. Assume that the number of letters cancelled by the machine is approximately normally distributed.
   a. What percentage of the time does the machine cancel more than 90 letters per minute?
   b. What percentage of the time does the machine cancel less than 50 letters per minute?

16. The average (mean) length of a phone call from a certain pay phone is 5 minutes, with a standard deviation of 1 minute. Assume that the lengths of the phone calls are approximately normally distributed.
   a. What percentage of the calls last more than $6\frac{1}{2}$ minutes?
   b. What percentage of the phone calls last between $5\frac{1}{2}$ and $6\frac{1}{2}$ minutes?
   c. What percentage of the calls cost extra, if a caller has to pay extra for any call over 3 minutes in length?

17. A manufacturer has been awarded a government contract to manufacture bleeps, which must be between 1.45 and 1.55 inches in length. Testing indicates that the manufacturer's machines produce bleeps whose lengths are normally distributed with a mean length of 1.5 inches and a standard deviation of 0.1 inch.
   a. What percentage of bleeps meet the government specifications?
   b. What percentage of bleeps do not meet the government specifications?

18. The results of an exam in a statistics class are approximately normally distributed with a mean of 40 and a standard deviation of 8.
   a. What percentage of the scores are above 52?
   b. What percentage of the scores are above 60?
   c. If 100 students took the exam and all students with grades from 52 to 60 received a B, how many students received a B?

19. Testing indicates that the lifetimes of a certain lot of light bulbs are approximately normally distributed with a mean of 100 hours and a standard deviation of 10.
   a. What is the probability that a bulb selected at random will last more than 100 hours?
   b. What is the probability that a bulb selected at random will last more than 105 hours?
   c. What is the probability that a bulb selected at random will last between 95 and 108 hours?

20. The results of an exam in a history class are normally distributed with a mean of 60 and a standard deviation of 6.
   a. What is the probability that a student who took the exam will have a score between 57 and 63?
   b. What is the probability that a student who took the exam will have a score between 45 and 75?

JUST FOR FUN

Can you connect all nine dots with only four connecting line segments and without raising your pencil from the paper?

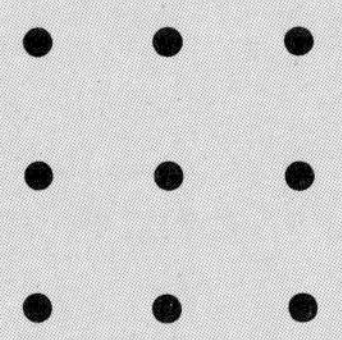

Summary

As we saw in this chapter, a *statistic* is a number derived from a set of data that is used in some way to represent or describe the data. The *measures of central tendency* and *measures of dispersion* are devices used to acquire knowledge from the data.

Measures of central tendency tell us where the data tend to center or cluster. The *mean* for a set of data is found by determining the sum of the data and dividing this sum by the total number of elements in the set. The *median* for a set of data is found by arranging the data in sequential order and finding the middle number. When we have an even number of data, it is customary to use the number halfway between the two middle numbers, that is, the mean of the two middle numbers. The median is the measure of central tendency that determines the middle of a given set of data. The *mode* for a given set of data is the item or value that occurs most frequently. There can be times when a set of data has no mode—that is, when each value occurs an equal number of times. A set of data can also have more than one mode. The midrange is found by adding the least value in the given set of data to the greatest value and dividing the sum by 2.

Measures of dispersion tell us about the spread of the data. The *range* for a set of data is found by subtracting the smallest value in the set from the largest value in the set. The range only involves the use of two extreme values; it does not tell us anything about the remaining values in the set of data.

A measure of dispersion that depends on all of the values in a given set of data is the *standard deviation.* To find the standard deviation for a given set of data, we must follow these steps.

1. Find the *mean* of the given set of data.
2. Find the *deviation from the mean* for each value in the set of data.
3. Square each deviation.
4. Find the mean of the squared deviations.
5. Find the principal square root of this number.

After determining where data tend to center or cluster and then where data tend to disperse or scatter, we next examined the location of data in relation to the whole set or to other data. *Percentiles* and *quartiles* are *measures of location* of data. *Percentiles* divide sets of data into 100 equal parts; hence 100% is the basis of the measure. In order to determine the percentile at which a datum lies, we need to find the number of data that are ranked below the given datum and divide that number by the total number of data in the given set. The 25th percentile, 50th percentile, and 75th percentile are unique because they divide sets of data into quarters; hence they are referred to as *quartiles*. The *first quartile* for a set of data is the value that has 25% of the data below it. The *second quartile* is the value that has 50% of the data below it, and the *third quartile* is the value that has 75% of the data below it.

Pictures of data are often used to transmit or summarize statistical information. In this chapter we studied the *circle* or *pie graph,* the *histogram,* and the *frequency polygon (line graph)*. A *circle graph* is used to show how a whole quantity is divided into parts. A *histogram* consists of a series of bars that are all the same width. Uniform units are used on the vertical axis to show the frequency of the data. A *frequency distribution table* is used to tabulate the given data and to determine the number of bars in the histogram. A *frequency polygon (line graph)* can also be used to graph the information in a frequency distribution table. It is constructed by connecting the midpoints of the tops of each bar in a histogram.

When we have a *normal distribution,* the area under the curve represents the entire set of data, or population. If we have data that are approximately normally distributed, then the mean, median, and mode all have approximately the same value. It can also be shown that approximately 68% of the data will be included within an interval of one standard deviation about the mean, and approximately 95% of the data will be included in the interval of two standard deviations about the mean. Practically all of the data (99.7%) will be included in the interval from three standard deviations below the mean to three standard deviations above the mean. These proportions will remain the same, regardless of the mean and standard deviation for a specific experiment, so long as the data are normally distributed.

The *Z-score,* or *standard score,* is the position of a piece of data in terms of the number of standard deviations it is located from the mean. The *Z*-score is found by the formula

$$Z = \frac{\text{piece of data} - \text{mean}}{\text{standard deviation}}$$

Vocabulary Check

mean	bimodal	histogram	frequency distribution
median	range	*Z-score*	frequency polygon
mode	percentile	standard deviation	normal distribution
midrange	quartile	central tendency	

Review Exercises for Chapter 4

1. a. Name four measures of central tendency.
   b. Which measure of central tendency gives the value that occurs most frequently?
   c. Which measure of central tendency gives the middle value of the given data?
   d. Which measure of central tendency gives the average value of the given data?

In Exercises 2–5, find the mean, median, mode, and midrange for each set of data. (Round off any decimal answer to the nearest tenth.)

2. 12, 18, 16, 12, 19
3. 2, 3, 4, 5, 5, 6, 10, 13
4. 2, 3, 5, 7, 11, 13, 17, 19
5. 99, 11, 88, 22, 33, 77
6. In order to receive a grade of C in her statistics class, Susan needs a mean score of 72 on five tests. If Susan had scores of 62, 78, 80, and 68 on her first four tests, what is the lowest score that Susan can get on her last test and still receive a grade of C in the course?
7. The mean of four of a set of five scores is 74. The fifth score is 88. What is the sum of the five scores? What is the mean of the five scores?
8. Find the range, midrange, and mean for the following set of test scores: {70, 70, 67, 66, 65, 65, 65, 62, 60, 60}.
9. Find the range, midrange, and mean for the following set of data: {58, 76, 76, 79, 84, 85, 100}.
10. Find the standard deviation for the following set of data: {10, 11, 13, 13, 14, 17}. (Round your answer to the nearest tenth.)

11. Find the standard deviation for the following set of data: {10, 14, 16, 26, 14, 16}. (Round your answer to the nearest tenth.)

12. In a senior class of 300 students, Joe has the rank of 45th. What is Joe's percentile rank in the class?

13. Andy is ranked at the third quartile in his statistics class of 32 students. Julie is in the same statistics class, and she is ranked sixth. Who ranks higher, Andy or Julie?

14. A student tossed a die 50 times. His results are shown below. Construct (a) a frequency distribution, (b) a histogram, and (c) a frequency polygon that represent these data.

2, 5, 4, 4, 1, 5, 3, 5, 6, 6
2, 5, 2, 5, 5, 1, 1, 5, 4, 6
2, 4, 4, 6, 6, 3, 2, 6, 1, 1
6, 1, 6, 4, 5, 4, 1, 6, 3, 2
3, 4, 4, 5, 6, 4, 6, 1, 2, 2

15. Tom's Tobacco Shop sells tobacco in many forms. All tobacco sales for the year were distributed as follows.

Cigarettes	40%
Cigars	30%
Pipe tobacco	20%
Chewing tobacco	6%
Snuff	4%

Construct a circle graph representing this information.

16. The heights (in inches) of 30 students in a physical education class were recorded as shown below.

65, 69, 64, 65, 70, 71, 75, 69, 70, 64
72, 70, 66, 68, 70, 72, 64, 68, 70, 69
67, 70, 67, 66, 67, 71, 70, 75, 69, 70

a. Construct a frequency distribution for these data.
b. Construct a histogram for these data.
c. Construct a frequency polygon for these data.
d. Does the frequency polygon obtained in part *c* approximate a normal curve? Why?

17. One thousand test scores are approximately normally distributed with a mean of 60 and a standard deviation of 8.
a. What percentage of the scores are above 68?
b. What percentage of the scores are above 76?
c. What percentage of the scores are below 52?
d. What percentage of the scores are between 52 and 76?
e. If a grade of C is assigned to the scores between 52 and 68, how many scores will receive a grade of C?

18. On the first day of class last semester, a group of students was asked for the number of hours of sleep they obtained the previous night. The data were approximately normally distributed with a mean of 5.0 and a standard deviation of 0.5. Find
a. the percentage of students who slept more than 5.8 hours
b. the percentage of students who slept less than 4.8 hours
c. the percentage of students who slept between 4.8 and 5.8 hours.
d. the percentage of students who slept more than 5.4 hours
e. the percentage of students who slept between 5.4 and 5.8 hours

19. Two exams were given to a certain mathematics class. On the first exam, the mean was 64 and the standard deviation was 8. On the second exam, the mean was 48 and the standard deviation was 12. The scores of four students who took both exams were

	Exam 1	Exam 2
Bill	64	63
Louie	70	60
Jack	60	54
Steve	80	66

a. Who improved on the second exam?
b. Who did not do as well on the second exam as on the first?
c. Who performed the same on both exams?
d. Who has the best grade for the two exams combined?

20. Babe Ruth was one of the first inductees into the Baseball Hall of Fame. He played in the major leagues from 1914 to 1935. The following is a list of the number of home runs he hit in each season: 0, 4, 3, 2, 11, 29, 54, 59, 35, 41, 46, 25, 47, 60, 54, 46, 49, 46, 41, 34, 22, 6. For this set of data, find the
   a. mean b. median c. mode
   d. midrange e. range

21. Use the set of data {20, 22, 26, 26, 28, 34} to find the following.
   a. mean b. median c. mode
   d. range e. midrange
   f. standard deviation (to the nearest tenth)

22. Given the following set of 12 scores: {29, 22, 24, 27, 22, 20, 20, 20, 15, 24, 22, 34}.
   a. What is the rank (from the top) of 27?
   b. What is the percentile rank of 27?
   c. What score is at the third quartile?
   d. What is the median?
   e. What is the mode?

23. After the midterm exam in a certain mathematics class, 25 students were asked for the amount of time (to the nearest hour) that they spent studying for the midterm exam. The table below provides the resulting data. Using the information provided in the table, find the following (to the nearest tenth).
   a. the mean time
   b. the median time
   c. the mode time
   d. the midrange time

Time (hours)	Number of People
12	2
10	3
8	3
6	4
4	10
2	2
0	1

Chapter Quiz

State whether the following are true or false.

1. Each of the measures of central tendency—mean median, mode, and midrange—is an "average."

2. The *median* for a set of data is found by arranging the data in sequential order and finding the middle value.

3. A given set of data may have more than one *mode*.

4. The *range* for a set of data is also a measure of central tendency.

5. The *midrange* is a measure of dispersion.

6. The standard deviation for the set of data {70, 70, 70, 65, 60} is 4.

7. The mean of the squared deviations will always be a nonnegative value.

8. When a student is told that he scored at the 77th percentile on an exam, he then knows that he received a score of 77% on the exam.

9. In a class of 30 students, Larry has a percentile rank of 70. Therefore, Larry's rank in the class is 21st.

10. The first quartile for a set of data is that value which has 75% of the data below it.

11. For a certain exam, Irene scored at the 30th percentile, whereas Janie scored at the 60th percentile on the same exam. Therefore, Janie's score was twice Irene's score.

12. Given the set of data {30, 35, 36, 37, 38, 39, 40, 41}, the first quartile is 40.

13. Sandra is ranked 50th in her senior class of 200 students. Eileen is in the same senior class and

Eileen has a percentile rank of 72. Of the two seniors, Eileen has the higher standing in the class.

14. Circle graphs are used in showing how a whole quantity is divided into parts.

15. A histogram consists of a series of bars that are drawn all with the same widths on the vertical axis, and uniform units on the horizontal axis.

16. A frequency polygon is a type of histogram.

17. A normal bell-shaped distribution will have a range that is approximately equal to six standard deviations.

18. For normally distributed data, approximately 68% of the data will be included within an interval of one standard deviation about the mean.

19. If we have data that are approximately normally distributed, then the mean, median, and mode all have approximately the same value.

20. All normal curves appear the same.

21. The *range* is a measure of central tendency.

22. The position of a piece of data, in terms of standard deviations from the mean, is called the *Z-score*.

23. If Sam scored at the zero percentile, then he scored the highest in his class.

24. Grouped frequency distributions representing the same set of data may be different.

25. The tails of a normal curve must touch the horizontal axis.

NOTE OF INTEREST

Statistics can be presented in a manner that misleads you. Averages, percentages, charts are all used to present statistics. But, by accident or on purpose, essentially accurate information may be used to present distorted or biased information.

For example, consider the production figures of two companies that manufacture passenger airplanes. The use of the percentage figures alone gives the impression that the Sullavin Air Company was the more successful of the two. But, in fact, the increase in production by Ronolog Plane Corporation was more than twice the number of Sullavin Air Company.

	1984	1985	Increase
Sullavin Air Co.	60	120	100%
Ronolog Plane Corp.	312	468	50%

If it is true that leading dentists recommend brand A toothpaste, it may also be true that they recommend most, if not all, brands of toothpaste available today.

If four out of five doctors recommend brand B aspirin, it is not known how large a sample was used or how many times the survey was taken until the desired result was obtained. Also, consider the possibility that many doctors do not recommend aspirin at all.

Statistics are often given by means of graphs or charts. These too can be misleading and deceptive. The graphs in Figure 19 show the total daily sales for two sales people in a store. Which one appears to be more successful?

Both graphs present identical information; the difference is that the vertical scale of Smith has been exaggerated.

Consider your average speed for a trip from town A to town B, a distance of 200 miles. If you took five hours to go from A to B, your average speed was 40 mph. If you took four hours to return to A from B, then your average speed was 50 mph. What was your average speed for the entire round trip? Many people would answer 45 mph. This is not correct! They have made the error of averaging averages. To find the average speed for the entire round trip, divide the total distance (400 miles) by the total time (9 hours) and we obtain an average speed of 44.4 mph, not 45.

Statistics provide valuable information when dealt with correctly. The misuse of them by some does not detract from their value when used properly.

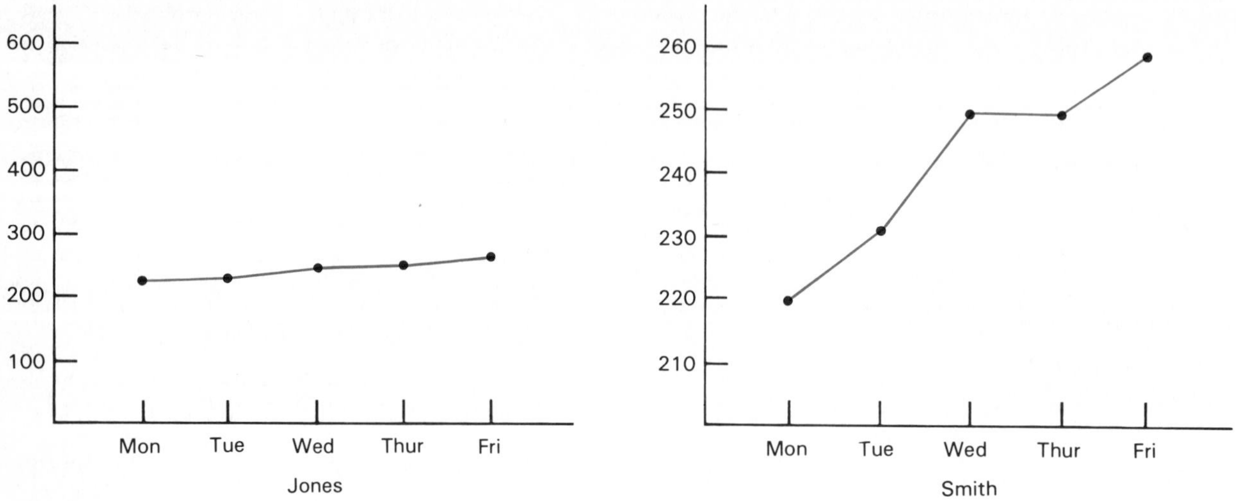
600
500
400
300
200
100
Mon
Tue
Wed
Thur
Fri
Jones
260
250
240
230
220
210
Mon
Tue
Wed
Thur
Fri
Smith

FIGURE 19

CHAPTER 5

PEPPERIDGE FARM
MILK
Chocolate Macadamia
SPECIAL COLLECTION COOKIES
NET WT. 6½ OZ (184g)
DANNON
FRESH FLAVORS
Vanilla
LOWFAT YOGURT
1.5% MILKFAT NET WT. 16 OZ. (1 LB.) 454g
1 LITER
diet Coke
Trade-mark
Dissolved Mineral Solids: 2,851 mg per Liter
Apollinaris
NATURALLY SPARKLING MINERAL WATER
Federal Republic of Germany
1 LITER · 33.8 FL. OZ.
BIRDS EYE
Green Peas
NET WT 10 OZ. (283 G)
Bring Out The Best
HELLMANN'S
REAL
MAYONNAISE
32 FL. OZ. (1 QT.)/.95 L

AN INTRODUCTION TO THE METRIC SYSTEM

AFTER STUDYING THIS CHAPTER YOU WILL BE ABLE TO DO THE FOLLOWING:

1. Describe the basic characteristics and advantages of the **metric system of measurement.**
2. Identify and use the following prefixes in the metric system: **kilo, hecto, deka, deci, centi,** and **milli.**
3. Name the basic unit of length in the metric system and describe its origin.
4. Convert between the various metric units for measuring length, and use conversion tables to convert our customary units of length to metric units and vice versa.
5. Name the basic unit of volume in the metric system; and describe the relationship between this unit and the basic unit of length.
6. Convert between the various metric units for measuring volume, and use conversion tables to convert our customary units of volume to metric units and vice versa.
7. Name the basic unit of weight in the metric system, and describe the relationship between this unit and the basic unit of volume.
8. Convert between the various metric units for weight, and use conversion tables to convert our customary units of weight to metric units and vice versa.
9. Describe the **Celsius** thermometer, and convert Fahrenheit degree readings to Celsius and vice versa.
10. Make approximate conversions from metric measurements to customary measurements and vice versa, without computation.
11. Develop an understanding of the metric measurements and be able to determine which metric unit best fits the given item.

SYMBOLS FREQUENTLY USED IN THIS CHAPTER

Basic Metric Unit	Symbol
meter	m
liter	L
gram	g
Celsius degree	°C

Prefix	Symbol	Meaning
kilo	k	1000
hecto	h	100
deka	da	10
deci	d	0.1
centi	c	0.01
milli	m	0.001

5.1 INTRODUCTION

> It is therefore declared that the policy of the United States shall be to coordinate and plan the increasing use of the metric system in the United States and to establish a United States Metric Board. . . .—*Metric Conversion Act of 1975*

The **metric system of weights and measures** is an international system that has been formalized as the International System of Units (SI). Many books and pamphlets dealing with the metric system approach the topic from strictly a metric viewpoint. We have not used this approach here. The purpose of this chapter is to introduce you to the metric system and its advantages and to help you understand how the system works.

Our customary system of measures is a "little of this and a little of that," a hodgepodge. It probably does not seem like a bad system, or one that is awkward to work with, but that is simply because you have used it all your life. However, it can be a confusing system. If you buy a quart of strawberries, you are buying a different amount than if you buy a quart of milk. The reason is that one is a dry quart and the other is a liquid quart. A dry quart is 16% greater in volume than a liquid quart. Ounces are also confusing because an ounce can mean volume, for example, the number of ounces in a quart, or weight, as the number of ounces in a pound.

Pound measurement is even more ambiguous. Pound can mean weight, as in a pound of coffee, or it can mean force as in pounds required to snap a cable or rope. Which is heavier, a pound of hamburger or a pound of gold? This question is not as silly as it seems. A pound of hamburger is weighed by *avoirdupois* weight and contains 7,000 grains, but a pound of gold is weighed by *troy* weight and contains 5,760 grains. Therefore, a pound of hamburger is heavier. These inconsistencies do not occur in the metric system because each quantity has its own unit of measurement and no unit is used to express more than one quantity.

The metric system is a system of measurement that has basic units of measure for length, width, area, volume, and weight that are in a decimal relationship to each other. For example, 1 meter = 10 decimeters = 100 centimeters. On the other hand, the United States uses a system of measurement that does not relate units in a decimal manner. For example, 12 inches = 1 foot, 3 feet = 1 yard, 16 ounces = 1 pound, and so on. You will find that it is much simpler to multiply or divide by 10 than it is to use 12 for one calculation, 3 for another, 16 for another, 5,280 for yet another (5,280 feet = 1 mile).

NOTE OF INTEREST

Measurements of length, weight, volume, and temperature are important to all of us. But many other kinds of measurement are also important in today's society. For example, measurements of force, work, power, and energy all provide necessary information to scientists and manufacturers. Other measurements that may be of interest to someone are related to heat, light, sound, and electricity. Most of these are measured with special units.

A *calorie* is the amount of heat needed to raise the temperature of one gram of water one degree Celsius. A *large calorie* (the nutritionist's calorie) is the amount of heat needed to raise the temperature of 1000 grams (1 kg) of water one degree Celsius.

The brightness or intensity of a light is measured in *foot-candles*. One *foot-candle* is the intensity of light of a standard one-inch-thick candle at a distance of one foot from the flame. One of the smallest units of length is the *angstrom*. It measures the *wavelength* of light. An *angstrom* is a metric unit equivalent to $\frac{1}{100,000,000}$ of a centimeter.

A *light year* is the astronomer's unit of length. It is the distance light travels in a year (about six trillion miles).

The intensity of sound (loudness) is measured in units called *decibels*. A sound that can hardly be heard by a normal ear has an intensity of zero decibel. A very loud sound, one that usually hurts a normal ear, has an intensity of approximately 120 decibels.

Electricity involves a variety of measuring units. The rate of flow of electric charges is measured in *amperes*, the difference in the pressure of electricity between two points in a closed electrical circuit in *volts*. Resistance to the flow of current is measured in *ohms*, and electrical power in *watts*.

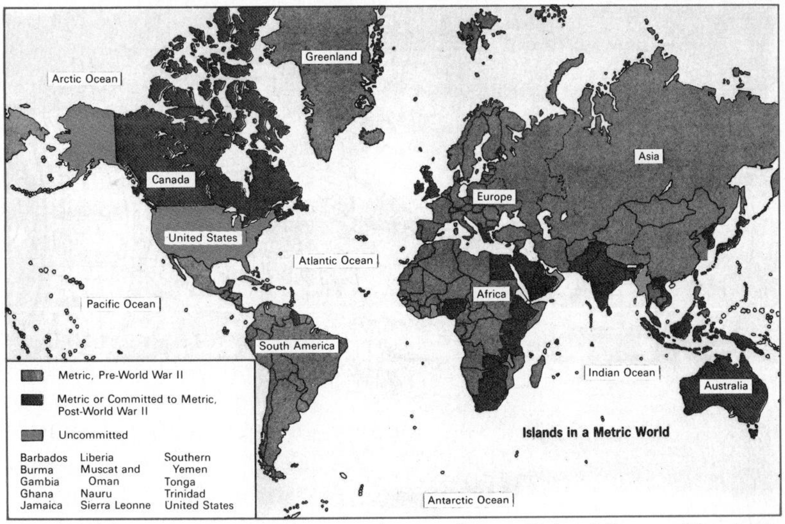

Source: National Bureau of Standards.

FIGURE 1

Why is the United States going metric after having lived with the English system for over 200 years? The answer is that the United States cannot afford not to. The Common Market adopted a directive that requires all exporters to the Common Market nations to indicate the dimensions of their products in metric units. Practically every country in the world presently uses the metric system or is in the process of converting to it. All of our major trading partners belong in one category or the other.

Figure 1 is a map of the world that shows the present status of the metric system.

As you can see from the map, only a few countries, such as Liberia, Burma, South Yemen (and the United States), are not committed to the metric system. Our neighbors to the north and south, Canada and Mexico, are "metric" countries. Canada announced a commitment to the metric system, and on April 1, 1975, began its conversion to the metric system by giving weather reports with Celsius temperature readings. Today these countries have gone completely metric. Making the conversion to metric on a timetable, that is, doing it abruptly, is known as a "hard conversion." The method that the United States is using, a slow and steady approach, is known as a "soft conversion." In 1981 President Reagan informed the American National Metric Council that he welcomed their efforts to expand metric usage, but emphasized that this was an area for private initiative, not government pressure.

Originally, the meter was not immediately welcomed by the general public for some of the same reasons that apply today. It was inconvenient for people to learn a new system. But it was gradually accepted, and by 1875 all of Europe (except England) had adopted the metric system. Throughout the history of the United States many people, including Benjamin Franklin, have recommended that we adopt the metric system. In 1790, Thomas Jefferson recommended that the United States adopt a uniform system of weights and measures that uses a decimal system. Alexander Graham Bell spoke before a Congressional committee in 1906 and stated that the metric system should be provided for the whole of the United States. On December 23, 1975, President Gerald Ford signed the Metric Conversion Act into law. This act established a national policy in support of metric measurement in the United States. Shortly after this, the Bureau of Alcohol, Tobacco, and Firearms of the Treasury Department established a policy that all distilled spirits will be sold in standard metric sizes. This took effect on January 1, 1980, and now the labels on wine and liquor bottles state metric measures only. One immediate economic benefit was to reduce the number of standardized bottles from 16 to 7. Hence, manufacturers do not have to keep on hand as many bottle sizes (inventories) to correspond with various products. The United States fastener industry is also in the process of replacing 59 customary sizes with 25 metric sizes for threaded fasteners.

We are already using the metric system in many ways, and it is in greater usage than is commonly realized. Movie cameras use 8- or 16-millimeter film, and 35 millimeters is a popular size for photographic film. Skis are sold in centimeter lengths. Most of the items that we purchase at the supermarket are now labeled in both customary and metric units. For example, a can of pepper whose net weight is 4 ounces is also labeled 113 grams, and a jar of mayonnaise contains 32 fluid ounces, 1 quart, or 0.95 liter. Soft drinks are now sold in liter, 2-liter, and 3-liter bottles.

The automobile industry has been one of the largest industries to use the metric system of measurement. Foreign cars are constructed using the metric system, and American-made autos have most parts made using the metric system. Approximately one-half of the cars being used in this country require some metric tools. Engine displacement is now being described in metric terms; for example, a car might have a 1.6-liter engine. Speedometers on new cars contain gauges that give readings in miles per hour and kilometers per hour. Some oil companies are now selling gasoline by the liter. Many other automotive items are sold using metric measurements, such as tires and spark plugs. Road signs give both metric and English information.

H. Armstrong Roberts

Skis are sold in centimeter lengths.

NOTE OF INTEREST

Japan converted to the metric system between 1951 and 1962. In fact, by 1962 the usage of the metric system was so widespread in Japan that the government established a fine of $140 for needless use of the old system of weights and measures.

Temperatures are often given in terms of degrees Celsius. Precipitation figures will be given in centimeters (or millimeters), and wind speeds will soon be given in kilometers per hour instead of miles per hour, as the National Weather Service has a conversion plan for weather reports and anticipates its use in the near future. International track events, such as the Olympic Games, are now conducted using the metric system. The 100-yard dash is changed to a slightly longer race, the 100-meter dash. Similarly, the 220-yard dash becomes the 200-meter dash, and the mile run becomes the 1500-meter run. Many American high schools and colleges now conduct their track meets using metric measurements.

Obviously, we will not stop using all of our customary units of measure. Even when the metric system is adopted, the height of horses will probably still be measured in terms of *hands* and depths at sea will still be measured in *fathoms*. Wood may still be sold by the *cord*. In a game like football, distances will probably always be given in yards; it will never be "first down, 9.14 meters to go." It is also doubtful that the Indianapolis 500-mile race will become the Indianapolis 804.7-kilometer race.

Conversion to the metric system will be made in those areas where it is advantageous to do so. Conversion to the metric system will make many computations easier; the metric system is also easier to learn. Conversion will help our economic situation in world markets because countries that already use the metric system will find American products more acceptable; this will benefit the entire population of the United States.

Dr. George Gerster/Photo Researchers, Inc.

Road signs like this one, showing both miles and kilometers, will soon be a common sight.

JUST FOR FUN The United States system of measurement uses two different miles, the statute mile and the nautical mile. Which is longer and by how much?

5.2 HISTORY OF SYSTEMS OF MEASUREMENT

Regardless of what period of time we examine, man has always developed some method for weighing and measuring things. History tells us that early man used a measuring system for making weapons, building places to live in, and even making clothing. The **cubit** is one of the first recorded units of measurement. Noah supposedly built his ark 300 cubits long, 50 cubits wide, and 30 cubits high. According to Egyptian records (4000 B.C), a cubit was the distance from the tip of the middle finger of the outstretched hand to the point of the elbow (approximately 19 inches). It is interesting to note that the side of one of the pyramids is 500 cubits long. Volume was measured by filling a container with seeds and then counting the number of seeds in the container. Stones were commonly used for determining the weight of heavy objects.

Man first used parts of his body or things that he could easily obtain, such as seeds and stones, as measuring instruments. A common brick is supposed to be a span long and one-half span wide. A **span** is the distance from the tip of the little finger to the tip of the thumb of an outstretched hand. A **palm** is the distance across the base of the four fingers that form your palm. A **digit** is the thickness or width of the middle of the middle finger, approximately three-quarters of an inch. The Romans used the idea of the digit to invent the inch. According to the Romans, an **inch** was the thickness or width of a thumb. There are no reliable facts on how the **foot** was invented.

According to most accounts, King Henry I of England decreed that a **yard** was the distance from the tip of his nose to the end of the thumb on his outstretched hand. It was by means of such royal commands that many standards of measurement were determined. For example, Queen Elizabeth I changed the measure of the mile from 5,000 feet to 5,280 feet. She did this because 1 furlong equaled 220 yards (660 feet), and if 1 mile equaled 5,280 feet, then a mile would equal 8 furlongs. Thus, a partial list of English measures about 1500 A.D. was

12 inches	= 1 foot
3 feet	= 1 yard

HISTORICAL NOTE

In 1821 John Quincy Adams finished a 4-year investigation on the metric question and published a comprehensive report for the Congress. It was the first United States Metric Study. Adams' report considered the metric system as an alternative for adoption because he believed it approached "the ideal perfection of uniformity applied to weights and measures." But the time was not right; then, most of our trade was with inch-pound England, and the metric system had not yet established itself as the international measurement system. Today the world has committed itself to the metric system, and even in the United States its use is increasing.

Barbara Hadley/courtesy of U.S. Department of Health and Human Services

5 feet	=	1 pace
125 paces	=	1 furlong
8 furlongs	=	1 mile
12 furlongs	=	1 league

England became a world power, and by means of trade and colonization, the English system became established in many parts of the world. However, the need for greater uniformity and a single, worldwide, coordinated measurement system was recognized over 300 years ago. As a result, other systems were also being developed in other countries.

Bariel Mouton, vicar of St. Paul's Church in Lyons, France, proposed in 1670 that a standard unit of length be one minute of arc of a great circle of the earth. (A great circle is any circumference of the earth; the meridians that pass through the poles of the earth are examples of great circles.) Another proposal was made by Jean Picard, a French astronomer, who proposed a unit of length that was the length of a pendulum that takes 1 second to swing back and forth. But since a pendulum swings faster at the north and south poles than it does at the equator, nothing ever became of Picard's proposal.

After Mouton's proposal, not much was done toward developing a standard unit of measurement for over 100 years. In 1790, at the request of the French government, the French Academy of Sciences devised a new system of measurement. The new basic unit of length was a portion of a meridian of the earth, similar to the unit proposed by Mouton. The new unit was called a *meter,* which was taken from the Greek word *metron,* "to measure."

Since the scientists wanted a unit similar in length to a yard, they chose a portion of the meridian that was approximately the same length. But they also wanted the unit to be part of a base 10, or decimal, system. Therefore they calculated the distance from the north pole to the equator along the meridian that runs through Dunkirk and Paris, and then took one ten-millionth ($\frac{1}{10,000,000}$) of that distance as the standard unit of measure, the **meter.** The French Academy of Sciences recommended this unit because all future calculations could be done using the decimal system. There would be no need to divide by 5,280, multiply by 16, and so on; all quantities that were larger or smaller than the meter could be converted by multiplying or dividing by 10 or powers of 10. Recall that it is quite easy to multiply or divide a number by 10: we simply move the decimal point to the right or left.

A meter is about 39.37 inches long—a little longer than a yard. By keeping this in mind, you will be able to visualize how long a meter is. It will also help to give you an idea of the size of other metric measurements. The metric unit used for determining mass (weight) is called a *gram,* and the metric unit used for determining volume is called a *liter*. We shall examine grams and liters in greater detail later in this chapter.

Before we proceed with a study of the metric system, we must first develop a familiarity with a set of prefixes used throughout the system. Listed below are the most common prefixes and their meanings.

Prefix	Symbol	Meaning
kilo	k	1000
hecto	h	100
deka	da	10
deci	d	$\frac{1}{10}$ or 0.1
centi	c	$\frac{1}{100}$ or 0.01
milli	m	$\frac{1}{1000}$ or 0.001

For example, a **kilometer** is 1000 meters and a **centimeter** is $\frac{1}{100}$ of a meter.

To help you remember these prefixes, observe that prefixes containing an *i (deci, centi, milli)* are all fractional parts of one unit. The prefix *deci* should remind you of a word involving 10, such as *decade* or *decimal. Centi* should remind you of a *century,* which is 100 years. The prefix *milli* should remind you of *millennium,* a period of 1,000 years.

The prefixes *deka, hecto,* and *kilo* are prefixes that indicate multiplication by 10, 100, and 1000 respectively. A *kilowatt* is a unit of measure used in electricity and is equivalent to 1000 watts. An easy way to remember the correct multiple for *hecto* is to notice that *hecto* and *hundred* both begin with the same letter, *h.* The prefix *deka* is similar to *deci* and should also remind you of words involving 10.

We did not list two prefixes, *micro* and *mega,* as they are not frequently used. These prefixes refer to 1 million. A *megaton* is 1 million tons. A *micrometer* is one-millionth of a meter. The prefix *micro* may remind you of *microfilm,* the very small film that libraries use to keep copies of printed matter.

Regardless of the fact that we do not yet know the meaning of gram or liter, we should be able to give an interpretation of the following terms: kilogram, decigram, hectoliter, and milliliter.

The key clues are the prefixes. The prefixes have the same meaning for all metric measures. Therefore, a kilogram is 1000 grams, whereas a decigram is $\frac{1}{10}$ of a gram or 0.1 gram. Similarly a hectoliter is 100 liters, and a milliliter is $\frac{1}{1000}$ of a liter or 0.001 liter.

EXAMPLE 1 Using the prefixes as a guide, find equivalent expressions in grams, meters, or liters for each of the following.

a. centigram b. kilometer c. deciliter d. hectometer

SOLUTION a. The prefix *centi* tells us that 1 centigram = 0.01 gram.
b. The prefix *kilo* indicates that 1 kilometer = 1000 meters.
c. The prefix *deci* indicates *ten* and the *i* tells us that it is a fractional part; that is, 1 deciliter = 0.1 liter.
d. The prefix *hecto* indicates that 1 hectometer = 100 meters.

EXAMPLE 2 What prefix can be used to indicate each number?

a. 0.001 b. 10 c. 0.1 d. 0.01

SOLUTION a. 0.001 is the same as $\frac{1}{1000}$ and is indicated by *milli.*
b. 10 is indicated by *deka.*
c. 0.1 is the same as $\frac{1}{10}$ and is indicated by *deci.*
d. 0.01 is the same as $\frac{1}{100}$ and is indicated by *centi.*

EXERCISES FOR SECTION 5.2

1. List two reasons why the United States should convert to the metric system.

2. a. Where did the metric system originate?
   b. What is the basic unit of length of the metric system?
   c. How was this basic unit of length obtained (what was it taken from)?

3. Name four units of measurement used by ancient civilizations.

4. Name the three basic units of measure in the metric system.

5. What prefix can be used to indicate each number?
   a. 10 b. 0.1 c. 1000
   d. 0.01 e. 100 f. 0.001

6. Name the prefix that means
   a. one thousand times b. one-tenth of
   c. ten times d. one-hundredth of
   e. a hundred times f. one-thousandth of

7. Given the fact that 1 meter is approximately equal to 39.37 inches, how many inches are contained in each of the following?
   a. 1 kilometer b. 1 hectometer
   c. 1 dekameter d. 1 decimeter
   e. 1 centimeter f. 1 millimeter

8. Given the fact that 1 liter is approximately equal to 1.06 liquid quarts, how many quarts are contained in each of the following?
   a. 1 kiloliter b. 1 hectoliter
   c. 1 dekaliter d. 1 deciliter
   e. 1 centiliter f. 1 milliliter

9. Complete each of the following:
   a. 1 kilometer = ______ meters
   b. 1 dekameter = ______ meters
   c. 1 decimeter = ______ meters
   d. 1 millimeter = ______ meters
   e. 1 centimeter = ______ meters
   f. 1 hectometer = ______ meters

10. Complete each of the following:
   a. 1 milligram = ______ grams
   b. 1 centigram = ______ grams
   c. 1 hectogram = ______ grams
   d. 1 kilogram = ______ grams
   e. 1 decigram = ______ grams
   f. 1 dekagram = ______ grams

11. Complete each of the following:
   a. 1 meter = ______ centimeters
   b. 1 liter = ______ milliliters
   c. 1 meter = ______ decimeters
   d. 1 gram = ______ centigrams
   e. 1 gram = ______ milligrams
   f. 1 liter = ______ deciliters

12. Complete each of the following:
   a. 1 kilometer = ______ hectometers
   b. 1 hectometer = ______ dekameters
   c. 1 dekagram = ______ decigram
   d. 1 kiloliter = ______ dekaliters
   e. 1 deciliter = ______ liters
   f. 1 kilogram = ______ hectograms

13. Complete each of the following:
   a. 2000 meters = ______ kilometers
   b. 32 kilograms = ______ grams
   c. 3 grams = ______ milligrams
   d. 40 dekaliters = ______ liters
   e. 3 hectoliters = ______ liters
   f. 5 kilometers = ______ meters

14. Complete each of the following:
   a. 1000 grams = ______ kilograms
   b. 1000 grams = ______ hectograms
   c. 20 liters = ______ dekaliters
   d. 50 liters = ______ hectoliters
   e. 40 meters = ______ dekameters
   f. 20 kilometers = ______ meters

15. Determine whether each statement is true or false.
   a. A liter is larger than a kiloliter.
   b. A liter is smaller than a milliliter.
   c. There are 500 milligrams in a gram.
   d. There are 100 grams in a kilogram.
   e. A centimeter is one-hundredth of a meter.

16. Determine whether each statement is true or false.
   a. There are 1000 liters in a milliliter.
   b. There are 100 liters in a centiliter.
   c. A gram is larger than a milligram.
   d. A gram is smaller than a kilogram.
   e. A kilometer is one-thousandth of meter.

JUST FOR FUN

Can you unscramble these 10 words to make the correct metric units of measurement?

1. ETMER
2. MAGR
3. TERIL
4. AGKDMRAE
5. OLTERILKI
6. ICERMTEED
7. OLMAGRKI
8. RTLLIEIILM
9. NITREMECTE
10. OERLMTEKI

5.3 LENGTH AND AREA

The basic unit of length in the metric system is the meter. A meter is slightly longer than a yard; 1 meter $\approx$ 39.37 inches. (Recall that the symbol $\approx$ means approximately equal.) The symbol for meter is a lowercase m. Most likely there is a *meter stick* in your classroom. If there is, examine it in order to get an idea of how long 1 meter is.

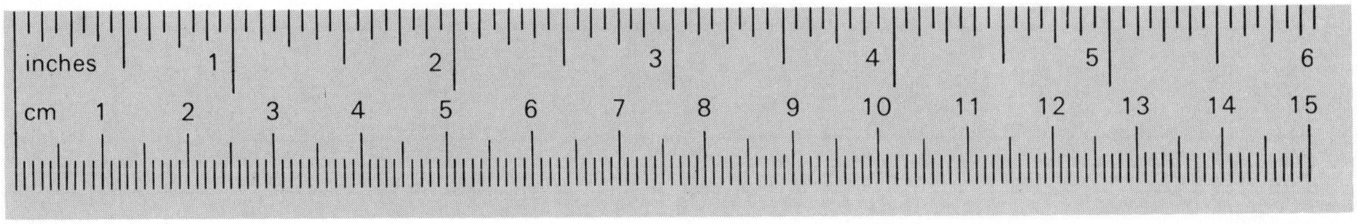

FIGURE 2

Figure 2 shows a ruler. Note that one edge is marked in inches and the other edge is marked in centimeters. A centimeter is 0.01 meter; that is, 1 meter = 100 centimeters. Each small division on the metric edge of the ruler is 1 millimeter; 10 millimeters = 1 centimeter. The symbol for centimeters is *cm,* and the symbol for millimeter is *mm.*

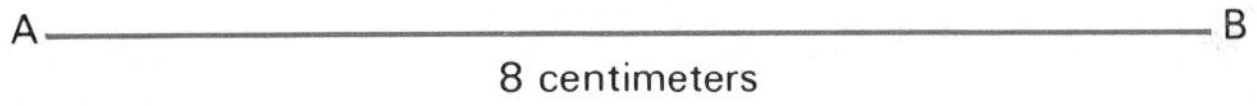

FIGURE 3

Suppose we want to find the length of line segment AB in Figure 3. Using inches, segment AB is approximately $3\frac{3}{16}$ inches long. But if you measure it using the metric edge, you will find that its length is 8 centimeters. This illustrates another reason why the metric system is favored over our customary units of measurement: the metric system eliminates fractions such as $\frac{1}{4}$, $\frac{3}{8}$, $\frac{5}{16}$, and $\frac{1}{32}$. Granted, a line segment that is 2 inches long (see Fig. 4) is easy to measure in terms of inches, but it is also not difficult to measure

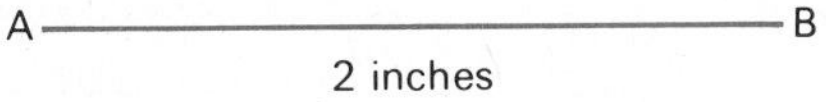

FIGURE 4

using the metric system. A line segment 2 inches long is a little longer than 5 centimeters; in fact, it is 51 millimeters in length. Note that we do not have to use a fraction to express this length.

EXAMPLE 1 Measure each of the given line segments to the nearest millimeter (mm).

a. ____________________________

b. _______

c. ________________

d. ___________________________

SOLUTION a. 55 mm b. 15 mm c. 32 mm d. 50 mm

Although the meter is the basic unit of measurement in the metric system, many lengths (and thicknesses) are measured in terms of parts of a meter, such as decimeter (dm), centimeter (cm), and millimeter (mm). In measuring greater lengths and distances, we can use the dekameter and hectometer, but the kilometer is the most commonly used unit for longer lengths. For example, all of the races in the Olympic Games are described in meters—100-meter dash, 200-meter dash, 400-meter dash, 800-meter run, 1500-meter run, 5000-meter run, and so on—but the distance between two cities is measured in kilometers. The symbol for kilometer is *km*. The distance from Los Angeles to New York is 4690 kilometers; from Dallas to Chicago, 1506 kilometers.

The following table summarizes the units related to the meter. Note that the first part of the symbol indicates the prefix and the second part (m) indicates *meter*.

Symbol	Word	Meaning
km	kilometer	1000 meters
hm	hectometer	100 meters
dam	dekameter	10 meters
m	meter	1 meter
dm	decimeter	0.1 meter
cm	centimeter	0.01 meter
mm	millimeter	0.001 meter

Now that we are somewhat familiar with the metric prefixes, we can use the handy chart below to find equivalent measures of length.

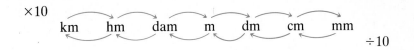

$\times 10$ km hm dam m dm cm mm $\div 10$

The chart gives a technique for changing from one metric length measure to another. To change from one unit to another, for each step to the right, multiply by 10 to obtain units that are smaller. You can also think of this as moving the decimal point one place to the right for each step to the right. For each step to the left, divide by 10 to obtain fewer units that are larger. You can also think of this as moving the decimal point one place to the left for each step to the left.

Suppose we want to convert 1.5 kilometers to dekameters. That is 1.5 km = ______ dam. Using the chart we start at kilometers and move two places to the right to obtain dekameters. Hence, we multiply by 10×10 (100) or simply move the decimal point two places to the right. That is,

$$1.5 \text{ km} = 150. \text{ dam}$$

To convert 42.7 decimeters to hectometers, that is, 42.7 dm = ______ hm, we start at decimeters and move three places to the left to obtain hectometers. Therefore, we divide by 10 for each step to the left or in other words move the decimal point three places to the left. That is,

$$42.7 \text{ dm} = 0.0427 \text{ hm}$$

EXAMPLE 2 Complete each of the following.

a. 22 m = ______ cm b. 31.4 mm = ______ cm
c. 15 mm = ______ dm d. 4.1 m = ______ mm
e. 4.2 km = ______ m f. 77 dm = ______ hm

SOLUTION a. Using the chart in this section, we start at meters and move two places to the right to obtain centimeters. Hence, we move the decimal point two places to the right. That is

$$22 \text{ m} = 2200. \text{ cm}$$

Note: If a decimal point is not indicated, it is understood to be immediately to the right of the last digit in the given whole number.

b. Using the chart, we start at millimeters and move one place to the left to centimeters. Hence, we move the decimal point one place to the left. That is,

$$31.4 \text{ mm} = 3.14 \text{ cm}$$

c. 15 mm = 0.15 dm   d. 4.1 m = 4100 mm

e. 4.2 km = 4200 m   f. 77 dm = 0.077 hm

How long is a kilometer? From the prefix *kilo,* we know that 1 kilometer = 1000 meters. But does that give you any idea how long it is? Let's compare it with something that is familiar. How does it compare with a mile? A meter is 39.37 inches long; therefore, a kilometer is 39,370 inches long. A mile contains 5280 feet and 1 foot = 12 inches. Therefore, 1 mile = 5280 × 12 = 63,360 inches. A mile is longer than a kilometer. In fact, a mile is approximately 1.6 kilometers.

As another illustration, a football field is 120 yards long (including the end zones); therefore a football field is 120 × 36 = 4,320 inches long. Nine football fields have a total length of 38,880 inches. Because a kilometer is 39,370 inches in length, a kilometer is 490 inches longer than nine football fields placed end to end. A kilometer is approximately 0.6 mile in length.

Keeping in mind that 1 mile ≈1.6 kilometers, we can also discuss speed in terms of kilometers per hour. For example, if a cyclist is pedaling his bike at the rate of 5 miles per hour (mi/hr), his rate of speed is 8 kilometers per hour (km/hr): because 1 mile = 1.6 kilometers, we have 5 × 1.6 = 8 kilometers per hour. If an automobile travels at the rate of 40 miles per hour, then it is also traveling at 64 (40 × 1.6) kilometers per hour.

If a person never drives faster than 90 kilometers per hour, how fast would this be in miles per hour? In order to convert kilometers to miles, we simply multiply the number of kilometers by 0.6 (1 kilometer ≈ 0.6 mile). Therefore, 90 kilometers per hour is approximately the same as 90 × 0.6 = 54 miles per hour. (*Note:* Although most of our conversion factors are approximate, we will use the equals sign for convenience.) Also, we shall use Table 1 in this section to obtain the conversion factors.

EXAMPLE 3 Convert each given measurement to the indicated measurement.

a. 10 mi = ______ km   b. 25 mi = ______ km

c. 150 km = ______ mi   d. 25 km = ______ mi

SOLUTION (USE TABLE 1) a. Since 1 mi = 1.6 km, 10 mi = 10 × 1.6 = 16 km. (*Note:* A better approximation of 1 mile is 1.61 kilometers, but we shall use 1 mi = 1.6 km.)

b. 25 mi = 25 × 1.6 = 40 km
c. Because 1 km = 0.6 mi, 150 km = 150 × 0.6 = 90 mi
d. 25 km = 25 × 0.6 = 15 mi

EXAMPLE 4 Convert each speedometer reading to the indicated measurement.

a. 30 mi/hr = ______ km/hr b. 45 mi/hr = ______ km/hr
c. 100 km/hr = ______ mi/hr d. 120 km/hr = ______ mi/hr

SOLUTION (USE TABLE 1) This example is similar to Example 3. To convert miles to kilometers, multiply the number of miles by 1.6. To convert kilometers to miles, multiply the numbers of kilometers by 0.6.

a. 30 mi/hr = 30 × 1.6 = 48 km/hr
b. 45 mi/hr = 45 × 1.6 = 72 km/hr
c. 100 km/hr = 100 × 0.6 = 60 mi/hr
d. 120 km/hr = 120 × 0.6 = 72 mi/hr

NOTE OF INTEREST

Because the meter is the basis of the metric system, this meter length was marked on a platinum rod. The rod was located in Paris, France and kept in a chamber at a constant temperature and pressure to keep the rod from expanding or contracting. In 1960 this procedure was changed; it was found that 1,650,763.73 wavelengths of the orange-red light of the heated gas krypton equaled a metric meter. Today the unit of length corresponding to 1 meter is defined as a length equal to the distance traveled by light in a vacuum during a time interval of 1/299,792,458 of a second. Now anyone with a properly equipped science laboratory can check the length of the meter—it is no longer necessary to go to France to check against the international standard.

Area is measured in square units. The floor of a room that measures 8 feet by 10 feet has an area of 80 square feet. This means that there are 80 squares, each measuring 1 foot by 1 foot, that will cover the surface of the floor. Area is sometimes referred to as *surface area.*

A square whose measurements are 1 centimeter by 1 centimeter is said to have an area of 1 square centimeter; this is denoted by 1 cm^2 (see Fig. 5). The square centimeter is used to find the area of relatively small regions, such as the area of this page. Larger regions are measured in terms of square meters (m^2). Since 1 meter = 100 centimeters, a square meter contains 10 000 square centimeters (100 × 100 = 10 000). (*Note:* We use a space instead of a comma in numbers of 10,000 or more when discussing metric measurements.)

The area of very large regions is measured in *hectares* (10 000 square meters). Land that is measured in terms of acres can also be measured in hectares. Since 1 hectare = 10 000 square meters, a hectare is the area of a square that measures 100 meters on each side. It is highly unlikely that the hectare will be used in measuring the area of land when the United States converts to the metric system. Land will probably continue to be sold in terms of acres for two reasons: (1) land cannot be shipped overseas, that is, we will not export land to metric countries as we do machinery and other products; and (2) it would be impractical to change all of the property deeds in the United States so that the area would be in terms of hectares. However, the other metric measures of area, such as the square centimeter and the square meter, will be

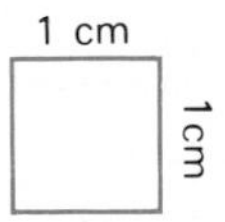

FIGURE 5
Area = 1 cm^2.

TABLE 1
METRIC CONVERSION FACTORS—LENGTH

Symbol	When You Know	Multiply by	To Find	Symbol
LENGTH		*To Metric*		
in	inches	2.5	centimeters	cm
ft	feet	30	centimeters	cm
yd	yards	0.9	meters	m
mi	miles	1.6	kilometers	km
		From Metric		
mm	millimeters	0.04	inches	in
cm	centimeters	0.4	inches	in
m	meters	3.3	feet	ft
m	meters	1.1	yards	yd
km	kilometers	0.6	miles	mi
AREA		*To Metric*		
in^2	square inches	6.5	square centimeters	cm^2
ft^2	square feet	0.09	square meters	m^2
yd^2	square yards	0.8	square meters	m^2
mi^2	square miles	2.6	square kilometers	km^2
	acres	0.4	hectares	ha
		From Metric		
cm^2	square centimeters	0.16	square inches	in^2
m^2	square meters	1.2	square yards	yd^2
km^2	square kilometers	0.4	square miles	mi^2
ha	hectares	2.5	acres	

used. Another metric area measure is the *are* (pronounced *air*). It is a square measuring 10 meters on each side; therefore its area is 100 square meters. *Note:* 100 ares = 1 hectare.

You may find the metric conversion factors in Table 1 helpful. *Remember, these are all approximate conversions.*

EXERCISES FOR SECTION 5.3

1. Measure each of the given line segments to the nearest millimeter.
   a. ______________
   b. ______________________________
   c. __________________
   d. ____________________________________
   e. ________________________
   f. _________________

2. Complete each of the following.
   a. 4.2 m = ______ mm
   b. 51 cm = ______ mm
   c. 15 mm = ______ cm
   d. 13.2 km = ______ m
   e. 40.3 dam = ______ m
   f. 36 m = ______ cm

3. Complete each of the following.
   a. 18 cm = ______ dm
   b. 37 dam = ______ dm
   c. 3.2 km = ______ cm
   d. 702 km = ______ dam
   e. 423 cm = ______ hm
   f. 8.14 cm = ______ mm

4. Complete each of the following.
   a. 43 dm = ______ dam
   b. 213.4 cm = ______ m
   c. 333 cm = ______ km
   d. 43.9 hm = ______ km
   e. 714 mm = ______ cm
   f. 0.7 m = ______ cm

5. Complete each of the following.
   a. 413.7 cm = ______ dm
   b. 0.45 cm = ______ mm
   c. 4.7 dam = ______ cm
   d. 4378 cm = ______ km
   e. 30.2 dam = ______ km
   f. 985 mm = ______ cm

6. If the distance from Miami to Atlanta is 1070 kilometers, how many miles is it?

7. If the distance from Seattle to New Orleans is 2,625 miles, how many kilometers is it?

8. If the distance from Mexico City to Chicago is 2,082.5 miles, how many kilometers is it?

9. If the distance from Los Angeles to New York City is 4690 kilometers, how many miles is it?

10. The speed limit in New York State is 55 miles per hour. What is the speed limit in New York State in kilometers per hour?

11. The speed limit in a certain town is 75 kilometers per hour. If radar records Larry's speed as 50 miles per hour, should he get a ticket?

12. The distance from the earth to the sun is approximately 93 million miles. How many kilometers is it?

13. Mercury is the closest planet to the sun; its distance from the sun is 58 million kilometers. How many miles is it?

14. Find the perimeter of each polygon below. Your answer should be in terms of the indicated unit.

a.

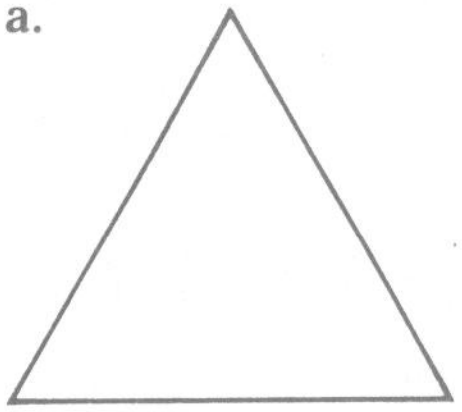

______ mm

b.

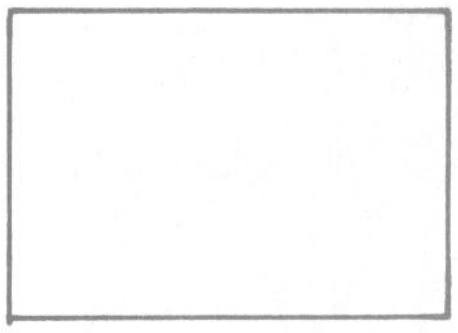

______ cm

c.

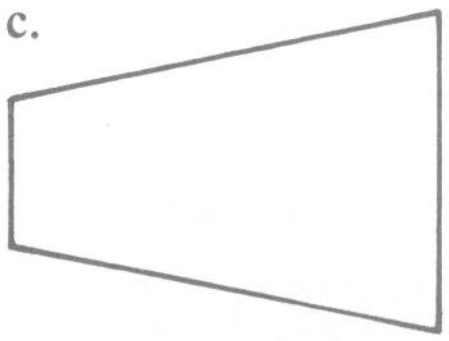

______ mm

d.

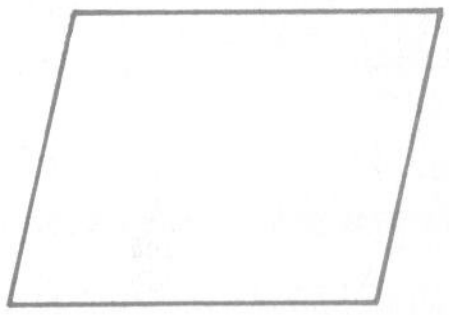

______ cm

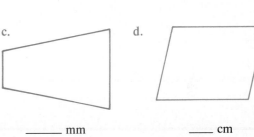

15. A square that measures 2 centimeters by 2 centimeters has what area?

16. A hectare is the area of a square that measures ______ meters on each side.

17. One mile is approximately 1.6 kilometers. Approximately how many square kilometers are in 1 square mile?

18. Which is the longer measurement in each case?
   a. 1 inch or 1 centimeter
   b. 1 yard or 1 meter
   c. 1 mile or 1 kilometer
   d. 6 inches or 13 centimeters
   e. 6 feet or 2 meters
   f. 1 foot or 35 centimeters

19. Which measurement of area is greater in each case?
   a. 1 square inch or 1 square centimeter
   b. 1 square foot or 1 square meter
   c. 1 square yard or 1 square meter
   d. 1 square mile or 1 square kilometer
   e. 1 acre or 1 hectare
   f. 1 square meter or 1 are

For Exercises 20–25, choose the most sensible answer.

20. Width of a newspaper:
a. 0.38 m b. 3.8 m c. 0.38 km

21. Length of a shovel:
a. 0.6 m b. 1.6 m c. 2.6 m

22. Height of a bowling pin:
a. 0.39 mm b. 3.9 cm c. 0.39 m

23. Length of a new pencil:
a. 19 mm b. 19 cm c. 30 cm

24. Length of a watch band:
a. 20 cm b. 30 mm c. 30 cm

25. Length of a shopping trip:
a. 5 mm b. 5 cm c. 5 km

26. What are your metric measurements?
a. height: feet ______ inches ______; meters ______ centimeters ______
b. waist: inches ______; centimeters ______
c. neck: inches ______; centimeters ______
d. wrist: inches ______; centimeters ______
e. biceps: inches ______; centimeters ______
f. foot length: inches ______; centimeters ______

JUST FOR FUN

Given the following equivalences, how many inches are contained in a distance that measures 2 miles 3 furlongs 4 rods 5 yards 2 feet?

12 inches = 1 foot
3 feet = 1 yard
$5\frac{1}{2}$ yards = 1 rod
40 rods = 1 furlong
8 furlongs = 1 statute mile

5.4 VOLUME

Volume is the measure of how much a container can hold, that is, its capacity. Unfortunately, our system of measuring volumes and weights is quite confusing. For example, some soft-drink bottles

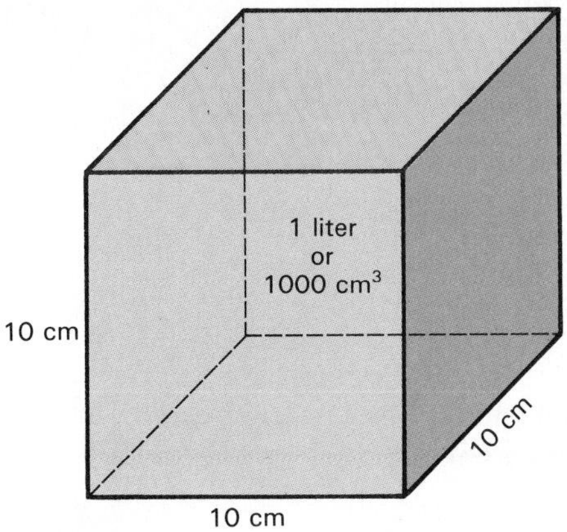

FIGURE 6

contain 16 ounces, and some cans of coffee contain 16 ounces, but these are two different kinds of ounces. The soft-drink bottle contains 16 **fluid ounces,** which is equivalent to 1 pint. The 16 ounces in the can of coffee are units of weight, equivalent to 1 pound.

In the metric system, the *liter* is the basic unit used to measure capacity. A **liter** is defined as the volume of a cubic decimeter. In other words, a liter is the capacity of a cube (box) that is 1 decimeter long, 1 decimeter wide, and 1 decimeter high (see Fig. 6). One liter, 1 cubic decimeter, and 1000 cubic centimeters all represent the same volume.

One of the reasons that volume is easier to work with in the metric system than in our present system is that one set of units is used for all volume measures, whether liquid or dry. Using our present system, the volume of a pint of blueberries is different from the volume of a pint of cream. But in the metric system, cream is sold in liter containers and blueberries are sold in liter boxes.

We shall compare a liter to a unit of volume with which you are already familiar, a quart (Fig. 7). A liter contains a little more than a quart. One liter is approximately 1.06 liquid quarts, and one liquid quart is approximately 0.95 liter.

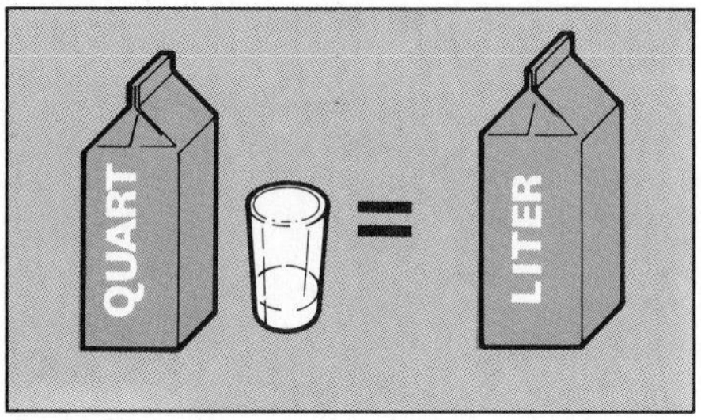

FIGURE 7
One liter of milk contains slightly more than a quart of milk.

Two other common units of volume measure in the metric system are the **milliliter** and the **cubic meter.** Recall that the prefix *milli* means $\frac{1}{1000}$ or 0.001; hence, a milliliter is $\frac{1}{1000}$ of a liter. Because a liter contains 1000 cubic centimeters, a milliliter is 1 cubic centimeter. We can think of a milliliter as a cube whose length, width, and height each measure 1 centimeter. The milliliter and the liter are the two most commonly used units of volume.

The milliliter is a small unit of volume measure (Fig. 8). For example, 5 milliliters = 1 teaspoon, and 15 milliliters = 1 tablespoon. Most liquid prescriptions obtained at pharmacies are sold in milliliters.

A cubic meter (m^3) is used to measure large volumes. One cubic meter is equivalent to 1.3 cubic yards. Items such as large amounts of sand and concrete are sold by the cubic yard at present. In the metric system, they would be sold by the cubic meter. Extremely large quantities of liquids would be measured in terms of the **kiloliter,** which is equivalent to 1000 liters. The capacity of fuel oil trucks and gasoline trucks, for example, would be expressed in kiloliters.

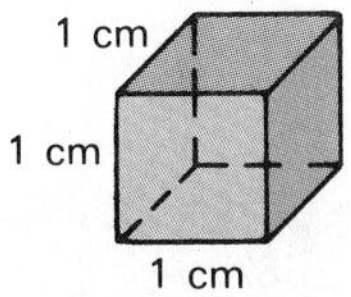

FIGURE 8
One milliliter (1 cubic centimeter).

Remember that a liter is defined as the volume of a cube that is 10 centimeters long, 10 centimeters wide, and 10 centimeters high. The volume of 1 liter is 1000 cubic centimeters. The following table lists metric volume measures and their equivalent measure in liters.

Symbol	Word	Meaning
kL	kiloliter	1000 liters
hL	hectoliter	100 liters
daL	dekaliter	10 liters
L	liter	1 liter
dL	deciliter	0.1 liter
cL	centiliter	0.01 liter
mL	milliliter	0.001 liter

We can use a conversion chart, similar to the one in the previous section, to find equivalent measures of volume.

$\times$ 10

kL hL daL L dL cL mL

$\div$ 10

Recall that for each step to the right we multiply by 10, or move the decimal point one place to the right. For each step to the left, divide by 10, or move the decimal point one place to the left.

EXAMPLE 1 Complete each of the following:

a. 642 mL = ____ L b. 74.3 L = ____ cL
c. 5 kL = ____ daL d. 15 L = ____ kL
e. 4000 mL = ____ L f. 253 L = ____ hL

SOLUTION a. To change milliliters to liters, move the decimal point three places to the left. That is,

$$642 \text{ mL} = 0.642 \text{ L}$$

b. To change liters to centiliters, move the decimal point two places to the right. That is,

$$74.3 \text{ L} = 7430 \text{ cL}$$

c. 5 kL = 500 daL d. 15 L = 0.015 kL

e. 4000 mL = 4.000 L = 4.0 L f. 253 L = 2.53 hL

You may find the metric conversion factors in Table 2 helpful. *Remember, these are all approximate conversions.* It should be noted that some texts list a cup as equivalent to 250 milliliters, but according to the National Bureau of Standards, 1 cup = 0.24 liter, which is 240 milliliters.

NOTE OF INTEREST

How much does a barrel contain? Several different sized barrels have been established by law or usage. For example, the taxes on fermented liquors are based on a barrel of 31 gallons. But some states recognize a liquid barrel as 31.5 gallons. Federal law recognizes a 40-gallon barrel for "proof spirits." A barrel of crude oil or petroleum products contains 42 gallons. Hence, a barrel may contain from 31 to 42 gallons depending upon where it is used and how it is used.

TABLE 2
METRIC CONVERSION FACTORS—VOLUME

Symbol	When You Know	Multiply by	To Find	Symbol
		To Metric		
tsp	teaspoons	5	milliliters	mL
tbsp	tablespoons	15	milliliters	mL
fl oz	fluid ounces	30	milliliters	mL
c	cups	0.24	liters	L
pt	pints	0.47	liters	L
qt	quarts	0.95	liters	L
gal	gallons	3.8	liters	L
ft^3	cubic feet	0.03	cubic meters	m^3
yd^3	cubic yards	0.76	cubic meters	m^3
		From Metric		
mL	milliliters	0.03	fluid ounces	fl oz
L	liters	2.1	pints	pt
L	liters	1.06	quarts	qt
L	liters	0.26	gallons	gal
m^3	cubic meters	35	cubic feet	ft^3
m^3	cubic meters	1.3	cubic yards	yd^3

One of the areas where the metric system can provide some help is in the kitchen. For example, a cake recipe that calls for

$2\frac{1}{4}$ cups of flour	$\frac{1}{3}$ cup of shortening
$1\frac{1}{2}$ cups of sugar	1 cup of milk
3 teaspoons of baking powder	1 tablespoon of flavoring
1 teaspoon of salt	2 eggs

would call for the following in the metric system.

540 milliliters of flour	80 milliliters of shortening
360 milliliters of sugar	240 milliliters of milk
15 milliliters of baking powder	15 milliliters of flavoring
5 milliliters of salt	2 eggs

Note that in the first recipe we have to use fractions, as well as different units of measure. However, in the metric recipe, everything (except the eggs) is measured in the same unit, milliliters. If the metric recipe did call for fractional amounts, they would indeed be in decimal notation, because the metric system only uses powers of 10.

Using the metric conversions in Table 2, we can convert any

recipe to a metric recipe. Since 1 cup = 0.24 liter = 240 milliliters, $2\frac{1}{4}$ cups = $2\frac{1}{4} \times 240$ = 540 milliliters. The other conversions are done in a similar manner, using Table 2.

EXAMPLE 2 Convert the given recipe to a metric recipe.

Blueberry Pie Filling

$\frac{1}{2}$ cup of sugar	$2\frac{1}{2}$ tablespoons of flour
1 tablespoon of lemon juice	$\frac{1}{2}$ teaspoon of cinnamon
$2\frac{1}{3}$ cups of drained blueberries	1 tablespoon of shortening
$\frac{1}{3}$ cup of blueberry juice	

SOLUTION $\frac{1}{2}$ cup of sugar = $\frac{1}{2} \times 240$ = 120 milliliters of sugar
1 tablespoon of lemon juice = 15 milliliters of lemon juice
$2\frac{1}{3}$ cups of blueberries = $2\frac{1}{3} \times 240$ = 560 milliliters of blueberries
$\frac{1}{3}$ cup of blueberry juice = $\frac{1}{3} \times 240$ = 80 milliliters of blueberry juice
$2\frac{1}{2}$ tablespoons of flour = $2\frac{1}{2} \times 15$ = 37.5 milliliters of flour
$\frac{1}{2}$ teaspoon of cinnamon = $\frac{1}{2} \times 5$ = 2.5 milliliters of cinnamon
1 tablespoon of shortening = 15 milliliters of shortening

Remember that in the metric system one type of unit, the liter or some multiple of the liter, is used to measure both liquid and dry volume. The liter and the milliliter are the two most commonly used units of volume. One liter is defined to be the volume of a cube that is 10 centimeters long, 10 centimeters wide, and 10 centimeters high. A milliliter is the volume of a cube that is 1 centimeter long, 1 centimeter wide, and 1 centimeter high.

EXAMPLE 3 How many liters will an aquarium hold if it is 70 centimeters long, 0.5 meter wide, and 500 millimeters high?

SOLUTION We must convert all measurements to the same unit before finding the volume. Recall that volume = length × width × height. Because we want to express the answer in terms of liters, we shall convert all measurements to centimeters because the volume of 1 liter is 1000 cubic centimeters.

length = 70 cm width = 0.5 m = 50 cm
height = 500 mm = 50 cm

Therefore,

$V = l \times w \times h$

$V = 70 \times 50 \times 50$

$V = 175\ 000\ \text{cm}^3$ *Note:* $1\ \text{L} = 1000\ \text{cm}^3$

$V = 175\ \text{L}$

It should be noted that in the United States a common abbreviation for cubic centimeters is cc, as opposed to the metric notation cm^3. This is particularly true in medicine and other areas where cubic centimeters has been used as a unit of measurement. But because 1 cubic centimeter is the same as 1 milliliter, the milliliter usage is becoming more and more common, particularly with regard to prescriptions.

EXERCISES FOR SECTION 5.4

1. Complete each of the following:
   a. 12 hL = ___ L b. 25 dL = ___ mL
   c. 3500 liters = ___ kL d. 100 mL = ___ cL
   e. 650 cL = ___ L f. 45 hL = ___ daL

2. Complete each of the following:
   a. 33 liters = ___ mL b. 560 mL = ___ L
   c. 38 kL = ___ L d. 4kL = ___ hL
   e. 25 daL = ___ hL f. 58 dL = ___ daL

3. Complete each of the following:
   a. 73.2 mL = ___ dL b. 2314 mL = ___ L
   c. 3.14 kL = ___ daL d. 14.95 mL = ___ L
   e. 0.49 daL = ___ cL f. 7.2 liters = ___ mL

4. Complete each of the following:
   a. 432 mL = ___ L b. 31.7 daL = ___ L
   c. 44 liters = ___ mL d. 6.4 L = ___ mL
   e. 73.4 kL = ___ daL f. 0.7 kL = ___ L

5. Arrange the following measurements of volume in descending order beginning with the largest: dekaliter, liter, hectoliter, milliliter, kiloliter, deciliter, centiliter.

6. Which has the greatest volume in each pair?
   a. 1 qt or 1 liter b. 1 gal or 3 L
   c. 2 pt or 1 liter d. 1 tsp or 2 mL
   e. 1 c or 1 liter f. 2 fl oz or 20 mL

7. Which has the greatest volume in each pair?
   a. 2 gal or 2 kL b. 3 qt or 3 liters
   c. 3 pt or 50 cL d. 2 c or 2 liters
   e. 3 tsp or 3 mL f. 5 fl oz or 5 mL

8. Find the volume of a box that is 1 meter long, 40 centimeters wide, and 50 centimeters high. (*Hint:* Convert all measurements to the same unit before finding the volume, where volume = length × width × height. Express your answer in liters.)

9. Find the volume of a box that is 80 centimeters long, 0.5 meter wide, and 50 millimeters high. Express your answer in liters.

10. How much water will an aquarium hold if it is 1 meter long, 60 centimeters wide, and 600 millimeters high? Express your answer in cubic meters.

11. What is the storage capacity of a food freezer whose inside measurements are 1.5 meters by 1 meter by 80 centimeters? Express your answer in cubic meters.

12. Convert the given recipe to a metric recipe.

Clam Chowder

1 teaspoon salt	$\frac{1}{2}$ cup water
$\frac{1}{4}$ cup butter	1 pint minced clams
2 cups milk	2 cups diced potatoes
$\frac{1}{4}$ cup minced onions	

13. Convert the given recipe to a metric recipe.

Rice Pudding

$\frac{1}{2}$ cup uncooked rice	$\frac{1}{3}$ cup seedless raisins
$2\frac{1}{2}$ cups milk	$\frac{1}{2}$ tablespoon cinnamon
$\frac{1}{4}$ cup sugar	$\frac{1}{2}$ teaspoon salt

For Exercises 14–20, choose the most sensible answer.

14. Automobile fuel tank:
a. 800 L b. 8 kL c. 80 L

15. Bottle of soda pop:
a. 36 mL b. 360 mL c. 36 L

16. Can of paint:
a. 4 mL b. 4 cL c. 4 L

17. Coffee cup:
a. 20 mL b. 25 L c. 250 mL

18. Measuring cups:
a. 5 mL b. 50 mL c. 500 mL

19. Soupspoon:
a. 15 cL b. 15 dL c. 15 mL

20. Teakettle:
a. 1 L b. 10 L c. 100 L

JUST FOR FUN

In the late 1880s, liquid measure was also known as wine measure because it was used to measure liquors and wines. Given the following equivalences, how many gills are contained in 1 hogshead 1 barrel 20 gallons 3 quarts 1 pint?

4 gills	= 1 pint
2 pints	= 1 quart
4 quarts	= 1 gallon
$31\frac{1}{2}$ gallons	= 1 barrel
2 barrels	= 1 hogshead

DARLING, LET'S JUST SPLIT A LITTLE HOGSHEAD OF CHAMPAGNE.

5.5 MASS (WEIGHT)

Weight is a measure of the earth's gravitational pull. (Gravity is the force that holds you on earth.) **Mass** is the measure of the amount of matter—that is, atoms and molecules—that objects are made of. In space, the mass of an object does not change, but its weight does. *Weight* and *mass* are not the same thing, but on earth the mass of an object is always proportional to the weight of the object. Therefore, for this course, we shall assume that weight and mass mean the same thing.

In the metric system, the most common measures of weight (mass) are the kilogram (kg), the gram (g), and milligram (mg). The basic unit of mass in the metric system is the gram. The weight of a common paperclip is approximately 1 gram, whereas a nickel weighs 5 grams (Fig. 9).

FIGURE 9
The weight of one nickel is 5 grams.

Imagine constructing a leakproof cubic centimeter out of weightless material (see Fig. 8 in Sec. 5.4) and filling it with very cold water. The mass (weight) of the water in such a container is 1 gram. Recall that one of the advantages of the metric system is that all of the measures (distance, volume, weight) are related. The mass (weight) of 1 milliliter of water is 1 gram, and a milliliter is the volume of a cube whose length, width, and height each measure 1 centimeter.

Now we can list some metric weight measures and their equivalent measure in grams.

Symbol	Word	Meaning
kg	kilogram	1000 grams
hg	hectogram	100 grams
dag	dekagram	10 grams
g	gram	1 gram
dg	decigram	0.1 gram
cg	centigram	0.01 gram
mg	milligram	0.001 gram

We can use a conversion chart, similar to those in the previous sections, to find equivalent measures of weight (mass).

× 10

kg hg dag g dg cg mg

÷ 10

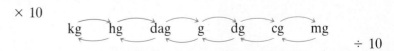

For each step to the right, we multiply by 10, or move the decimal point one place to the right. For each step to the left, divide by 10, or move the decimal point one place to the left.

EXAMPLE 1 Complete each of the following:

a. 3.5 kg = ____ g b. 2000 g = ____ hg
c. 250 mg = ____ g d. 3.7 g = ____ dg
e. 4.5 dg = ____ mg f. 380 mg = ____ dag

SOLUTION a. To change kilograms to grams, move the decimal point three places to the right. That is

$$3.5 \text{ kg} = 3500 \text{ g}$$

b. To change grams to hectograms, move the decimal point two places to the left. That is

$$2000 \text{ g} = 20.00 \text{ hg} = 20 \text{ hg}$$

c. 250 mg = 0.250 g d. 3.7 g = 37 dg
e. 4.5 dg = 450 mg f. 380 mg = 0.0380 dag

A kilogram is equivalent to 1000 grams. Therefore 1000 milliliters—that is, 1 liter—filled with very cold water will approximate the weight of 1 kilogram. One pound is approximately 0.45 kilogram and 1 kilogram is approximately 2.2 pounds. Larger foodstuffs such as meats and fish are weighed in terms of kilograms. A piece of meat that weighs 1 kilogram is ample for four people.

A milligram is equivalent to 0.001 gram. It is an extremely small unit of weight and is used only in medical prescriptions and some areas of science. For example, if you examine the label on a bottle of cold tablets, you will note that the amounts of different substances that each tablet contains are given in terms of milligrams. A common cold tablet contains, among other things, 225 milligrams of aspirin, 30 milligrams of caffeine, and 50 milligrams of ascorbic acid.

The weights of objects that are quite heavy, such as automobiles, are given in *metric tonnes.* One **metric tonne** is equivalent to 1000 kilograms. Because 1 kilogram is approximately 2.2 pounds, a metric tonne (1000 kilograms) is equivalent to 2,200 pounds, or 1.1 tons (1 ton = 2,000 pounds).

It is interesting to note that 1 gram is the weight of 1 milliliter of water, 1 kilogram is the weight of 1 liter of water, and a metric tonne is the weight of 1 cubic meter of water.

In order to develop a sense of the metric weights discussed, remember that a common paperclip weighs approximately 1 gram and a nickel weighs about 5 grams. A kilogram is equivalent to 2.2 pounds, so a man who weighs 220 pounds weighs 100 kilograms.

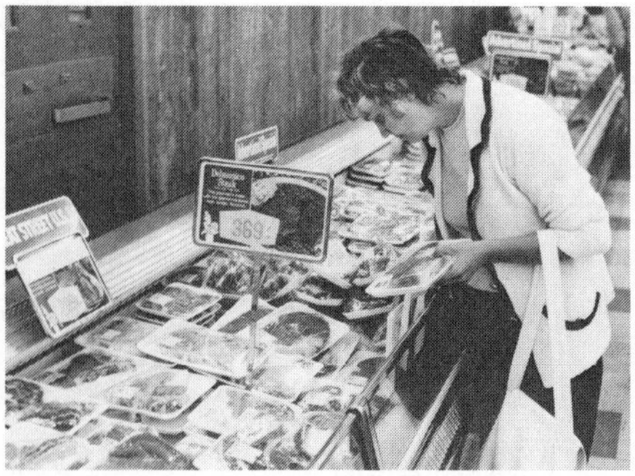

Mark Antman/The Image Works

How many kilograms does a 2-pound steak weigh?

TABLE 3
METRIC CONVERSION FACTORS—MASS (Weight)

Symbol	When You Know	Multiply by	To Find	Symbol
To Metric				
oz	ounces	28	grams	g
lb	pounds	0.45	kilograms	kg
T	tons	0.9	tonnes	t
From Metric				
g	grams	0.035	ounces	oz
kg	kilograms	2.2	pounds	lb
t	tonnes	1.1	tons	T

A milligram ($\frac{1}{1000}$ of a gram) is a minute quantity and is difficult to approximate. A small grain of sand weighs about 1 milligram. The metric tonne is used in measuring the weight of very heavy objects. A metric tonne (1000 kilograms) is equivalent to 2,200 pounds.

You may find the metric conversion factors in Table 3 helpful. *These are all approximate conversions.*

EXAMPLE 2 Recall that 1 gram is the mass (weight) of 1 milliliter of water whose temperature is approximately 39° Fahrenheit. Give the mass (weight) in grams of

a. 1 liter of water b. 1 cubic meter of water

SOLUTION a. The volume of 1 liter is 1000 cubic centimeters. A milliliter is the volume of 1 cubic centimeter. Therefore, 1 liter contains 1000 milliliters and 1 liter of water weighs 1000 grams.

As an alternate solution note that a milliliter is $\frac{1}{1000}$ of a liter (recall the meaning of the prefix *milli*). Therefore, 1 liter contains 1000 milliliters and 1 liter of water weighs 1000 grams.

b. A meter contains 100 centimeters, so a cubic meter contains 1 000 000 cubic centimeters. Thus, a cubic meter of water contains 1 000 000 milliliters of water and weighs 1 000 000 grams. Note that $1\ 000\ 000 = 1000 \times 1000$; therefore, 1 000 000 grams is the same as 1000 kilograms. But 1000 kilograms is equivalent to 1 metric tonne. The weight of 1 cubic meter of very cold water is 1 metric tonne (1 Mg).

EXAMPLE 3 Convert each of the following to the indicated weight.

a. 16 oz = ____ g b. 10 lb = ____ kg
c. 50 kg = ____ lb d. 100 g = ____ oz
e. 4,000 lb = ____ t f. 180 lb = ____ kg

SOLUTION

a. In order to convert ounces to grams, we must multiply the number of ounces by 28. Therefore 16 oz = 16 × 28 = 448 g.
b. One pound is equivalent to 0.45 kilogram. Therefore 10 lb = 10 × 0.45 = 4.5 kg.
c. One kilogram is equivalent to 2.2 pounds. Therefore 50 kg = 50 × 2.2 = 110 lb.
d. One gram is equivalent to 0.035 ounce. Therefore 100 g = 100 × 0.035 = 3.5 oz.
e. 2,000 pounds = 1 ton, so 4,000 pounds = 2 tons. Using Table 3, 1 T = 0.9 t and 4,000 lb = 2 T = 2 × 0.9 = 1.8 t.
f. One pound is equivalent to 0.45 kilogram. Therefore, 180 lb = 180 × 0.45 = 81 kg.

The metric system is more consistent than our customary system of weights and measures. In the metric system, length, volume, and weight are directly related to each other. For example, the volume of a cube that has length, width, and height of 1 centimeter (0.01 meter) is called a milliliter, and if we fill the milliliter with very cold water, it will have a weight of 1 gram.

A liter is the volume of a cube that has a length, width, and height of 10 centimeters. If we fill the liter with very cold water, it will have a weight of 1 kilogram. Similarly, the volume of a cube that has length, width, and height of 1 meter is called a kiloliter, and if we fill the kiloliter with very cold water, it will have a weight of 1,000 kilograms or 1 tonne.

EXERCISES FOR SECTION 5.5

1. Complete each of the following:
   a. 52 g = ____ mg b. 3.5 kg = ____ g
   c. 250 mg = ____ g d. 4.3 dg = ____ g
   e. 7.2 hg = ____ dg f. 4.7 g = ____ mg

2. Complete each of the following:
   a. 4200 mg = ____ g b. 5 dag = ____ g
   c. 60 cg = ____ g d. 10.7 mg = ____ cg
   e. 36.7 dag = ____ hg f. 7.3 kg = ____ mg

3. Complete each of the following:
   a. 417 mg = ____ g b. 342 kg = ____ g
   c. 14.3 kg = ____ g d. 885.9 mg = ____ g
   e. 14 359 mg = ____ kg
   f. 2.71 kg = ____ g

4. Complete each of the following:
   a. 483 g = ____ mg b. 432 g = ____ kg
   c. 813.2 g = ____ kg d. 487.3 g = ____ mg
   e. 819 dg = ____ dag f. 498 dag = ____ dg

5. Convert each of the following to the indicated weight:
   a. 8 oz = ____ g b. 100 kg = ____ lb
   c. 500 g = ____ oz d. 10 T = ____ t
   e. 200 lb = ____ kg f. 6,000 lb = ____ t

6. Convert each of the following to the indicated weight:
   a. 4 lb = ____ kg b. 10 kg = ____ lb
   c. 100 g = ____ oz d. 32 oz = ____ g
   e. 10 mg = ____ oz f. 5,000 lb = ____ t

7. A certain vitamin tablet contains 150 mg of vitamin A, 225 mg of vitamin B, and 125 mg of vitamin C. What is the weight of the vitamin tablet in grams?

8. A certain antacid tablet contains 248 mg of aluminum hydroxide, 75 mg of magnesium hydroxide, and 41 mg of sodium. What is the weight of the antacid tablet in grams?

9. A container is 60 centimeters long, 40 centimeters wide, and 50 centimeters high. It weighs 3 kilograms when it is empty. What will it weigh when it is filled with cold water? Express your answer in kilograms.

10. How many liters will an aquarium hold if it is 1 meter long, 60 centimeters wide, and 60 centimeters high? What is the mass of the water in kilograms if it is approximately 39° Fahrenheit?

11. A swimming pool is 20 meters long and 9 meters wide. It has a uniform depth of 2 meters. How many kiloliters of water will it hold? What is the mass of the water if it is approximately 39° Fahrenheit?

12. Assuming that each of the following is filled with very cold water, give the mass (in grams) of
    a. 2 liters b. 1 dekaliter
    c. 5 milliliters d. 2 kiloliters
    e. 3 hectoliters f. 5 centiliters

13. Which has the greater weight?
    a. 1 lb or 1 kg b. 2 oz or 20 g
    c. 10 lb or 5 kg d. 4,000 lb or 3 t
    e. 40 oz or 2 kg f. 280 g or 9 oz

For Exercises 14–19, choose the most sensible answer.

14. Mass of a liter of wine:
    a. 10g b. 100 g c. 1000 g

15. Mass of a newborn baby:
    a. 3 kg b. 30 kg c. 300 kg

16. Mass of a grain of sand:
    a. 1 kg b. 1 g c. 1 mg

17. Mass of an adult person:
    a. 7.5 kg b. 75 kg c. 750 kg

18. Mass of a nickel:
    a. 5 g b. 5 mg c. 5 kg

19. Mass of a paperclip:
    a. 1 mg b. 1 g c. 1 dag

JUST FOR FUN

In the late 1800s, apothecaries (pharmacists) used the standard apothecaries' weight to make up different medicines. Given the following equivalences, how many grains are contained in 2 pounds 3 ounces 2 drams 2 scruples?

20 grains = 1 scruple
3 scruples = 1 dram
8 drams = 1 ounce
12 ounces = 1 pound

5.6 TEMPERATURE

If the temperature is 30 °C, it's hot!

One type of measure with which we are very familiar is temperature. All of us are concerned about weather on any given day, and one of the first questions we ask is "What's the temperature?" Wherever we go, there are time-temperature clocks that indicate how warm or how cold it is. It is even possible to dial a number and hear a recorded message stating the current time and temperature.

Gabriel Fahrenheit (1686–1736) used a brine solution (salt, water, and ice) to devise a scale for the mercury thermometer such that the boiling point of water was 212° and the freezing point of water was 32°. Zero degrees was the lowest point on the thermometer because that was the coldest temperature he could get with the brine solution. However, there are colder temperatures. A temperature of −81° was once recorded in Canada, −90° in Russia, and −127° in Antarctica.

Shortly after Fahrenheit developed his scale, Anders Celsius (1701–1744), a Swedish astronomer, developed another scale. Celsius developed his scale so that the boiling point of water was 100° and the freezing point of water was 0°. You may be familiar with the Celsius thermometer as the centigrade thermometer. Recall that *centi* means $\frac{1}{100}$ or 0.01; there are 100 intervals between 0° centigrade and 100° centigrade. Because the scale that Celsius developed was so convenient, the centigrade thermometer was adopted by scientists, and because 100 is a power of 10 ($100 = 10^2$), all of the countries using the metric system have also adopted it. This thermometer was formerly known as the centigrade thermometer, but, in honor of the man who developed the scale, it is now officially known as the Celsius thermometer, and each unit is called a degree Celsius (°C).

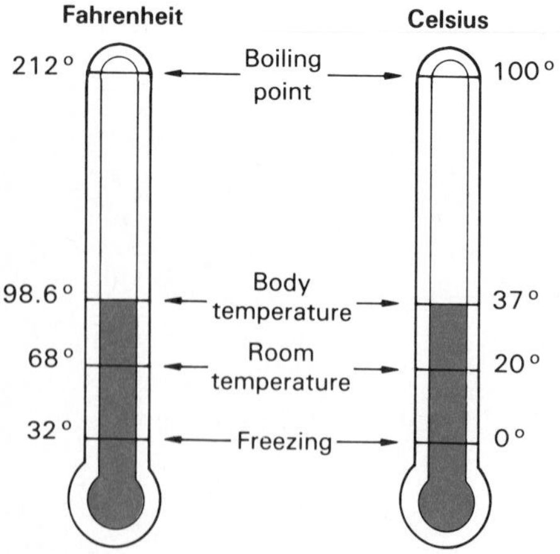

FIGURE 10

Pictured in Figure 10 are a Celsius thermometer and a Fahrenheit (°F) thermometer with indicated temperatures.

As shown in Figure 10, the boiling point of water is 212 °F or 100 °C, and the freezing point of water is 32 °F or 0 °C. Most of us will be able to remember these comparative temperatures on the Celsius thermometer. But we would like to be able to interpret other Celsius thermometer readings as well. When someone tells us that their temperature is 37 °C, we should be aware that this is normal, or 98.6 °F. Note also that Figure 10 indicates that a room temperature of 68 °F is the same as a room temperature of 20 °C. It is obviously important to identify a degree measurement as degrees Celsius or degrees Fahrenheit.

The following two formulas will enable you to convert from Fahrenheit to Celsius and vice versa. When you know the Fahrenheit temperature and wish to convert to Celsius, you

HISTORICAL NOTE

Gabriel Fahrenheit believed that the lowest temperature that could be reached was the temperature of a mixture of ice and salt. He immersed the end of his thermometer in such a mixture. When the mercury would go no lower, he marked the place on the thermometer and called it zero.

Fahrenheit supposedly put the end of the thermometer in his assistant's mouth and when the mercury stopped rising, he called this point 100°. The space between 0° and 100° was then divided into 100 units. A freezing mark of 32° was obtained because that is where the mercury stopped when the thermometer was put into freezing water. Using the same scaling technique, Fahrenheit discovered that 212° was the temperature of boiling water.

NOTE OF INTEREST

This chapter has provided an introduction to the metric system. Following is a list of prefixes that provide the multiples in the International System.

Prefix	Symbol	Multiple	Equivalent
exa	E	10^{18}	quintillion
peta	P	10^{15}	quadrillion
tera	T	10^{12}	trillion
giga	G	10^{9}	billion
mega	M	10^{6}	million
kilo	k	10^{3}	thousand
hecto	h	10^{2}	hundred
deka	da	10	ten
deci	d	10^{-1}	tenth
centi	c	10^{-2}	hundredth
milli	m	10^{-3}	thousandth
micro	μ	10^{-6}	millionth
nano	n	10^{-9}	billionth
pico	P	10^{-12}	trillionth
femto	f	10^{-15}	quadrillionth
atto	a	10^{-18}	quintillionth

should use the formula

$$C = \frac{5}{9}(F - 32) \qquad (C = °\text{C}; F = °\text{F})$$

When you know the Celsius temperature and wish to convert to Fahrenheit, you should use

$$F = \frac{9}{5}C + 32$$

Let's work out a problem using the first formula. Suppose we wish to find the temperature on the Celsius thermometer equivalent to 32 °F. From our previous discussion, you should know that the answer will be 0 °C. Using the formula,

$$C = \frac{5}{9}(F - 32)$$

$$C = \frac{5}{9}(32 - 32) \qquad \text{substituting 32 for } F$$

$$C = \frac{5}{9}(0) \qquad \text{subtracting 32 from 32}$$

$$C = 0 \qquad \text{multiplying by 0}$$

Therefore, 32 °F = 0 °C.

Let's try another example using the same formula. Suppose the given temperature is 212 °F and we wish to find the equivalent temperature on the Celsius thermometer.

$$C = \frac{5}{9}(F - 32)$$

$$C = \frac{5}{9}(212 - 32) \qquad \text{substituting 212 for } F$$

$$C = \frac{5}{9}(180) \qquad \text{subtracting 32 from 212}$$

$$C = \frac{900}{9} \qquad \text{multiplying 5 times 180}$$

$$C = 100 \qquad \text{dividing 900 by 9}$$

Therefore, 212 °F = 100 °C.

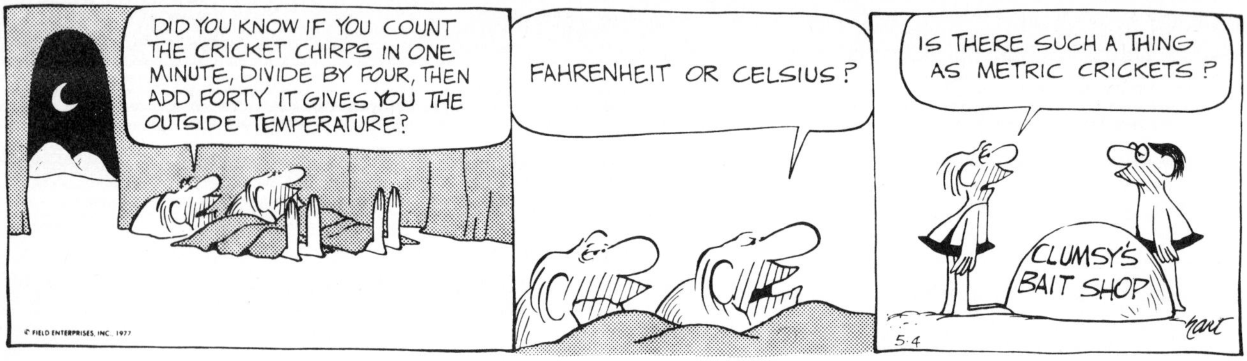

B.C. by permission of Johnny Hart and Creators Syndicate, Inc.

Suppose the given Celsius temperature is 100 °C and we wish to find the equivalent temperature on the Fahrenheit thermometer. We use the formula $F = \frac{9}{5}C + 32$:

$F = \frac{9}{5}(100) + 32$	substituting 100 for C
$F = \frac{900}{5} + 32$	multiplying 9 times 100
$F = 180 + 32$	dividing 900 by 5
$F = 212$	adding 180 + 32

Therefore, 100 °C = 212 °F.

EXAMPLE 1 Most people set their thermostats at 68 °F. Find the equivalent temperature on the Celsius thermometer.

SOLUTION In order to convert Fahrenheit to Celsius, we use the formula $C = \frac{5}{9}(F - 32)$.

$C = \frac{5}{9}(68 - 32)$	substituting 68 for F
$C = \frac{5}{9}(36)$	subtracting 32 from 68

$$C = \frac{180}{9} \quad \text{multiplying 5 times 36}$$

$$C = 20 \quad \text{dividing 180 by 9}$$

Therefore, 68 °F = 20 °C.

EXAMPLE 2 Usually a person in good health has a body temperature of 98.6 °F. Find the equivalent temperature on the Celsius thermometer.

$$C = \frac{5}{9}(F - 32)$$

$$C = \frac{5}{9}(98.6 - 32) \quad \text{substituting 98.6 for } F$$

$$C = \frac{5}{9}(66.6) \quad \text{subtracting 32 from 98.6}$$

$$C = \frac{333.0}{9} \quad \text{multiplying 5 times 66.6}$$

$$C = 37 \quad \text{dividing 333 by 9}$$

Therefore, 98.6 °F = 37 °C.

EXAMPLE 3 A recipe in a metric cookbook calls for an oven setting of 175 °C. Find the equivalent temperature on the Fahrenheit thermometer.

SOLUTION In order to convert Celsius to Fahrenheit, we use the formula $F = \frac{9}{5}C + 32$.

$$F = \frac{9}{5}(175) + 32 \quad \text{substituting 175 for C}$$

$$F = \frac{1575}{5} + 32 \quad \text{multiplying 9 times 175}$$

$$F = 315 + 32 \quad \text{dividing 1575 by 5}$$

$$F = 347 \quad \text{adding 315 and 32}$$

Therefore, 175 °C = 347 °F. (*Note:* We could also divide 175 by 5, which equals 35, and then multiply 35 by 9, which also equals 315 and brings us to the last step.)

EXAMPLE 4 Convert 37 °C to Fahrenheit.

SOLUTION

$F = \frac{9}{5}(37) + 32$	substituting 37 for C
$F = \frac{333}{5} + 32$	multiplying 9 times 37
$F = 66.6 + 32$	dividing 333 by 5
$F = 98.6$	adding 66.6 and 32

Therefore, 37 °C = 98.6 °F.

Beside being able to convert from Fahrenheit to Celsius and vice versa, you should develop a "feel" for temperatures that are given in terms of the Celsius scale. For example, a temperature of 30 °C is a hot day in most parts of the country and a temperature of 40 °C is a scorcher anywhere. Similarly, if someone has a body temperature of 40 °C then that person is very sick, and if his temperature rises to 41 °C, then he is near death.

The following may help you to remember what it is like when you are given certain Celsius temperatures.

Thirty makes it hot.
Twenty makes it nice.
Ten gives us a cool spot.
Zero gives us ice.

EXERCISES FOR SECTION 5.6

1. Convert each Fahrenheit temperature to Celsius.
   a. 104 °F b. 50 °F c. 41 °F
   d. 95 °F e. 86 °F f. 23 °F

2. Convert each Fahrenheit temperature to Celsius.
   a. 59 °F b. 77 °F c. 5 °F
   d. 113 °F e. 32 °F f. 14 °F

3. Convert each Celsius temperature to Fahrenheit.
   a. 20 °C b. 50 °C c. 85 °C
   d. 10 °C e. 0 °C f. 65 °C

4. Convert each Celsius temperature to Fahrenheit.
   a. 95 °C b. 30 °C c. 5 °C
   d. 15 °C e. 25 °C f. −5 °C

5. Convert each temperature to the indicated scale.
   a. 14 °F = _____ °C b. 15 °C = _____ °F
   c. 25 °C = _____ °F d. 86 °F = _____ °C

6. Convert each temperature to the indicated scale.
   a. 95 °F = _____ °C b. 35 °C = _____ °F
   c. −10 °C = _____ °F d. −4 °F = _____ °C

7. A sick person has a high fever if his temperature is 40 °C. Find the equivalent temperature on the Fahrenheit thermometer.

8. A recipe in a metric cookbook calls for an oven setting of 200 °C. Find the equivalent temperature on the Fahrenheit thermometer.

9. Iron will melt at 5432 °F. Find the equivalent temperature on the Celsius thermometer.

10. The highest temperature ever recorded in the state of Ohio is 113 °F. Find the equivalent temperature on the Celsius thermometer.

11. The lowest temperature ever recorded in the state of South Dakota is −58 °F. Find the equivalent temperature on the Celsius thermometer.

12. The lowest temperature ever recorded in the United States is −80 °F at Prospect Creek, Alaska. Find the equivalent temperature (to the nearest whole degree) on the Celsius thermometer.

13. A temperature of 136 °F is generally accepted as the world's highest temperature ever recorded under standard conditions. This occurred in Northern Africa in 1922. Find the equivalent temperature (to the nearest whole degree) on the Celsius thermometer.

14. In Montana, the temperature once dropped from 44 °F to −56 °F in a period of 24 hours. This is a change in temperature of 100 °F. Approximately how much of a change is it on the Celsius scale? (Think carefully!)

15. The highest temperature ever recorded in the United States is 134 °F, recorded in Death Valley. Find the equivalent temperature (to the nearest whole degree) on the Celsius thermometer.

For Exercises 16–22, choose the most sensible answer.

16. Normal body temperature:
a. 98 °C b. 37 °C c. 45 °C

17. Room temperature:
a. 70 °C b. 20 °C c. 30 °C

18. Temperature of freezing rain:
a. 32 °C b. 0 °C c. 10 °C

19. Temperature of melting wax:
a. 100 °C b. 212 °C c. 80 °C

20. Temperature of a frozen TV dinner:
a. 0 °C b. −10 °C c. 32 °C

21. Temperature of an oven when set for baking:
a. 100 °C b. 200 °C c. 300 °C

22. Ice-skating weather:
a. −5 °C b. 0 °C c. 5 °C

JUST FOR FUN

Replace the usual measures (underlined) in the given expressions with the appropriate metric measures.

a. It's third down and there are <u>10 yards</u> to go.
b. Give them an <u>inch</u> and they will take a <u>mile</u>.
c. The cowboy wore a 10-<u>gallon</u> hat.
d. Did you see the <u>foot</u>note?
e. A miss is as good as a <u>mile</u>.
f. He climbed the rope <u>inch</u> by <u>inch</u>.
g. An <u>ounce</u> of prevention is worth a <u>pound</u> of cure.
h. Have you read <u>"Fahrenheit 451"</u>?
i. Erskine Caldwell wrote "God's Little <u>Acre</u>."
j. A <u>pint's</u> a <u>pound</u> the world around.

Summary

The question is not *if* the United States will convert to the metric system, but *when*. Practically every country in the world uses the metric system or is presently converting to the metric system. The United States has not yet made an official act to convert to the metric system, but it is slowly inching its way in that direction.

The advantages of the metric system compared to our customary system of weights and measures are many. The metric system is a well-planned, logical system, based on powers of 10. Once we have learned the meaning of the various prefixes, the system becomes quite simple. The common prefixes in the metric system are:

Prefix	Symbol	Meaning
kilo	k	1000
hecto	h	100
deka	da	10
deci	d	$\frac{1}{10}$ or 0.1
centi	c	$\frac{1}{100}$ or 0.01
milli	m	$\frac{1}{1000}$ or 0.001

There are other reasons for converting to the metric system. The United States will probably be able to export more manufactured goods if the products are made according to metric specifications. Countries that are already metric (most of them are) give preference to metric products. Also, the metric system would eliminate many of the different kinds of measurements that we now use. For example, consider shoe sizes: In our current system there are babies' shoe sizes, children's shoe sizes, men's shoe sizes, and women's shoe sizes. In the metric system, people's feet are measured in centimeters.

You should know that a kilogram is 1000 grams, and a milliliter is 0.001 liter. For reference, remember that a meter is a little longer than 1 yard, a liter is a little more than 1 quart, and a gram is approximately the weight of a paperclip; also, a nickel weighs about 5 grams.

A room temperature of 20 °C is the same as 68 °F. Other important Celsius temperatures to keep in mind are: 0 °C = 32 °F (freezing) and 100 °C = 212 °F (boiling).

We can make metric–metric conversions by noting the prefixes and using the chart below:

× 10

kilo hecto deka base unit deci centi milli

÷ 10

The chart illustrates a technique for changing from one metric measure to another. For each step to the right, we multiply by 10 or move the decimal point one place to the right. For each step to the left, divide by 10, or move the decimal point one place to the left. Table 3 in the Appendix is a table of metric conversion factors. It is a guide for converting to and from metric measurements. Remember, these are all approximate conversions.

For more information (free) on the metric system, write to

Office of Metric Programs
U.S. Department of Commerce
Room 4082 Hoover Building
Washington, DC 20230

Vocabulary Check

Celsius	metric ton	meter	kilo
deka	second	milli	linear
gram	centi	deci	metric system
kilogram	foot	decimal system	SI
liter	hecto	going metric	

Review Exercises for Chapter 5

1. Describe the basic characteristics of the metric system and list some of its advantages.

2. Name the prefix that indicates each of the following.
   a. 1000 b. 10 c. 0.1
   d. 100 e. 0.01 f. 0.001

3. How was the length of a meter determined?

4. Complete each of the following.
   a. 10 dam = ______ m
   b. 20 hm = ______ km
   c. 40 mm = ______ cm
   d. 35 m = ______ dm
   e. 32 km = ______ m
   f. 15 m = ______ dam

5. If 1 meter = 39.37 inches, then how many inches are contained in each of the following?
   a. 1 kilometer b. 1 millimeter
   c. 1 hectometer d. 1 centimeter
   e. 1 dekameter f. 1 decimeter

6. If the distance from New York City to Seattle is 4672 kilometers, approximately how many miles is it?

7. If the distance from New Orleans to Montreal is 1,583 miles, approximately how many kilometers is it?

8. Convert each of the following to the indicated unit.
   a. 150 cm = ______ in. b. 300 ft = ______ m
   c. 5 mi = ______ km d. 10 m = ______ yd
   e. 5 m^2 = ______ yd^2 f. 10 $in.^2$ = ______ cm^2

9. How is the volume of a liter determined?

10. Complete each of the following.
    a. 2 L = ______ mL b. 3 hL = ______ daL
    c. 5 kL = ______ hL d. 100 cL = ______ L
    e. 20 cL = ______ mL f. 200 dL = ______ daL

11. Find the volume of a box (in liters) that is 1 meter long, 30 centimeters wide, and 30 centimeters high.

12. How many liters of water will an aquarium hold if it is 80 centimeters long, 0.5 meter wide, and 600 millimeters high?

13. How is the weight of a gram determined?

14. Complete each of following.
    a. 4 kg = ____ g b. 8 g = ____ mg
    c. 30 dag = ____ hg d. 20 cg = ____ mg
    e. 40 hg = ____ kg f. 18 g = ____ cg

15. Assuming that each of the following containers is filled with very cold water, give the mass (in grams) of
    a. 1 liter b. 1 cubic meter
    c. 1 centiliter d. 1 hectoliter
    e. 45 milliliters f. 50 liters

16. The lowest temperature ever recorded was −127 °F, recorded in Antarctica. Find the equivalent temperature (to the nearest degree) on the Celsius thermometer.

17. The highest temperature ever recorded was 58 °C, recorded in Africa. Find the equivalent temperature (to the nearest degree) on the Fahrenheit thermometer.

For Exercises 18–25, choose the best answer.

18. The height of a basketball player 6 feet 6 inches tall is approximately
    a. 1 km b. 100 cm c. 2 m d. 3000 mm

19. The weight of a football player weighing 200 pounds is approximately
    a. 1 t b. 1000 g c. 90 kg d. 100 kg

20. Five gallons of apple cider is approximately equal to
    a. 30 L b. 19 L c. 100 cL d. 5 L

21. On a very hot August day in Phoenix, Arizona, the temperature is likely to be
    a. 60 °C b. 20 °C c. 25 °C d. 40 °C

22. The height of a flagpole is likely to be
    a. 10 m b. 100 m c. 1000 m

23. The distance from home plate to first base on a baseball field is approximately
a. 27 cm b. 2.7 km c. 27 m

24. The area of a floor tile is likely to be
a. 9 m^2 b. 90 mm^2 c. 900 cm^2

25. If the weather forecast calls for sunny skies with a high of 20 °C, you may plan on
a. swimming b. hiking c. ice skating

26. If the temperature outside is 35 °C, what should you wear?
a. a heavy sweater
b. a ski parka
c. a T-shirt

27. If you purchased 10 kilograms of groceries at the supermarket, in which of the following should you carry the groceries?
a. a shopping cart
b. a small bag
c. a trailer

28. What would you be most likely to do if you purchased 400 milliliters of soda pop at the store?
a. Call your friends to have a party.
b. Drink it and satisfy your thirst.
c. Store all the cases in the garage.

29. If your home was located 1 kilometer from your college, how would you get home?
a. Take a bus.
b. Take a plane.
c. Walk.

30. Convert each of the following to the indicated unit.
a. 2 tsp = ____ mL b. 1 qt = ____ L
c. 100 liters = ____ qt d. 2 oz = ____ g
e. 200 lb = ____ kg f. 95 °F = ____ °C

31. Convert the given recipe to a metric recipe.

Sour Cream Cookies

$\frac{1}{2}$ cup sour cream	$\frac{1}{2}$ pound butter
1 teaspoon vanilla	1 teaspoon baking soda
1 cup brown sugar	$2\frac{1}{2}$ cups flour
2 eggs	Bake at 350 °F

Chapter Quiz

Indicate whether each statement is true or false.

1. A dekagram is one-tenth of a gram.
2. A liter is larger than a quart.
3. One hundred degrees Celsius is the boiling point of water.
4. A yard is longer than a meter.
5. A kilometer is shorter than a mile.
6. A liter contains 500 milliliters.
7. The abbreviation for dekagram is dg.
8. Water freezes at 32 °C.
9. A newborn baby weighs about 30 kg.
10. Normal body temperature is 40 °C.
11. Metric conversion is voluntary in the United States.
12. A centimeter is equivalent to 100 meters.
13. A hectare is a measure of volume.
14. A liter is defined as the volume of a cube that is 10 cm long, 10 cm wide, and 10 cm high.
15. The liter is used to measure both liquid and dry volume.
16. A milligram is equivalent to 1000 grams.
17. A common paperclip weighs approximately 1 gram.
18. It is most likely to snow at a temperature of 25 °C.

19. The height of a basketball player is likely to be 200 cm.

20. The thickness of a dime is about 1 mm.

21. A gallon is larger than 3 liters.

22. A hectare is the area of a square that measures 100 meters on each side.

23. The volume of 1 liter is 1000 cubic centimeters, or 1000 milliliters.

24. A small grain of sand weighs about 1 milligram.

25. A liter is the volume of a cube that has a length, width, and height of 100 centimeters.

JUST FOR FUN

Can you find at least 14 metric terms hidden in this puzzle?

A	K	E	I	B	C	U	B	I	C	M	E	T	E	R
M	I	L	L	I	G	R	A	M	A	B	C	D	R	E
I	L	O	U	L	A	E	I	O	U	H	F	M	G	T
L	O	C	E	L	S	I	U	S	H	E	I	R	J	E
L	L	E	K	L	M	N	O	P	Q	C	E	T	R	M
I	I	N	J	S	T	U	S	C	O	T	T	R	V	A
M	T	T	O	N	N	E	W	X	E	A	Y	I	Z	K
E	E	I	E	A	G	R	A	M	M	R	M	C	C	E
T	R	M	B	C	D	E	O	A	F	E	G	H	I	D
E	J	E	K	L	M	L	R	D	S	H	T	A	M	E
R	O	T	P	Q	I	G	I	R	N	E	T	E	S	R
T	U	E	V	K	A	W	X	T	Y	A	T	A	R	A
Z	T	R	E	K	L	F	O	U	E	B	L	E	M	U
O	H	T	E	N	O	A	D	D	I	R	E	S	K	Q
T	A	D	C	U	B	I	C	M	E	T	E	R	I	S

CHAPTER 6

MATHEMATICAL SYSTEMS

AFTER STUDYING THIS CHAPTER YOU WILL BE ABLE TO DO THE FOLLOWING:

1. Add, subtract, and multiply in the 12-hour clock system.
2. Identify the basic parts of a **mathematical system.**
3. Determine whether a set is **closed** with respect to a given operation.
4. Determine whether an **identity element** exists for a given operation, and whether each element in the set has an **inverse** for a given operation.
5. Determine whether a set is **associative** with respect to a given operation, and whether the set is **commutative** with respect to a given operation.
6. Determine whether a mathematical system is a **group.**
7. Add, subtract, and multiply in other **modular systems.**
8. Evaluate problems in an abstract system, given a table that defines an operation for the elements in the system, and determine whether the properties of a group are satisfied by the system.
9. Identify the basic parts of an **axiomatic system.**
10. Construct a **diagram** (model) for which all of the axioms of a system are satisfied, and prove a **theorem** given the undefined terms, defined terms, and axioms for the system.

6.1 INTRODUCTION

In this chapter, we shall examine mathematical systems and their properties. That is, we shall study the nature and structure of mathematical systems. Regardless of what area of mathematics is examined (sets, logic, etc.), there are certain basic common characteristics that these topics possess. For the present time, we shall think of a **mathematical system** as a set of elements together with one or more operations (rules) for combining elements of the set. We shall expand upon this idea later in the chapter, but for now this concept of a system is sufficient.

One of the first mathematical systems to which we are exposed in school is the set of counting numbers {1, 2, 3, . . . } together with the operation of addition. This system is considered to be an infinite system, because there are an infinite number of elements in the set. We shall begin our study of mathematical systems by examining a finite system, one that has some unusual properties.

6.2 CLOCK ARITHMETIC

Clock arithmetic is an example of a finite mathematical system that will enable us to understand the nature and structure of mathematical systems.

Consider the following addition problems.

$$1 + 2 = 3 \qquad 5 + 7 = 12 \qquad 9 + 10 = 7$$
$$3 + 4 = 7 \qquad 5 + 8 = 1 \qquad 11 + 11 = 10$$
$$5 + 6 = 11 \qquad 6 + 12 = 6 \qquad 9 + 9 = 6$$

Each addition problem listed is correct; there are no mistakes. The reason all of these examples are correct is that they come from a system called clock arithmetic. The first four examples in the list look exactly like examples from ordinary arithmetic, but $5 + 8 = 1$ and $9 + 9 = 6$ do not. In clock arithmetic, $5 + 8 = 1$ because 1:00 comes 8 hours after 5:00. Similarly, 6:00 comes 9 hours after 9:00, so $9 + 9 = 6$. It also follows that, if it is 6:00 now, then 12 hours from now it will be 6:00, so $6 + 12 = 6$. Hence, we see that our mathematical system, clock arithmetic, has a set of elements, {1, 2, 3, 4, 5, 6, 7, 8, 9, 10, 11, 12}, the numerals 1 through 12 on the face of a clock, and it also has an operation (addition) which consists of counting hours in a clockwise direction.

FIGURE 1

Using the clock face in Figure 1, we can see that 9 + 6 = 3. We start at 9 and count 6 units (hours) in a clockwise direction. We complete the counting at 3. Therefore, 9 + 6 = 3 in clock arithmetic. Any of the examples listed earlier can be figured out in this manner.

Using this technique, we can also verify that 6 + 12 = 6, 9 + 10 = 7, 11 + 11 = 10, and 9 + 9 = 6. In fact, we can construct a table of addition facts for a 12-hour clock using the set of elements {1, 2, 3, 4, 5, 6, 7, 8, 9, 10, 11, 12} and the operation of addition. (See Table 1).

Table 1 gives us the answer when we add any two numbers on a 12-hour clock. All answers are included, since we have combined each element in the set with every other element in the set. This underscores the fact that we are working with a finite mathematical system. Every answer in the table is a member of the original set {1, 2, 3, 4, 5, 6, 7, 8, 9, 10, 11, 12}, so there are no new elements in the set of answers. This is a special characteristic for some systems and is called **closure.**

A system is said to be *closed* with respect to an operation (in this case addition) if, when we operate on any two elements in the system, the result is also an element in the system. In the case of addition of clock numbers, when we add any two clock numbers, the sum is also a clock number.

More formally, we can say that

A system consisting of a set of elements $\{a, b, c, \ldots\}$ and an operation $*$ is *closed* if for any two elements a and b in the set, $a * b$ (read "a operation b") is also a member of the set.

TABLE 1

+	1	2	3	4	5	6	7	8	9	10	11	12
1	2	3	4	5	6	7	8	9	10	11	12	1
2	3	4	5	6	7	8	9	10	11	12	1	2
3	4	5	6	7	8	9	10	11	12	1	2	3
4	5	6	7	8	9	10	11	12	1	2	3	4
5	6	7	8	9	10	11	12	1	2	3	4	5
6	7	8	9	10	11	12	1	2	3	4	5	6
7	8	9	10	11	12	1	2	3	4	5	6	7
8	9	10	11	12	1	2	3	4	5	6	7	8
9	10	11	12	1	2	3	4	5	6	7	8	9
10	11	12	1	2	3	4	5	6	7	8	9	10
11	12	1	2	3	4	5	6	7	8	9	10	11
12	1	2	3	4	5	6	7	8	9	10	11	12

EXAMPLE 1 Using 12-hour clock addition, evaluate each of the following:

a. $4 + 7$ b. $7 + 7$ c. $10 + 9$
d. $5 + 8$ e. $9 + 7$ f. $8 + 12$

SOLUTION We use the table of addition facts for a 12-hour clock (Table 1) to find the answer to each problem.

a. $4 + 7 = 11$ b. $7 + 7 = 2$ c. $10 + 9 = 7$
d. $5 + 8 = 1$ e. $9 + 7 = 4$ f. $8 + 12 = 8$

It may have occurred to you that the technique of using a table to solve the problems in Example 1 is not the only way that these problems could be done. There is a more efficient way that we shall now explore.

The armed forces and many factories operate on a 24-hour clock. These clocks begin the same as the 12-hour clock, but once it becomes noon, the 12-hour system starts over, whereas the 24-hour system continues. For example, 1:00 P.M. becomes 1300 hours, or 13:00. Similarly, 2:00 P.M. is the same as 14:00, 3:00 P.M. is the same as 15:00, and so on. This idea can be used to express any number as one of the numbers on a 12-hour clock, that is, as one of the numbers in the set {1, 2, 3, 4, 5, 6, 7, 8, 9, 10, 11, 12}. For example, 13 can be expressed as 1, since 13 hours is the same as 1 rotation around the clock (12 hours) plus 1 additional hour, that is, $13 = 12 + 1$. The number 15 can be expressed as 3, since $15 = 12 + 3$.

The number 12 has a special property in the 12-hour clock system: whenever we add 12 to a number, we obtain that number as a solution ($8 + 12 = 8$, $2 + 12 = 2$, and so on). The number 12 is the *identity element* in this system.

Courtesy Andrews Air Force Base. PAR/NYC photo.

A 24-hour clock.

A system consisting of a set of elements $\{a, b, c, \ldots\}$ and an operation $*$ has an *identity element* (we will call it e) if for every element a in the system,

$$a * e = a \qquad \text{and} \qquad e * a = a$$

The identity element does not change any element when it is operated on together with that element. In ordinary arithmetic, the identity element for the operation of addition is 0 (zero), because $4 + 0 = 4$ and $0 + 4 = 4$. The sum of any number and zero is that number. This is known as the addition property of zero. The identity element for the operation of multiplication is 1 (one), because $5 \times 1 = 5$ and $1 \times 5 = 5$. The product of any number and one is that number. This is known as the multiplication property of one.

We have already changed some numbers to one of the numbers in the system, but suppose we want to change a number such as 55. It can be expressed as 7 in the 12-hour system as follows: starting at 12, we complete four rotations around the clock (48 hours), plus 7 more hours, to get 55 hours. That is, $55 = 12 + 12 + 12 + 12 + 7$. Hence, the number 55 can be expressed as 7 in our system, because 12 is the identity element and does not change the identity of the number 7. Therefore 55 and 7 are in the same position on a 12-hour clock.

Another way to show that 55 can be expressed as 7 is by means of division. If we divide 55 by 12, the remainder is 7.

$$\begin{array}{r} 4 \\ 12\overline{)55} \\ \underline{48} \\ 7 \end{array}$$

This division indicates that there are four 12s in 55. The 12s do not affect the value of the number in the 12-hour clock system; therefore the remainder, 7, is our answer.

In order to convert any number into a number in the 12-hour clock system, we divide it by 12 and record the remainder. The number 116, for example, can be expressed as 8.

$$\begin{array}{r} 9 \\ 12\overline{)116} \\ \underline{108} \\ 8 \end{array}$$

The nine 12s contained in 116 do not affect the value of the number in the 12-hour system, so the remainder, 8, is our answer.

EXAMPLE 2 Find the equivalent of each of the following on a 12-hour clock.

a. 124 b. 258 c. 2,000 d. 300

SOLUTION a. We divide 124 by 12 and record the remainder.

$$\begin{array}{r} 10 \\ 12\overline{)124} \\ \underline{12} \\ 04 \\ \underline{0} \\ 4 \end{array}$$

Hence 124 is equivalent to 4 on a 12-hour clock.

b. Using the same technique we used in part *a,* we divide 258 by 12. The remainder is 6.

c. The number 2,000 is equivalent to 8 on a 12-hour clock, because the remainder when 2,000 is divided by 12 is 8.

d. The number 300 is equivalent to 12 on a 12-hour clock. Recall that 12 is the identity element in our system under the operation of addition; it has the same property as zero for the operation of addition in ordinary arithmetic. Thus 300 hours would take 25 complete rotations around the clock and would stop at the same place it began, at 12. Hence, after 300 hours, the clock is again at the beginning position.

EXAMPLE 3 Evaluate each of the following on a 12-hour clock.

a. $8 + 4$ b. $9 + 3$ c. $6 + 6$

SOLUTION a. $8 + 4 = 12$ b. $9 + 3 = 12$ c. $6 + 6 = 12$

The problems presented in Example 3 all have the answer 12. Recall that 12 is the identity element in our system, and that when 12 is added to a number the identity or position of the number will not be changed. That is, $2 + 12 = 2$, $3 + 12 = 3$, and so on. In the problems in Example 3, we have added one number to another number and obtained the identity element as the result. These problems illustrate another property found in mathematical systems. In the problem $8 + 4 = 12$, 4 is called the *inverse* or *additive inverse* of 8 because when we add 4 to 8 we obtain the identity element 12. Given any clock number in the 12-hour system {1, 2, 3, 4, 5, 6, 7, 8, 9, 10, 11, 12}, we can find another clock number such that the sum of the two numbers is the identity element.

FIGURE 2

Each element in a system consisting of a set of elements $\{a, b, c, \ldots\}$ and an operation $*$ has an *inverse* if for every element a in the system there exists an element b (also in the system) such that

$$a * b = e \quad \text{and} \quad b * a = e$$

where e is the identity element of the system.

Note that if a system has no identity element, then the inverses of elements cannot occur. In the 12-hour clock system, every element has an inverse. For example, 11 is the inverse of 1, because $1 + 11 = 12$ and $11 + 1 = 12$.

EXAMPLE 4 Find the additive inverse of each number in the 12-hour clock system.

a. 7 b. 2 c. 12

SOLUTION a. The additive inverse of 7 is 5. We have $5 + 7 = 12$ and $7 + 5 = 12$.
b. The additive inverse of 2 is 10: $2 + 10 = 12$ and $10 + 2 = 12$.
c. The additive inverse of 12 is 12 because in a 12-hour clock system $12 + 12 = 12$. Thus 12 is its own inverse.

Let us next consider subtracting numbers in our system. What is the answer to the problem $2 - 3$ on a 12-hour clock? Using the clock face in Figure 2, we can see that $2 - 3 = 11$: we start at 2 and count 3 units (hours) in a counterclockwise direction; we complete the counting at 11. Therefore $2 - 3 = 11$ in clock arithmetic.

This problem may also be solved in another manner. Recall that 12 is the identity element in our system, so if we add 12 to a number we will not change its value in the system. Therefore when we add 12 to a number we will not change its position on the clock face. Hence 2:00 is the same as 14:00, and we can think of the problem $2 - 3$ as $14 - 3$. Therefore $2 - 3 = 14 - 3 = 11$ in the 12-hour clock system.

EXAMPLE 5 Evaluate each of the following on a 12-hour clock.

a. $4 - 7$ b. $3 - 8$ c. $4 - 12$

SOLUTION a. In the 12-hour clock system, we can add 12 to a number and not change its identity. Hence $4 - 7 = 16 - 7 = 9$.
b. $3 - 8 = 15 - 8 = 7$
c. $4 - 12 = 16 - 12 = 4$ is one way of solving this problem. An alternate method is to recall that 12 may also be thought of as zero in our system. Therefore we have $4 - 12 = 4 - 0 = 4$.

NOTE OF INTEREST

Evariste Galois was a brilliant French mathematician who was killed in a duel (in 1832) when he was 20 years old. The night before the duel he worked furiously on a set of notes and proofs that solved a long-standing mathematical problem. His calculations laid the basis for what has become the Theory of Groups.

This type of mathematical abstraction is considered to be a powerful tool because it has provided a structural link between arithmetic, algebra, geometry, coding theory, crystallography, quantum mechanics, and elementary-particle physics. There are physicists who believe that they will obtain a most intimate insight into the basic structures of the universe because of group theory.

Thus far in our discussion of clock arithmetic, we have encountered three properties of a mathematical system: the clock system has the *closure* property with respect to the operation of addition, it has an *identity element* for addition, and each element has an *inverse* with respect to addition. This system also has a property that has not been mentioned yet: the *associative property for addition.*

An operation is associative if the location of parentheses in a problem does not affect the answer. To be more specific, consider the following addition problem in ordinary arithmetic.

$$4 + 6 + 9$$

We can find the sum by adding 4 and 6 first, obtaining 10, and then adding 9 to get an answer of 19; that is, $(4 + 6) + 9 = 10 + 9 = 19$. Or we might add 6 and 9 first, obtaining 15, and then add 4 to get an answer of 19; that is, $4 + (6 + 9) = 4 + 15 = 19$. Regardless of which two numbers are added first, the answer is the same. Hence we can say that

$$(4 + 6) + 9 = 4 + (6 + 9)$$

If a system consists of a set of elements $\{a, b, c, \ldots\}$ and an operation $*$, we say that the operation is *associative* (or has the *associative property*) if, for all of the elements in the system,

$$(a * b) * c = a * (b * c)$$

The associative property does not hold for all operations. We have seen that $(4 + 6) + 9 = 4 + (6 + 9)$ in ordinary arithmetic. But, consider the operation of subtraction in ordinary arithmetic. Let us see if

$$(7 - 4) - 3 = 7 - (4 - 3)$$

is true. In computing the answer, we always operate inside the parentheses first. Therefore $(7 - 4) - 3 = 3 - 3 = 0$, while $7 - (4 - 3) = 7 - 1 = 6$, and we have shown that

$$(7 - 4) - 3 \neq 7 - (4 - 3)$$

Hence the associative property does not hold for the operation of subtraction in ordinary arithmetic, because one example showing that a property does not hold is sufficient to illustrate that a property does not work for all elements in the system. Such an example is sometimes called a **counterexample.**

On a 12-hour clock, we have the set of elements {1, 2, 3, 4, 5, 6, 7, 8, 9, 10, 11, 12}. We can illustrate the associative property for addition by considering the following example.

$$(7 + 8) + 10 \stackrel{?}{=} 7 + (8 + 10)$$

Working with the left side of the equation first, we have $(7 + 8) + 10 = 15 + 10$. But on a 12-hour clock, 15 is the same as 3; hence $15 + 10 = 3 + 10 = 13$, which is the same as 1 on a 12-hour clock. Hence, $(7 + 8) + 10 = 1$ on a 12-hour clock. Now working with the right side, we have $7 + (8 + 10) = 7 + 18$. But 18 on a 12-hour clock is the same as 6, so $7 + 18 = 7 + 6 = 13$, which is the same as 1 on a 12-hour clock. Hence $7 + (8 + 10) = 1$ on a 12-hour clock. We have verified that the associative property holds for the example $(7 + 8) + 10 = 7 + (8 + 10)$.

Space does not permit the verification of the associative property for addition for all of the elements in the 12-hour clock system, but the associative property for addition does hold for this system.

Thus far, the 12-hour clock system, consisting of the set of elements {1, 2, 3, 4, 5, 6, 7, 8, 9, 10, 11, 12} and the operation of addition

1. Is *closed* with respect to addition
2. Contains an *identity element* with respect to addition
3. Contains an *inverse* for each of its elements with respect to addition
4. Is *associative* with respect to addition.

When a set of elements and an operation satisfy these properties, we say that the elements form a **group** under the operation (in this case addition). The group operation must be a **binary operation,** that is, one which combines two elements to produce a third element. Addition is a binary operation since it acts on two numbers to produce a third. Squaring a number is not a binary operation, since it acts on only one number.

EXAMPLE 6 Consider the set of counting numbers, {1, 2, 3, 4, . . .}, and the operation of addition. Does this system form a group? Why or why not?

SOLUTION No; in order to form a group, a system must satisfy the closure property, identity property, inverse property, and associative property. The set of counting numbers {1, 2, 3, 4, . . .} does not have an identity element under the operation of addition, because zero is not included in the given set of elements. In addition, the elements have no additive inverses.

EXAMPLE 7 Consider the set of counting numbers, {1, 2, 3, 4, . . .}, and the operation of multiplication. Does this system form a group? Why or why not?

SOLUTION In order to form a group under the operation of multiplication, the system must satisfy the four properties: closure, identity, inverse, and associative. We shall check to see if these properties hold under the operation of multiplication.

a. The system is closed under multiplication. Whenever we multiply two counting numbers, the product is a counting number.
b. The system does contain an identity element under multiplication, the number 1. The number 1 does not change the identity of a number when the two are multiplied together. That is, $2 \times 1 = 1 \times 2 = 2$, $3 \times 1 = 1 \times 3 = 3$, and so on.
c. The system does *not* contain an inverse element under multiplication for each element. The number 1 is the only element in the set that has an inverse, and it is its own inverse: $1 \times 1 = 1$. Two does not have an inverse in the set of elements: there is no number b in the set such that $2 \times b = 1$. This counterexample shows that the set of counting numbers {1, 2, 3, 4, . . .} under the operation of multiplication *does not* form a group.

When you find the sum of 4 and 5, whether you add them as $4 + 5$, or as $5 + 4$, the answer is 9. That is, $4 + 5 = 5 + 4$. Similarly, if you multiply 4 and 5 together, you can multiply 4×5 or 5×4, and the answer is 20. So $4 \times 5 = 5 \times 4$. No matter what order you do the operation in, the answer is the same. In other words, we can switch the elements around; we can *commute* them. When we can do this for all elements in a system using a given operation, we say that the operation is *commutative.*

Given a system consisting of a set of elements $\{a, b, c, \ldots\}$ and an operation $*$, we say that the operation is *commutative* if for all elements a and b in the system

$$a * b = b * a$$

Not all operations are commutative. We have seen that addition and multiplication of counting numbers are commutative operations. But consider the operation of subtraction: does $7 - 6$

$= 6 - 7$? The answer is no, because $7 - 6 = 1$, whereas $6 - 7 = -1$. This verifies that $7 - 6 \neq 6 - 7$, and that subtraction is not commutative. A group with operation $*$ is called a **commutative group** if $a * b = b * a$ for all elements a and b in the group.

Thus far in this section, we have performed the operations of addition and subtraction on the numbers in a 12-hour clock system. Now let's examine the operation of multiplication. Consider the problem 3×9 on a 12-hour clock. We can consider this as an addition problem, because $3 \times 9 = 9 + 9 + 9$. On a 12-hour clock, $9 + 9 = 18 = 6$, so $(9 + 9) + 9$ becomes $6 + 9$, and $6 + 9 = 3$. Therefore, on a 12-hour clock, $3 \times 9 = 3$. This process is quite tedious. We can multiply two numbers on a 12-hour clock in a more efficient manner: first multiply the numbers as you would in ordinary arithmetic, and then convert your answer to its equivalent on the 12-hour clock. Using this technique for 3×9, we have $3 \times 9 = 27$, which is equivalent to 3 on the 12-hour clock. Hence $3 \times 9 = 3$ on a 12-hour clock.

EXAMPLE 8 Evaluate each of the following on a 12-hour clock.

a. 4×5 b. 6×7 c. 8×10
d. 11×11 e. 10×12

SOLUTION
a. $4 \times 5 = 20$ and 20 is equivalent to 8; hence $4 \times 5 = 8$
b. $6 \times 7 = 42$ and 42 is equivalent to 6; hence $6 \times 7 = 6$
c. $8 \times 10 = 80$ and 80 is equivalent to 8; hence $8 \times 10 = 8$
d. $11 \times 11 = 121$ and 121 is equivalent to 1; hence $11 \times 11 = 1$
e. $10 \times 12 = 120$ and 120 is equivalent to 12; hence $10 \times 12 = 12$

We shall not explore in detail the operation of division in a 12-hour clock system because of the problems that arise in doing division problems in this system. Consider the problem $8 \div 4$. In ordinary arithmetic, the answer is 2; we can verify this because $2 \times 4 = 8$. But in a 12-hour clock system, $8 \div 4 = 2$, $8 \div 4 = 5$, $8 \div 4 = 8$, and $8 \div 4 = 11$, because $4 \times 2 = 8$, $4 \times 5 = 8$, $4 \times 8 = 8$, and $4 \times 11 = 8$ on a 12-hour clock. Thus some problems have more than one answer, whereas other problems such as $8 \div 3$, have no answer. In ordinary arithmetic, the answer to this is $2\frac{2}{3}$, but in a 12-hour clock system there is no answer. In order for 3 to divide 8 in a 12-hour clock system, the answer must be one of the elements in the set $\{1, 2, 3, 4, 5, 6, 7, 8, 9, 10, 11, 12\}$. But not one of these numbers will satisfy the statement $3 \times a = 8$. Hence there is no number in this system that when multiplied by 3 will yield 8 as an answer. In other words, $8 \div 3$ has no answer in a 12-hour clock system.

EXERCISES FOR SECTION 6.2

1. Evaluate each sum on a 12-hour clock.
   a. $2 + 4$ b. $7 + 6$
   c. $9 + 2$ d. $11 + 11$
   e. $10 + 11$ f. $9 + 10$

2. Evaluate each sum on a 12-hour clock.
   a. $8 + 8$ b. $8 + 9$
   c. $8 + 10$ d. $8 + (9 + 11)$
   e. $(9 + 7) + 6$ f. $9 + (8 + 7)$

3. Find the equivalent of each number on a 12-hour clock.
   a. 33 b. 44 c. 55
   d. 66 e. 277 f. 188

4. Find the equivalent of each number on a 12-hour clock.
   a. 342 b. 201 c. 400
   d. 1,984 e. 2,001 f. -17

5. Evaluate each difference on a 12-hour clock.
   a. $5 - 7$ b. $6 - 8$ c. $7 - 11$
   d. $9 - 10$ e. $2 - 5$ f. $3 - 7$

6. Evaluate each difference on a 12-hour clock.
   a. $2 - 6$ b. $4 - 9$
   c. $3 - 11$ d. $2 - 12$
   e. $2 - (3 - 8)$ f. $4 - (7 - 9)$

7. Evaluate each product on a 12-hour clock.
   a. 6×8 b. 4×6 c. 3×7
   d. 4×10 e. 5×11 f. 9×9

8. Evaluate each product on a 12-hour clock.
   a. 7×11 b. 8×11
   c. 10×10 d. $3 \times (4 + 7)$
   e. $2 \times (9 + 10)$ f. $5 \times (6 - 10)$

9. Evaluate each product on a 12-hour clock.
   a. $2 \times (5 + 8)$ b. $3 \times (9 - 11)$
   c. $8 \times (7 + 6)$ d. $9 \times (11 + 10)$
   e. $10 \times (2 - 11)$ f. $11 \times (1 - 11)$

10. State the property of the 12-hour clock system that is illustrated by each of the following.
    a. $(4 + 3) + 5 = 4 + (3 + 5)$
    b. $7 + 6 = 1$, a number in the 12-hour clock system
    c. $9 + 12 = 9$
    d. $8 + 9 = 9 + 8$
    e. $7 \times 7 = 1$
    f. $4 \times 5 = 8$, a number in the 12-hour clock system

11. State the property of the 12-hour clock system that is illustrated by each of the following:
    a. $7 + 6 = 1$
    b. $7 \times 5 = 5 \times 7$
    c. $(7 + 6) + 3 = (6 + 7) + 3$
    d. $(2 \times 4) \times 3 = 2 \times (4 \times 3)$
    e. $5 \times 5 = 1$
    f. $12 + 7 = 7$

12. Construct a complete table of multiplication facts for the 12-hour clock system. Use your table to answer each question.
    a. What is the identity element for multiplication in this system?
    b. Does the closure property hold for this system?
    c. Does the commutative property hold for this system?
    d. Verify that the associative property holds for one specific instance in this system.
    e. What elements in this system have an inverse?

13. Determine whether each statement is true or false. Given the set of counting numbers {1, 2, 3, 4, . . .}, it
    a. is closed with respect to addition.
    b. is closed with respect to subtraction.
    c. is associative with respect to addition.
    d. is commutative with respect to division.
    e. contains an identity element for addition.
    f. contains an identity element for multiplication.

14. Determine whether each statement is true or false for the set of all integers, {. . ., -2, -1, 0, 1, 2, . . .}.
    a. The set is closed with respect to addition.
    b. The set is closed with respect to subtraction.
    c. The set is closed with respect to division.
    d. The set is a system that contains an identity element for addition.
    e. The set contains an identity element for multiplication.

f. The set contains an additive inverse element for each element in the set.

15. Consider the set of integers $\{\ldots, -2, -1, 0, 1, 2, \ldots\}$ and the operation of addition. Does this system form a group? Why or why not?

16. Consider the set of whole numbers, $\{0, 1, 2, 3, \ldots\}$, and the operation of addition. Does this system form a group? Why or why not?

17. Replace each question mark with a number from the 12-hour clock system to give a true statement.
a. $6 + ? = 1$ b. $? - 4 = 9$
c. $5 \times ? = 1$ d. $5 \times (8 + 7) = ?$
e. $? \times 7 = 9$ f. $? \times (2 + 11) = 2$

18. Replace each question mark with a number from the 12-hour clock system to give a true statement.
a. $3 + ? = 1$ b. $4 - ? = 9$
c. $4 - 7 = ?$ d. $3 \times (4 + ?) = 9$
e. $8 + ? = 2$ f. $4 + ? = 8 - ?$

19. Is the following statement ever true in the 12-hour clock system? If so, give an example. For any numbers a, b, and c, $(a - b) - c = a - (b - c)$.

20. Is the following statement ever true in the 12-hour clock system? If so, give an example. For any numbers a and b, $a - b = b - a$.

"Daddy, 28 plus (36 plus 49) equals (28 plus 36) plus what--using the associative principle?"

The Family Circus reprinted with special permission of Cowles Syndicate, Inc.

JUST FOR FUN Are the horizontal lines straight?

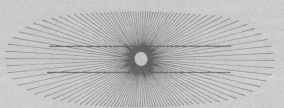

6.3 MORE NEW SYSTEMS

In the preceding section, we examined the nature of clock arithmetic. Using the set of numbers {1, 2, 3, 4, 5, 6, 7, 8, 9, 10, 11, 12} and the properties of a 12-hour clock, we constructed a new system of arithmetic.

There are many other systems of arithmetic that can be created. For another example, let

1 = spring 3 = fall
2 = summer 4 = winter

If it is now spring, then three seasons from now it will be winter. That is, 1 + 3 = 4, or spring + fall = winter. Suppose it is now winter. What season will it be five seasons from now? Starting with winter and counting off five seasons, we wind up at spring. This suggests that 4 + 5 = 1 in our new system. That is, we have a cycle of four seasons, and so once we reach 4 in our counting, we start over again. For the previous problem we have 4 + 5 = 9 in ordinary arithmetic, but 9 = 4 + 4 + 1, so in our new system 9 is equivalent to 1. To help us in our calculations, we could think of this new system in terms of a 4-hour clock. Tables 2 and 3 show the addition and multiplication facts for the seasons system.

TABLE 2

+	1	2	3	4
1	2	3	4	1
2	3	4	1	2
3	4	1	2	3
4	1	2	3	4

TABLE 3

×	1	2	3	4
1	1	2	3	4
2	2	4	2	4
3	3	2	1	4
4	4	4	4	4

EXAMPLE 1 Let spring = 1, summer = 2, fall = 3, and winter = 4. Evaluate the following.

a. summer + summer b. fall + summer
c. summer × fall d. fall × fall

SOLUTION a. Since summer = 2, we have 2 + 2 = 4, and 4 = winter, so summer + summer = winter.

b. Fall = 3 and summer = 2; therefore, we have 3 + 2 = 1, according to table 2. Since 1 = spring, fall + summer = spring.

As an alternate solution, $3 + 2 = 5$, which is equivalent to 1 in this system, and $1 =$ spring.

c. Summer $= 2$ and fall $= 3$; therefore, we have $2 \times 3 = 2$, according to Table 3, because $2 =$ summer, summer $\times$ fall $=$ summer.

Alternately, we have $2 \times 3 = 6$, which is equivalent to 2 in this system, and $2 =$ summer.

d. Fall $= 3$; therefore, we have $3 \times 3 = 1$, according to Table 3, because $1 =$ spring, fall $\times$ fall $=$ spring.

As an alternate solution, we have $3 \times 3 = 9$, which is equivalent to 1 in this system, and $1 =$ spring.

Recall that Table 2 illustrates the addition facts for the seasons system. If we substitute the names of the seasons for the numbers in Table 2, we would have Table 4, which may appear strange at first glance. But remember, the only thing that we have done is substitute the names of the seasons in place of the numbers in Table 2. This system may seem more abstract, but it is still a system.

TABLE 4

+	**Spring**	**Summer**	**Fall**	**Winter**
Spring	summer	fall	winter	spring
Summer	fall	winter	spring	summer
Fall	winter	spring	summer	fall
Winter	spring	summer	fall	winter

Note that we have closure in this system, because whenever we combine any two seasons the result is a member of the set {spring, summer, fall, winter}. What is the identity element in this system? Recall that the identity element does not change any element when it is operated on together with that element. Upon examination of Table 4, we see that winter is the identity element in our seasons system for the operation of addition. We can verify this by checking each season's addition with winter: for example, spring + winter = spring, and winter + spring = spring.

EXAMPLE 2 Using Table 4, find the additive inverse of each of the following:
a. spring b. summer c. fall d. winter

SOLUTION Recall that each element a in a system has an *inverse* element if there exists an element b such that $a * b = e$ and $b * a = e$, where e is the identity element of the system and $*$ is the operation of the system. Since winter is the identity element in the seasons system, we examine the table to see what season added to a given season yields winter as the answer.

a. The additive inverse of spring is fall.
b. The additive inverse of summer is summer.
c. The additive inverse of fall is spring.
d. The additive inverse of winter is winter.

Suppose that today is Wednesday, the fourth day of the month. What day of the week is the 27th? One way we could determine this would be to examine a calendar. Another interesting technique is the following: Because Wednesday corresponds to 4 (it is the fourth day of the month), then Thursday corresponds to 5, Friday corresponds to 6, and so on. Twenty-one days from now it will be Wednesday again (the days of the week are in a cycle of 7), and the date will be the 25th; hence, the 27th will fall on a Friday. We can also solve this problem in another way: Since every seven days brings us back to the same day of the week, 27 is equivalent to 6 in a week system, because $27 = 7 + 7 + 7 + 6$. Recall that Wednesday corresponded to 4; hence Friday must correspond to 6, and the 27th must be a Friday. This discussion suggests that we can use the days of the week to create another new system of arithmetic.

Let

1 = Sunday	5 = Thursday
2 = Monday	6 = Friday
3 = Tuesday	7 = Saturday
4 = Wednesday	

Photo by PAR/NYC

If today is Sunday, then three days from now it will be Wednesday; that is, $1 + 3 = 4$. If today is Tuesday, then four days from now it will be Saturday; that is, $3 + 4 = 7$. Suppose that today is Monday; what day will it be eight days from now? We know that eight days from now it will be Tuesday, because seven days from now it will be Monday again, and one more day will bring us to Tuesday. But we could also say Monday $= 2$, and $2 + 8 = 10$. Ten is equivalent to 3 in the week system, because 7 acts as an identity element in this system for the operation of addition and $10 = 3 + 7$.

Table 5 shows the addition facts for the week system. Note that whenever 7 is added to another number in the system, the result is always the original number. We also have closure in this new system, because the results of adding elements in the system are always in the system. There is an identity element (7), so we can determine the additive inverse for an element in this system. For example, the additive inverse of 2 is 5, since $2 + 5 = 7$ and $5 + 2 = 7$.

TABLE 5

+	1	2	3	4	5	6	7
1	2	3	4	5	6	7	1
2	3	4	5	6	7	1	2
3	4	5	6	7	1	2	3
4	5	6	7	1	2	3	4
5	6	7	1	2	3	4	5
6	7	1	2	3	4	5	6
7	1	2	3	4	5	6	7

EXAMPLE 3 Given that Sunday = 1, Monday = 2, Tuesday = 3, Wednesday = 4, Thursday = 5, Friday = 6, and Saturday = 7, evaluate the following.

a. Sunday + Tuesday b. Monday + Wednesday
c. Friday + Tuesday d. Wednesday + Friday

SOLUTION a. Because Sunday = 1 and Tuesday = 3, we have $1 + 3 = 4$, and 4 = Wednesday; Sunday + Tuesday = Wednesday.

b. Monday = 2 and Wednesday = 4; therefore we have $2 + 4 = 6$, and 6 = Friday; Monday + Wednesday = Friday.

c. Friday = 6 and Tuesday = 3; therefore, we have $6 + 3 = 2$ according to the table, and 2 = Monday. Hence, Friday + Tuesday = Monday.

As an alternate solution, $6 + 3 = 9$, which is equivalent to 2 in this system, and 2 = Monday.

d. Wednesday = 4 and Friday = 6; hence, $4 + 6 = 3$ according to the table, and 3 = Tuesday. Therefore, Wednesday + Friday = Tuesday.

We could also note that $4 + 6 = 10$, which is equivalent to 3 in this system, and 3 = Tuesday.

EXAMPLE 4 Using the week system, evaluate the following.

a. Wednesday − Thursday
b. Monday × Tuesday
c. Wednesday × Thursday
d. (Friday × Monday) − Wednesday

SOLUTION a. Wednesday = 4 and Thursday = 5; therefore we have 4 − 5. In order to subtract 5 from 4, we must change 4 to an equivalent number: 4 + 7 = 11, so 4 is equivalent to 11. Hence, in our system, 4 − 5 = 11 − 5 = 6, and 6 = Friday. Hence Wednesday − Thursday = Friday.

b. Monday = 2 and Tuesday = 3; 2 × 3 = 6 = Friday. Hence Monday × Tuesday = Friday.

c. Wednesday = 4 and Thursday = 5; 4 × 5 = 20, which is equivalent to 6, because in the week system, 20 = 7 + 7 + 6, and 6 = Friday. Hence Wednesday × Thursday = Friday.

d. Friday = 6, Monday = 2, and Wednesday = 4. Working inside the parentheses first, we have (6 × 2), which equals 12, and 12 − 4 = 8, which is equivalent to 1 in the week system. Hence (Friday × Monday) − Wednesday = Sunday.

EXERCISES FOR SECTION 6.3

In Exercises 1–12, assume that 1 = spring, 2 = summer, 3 = fall, and 4 = winter. Evaluate each of the following, giving your answer in terms of a season.

1. summer + fall
2. fall + winter
3. winter + summer
4. winter − spring
5. spring − summer
6. fall − winter
7. summer × summer
8. spring × summer
9. fall × winter
10. fall × (winter + spring)
11. fall × (winter − spring)
12. winter × winter
13. What is the identity element in the seasons system for the operation of multiplication?
14. What is the multiplicative inverse of winter?
15. What is the multiplicative inverse of fall?
16. Does the season system form a group under the operation of addition?
17. Does the season system form a group under the operation of multiplication?

In Exercises 18–34 assume that 1 = Sunday, 2 = Monday, 3 = Tuesday, 4 = Wednesday, 5 = Thursday, 6 = Friday, and 7 = Saturday. Evaluate each of the following, giving your answer in terms of a day of the week.

18. Sunday + Friday
19. Monday + Saturday
20. Tuesday + Monday
21. Thursday + Friday
22. Friday + Friday
23. Sunday + Sunday
24. Friday − Wednesday
25. Thursday − Friday
26. Friday − Saturday
27. Thursday − Saturday
28. Wednesday − Saturday

29. Tuesday − Thursday

30. Tuesday × Friday

31. Wednesday × Saturday

32. Saturday × Saturday

33. Tuesday × (Monday + Friday)

34. Friday × (Friday − Saturday)

35. What is the identity element in the week system for the operation of addition?

36. What is the identity element in the week system for the operation of multiplication?

37. What is the additive inverse of
a. Monday b. Tuesday c. Wednesday

38. What is the multiplicative inverse of
a. Monday b. Friday

39. Verify that Sunday × (Tuesday × Friday) = (Sunday × Tuesday) × Friday.

40. Does the week system form a group under the operation of multiplication?

41. Does the week system form a group under the operation of addition?

JUST FOR FUN

Take any size piece of paper and fold it in half, then fold it in half again, and keep doing this as many times as possible. You probably cannot do it eight times.

6.4 MODULAR SYSTEMS

FIGURE 3

Thus far in this chapter, we have examined various new systems of arithmetic. One of the systems discussed was the 12-hour clock system, since everyone is familiar with the 12-hour clock. We will next consider a mathematical system that is also based on a clock, but this time we will consider a 5-hour clock.

A 5-hour clock might be one like the one shown in Figure 3, but to make our new clock a little easier to understand, we will make it look like a timer. You have probably seen some sort of a timer before—for example, an egg timer or a stop watch. These timers have zero at the top of the face, rather than some other

FIGURE 4

number. Figure 4 shows our new 5-hour clock. This 5-hour clock system contains the elements 0, 1, 2, 3, 4. If we begin to count in this system (clockwise), we have 1, 2, 3, 4, 0, and then the system starts to repeat itself; that is, 1, 2, 3, 4, 0, 1, 2, 3, 4, 0, 1, 2, ...

If it is now 1:00 on our 5-hour clock, then 3 hours from now it will be 4:00. Thus $1 + 3 = 4$. Suppose it is now 2:00 on the 5-hour clock. What time will it be 4 hours from now? To find out, we begin at 2 and count 4 units in a clockwise direction, ending up at 1. Therefore in this new system, $2 + 4 = 1$.

Similarly, we can verify that $4 + 3 = 2$ using the 5-hour clock. Continuing with this technique, we can construct a table of addition facts for a 5-hour clock using the set of elements {0, 1, 2, 3, 4} (see Table 6).

TABLE 6

+	**0**	**1**	**2**	**3**	**4**
0	0	1	2	3	4
1	1	2	3	4	0
2	2	3	4	0	1
3	3	4	0	1	2
4	4	0	1	2	3

When we have a system such as the 5-hour or 12-hour clock that repeats itself in a cycle, we call it a **modular system.** The 5-hour clock is called the *modulo 5 system.* We abbreviate this to *mod 5 system.*

Recall that in order to convert a given number into a number in the 12-hour clock system, we divided it by 12 and recorded the remainder. Twelve did not affect the answer because it was the identity element. In order to convert a given number into a number in the 5-hour clock system (modulo 5), we divide it by 5 and record the remainder. For example, from Table 6 we know that $4 + 4 = 3$ in the modulo 5 system, but we can also do this by finding the sum of 4 and 4 in ordinary arithmetic, 8, and converting it to a number in the modulo 5 system.

$$\begin{array}{r} 1 \\ 5\overline{)8} \\ \underline{5} \\ 3 \end{array}$$

When we divide 8 by 5, the remainder is 3, our desired result.

Therefore, we can write

$$4 + 4 \equiv 3 \pmod 5$$

This is read "4 + 4 is equivalent to 3, mod 5."

Similarly, we can write $6 \equiv 1 \pmod 5$, $21 \equiv 1 \pmod 5$, and $36 \equiv 1 \pmod 5$. In each case, this indicates that 1 is the remainder when the numbers 6, 21, and 36 are divided by 5. In general terms,

$$a \equiv b \pmod m$$

means that a and b both have the same remainder when they are divided by m. We say that *a is equivalent to b mod m.*

EXAMPLE 1 Evaluate the following in the modulo 5 system.

a. $4 + 4$ b. $2 + 4$ c. $3 + 2$

SOLUTION

a. Using Table 6, we note that the sum of 4 and 4 is 3; that is, $4 + 4 \equiv 3 \pmod 5$.
b. According to the table, the sum of 2 and 4 is 1, or $2 + 4 \equiv 1 \pmod 5$.
c. Since $3 + 2$ is equivalent to 0, $3 + 2 \equiv 0 \pmod 5$.

ALTERNATE SOLUTION

a. $4 + 4 = 8$; divide 5 into 8, and the remainder is 3. Hence $4 + 4 \equiv 3 \pmod 5$.
b. $2 + 4 = 6$; divide 5 into 6, and the remainder is 1. Hence $2 + 4 \equiv 1 \pmod 5$.
c. $3 + 2 = 5$; divide 5 into 5, and the remainder is 0. Hence $3 + 2 \equiv 0 \pmod 5$.

EXAMPLE 2 Evaluate the following in the modulo 7 system.

a. $4 + 5$ b. $4 + 4$ c. $6 + 5$ d. $4 + 3$

SOLUTION

a. $4 + 5 = 9$; divide 7 into 9, and the remainder is 2. Therefore $4 + 5 \equiv 2 \pmod 7$.
b. $4 + 4 = 8$; divide 7 into 8, and the remainder is 1. Therefore $4 + 4 \equiv 1 \pmod 7$.
c. $6 + 5 = 11$; divide 7 into 11, and the remainder is 4. Therefore $6 + 5 \equiv 4 \pmod 7$.
d. $4 + 3 = 7$; divide 7 into 7, and the remainder is 0. Therefore $4 + 3 \equiv 0 \pmod 7$.

EXAMPLE 3 Evaluate the following in the modulo 5 system.

a. $2 - 4$ b. $1 - 3$ c. $3 - 4$ d. $4 - 0$

SOLUTION a. In the mod 5 system, we can add 5 to a number and not change its identity. Therefore, $2 - 4 = 7 - 4 = 3$, so $2 - 4 \equiv 3 \pmod{5}$.

b. $1 - 3 = 6 - 3 = 3$, so $1 - 3 \equiv 3 \pmod{5}$

c. $3 - 4 = 8 - 4 = 4$, so $3 - 4 \equiv 4 \pmod{5}$

d. $4 - 0 \equiv 4 \pmod{5}$

Let's examine the properties of the modulo 5 mathematical system with the operation of addition. This system consists of the set of elements {0, 1, 2, 3, 4}. The addition operation on this set is shown in Table 6 (page 302).

1. **Closure property:** All of the entries in the table are elements of the given set; there are no new elements appearing in the table. Therefore, the system satisfies the closure property.
2. **Identity property:** There is an element, 0, in the set of elements that does not change any element when it is added to that element.

$$0 + 0 = 0, \quad 1 + 0 = 1, \quad 2 + 0 = 2,$$
$$3 + 0 = 3, \quad 4 + 0 = 4$$

3. **Inverse property:** Each element in this system has an inverse, an element which when added to the given element results in the identity element, 0. Note that

$$0 + 0 = 0 \quad 1 + 4 = 0, \quad 2 + 3 = 0,$$
$$3 + 2 = 0, \quad 4 + 1 = 0$$

Therefore, the inverse of 0 is 0, the inverse of 1 is 4, the inverse of 2 is 3, the inverse of 3 is 2, and the inverse of 4 is 1.

4. **Associative property:** The addition operation is associative if $(a + b) + c = a + (b + c)$. Trying an example, we have $(1 + 2) + 3 = 3 + 3 = 1$ and $1 + (2 + 3) = 1 + 0 = 1$, so that

$$(1 + 2) + 3 = 1 + (2 + 3)$$

Any other example for this system also works; hence, the system satisfies the associative property.

NOTE OF INTEREST

One of the more popular toys and puzzles of the 1980s is Rubik's Cube. It was developed in 1974 by Erno Rubik, a Hungarian design professor who wanted to give his students more experience in understanding three-dimensional solids.

Each side of a Rubik's Cube is made up of nine squares. In the beginning, all nine squares on each face are of one color, that is there are six faces and six colors. The cube has a system of axles such that it can rotate about its center so that each corner's small cube can rotate in any of the three dimensions. Also, with as few as four twists of the cube the solid faces of the cube will become completely scrambled. The challenge is to manipulate the cube back to its original state. Most people cannot do it, even with a solutions manual.

One of the reasons for the difficulty in solving it is that the cube has approximately 43,252,003,274,489,856,000 (43 quintillion) positions.

In addition to the fascination of trying to solve Rubik's Cube, it should be noted that it satisfies the properties of a group. That is, the moves are closed: No matter how you twist it, the result is a member of the original set. Each move has an inverse, and identity moves exist for each one as well. Similarly, the moves are associative, and in fact the commutative property holds true as well.

5. **Commutative property:** Addition is commutative if $a + b = b + a$. Trying an example, we have $2 + 4 = 1$ and $4 + 2 = 1$; therefore, $2 + 4 = 4 + 2$. Any other example for this system also works; hence, the system satisfies the commutative property.

The mathematical system (modulo 5) that consists of the set of elements {0, 1, 2, 3, 4} and the operation of addition satisfies all of the properties listed. Since a group is composed of a set of elements together with a binary operation that satisfies the closure property, identity property, inverse property, and associative property, the modulo 5 system with the operation of addition forms a group. The fact that the system also satisfies the commutative property means that we have a commutative group under the operation of addition.

Now let's consider the modulo 5 system with the operation of multiplication. This system consists of the set of elements {0, 1, 2, 3, 4} and the operation of multiplication. We first construct a multiplication table for mod 5 (see Table 7). We next determine what properties are satisfied under this system. The system is closed with respect to multiplication. It has an identity element, 1; that is, $a \times 1 = a$ for any a in the system. But not every element has an inverse. In particular, there is no number in the system which when multiplied by zero will yield the identity element 1. Zero times a number is zero for every number in the system.

TABLE 7

×	**0**	**1**	**2**	**3**	**4**
0	0	0	0	0	0
1	0	1	2	3	4
2	0	2	4	1	3
3	0	3	1	4	2
4	0	4	3	2	1

EXAMPLE 4 Using the elements of the mod 5 system, find replacements for the question marks so that each of the following is true.

a. $4 + 3 \equiv ? \pmod 5$ b. $? + 4 \equiv 1 \pmod 5$
c. $3 - ? \equiv 4 \pmod 5$ d. $3 \times ? \equiv 1 \pmod 5$

SOLUTION a. $4 + 3 = 7$; divide 5 into 7, and the remainder is 2. Hence, $4 + 3 \equiv 2 \pmod{5}$.

b. Because the mod 5 system contains the set of numbers $\{0, 1, 2, 3, 4\}$, we can try each one of these in place of the question mark. Starting with 0, we have: $0 + 4 = 4$; $1 + 4 = 0$; $2 + 4 = 6$, but $6 \equiv 1 \pmod{5}$. Therefore, our answer is 2.

c. Using the same technique, we have $3 - 0 = 3$, $3 - 1 = 2$, $3 - 3 = 0$, and $3 - 4 = 8 - 4 = 4$. Therefore the solution is 4.

d. We see that $3 \times 0 = 0$, $3 \times 1 = 3$, and $3 \times 2 = 6$. Because $6 \equiv 1 \pmod{5}$, the solution is 2.

EXAMPLE 5 Joe, an avid sports fan, collects baseball cards. Each day he studies his collection of special cards. On Monday, he divides his collection into piles of 5 with 2 left over; on Tuesday, he divides his set of cards into piles of 4 with 2 left over; and on Wednesday, he divides his set into piles of 7 with 0 left over. If Joe's collection of special baseball cards consists of less than 50 cards, how many does he have?

SOLUTION There is no information given as to exactly how many cards Joe has, but we do know how many are in each pile on each day. Let x represent the number of cards. On Monday each pile contained 5 cards with 2 left over. Therefore if 5 is divided into x, there is a remainder of 2; that is, $x \equiv 2 \pmod{5}$. On Tuesday, each pile contained 4 cards with 2 left over. Therefore if 4 is divided into x, there is a remainder of 2; so $x \equiv 2 \pmod{4}$. On Wednesday, each pile contained 7 cards with 0 left over. Hence, if 7 is divided into x, there is a remainder of 0; so $x \equiv 0 \pmod{7}$.

Now we must find a value that satisfies all three statements:

$$x \equiv 2 \pmod{5}, \qquad x \equiv 2 \pmod{4}, \qquad x \equiv 0 \pmod{7}$$

For the first statement, $x \equiv 2 \pmod{5}$, the set of possible replacements for x is $\{7, 12, 17, 22, 27, 32, 37, 42, 47\}$. We stop at 47 because Joe has less than 50 cards. Now which of these numbers also satisfies the second statement, $x \equiv 2 \pmod{4}$? That is, which of these numbers has a remainder of 2 when divided by 4? These numbers are 22 and 42. Now which of these two numbers also satisfies the third statment, $x \equiv 0 \pmod{7}$? The remainder is zero when 42 is divided by 7, and therefore Joe has 42 cards in his collection.

Courtesy Topps Chewing Gum, Inc. Used by permission.

EXERCISES FOR SECTION 6.4

1. Evaluate each sum on a 5-hour clock, that is, in the modulo 5 system.
   a. $1 + 3$ b. $2 + 3$
   c. $4 + 2$ d. $3 + 4$
   e. $(3 + 2) + 4$ f. $(4 + 3) + 4$

2. Evaluate each sum in the modulo 5 system.
   a. $3 + 3$ b. $4 + 4$
   c. $2 + 2$ d. $1 + 4$
   e. $3 + (3 + 4)$ f. $(2 + 2) + 2$

3. Find the equivalent of each number in the modulo 5 system.
   a. 33 b. 44 c. 55
   d. 342 e. 780 f. -8

4. Find the equivalent of each number in the modulo 5 system.
   a. 32 b. 41 c. 53
   d. 287 e. 2001 f. -17

5. Evaluate each difference in the modulo 5 system.
   a. $2 - 4$ b. $1 - 4$
   c. $1 - 3$ d. $3 - 4$
   e. $3 - (2 - 4)$ f. $2 - (3 - 4)$

6. Evaluate each difference in the modulo 5 system.
   a. $2 - 3$ b. $1 - 2$
   c. $4 - 4$ d. $(2 - 3) - 4$
   e. $(3 - 4) - 1$ f. $1 - (2 - 4)$

7. Evaluate each product in the modulo 5 system.
   a. 4×3 b. 2×4
   c. 3×2 d. $2 \times (4 \times 3)$
   e. $3 \times (4 + 2)$ f. $2 \times (3 - 4)$

8. Evaluate each product in the modulo 5 system.
   a. 4×4 b. 3×3
   c. 4×2 d. $3 \times (3 \times 3)$
   e. $2 \times (3 + 4)$ f. $3 \times (2 - 4)$

9. Given the set of elements $\{0, 1, 2, 3, 4, 5, 6\}$, construct a complete table of addition facts for the modulo 7 system.
   a. Does the closure property hold for this system?
   b. What is the identity element (if any) for the operation of addition in this system?
   c. Does the commutative property hold for this system?
   d. Verify that the associative property holds for one specific instance in this system.
   e. List the elements in this system that have an additive inverse, and list their inverses.
   f. Does this system form a group under the operation of addition?

10. Given the set of elements $\{0, 1, 2, 3, 4, 5, 6\}$, construct a complete table of multiplication facts for the modulo 7 system.
    a. Does the closure property hold for this system?
    b. What is the identity element (if any) for the operation of multiplication?
    c. Does the commutative property hold for this system?
    d. List the elements in this system that have a multiplicative inverse, and list their inverses.
    e. Does this system form a group under the operation of multiplication?

11. Evaluate each of the following in the modulo 7 system.
    a. $5 + 6$ b. $3 - 6$
    c. $2 \times (5 + 3)$ d. $3 \times (1 - 6)$
    e. $2 - (3 - 4)$ f. $4 \times (3 - 5)$

12. Evaluate each of the following in the modulo 7 system.
    a. $6 + 6$ b. $3 - 5$
    c. 5×5 d. $4 - (1 - 3)$
    e. $2 \times (5 + 4)$ f. $3 \times (4 - 5)$

13. Determine whether each statement is true or false.
    a. $18 \equiv 3 \pmod 5$ b. $22 \equiv 1 \pmod 5$
    c. $144 \equiv 4 \pmod 5$ d. $33 \equiv 5 \pmod 7$
    e. $49 \equiv 0 \pmod 7$ f. $99 \equiv 2 \pmod 7$

14. Determine whether each statement is true or false.
    a. $44 \equiv 9 \pmod 5$ b. $27 \equiv 2 \pmod 5$

c. $17 \equiv 3 \pmod 5$ d. $140 \equiv 88 \pmod 7$
e. $213 \equiv 12 \pmod 7$ f. $1000 \equiv 55 \pmod 5$

15. Determine whether each statement is true or false.
a. $10 \equiv 3 \pmod 7$ b. $122 \equiv 1 \pmod 3$
c. $1234 \equiv 27 \pmod 5$ d. $0 \equiv 171 \pmod 2$
e. $184 \equiv 32 \pmod 8$ f. $121 \equiv 22 \pmod{11}$

16. Using the elements of the indicated modular system, find a replacement for each question mark so that each statement is true.
a. $4 + ? \equiv 1 \pmod 5$ b. $3 + ? \equiv 2 \pmod 5$
c. $? + 4 \equiv 3 \pmod 5$ d. $? + 3 \equiv 2 \pmod 7$
e. $2 - ? \equiv 4 \pmod 6$ f. $3 - ? \equiv 4 \pmod 7$

17. Using the elements of the indicated modular system, find a replacement for each question mark so that each statement is true.
a. $2 \times ? \equiv 3 \pmod 7$ b. $? \times 4 \equiv 1 \pmod 5$
c. $? - 6 \equiv 2 \pmod 7$ d. $1 - ? \equiv 4 \pmod 5$
e. $2 \times ? \equiv 3 \pmod 9$ f. $2 - ? \equiv 3 \pmod{12}$

18. Stan, a stock boy in a local supermarket, had to change the price on a set of soup cans on a particular shelf. In order to make the job easier, Stan decided to arrange the cans in stacks of 10, but when he did this he had 1 left over. Trying again, he arranged them in stacks of 7, but then he had 4 left over. Finally, he arranged them in stacks of 3, and there were none left over. If the shelf can only hold 100 cans, on how many cans did Stan have to change the price?

19. Irene is a cashier in a restaurant. After the rush hour, she began to tabulate the customers' checks. First, she arranged the checks in stacks of 5, and there was 1 left over. Next, Irene arranged the checks in stacks of 7, and there were 2 left over. Finally, she arranged them in stacks of 4, and there were 3 left over. If Irene had less than 100 checks, how many did she have?

20. A girl was carrying a basket of tomatoes, and a man on a bicycle hit the basket and smashed all the tomatoes. Wishing to pay for the damages, he asked the girl how many tomatoes she had. The girl replied that she didn't know, but she remembered that when she counted them by 2s, there was 1 tomato left over; when she counted them by 3s, there was also 1 tomato left over; when she counted them by 4s, there was also 1 left over; but when she counted them by 5s, there were no tomatoes left over. How many tomatoes do you think the girl had in her basket? Why?

JUST FOR FUN

Can you draw a straight line that intersects all three sides of triangle ABC?

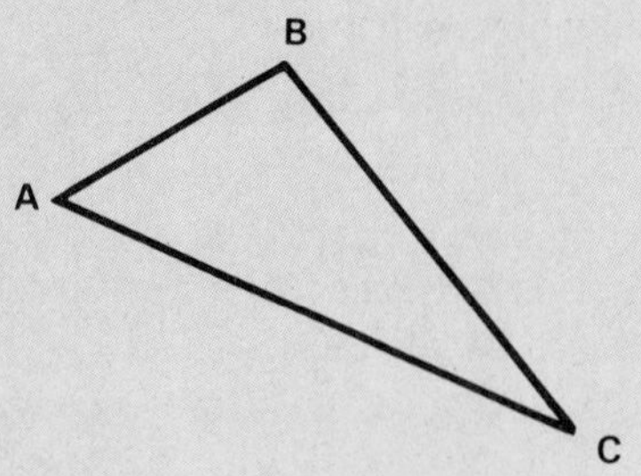

6.5 MATHEMATICAL SYSTEMS WITHOUT NUMBERS

Thus far in this chapter, we have examined mathematical systems such as clock arithmetic, the system of months, the week system, and various modular systems. Recall that a mathematical system consists of a set of elements together with one or more operations (rules) for combining elements of the set. The systems that we have considered so far have been based on the use of numbers, but now we want to consider systems that are more abstract.

TABLE 8

$*$	A	B	C	D
A	A	B	C	D
B	B	C	D	A
C	C	D	A	B
D	D	A	B	C

Consider the set of elements $\{A, B, C, D\}$, together with the operation $*$. We define this operation by means of Table 8. The operation $*$ is a binary operation because we combine two elements of the given set to obtain each result. For example, to find the answer to $B * C$, we find the first element, B, in the vertical column under $*$, and the second element, C, in the top horizontal row following $*$. The answer is found where the row containing B and the column containing C intersect, at D. Hence, $B * C = D$. Similarly, $C * D = B$ and $D * C = B$.

Let's examine the mathematical properties of this particular system, the set of elements $\{A, B, C, D\}$ and the operation $*$.

1. **Closure property:** All of the entries in the table are elements in the given set; there are no new elements appearing in the table. Therefore the system satisfies the closure property.
2. **Identity property:** There is an element, A, in the set that does not change any element when it is operated on together with that element. Note that $A * A = A$, $B * A = B$, $C * A = C$, $D * A = D$.
3. **Inverse property:** Each element in this system has an inverse, an element which when operated on with the given element results in the identity element A. Note that $A * A = A$, $B * D = A$, $C * C = A$, $D * B = A$. From these equations, we can see that the inverse of A is A, the inverse of B is D, the inverse of C is C, and the inverse of D is B.
4. **Associative property:** An operation is associative if the location of parentheses does not affect the answer. Trying

HISTORICAL NOTE

A group is Abelian or commutative if, in addition to satisfying the four basic properties of a group, it also satisfies the commutative property. That is, $a * b = b * a$, where a and b are any two members of the group.

The Abelian group is named after a famous Norwegian mathematician, Neils Henrik Abel (1802–1829). As a child he was considered a mathematical genius. He thoroughly studied the works of many great mathematicians and began producing original work of his own while still in his teens.

He received little recognition during his lifetime, lived in poverty, and died at the age of 27 from a variety of ills.

an example, we have $(B * C) * D = D * D = C$, and $B * (C * D) = B * B = C$; hence $(B * C) * D = B * (C * D)$. Any other example for this system also works, and therefore the system satisfies the associative property.

5. **Commutative property:** An operation is commutative if it does not matter in what order you perform the operation. For an example in this system, we find that $B * C = D$ and $C * B = D$; therefore $B * C = C * B$. Any other example for this system also works, so the system satisfies the commutative property.

Note: A quick check for the commutative property may be performed by examining the table itself. Look at the "diagonal" from the upper left hand corner to the lower right hand corner. If there is "symmetry" along the diagonal, then the system is commutative. In Table 8 we examine the diagonal consisting of $A—C—A—C$. We see that there exists a B on each side of the diagonal, similarly we have a C on each side, $D—D$ on each side, A on each side, and another B on each side. This is known as symmetry or a mirror image. Hence, the system satisfies the commutative property.

The abstract mathematical system consisting of the set of elements $\{A, B, C, D\}$ and the binary operation $*$ satisfies all of the properties listed. Therefore this system forms a group, and because it satisfies the commutative property, it forms a commutative group.

EXAMPLE 1 The following table defines the operation $\times$ for the set of elements {*odd*, *even*}.

$\times$	*odd*	*even*
odd	*odd*	*even*
even	*even*	*even*

Find each product using this table.

a. *odd* $\times$ *even* b. *odd* $\times$ *odd* c. *even* $\times$ *even*

SOLUTION a. *odd* $\times$ *even* = *even*
b. *odd* $\times$ *odd* = *odd*
c. *even* $\times$ *even* = *even*

EXAMPLE 2 Using the table in Example 1, answer the following.

a. Is the set closed with respect to the operation $\times$? Why or why not?

b. What is the identity element (if any)?
c. Which elements of the set have an inverse? Name the inverse of each of these elements.
d. Does the commutative property hold for this system? Verify your answer.
e. Does this set form a group under the operation of $\times$?

SOLUTION a. Yes, the set is closed with respect to the operation $\times$. There are no new elements appearing in the table.
b. The identity element is *odd*. It does not change any of the elements in the given set when it is operated on with them: *odd* $\times$ *odd* = *odd*, *even* $\times$ *odd* = *even*.
c. The element *even* does not have an inverse. However, *odd* $\times$ *odd* = *odd*, so *odd* has an inverse, itself.
d. Yes, the commutative property does hold for this system. Checking the elements, we have *odd* $\times$ *even* = *even* and *even* $\times$ *odd* = *even*. Therefore, *odd* $\times$ *even* = *even* $\times$ *odd*. Note that *odd* $\times$ *odd* = *odd* $\times$ *odd* and *even* $\times$ *even* = *even* $\times$ *even*.
e. No, the set of elements {*odd, even*} with the operation $\times$, does not form a group. Not every element has an inverse, because *even* does not have an inverse.

EXAMPLE 3 Sergeant Gig, a drill instructor, drills his drill team daily. During the drills, he issues four commands: *right face, left face, about face,* and *as you were*. Let

r = *right face* a = *about face*
l = *left face* y = *as you were*

Sometimes Sergeant Gig gives two commands in succession, such as *right face* followed by *left face*. If a person followed these commands he would wind up in the original position, the position of *as you were*. If we let $\otimes$ stand for the operation *followed by,* then $r \otimes l = y$. From this information, evaluate the following:

a. $r \otimes r$ b. $a \otimes r$ c. $l \otimes a$

SOLUTION a. $r \otimes r$ means *right face* followed by *right face,* which would be the same as *about face*. Therefore $r \otimes r = a$.
b. $a \otimes r$ means *about face* followed by *right face,* which would be the same as *left face*. Therefore $a \otimes r = l$.
c. $l \otimes a$ means *left face* followed by *about face,* which would be the same as *right face*. Therefore $l \otimes a = r$.

Using the idea of drill commands in Example 3, we can create a table to illustrate the results when any two commands are given. Using the notation r = *right face,* l = *left face,* a = *about face,* and y = *as you were,* and letting $\otimes$ stand for the operation *followed by,* we have a system consisting of a set of elements $\{r, l, a, y\}$ and an operation $\otimes$. Table 9 illustrates the operation $\otimes$ in this system.

TABLE 9

$\otimes$	r	l	a	y
r	a	y	l	r
l	y	a	r	l
a	l	r	y	a
y	r	l	a	y

From Table 9, we can verify the exercises in Example 3: $r \otimes r = a$, $a \otimes r = l$, and $l \otimes a = r$. Examining the entries in the table, we see that the set is closed with respect to the operation $\otimes$. The set also contains an identity element for the operation $\otimes$, the element y. Since y is the identity element, we can find the inverse of each element. The inverse of r is l, the inverse of l is r, a is its own inverse, and y is its own inverse. It can also be shown that the set satisfies the associative property for the operation $\otimes$. Hence, the set of element $\{r, l, a, y\}$ with the operation $\otimes$ satisfies the closure property, identity property, inverse property, and associative property, and therefore forms a group.

It is interesting to note that many of the examples discussed in this chapter are different, yet similar. We have discussed examples that contain different sets of elements which are combined using different operations, but they still satisfy the same properties of closure, identity, and so on. This is one of the unique characteristics of mathematical systems.

EXERCISES FOR SECTION 6.5

For Exercises 1–14, use Table 10, which defines an operation for the set of elements $\{P, Q, R, S\}$.

TABLE 10

:	P	Q	R	S
P	P	Q	R	S
Q	Q	R	S	P
R	R	S	P	Q
S	S	P	Q	R

Evaluate the following.

1. $P:Q$
2. $R:S$
3. $S:S$
4. $R:R$
5. $Q:R$
6. $R:P$
7. $R:(Q:S)$
8. $(S:Q):R$
9. $(P:Q):R$
10. $P:(Q:R)$
11. Is the set closed with respect to the operation :? Why or why not?

12. What is the identity element (if any)?

13. Which elements of the set have an inverse? Name the inverse of each of these elements.

14. Does this set form a group under the operation :?

Answer Exercises 15–29 by using Table 11, which defines an operation ⊙ for the elements of the set {$, ?, ¢}.

TABLE 11

⊙	$	?	¢
$	$	!	?
?	?	¢	!
¢	¢	?	$

Evaluate the following.

15. $ ⊙ ?

16. ? ⊙ $

17. ? ⊙ ¢

18. ¢ ⊙ ¢

19. ¢ ⊙ ?

20. $ ⊙ $

21. $ ⊙ (¢ ⊙ ?)

22. ($ ⊙ ¢) ⊙ ?

23. ¢ ⊙ (? ⊙ $)

24. (¢ ⊙ ?) ⊙ $

25. Is the set closed with respect to the operation ⊙? Why or why not?

26. What is the identity element (if any)?

27. Which elements of the set have an inverse? Name the inverse of each of these elements.

28. Is the commutative property satisfied for this system? Why or why not?

29. Does this set form a group under the operation ⊙?

Answer Exercises 30–46 by using Table 12, which defines an operation $*$ for the elements of the set $\{a, b, c, d, e\}$.

TABLE 12

*	**a**	**b**	**c**	**d**	**e**
a	a	b	c	d	e
b	b	c	d	e	a
c	c	d	e	a	b
d	d	e	a	b	c
e	e	a	b	c	d

Evaluate the following.

30. $b * a$

31. $c * d$

32. $a * d$

33. $d * a$

34. $e * e$

35. d * d

36. $c * a$

37. $c * c$

38. $c * (d * e)$

39. $(c * d) * e$

40. $(b * a) * d$

41. $b * (a * d)$

42. Is the set closed with respect to the operation $*$? Why or why not?

43. What is the identity element (if any)?

44. Which elements of the set have an inverse? Name the inverse of each of these elements.

45. Is the commutative property satisfied for this system? Why or why not?

46. Does this set form a group under the operation of $*$?

47. Given the set of counting numbers $\{1, 2, 3, 4, \ldots\}$ and the operation of addition, answer the following.
   a. Is this set closed with respect to the operation of addition?
   b. What is the identity element (if any)?
   c. Which elements of the set have an inverse with respect to addition?
   d. Does this set form a group under the operation of addition?

48. Given the set of counting numbers $\{1, 2, 3, 4, \ldots\}$ and the operation of multiplication, answer the following.
   a. Is the set closed with respect to the operation of multiplication?
   b. What is the identity element (if any)?
   c. Which elements of the set have an inverse with respect to the operation of multiplication?
   d. Does this set form a group under the operation of multiplication?

49. Given the equilateral triangle ABC (all sides equal), with a point in the center of the triangle so

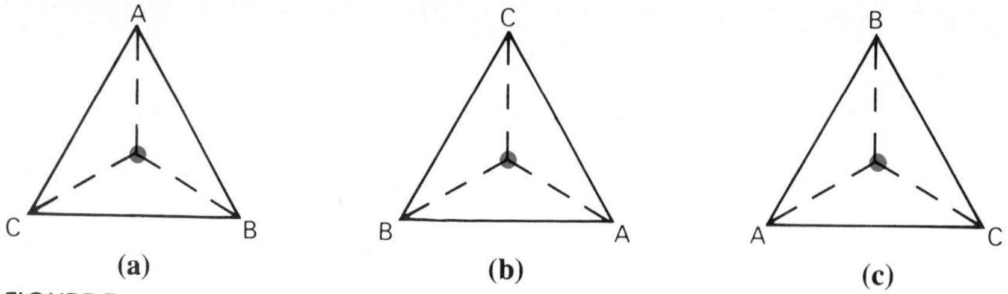

FIGURE 5

that the triangle may be rotated, as in Figure 5a. If we rotate the triangle 120° clockwise (call it rotation a), the triangle changes position: A goes to B, B goes to C, and C goes to A (see Fig. 5b). If we rotate the triangle 240° clockwise (call it rotation b), the triangle again changes position: A goes to C, B goes to A, and C goes to B (see Fig. 5c). If we rotate the triangle 360° clockwise (call it rotation c), the triangle returns to its original position: A goes to A, B goes to B, C goes to C (see Fig. 5a).

Let's define an operation $*$ to mean "followed by." Therefore, $a * b$ would mean rotation a followed by rotation b, which is the same as a rotation of 360°, or rotation c. Hence, for this system $a * b = c$.

a. Complete the following table for this system.

$*$	a	b	c
a		c	
b			b
c	a		

b. Is this set closed with respect to the operation $*$?
c. What is the identity element (if any)?
d. Which elements of the set have an inverse? Name the inverse of each of these elements.
e. Does this set form a group under the operation of $*$?

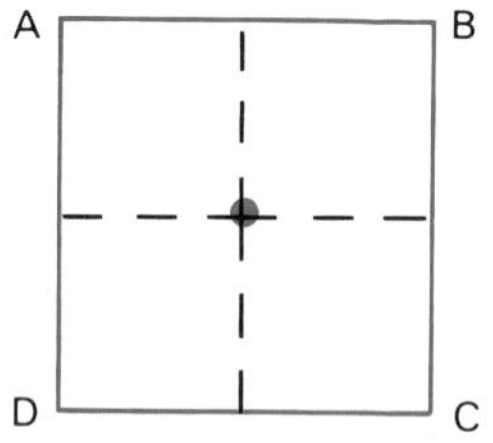

FIGURE 6

50. Given the square $ABCD$ (all sides equal), with a point in the center of the square so that the square can be rotated (see Fig. 6). Define the following rotations.

a = a clockwise rotation of 90°
b = a clockwise rotation of 180°
c = a clockwise rotation of 270°
d = a clockwise rotation of 360°

Let the operation $*$ mean "followed by."

a. Complete the following table for this system.

$*$	a	b	c	d
a		c		
b		d		
c				b
d	a			

b. Is this set closed with respect to the operation $*$?
c. What is the identity element (if any)?
d. Which elements of the set have an inverse? Name the inverse of each of these elements.
e. Does the set form a group under the operation $*$?

JUST FOR FUN If a certain kind of egg sells for $1 per dozen, which costs more, $\frac{1}{2}$ dozen dozen eggs, or 6 dozen dozen eggs?

6.6 AXIOMATIC SYSTEMS

Throughout this chapter we have examined the basic structure of mathematical systems. We have considered a mathematical system to be a set of elements together with one or more operations (rules) for combining any two elements of the set. In this section we will examine more closely the structure of mathematics itself.

Regardless of the branch of mathematics we choose to examine, the various branches are similar in the way that they are constructed. We can compare the basic characteristics of mathematics to the basic characteristics of a game.

Most games have a vocabulary of special terms, some defined and some undefined. After a player acquires this vocabulary, he learns the rules of the games—that is, what moves he can make, and what moves he cannot make. Normally, he accepts these rules without question. For instance, in the game of baseball, one rule says that a runner must run the bases in a counterclockwise direction. When a youngster is first learning how to play baseball, he sometimes wants to run the bases in a clockwise direction. But when he is told that the rules state that a runner must run the bases the other way (counterclockwise), he accepts this. Similarly, we all accept the fact that a queen ranks higher than a jack in card games. Why? Because the rules say so!

In mathematics, the rules are called *axioms*. An **axiom** is a statement that is accepted as true without proof. Each mathematical system must have axioms that are consistent. They must not contradict one another, just as the rules of a game should not contradict one another.

After learning the undefined terms, defined terms, and rules of a game, we are ready to play. Once we have mastered the elementary moves of the game, we usually try more complicated moves using the rules (as in the game of chess). In mathematics, these new results that have evolved from the undefined terms, defined terms, and axioms (rules) are called *theorems*. **Theorems are logical deductions that are made from undefined terms, defined terms, and axioms. Some theorems are even logical deductions from other theorems.**

In essence, an *axiomatic* system consists of four main parts.

1. **Undefined terms**
2. **Defined terms (definitions)**
3. **Axioms**
4. **Theorems**

Undefined terms are necessary in an axiomatic system, as they are used to form a fundamental vocabulary with which other terms can be defined. Even though a term may be undefined, that does not mean that we do not know what it is. In a high school geometry course, the terms *point* and *line* are not defined terms, yet we know what they are. In Chapter 1 of this text, *set* is an undefined term, but that does not prevent us from having an intuitive idea of what *set* means.

Definitions of defined terms may use undefined terms or terms that have been previously defined. Definitions should be concise, consistent, and not circular.

As we noted earlier, axioms are statements that are accepted as true without proof. In a given system, no axiom should contradict another; that is, the axioms must be consistent. Axioms are necessary in a system because not everything can be proved. Axioms are needed to derive other statements. The derived statements are the theorems.

Let us now examine a proved theorem for a system that uses undefined terms, defined terms, and axioms. The following proof is found in a high school geometry course. The statement (theorem) to be proved is "If two straight lines intersect, then the vertical angles thus formed are equal."

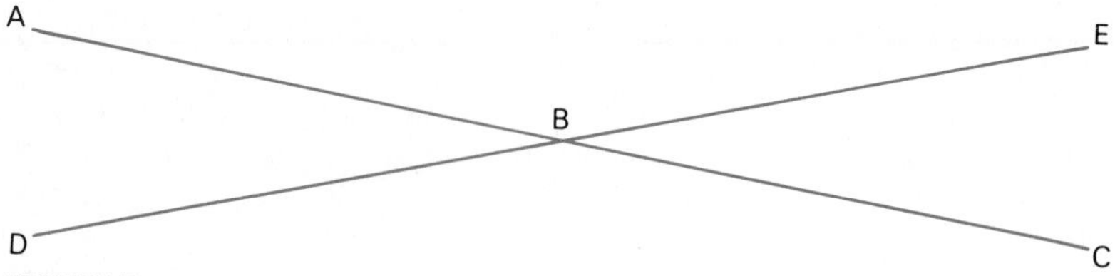

FIGURE 7

We first construct a diagram representing this situation (see Fig. 7).

This diagram shows straight lines *AC* and *DE* intersecting at *B* and forming vertical angles *ABD* and *EBC*. We want to prove that the measure of angle *ABD* equals the measure of angle *EBC*, that is, $m\angle ABD = m\angle EBC$. The following is a formal proof.

Statements	Reasons
1. *AC* and *DE* are straight lines	1. Given
2. Angles *ABC* and *DBE* are straight angles	2. Definition of a straight angle
3. $m\angle ABC = m\angle DBE$	3. All straight angles are equal
4. $m\angle ABE = m\angle ABE$	4. Identity
5. $m\angle ABD = m\angle EBC$	5. If the same quantity is subtracted from two equal quantities, the remainders are equal

This proof uses undefined terms (straight lines), definitions (the definitions of a straight angle), and axioms. The axioms appear in reason 3 (All straight angles are equal) and reason 5 (If the same quantity is subtracted from two equal quantities, the remainders are equal). The axioms used are consistent; that is, they do not contradict each other. The diagram in Figure 7 could be considered a *model*—that is, a physical interpretation of the undefined terms which satisfies the axioms.

Let's examine another example, but one that is less familiar. We start with the following axioms, and the undefined terms *road, town,* and *stop sign.*

1. There is at least one road in the town.
2. Every stop sign is on exactly two roads.
3. Every road has exactly two stop signs on it.

We wish to prove that there is at least one stop sign in the town. It is helpful to construct a diagram (model) that satisfies all of the axioms (see Fig. 8).

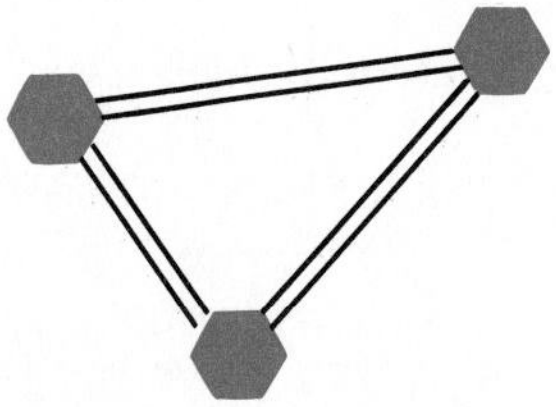

FIGURE 8

We want to prove that there is at least one stop sign in the town. Axiom 1 states that there must be at least one road in the town, and axiom 3 says that every road has exactly two stop signs on it. Hence, there must be at least one stop sign in the town.

Note that from this set of axioms we could have derived other conclusions, but we only derived the desired conclusion.

EXAMPLE 1 Given:

Axiom 1: There is exactly one road between any two traffic lights.
Axiom 2: For every road there exists a traffic light not on that road.
Axiom 3: There exist at least two traffic lights.

Prove: There exist at least three roads.

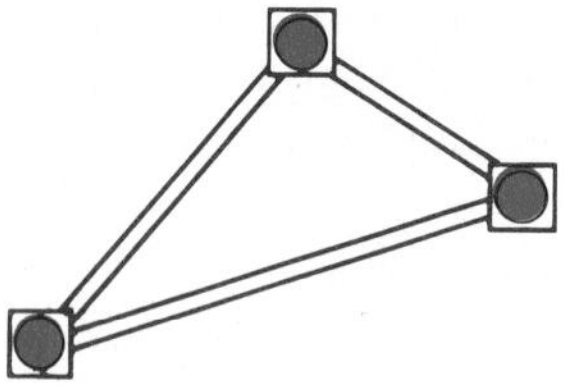

FIGURE 9

SOLUTION First we construct a model that satisfies all of the given axioms Let  represent a traffic light and ═══ represent a road. Our model appears in Figure 9; this helps us understand the axioms.

Axiom 3 tells us that there are at least two traffic lights, and axiom 1 states that there is exactly one road between any two traffic lights. So far we have one road. Now axiom 2 tells us that there has to be another traffic light not on the given road. But if there is another traffic light, then there must be two more roads: axiom 1 tells us that there is exactly one road between any two traffic lights, and since there is a third traffic light, it must be connected to the other two lights by means of two roads. Therefore we have at least three roads.

EXERCISES FOR SECTION 6.6

1. Consider the following set of axioms.

   I. There are at least two buildings on campus.
   II. There is exactly one sidewalk between any two buildings.
   III. Not all of the buildings are on the same sidewalk.

   If the capital letters *A*, *B*, *C*, and so on represent buildings and the lines represent sidewalks, which of the following models represent the given axiomatic system?

   a. 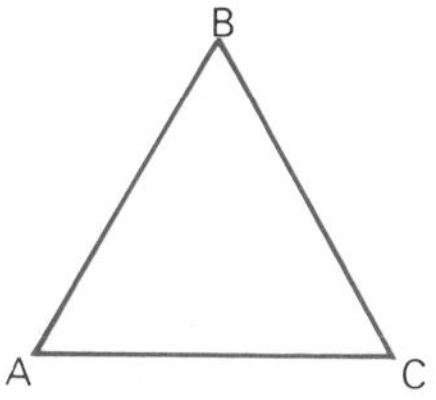

   b. 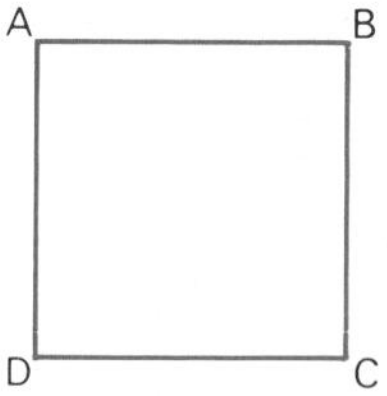

c.
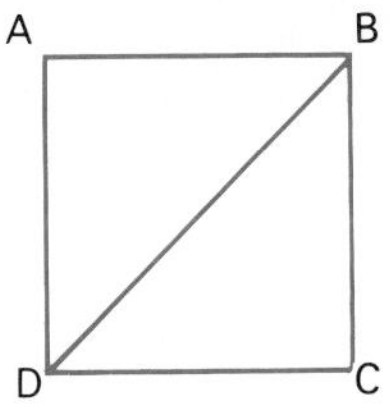

d.
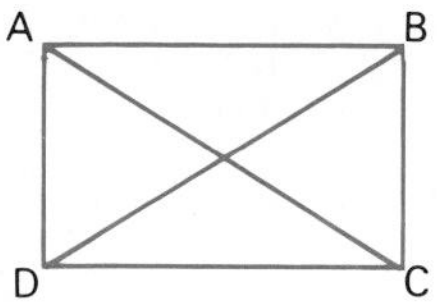

e.
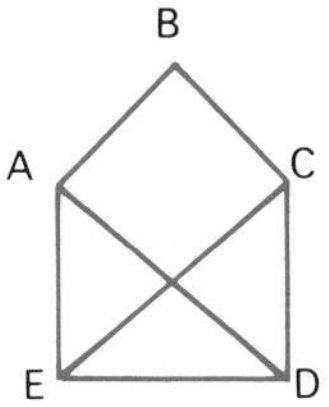

f.
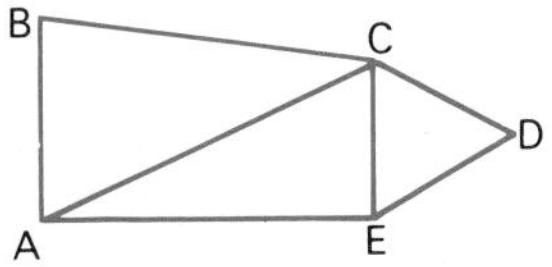

2. Consider the following set of axioms.
   I. There are exactly four bleeps.
   II. Each bleep is on a cleep.
   III. No bleep is on a cleep by itself.

   If the capital letters, *A, B, C,* and so on, represent bleeps and the lines represent cleeps, which of the following models represent the given axiomatic system?

a.
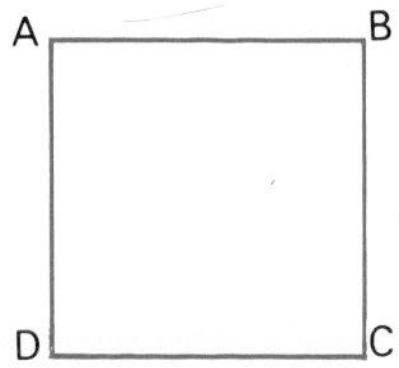

b.
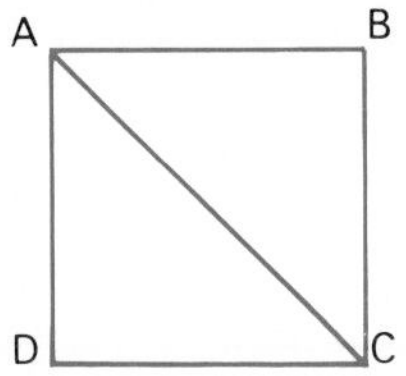

c.
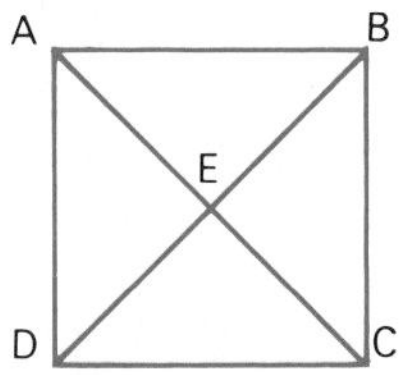

d.
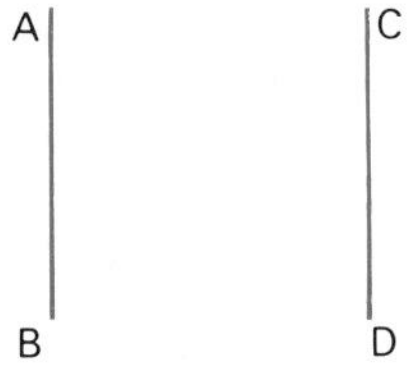

e.
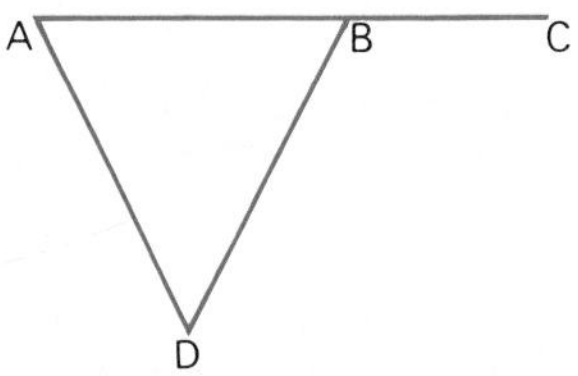

f.
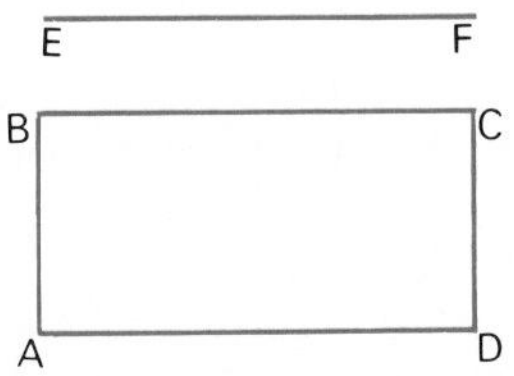

3. What is the basic structure of an axiomatic system?

4. Given the following axioms:
   I. There are at least two buildings on campus.
   II. There is exactly one sidewalk between any two buildings.
   III. Not all of the buildings are on the same sidewalk.

   Prove: There exist at least three buildings on campus.

5. Given the following axioms:
   I. There are exactly two points on each line.
   II. There is at least one line.
   III. For each pair of points, there is one and only one line containing them.
   IV. Corresponding to each line there is exactly one other line which has no point in common with it.

   Prove: There are at least four points.

6. Given the following axioms:
   I. Not all cars are in the same garage.
   II. There exist at least two cars.
   III. There is exactly one garage for any two cars.

   Prove: There exist at least two garages.

7. Given the following axioms:
   I. If equal quantities are added to equal quantities, then the sums are equal.
   II. If equal quantities are subtracted from equal quantities, then the differences are equal.
   III. If equal quantities are multiplied by equal quantities, then the products are equal.
   IV. If equal quantities are divided by equal quantities (except zero), then the quotients are equal.
   V. A quantity may be substituted for its equal in any process.

   State which axiom is used in each step in solving the following equations.
   a. Given: $3x + 4 = 25$
      1. $3x = 21$
      2. $x = 7$

b. Given: $2x - 7 = 15$
   1. $2x = 22$
   2. $x = 11$

c. Given: $\frac{x}{2} - 4 = 3$
   1. $\frac{x}{2} = 7$
   2. $x = 14$

d. Given: $\begin{cases} x + 3y = 7 \\ x + y = 5 \end{cases}$
   1. $2y = 2$
   2. $y = 1$
   3. $x + 1 = 5$
   4. $x = 4$

e. Given: $\begin{cases} 2x + y = 8 \\ x - y = 4 \end{cases}$
   1. $3x = 12$
   2. $x = 4$
   3. $4 - y = 4$
   4. $4 = 4 + y$
   5. $0 = y$

8. Using the axioms given in Exercise 7, prove the desired conclusion for each of the following.
   a. Given: $\overline{AC} = \overline{BC}$
      $\overline{DC} = \overline{EC}$
      Prove: $\overline{AE} = \overline{BD}$

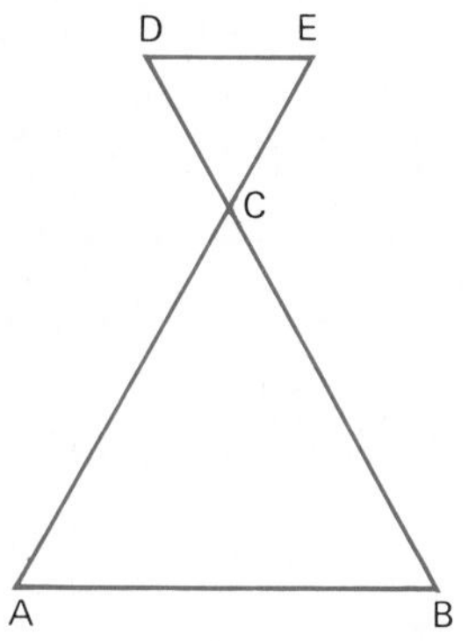

b. Given: $m\angle EBC = m\angle DBA$
   Prove: $m\angle 1 = m\angle 2$

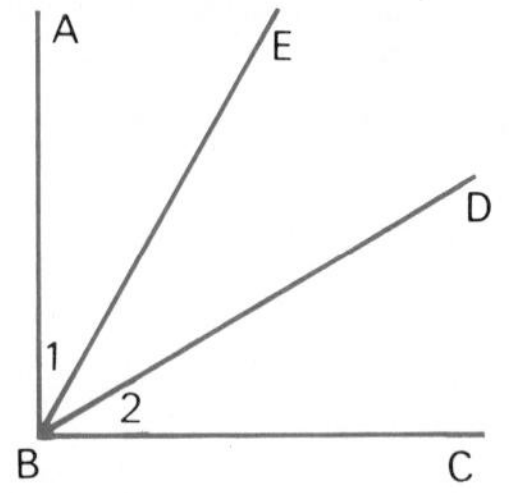

c. Given: $m\angle ABC = m\angle EFG$
   $m\angle 1 = m\angle 3$
   Prove: $m\angle 2 = m\angle 4$

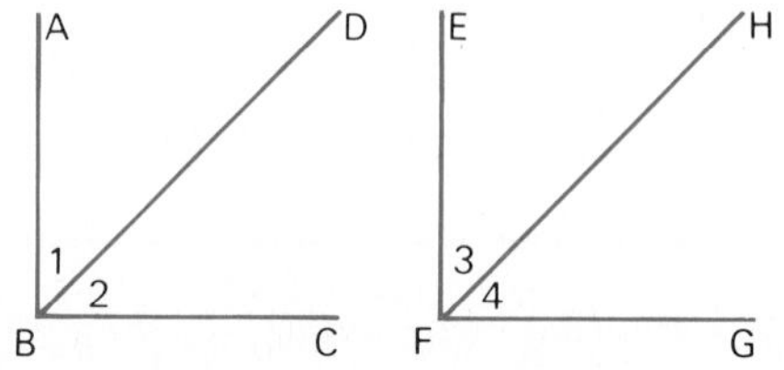

d. Given: $m\angle 1 = m\angle 2$
   $m\angle 1 = m\angle 3$
   $m\angle 3 = m\angle 4$
   Prove: $m\angle 4 = m\angle 2$

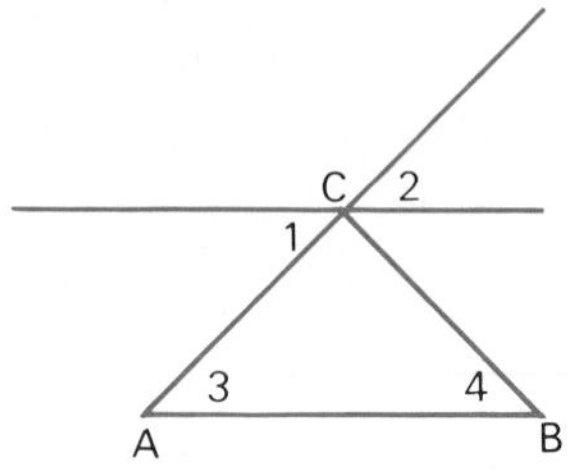

e. Given: $\overline{AB}$, $\overline{CD}$ and EF are straight lines.
   $m\angle b = m\angle c$
   $m\angle a = m\angle b$
   Prove: $m\angle a = m\angle c$

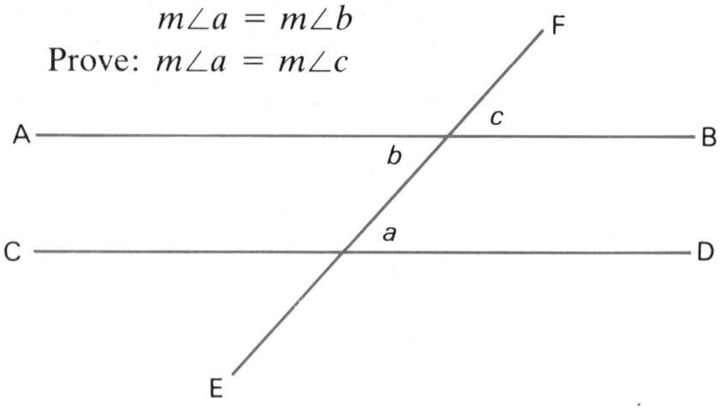

JUST FOR FUN

Benny wanted to plant some tomato plants in his garden. Altogether he had 12 tomato plants. He wanted to plant these in six rows, but with four tomato plants in each row. How did Benny do it?

Summary

A mathematical system is composed of a nonempty set of elements together with one or more operations (rules) for combining any two elements of the set. A system is *closed* if when we operate on any two elements in the given set, the result is also a member of the given set. If there is an element e in the set that does not change any element when it is operated on together with that element, then e is the *identity element.* Each element in the system has an *inverse* if there exists an element which when operated on together with the given element results in the identity element. An operation is *associative* if the location of parentheses in a problem does not affect the answer. When a set of elements and an operation satisfy the closure property, identity property, inverse property, and associative property, we say that the set of elements forms a *group* under the given operation. A group is called a *commutative group* if it is also satisfies the *commutative property*—that is, if $a * b = b * a$ for all elements a and b in the group.

Clock arithmetic is an example of a mathematical system, and can be used to illustrate the properties mentioned. A system that repeats itself or is cyclical is called a *modular system.* If we have $a \equiv b \pmod{m}$, we say that *a is equivalent to b mod m.* This means that a and b both have the same remainder when they are divided by m.

Abstract mathematical systems can be thought of as mathematical systems without numbers. This is just an extension of our basic notion of a system. The elements in the set can be anything (they are usually letters), and the operation can be defined in any way (e.g., "followed by" in Example 3 of Sec. 6.5). An abstract system can also form a group, providing it satisfies the properties of a group.

An *axiomatic system* consists of *undefined terms, defined terms, axioms,* and *theorems.* A *model* for an axiomatic system is a physical interpretation or diagram of the undefined terms; it must be constructed so that all of the axioms of the system are satisfied. The axioms in a system must also be consistent; that is, they must not contradict each other. As we continue our study of mathematics, we will see that all branches of mathematics are similar in that they are based upon an axiomatic system.

Vocabulary Check

group	undefined terms	model	axiomatic system
modulo	axiom	inverse element	commutative
closure	theorem	mathematical system	defined terms
identity element	associative		

Review Exercises for Chapter 6

1. Evaluate each of the following on a 12-hour clock.
   a. $7 + 7$ b. $9 + 4$
   c. $8 + 7$ d. $11 + 12$
   e. $10 + 10$ f. $9 + 8$

2. Evaluate each of the following on a 12-hour clock.
   a. $6 - 8$ b. $7 - 11$
   c. $3 - 7$ d. $2 - 12$
   e. $6 - 10$ f. $8 - 9$

3. Evaluate each of the following on a 12-hour clock.
   a. 4×9 b. 8×4
   c. 6×7 d. 9×3
   e. 6×5 f. 6×4

4. Evaluate each of the following on a 12-hour clock.
   a. $7 \times (8 + 6)$ b. $4 \times (3 - 6)$
   c. $4 - (6 - 8)$ d. $(4 - 6) - 8$
   e. $2 \times (6 - 8)$ f. $(2 \times 6) - 8$

5. Describe in your own words what a *mathematical system* is.

6. Given the set of whole numbers $\{0, 1, 2, 3, \ldots\}$, tell whether each sentence is true or false.
   a. The set is closed with respect to addition.
   b. The set is closed with respect to subtraction.
   c. The set is closed with respect to multiplication.
   d. The set is closed with respect to division.

7. True or false?
   a. The set of whole numbers $\{0, 1, 2, 3, \ldots\}$ contains an identity element for the operation of addition.
   b. The set of whole numbers contains an additive inverse element for each element in the set.
   c. The 12-hour clock system contains an identity element for the operation of multiplication.
   d. The 12-hour clock system contains a multiplicative inverse element for each element.

8. True or false: The set of whole numbers
   a. is associative with respect to addition
   b. is commutative with respect to addition.

9. True or false: The set of whole numbers
   a. is a group with respect to addition
   b. is a group with respect to multiplication.

For Questions 10 and 11, evaluate the following, given that 1 = spring, 2 = summer, 3 = fall, and 4 = winter. (Your answer should be in terms of a season.)

10. a. summer + summer
    b. winter + summer
    c. fall + winter
    d. summer − fall
    e. fall − winter
    f. spring − summer

11. a. fall × fall
    b. summer × fall
    c. fall × (fall + winter)
    d. fall × (spring − winter)
    e. winter × (fall − winter)
    f. summer × (winter + spring)

12. Find the equivalent to each of the following in the modulo 5 system.
    a. 42 b. 61 c. 89
    d. −32 e. −108 f. 2003

13. Using the elements of the indicated modular systems, find a replacement for each question mark so that the statement is true.
    a. $4 + ? \equiv 2 \pmod 5$
    b. $2 - ? \equiv 4 \pmod 7$

c. $3 \times ? \equiv 2 \pmod 7$
d. $? \times 2 \equiv 3 \pmod 7$
e. $4 \times (3 + ?) \equiv 1 \pmod 5$
f. $2 \times (3 - ?) \equiv 4 \pmod 7$

14. Carl the coin collector obtained two rolls of pennies from the bank. After examining the coins, he divided the pennies into two groups: those with mint marks, and those without mint marks. Working with the coins that had mint marks, Carl arranged these pennies in stacks of 5, with 2 left over. Next he arranged these coins in stacks of 7, with 1 left over. Finally, he arranged them in stacks of 3, with 0 left over. How many of the pennies had a mint mark?

15. The speedometer on a car indicates how fast a car travels; the odometer indicates how many miles a car travels. The odometer is an example of a modular system. Why? What modular system is used on the odometer of an ordinary car?

16. Answer the following by using Table 13, which defines an operation * for the elements of the set {$, ¢, &, ?}.

TABLE 13

*	$	¢	&	?
$	¢	$	?	&
¢	$	¢	&	?
&	?	&	π	$
?	&	?	$	π

a. $ * ¢
b. ¢ * ?
c. & * &
d. ? * ?
e. ¢ * &
f. & * ?
g. ¢ * (& * $)
h. & * (? *$)
i. Is this set closed with respect to the operation * ? Why or why not?
j. What is the identity element (if any)?
k. Which elements of the set have an inverse?
l. Does this set form a group under the operation *?

17. Name the basic parts of an axiomatic system.

18. Given the following axioms:

I. There are at least three squirrels.
II. Each squirrel is in exactly one tree.
III. No squirrel is in a tree by itself.
IV. For every tree, there is a squirrel that is not in that tree.

Prove: There exist at least four squirrels.

19. The sides of a triangle are represented by $3x + 4$, $2x + 8$, and $5x - 4$. If the perimeter of the triangle is 48, prove that the triangle is equilateral, that is, that all three sides are equal.

20. Given: $\overline{EOA}$ is a straight line
Prove: $\overline{OC}$ bisects $\angle BOD$

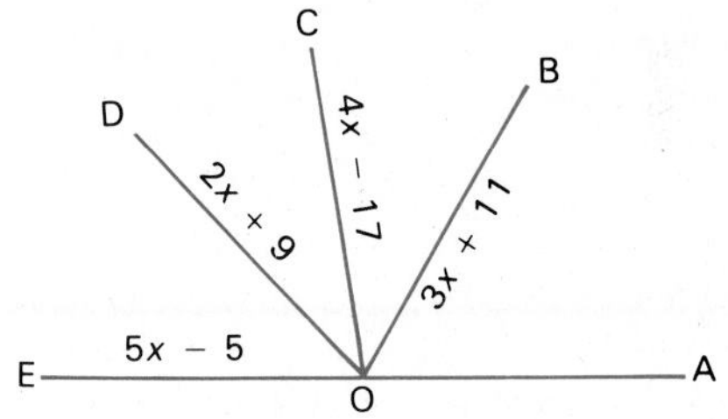

Chapter Quiz

Determine whether the following statements are true or false.

1. The set of whole numbers is closed for the operation of addition.
2. The set of whole numbers contains an identity element for the operation of multiplication.
3. For every element in a group, there must be an

element such that combining the two produces the identity element.

4. $(a * b) * c = (b * a) * c$ is an example of the associative property.

5. In the 12-hour clock system $9 + 7 = 4$ and $10 + 9 = 7$.

6. If a system consisting of a set of elements $\{a, b, c, \ldots\}$ and an operation $*$ forms a group, then we may conclude that the operation is commutative.

7. The set of integers is closed with respect to division.

8. $2{,}001 \equiv 111 \pmod{5}$

9. $3 \times (4 - 2) \equiv 2 \times (5 - 6) \pmod{7}$

10. The set of counting numbers forms a group under the operation of additon.

11. The identity in the "week" system for the operation of addition is Sunday.

12. A mathematical system consists of a set of elements together with one or more rules for combining elements of the set.

13. An axiomatic system consists of undefined terms, definitions, axioms, and theorems.

14. The 12-hour clock system under the operation of addition contains an inverse for each of its elements.

15. The 12-hour clock system under the operation of multiplication contains an inverse for each of its elements.

16. If $4 \times y \equiv 1 \pmod{5}$, then $y = \frac{1}{4}$

17. If $x - 6 \equiv 2 \pmod{7}$, then $x = 2$

18. The set of odd counting numbers is closed with respect to addition.

19. The set of odd whole numbers is closed with respect to multiplication.

20. Axioms are necessary in an axiomatic system because not everything can be proved.

21. If a system consisting of a set of elements $\{a, b, c, \ldots\}$ and an operation $*$ satisfies the commutative property, then the system forms a group.

22. $99 \equiv 44 \pmod{7}$

23. The set of counting numbers and the operation of additon form a group.

24. If $a \equiv b \pmod{m}$, then a and b both have the same remainder when they are divided by m.

25. Theorems are logical deductions that are made from undefined terms, defined terms, and axioms.

JUST FOR FUN

Most proofs are done by means of deduction; that is, we proceed from the premises, step by step, to the conclusion. As we go from one step to the next, we must have a reason for each step to show that it follows logically. The following is an example of a proof that does not obey the rules: even though the derivation appears to be correct, it is not. Can you find the error?

Statements	Reasons
1. $a = b$	Given
2. $a^2 = ab$	Multiplying both sides by a
3. $a^2 - b^2 = ab - b^2$	Substracting b^2 from both sides
4. $(a + b)(a - b) = b(a - b)$	Factoring both sides
5. $\frac{(a + b)(a - b)}{(a - b)} = \frac{b(a - b)}{(a - b)}$	Dividing both sides by $(a - b)$
6. $(a + b) = b$	Result of step 5
7. $b + b = b$	Recall that $a = b$ (step 1), so we may substitute b for a
8. $2b = b$	Combining $b + b$
9. $\frac{2b}{b} = \frac{b}{b}$	Dividing both sides by b
10. Therefore, $2 = 1$	Result of step 9 (this is the conclusion)

CHAPTER 7

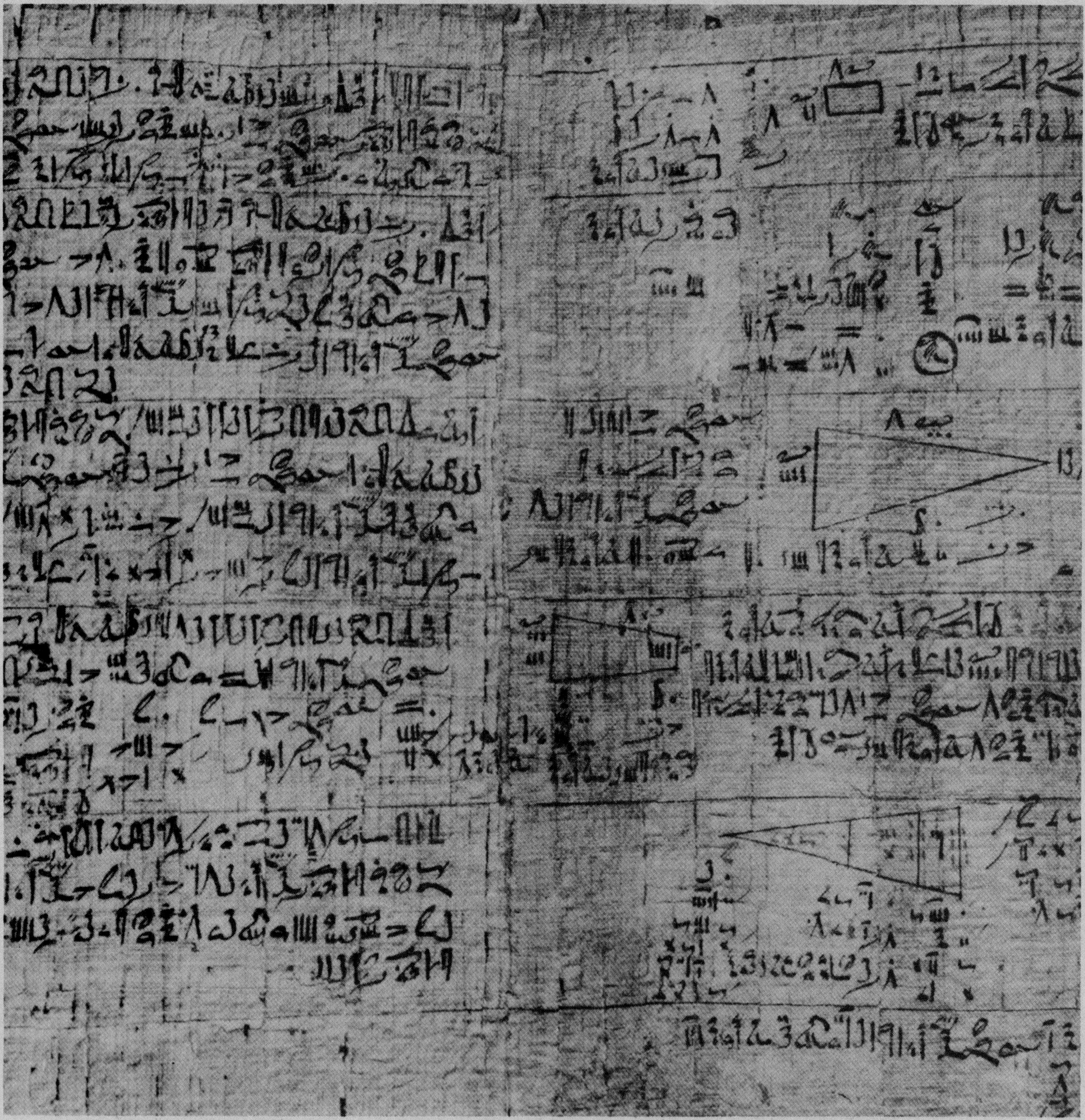

SYSTEMS OF NUMERATION

AFTER STUDYING THIS CHAPTER YOU WILL BE ABLE TO DO THE FOLLOWING:

1. Express a counting number as an **Egyptian numeral,** and express an Egyptian numeral as a **base 10 numeral.**
2. Add and subtract Egyptian numerals.
3. List the distinguishing characteristics of systems of numeration that use a **simple grouping, multiplicative grouping,** or **place-value system.**
4. Identify systems of numeration that use simple grouping, multiplicative grouping, and place-value systems.
5. Convert a numeral in any base other than 10 to a base 10 numeral by means of **expanded notation.**
6. Convert a base 10 numeral to any desired base by means of the **division algorithm.**
7. Identify systems of numeration other than base 10 to which we are exposed in everyday life.
8. Add, subtract, and multiply in base 5.
9. Add, subtract, and multiply in base 2.

7.1 INTRODUCTION

Regardless of the human culture we examine, the people in that culture need a system of counting. Man needs numbers: he must be able to count. No matter what type of job a person has, he or she must be able to count in order to cope with situations encountered in everyday living, even in such common activities as paying bills and making change. Man has needed some form of counting (numeration) system ever since he began to reason and to develop a civilization.

Anthropologists maintain that many of the primitive tribes of prehistoric times had some system for counting. The most primitive counting systems went only as high as three or four, with anything greater being described as "many"; however, many so-called primitive peoples had much more sophisticated systems.

Probably the first form of counting was done by matching the things to be counted (such as animals) with something else, such as fingers, toes, stones, or sticks. A pile of pebbles may have been matched, one-to-one, to the number of animals in a certain herd. A herdsman might have placed the pebbles in a pile as he was letting the animals out to graze, one pebble corresponding to each animal. When the animals returned, the herdsman would match a pebble from the pile with each animal. If, after all the animals had passed by, there were some pebbles still remaining that had not been matched, then the herdsman would know that there were some animals not accounted for, perhaps one, two, three, or even "many."

Man has invented many different ways to record numbers. The ancient Chinese recorded numbers by tying knots on a string, as did the Incas. Another way of recording numbers used in early history was to draw pictures or slashes in the dirt. Early man made slashes (marks) in the dirt; each mark represented an animal that he owned. Other civilizations recorded these marks on stones or pieces of clay, or made notches in a stick. Even today we still sometimes use tally marks to record numbers.

Tally marks such as /// are used to represent numbers. The mark /// represents the number three, whereas ////// represents the number six. Note that the tally marks are symbols used to represent numbers. All systems of numeration use symbols to represent numbers. In the Roman system of numeration, V represents the number five. You may recall that the symbols, I, II, III, IV, V, X, L, D, and so on, are called *Roman numerals.* Any symbol for a number is called a **numeral.** Symbols like V or 5 represent the number five; they are numerals, not numbers. Numbers exist in your mind, but when you put the symbol V or 5 on a piece of paper you are using a written symbol (numeral) to represent the number five.

The distinction between a number and a numeral is pointed out here because of the many different types of numerals used throughout history by different cultures. Do not become overly concerned with this distinction; we shall make use of it only when it is helpful to do so. Many times we shall use numerals just as we would numbers.

The distinction between a number and a numeral can be compared to the distinction between the letter O and the number zero (0). There is distinct difference of meaning between the two, but we all know what a person means when he tells us that his phone number is "two-oh-eight-five," that is, 2085. Similarly, there is a distinction between a number and a numeral, but we shall only emphasize this distinction when it is helpful to do so.

7.2 SIMPLE GROUPING SYSTEMS

One of the oldest known systems of numeration is that of the Egyptians. The Egyptian culture was a fairly advanced one, and consequently they had an advanced system of numeration. The pyramids are proof of their technical ingenuity. These pyramids were tombs for Egyptian kings. Pictures called *hieroglyphics* were painted on the walls of these tombs to represent numbers.

The Egyptians used hieroglyphics as early as 3400 B.C. A single line or stroke represented items up to 10. After reaching 10, a new symbol is used to indicate a set of 10 things. This is an example of *simple grouping.*

A *simple grouping system* is a system in which the position of a symbol does not affect the number represented, and in which a different symbol is used to indicate a certain number or group of things.

In the Egyptian system, special symbols were used to represent tens, hundreds, thousands, ten-thousands, hundred-thousands, and millions. That is, the Egyptians used a different symbol for each power of 10. A **power of 10** is 10 raised to a number. For example, $10^1 = 10$ is the first power of 10, $10^2 = 10 \times 10 = 100$ is the second power of 10, and $10^3 = 10 \times 10 \times 10 = 1{,}000$ is the the third power of 10. Because the Egyptians used a different symbol for each power of 10, we say that their system had a **base** of 10. In the expression 10^3, 10 is the base and 3 is the exponent, and 1,000, or 10^3, is the third power of 10.

TABLE 1
EGYPTIAN NUMERALS AND THEIR VALUES

Egyptian Numerals	Name	Value	Power of 10
𓏺	Stroke	1	10^0
𓎆	Heelbone	10	10^1
𓍢	Coiled Rope	100	10^2
𓆼	Lotus Flower	1,000	10^3
𓂭	Pointed Finger	10,000	10^4
𓆐	Polywog	100,000	10^5
𓁨	Astonished Man	1,000,000	10^6

Some of the Egyptian hieroglyphic numerals and their corresponding values are given in Table 1. It should be noted that the order and shape of the figures may vary from one reference to another. This is because different examples were obtained from different sources (inscriptions) and are a reflection of the individuals who made the original inscriptions.

The only fractions that the ancient Egyptians considered were those they called *unit fractions.* That is, the numeral 1 over an ordinary number. These unit fractions were denoted by placing the symbol 𓂋 over the number for the denominator. For example,

𓂋 over 𓏺𓏺𓏺 $= \frac{1}{3}$ and 𓂋 over 𓎆𓎆 $= \frac{1}{20}$

Fractions that were not unit fractions were expressed as sums of unit fractions. In the Egyptian system $\frac{3}{4}$ would be expressed as the sum of $\frac{1}{2}$ and $\frac{1}{4}$. That is, $\frac{3}{4}$ would be written as

$\frac{1}{2} + \frac{1}{4} =$ 𓂋 over 𓏺𓏺 + 𓂋 over 𓏺𓏺𓏺𓏺

Because the Egyptians used a simple grouping system, the position of a symbol does not affect the number represented. That is, the numeral 𓎆𓏺𓏺 represents 12, as do 𓏺𓏺𓎆 and 𓎆 over 𓏺𓏺. This is certainly not the case for other systems of numeration. For example, 23 does not equal 32 and 128 does not equal 821 in our system of numeration.

NOTE OF INTEREST

One of the Great Pyramids in Egypt that is considered to have an architectural design that is technically perfect is Cheops' pyramid. The four sides of the pyramid are almost perfectly aligned with true north, south, east, and west. The base of the pyramid covers an area that is over 13 acres! Yet each side of the pyramid joins the other at almost perfect right angles. The blocks that cover the exterior of the pyramid are fitted together with joints that measure approximately $\frac{1}{50}$ of an inch. Truly, the ancient Egyptians were masters of the mathematical tools they possessed.

NOTE OF INTEREST

Following is an illustrative example of multiplication using a technique employed by the ancient Egyptians. Consider 12 × 13. Using the Egyptians' method, prepare two columns. In the first column successively double the number 1; in the second column successively double the multiplicand (13). Stop when you have obtained numbers in the first column whose sum equals the multiplier (12). For example:

1	13
2	26
*4	→ 52
*8	→ 104
STOP (4 + 8 = 12)	156

Add the corresponding "doubles" and the result is the desired answer. Therefore,

$$12 \times 13 = 156$$

A simple grouping system can also be thought of as an "additive" system, because we must add the values of the symbols rather than concern ourselves with the position of the symbols. Sometimes the Egyptians wrote their numerals in left to right order, from largest to smallest; at other times the numerals are written in a right to left order. Regardless of the arrangement of the symbols, the value of the number represented is not affected. Both of the following pictures represent 2304.

In order to evaluate the Egyptian numeral represented, we just add the values of the hieroglyphics, regardless of their position.

To add numbers that are expressed as Egyptian numerals, we group the same symbols together, and then simplify (rewrite) the expression. Consider the following addition problem.

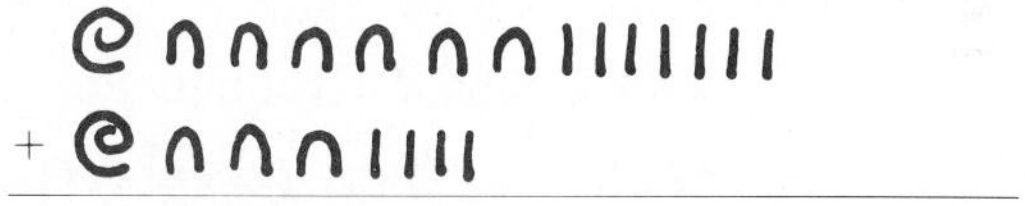

In order to add these two numbers, we first group the same symbols together. Then to obtain our final answer, we rewrite 10 of the strokes as a heelbone. This results in a string of 10 heelbones, which we can rewrite as a coiled rope.

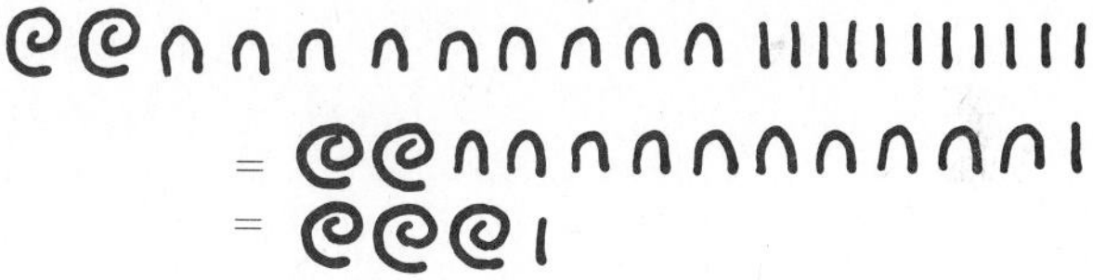

In order to subtract two numbers that are expressed as Egyptian numerals, we simply take away those symbols that are contained in both numbers. It may be necessary to rewrite some numerals in terms of other symbols, such as ∩ = ||||||||||. Consider the following subtraction problem.

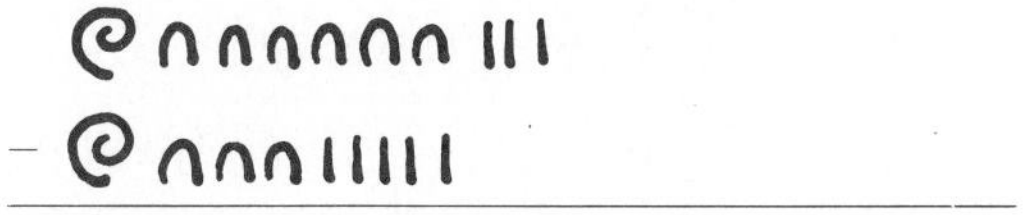

As we cannot subtract five strokes from three strokes, we rewrite one of the heelbones as ten strokes, and we have

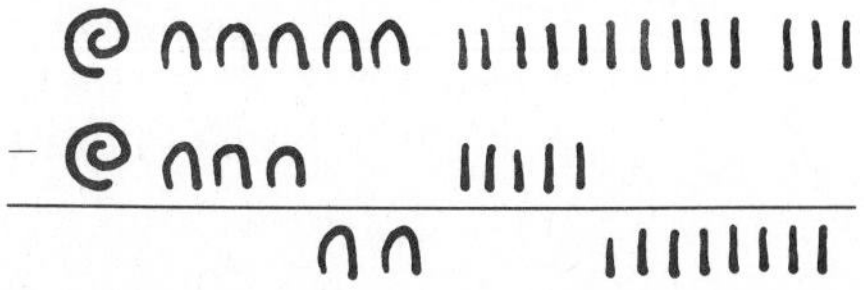

Remember that in the Egyptian system of computation, it is necessary to group the symbols together, and it is also sometimes necessary to rewrite some numerals in terms of other symbols.

EXAMPLE 1 Express the following as Egyptian numerals.

a. 13   b. 231   c. 13,423

SOLUTION a. 13 is one 10 and three 1s; that is, 𓎆 𓏺𓏺𓏺

b. 231 is two 100s, three 10s, and one 1; that is, 𓍢𓍢 𓎆𓎆𓎆 𓏺

c. 13,423 is one 10,000, three 1,000s, four 100s, two 10s, and three 1s; that is, 𓂭𓆼𓆼𓆼𓍢𓍢𓍢𓍢𓎆𓎆𓏺𓏺𓏺

EXAMPLE 2 Evaluate the following Egyptian numerals.

a. 𓍢𓍢𓎆𓎆𓎆𓎆 𓏺𓏺𓏺𓏺   b. 𓆼𓆼𓍢𓎆𓎆𓎆𓏺

c. 𓁨 𓂭𓍢𓎆𓏺𓏺

SOLUTION a. 𓍢𓍢𓎆𓎆𓎆𓎆𓏺𓏺𓏺𓏺 = two 100s, four 10s, four 1s; that is, $(2 \times 100) + (4 \times 10) + (4 \times 1) = 200 + 40 + 4 = 244$.

b. 𓆼𓆼𓍢𓎆𓎆𓎆𓏺 = two 1,000s, one 100, three 10s, and one 1; that is, $(2 \times 1000) + (1 \times 100) + (3 \times 10) + (1 \times 1) = 2000 + 100 + 30 + 1 = 2131$.

c. 𓁨𓂭𓍢𓎆𓏺𓏺 = one 1,000,000, one 10,000, one 100, one 10, and two 1s; that is, $(1 \times 1{,}000{,}000) + (1 \times 10{,}000) + (1 \times 100) + (1 \times 10) + (2 \times 1) = 1{,}000{,}000 + 10{,}000 + 100 + 10 + 2 = 1{,}010{,}112$.

EXAMPLE 3 Add.

a. 𓎆𓎆𓎆𓏺𓏺 + 𓎆𓎆𓎆𓎆𓏺𓏺𓏺

b. 𓆼𓎆𓎆𓎆𓎆𓎆𓎆𓏺𓏺𓏺𓏺𓏺𓏺 + 𓆼𓎆𓎆𓎆𓏺𓏺𓏺𓏺𓏺

SOLUTION a. In order to add these two numbers, we simply combine the symbols.

𓎆𓎆𓎆𓏺𓏺 + 𓎆𓎆𓎆𓎆𓏺𓏺𓏺 = 𓎆𓎆𓎆𓎆𓎆𓎆𓎆𓏺𓏺𓏺𓏺𓏺

b. In order to add these two numbers, we first group the same symbols together and then simplify.

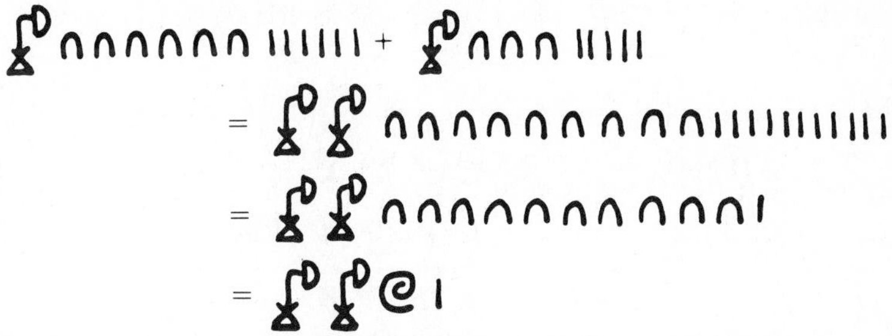

EXAMPLE 4 Subtract.

a. 

b. 

SOLUTION a. In order to subtract two numbers that are expressed as Egyptian numerals, we subtract (or take away) those symbols that are contained in both numerals.

b. This subtraction cannot be performed immediately; we must first rewrite the first number, since it does not contain 7 ones or 4 tens as such.

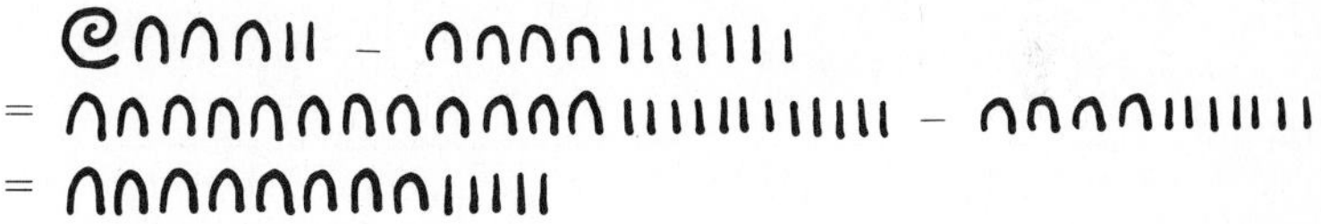

EXERCISES FOR SECTION 7.2

1. Express each number with an Egyptian numeral.
   a. 18 b. 23 c. 124
   d. 102 e. 201 f. 1,132

2 Express each number with an Egyptian numeral.
   a. 13 b. 22 c. 124
   d. 1,492 e. 13,213 f. 22, 123

3. Express each number with an Egyptian numeral.
   a. 14 b. 21 c. 132
   d. 1,776 e. 1,984 f. 10,001

4. Evaluate each Egyptian numeral.
   a.  b.  c. 
   d. 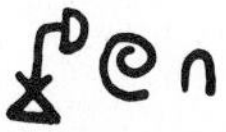 e.  f. 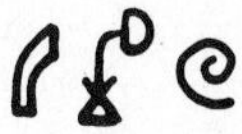

5. Evaluate each Egyptian numeral.
   a.  b. 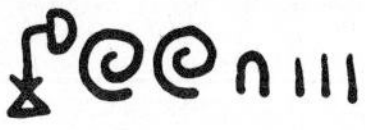

c. 𓆼𓍢𓏺
d. 𓂭𓆼𓆼𓍢𓍢𓎆𓏺𓏺
e. 𓎆𓎆𓆼𓍢𓍢𓏺𓏺
f. 𓏺𓎆𓏺𓆼𓍢𓆼𓍢

6. Evaluate each Egyptian numeral.
a. 𓎆𓎆𓏺𓏺𓎆
b. 𓍢𓍢𓎆𓏺𓎆
c. 𓆼𓏺𓏺𓎆
d. 𓆼𓆼𓍢𓍢𓏺𓏺𓏺𓏺
e. 𓂭𓆼𓏺𓍢𓆼
f. 𓏺𓎆𓍢𓆼𓂭

7. Add the following.
a. 𓎆𓏺 + 𓏺𓎆𓏺
b. 𓍢𓎆𓎆𓎆 + 𓎆𓎆𓎆𓎆𓎆
c. 𓏺𓏺𓏺𓏺𓏺𓏺 + 𓏺𓏺𓏺𓏺
d. 𓎆𓎆𓎆𓎆𓎆𓎆𓎆 + 𓎆𓎆𓎆

8. Add the following.
a. 𓎆𓎆𓎆𓏺 + 𓎆𓎆𓏺𓏺𓏺
b. 𓍢𓍢𓎆𓎆𓎆𓏺𓏺𓏺𓏺𓏺 + 𓍢𓎆𓏺𓏺𓏺𓏺𓏺
c. 𓍢𓍢𓎆𓎆𓎆𓏺𓏺 + 𓍢𓎆𓎆𓎆𓎆𓏺𓏺
d. 𓆼𓆼𓍢𓍢𓍢
+ 𓍢𓍢𓍢𓍢𓍢𓍢𓍢

9. Add the following.
a. 𓏺𓏺𓏺𓏺𓏺𓏺 + 𓏺𓏺𓏺𓏺𓏺
b. 𓍢𓎆𓎆𓎆𓎆𓎆𓏺𓏺𓏺𓏺 + 𓏺𓏺𓏺𓏺𓏺𓎆
c. 𓎆𓎆𓎆𓎆 + 𓎆𓎆𓎆𓎆𓎆𓎆𓎆
d. 𓆼𓍢𓍢𓍢𓍢𓍢𓍢 + 𓆼𓍢𓍢𓍢𓍢

10. Perform each indicated subtraction.
a. 𓎆𓎆𓎆𓏺 − 𓎆𓎆𓏺𓏺𓏺𓏺𓏺
b. 𓍢𓎆𓎆𓏺𓏺 − 𓎆𓎆𓎆𓏺𓏺𓏺
c. 𓍢𓍢𓎆𓎆𓎆𓎆𓏺𓏺 − 𓍢𓎆𓎆𓏺
d. 𓆼𓆼𓍢𓍢 − 𓍢𓎆𓎆𓏺𓏺𓏺

11. Perform each indicated subtraction.
a. 𓎆𓏺 − 𓏺𓏺𓏺𓏺𓏺
b. 𓎆𓎆 − 𓎆𓏺𓏺𓏺𓏺𓏺𓏺𓏺
c. 𓍢𓎆𓏺 − 𓎆𓎆𓎆𓎆𓎆𓎆𓏺𓏺
d. 𓆼𓍢 − 𓍢𓍢𓎆𓎆𓎆𓎆𓏺𓏺𓏺𓏺𓏺

12. Perform each indicated subtraction.
a. 𓎆𓎆 − 𓏺𓏺𓏺𓏺𓏺𓏺
b. 𓍢 − 𓎆𓎆
c. 𓎆𓎆𓏺 − 𓎆𓏺𓏺𓏺𓏺
d. 𓍢𓍢 − 𓎆𓎆𓎆𓏺𓏺𓏺

JUST FOR FUN

Fill in the blanks correctly to obtain the sum shown.

1. Fahrenheit ______ (   )
2. House of ______ Gables (   )
3. Life begins at ______ (   )
4. The Indianapolis ______ (   )
5. Ali Baba and the ______ Thieves (   )
6. ______ Horsemen of the Apocalypse (   )
7. ______ Downing Street (   )
8. The boiling point of water is ______ Celsius (   )
9. ______ Leagues under the Sea (   )
10. Into the valley of death rode the ______ (   )

21,752

7.3 MULTIPLICATIVE GROUPING SYSTEMS

One of the numeration systems developed by the Greeks used letters to represent numbers; that is, letters were used as numerals. Listed in Table 2 are the Greek numerals together with their corresponding values.

TABLE 2
GREEK NUMERALS AND THEIR VALUES

Greek Numerals	Values
Ι (iota)	1
Γ (gamma)	5
Δ (delta)	10
Η (eta)	100
Χ (chi)	1,000
Μ (mu)	10,000

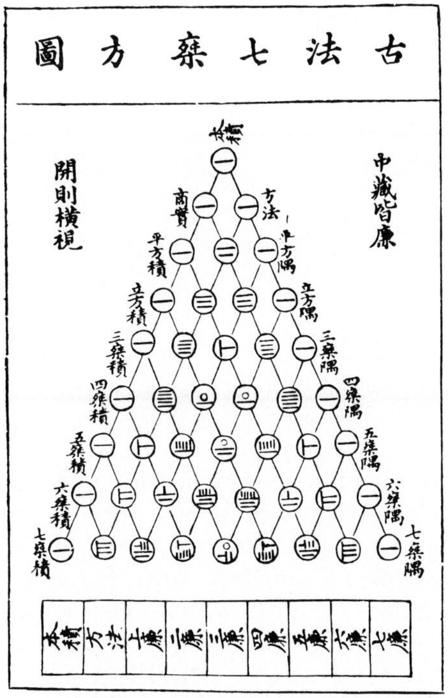

Joseph Needham, Cambridge University Press

This excerpt from a fourteenth-century Chinese mathematical treatise shows an array of numbers known in the West as *Pascal's triangle*.

One thing that the Greeks did differently from the other systems of numeration is to create symbols for multiples of 5. Since the Greeks had no symbol for 50, they thought of 50 as five 10s and wrote it as ΓΔ Similarly, 500 was thought of as five 100s and written as ΓΗ. It follows that ΓΧ = 5,000 and ΓΜ = 50,000.

In order to express 1,984 in Greek numerals, we first think of 1,984 as 1,000 + 900 + 80 + 4. But 900 = 500 + 400 and 80 = 50 + 30. Therefore, 1,984 = 1,000 + 500 + 400 + 50 + 30 + 4, or

1,984 = ΧΓΗΗΗΗΗΓΔΔΔΔΙΙΙΙ

Note that the number is represented by using multiples of 5; this enabled the Greeks to use fewer symbols to express a number. The use of multiples of 5 is an example of *multiplicative grouping*.

A *multiplicative grouping system* is a system that uses certain symbols for numbers in a basic group, together with a second symbol or notation to represent numbers that are multiples of the basic group.

In the Greek system, 5 is the basic group. The symbol for 5, Γ, is used together with other symbols to represent numbers that are multiples of 5.

EXAMPLE 1 Express each number with a Greek numeral.

a. 12 b. 56 c. 88 d. 167 e. 1776

SOLUTION

a. 12 is one 10 and two 1s: ΔII
b. 56 is five 10s, one 5, and one 1: $\Gamma_{\Delta}\Gamma I$
c. 88 is one 50, three 10s, one 5, and three 1s: $\Gamma_{\Delta}\Delta\Delta\Delta\Gamma III$
d. 167 is one 100, one 50, one 10, one 5, and two 1s: $H\Gamma_{\Delta}\Delta\Gamma II$
e. 1,776 is one 1,000, one 500, two 100s, one fifty, two 10s, one 5, and one 1: $X\Gamma_{H}HH\Gamma_{\Delta}\Delta\Delta\Gamma I$

EXAMPLE 2 Evaluate each Greek numeral.

a. $\Gamma_{\Delta}\Gamma II$
b. $\Gamma_{H}HH\Gamma_{\Delta}I$
c. $\Gamma_{X}XX\Gamma_{H}H\Gamma_{\Delta}\Delta II$

SOLUTION

a. Γ_{Δ} is five 10s, or 50, Γ is 5, and II is 2; hence, $\Gamma_{\Delta}\Gamma II = 50 + 5 + 2 = 57$.
b. $\Gamma_{H} = 500$, $HH = 200$, $\Gamma_{\Delta} = 50$, and $I = 1$; hence, $\Gamma_{H}HH\Gamma_{\Delta}I = 500 + 200 + 50 + 1 = 751$.
c. $\Gamma_{X} = 5{,}000$, $XX = 2{,}000$, $\Gamma_{H} = 500$, $H = 100$, $\Gamma_{\Delta} = 50$, $\Delta = 10$, and $II = 2$; hence, $\Gamma_{X}XX\Gamma_{H}H\Gamma_{\Delta}\Delta II = 5{,}000 + 2{,}000 + 500 + 100 + 50 + 10 + 2 = 7662$

The Greek system of numeration uses six symbols. The system is repetitive and it uses multiples of 5. The use of multiples of 5 in the Greek numeration system is an example of multiplicative grouping.

The Chinese-Japanese system of numeration also involves multiplicative grouping. But this system differs from the Greek system in that it uses multiples of 10, 100, and 1,000 in its grouping system. Before we examine the multiplicative grouping of this system, we must first acquaint ourselves with the characteristics of the Chinese-Japanese system. One of the most important things to remember is that the system uses vertical instead of horizontal writing. Listed in Table 3 are the Chinese-Japanese numerals together with their corresponding values.

Because the Chinese-Japanese system of numeration is a system that uses multiplicative grouping, a number such as 2,347 is thought of as two 1,000s, three 100s, four 10s, and 7. Two

TABLE 3
CHINESE-JAPANESE NUMERALS AND THEIR VALUES

Chinese-Japanese Numerals	Values	Chinese-Japanese Numerals	Values
一	1	七	7
二	2	八	8
三	3	九	9
四	4	十	10
五	5	百	100
六	6	千	1,000

二 千	two 1,000s
三 百	three 100s
四 十	four 10s
七	7

1,000s is written as 二千, three 100s is written as 三百, and four 10s is written as 四十. Hence 2,347 is written as shown in the margin.

EXAMPLE 3 Express each number with a Chinese-Japanese numeral.

a. 12 b. 56 c. 88 d. 167 e. 2,776

SOLUTION a. 十二 b. 五十六 c. 八十八 d. 百六十七 e. 二千七百七十六

EXAMPLE 4 Evaluate each Chinese-Japanese numeral.

a. 二百三十四 b. 百六十七 c. 五千百五十四

SOLUTION a. Because the Chinese-Japanese system of numeration uses multiplicative grouping, we have two 100s, three 10s, and 4: $(2 \times 100) + (3 \times 10) + 4 = 234$.

b. We have one 100, six 10s, and 7: $100 + (6 \times 10) + 7 = 167$.

c. This numeral contains five 1,000s, one 100, five 10s, and 4: $(5 \times 1000) + 100 + (5 \times 10) + 4 = 5154$.

EXERCISES FOR SECTION 7.3

1. Express each number with a Greek numeral.
   a. 18 b. 23 c. 34
   d. 44 e. 187 f. 598

2. Express the following as Greek numerals.
   a. 16 b. 54 c. 147
   d. 1492 e. 2194 f. 6875

3. Express the following as Greek numerals.
   a. 21 b. 57 c. 137
   d. 1,776 e. 1,984 f. 2,001

4. Evaluate each Greek numeral.
   a. ΓII b. ΔI c. ΔΓIII
   d. HΔΔΓII e. H𐅄ΓI f. 𐅅H𐅄ΔΓI

5. Evaluate the following Greek numerals.
   a. ΔΓII b. HΔΓII c. MXΔ
   d. 𐅅H𐅄ΓI e. 𐅇X𐅅HI f. 𐅇𐅅𐅄Γ

6. Evaluate each Greek numeral.
   a. ΓIII b. ΔIII c. ΔΓIII
   d. HΔΓI e. H𐅄I f. 𐅅ΔI

7. Express each number with a Chinese-Japanese numeral.
   a. 21 b. 57 c. 137
   d. 1,776 e. 1,984 f. 2,001

8. Express each number with a Chinese-Japanese numeral.
   a. 18 b. 23 c. 34
   d. 46 e. 234 f. 477

9. Express the following as Chinese-Japanese numerals.
   a. 16 b. 54 c. 147
   d. 897 e. 3473 f. 4176

10. Evaluate each Chinese-Japanese numeral.
    a. 二十 b. 七百五 c. 十三 d. 四十九

11. Evaluate each Chinese-Japanese numeral.
a. 五百二十七
b. 三百十五
c. 千百十
d. 八千二百三十六

12. Evaluate each Chinese-Japanese numeral.
a. 二千二
b. 五十八
c. 百五十五
d. 八千二

JUST FOR FUN

The following is an addition problem in which each letter represents a number. Two different letters cannot represent the same number. What numbers do the letters represent?

```
  SEND
+ MORE
 MONEY
```

7.4 PLACE-VALUE SYSTEMS

Thus far in our discussion of systems of numeration, we have examined a simple grouping system of numeration—the Egyptian system—and two multiplicative grouping systems—the Greek and the Chinese-Japanese. We shall now examine two systems of numeration that use a *place-value system*—the Babylonian system and the Hindu-Arabic system.

A *place-value system* is a system in which the position of a symbol matters; that is, the value that any symbol represents depends on the position it occupies within the numeral.

The Babylonian system of numeration uses only two symbols to represent numbers, ▼ to represent 1 and ◀ to represent 10.

Because the signs looked like little wedges, they were called *cuneiform* signs, a word meaning "wedge-shaped." The Babylonians pressed the end of a stick into a clay tablet in order to write their numbers. They used these two symbols, ▼ = 1, and ◀ = 10, to write any number up to 60. That is, ▼ = 1, ▼▼ = 2, ▼▼▼ = 3, and so on. They used the principle of addition to write numbers. For example, to record the number 7, the Babylonians would write seven as

7 = ▼▼▼▼ ▼▼▼

Since ◀ = 10, we can express 11 as one 10 and one 1: 11 = ◀ ▼, 12 = ◀ ▼▼, 13 = ◀▼▼▼, and so on. Forty-three can be expressed as four 10s and three 1s.

43 = ◀◀ ◀◀ ▼▼▼

The Babylonians used a place-value system, in which the position of a symbol is important. The symbols for 10 were always placed to the left of the symbols for 1. Therefore to represent the number 57 in the Babylonian system, we use five 10s and seven 1s arranged as follows.

57 = ◀◀◀ ◀◀ ▼▼▼▼ ▼▼▼

In order to represent a number greater than 60, such as 85, the Babylonians used a *sexagesimal* system, a system based on 60. The number 85 was thought of as one 60 and 25, that is, as $(1 \times 60) + 25$. To indicate this, the Babylonians placed a symbol for one, ▼, to the left of the numeral for 25.

85 = ▼ ◀◀ ▼▼▼ ▼▼

one 60 two 10s 5

In our system, a number such as 1,984 is read "one thousand, nine hundred, eighty-four." In other words, 1984 is composed of one 1,000, nine 100s, eight 10s, and four 1s. Symbolically, we can write this as $(1 \times 10^3) + (9 \times 10^2) + (8 \times 10^1) + (4 \times 1)$. Our system is based on powers of 10; hence it is called a **decimal** system of numeration. (The word *decimal* is derived from the Latin word *decem,* which means "ten.") The Babylonian system is based on powers of 60; hence it is called a **sexagesimal** system. Therefore, a Babylonian numeral such as

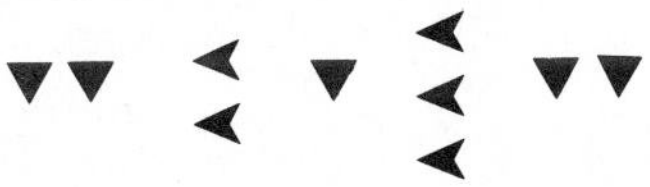

is interpreted as $(2 \times 60^2) + (21 \times 60^1) + 32$, or $(2 \times 3{,}600) + (21 \times 60) + 32 = 7{,}200 + 1{,}260 + 32 = 8{,}492$.

EXAMPLE 1 Express the following as Babylonian numerals.

a. 12 b. 42 c. 56 d. 88 e. 147

SOLUTION a. 12 is one 10 and two 1s; hence 12 = 

b. 42 is four 10s and two 1s; hence 42 = 

c. 56 is five 10s and six 1s; hence 56 = 

d. 88 is one 60 and 28, or one 60, two 10s, and eight 1s; hence

88 = 

e. 147 is two 60s, two 10s, and seven 1s; hence 147 =

EXAMPLE 2 Evaluate each Babylonian numeral.

a.  b.  c. 

SOLUTION a. In this expression, we have one 10 and three 1s; $10 + 3 = 13$
b. Here we have four 10s and four 1s; $(4 \times 10) + 4 = 44$
c. In this expression, we have one 60; three 10s, and two 1s; $(1 \times 60) + (3 \times 10) + 2 = 60 + 30 + 2 = 92$.

The Babylonian system of numeration uses only two symbols, and the symbols can be repeated for numbers up to 60. After that, the numbers are expressed in powers of 60. The numeral

represents one 60, two 10s, and two 1s; that is, $(1 \times 60) + (2 \times 10) + 2 = 82$.

The system of numerals we use today is called the **Hindu-Arabic system.** That is, we use Hindu-Arabic numerals to express numbers. The symbols we use are 0, 1, 2, 3, 4, 5, 6, 7, 8, 9. These symbols had their beginning in India; it was the Arabs who were responsible for making their existence known in Europe. The Arabs did their calculations with these new numerals and then exposed the Europeans to their new techniques.

One of the most significant contributions of the Hindu numerals was zero. Zero evolved from a need for a placeholder, because it was important to distinguish between numerals such as 501 and 51. The transition to Hindu-Arabic numerals was a slow process, and it was not until the end of the sixteenth century that the changeover was fairly complete.

Recall that the Egyptian system of numeration uses a different symbol for each power of 10, but it has no place value. That is, the positions of the symbols do not affect the value of the number. Our system of numeration also uses powers of 10, but it does have **place value:** the position that a symbol has within a numeral is important.

The Hindu-Arabic system of numeration uses multiplicative grouping based on powers of 10. For example, in the numeral 1,978, the 8 represents eight 1s, the 7 represents seven 10s, the 9 represents nine 100s, and the 1 represents one 1,000. One, 10, 100, and 1,000 are all powers of 10.

$$\begin{aligned} 1 &= 10^0 \\ 10 &= 10^1 \\ 100 &= 10 \times 10 = 10^2 \\ 1{,}000 &= 10 \times 10 \times 10 = 10^3 \end{aligned}$$

The number 1,978 is the result of combining multiples of these powers of 10.

$$\begin{aligned} 8 &= 8 \times 10^0 \\ 70 &= 7 \times 10^1 \\ 900 &= 9 \times 10^2 \\ 1{,}000 &= 1 \times 10^3 \\ \hline 1{,}978 & \end{aligned}$$

Notice that the positions of the numerals are important. The only way that we know that the 8 in 1978 represents eight 1s is by its position; in 1,987, the 8 represents eight 10s. Thus we say that the Hindu-Arabic system of numeration is a place-value system.

HISTORICAL NOTE

In ancient times people wrote in the dirt and on walls, pieces of stone, bricks, pottery, and clay tablets. The Egyptians improved upon these methods with the discovery of a large water plant with peculiar properties. By trial and error the Egyptians learned to cut this plant into thin strips, lay the strips together, and then place a similar layer of strips crosswise over the first. Next they pressed the layers together and allowed them to dry, after which they had something to write on. The coarse, brown writing material was called *papyrus* because it was obtained from the papyrus reed. It is believed that the word *paper* is derived from the term for this ancient writing material.

When we write a number in terms of powers of 10, we are writing it in **expanded notation.** In expanded notation,

$$1{,}978 = (1 \times 10^3) + (9 \times 10^2) + (7 \times 10^1) + (8 \times 10^0)$$

When expressing a number in expanded notation, it is convenient to start with the 1s place and proceed from right to left.

EXAMPLE 3 Write each number in expanded notation.

a. 123 b. 2,347 c. 2,003

SOLUTION
a. The Hindu-Arabic numeral 123 is composed of one 100, two 10s, and three 1s. Therefore, we have $(1 \times 100) + (2 \times 10) + (3 \times 1)$. Rewriting this using powers of 10, we have $(1 \times 10^2) + (2 \times 10^1) + (3 \times 10^0)$. (Recall that $10^0 = 1$. Any number raised to the zero power, except zero, equals 1.)
b. $2{,}347 = (2 \times 1{,}000) + (3 \times 100) + (4 \times 10) + (7 \times 1) = (2 \times 10^3) + (3 \times 10^2) + (4 \times 10^1) + (7 \times 10^0)$
c. $2{,}003 = (2 \times 10^3) + (0 \times 10^2) + (0 \times 10^1) + (3 \times 10^0)$. Note that there are no 10s or 100s, but we still must indicate this in expanded notation, as the zeros are placeholders, and 2,003 is not the same as 23.

EXAMPLE 4 Write each of the following as Hindu-Arabic numerals in base 10 (decimal) notation. *Note:* This is also called *standard notation,* or **standard form.**
a. one hundred eighty-seven
b. two thousand three hundred forty-one
c. one thousand two
d. $(3 \times 10^3) + (2 \times 10^2) + (1 \times 10^1) + (0 \times 10^0)$
e. $(4 \times 10^3) + (2 \times 10^0)$

SOLUTION
a. 187
b. 2,341
c. 1,002 (Note that we did not write "one thousand *and* two." In mathematics, the word *and* is used to indicate the position of the decimal point, as in one hundred three and two-tenths, which is 103.2.)
d. $(3 \times 10^3) + (2 \times 10^2) + (1 \times 10^1) + (0 \times 10^0) = (3 \times 1000) + (2 \times 100) + (1 \times 10) + (0 \times 1) = 3000 + 200 + 10 + 0 = 3210$
e. $(4 \times 10^3) + (2 \times 10^0) = (4 \times 1000) + (2 \times 1) = 4000 + 2 = 4{,}002$. (Note that the 100s and 10s places were omitted in the expanded notation, but we were able to obtain the correct result by proceeding in an orderly manner.)

EXERCISES FOR SECTION 7.4

1. Express the following as Babylonian numerals.
   a. 18 b. 23 c. 82
   d. 102 e. 349 f. 864

2. Express the following as Babylonian numerals.
   a. 34 b. 44 c. 93
   d. 201 e. 423 ⋆f. 3,674

3. Express the following as Babylonian numerals.
   a. 42 b. 58 c. 65
   d. 132 e. 193 ⋆f. 3,682

4. Evaluate each Babylonian numeral.
   a. ◀▼▼▼ b. ◀◀▼▼
   c. ◀◀◀▼ d. ▼◀◀▼▼▼

5. Evaluate the following Babylonian numerals.
   a. ▼◀◀◀▼ b. ▼▼◀◀▼▼
   c. ▼◀▼ d. ▼◀◀▼

6. Evaluate the following Babylonian numerals.
   a. ◀▼▼ b. ◀◀▼
   c. ◀◀◀▼ d. ▼◀◀▼

7. Write each number in expanded notation.
   a. 243 b. 378 c. 1,234
   d. two thousand fifty-one
   e. ten thousand four hundred one

8. Write each number in expanded notation.
   a. 345 b. 1,776 c. 19,876
   d. three thousand four hundred fifty-six
   e. twelve thousand nine hundred three

9. Write each number in expanded notation.
   a. 402 b. 1,476 c. 20,182
   d. two thousand three hundred eighty-one
   e. ten thousand fifty

10. Write each of the following in base 10 (decimal) notation.
    a. one thousand four
    b. two billion two
    c. seventy-five thousand seventy-five
    d. $(4 \times 10^3) + (2 \times 10^1) + (3 \times 10^0)$
    e. $(1 \times 10^4) + (1 \times 10^2) + (1 \times 10^0)$
    f. $(5 \times 10^5) + (2 \times 10^4) + (7 \times 10^3) + (6 \times 10^0)$

11. Write each of the following in base 10 (decimal) notation.
    a. two hundred forty
    b. two thousand three hundred eleven
    c. one thousand seven hundred seventy-six
    d. $(4 \times 10^3) + (2 \times 10^2) + (1 \times 10^1) + (3 \times 10^0)$
    e. $(2 \times 10^2) + (0 \times 10^1) + (4 \times 10^0)$
    f. $(4 \times 10^4) + (3 \times 10^2) + (1 \times 10^0)$

12. Write each of the following in base 10 (decimal) notation.
    a. three hundred forty-five
    b. forty thousand two
    c. one million one
    d. $(5 \times 10^3) + (3 \times 10^2) + (2 \times 10^1) + (1 \times 10^0)$
    e. $(2 \times 10^4) + (3 \times 10^2) + (4 \times 10^1) + (5 \times 10^0)$
    f. $(7 \times 10^5) + (8 \times 10^3) + (9 \times 10^1)$

JUST FOR FUN Look at the equation

XI + I = X

Can you make it a correct statement without adding, crossing out, or changing anything?

7.5 NUMERATION IN BASES OTHER THAN 10

It is common practice to group items by 10s. For example, a decade is a period of 10 years. A dime is equal to 10 pennies and 10 dimes make one dollar. A decathlon is a famous Olympic athletic contest in which each contestant participates in 10 events. But it is not uncommon to group items in some other manner. We group things such as socks and mittens by 2s. Another common grouping is by 12s. How do you purchase doughnuts and eggs? We buy items like these by the dozen, that is, in groups of 12. Three dozen doughnuts is three 12s, or 36 doughnuts.

Other common groupings of 12 include: 12 inches in a foot, 12 hours in one complete cycle of the clock, and 12 months in a year. Consider the instructor who is ordering supplies and orders a *gross* of chalk. One *gross* is a dozen dozen, or twelve 12s. Therefore, the instructor ordered 144 pieces of chalk. There also is another type of *gross* grouping; it is called a *great gross,* 1 great gross = 12 gross. Hence a great gross is a dozen gross or 12×144 or 1,728 units. Gross and great gross are common units when ordering items in bulk; usually wholesalers order items in this manner.

© Brent Jones

Suppose we have $2\frac{1}{2}$ dozen doughnuts. How many doughnuts do we have? We have 2 dozen and $\frac{1}{2}$ of a dozen, or $(2 \times 12) + 6 = 24 + 6 = 30$. Because we are grouping by dozens, we could have written this as two dozen + 6. But, we also could have said 2 dozen + six 1s, and this is the same as $(2 \times 12) + (6 \times 1)$. Since any number (except zero) raised to the zero power is 1, we can write this as $(2 \times 12^1) + (6 \times 12^0)$. This expanded notation indicates that we are grouping by 12s. In decimal notation, we grouped by 10s and hence were working in base 10. Now we are grouping by 12s and therefore we can say that we are working in base 12. Hence we have

$$(2 \times 12^1) + (6 \times 12^0) = 26_{\text{twelve}}$$

When we write numerals in some base other than base 10, we must indicate what base we are working with. We do this by using a subscript. The numeral

$$47_{\text{twelve}}$$

indicates that we are grouping by 12s, and for this particular example we have four 12s and seven 1s—that is, $(4 \times 12^1) + (7 \times 12^0)$, or $(4 \times 12) + (7 \times 1) = 55$.

EXAMPLE 1 Change to base 10 notation.

a. 42_{twelve} b. 30_{twelve} c. 234_{twelve}

SOLUTION a. In order to change a numeral to base 10 notation, we first write the numeral in expanded notation.

$$42_{\text{twelve}} = (4 \times 12^1) + (2 \times 12^0) = (4 \times 12) + (2 \times 1)$$
$$= 48 + 2 = 50$$

b. $30_{\text{twelve}} = (3 \times 12^1) + (0 \times 12^0) = (3 \times 12) + (0 \times 1)$

$$= 36 + 0 = 36$$

c. To write numerals in expanded notation, we start from the right (note that this is the units place), and proceed to the left in successive higher powers of the indicated base. Therefore,

$$234_{\text{twelve}} = (2 \times 12^2) + (3 \times 12^1) + (4 \times 12^0)$$
$$= (2 \times 144) + (3 \times 12) + (4 \times 1) = 288 + 36 + 4$$
$$= 328$$

Many people believe that the reason we normally group items by 10s is that humans have 10 fingers. But we could just as easily group items by 5s because we have five fingers on one hand. The *base 5 system* is a system of numeration that groups items by 5s. Consider the numeral 13: in base 10, this is one 10 and three 1s, but in base 5, it is two 5s and three 1s. Therefore $13 = 23_{\text{five}}$. If we are given a numeral in base 5 notation, we can convert it to base 10 by writing the base 5 numeral in expanded notation. Recall that in order to write numerals in expanded notation we start from the right, which is the units place. This represents the zero power of the indicated base. Then we proceed to the left in successive higher powers of the indicated base. Therefore we have

$$23_{\text{five}} = (2 \times 5^1) + (3 \times 5^0) = (2 \times 5) + (3 \times 1)$$
$$= 10 + 3 = 13$$

Next, let us convert 342_{five} to base ten notation.

$$342_{\text{five}} = (3 \times 5^2) + (4 \times 5^1) + (2 \times 5^0)$$
$$= (3 \times 25) + (4 \times 5) + (2 \times 1)$$
$$= 75 + 20 + 2 = 97$$

Remember that when we want to convert from a given base to base 10, we use expanded notation.

EXAMPLE 2 Write each number in base 10 notation.

a. 40_{five} b. 121_{five} c. 203_{five}

SOLUTION a. We first write the numeral in expanded notation, then simplify.

$$\begin{aligned} 40_{\text{five}} &= (4 \times 5^1) + (0 \times 5^0) = (4 \times 5) + (0 \times 1) \\ &= 20 + 0 = 20 \end{aligned}$$

b.
$$\begin{aligned} 121_{\text{five}} &= (1 \times 5^2) + (2 \times 5^1) + (1 \times 5^0) \\ &= (1 \times 25) + (2 \times 5) + (1 \times 1) \\ &= 25 + 10 + 1 = 36 \end{aligned}$$

c.
$$\begin{aligned} 203_{\text{five}} &= (2 \times 5^2) + (0 \times 5^1) + (3 \times 5^0) \\ &= (2 \times 25) + (0 \times 5) + (3 \times 1) \\ &= 50 + 0 + 3 = 53 \end{aligned}$$

Thus far we have only considered changing numerals in a given base to base 10 notation. We should also be able to express base 10 numerals in terms of other bases. Suppose we want to express 13 as a numeral in base 5. Because base 5 groups items by 5s, we determine how many 5s are contained in 13. There are two 5s and three 1s. Therefore,

$$13 = 23_{\text{five}}$$

There is a convenient rule that enables us to convert from base 10 to any given base. We can convert any number from base 10 to another base by recording the remainders of successive divisions. We stop dividing when we obtain a quotient of zero. Using the last example, we can illustrate the procedure involved. We wish to convert 13 to base 5, so we divide 13 by 5 and record the remainders.

5⌊13
5⌊2 3 } remainders ↑
→0 2
STOP (a zero quotient)

The answer is determined by reading the remainders from *bottom to top*. Therefore, $13 = 23_{\text{five}}$.

Consider the number 97: let's express it with a base 5 numeral. Performing the successive divisions, we have

```
 5|97
 5|19   2 ↑
  5|3   4 |
    0   3 |
   ↗
STOP
```

Reading the remainders from bottom to top, we have

$$97 = 342_{\text{five}}$$

We can check the answer by converting the base 5 numeral to base 10: $342_{\text{five}} = (3 \times 5^2) + (4 \times 5^1) + (2 \times 5^0) = (3 \times 25) + (4 \times 5) + (2 \times 1) = 75 + 20 + 2 = 97$. The answer checks.

EXAMPLE 3 Express each number with base 5 notation.

a. 43 b. 147 c. 520

SOLUTION In each case, we perform successive divisions by 5, the new base, and record the remainders for each division. We determine the answer by reading the remainders from bottom to top.

```
a. 5|43          b. 5|147          c. 5|520
    5|8   3 ↑        5|29   2 ↑        5|104   0 ↑
    5|1   3 |         5|5   4 |         5|20   4 |
      0   1 |         5|1   0 |          5|4   0 |
     ↗                  0   1 |            0   4 |
  STOP                 ↗                  ↗
                    STOP               STOP
```

$43 = 133_{\text{five}}$ $147 = 1042_{\text{five}}$ $520 = 4040_{\text{five}}$

EXAMPLE 4 Change the following to base 12 notation.

a. 43 b. 100 c. 520

SOLUTION In each case, we shall apply our handy rule and perform successive divisions. Remember that the answer is determined by reading the remainders from bottom to top. Since we are converting to base 12, we shall divide by 12.

a.		b.		c.	
$12\overline{)43}$		$12\overline{)100}$		$12\overline{)520}$	
$12\overline{)3}$	7 ↑	$12\overline{)8}$	4 ↑	$12\overline{)43}$	4 ↑
0 ↙ STOP	3	0 ↙ STOP	8	$12\overline{)3}$	7
				0 ↙ STOP	3
$43 = 37_{\text{twelve}}$		$100 = 84_{\text{twelve}}$		$520 = 374_{\text{twelve}}$	

Suppose you buy 22 of something (eggs, doughnuts, or bagels, for example). You have purchased 1 dozen plus 10 more. How can we express this in base 12 notation? We cannot say $22 = 110_{\text{twelve}}$. Why? Because if we evaluate 110_{twelve}, we have $(1 \times 12^2) + (1 \times 12^1) + (0 \times 12^0)$, or $(1 \times 144) + (1 \times 12) + (0 \times 1) = 144 + 12 + 0 = 156$, which is not 22!

In our decimal system of numeration we use 10 symbols, 0, 1, 2, 3, 4, 5, 6, 7, 8, 9. Consequently we should use 12 symbols in the base 12 numeration system. Because we already have 10 symbols we can borrow from the base 10 system, let us agree to use the additional symbols T and E in the base 12 system of numeration. Let T stand for 10 and E stand for 11. Then the 12 symbols that we shall use in base 12 are 0, 1, 2, 3, 4, 5, 6, 7, 8, 9, T, E. Keep in mind that, in the base 10 system of numeration, 11 represents one 10 and one 1, or eleven. But in the base 12 system of numeration, T represents 10, and 11_{twelve} represents one 12 and one 1, or 13.

EXAMPLE 5 Change to base 10 notation.

a. 40_{twelve} b. $4T_{\text{twelve}}$ c. $ET2_{\text{twelve}}$

SOLUTION

a. $40_{\text{twelve}} = (4 \times 12^1) + (0 \times 12^0) = (4 \times 12) + (0 \times 1)$
$= 48 + 0 = 48$

b. $4T_{\text{twelve}} = (4 \times 12^1) + (T \times 12^0) = (4 \times 12) + (10 \times 1)$
$= 48 + 10 = 58$

c. $ET2_{\text{twelve}} = (E \times 12^2) + (T \times 12^1) + (2 \times 12^0)$
$= (E \times 144) + (T \times 12) + (2 \times 1)$
$= (11 \times 144) + (10 \times 12) + (2 \times 1)$
$= 1{,}584 + 120 + 2 = 1{,}706$

The base 12 system of numeration is a fairly common system. It is also called the **duodecimal** system of numeration, which indicates that it is a system of numeration with 12 as its base, as opposed to the decimal system of numeration, which has 10 as its base.

EXERCISES FOR SECTION 7.5

1. In each example, items are grouped by 12. Perform the indicated operations.
   a. 4 years 7 months
   + 2 years 9 months

   b. 7 years 3 months
   − 4 years 10 months

   c. 13 feet 11 inches
   + 11 feet 10 inches

   d. 14 feet 6 inches
   − 10 feet 7 inches

   e. 2 gross 3 dozen 8 units
   + 4 gross 11 dozen 6 units

   f. 5 gross 9 dozen 3 units
   − 2 gross 11 dozen 7 units

2. In each example, items are grouped by 12. Perform the indicated operations. (*Note:* 1 great gross = 12 gross.)
   a. 3 years 9 months
   + 2 years 7 months

   b. 8 years 2 months
   − 3 years 9 months

   c. 3 gross 3 dozen 9 units
   + 5 gross 8 dozen 4 units

   d. 7 gross 2 dozen 5 units
   − 3 gross 9 dozen 7 units

   e. 3 great gross 2 gross 4 dozen
   + 4 great gross 10 gross 8 dozen

   f. 3 great gross 2 gross 4 dozen
   − 1 great gross 3 gross 7 dozen

3. In each example, items are grouped by 12. Perform the indicated operations. (*Note:* 1 great gross = 12 gross.)
   a. 4 feet 7 inches
   + 5 feet 8 inches

   b. 6 feet 2 inches
   − 3 feet 9 inches

   c. 2 years 4 months
   + 3 years 11 months

   d. 8 years 5 months
   − 7 years 7 months

   e. 4 gross 10 dozen 8 units
   + 9 gross 5 dozen 7 units

   ★f. 4 great gross
   − 2 gross 3 dozen 2 units

4. Change to base 10 notation.
   a. 32_{five} b. 12_{five} c. 14_{five}
   d. 123_{five} e. 203_{five} f. 2031_{five}

5. Change to base 10 notation.
   a. 13_{five} b. 44_{five} c. 231_{five}
   d. 304_{five} e. 100_{five} f. 4021_{five}

6. Change to base 10 notation.
   a. 42_{five} b. 43_{five} c. 102_{five}
   d. 301_{five} e. 4011_{five} f. 3001_{five}

7. Change to base 5 notation.
   a. 6 b. 19 c. 38
   d. 3 e. 121 f. 497

8. Change to base 5 notation.
a. 9 b. 27 c. 4
d. 243 ⋆e. 2003 ⋆f. 3421

9. Change to base 5 notation.
a. 6 b. 2 c. 11
d. 48 e. 101 ⋆f. 1001

10. Change to base 10 notation.
a. 47_{twelve} b. 28_{twelve} c. 59_{twelve}
d. 124_{twelve} e. 347_{twelve} f. 1001_{twelve}

11. Change to base 10 notation.
a. 42_{twelve} b. 54_{twelve} c. 99_{twelve}
d. 137_{twelve} e. 243_{twelve} f. 2001_{twelve}

12. Change to base 10 notation.
a. 11_{twelve} b. 21_{twelve} c. 42_{twelve}
d. 101_{twelve} e. 123_{twelve} f. 1009_{twelve}

13. Change to base 12 notation.
a. 42 b. 53 c. 60
d. 137 e. 234 f. 876

14. Change to base 12 notation.
a. 49 b. 66 c. 118
d. 341 ⋆e. 3421 ⋆f. 5736

15. Change to base 12 notation.
a. 43 b. 61 c. 112
d. 201 e. 432 ⋆f. 4101

16. Change to base 10 notation.
a. $2E_{\text{twelve}}$ b. TE_{twelve}
c. $T2E_{\text{twelve}}$ d. $E2T_{\text{twelve}}$

17. Change to base 10 notation.
a. $T3_{\text{twelve}}$ b. ET_{twelve}
c. $3E4_{\text{twelve}}$ d. $TE5_{\text{twelve}}$

18. Change to base 10 notation.
a. $3E_{\text{twelve}}$ b. $E3_{\text{twelve}}$
c. $1ET_{\text{twelve}}$ d. $1TE_{\text{twelve}}$

⋆19. Change to base 10 notation.
a. 47_{nine} b. 52_{nine} c. 34_{six}
d. 52_{six} e. 25_{seven} f. 462_{seven}

⋆20. Change to base 10 notation.
a. 21_{four} b. 301_{four} c. 56_{eight}
d. 712_{eight} e. 22_{three} f. 201_{three}

⋆21. Change to base 9 notation.
a. 34 b. 60 c. 102
d. 135 e. 234 f. 716

⋆22. Change to base 4 notation.
a. 21 b. 65 c. 83
d. 137 e. 260 f. 301

23. An instructor ordered the following classroom supplies: a gross of pencils, a great gross of chalk (1 great gross is 12 gross), 3 dozen notebooks, $\frac{1}{2}$ dozen pens, and one grade book. What is the total number of items ordered?

24. Jake the bagel maker prepared the following order: He made 6 dozen salt bagels, 3 dozen poppy bagels, 6 dozen rye bagels, a gross of plain bagels, and a gross of onion bagels. How many bagels did Jake make?

25. a. What numeral is 47_{twelve} in base 5?
b. What numeral is 34_{five} in base 12?

JUST FOR FUN

Following are some units of measure that are probably not familiar to you. See if you can find equivalent measures (for example, 1 yard = 3 feet).

1 fathom = ? (A fathom is used in measuring depths at sea.)

1 hand = ? (A hand is used in measuring the height of a horse.)

3 barleycorns = ? (A barleycorn is used by shoe manufacturers in measuring the length of a foot.)

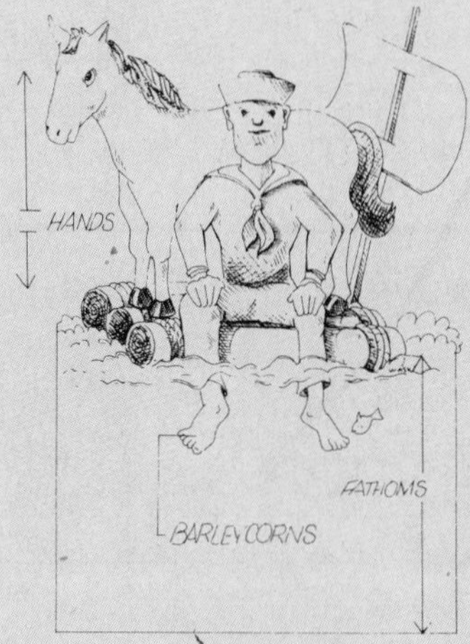

7.6 BASE 5 ARITHMETIC

Thus far we have converted base 10 numerals to base 12 numerals and also to base 5 numerals. We have also converted base 12 and base 5 numerals to base 10 numerals. In a manner of speaking, we have learned to count in some bases other than base 10. The next step is to perform some arithmetic operations in a base system other than base 10. A convenient base to work with is base 5. Using conventional Hindu-Arabic numerals, the base 5 system involves the numerals 0, 1, 2, 3, 4. Recall that the numeral 14_{five} is composed of one 5 and four 1s; hence $14_{\text{five}} = 9$ in base 10. For the sake of convenience, we shall now write all numerals in base five with the numeral 5 as a subscript. Therefore, 14_{five} is the same as 14_5. Recall that if no subscript is written—that is, if no base is indicated—then it is understood that the numeral is written in base 10 notation.

Now suppose we want to perform the following addition: $14_5 + 24_5$. At first glance, you might want to say that the answer is 38. But 38 is an impossible answer! Why? For one reason, the base 5 system of numeration uses only the digits 0, 1, 2, 3, 4, and therefore we cannot have an 8 in our answer.

Let us try again. One way to find an answer to the problem $14_5 + 24_5$ is to convert each of the base 5 numerals to base 10, add the base 10 numerals, and then convert the answer back to base 5.

$$14_5 = 9 \quad \text{and} \quad 24_5 = 14$$
$$9 + 14 = 23 \quad \text{and} \quad 23 = 43_5$$

Therefore,

$$14_5 + 24_5 = 43_5$$

This may seem like the best method, but it isn't—especially if we want to add three- and four-digit numerals such as 1223_5 and 4223_5. One thing that would help us to add numerals in base 5 is a table of addition facts. Table 4 is a table of addition facts for base 5.

TABLE 4
BASE 5 ADDITION TABLE

+	0_5	1_5	2_5	3_5	4_5
0_5	0_5	1_5	2_5	3_5	4_5
1_5	1_5	2_5	3_5	4_5	10_5
2_5	2_5	3_5	4_5	10_5	11_5
3_5	3_5	4_5	10_5	11_5	12_5
4_5	4_5	10_5	11_5	12_5	13_5

The entries in this table may seem odd at first. Let us check a few of them so we can see that they do make sense. How does $3_5 + 4_5 = 12_5$? If we add 3 + 4 in base 10, we get 7. Since 7 is composed of one 5 and two 1s, $7 = 12_5$, and so $3_5 + 4_5 = 12_5$. Similarly, $4_5 + 4_5 = 13_5$, because 4 + 4 = 8 in base 10, and 8 is composed of one 5 and three 1s; hence $8 = 13_5$. Let us solve an addition problem.

$$\begin{array}{r} 23_5 \\ +\ 34_5 \\ \hline \end{array}$$

We start with the units place: $3_5 + 4_5 = 12_5$. Writing the 2 and carrying the 1, we have

$$\begin{array}{r} {}^{1} \\ 23_5 \\ +\ 34_5 \\ \hline 2_5 \end{array}$$

Next we add the 5s: $1_5 + 2_5 + 3_5 = 3_5 + 3_5 = 11_5$. Therefore the completed problem is

$$\begin{array}{r} 23_5 \\ +\ 34_5 \\ \hline 112_5 \end{array}$$

We can check our work in base 10.

$$\begin{array}{rl} 23_5 & = (2 \times 5^1) + (3 \times 5^0) = (2 \times 5) + (3 \times 1) = 10 + 3 = 13 \\ +\ 34_5 & = (3 \times 5^1) + (4 \times 5^0) = (3 \times 5) + (4 \times 1) = 15 + 4 = \underline{19} \\ \hline 112_5 & = (1 \times 5^2) + (1 \times 5^1) + (2 \times 5^0) \\ & = (1 \times 25) + (1 \times 5) + (2 \times 1) = 25 + 5 + 2 \qquad = 32 \end{array}$$

Let us try the preceding problem without using the addition table. In order to add 23_5 and 34_5, we can combine 3 + 4 as we normally would; that is, 3 + 4 = 7. Remember that the numerals 0, 1, 2, 3, 4 have the same meaning in base 5 as they do in base 10. Our only problem is the 7, but 7 is composed of one 5 and two 1s, and therefore $3_5 + 4_5 = 12_5$. We write the 2 and carry the 1 as before.

$$\begin{array}{r} {\scriptstyle 1} \\ 23_5 \\ +\ 34_5 \\ \hline 2_5 \end{array}$$

We now add 1 + 2 + 3 = 6, but in base 5 notation $6 = 11_5$ (one 5 and one 1). Hence $23_5 + 34_5 = 112_5$.

EXAMPLE 1 Find the sum of 234_5 and 341_5.

SOLUTION Starting with the units place, 4 + 1 = 5, but in base 5, $5 = 10_5$ (one 5 and no 1s). Placing the 0 and carrying the 1, we have 1 + 3 + 4 = 8; but in base 5, $8 = 13_5$ (one 5 and three 1s). Placing the 3 and carrying the 1, we now have 1 + 2 + 3 = 6, and in base 5, $6 = 11_5$ (one 5 and one 1). Our addition is complete.

$$\begin{array}{r} 234_5 \\ +\ 341_5 \\ \hline 1130_5 \end{array}$$

EXAMPLE 2 Find the sum of 133_5 and 341_5.

SOLUTION Starting with the units place, 3 + 1= 4. This is the same for base 5 as for base 10, because the numerals 0, 1, 2, 3, 4 have the same meaning in base 5 as they do in base 10. Next we proceed to the 5s place: 3 + 4 = 7, but in base 5, $7 = 12_5$ (one 5 and two 1s). Placing the 2 and carrying the 1, we now have 1 + 1 + 3 = 5, and in base 5, 5 = 10_5 (one 5 and no 1s). Our addition is complete.

$$\begin{array}{r} 133_5 \\ +\ 341_5 \\ \hline 1024_5 \end{array}$$

EXAMPLE 3 Find the sum of 342_5 and 324_5.

SOLUTION

$$\begin{array}{r} 342_5 \\ +\ 324_5 \\ \hline 1221_5 \end{array}$$

Check:

$$\begin{aligned}
342_5 &= (3 \times 5^2) + (4 \times 5^1) + (2 \times 5^0) \\
&= (3 \times 25) + (4 \times 5) + (2 \times 1) = 75 + 20 + 2 = \quad 97 \\
+\ 324_5 &= (3 \times 5^2) + (2 \times 5^1) + (4 \times 5^0) \\
&= (3 \times 25) + (2 \times 5) + (4 \times 1) = 75 + 10 + 4 = \quad \underline{89} \\
1221_5 &= (1 \times 5^3) + (2 \times 5^2) + (2 \times 5^1) + (1 \times 5^0) \\
&= (1 \times 125) + (2 \times 25) + (2 \times 5) + (1 \times 1) \\
&= 125 + 50 + 10 + 1 \qquad = 186
\end{aligned}$$

Another arithmetic operation that goes hand in hand with addition is the operation of subtraction. Before we try a subtraction problem in base 5, let's review subtraction in base 10. Suppose we wish to subtract 248 from 735, that is 735 − 248. We set the problem up as shown below. Note that the parts of the problem have been labeled.

$$\begin{array}{rl} 735 & \textit{minuend} \\ -\ 248 & \textit{subtrahend} \\ \hline \end{array}$$

The answer is usually called the *difference,* but it is also sometimes called the *remainder*. Performing the subtraction, we have

$$\begin{array}{rrr} {\scriptstyle 6} & {\scriptstyle 12} & {\scriptstyle 15} \\ 7 & 3 & 5 \\ -\ 2 & 4 & 8 \\ \hline 4 & 8 & 7 \end{array}$$

Notice that we must rename three 10s, five 1s as two 10s, fifteen 1s. Seven 100s, two 10s are then renamed as six 100s, twelve 10s. The numbers in small type over the minuend are shown only to indicate the new arrangement of the number in order to facilitate the subtraction.

Now let us try a subtraction problem in base 5. Consider the following.

$$\begin{array}{r} 42_5 \\ -\ 13_5 \\ \hline \end{array}$$

We note that we cannot subtract 3 from 2, so we must borrow from the 4. But what do we borrow? Since we are in base 5, we borrow one 5. So we now have one 5 and two 1s in the units place, that is, 12_5. But this is the same as 7, and 3 from 7 is 4. Now we have only 3 in the 5s place, and 1 from 3 is 2. This process is illustrated below.

$$\begin{array}{rcc} & \scriptstyle 3 & \scriptstyle 1 \\ & \not{4} & 2_5 \\ - & 1 & 3_5 \\ \hline & 2 & 4_5 \end{array}$$

Remember to indicate the base with which you are working in your answer. If no subscript is written, then it is understood that the numeral is written in base 10 notation.

Let us check our answer to the preceding problem by translating it into base 10.

$$\begin{array}{rl} 42_5 &= (4 \times 5^1) + (2 \times 5^0) = (4 \times 5) + (2 \times 1) = 20 + 2 = \quad 22 \\ -\ 13_5 &= (1 \times 5^1) + (3 \times 5^0) = (1 \times 5) + (3 \times 1) = \ \ 5 + 3 = -\ \ 8 \\ \hline 24_5 &= (2 \times 5^1) + (4 \times 5^0) = (2 \times 5) + (4 \times 1) = 10 + 4 = \quad 14 \end{array}$$

Since $22 - 8 = 14$, the answer checks.

Now let us try another problem. Consider

$$\begin{array}{r} 431_5 \\ -\ 132_5 \\ \hline \end{array}$$

Since we cannot subtract 2 from 1, we borrow 1 from the 3 in the 5s place, leaving a 2 in the 5s place and giving us 11_5 in the units place. We know 11_5 is the same as 6, and 2 from 6 is 4. So far we have

$$\begin{array}{rccc} & & \scriptstyle 2 & \scriptstyle 1 \\ & 4 & \not{3} & 1_5 \\ - & 1 & 3 & 2_5 \\ \hline & & & 4_5 \end{array} \qquad 11_5 = 6 \quad \text{so} \quad 11_5 - 2_5 = 4_5$$

We cannot subtract 3 from 2 in the 5s place, so again we borrow. This time we borrow 1 from the 4 in the next place (the 25s). This means we have borrowed one 25, or five 5s, which gives us a total of seven 5s, and three 5s from seven 5s is four 5s. We can also think of this as subtracting 3_5 from 12_5, or 3 from 7, which gives us a 4 in the 5s place. Now we have only 3 in the 25s place, and 1 from 3 is 2. The completed subtraction process is

$$\begin{array}{rrr} {}^{3} & {}^{12} & {}^{1} \\ 4 & 3 & 1_5 \\ -\,1 & 3 & 2_5 \\ \hline 2 & 4 & 4_5 \end{array} \qquad 12_5 = 7 \quad \text{so} \quad 12_5 - 3_5 = 4_5$$

Table 1, the base 5 addition table, may be helpful to you in subtraction problems. Remember that when you borrow, you are borrowing a number in the indicated base, not a 10. In these examples we are borrowing 5s and 25s. For problems in other bases, you may be borrowing 7s, 4s, and so on.

EXAMPLE 4 Subtract 24_5 from 33_5.

SOLUTION Since we cannot subtract 4 from 3, we borrow 1 from the 5s place, which gives us one 5 and three 1s in the units place, or $13_5 = 8$; hence 4_5 from 13_5 is 4_5. We are left with a 2 in the 5s place, and 2 from 2 is zero.

$$\begin{array}{rr} {}^{2} & {}^{1} \\ 3 & 3_5 \\ -\,2 & 4_5 \\ \hline & 4_5 \end{array} \qquad \text{or} \qquad \begin{array}{r} 33_5 \\ -\,24_5 \\ \hline 4_5 \end{array}$$

EXAMPLE 5 Subtract 234_5 from 433_5.

SOLUTION We cannot subtract 4 from 3, so we borrow 1 from the 5s place. This gives one 5 and three 1s in the units place, or $13_5 = 8$, and 4 from 8 is 4. In the 5s place, we cannot subtract 3 from 2, so we borrow 1 from the 4 in the 25s place. One 25 and two 5s gives us seven 5s, and three 5s from seven 5s is four 5s. We are left with a 3 in the 25s place, and 2 from 3 is 1.

$$\begin{array}{rrr} {}^{3} & {}^{12} & {}^{1} \\ 4 & 3 & 3_5 \\ -\,2 & 3 & 4_5 \\ \hline 1 & 4 & 4_5 \end{array} \qquad \text{or} \qquad \begin{array}{r} 433_5 \\ -\,234_5 \\ \hline 144_5 \end{array}$$

We shall also examine the operation of multiplication in base 5. The procedure for multiplication in base 5 (or in any other base) is the same as that in base 10. Table 5 is a table of multiplication facts for multiplying numerals in base 5.

TABLE 5
BASE 5 MULTIPLICATION TABLE

$\times$	0_5	1_5	2_5	3_5	4_5
0_5	0_5	0_5	0_5	0_5	0_5
1_5	0_5	1_5	2_5	3_5	4_5
2_5	0_5	2_5	4_5	11_5	13_5
3_5	0_5	3_5	11_5	14_5	22_5
4_5	0_5	4_5	13_5	22_5	31_5

We see that $3_5 \times 3_5 = 14_5$ because $3 \times 3 = 9$ and 9 is equal to one 5 and four 1s. Similarly, $4_5 \times 3_5 = 22_5$; $4 \times 3 = 12$, but 12 is two 5s and two 1s.

We must master single-digit multiplication before we can proceed to other examples. To help in an example such as 4_5 times 4_5, just remember that $4 \times 4 = 16$, and 16 equals three 5s and one 1; therefore $4_5 \times 4_5 = 31_5$.

Consider the following multiplication problem.

$$\begin{array}{r} 21_5 \\ \times \quad 31_5 \\ \hline 21_5 \\ 113_5 \\ \hline 1201_5 \end{array}$$

As the procedure for multiplication is the same for any base, we first multiply 21_5 by 1_5: $21_5 \times 1_5 = 21_5$. Now we multiply by the next digit, that is, $21_5 \times 3_5$. To do this, we multiply each digit of 21_5 by 3_5: $3 \times 1 = 3$ and $3 \times 2 = 6$. However, in base 5, $6 = 11_5$; hence $21_5 \times 3_5 = 113_5$. Note that the partial product is indented just as in base 10. Next we find the sum of the partial products.

$$\begin{array}{r} 21_5 \\ +\ 113_5 \\ \hline 1201_5 \end{array}$$

Note that in the 5s place $2 + 3 = 5$, but in base 5, $5 = 10_5$, so we write the 0 and carry the 1 to the next place, adding it to the 1 already there.

Let us try another example. Consider

$$\begin{array}{r} 433_5 \\ \times \quad 2_5 \\ \hline \end{array}$$

One way to do this problem is shown below. Note that partial products are used here, and each one is indented to indicate the powers of 5 involved.

$$\begin{array}{r} 433_5 \\ \times \quad 2_5 \\ \hline 11_5 \\ 11_5 \\ 13_5 \\ \hline 1421_5 \end{array}$$

We can do this problem in another way, but we will have to do some work mentally. The solution to the problem could have appeared as

$$\begin{array}{r} 433_5 \\ \times \quad 2_5 \\ \hline 1421_5 \end{array}$$

To do the multiplication in the shortened form, we first multiply the 3 in the units place by 2 : $2 \times 3 = 6 = 11_5$. Therefore we write 1 and carry 1. In the 5s place, we again have 2×3 which equals 11_5, but we carried 1 from the units place, so we have $11_5 + 1_5 = 12_5$. Hence we write the 2 and carry the 1. In the 25s place, we have 2×4 which equals 13_5, and we carried a 1 from the 5s place, so $13_5 + 1_5 = 14_5$ and our multiplication is complete.

EXAMPLE 6 Find the product

$$\begin{array}{r} 342_5 \\ \times \quad 23_5 \\ \hline \end{array}$$

SOLUTION First, we multiply 342_5 by 3_5 : $3 \times 2 = 6$, but $6 = 11_5$, so we write 1 and carry 1. We have $3 \times 4 = 12 = 22_5$, and because we carried a 1, we have $22_5 + 1_5 = 23_5$. We write the 3 and carry the 2. $3 \times 3 = 9 = 14_5$, and because we carried a 2, we have $14_5 + 2_5 = 21_5$. Therefore our partial product ($342_5 \times 3_5$) appears in the problem as

$$\begin{array}{r} 342_5 \\ \times \quad 23_5 \\ \hline 2131_5 \end{array}$$

Now we are ready to multiply by the next digit, 2: $2 \times 2 = 4 = 4_5$; we write the 4 and proceed. We have $2 \times 4 = 8$, but $8 = 13_5$, so we write the 3 and carry the $1: 2 \times 3 = 6 = 11_5$ and because we carried a 1, we have $11_5 + 1_5 = 12_5$. Now the problem looks like

$$\begin{array}{r} 342_5 \\ \times \quad 23_5 \\ \hline 2131_5 \\ 1234_5 \\ \hline \end{array}$$

Note that the partial product is indented to indicate the powers of 5 involved. Next, we find the sum of the partial products, and the completed problem is

$$\begin{array}{r} 342_5 \\ \times \quad 23_5 \\ \hline 2131_5 \\ 1234_5 \\ \hline 20021_5 \end{array}$$

EXAMPLE 7 Multiply 44_5 by 23_5.

SOLUTION According to Table 5, $3_5 \times 4_5 = 22_5$; we write 2 and carry 2. Again, we have $3_5 \times 4_5 = 22_5$ and because we carried a 2, we have $22_5 + 2_5 = 24_5$. Therefore the partial product is 242_5. We find the next partial product as follows: $2_5 \times 4_5 = 13_5$, so we write the 3 and carry the 1; again, we have $2_5 \times 4_5 = 13_5$, and because we carried a 1, we have $13_5 + 1_5 = 14_5$. This partial product is 143_5. We next find the sum of the partial products and obtain the final result, 2222_5.

$$\begin{array}{r} 44_5 \\ \times \quad 23_5 \\ \hline 242_5 \\ 143_5 \\ \hline 2222_5 \end{array}$$

A true test of understanding of the process of computation in base 5 is the operation of division. We divide in base 5 in the same way that we divide in base 10. In fact, we can think in base 10, but we must write our computation and answer in base 5. Consider the problem 123_5 divided by 2_5. We first set up this problem just as we would in base 10, but remember that we are working in base 5. Therefore we have

$$2_5\overline{)123_5}$$

At first glance, we might want to say that 2 divides 12 six times; but this is not the case. We are dividing 2_5 into 12_5. Besides, we cannot have a 6 for an answer because base 5 only uses the numerals 0, 1, 2, 3, 4. Since we are dividing 2_5 into 12_5, we can think of this as dividing 2 into $7 = 12_5$. Two divides 7 three times, and $2_5 \times 3_5 = 11_5$. Thus we have

$$\begin{array}{r} 3 \\ 2_5\overline{)123_5} \\ \underline{11_5} \end{array}$$

Next, we subtract and bring down the next digit. The problem now appears as

$$\begin{array}{r} 3 \\ 2_5\overline{)123_5} \\ \underline{11_5} \\ 13_5 \end{array}$$

Now we must divide 2_5 into 13_5. We can think of this as dividing 2 into 8, which equals 4, and $2_5 \times 4_5 = 13_5$. We have completed the division, and there is no remainder. The completed problem and check appear as follows.

$$\begin{array}{rl} 34_5 & \longleftarrow \text{quotient} \\ \text{divisor} \longrightarrow 2_5\overline{)123_5} & \longleftarrow \text{dividend} \\ \underline{11_5} & \\ 13_5 & \\ \underline{13_5} & \\ 0_5 & \end{array} \qquad \textit{Check:} \quad \begin{array}{rl} 34_5 & \longleftarrow \text{quotient} \\ \underline{\times\ 2_5} & \longleftarrow \text{divisor} \\ 123_5 & \longleftarrow \text{dividend} \end{array}$$

EXAMPLE 8 Divide 141_5 by 2_5.

SOLUTION First we divide 2_5 into 14_5 (think of this as 2 into 9), which equals 4. As we do in base 10, we now multiply $4_5 \times 2_5 = 13_5$; we then subtract this from 14_5 in the dividend. We bring down the next digit, 1. Now we must divide 2_5 into 11_5 (think of this as 2 into 6), which equals 3, and $3_5 \times 2_5 = 11_5$. We have completed the division, and there is no remainder.

$$\begin{array}{r} 43_5 \\ 2_5\overline{)141_5} \\ \underline{13_5} \\ 11_5 \\ \underline{11_5} \\ 0_5 \end{array} \qquad \text{Recall that } \begin{cases} 1_5 \times 2_5 = 2_5 \\ 2_5 \times 2_5 = 4_5 \\ 3_5 \times 2_5 = 11_5 \\ 4_5 \times 2_5 = 13_5 \end{cases}$$

EXAMPLE 9 Divide 234_5 by 4_5.

SOLUTION Dividing 4_5 into 23_5 (think of this as 4 into 13), we get 3. Multiplying $3_5 \times 4_5$, we obtain 22_5, and subtract this from 23_5 in the dividend. After bringing down the next digit, we divide 4_5 into 14_5 (think of this as 4 into 9), which equals 2, and $2_5 \times 4_5 = 13_5$. Subtracting this from 14_5, we obtain a remainder of 1_5, and the division is complete.

$$\begin{array}{r} 32_5 \\ 4_5\overline{)234_5} \\ \underline{22_5} \\ 14_5 \\ \underline{13_5} \\ 1_5 \end{array} \qquad \text{remainder: } 1_5$$

Remember that we can always check our work in division. In order to check, we multiply the quotient by the divisor and add the remainder, if any, to the resulting product. If the result is equal to the dividend, then the work is correct.

EXERCISES FOR SECTION 7.6

1. Perform the following additions in base 5. Check your work by converting to base 10.

a. $\begin{array}{r} 13_5 \\ +\ 23_5 \\ \hline \end{array}$ b. $\begin{array}{r} 14_5 \\ +\ 22_5 \\ \hline \end{array}$ c. $\begin{array}{r} 23_5 \\ +\ 32_5 \\ \hline \end{array}$

d. $\begin{array}{r} 123_5 \\ +\ 124_5 \\ \hline \end{array}$ e. $\begin{array}{r} 231_5 \\ +\ 222_5 \\ \hline \end{array}$ f. $\begin{array}{r} 343_5 \\ +\ 112_5 \\ \hline \end{array}$

2. Perform the following additions in base 5. Check your work by converting to base 10.

a. $\begin{array}{r} 34_5 \\ +\ 34_5 \\ \hline \end{array}$ b. $\begin{array}{r} 42_5 \\ +\ 24_5 \\ \hline \end{array}$ c. $\begin{array}{r} 34_5 \\ +\ 11_5 \\ \hline \end{array}$

d. $\begin{array}{r} 1213_5 \\ +\ 2312_5 \\ \hline \end{array}$ e. $\begin{array}{r} 3142_5 \\ +\ 2233_5 \\ \hline \end{array}$ f. $\begin{array}{r} 4241_5 \\ +\ 1204_5 \\ \hline \end{array}$

3. Perform the following additions in base 5. Check your work by converting to base 10.

a. $\begin{array}{r} 31_5 \\ +\ 32_5 \\ \hline \end{array}$ b. $\begin{array}{r} 41_5 \\ +\ 44_5 \\ \hline \end{array}$ c. $\begin{array}{r} 32_5 \\ +\ 14_5 \\ \hline \end{array}$

d. $\begin{array}{r} 134_5 \\ +\ 122_5 \\ \hline \end{array}$ e. $\begin{array}{r} 342_5 \\ +\ 234_5 \\ \hline \end{array}$ f. $\begin{array}{r} 1234_5 \\ +\ 1441_5 \\ \hline \end{array}$

4. Perform the following subtractions in base 5. Check your work by converting to base 10.

a. $23_5 - 14_5$ b. $22_5 - 13_5$ c. $12_5 - 4_5$

d. $321_5 - 231_5$ e. $231_5 - 132_5$ f. $411_5 - 122_5$

5. Perform the following subtractions in base 5. Check your work by converting to base 10.

a. $32_5 - 23_5$ b. $22_5 - 14_5$ c. $11_5 - 3_5$

d. $434_5 - 332_5$ e. $4211_5 - 1232_5$ f. $3212_5 - 2233_5$

6. Perform the following subtractions in base 5. Check your work by converting to base 10.

a. $43_5 - 24_5$ b. $31_5 - 12_5$ c. $12_5 - 4_5$

d. $341_5 - 43_5$ e. $421_5 - 122_5$ f. $4121_5 - 1213_5$

7. Perform the following multiplications in base 5. Check your work by converting to base 10.

a. $231_5 \times 3_5$ b. $432_5 \times 2_5$ c. $432_5 \times 4_5$

d. $231_5 \times 21_5$ e. $324_5 \times 23_5$ f. $432_5 \times 34_5$

8. Perform the following multiplications in base 5. Check your work by converting to base 10.

a. $343_5 \times 2_5$ b. $434_5 \times 3_5$ c. $231_5 \times 21_5$

d. $234_5 \times 42_5$ e. $434_5 \times 34_5$ f. $434_5 \times 234_5$

9. Perform the following multiplications in base 5. Check your work by converting to base 10.

a. $342_5 \times 2_5$ b. $123_5 \times 3_5$ c. $121_5 \times 31_5$

d. $312_5 \times 13_5$ e. $434_5 \times 23_5$ f. $132_5 \times 213_5$

10. Perform the following divisions in base 5. Check your work by converting to base 10.

a. $2_5\overline{)11_5}$ b. $3_5\overline{)22_5}$ c. $4_5\overline{)31_5}$

d. $2_5\overline{)32_5}$ e. $3_5\overline{)212_5}$ f. $4_5\overline{)103_5}$

11. Perform the following divisions in base 5. Check your work by converting to base 10.

a. $2_5\overline{)124_5}$ b. $3_5\overline{)343_5}$ c. $4_5\overline{)342_5}$

d. $3_5\overline{)1234_5}$ e. $11_5\overline{)243_5}$ f. $13_5\overline{)341_5}$

12. Perform the following divisions in base 5. Check your work by converting to base 10.

a. $2_5\overline{)13_5}$ b. $3_5\overline{)14_5}$ c. $4_5\overline{)103_5}$

d. $3_5\overline{)113_5}$ e. $12_5\overline{)223_5}$ f. $14_5\overline{)311_5}$

JUST FOR FUN What three words in the English language are pronounced the same as the numeral 4?

7.7 BINARY NOTATION AND OTHER BASES

Arithmetic operations in other bases are performed in the same manner as in base 10 or base 5. The base 2 system of numeration (also called *binary notation*) is of particular interest because it has some useful applications, particularly in the area of computers. **Binary notation** is a system of numeration that groups items by 2s. Therefore, the only numerals used in this system are 0 and 1. We can illustrate this if we examine a numeral in base 2 notation. Consider 10_2: writing this in expanded notation, we have $10_2 = (1 \times 2^1) + (0 \times 2^0) = (1 \times 2) + (0 \times 1) = 2 + 0 = 2$. How would we express 3 in base 2? Three is composed of one 2 and one 1; hence $3 = 11_2$.

The binary system is unique in that it uses only two symbols, 0 and 1, to represent any number. This is important for computers, because these machines consist of a large number of electrical switches. Each of these switches, like a light switch, has only two possible positions: on or off. The position of a switch can be changed in nanoseconds. (A *nanosecond* is one-billionth of a second.) This is because electricity travels at a rate of approximately 186,000 miles per second, or about 1 foot per nanosecond. If a switch is on, then it represents the numeral 1; if it is off, it represents the numeral 0. Using only these two symbols, a computer can perform thousands of calculations per second, thereby saving thousands of hours of human worktime. Informa-

Courtesy AT & T

This computer performs all of its internal calculations in base 2. The results are converted to base 10 for display on the monitor screen.

tion is fed into most computers in base 10 notation and then converted to base 2. The machine performs the necessary base 2 calculations by turning many switches on and off.

TABLE 6
ADDITION TABLE FOR BASE 2

+	0_2	1_2
0_2	0_2	1_2
1_2	1_2	10_2

TABLE 7
MULTIPLICATION TABLE FOR BASE 2

×	0_2	1_2
0_2	0_2	0_2
1_2	0_2	1_2

To convert a base 10 numeral to base 2, we perform successive divisions by 2, the new base, and record the remainders for each separate division. The answer is determined by reading the remainders from the bottom to the top.

To express 7 as a base 2 numeral, we perform the division in the following manner.

$$\begin{array}{r|l l} 2 & 7 & \\ 2 & 3 & 1 \uparrow \\ 2 & 1 & 1 \\ & 0 & 1 \end{array} \qquad 7 = 111_2$$

We can check our answer by converting 111_2 back to base 10. We can do this by writing 111_2 in expanded notation: $111_2 = (1 \times 2^2) + (1 \times 2^1) + (1 \times 2^0) = (1 \times 4) + (1 \times 2) + (1 \times 1) = 4 + 2 + 1 = 7$. Note that because we are now working in base 2, each place value is a power of 2: 2^0, 2^1, 2^2, 2^3, and so on.

EXAMPLE 1 Change to base 10 notation.

a. 101_2 b. 1101_2

SOLUTION a. In order to change a numeral to base 10 notation, we must write the numeral in expanded notation.

$$\begin{aligned} 101_2 &= (1 \times 2^2) + (0 \times 2^1) + (1 \times 2^0) \\ &= (1 \times 4) + (0 \times 2) + (1 \times 1) \\ &= 4 + 0 + 1 \\ &= 5 \end{aligned}$$

b. $1101_2 = (1 \times 2^3) + (1 \times 2^2) + (0 \times 2^1) + (1 \times 2^0)$
$= (1 \times 8) + (1 \times 4) + (0 \times 2) + (1 \times 1)$
$= 8 + 4 + 0 + 1$
$= 13$

EXAMPLE 2 Change to base 2 notation.
a. 9 b. 15

SOLUTION We perform successive divisions by 2 and record the remainders for each division. The answer is determined by reading the remainders from bottom to top.

a.		b.	
2⌊9		2⌊15	
2⌊4	1 ↑	2⌊7	1 ↑
2⌊2	0	2⌊3	1
2⌊1	0	2⌊1	1
0	1	0	1

$9 = 1001_2$ $15 = 1111_2$

In order to perform addition in base 2, we must remember the four addition facts listed in Table 6. We must also remember the "carrying" process. Consider the following addition problem.

$$\begin{array}{r} 111_2 \\ +\ 110_2 \\ \hline \end{array}$$

We start with the units place: $0 + 1 = 1$, and this is the answer for both base 2 and base 10, because the numerals 0 and 1 have the same meaning in base 2 and base 10. The sum of 0 and 1 is 1, and there is nothing to carry. We proceed to the 2s place: $1 + 1 = 2$, but in base 2, $2 = 10_2$. We write the 0 and carry the 1. Next we have $1 + 1 + 1$, which equals 3; in base 2, $3 = 11_2$ (one 2 and one 1). The addition is complete, as shown below.

$$\begin{array}{r} 111_2 \\ +\ 110_2 \\ \hline 1101_2 \end{array}$$

EXAMPLE 3 Find the sum of 1010_2 and 1011_2.

SOLUTION

$$\begin{array}{r} 1010_2 \\ +\ 1011_2 \\ \hline 10101_2 \end{array}$$

Check:

$$\begin{aligned}
1010_2 &= (1 \times 2^3) + (0 \times 2^2) + (1 \times 2^1) + (0 \times 2^0) \\
&= (1 \times 8) + (0 \times 4) + (1 \times 2) + (0 \times 1) \\
&= 8 + 0 + 2 + 0 && = 10 \\
+\ 1011_2 &= (1 \times 2^3) + (0 \times 2^2) + (1 \times 2^1) + (1 \times 2^0) \\
&= (1 \times 8) + (0 \times 4) + (1 \times 2) + (1 \times 1) \\
&= 8 + 0 + 2 + 1 && = \underline{11} \\
10101_2 &= (1 \times 2^4) + (0 \times 2^3) + (1 \times 2^2) + (0 \times 2^1) + (1 \times 2^0) \\
&= (1 \times 16) + (0 \times 8) + (1 \times 4) + (0 \times 2) + (1 \times 1) \\
&= 16 + 0 + 4 + 0 + 1 && = 21
\end{aligned}$$

EXAMPLE 4 Subtract 111_2 from 1011_2.

SOLUTION In the units and 2s place, we subtract 1 from 1 and obtain 0. But in the 4s place, we cannot subtract 1 from 0. Therefore, we borrow 1 from the 8s place. One 8 gives us two 4s, and one 4 from two 4s is one 4. The subtraction is complete.

$$\begin{array}{r} 1011_2 \\ -\ \ 111_2 \\ \hline 100_2 \end{array}$$

Multiplication in base 2 does not present much of a problem, providing we can add the partial products, because we only have to remember the four multiplication facts listed in Table 7: $0 \times 0 = 0$, $0 \times 1 = 0$, $1 \times 0 = 0$, and $1 \times 1 = 1$. Consider the following multiplication problem.

$$\begin{array}{r} 101_2 \\ \times\ \ 11_2 \\ \hline 101_2 \\ 101_2\ \ \\ \hline 1111_2 \end{array}$$

Because the procedure for multiplication is the same for any base, we first multiply 101_2 by 1_2, which equals 101_2. Now we multiply by the next digit, that is, $101_2 \times 1_2$, which equals 101_2. Note that the partial product is indented, as in base 10. Next we find the sum of the partial products. For this example, we do not have to carry anything.

EXAMPLE 5 Multiply 110_2 by 11_2.

SOLUTION

$$\begin{array}{r} 110_2 \\ \times \;\; 11_2 \\ \hline 110_2 \\ 110_2\;\, \\ \hline 10010_2 \end{array}$$

The only problem that occurs here is in the adding of the partial products.

$$\begin{array}{r} 110_2\;\, \\ +\; 110_2\;\, \\ \hline 10010_2 \end{array}$$

Note that in the 4s place we have 1 + 1, which equals 10_2, so we place the 0 and carry the 1. This again gives 1 + 1, which is 10_2. Since our addition is completed, we write down this sum to give us the final answer, 10010_2.

EXAMPLE 6 Multiply 101_2 by 101_2.

SOLUTION The only difference between the two solutions below is that the first solution indicates the multiplication by 0 in the second partial product, whereas the second solution actually shows the multiplication by 0.

$$\begin{array}{r} 101_2 \\ \times\, 101_2 \\ \hline 101_2 \\ 1010_2\;\, \\ \hline 11001_2 \end{array} \qquad \text{or} \qquad \begin{array}{r} 101_2 \\ \times\, 101_2 \\ \hline 101_2 \\ 000_2\;\, \\ 101_2\;\;\;\, \\ \hline 11001_2 \end{array}$$

HISTORICAL NOTE

The earliest known use of coins occurred around 550 B.C. Before this time, merchants traded their wares for other items and services they needed. Next, they began to sell their merchandise for silver and copper coins. The significance of this is that the invention of "money" made it necessary for people to know more about counting and working with numbers.

If we wish to convert a base 10 numeral to a different base, we perform successive divisions by the numeral of the new base, and record the remainders for each separate division. The answer is determined by reading the remainders from the bottom to the top. For example, if we wanted to express 34 as a numeral in base 6, we would divide 34 by 6.

$$\begin{array}{r l}
6\,\underline{|34} & \uparrow \\
6\,\underline{|5}\quad 4 & | \\
0\quad 5 & | \quad 34 = 54_6 \quad \text{(five 6s and four 1s)}
\end{array}$$

If we want to convert a numeral in some other base to a base 10 numeral, then we use the concept of expanded notation. To convert 342_8 to a base 10 numeral, for example, we would write 342_8 as $(3 \times 8^2) + (4 \times 8^1) + (2 \times 8^0)$ and proceed to evaluate it as $(3 \times 64) + (4 \times 8) + (2 \times 1) = 192 + 32 + 2 = 226$.

In order to convert numerals from one base to another base, where neither base is base 10, it is best to convert first to base 10 and then convert this result to the desired base. For example, suppose we want to convert 54_6 to base 8. First we convert 54_6 to base 10.

$$54_6 = (5 \times 6^1) + (4 \times 6^0) = (5 \times 6) + (4 \times 1) = 30 + 4 = 34$$

Now we convert 34 to base 8 by performing successive divisions.

$$\begin{array}{r l}
8\,\underline{|34} & \\
8\,\underline{|4}\quad 2 & \uparrow \\
0\quad 4 & | \qquad 34 = 42_8
\end{array}$$

Therefore, $54_6 = 42_8$.

EXAMPLE 7 Convert 354_6 to base 5.

SOLUTION First we convert 354_6 to base 10.

$$\begin{aligned}
354_6 &= (3 \times 6^2) + (5 \times 6^1) + (4 \times 6^0) \\
&= (3 \times 36) + (5 \times 6) + (4 \times 1) \\
&= 108 + 30 + 4 = 142
\end{aligned}$$

Now we have $354_6 = 142$, and we can convert 142 to a base 5 numeral.

$$\begin{array}{r|l l}
5 & 142 & \\
\hline
5 & 28 & 2 \uparrow \\
\hline
5 & 5 & 3 \\
\hline
5 & 1 & 0 \\
\hline
 & 0 & 1
\end{array}$$

Therefore, $354_6 = 1032_5$.

Check:

$$\begin{aligned}
354_6 &= (3 \times 6^2) + (5 \times 6^1) + (4 \times 6^0) \\
&= (3 \times 36) + (5 \times 6) + (4 \times 1) \\
&= 108 + 30 + 4 = 142 \\
1032_5 &= (1 \times 5^3) + (0 \times 5^2) + (3 \times 5^1) + (2 \times 5^0) \\
&= (1 \times 125) + (0 \times 25) + (3 \times 5) + (2 \times 1) \\
&= 125 + 0 + 15 + 2 = 142
\end{aligned}$$

EXAMPLE 8 Which is greater, 211_3 or 10111_2?

SOLUTION We will convert each of the numerals to base 10 and compare the results.

$$\begin{aligned}
211_3 &= (2 \times 3^2) + (1 \times 3^1) + (1 \times 3^0) \\
&= (2 \times 9) + (1 \times 3) + (1 \times 1) = 18 + 3 + 1 = 22 \\
10111_2 &= (1 \times 2^4) + (0 \times 2^3) + (1 \times 2^2) + (1 \times 2^1) + (1 \times 2^0) \\
&= (1 \times 16) + (0 \times 8) + (1 \times 4) + (1 \times 2) + (1 \times 1) \\
&= 16 + 0 + 4 + 2 + 1 = 23
\end{aligned}$$

Because $211_3 = 22$ and $10111_2 = 23$, $10111_2 > 211_3$.

CONCHY by James Childress.

EXERCISES FOR SECTION 7.7

1. Change to base 10 notation.
   a. 10_2 b. 11_2 c. 101_2
   d. 111_2 e. 1011_2 f. 11011_2

2. Change to base 10 notation.
   a. 100_2 b. 110_2 c. 1101_2
   d. 1010_2 e. 10110_2 f. 11111_2

3. Change to base 10 notation.
   a. 1000_2 b. 1001_2 c. 1100_2
   d. 1111_2 e. 10000_2 f. 10001_2

4. Change to binary notation.
   a. 5 b. 7 c. 8
   d. 11 e. 17 f. 23

5. Change to binary notation.
   a. 6 b. 9 c. 13
   d. 21 e. 25 f. 33

6. Change to binary notation.
   a. 4 b. 10 c. 12
   d. 16 e. 22 f. 36

7. Perform the following additions in base 2. Check your work by converting to base 10.
   a. $10_2 + 11_2$ b. $11_2 + 11_2$ c. $10_2 + 10_2$
   d. $110_2 + 10_2$ e. $100_2 + 101_2$ f. $110_2 + 110_2$

8. Perform the following additions in base 2. Check your work by converting to base 10.
   a. $110_2 + 11_2$ b. $101_2 + 10_2$ c. $111_2 + 110_2$
   d. $111_2 + 111_2$ e. $1110_2 + 1011_2$ f. $1011_2 + 1111_2$

9. Perform the following additions in base 2. Check your work by converting to base 10.
   a. $1000_2 + 101_2$ b. $1001_2 + 111_2$ c. $1110_2 + 111_2$
   d. $1110_2 + 1110_2$ e. $1001_2 + 1011_2$ f. $1111_2 + 1111_2$

10. Perform the following subtractions in base 2. Check your work by converting to base 10.
   a. $10_2 - 1_2$ b. $11_2 - 10_2$ c. $101_2 - 10_2$
   d. $101_2 - 11_2$ e. $111_2 - 101_2$ f. $1011_2 - 101_2$

11. Perform the following subtractions in base 2. Check your work by converting to base 10.
   a. $11_2 - 1_2$ b. $111_2 - 10_2$ c. $111_2 - 101_2$
   d. $1010_2 - 101_2$ ⋆e. $1001_2 - 11_2$ ⋆f. $1001_2 - 110_2$

12. Perform the following subtractions in base 2. Check your work by converting to base 10.
   a. $111_2 - 110_2$
   b. $1011_2 - 100_2$
   c. $1011_2 - 110_2$
   d. $1100_2 - 10_2$
   ★e. $1100_2 - 110_2$
   ★f. $1100_2 - 111_2$

13. Perform the following multiplications in base 2. Check your work by converting to base 10.
   a. $11_2 \times 11_2$
   b. $11_2 \times 10_2$
   c. $101_2 \times 11_2$
   d. $110_2 \times 11_2$
   e. $101_2 \times 111_2$
   f. $101_2 \times 101_2$

14. Perform the following multiplications in base 2. Check your work by converting to base 10.
   a. $111_2 \times 10_2$
   b. $101_2 \times 10_2$
   c. $111_2 \times 100_2$
   d. $1001_2 \times 11_2$
   e. $1011_2 \times 101_2$
   ★f. $111_2 \times 111_2$

15. Perform the following multiplications in base 2. Check your work by converting to base 10.
   a. $111_2 \times 11_2$
   b. $100_2 \times 10_2$
   c. $100_2 \times 11_2$
   d. $100_2 \times 101_2$
   e. $1100_2 \times 101_2$
   f. $1001_2 \times 101_2$

16. Convert 321_6 to base 5.

17. Convert 434_5 to base 4.

18. Convert 101101_2 to base 7.

19. Convert 243_5 to base 8.

20. Convert 111_5 to base 2.

21. Indicate whether each statement is true or false.
   a. $111_2 > 12_5$
   b. $321_4 < 10111_2$
   c. $1011_2 = 21_5$
   d. $110_2 = 11_5$

22. Indicate whether each statement is true or false.
   a. $41_8 < 46_7$
   b. $32_8 > 1111_2$
   c. $421_5 = 111$
   d. $213_4 < 211_5$

★23. Perform each operation in the indicated base.
   a. $23_6 + 34_6$
   b. $352_7 + 405_7$
   c. $41_7 - 23_7$
   d. $241_6 - 42_6$

★24. Perform each operation in the indicated base.
   a. $21_4 + 33_4$
   b. $35_6 + 42_6$
   c. $32_4 - 13_4$
   d. $44_6 - 25_6$

★25. Perform each operation in the indicated base.
   a. $32_7 \times 43_7$
   b. $24_6 \times 43_6$
   c. $46_8 \times 23_8$
   d. $56_9 \times 34_9$

★26. Perform each operation in the indicated base.
   a. $121_3 + 122_3$
   b. $234_5 + 141_5$
   c. $441_6 - 212_6$
   d. $325_7 - 146_7$

★27. Perform each operation in the indicated base.
   a. $212_3 \times 22_3$
   b. $424_6 \times 32_6$
   c. $423_5 \times 22_5$
   d. $325_7 \times 46_7$

JUST FOR FUN

Three Yankee fans and three Met fans have to ride an elevator up to the top floor (there are no steps between the ground floor and the top), but the elevator will hold no more than two people. The Met fans always start an argument if they are left in a situation where they outnumber the Yankee fans, but they are fine if they are left alone or if they are with the same or a greater number of Yankee fans. How do they all get to the top floor, using the elevator, without any arguments?

Summary

One of the oldest known systems of numeration is that of the Egyptians (approximately 3400 B.C.). The Egyptians used a simple grouping system, grouping items by 10s. They used a different symbol for each power of 10.

When we say that the Egyptians used a simple grouping system, we mean that the position of a symbol in an Egyptian numeral does not affect the number represented.

The Greek system of numeration differs from other systems of numeration in that it uses special symbols for multiples of 5. For example, since they had no symbol for 50, they thought of 50 as five 10s, and similarly, 500 was thought of as five 100s. This concept enabled the Greeks to use fewer symbols to express a number. The use of multiples of 5 is an example of *multiplicative grouping*. Another system of numeration that involves multiplicative grouping is the Chinese-Japanese system of numeration. This system arranges numbers vertically instead of horizontally. It also uses positional notation and groups items in powers of ten.

The Babylonian system of numeration uses only two symbols to represent numerals. This system also uses a *place-value* system: the position of a symbol matters. The Babylonian system of numeration is based on powers of 60, and is called a sexagesimal system.

Our decimal system of numeration uses *Hindu-Arabic* numerals to express numbers. The decimal system of numeration groups by powers of 10. It also uses a place-value system: the position that a symbol has in a decimal numeral affects its meaning.

Even though we use the decimal system of numeration, it is not uncommon for us to group items in some other manner, such as pairs (2s) or dozens. When we group items by some number other than 10, we say that we are working with a *base* different from 10. When we write numerals in a base other than base 10, we must indicate what base we are working with. We do this by using a subscript to indicate the base. An example is 23_5, which indicates that we are grouping by 5s.

In order to change a numeral in some other base to base 10 notation, we write the numeral in expanded notation. For example, $23_5 = (2 \times 5^1) + (3 \times 5^0) = (2 \times 5) + (3 \times 1) = 10 + 3 = 13$. We can also convert any number from base 10 to another base by performing successive divisions by the new base and recording the remainders for each division; we stop dividing when we get a quotient of zero. The answer in the new base is found by reading the remainders from bottom to top.

In this chapter, we performed arithmetic operations in base 5 and base 2.

When we perform any operation in any base other than base 10, we must write the numerals in the indicated base; however, it is sometimes helpful to think in terms of base 10, because the arithmetic operations are performed in the same manner in every base.

Vocabulary Check

number	place-value system	expanded notation	gross
decimal system	simple grouping system	binary notation	great gross
duodecimal system	numeral	sexagesimal system	algorithm
decimal notation	base 5 system	multiplicative grouping system	

Review Exercises for Chapter 7

1. Express 78 as
   a. an Egyptian numeral
   b. a Greek numeral
   c. a Chinese-Japanese numeral
   d. a Babylonian numeral
2. Express the following as Egyptian numerals.
   a. 32 b. 211 c. 1,111
3. Express the following Egyptian numerals as base 10 numerals.
   a. 
   b. 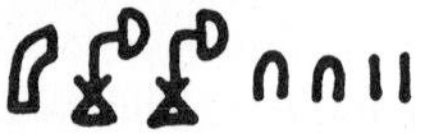
4. Perform the indicated operations. Express your answers in Egyptian numerals.
   a. 
   b. 
   c. 
   d. 
5. List the distinguishing characteristics of
   a. a simple grouping system
   b. a multiplicative grouping system
   c. a place-value system.

6. Name a system of numeration that uses
   a. simple grouping
   b. multiplicative grouping
   c. place value.

7. Write in expanded notation.
   a. 345 b. 342_5 c. 10111_2

8. Convert to base 10.
   a. 47_{12} b. 34_5 c. 241_5
   d. 36_8 e. 10101_2 f. $2TE_{12}$

9. Convert to the indicated base.
   a. $45 = \_\_\_\__5$ b. $51 = \_\_\_\__5$
   c. $32 = \_\_\_\__{12}$ d. $33 = \_\_\_\__2$
   e. $63 = \_\_\_\__2$ f. $34 = \_\_\_\__{12}$

10. Name at least two systems of numeration, other than base 10, to which we are exposed in everyday life and give examples of how each is used.

11. Perform the indicated operations in base 5.
   a. $23_5 + 22_5$ b. $314_5 + 221_5$ c. $22_5 - 13_5$ d. $231_5 - 132_5$

12. Perform the indicated operations in base 5.
   a. $231_5 \times 3_5$ b. $123_5 \times 4_5$ c. $231_5 \times 34_5$ d. $2_5\overline{)124_5}$

13. Perform the indicated operations in base 2.
   a. $101_2 + 110_2$ b. $101_2 + 101_2$ c. $1101_2 + 1101_2$ d. $101_2 - 10_2$

14. Perform the indicated operations in base 2.
   a. $111_2 - 11_2$ b. $1011_2 - 101_2$ c. $111_2 \times 10_2$ d. $1011_2 \times 101_2$

15. Convert each of the following to the indicated base.
   a. $321_5 = \_\_\_\__6$ b. $42_5 = \_\_\_\__3$
   c. $10111_2 = \_\_\_\__5$ d. $77_8 = \_\_\_\__2$

16. Express each of the following as a numeral in base 5 and base 2.
   a. 7 b. 13 c. 33

17. Convert to base 10.
   a. 58_9 b. 34_7 c. 65_8
   d. 402_9 e. 321_6 f. 465_7

18. Convert to the indicated base.
   a. $33 = \_\_\_\__9$ b. $46 = \_\_\_\__7$
   c. $65 = \_\_\_\__4$ d. $92 = \_\_\_\__3$
   e. $75 = \_\_\_\__6$ f. $135 = \_\_\_\__8$

19. Perform each operation in the indicated base.
   a. $432_6 + 324_6$ b. $321_4 + 233_4$
   c. $521_7 - 436_7$ d. $843_9 - 627_9$

20. Perform each operation in the indicated base.
   a. $314_6 \times 42_6$ b. $122_3 \times 12_3$
   c. $346_7 \times 34_7$ d. $843_9 \times 27_9$

JUST FOR FUN Three books stand in order on a shelf. The pages of each book take up 2 inches on the shelf, and their front and back covers are each $\frac{1}{4}$-inch thick. What is the minimum distance from the first page of Volume 1 to the last page of Volume 3?

NOTE OF INTEREST

Using marks or notches to represent numbers was a common method throughout history. Roman numerals seem to have evolved from this technique: consider the symbols I, II, III. Even today, we find the use of Roman numerals: on clocks, room numbers, conerstones, and so on. The date 1776 is recorded in Roman numerals on the base of the pyramid pictured on the back of a one-dollar bill as MDCCLXXVI.

The Roman system of numeration uses seven capital letters to represent numbers. They are listed below, together with their corresponding values.

Roman Numerals	Values
I	1
V	5
X	10
L	50
C	100
D	500
M	1000

These seven letters are used in the Roman system of numeration to represent numbers. Since there are more than seven numbers, we must have some rules for combining these letters together in order to represent other numbers. We shall adopt the following rules.

1. If we repeat a letter, we repeat its value.

   XX = 2 tens or 20
   CCC = 3 hundreds or 300

2. If a letter is positioned after another letter of greater value, we add its value to the value of the letter of greater value.

   XI = X + I = 11
   CL= C + L = 150
   XVI = X + V + I = 16

3. If a letter is positioned before another letter of greater value, we subtract its value from the value of the letter of greater value.

   IV = V − I = 4
   XL = L − X = 40
   CM = M − C = 900

4. If a letter is positioned between two letters, each of greater value, its value is subtracted from the sum of the values of those letters of greater value.

   XIX = (X + X) − I = 19
   LIX = (L + X) − I = 59
   XLIV = (L − X) + V − I = 44

   MCMLXXVI = 1976

Chapter Quiz

Indicate whether each statement is true or false.

1. Any symbol for a number is called a numeral.
2. The Egyptians used a multiplicative grouping system.
3. In the Egyptian system of numeration

   ∩II = I∩I

4. If the position of a symbol does not affect the number represented, then the system is a simple grouping system.
5. The Greek system of numeration is a simple grouping system.
6. One of the advantages of a multiplicative grouping system is that it enables us to use fewer symbols to express a number as compared with a simple grouping system.
7. The Chinese-Japanese system of numeration uses multiplicative grouping.
8. The Babylonian system of numeration uses only two symbols.
9. One of the most significant contributions of the Babylonian system was devising a numeral for zero.

10. $43_{\text{twelve}} = 51$

11. $32_{\text{twelve}} = 38$

12. $121_{\text{five}} = 37$

13. $ET1_{\text{twelve}} = 1{,}704$

14. $433_5 + 234_5 = 667_5$

15. $33_5 \times 44_5 = 3212_5$

16. $1011_2 + 111_2 = 110010_2$

17. $1001_2 - 111_2 = 110_2$

18. $111_2 \times 11_2 = 10111_2$

19. $43_5 > 11111_2$

20. $78_8 = 213_5$

21. $354_6 = 1032_5$

22. $134_7 + 265_7 = 432_7$

23. $432_6 \times 43_6 = 31300_6$

24. $47_9 < 11101_2$

25. $82_9 = 2202_3$

CHAPTER 8

SETS OF NUMBERS AND THEIR STRUCTURE

AFTER STUDYING THIS CHAPTER
YOU WILL BE ABLE TO DO THE FOLLOWING:

1. Determine whether a natural number is **prime** or **composite.**
2. Find the **prime factors** of a given composite number.
3. Find the **greatest common divisor** and the **least common multiple** for a given pair of numbers.
4. Add, subtract, and multiply integers.
5. Add, subtract, multiply, and divide rational numbers.
6. Express a rational number as a decimal.
7. Express a **terminating decimal** or a **repeating nonterminating decimal** as a quotient of integers.
8. Identify an **irrational number** as a nonterminating and nonrepeating decimal.
9. Identify the set of **real numbers** as the union of the sets of rational and irrational numbers.

★10. Use scientific notation to evaluate expressions that contain very large or very small numbers.

8.1 INTRODUCTION

Numbers, like many other things, can be described or classified in a variety of ways. It is not unusual to classify an item in a certain category and then later reclassify it differently to describe it better.

If a person is holding a playing card from a standard deck of cards, we don't know much about the card: it is one of 52 possibilities. If the person tells us that it is a heart, then we know that it is one of 13 possibilities. If we are told that it is also a picture card, then we know it is one of three possibilities—king, queen, or jack of hearts. Finally, the person may identify it completely by telling us that the card is the king of hearts.

Once an item has been placed in a general category, we can better describe it by placing it in more specific categories. This can also be done with numbers. That is, numbers can be classified in various ways. For example, in Chapter 1 we discussed cardinal numbers. A *cardinal number*—for example, 1, 2, 10, or 2,001—tells us "how many." In Chapter 1 we used cardinal numbers to determine the number of elements in a set.

An *ordinal number* refers to order—for example, first, second, 14th, and 123rd are ordinal numbers.

Your phone number, student number, and social security number are neither cardinal nor ordinal numbers: they are used strictly for identification purposes.

In this chapter we shall classify numbers in another manner and examine the properties of numbers that belong to these particular categories.

8.2 NATURAL NUMBERS—PRIMES AND COMPOSITES

When human beings first started to count they did not begin with zero. They began with 1 and then one more, 2, and proceeded in a like manner. They counted in this manner.

$$1, 2, 3, 4, 5, 6, 7, 8, 9, \ldots$$

That is they began with 1 and continued to count. The set of numbers $\{1, 2, 3, 4, \ldots\}$ is often referred to as the set of **counting numbers.** More formally, they are called the set of **natural numbers.** The first natural number is 1, sometimes called the unit number.

Any natural number can be expressed as the product of two or more natural numbers. For example, $6 = 3 \times 2$, $8 = 4 \times 2$, $3 = 3 \times 1$, and $1 = 1 \times 1$. The numbers that are multiplied together to form a number are the *factors* or the number. Since $6 = 3 \times 2$, 3 and 2 are factors of 6. They are also called *divisors* of 6. A factor or divisor of a number divides the given number with a zero remainder. Every natural number is divisible by itself and 1. Certain natural numbers are divisible only by themselves and 1. These natural numbers are called **prime numbers.**

A *prime number* is any natural number greater than 1 that is divisible only by itself and 1.

A prime number cannot be written as the product of natural numbers that are less than itself. A prime number has exactly two factors, 1 and the number itself.

Numerals 2, 3, 5, 7, and 11 are the first five prime numbers. Note that 1 is not a prime number. The first prime number, 2, is unique in that it is the only even prime number. Any other even natural number such as 10, 200, or 484 is divisible by 2 and therefore cannot be prime, because a prime number is divisible only by itself and 1. If a natural number greater than 2 is not prime, then it is called a **composite number** because it can be composed of other factors. A composite number can be written as the product of natural numbers that are less than itself.

Any composite number can be expressed as a product of prime factors. For example, $6 = 2 \times 3$, $12 = 2 \times 2 \times 3$, and $4 = 2 \times 2$. In fact, if we disregard order, every composite natural number can be expressed as a product of prime factors in one and only one way. ($6 = 2 \times 3$ or $6 = 3 \times 2$, but these are the same if we disregard order.) If we phrase this statement more formally, we have

Every natural number greater than 1 is either a prime or can be expressed as a product of prime factors. Except for the order of the factors, this can be done in one and only one way.

This statement is the *fundamental theorem of arithmetic*. Because every natural number except 1 is either a prime number or can be expressed as a product of primes, the concept of a prime number is an important one.

How can we tell if a number is prime? There is no quick and easy solution; no formula exists for finding primes. Suppose we want to determine if 29 is prime. How do we go about it? In order to determine whether a number is prime, we check divisors to see if they divide it. The first divisor to check is 2: 29 is not divisible by 2. Try 3: 29 is not divisible by 3. How about 4? We need not test 4 because it is a composite number ($4 = 2 \times 2$) and contains smaller divisors (2) which have already been tried. Next we try 5: 29 is not divisible by 5. We do not need to test 6. Why? Because $6 = 3 \times 2$ and neither 3 nor 2 divide 29. Try 7: 29 is not divisible by 7. How far do we keep testing? All the way to 29? No; in fact, we should have stopped at 6, because $6 \times 6 = 36$ and $36 > 29$. If 6 or a number greater than 6 divided 29, then the quotient would be less than 6 and would also be a factor of 29. But we have already tested all possible factors less than 6, and they all failed. Therefore, our conclusion is that 29 is prime.

To determine if a number is prime, we need only test the prime divisors $\{2, 3, 5, 7, 11, \ldots\}$ up to the largest natural number whose square is less than or equal to the number we are testing.

Remember, we do not have to check composite divisors, because a composite number can be expressed as a product of prime factors.

EXAMPLE 1 Is 43 prime?

SOLUTION Yes, 43 is not divisible by 2, 3, 5, or 7. We need not check any other divisors, since $7^2 = 49$ and $49 > 43$.

EXAMPLE 2 Is 91 prime?

SOLUTION No. We need not check past 10, since $10^2 = 100$ and $100 > 91$. In fact, we only have to check through 7, because 8, 9, and 10 are all composite numbers. Two does not divide 91, 3 does not divide 91, and 5 does not divide 91. But 7 does divide 91: $91 = 7 \times 13$. Hence, 91 is not prime.

EXAMPLE 3 Is 1,001 prime?

SOLUTION No. The primes less than 100 are 2, 3, 5, 7, 11, 13, 17, 19, 23, 29, 31, 37, 41, 43, 47, 53, 59, 61, 67, 71, 73, 79, 83, 89, and 97. We do not have to check past 37, since $37^2 = 1{,}369$ and $1{,}369 > 1{,}001$. In fact, we only have to check through 31, since the other natural numbers up to 37 are composite. We see that 2 does not divide 1,001, 3 does not divide 1,001, and 5 does not divide 1,001. But 7 does divide 1,001: $1{,}001 = 7 \times 143$. Therefore, 1,001 is not prime.

EXAMPLE 4 Is 2003 prime?

SOLUTION Yes. We need not check past 45, since $45^2 = 2{,}025$ and $2{,}025 > 2{,}003$. None of the prime numbers 2, 3, 5, 7, 11, 13, 17, 19, 23, 29, 31, 37, 41, or 43 divides 2,003. This is far enough to check, since 44 and 45 are composite numbers. Therefore, 2,003 is prime.

The following is a list of rules for divisibility by certain numbers. These rules may aid you in determining whether a given number is prime, or in finding the prime factors of a number.

1. A natural number is divisible by 2 if the natural number is an even number. (For example, 2 divides 2,754, since 2,754 ends in 4 and 4 is even.)
2. A natural number is divisible by 3 if the sum of the digits is divisible by 3. (For example, 3 divides 2,754, because $2 + 7 + 5 + 4 = 18$, and 18 is divisible by 3.)
3. A natural number is divisible by 4 if the number formed by its last two digits is divisible by 4. (For example, 4 divides 2,924 because the last two digits, 24, form a number divisible by 4.)
4. A natural number is divisible by 5 if the last digit on the right is 0 or 5. (For example, 5 divides 1,350 and 234,795.)
5. A natural number is divisible by 8 if the number formed by the last three digits is divisible by 8. (For example, 8 divides 4,560 because the last three digits, 560, form a number divisible by 8.)
6. A natural number is divisible by 9 if the sum of the digits is divisible by 9. (For example, 9 divides 2,754, because $2 + 7 + 5 + 4 = 18$, and 18 is divisible by 9.)
7. A natural number is divisible by 10 if its last digit (the units place) is 0. (For example, 10 divides 1,350 and 408,700.)
8. A natural number is divisible by 11 if the difference between the sum of the digits in the odd places and the sum of the digits in the even places is 0 or divisible by 11. (For example, 368,610 is divisible by 11 since the difference of the sum of the digits in the odd places, $6 + 6 + 0 = 12$, and the sum of the digits in the even places, $3 + 8 + 1 = 12$, is 0.)

EXAMPLE 5 Test each of the following numbers for divisibility by 2, 3, 4, 5, 8, 9, 10, and 11.

a. 330 b. 410 c. 369 d. 1,331

SOLUTION a. 330 is divisible by 2, 3, 5, 10, and 11.
b. 410 is divisible by 2, 5, and 10.
c. 369 is divisible by 3 and 9.
d. 1,331 is divisible by 11.

There exists an interesting and ancient technique for finding primes. It is called the **sieve of Eratosthenes** and was invented by a Greek scholar, Eratosthenes (276–194 B.C.), who was also head of the famous library in Alexandria. We shall use the numbers from 1 to 100 to illustrate Eratosthenes' method.

~~1~~	2	3	~~4~~	5	~~6~~	7	~~8~~	~~9~~	~~10~~
11	~~12~~	13	~~14~~	~~15~~	~~16~~	17	~~18~~	19	~~20~~
~~21~~	~~22~~	23	~~24~~	~~25~~	~~26~~	~~27~~	~~28~~	29	~~30~~
31	~~32~~	~~33~~	~~34~~	~~35~~	~~36~~	37	~~38~~	~~39~~	~~40~~
41	~~42~~	43	~~44~~	~~45~~	~~46~~	47	~~48~~	~~49~~	~~50~~
~~51~~	~~52~~	53	~~54~~	~~55~~	~~56~~	~~57~~	~~58~~	59	~~60~~
61	~~62~~	~~63~~	~~64~~	~~65~~	~~66~~	67	~~68~~	~~69~~	~~70~~
71	~~72~~	73	~~74~~	~~75~~	~~76~~	~~77~~	~~78~~	79	~~80~~
~~81~~	~~82~~	83	~~84~~	~~85~~	~~86~~	~~87~~	~~88~~	89	~~90~~
~~91~~	~~92~~	~~93~~	~~94~~	~~95~~	~~96~~	97	~~98~~	~~99~~	~~100~~

We exclude 1 because 1 is not prime. Eratosthenes determined that 2 is prime, and also that every second number (beginning with 2) is not prime because it has 2 as a factor. Therefore we cross out 4, 6, 8, 10, ... Similarly, 3 is a prime number, and every third number (beginning with 3) is not prime because it has 3 as a factor. Therefore, we cross out 6, 9, 12, ... Note that some of these numbers, such as 6, 12, and 18, have already been crossed out. The next number to consider is 5; it is prime, so we cross out every fifth number.

We can continue this process, but for how long? We do not have to go past 11, since $11^2 = 121$ and $121 > 100$. In fact, after determining that 7 is prime and crossing out every seventh number, we have a list of all the prime numbers less than 100. This process is called the sieve of Eratosthenes because instead of crossing out the numbers he punched them out with a sharp stick to form a "sieve."

Because every natural number (except 1) is either a prime number or a composite number, the concept of a prime number is an important one. If a number is not prime, then it is composite. A composite number can be expressed as a product of its prime factors in one and only one way, disregarding order. Is 60 prime? No, it is a composite number. What are its prime factors? To determine the prime factors of 60 in a systematic manner, we first try 2; 2 is a factor of 60 because $60 = 2 \times 30$. Next we try to factor 30, and 2 is a factor of 30: $30 = 2 \times 15$. We now have $60 = 2 \times 2 \times 15$. What are the factors of 15 (if any)? $15 = 3 \times 5$ and 3 and 5 are both prime. Hence, $60 = 2 \times 2 \times 3 \times 5$. We have expressed 60 as a product of its prime factors.

A *factor tree* can be used to express a composite number as a product of primes. The factor tree that follows illustrates the factoring process we have above for 60.

$$60$$
$$2 \times 30$$
$$2 \times 2 \times 15$$
$$2 \times 2 \times 3 \times 5$$

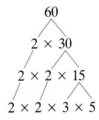

Write each composite number as a product of natural numbers. When a prime is determined, bring it down to the next line so it does not get "lost."

$$60 = 2 \times 2 \times 3 \times 5$$

Another way to determine the prime factors of 60 is by successive divisions by prime divisors. This division technique is the same as we used in Chapter 7, but now we must always have remainders of zero. In order to use the technique of successive divisions, we first determine the smallest prime that will divide into 60, which is 2. We divide 2 into 60 and continue in this manner until we reach a quotient that is prime. This indicates that no more divisions are necessary. This technique is illustrated as follows.

$$2 \underline{\lfloor 60}$$
$$2 \underline{\lfloor 30}$$
$$3 \underline{\lfloor 15}$$
$$5$$

NOTE OF INTEREST

Throughout the history of mathematics, many mathematicians have tried to find the largest prime number. In 1876, Anatole Lucas, a French mathematician, discovered a prime number that is 39 digits long. Since the invention of the computer, much larger prime numbers have been discovered. In 1952 a prime number with 386 digits was found with the aid of a computer. More recently, two college students with the assistance of a university computer discovered the largest known prime to be $2^{216,091} - 1$. This prime number contains 65,050 digits. The students spent approximately three years working on the project.

Because 5 is prime, no more divisions are necessary. Therefore,

$$60 = 2 \times 2 \times 3 \times 5$$

Note that we could also express this as $60 = 2 \times 3 \times 5 \times 2$. This is not a different factorization, but merely a rearrangement of the original one. A composite number can be expressed as a product of its prime factors in one and only one way, *disregarding order:* we still have two factors of 2, one factor of 3, and one factor of 5.

EXAMPLE 6 Determine the prime factors of 345.

SOLUTION 345 is not divisible by 2 since 345 is not even. But 345 is divisible by 3, since the sum of the digits $3 + 4 + 5 = 12$ is divisible by 3. So we have

$$\begin{array}{r|l} 3 & 345 \\ \hline & 115 \end{array}$$

Because 115 ends in 5, it is divisible by 5.

$$\begin{array}{r|l} 5 & 115 \\ \hline & 23 \quad \text{STOP: 23 is prime} \end{array}$$

Putting the two steps together, we have

$$\begin{array}{r|l} 3 & 345 \\ \hline 5 & 115 \\ \hline & 23 \end{array}$$

and $345 = 3 \times 5 \times 23$.

EXAMPLE 7 Determine the prime factors of 4,830.

SOLUTION

$$\begin{array}{r|l} 2 & 4,830 \\ \hline 3 & 2,415 \\ \hline 5 & 805 \\ \hline 7 & 161 \\ \hline & 23 \quad \text{STOP: 23 is prime} \end{array}$$

$$4,830 = 2 \times 3 \times 5 \times 7 \times 23$$

EXAMPLE 8 Determine the prime factors of 900.

SOLUTION

$$\begin{array}{r|l} 2 & 900 \\ 2 & 450 \\ 3 & 225 \\ 3 & 75 \\ 5 & 25 \\ & 5 \end{array}$$

5 STOP: 5 is prime

$$900 = 2 \times 2 \times 3 \times 3 \times 5 \times 5$$

(*Note:* We also could have said $900 = 2 \times 3 \times 5 \times 3 \times 2 \times 5$ and still be correct, as this is still the same set of prime factors. We can also write this as $900 = 2^2 \times 3^2 \times 5^2$.)

EXERCISES FOR SECTION 8.2

1. Determine whether the number used in each of the following is used as a cardinal number, an ordinal number, or an identification number. (*Hint:* See Sec. 8.1.)
   a. He shot 82 for a round of golf.
   b. Julie was third in line.
   c. My social security number is 089-20-4944.
   d. Ben's phone number is 727-3100.

2. Determine whether the number used in each of the following is used as a cardinal number, an ordinal number, or an identification number.
   a. Janie plays first singles on the tennis team.
   b. Pam bowled 157 for her only score.
   c. Mike has three television sets.
   d. Juan lives at 128 Washington Avenue.

3. Test each of the following numbers for divisibility by 2, 3, 4, 5, 8, 9, 10, and 11.
   a. 2,688 b. 73,440 c. 3,290,154

4. Test each of the following numbers for divisibility by 2, 3, 4, 5, 8, 9, 10, and 11.
   a. 7,128 b. 67,255 c. 4,194,168

5. List the first ten prime numbers.

6. What is the largest prime number less than 100?

7. Determine whether each number is prime or composite.
   a. 97 b. 89 c. 1
   d. 243 e. 741 f. 1,955

8. Determine whether each number is prime or composite.
   a. 101 b. 103 c. 323
   d. 2,007 e. 1,003 f. 4,159,731

9. Determine the prime factors for each number.
   a. 36 b. 72 c. 216
   d. 475 e. 625 f. 147

10. Find the prime factors of each number.
    a. 234 b. 213 c. 891
    d. 1,331 c. 902 f. 7,429

11. Two is the only even prime number; all of the other primes are odd. Hence any two consecutive odd primes must differ by at least 2, as do 3 and 5, and 5 and 7. Prime numbers that differ by exactly 2 are called *twin primes*. Find three pairs of twin primes other than those already mentioned.

12. Numerals 37 and 73 are prime numbers with reversed digits. What are the other pairs of two-digit primes with reversed digits that are less than 100?

JUST FOR FUN A perfect number is one that is equal to the sum of its divisors, excluding itself. An example of a perfect number is 6: its divisors are 1, 2, 3 and 6; we exclude 6, and 1 + 2 + 3 = 6. Therefore, 6 is perfect. Can you discover some more perfect numbers? (*Hint:* The next perfect number is less than 50.)

8.3 GREATEST COMMON DIVISOR AND LEAST COMMON MULTIPLE

Now that we are able to determine the prime factors of a given natural number, we shall use this process in determining some other properties of natural numbers. One important concept we will need is that of the *greatest common divisor.*

The *greatest common divisor* (GCD) of two natural numbers is the greatest (largest) natural number that divides a given pair of natural numbers with remainders of zero.

It should be noted here that the greatest common divisor is also called the *greatest common factor* (GCF) and either of these two names may be used.

Consider the two natural numbers 32 and 40. We list the set of divisors of each number.

$$32: \{1, 2, 4, 8, 16, 32\}$$
$$40: \{1, 2, 4, 5, 8, 10, 20, 40\}$$

If we find the intersection of these two sets of divisors, $\{1, 2, 4, 8, 16, 32\} \cap \{1, 2, 4, 5, 8, 10, 20, 40\}$, we have $\{1, 2, 4, 8\}$. As 8 is the greatest number in the intersection, it is the greatest common divisor of 32 and 40.

When we are given two prime numbers, then their greatest common divisor is 1. But there are pairs of composite numbers whose greatest common divisor is also 1. Consider the two natural numbers 24 and 25. The sets of divisors of these numbers are

$$24: \{1, 2, 3, 4, 6, 8, 12, 24\}$$
$$25: \{1, 5, 25\}$$

The intersection of these two sets of divisors is $\{1, 2, 3, 4, 6, 8, 12, 24\} \cap \{1, 5, 25\} = \{1\}$. Therefore 1 is the greatest common divisor of 24 and 25. Two numbers whose GCD is 1 are said to be **relatively prime.**

Do we always have to list the sets of divisors to determine the greatest common divisor? The answer is no. We can do it in a more efficient manner. Let's consider our original example, 32 and 40, and find the prime factors (divisors) of each number.

$$\begin{array}{r|l} 2 & 32 \\ 2 & 16 \\ 2 & 8 \\ 2 & 4 \\ & 2 \end{array} \qquad\qquad \begin{array}{r|l} 2 & 40 \\ 2 & 20 \\ 2 & 10 \\ & 5 \end{array}$$

$$32 = 2 \times 2 \times 2 \times 2 \times 2 \qquad 40 = 2 \times 2 \times 2 \times 5$$

Now examine the two sets of prime factors and determine the factors common to both sets.

$$\begin{aligned} 32 &= (2) \times (2) \times (2) \times 2 \times 2 \\ 40 &= (2) \times (2) \times (2) \times 5 \end{aligned}$$

Note that both sets of prime divisors contain $2 \times 2 \times 2$. Therefore the greatest common divisor of 32 and 40 is $2 \times 2 \times 2$, or 8.

Let's try another example. What is the greatest common divisor of 30 and 45? First we find the prime factors of each.

$$\begin{array}{r|l} 2 & 30 \\ \hline 3 & 15 \\ \hline & 5 \end{array} \qquad \begin{array}{r|l} 3 & 45 \\ \hline 3 & 15 \\ \hline & 5 \end{array}$$

$$30 = 2 \times \textcircled{3} \times \textcircled{5}$$
$$45 = 3 \times \textcircled{3} \times \textcircled{5}$$

Examining the two sets of prime factors (divisors), we see that the intersection is 3×5. Therefore, the greatest common divisor (or greatest common factor) of 30 and 45 is 3×5, or 15.

EXAMPLE 1 Find the greatest common divisor of 8 and 12.

SOLUTION First we find the prime factors of 8 and 12.

$$\begin{array}{r|l} 2 & 8 \\ \hline 2 & 4 \\ \hline & 2 \end{array} \quad 8 = 2 \times 2 \times 2 \qquad \begin{array}{r|l} 2 & 12 \\ \hline 2 & 6 \\ \hline & 3 \end{array} \quad 12 = 2 \times 2 \times 3$$

Next, we examine the intersection of the two sets of prime factors and we note that 2×2 is common to both sets; hence $2 \times 2 = 4$ is the GCD.

EXAMPLE 2 Find the greatest common divisor of 8 and 15.

SOLUTION We first find the prime factors of 8 and 15.

$$\begin{array}{r|l} 2 & 8 \\ \hline 2 & 4 \\ \hline & 2 \end{array} \quad 8 = 2 \times 2 \times 2 \qquad \begin{array}{r|l} 3 & 15 \\ \hline & 5 \end{array} \quad 15 = 3 \times 5$$

Note that the two sets of prime factors have no elements in common; their intersection is empty. When this occurs, the GCD for the two numbers is 1, so the numbers are relatively prime. Note that we did not list 1 as a prime factor for either number because 1 is not a prime number.

EXAMPLE 3 Find the greatest common divisor of 342 and 380.

SOLUTION Finding the prime factors of each number, we have

$$\begin{array}{r|l} 2 & 342 \\ \hline 3 & 171 \\ \hline 3 & 57 \\ \hline & 19 \end{array} \qquad \begin{array}{r|l} 2 & 380 \\ \hline 2 & 190 \\ \hline 5 & 95 \\ \hline & 19 \end{array}$$

$$342 = 2 \times 3 \times 3 \times 19 \qquad 380 = 2 \times 2 \times 5 \times 19$$

Examining the intersection of the two sets of prime factors, we see that 2×19 is common to both; hence the GCD of 342 and 380 is $2 \times 19 = 38$.

One application of the greatest common divisor is in the reduction of fractions. We can simplify, or reduce, fractions if we determine the greatest common divisor of both the numerator and denominator. Suppose we are asked to reduce the fraction

$$\frac{65}{91}$$

We can do this in the following manner. We find the greatest common divisor of both the numerator and denominator; that is, we find the GCD of 65 and 91.

$$\begin{array}{r|l} 5 & 65 \\ \hline & 13 \end{array} \quad 65 = 5 \times 13 \qquad \begin{array}{r|l} 7 & 91 \\ \hline & 13 \end{array} \quad 91 = 7 \times 13$$

The GCD of 65 and 91 is 13. We now rewrite the original fraction.

$$\frac{65}{91} = \frac{\not{13} \times 5}{\not{13} \times 7} = \frac{5}{7}$$

The numerator and denominator of the reduced fraction are relatively prime; that is, their GCD is 1.

EXAMPLE 4 Reduce $\frac{130}{455}$ to lowest terms.

SOLUTION We first find the GCD of 130 and 455.

$$\begin{array}{r|l} 2 & 130 \\ \hline 5 & 65 \\ \hline & 13 \end{array} \quad 130 = 2 \times 5 \times 13 \qquad \begin{array}{r|l} 5 & 455 \\ \hline 7 & 91 \\ \hline & 13 \end{array} \quad 455 = 5 \times 7 \times 13$$

The GCD of 130 and 455 is 5×13. Now we rewrite the original fraction.

$$\frac{130}{455} = \frac{\cancel{(5 \times 13)} \times 2}{\cancel{(5 \times 13)} \times 7} = \frac{2}{7}$$

EXAMPLE 5 Reduce $\frac{310}{460}$ to lowest terms.

SOLUTION We first find the GCD of 310 and 460.

$$\begin{array}{r|l} 2 & 310 \\ \hline 5 & 155 \\ \hline & 31 \end{array} \quad 310 = 2 \times 5 \times 31 \qquad \begin{array}{r|l} 2 & 460 \\ \hline 2 & 230 \\ \hline 5 & 115 \\ \hline & 23 \end{array} \quad 460 = 2 \times 2 \times 5 \times 23$$

The GCD of 310 and 460 is 2×5, and we can rewrite the original fraction.

$$\frac{310}{460} = \frac{\cancel{(2 \times 5)} \times 31}{\cancel{(2 \times 5)} \times 2 \times 23} = \frac{31}{2 \times 23} = \frac{31}{46}$$

If we want to find the greatest common divisor for three or more numbers, we extend the process for finding the GCD for two numbers. We find the prime factors that are common to all the sets of prime factors.

When we are given two natural numbers, we can now find their greatest common divisor. Another concept that goes hand in hand with the GCD is the *least common multiple*.

The *least common multiple* (LCM) of two natural numbers is the smallest (least) natural number that is a multiple of each of the two given numbers.

The least common multiple can also be thought of as the smallest (least) natural number that is divisible by both of the given numbers. Four is a multiple of 2, as is 6, 8, 10, and so on, because each of these numbers has 2 as a factor. The multiples of

3 are: {3, 6, 9, 12, 15, . . .}. Listing the sets of multiples for 2 and 3, we have

2: {2, 4, 6, 8, 10, . . .}
3: {3, 6, 9, 12, 15, . . .}

Upon inspection, we note that 6 is the least common multiple (LCM) of 2 and 3. Also, it is the smallest number that is divisible by both of the given numbers, 2 and 3.

Let's consider another example, the LCM of 10 and 12. Listing the sets of multiples for 10 and 12, we have

10: {10, 20, 30, 40, 50, 60, . . .}
12: {12, 24, 36, 48, 60, 72, . . .}

Inspection of these two sets of multiples indicates that 60 is the least common multiple of 10 and 12. It is the smallest number that is divisible by both of the given numbers, 10 and 12.

EXAMPLE 6 Find the least common multiple of 8 and 12.

SOLUTION Multiples of 8: {8, 16, 24, 32, 40, 48, . . . }
Multiples of 12: {12, 24, 36, 48, 60, . . . }

The common multiples are 24 and 48. The least common multiple (LCM) of 8 and 12 is 24.

EXAMPLE 7 Find the LCM of 48 and 72.

SOLUTION Multiples of 48: {48, 96, 144, 192, . . . }
Multiples of 72: {72, 144, 216, . . . }

The least common multiple of 48 and 72 is 144.

When do we use the least common multiple? It is usually used in combining fractions. Suppose we want to add $\frac{2}{5}$ and $\frac{1}{6}$.

$$\frac{2}{5} + \frac{1}{6}$$

We cannot add these two fractions as they are represented here. Before we can add or subtract two fractions, they must have a common denominator; when they have a common denominator, we can add or subtract the numerators. Because $\frac{2}{5}$ and $\frac{1}{6}$ do not have a common denominator, we must rewrite them as equivalent fractions that have the same denominator. In doing this, we usually use the **least common denominator,** which is the least common multiple of the given denominators. The least common multiple of 5 and 6 is 30. Therefore we now have

$$\frac{2}{5}+\frac{1}{6}=\frac{12}{30}+\frac{5}{30}=\frac{17}{30}$$

EXAMPLE 8 Add $\frac{1}{4}$ and $\frac{2}{9}$.

SOLUTION Before we can add these two fractions, they must have a common denominator. A common denominator is the LCM of 4 and 9. Because 4 and 9 are relatively prime, their GCD is 1. Therefore, the LCM is

$$\frac{4\times 9}{1}=36$$

$$\frac{1}{4}+\frac{2}{9}=\frac{9}{36}+\frac{8}{36}=\frac{17}{36}$$

EXAMPLE 9 Subtract $\frac{1}{6}$ from $\frac{2}{9}$.

SOLUTION We must first rewrite the fractions so that they have a common denominator, the LCM of 9 and 6. First we find the GCD of 9 and 6, which is 3, Therefore, the LCM is

$$\frac{9\times 6}{3}=\frac{54}{3}=18.$$

$$\frac{2}{9}-\frac{1}{6}=\frac{4}{18}-\frac{3}{18}=\frac{1}{18}$$

EXERCISES FOR SECTION 8.3

1. Find the greatest common divisor for each of the following:
   a. 8 and 14
   b. 14 and 28
   c. 15 and 24
   d. 52 and 78
   e. 111 and 267
   f. 24, 48, and 60

2. Find the greatest common divisor for each of the following:
   a. 8 and 15
   b. 21 and 25
   c. 48 and 72
   d. 234 and 470
   e. 801 and 999
   f. 60, 90, and 210

3. Reduce each fraction to lowest terms.
   a. $\frac{30}{36}$ b. $\frac{42}{54}$ c. $\frac{39}{65}$
   d. $\frac{120}{180}$ e. $\frac{294}{304}$ f. $\frac{195}{390}$

4. Reduce each fraction to lowest terms.
   a. $\frac{16}{24}$ b. $\frac{15}{28}$ c. $\frac{28}{72}$
   d. $\frac{50}{273}$ e. $\frac{213}{450}$ f. $\frac{115}{450}$

5. Find the least common multiple for each of the following:
   a. 8 amd 14 b. 14 and 28
   c. 15 and 24 d. 52 and 78
   e. 66 and 90 ⋆f. 111 and 267

6. Find the least common multiple for each of the following:
   a. 8 and 15 b. 21 and 25
   c. 48 and 72 d. 13 and 17
   ⋆e. 234 and 470 ⋆f. 801 and 999

7. Perform the indicated operations and reduce your answers to lowest terms.
   a. $\frac{3}{4}+\frac{2}{9}$ b. $\frac{5}{9}+\frac{1}{12}$ c. $\frac{1}{6}+\frac{3}{4}$
   d. $\frac{7}{8}-\frac{1}{12}$ e. $\frac{10}{11}-\frac{4}{5}$ f. $\frac{13}{15}-\frac{3}{20}$

8. Perform the indicated operations and reduce your answers to lowest terms.
   a. $\frac{7}{15}+\frac{1}{40}$ b. $\frac{4}{9}+\frac{5}{36}$ c. $\frac{2}{7}+\frac{3}{11}$
   d. $\frac{7}{9}-\frac{1}{5}$ e. $\frac{5}{18}-\frac{1}{24}$ f. $\frac{7}{11}-\frac{2}{5}$

9. Benny has two pieces of plywood. Both pieces have a height of 6 feet. But one piece is 54 inches wide whereas the other is 36 inches wide. If Benny is going to cut the pieces of plywood into 6-foot lengths to make book shelves, what is the widest shelf that can be cut from both pieces without any wood being left over?

10. Larry has adjustable shelves in his shoe store. He received a shipment of shoes and boots. The shoes come in boxes that are 9 inches high, whereas the boots come in boxes that are 12 inches high. Both types of boxes must fit exactly between the shelves. What height can Larry set the shelves at so that a stack of shoe boxes and a stack of boot boxes will fit on the same shelf?

11. Perform the indicated operations and reduce your answers to lowest terms.
   a. $\frac{3}{7}+\frac{4}{11}$ b. $\frac{5}{9}+\frac{7}{36}$ c. $\frac{8}{15}+\frac{3}{40}$
   d. $\frac{8}{11}-\frac{3}{5}$ e. $\frac{7}{18}-\frac{3}{24}$ f. $\frac{8}{9}-\frac{3}{5}$

12. Perform the indicated operations and reduce your answers to lowest terms.
   a. $\frac{5}{6}+\frac{1}{4}$ b. $\frac{7}{9}+\frac{5}{12}$ c. $\frac{4}{9}+\frac{3}{5}$
   d. $\frac{13}{15}-\frac{7}{20}$ e. $\frac{10}{11}-\frac{2}{5}$ f. $\frac{7}{8}-\frac{5}{12}$

13. Find the least common multiple for each of the following.
   a. 12 and 8 b. 3 and 10 c. 5 and 9
   d. 8 and 10 e. 14 and 21 f. 15 and 35

14. Find the least common multiple for each of the following.
   a. 12 and 30 b. 4 and 7 c. 12 and 16
   d. 9 and 11 e. 45 and 18 f. 36 and 12

15. Reduce each fraction to lowest terms.
   a. $\frac{14}{42}$ b. $\frac{13}{52}$ c. $\frac{16}{96}$
   d. $\frac{3}{57}$ e. $\frac{17}{51}$ f. $\frac{28}{280}$

16. Reduce each fraction to lowest terms.
   a. $\frac{45}{900}$ b. $\frac{72}{360}$ c. $\frac{35}{210}$
   d. $\frac{36}{324}$ e. $\frac{132}{165}$ f. $\frac{209}{323}$

17. Find the greatest common divisor for each of the following.
    a. 12 and 20 b. 22 and 68 c. 40 and 23
    d. 9 and 26 e. 45 and 63 f. 36 and 42

18. Find the greatest common divisor for each of the following.
    a. 24 and 60 b. 18 and 54 c. 5 and 23
    d. 32 and 80 e. 63 and 105 f. 80 and 240

JUST FOR FUN

It has been stated that "every even number greater than 2 can be expressed as the sum of two prime numbers." For example, $6 = 3 + 3$, $8 = 3 + 5$, $12 = 7 + 5$, and $14 = 11 + 3$. This statement is called Goldbach's conjecture. It is a conjecture because it has never been proved. Show that Goldbach's conjecture is true for all even numbers except 2, up to and including 30.

8.4 INTEGERS

In Section 8.2, we introduced the set of natural numbers {1, 2, 3, 4, . . . }. The first natural number is 1. This is the first number that man used in counting. The use of zero did not come about until approximately 700 A.D. Zero was first used to indicate position, distinguishing between numbers like 32 and 302. Later, zero began to be used as a starting point in counting. The set of natural numbers was expanded to include zero, forming the set {0, 1, 2, 3, 4, . . . }. This set is called the set of **whole numbers** to indicate that it is different from the set of natural numbers, which does not include zero.

$$\{1, 2, 3, 4, \ldots\} = \text{natural numbers}$$
$$\{0, 1, 2, 3, 4, \ldots\} = \text{whole numbers}$$

Recall that 0 is the identity element for the operation of addition: It does not change the identity of a number when it is added to that number. That is, $1 + 0 = 1$, $2 + 0 = 2$, $3 + 0 = 3$, and so on. Recall also that -3 is the additive inverse of 3, because $-3 + 3 = 0$, and -1 is the additive inverse of 1, as $-1 + 1 = 0$. The set of numbers consisting of the set of whole numbers and their additive inverses is called the set of **integers;** that is,

$$\{\ldots, -4, -3, -2, -1, 0, 1, 2, 3, 4, \ldots\} = \text{integers}$$

It should be noted that a number such as 2 may be classified as a natural number, a whole number, or an integer, because it is an element of all three of these sets. But a number such as −2 can only be classified as an integer, and not as a natural number or a whole number. We can also classify −2 as a negative integer, whereas 2 is a positive integer. Note that 0 is neither positive nor negative, but it is an integer.

It is possible to picture the set of integers, {. . . , −4, −3, −2, −1, 0, 1, 2, 3, 4, . . . }, on a **number line,** as shown in Figure 1. First, we draw a line, pick a point on the line, and label it 0. Next mark off equal units to the right and left of 0. Label the end points of the intervals to the right of 0 with the *positive integers,* and those to the left of 0 with the *negative integers.*

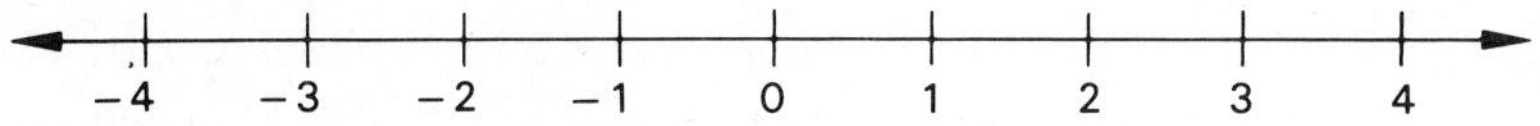

FIGURE 1

The number line may be extended indefinitely in either direction, that is to the left or right of zero. Remember that zero is an integer, but it is neither positive nor negative.

Using the number line, we can determine the order of the integers. Three is greater than 2 ($3 > 2$) because 3 is to the right of 2 on the number line. Similarly, 4 is greater than 1 ($4 > 1$) because 4 is to the right of 1 on the number line. For the same reason, 0 is greater than −1 ($0 > -1$) and −1 is greater than −4 ($-1 > -4$). Other observations that we might make are

$$4 > 0, \qquad 1 > 0, \qquad 1 > -1, \qquad 2 > -2$$

Instead of stating that 3 is greater than 2 ($3 > 2$), we could have said that 2 is less than 3 ($2 < 3$), because 2 is to the left of 3 on the number line. Similarly, $1 < 4$ (1 is less than 4) because 1 is to the left of 4 on the number line. We can also say that

$$0 < 4, \qquad 0 < 1, \qquad -1 < 1, \qquad -2 < 2$$

Prior to our discussion of the set of integers, we only used the minus sign (−) to indicate subtraction. Now we are also using it to label negative integers. The number 3 is three units to the right of 0 on the number line and its "opposite," negative 3 (−3), is three units to the left of 0 on the number line. The opposite of 4 is negative 4 (−4), and the opposite of 1 is negative 1 (−1). Note that the sum of any number and its opposite is 0: $3 + (-3) = 0$, $4 + (-4) = 0$, and $1 + (-1) = 0$. It should also be noted that the opposite of 0 is 0: $0 + 0 = 0$. Recall that we can also describe −3 as the additive inverse of 3, as $3 + (-3) = 0$. Similarly, −1 is the additive inverse of 1, because $1 + (-1) = 0$.

Wide World Photo

This desert area has a negative altitude because it is below sea level.

How do we combine integers? Consider the sum of the positive integers 2 and 3. We know that $2 + 3 = 5$. However, let us work this problem on the number line. To find the solution to $2 + 3$ on the number line, we start at 0 and proceed to the right two units to 2. Because we want to add 3, we proceed three more units to the right; this brings us to 5. Hence $2 + 3 = 5$. This process is illustrated in Figure 2.

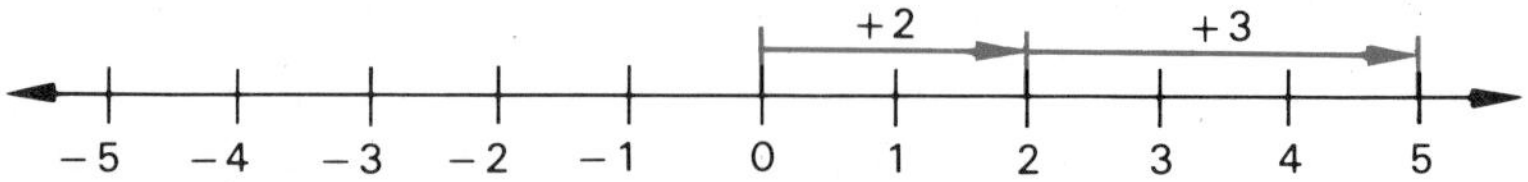

FIGURE 2
$2 + 3 = 5$

Now let's combine a positive and a negative integer, such as $2 + (-3)$, on a number line. We start at 0 and proceed two units to the right to 2. Because we want to add -3 to 2, we move three units to the *left* from 2. We end up at -1. Therefore $2 + (-3) = -1$. This is illustrated in Figure 3.

We could do this problem in another way: -3 can be expressed as $(-2) + (-1)$; then the original problem becomes

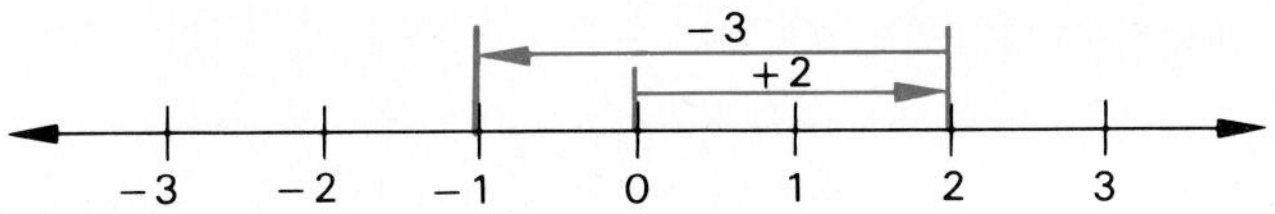

FIGURE 3
$2 + (-3) = -1$

$2 + (-3) = 2 + (-2) + (-1)$. But (-2) is the opposite, or additive inverse, of 2 and therefore $2 + (-2) + (-1) = 0 + (-1) = (-1)$. Summarizing these steps, we have

$$2 + (-3) = [2 + (-2)] + (-1) = 0 + (-1) = -1$$

We are not advocating one technique over the other. The method that you best understand is the one to use.

EXAMPLE 1 Evaluate $1 + (-4)$.

SOLUTION $1 + (-4) = -3$
On the number line in Figure 4, we start at 0 and proceed one unit to the right to 1. Next, we move four units in a negative direction—that is, to the left—from 1. We end up at -3.

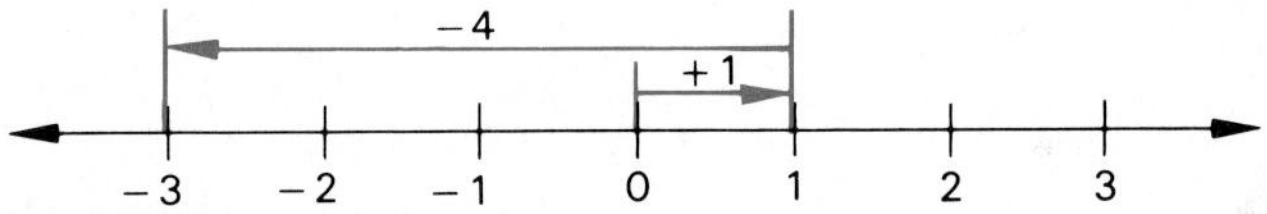

FIGURE 4
$1 + (-4) = -3$

ALTERNATE SOLUTION $1 + (-4) = [1 + (-1)] + (-3) = 0 + (-3) = -3$

EXAMPLE 2 Evaluate $-2 + 5$.

SOLUTION $-2 + 5 = 3$
We start at 0 on the number line in Figure 5 and move two units to the left to -2. Next we move five units to the right (positive direction) from -2; we end up at 3.

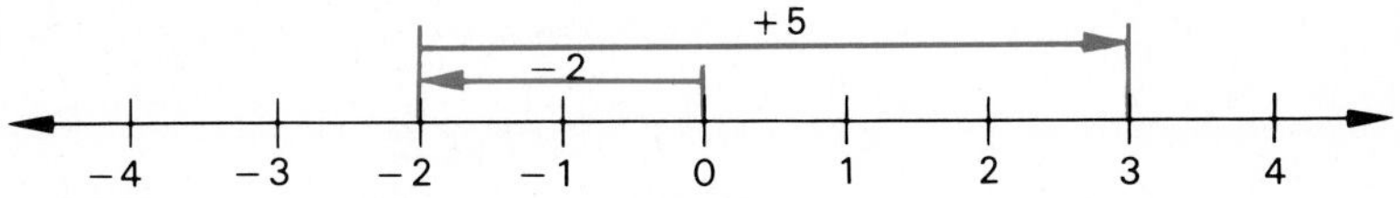

FIGURE 5
$-2 + 5 = 3$

ALTERNATE SOLUTION $-2 + 5 = [-2 + 2] + 3 = 0 + 3 = 3$

If we add two positive integers, then their sum will be a positive integer. If we add two negative integers, then their sum will be a negative integer. If we add a positive integer and a negative integer, then their sum may be a positive integer, a negative integer, or zero.

The problem $8 - 5$ has the same answer as $8 + (-5)$. Similarly, $7 - 4 = 7 + (-4)$ and $3 - 1 = 3 + (-1)$. This seems to indicate that for integers a subtraction problem can be thought of as an addition problem. That is, subtracting integers is the same as adding the opposite of the second integer to the first integer.

You probably do not need this rule for problems like $8 - 5$ and $7 - 4$. But what about $5 - 8$? According to the above rule, we can think of $5 - 8$ as $5 + (-8)$. Now we have a problem similar to the ones that we did in Examples 1 and 2. Using the number line, as in Figure 6, $5 + (-8) = -3$.

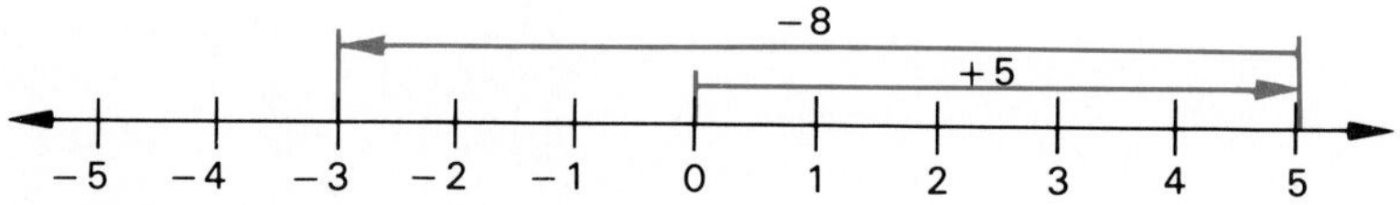

FIGURE 6
$5 + (-8) = -3$

Using an alternate method,

$$5 - 8 = 5 + (-8) = [5 + (-5)] + (-3) = 0 + (-3) = -3$$

EXAMPLE 3 Evaluate $3 - 4$.

SOLUTION $3 - 4 = -1$

$3 - 4$ is the same as $3 + (-4)$. Using the number line as in Figure 7, we have $3 + (-4) = -1$.

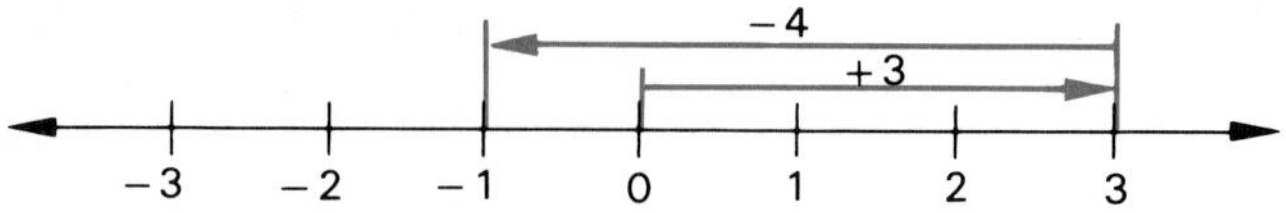

FIGURE 7
$3 + (-4) = -1$

ALTERNATE SOLUTION $3 - 4 = 3 + (-4) = [3 + (-3)] + (-1) = 0 + (-1) = -1$

EXAMPLE 4 Evaluate $-3 - 2$.

SOLUTION $-3 - 2 = -5$

$-3 - 2$ is the same as $-3 + (-2)$. Using the number line, as in Figure 8, we have $-3 + (-2) = -5$. Note that when we add two negative integers their sum is a negative integer.

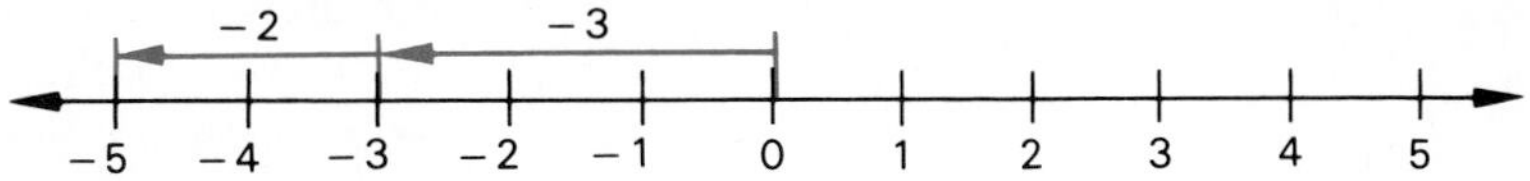

FIGURE 8
$-3 + (-2) = -5$

ALTERNATE SOLUTION $-3 - 2 = -3 + (-2) = -(3 + 2) = -5$

EXAMPLE 5 Evaluate $-2 - (-3)$.

SOLUTION $-2 - (-3) = 1$

Subtracting integers is the same as adding the opposite of the second integer to the first integer. Therefore, $-2 - (-3)$ is the same as $-2 + 3$. (Note that 3 is the opposite of -3). Using the number line, we see that $-2 + 3 = 1$, as shown in Figure 9.

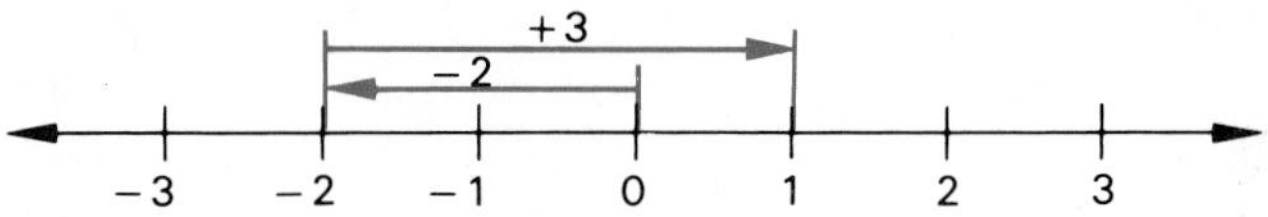

FIGURE 9
$-2 + 3 = 1$

ALTERNATE SOLUTION $-2 - (-3) = -2 + 3 = [-2 + 2] + 1 = 0 + 1 = 1$

EXAMPLE 6 Evaluate $-5 - (-3)$.

SOLUTION $-5 - (-3) = -2$

This problem is similar to Example 5: $-5 - (-3)$ is the same as $-5 + 3$. By means of the number line in Figure 10, we see that $-5 + 3 = -2$.

FIGURE 10
$-5 + 3 = -2$

ALTERNATE SOLUTION $-5 - (-3) = -5 + 3 = -2 + [(-3) + 3] = -2 + 0 = -2$

To add two integers with opposite signs, consider the distance each integer is from zero. Subtract the shorter distance from the longer distance, and use the sign of the longer distance in the answer.

To add integers with the same sign, add without regard to the signs. The answer contains the same sign as the given numbers.

To subtract two integers, add the opposite of the second integer to the first integer.

Once we have mastered addition and subtraction of integers, we can proceed to the operation of multiplication. We already know how to multiply the natural numbers: for example, $4 \times 2 = 8$, $5 \times 3 = 15$, and $7 \times 5 = 35$. These problems are also examples of multiplying positive integers. One observation that we can make from these examples is the following.

> A positive integer multiplied by a positive integer yields a positive integer.

But what happens when we multiply a positive integer by a negative integer? Consider the example $2 \times (-4)$. What is the answer? We can think of $2 \times (-4)$ as $(-4) + (-4)$, which equals -8. Therefore, $2 \times (-4) = -8$. Let's try another example, $3 \times (-5)$; this can also be expressed as $(-5) + (-5) + (-5)$, which equals -15. Therefore, $3 \times (-5) = -15$.

> Whenever we multiply a positive integer by a negative integer, the answer is always a negative integer.

Before we consider the product of two negative integers, let us consider the *distributive law.* It states that

$$a \times (b + c) = a \times b + a \times c$$

for all integers *a, b,* and *c*. In terms of a specific example,

$$3 \times (2 + 5) = 3 \times 2 + 3 \times 5.$$

More formally, we call this the **distributive law (or property) for multiplication over addition.** When we are given an expression such as $3 \times (2 + 5)$, we can evaluate it in two different ways.

$$3 \times (2 + 5) = 3 \times 7 = 21$$

or

$$3 \times (2 + 5) = 3 \times 2 + 3 \times 5 = 6 + 15 = 21$$

Similarly,

$$(-2) \times (5 + 6) = (-2) \times 11 = -22$$

or

$$(-2) \times (5 + 6) = (-2) \times 5 + (-2) \times 6 = -10 + (-12) = -22$$

Consider the example $2 \times (-4 + 4)$. We can also evaluate this in two ways.

$$2 \times (-4 + 4) = 2 \times (0) = 0$$

or

$$2 \times (-4 + 4) = 2 \times (-4) + 2 \times 4 = -8 + 8 = 0$$

Now consider the example $(-2) \times (-4 + 4)$. Evaluating this both ways, we have

$$(-2) \times (0) = 0 \qquad \text{and} \qquad (-2) \times (-4) + (-2) \times 4 = ?$$

Here we run into the problem of multiplying two negative integers. But now we can determine the answer. Because $a \times (b + c) = a \times b + a \times c$ for all integers a, b, and c, $(-2) \times (-4 + 4)$ must have the same answer regardless of which way we evaluate it. We already know that

$$(-2) \times (-4 + 4) = (-2) \times (0) = 0$$

Consequently,

$$(-2) \times (-4 + 4) = (-2) \times (-4) + (-2) \times 4$$

must also equal 0. We already know that $(-2) \times 4 = -8$; we must determine the answer to $(-2) \times (-4)$. We also know that when the answer to $(-2) \times (-4)$ is added to -8, the final answer must be 0. What number added to -8 will give an answer of 0? The opposite, or additive inverse, of -8: namely, 8. Therefore, $(-2) \times (-4)$ must equal 8. We list the steps again.

$$\begin{aligned} (-2) \times (-4 + 4) &= (-2) \times (-4) + \underbrace{(-2) \times 4} = 0 \\ &= \underbrace{(-2) \times (-4)} + \quad (-8) \quad = 0 \\ &= \qquad 8 \qquad + \quad (-8) \quad = 0 \end{aligned}$$

Whenever we multiply two negative integers, the product is always positive.

The following is a summary of the rules we have developed for multiplication of integers.

1. The product of two positive integers is positive. For example, $2 \times 3 = 6$.
2. The product of a positive integer and a negative integer is negative. For example, $2 \times (-3) = -6$.
3. The product of two negative integers is positive. For example, $(-2) \times (-3) = 6$.
4. The product of any integer and zero is zero. For example, $(-2) \times 0 = 0$, $100 \times 0 = 0$.

EXAMPLE 7 Evaluate each of the following.

a. 4×3 b. $4 \times (-3)$ c. $(-4) \times (-3)$ d. 4×0

SOLUTION a. $4 \times 3 = 12$. The product of two positive integers is positive.
b. $4 \times (-3) = -12$. The product of a positive integer and a negative integer is negative.
c. $(-4) \times (-3) = 12$. The product of two negative integers is positive.
d. $4 \times 0 = 0$. The product of any integer and zero is zero.

EXAMPLE 8 Evaluate each of the following.

a. $7 \times (-8)$ b. $(-3) \times 9$
c. $(-9) \times (-4)$ d. $(-6) \times (-6)$

SOLUTION a. $7 \times (-8) = -56$ b. $(-3) \times 9 = -27$
c. $(-9) \times (-4) = 36$ d. $(-6) \times (-6) = 36$

Sometimes a numerical expression may involve both addition and multiplication. For example, $6 + (3 \times 4)$, $9 + [(3 + 4) \times 5]$, and $4 \times 3 + 5 - 2 \times 9 - 11$ are three such expressions. To evaluate an expression that contains two or more grouping symbols, the expression in the innermost grouping symbol is evaluated first. Hence.

$$9 + [(3 + 4) \times 5] = 9 + [7 \times 5] = 9 + 35 = 44$$

If no grouping symbols appear in a numerical expression, the standard convention is to perform all multiplications and divisions in the order in which they occur (from left to right) and then do all the additions and subtractions in the order in which they occur (from left to right).

Therefore,

$$\begin{aligned} 4 \times 3 + 5 - 2 \times 9 - 11 &= 12 + 5 - 18 - 11 \\ &= 17 - 18 - 11 = 17 - 29 \\ &= -12 \end{aligned}$$

EXAMPLE 9 Evaluate each of the following.

a. $3 + (6 \times 4)$ b. $12 - (3 \times 4 + 5)$
c. $2 \times 4 + 5 - 6 \div 3$ d. $7 + [(3 \times 4) \div 6]$

SOLUTION

a. $3 + (6 \times 4) = 3 + 24 = 27$
b. $12 - (3 \times 4 + 5) = 12 - (12 + 5) = 12 - (17) = -5$
c. $2 \times 4 + 5 - 6 \div 3 = 8 + 5 - 2 = 13 - 2 = 11$
d. $7 + [(3 \times 4) \div 6] = 7 + [12 \div 6] = 7 + (2) = 9$

Every integer can be classified as either an even integer or an odd integer. An integer is even if it can be expressed as 2 times another integer. For example, 6 is an even integer because $6 = 2 \times 3$. Similarly, -2 is even because $-2 = 2 \times (-1)$, and 0 is even since $0 = 2 \times 0$. In general, any even integer may be expressed in the form $2 \times n$, $2 \cdot n$, or $2(n)$, where n represents some integer.

If an integer is not even, then it must be odd. If $2 \cdot n$ represents an even integer, then adding 1 more to it will make it odd. Therefore $2 \cdot n + 1$ represents an odd integer. Because 6 is even, adding 1 more will yield an odd integer, 7: $7 = 2 \cdot 3 + 1$. Another odd integer is -5: $-5 = 2 \cdot (-3) + 1$.

Some interesting observations can be made about odd and even integers. For example, what happens when you add two even integers? Considering a few examples, we see that $2 + 2 = 4$, $2 + 4 = 6$, $8 + 10 = 18$, and $12 + 6 = 18$. It appears that the sum of two even integers is also even. Are you sure? Considering some more examples, we see that $6 + 6 = 12$, $8 + 12 = 20$, and so on.

© *Topham/The Image Works*

You are probably convinced that the sum of two even integers is even, but have we proved it? No, we have just observed what happens for a certain set of examples. We have not examined all of the possibilities, and therefore we cannot be sure that it is always true. If we can show that it is true for *any* two even integers, then we can be certain that the sum of any two specific even integers is even. Any even integer can be expressed as 2 times another integer, so let's consider the even integers $2 \times n$

and $2 \times k$, where n and k are also integers. Then we have

$$(2 \times n) + (2 \times k)$$

But

$$(2 \times n) + (2 \times k) = 2 \times (n + k)$$

by means of the distributive property. Note that $(n + k)$ is an integer because the sum of any two integers is an integer. Therefore, $2 \times (n + k)$ is an even integer, since 2 times any integer is an even integer, and we have shown that the sum of any two even integers is also an even integer.

What happens when you add two odd integers? From $3 + 5 = 8$, $1 + 3 = 4$, and $5 + 7 = 12$, it appears that the sum of two odd integers is an even integer. Let's examine what happens when we add *any* two odd integers. Consider the odd integers $2 \times n + 1$ and $2 \times k + 1$: adding them, we have

$$\begin{aligned}(2 \times n + 1) + (2 \times k + 1 &= 2 \times n + 2 \times k + 2 \\ &= 2 \times (n + k + 1)\end{aligned}$$

by means of the distributive property. Note that $(n + k + 1)$ is an integer because $n + k$ is an integer and the sum of 1 and an integer is still an integer. Therefore, $2 \times (n + k + 1)$ is an even integer because 2 times any integer is an even integer.

EXAMPLE 10 Prove that the product of any two even integers is an even integer.

SOLUTION Any two even integers may be expressed as $2 \times n$ and $2 \times k$. Consequently, we have

$$(2 \times n) \times (2 \times k) = 2 \times (2 \times n \times k)$$

We see that $(2 \times n \times k)$ represents an integer because the product of three integers is an integer. Therefore, $2 \times (2 \times n \times k)$ is an even integer. Hence, the product of any two even integers is an even integer.

EXERCISES FOR SECTION 8.4

1. Evaluate each of the following.
   a. $2 + (-3)$ b. $4 + (-7)$
   c. $5 + (-8)$ d. $-3 + 5$
   e. $-7 + 8$ f. $-9 + 12$

2. Evaluate each of the following.
   a. $-13 + 14 + (-7)$ b. $-2 + (-1) + (-3)$
   c. $-2 + (-1) + 3$ d. $7 + 8 + 9$
   e. $3 + (-7) + (-4)$
   f. $5 + (-1) + (-4) + (-3)$

3. Evaluate each of the following.
   a. $3 - 5$ b. $6 - 9$
   c. $-8 - 6$ d. $6 - (-5)$
   e. $-2 - (-1)$ f. $-7 - (-3)$

4. Evaluate each of the following.
   a. $17 - (-1)$ b. $-18 - 2$
   c. $14 - [-6 - (-1)]$ d. $-19 - [-3 - (-2)]$
   e. $-12 - [-5 - (-8)]$ f. $[-16 - 9] - (-3)$

5. Evaluate each of the following.
   a. $-7 - (-2)$ b. $-10 - (-10)$
   c. $12 - (-13)$ d. $6 \times (-5 + 5)$
   e. $7 \times (-5)$ f. $(-8 + 5) \times (-3)$

6. Evaluate each of the following.
   a. $-8 + 8$ b. $-13 - (-12)$
   c. $-6 - (8 + 3)$ d. $(-8) \times 9$
   e. $(5 - 3) \times (-2)$ f. $(-2 - 1) \times (-3)$

7. Evaluate each of the following.
   a. $(-4) \times (-5)$ b. $(-3) \times 2 \times 4$
   c. $(-4) \times (-6) \times (-1)$ d. $(-2)(-2)(-2)(-2)$
   e. $4 \times (-5)(-2)$ f. $7 \times (-3) \times (-3) \times 2$

8. Evaluate each of the following.
   a. $4(3 + 2) - 5$ b. $(7 + 6 \times 4) - 12$
   c. $15 - (3 \times 2 - 1)$ d. $8 \times 2 - 5 + 6 \div 3$
   e. $8 \div 2 \times 4 + 5 - 3$ f. $7 \times [(3 + 15) \div 6]$

9. Evaluate each of the following.
   a. $3 \times (5 - 2) + 8$ b. $(6 \times 4 + 3) \div 9$
   c. $16 - 2(3 + 4 \div 2)$ d. $16 \div 4 - 2 + 6 \times 3$
   e. $12 \times 2 \div 4 + 8 - 2$ f. $24 \div [(5 - 3) \times 3]$

10. Replace each question mark with $=$, $>$, or $<$ to make the sentence true.
    a. $11 ? 3$ b. $0 ? 1$
    c. $0 ? -1$ d. $-2 + 3 ? 3 - 2$
    e. $2 + (-3) ? 1 \times (-1)$
    f. $(-2)(3) ? (-2)(-3)$

11. Replace each question mark with $=$, $>$, or $<$ to make the sentence true.
    a. $-1 ? -2$ b. $-10 ? -3$
    c. $-1 - (-1) ? 0$ d. $-3 + 2 ? 2 - (-3)$
    e. $-1 - (-2) ? -3 + 4$
    f. $(-1)(-3) ? -3 + (-1)$

12. Classify each integer as odd or even and express it in the form $2 \times n + 1$ or $2 \times n$. (*Example:* 7 is an odd integer; $7 = 2 \times 3 + 1$.)
    a. 10 b. 15 c. 21 d. -5

13. Classify each integer as odd or even and express it in the form $2 \times n + 1$ or $2 \times n$. (*Example:* 6 is an even integer; $6 = 2 \times 3$.)
    a. 12 b. 17 c. -6 d. 101

14. One of Goldbach's conjectures is that any odd number greater than 7 can be expressed as the sum of three odd primes (for example, $9 = 3 + 3 + 3$). Show that this conjecture is true for all positive integers from 11 up to and including 29.

15. On a particular Monday, the Dow Jones average opened at 842. At the close of trading on Monday, it had fallen 11 points. On Tuesday at the close of trading it had gained 5 points. On Wednesday at the close of trading it had gained 12 points. On Thursday at the close of trading it had fallen 15 points. What was the opening Dow Jones average on Friday?

16. An elevator in the Empire State Building started at the 35th floor, rose 7 floors, descended 12 floors, rose 4 floors, descended 11 floors, descended 3 floors, and then rose 15 floors and stopped. At what floor did the elevator stop?

Assuming that there are 15 feet between floors, how far did the elevator travel?

17. If Aristotle was born in 384 B.C. and Euclid was born in 365 B.C., who was born first?

18. Prove that the product of any two odd integers is an odd integer.

19. Prove that the sum of an odd integer and an even integer is an odd integer.

20. When you multiply an even integer by an odd integer, is the answer even or odd? Prove your answer.

B.C. by permission of Johnny Hart and Creaters Syndicate, Inc

JUST FOR FUN

A numismatist (coin collector) was examining a collection of coins. In this collection he discovered a coin dated 384 B.C. What is your conclusion regarding this coin?

8.5 RATIONAL NUMBERS

Thus far we have examined the set of natural numbers, the set of whole numbers, and the set of integers. All of these can be shown on a number line like that in Figure 11.

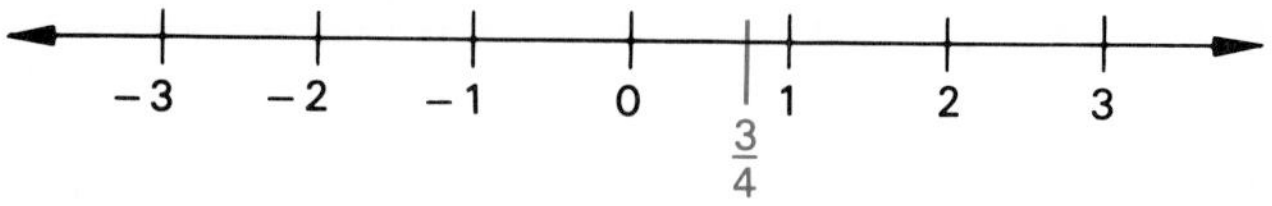

FIGURE 11

But what about the intervals between the numbers on the number line? Do any other numbers belong in these intervals? The answer is yes. Consider the number $\frac{3}{4}$: it is greater than 0 and less than 1, so it belongs in the interval between 0 and 1, as shown in Figure 11. What kind of number is it? It is not a natural number, it is not a whole number, and it is not an integer. It is a *rational number.*

A **rational number** is a number that can be expressed in the form a/b, where a and b are integers and $b \neq 0$ (we cannot divide by zero). In other words, a rational number is any number that can be expressed as the quotient of two integers. A rational number like $\frac{3}{4}$ is commonly referred to as a fraction. Remember that both the numerator and denominator of the fraction must be integers, and that the denominator of the fraction cannot be 0.

Is the number 4 a rational number? The answer is yes, because a rational number is any number that can be expressed as the quotient of two integers, and

$$4 = \frac{4}{1}$$

In fact, using this idea we can see that any integer can be expressed as a quotient of two integers.

$$7 = \frac{7}{1}, \quad 10 = \frac{10}{1}, \quad -4 = \frac{-4}{1}, \quad -8 = \frac{-8}{1}, \quad \text{and} \quad 0 = \frac{0}{1}$$

Every integer is also a rational number, but remember that not all rational numbers are integers. (The fraction $\frac{3}{4}$ is an example of a rational number that is not an integer.)

A rational number may be expressed in many different ways. Recall that 4 is a rational number since $4 = \frac{4}{1}$, but we could also express 4 in other ways.

$$4 = \frac{4}{1} = \frac{8}{2} = \frac{16}{4} = \frac{32}{8} = \cdots$$

Note that $\frac{4}{1} \times \frac{2}{2} = \frac{8}{2}$, $\frac{16}{4} = \frac{8}{2} \times \frac{2}{2}$, and so on.

For any fraction $\frac{a}{b}$, if k is any number other than zero,

$$\frac{a \times k}{b \times k} = \frac{a}{b}$$

This rule is helpful in reducing, or simplifying, fractions. Given the fraction $\frac{15}{25}$, we can reduce it by factoring both the numerator and denominator into prime factors.

$$\frac{15}{25} = \frac{3 \times 5}{5 \times 5}$$

Applying the rule, we have

$$\frac{15}{25} = \frac{3 \times \not{5}}{5 \times \not{5}} = \frac{3}{5}$$

Note that we eliminated the factors that were common to both the numerator and denominator.

How do we know that our answer is correct? Does $\frac{15}{25} = \frac{3}{5}$? An expression such as this can be verified by cross multiplying to see if the products are equal.

$$15 \times 5 \stackrel{?}{=} 25 \times 3$$
$$75 = 75$$

You may recall that the expression $\frac{15}{25} = \frac{3}{5}$ can be thought of as a proportion. For any proportion to be true, the product of the *means* must equal the product of the *extremes*. More formally, we say that

$$\frac{a}{b} = \frac{c}{d} \quad \text{if and only if} \quad \underbrace{a \times d}_{\text{extremes}} = \underbrace{b \times c}_{\text{means}}$$

EXAMPLE 1 Does $\frac{3}{11} = \frac{9}{33}$?

SOLUTION The answer is yes, because $3 \times 33 = 9 \times 11$, that is, $99 = 99$.

ALTERNATE SOLUTION We reduce $\frac{9}{33}$ and see if the result is $\frac{3}{11}$:

$$\frac{9}{33} = \frac{\not{3} \times 3}{\not{3} \times 11} = \frac{3}{11}$$

EXAMPLE 2 Reduce $\frac{42}{54}$.

SOLUTION $$\frac{42}{54} = \frac{\not{2} \times \not{3} \times 7}{\not{2} \times \not{3} \times 3 \times 3} = \frac{7}{3 \times 3} = \frac{7}{9}$$

Check:

$$\frac{42}{54} = \frac{7}{9} \quad \text{if and only if} \quad 42 \times 9 = 7 \times 54$$
$$378 = 378$$

How do we combine rational numbers? Because 2 and 3 are rational numbers and $2 + 3 = 5$, we already know how to combine some rationals. But consider the problem of adding the rational numbers $\frac{1}{5} + \frac{2}{3}$. In Section 8.3 we discussed the process of adding and subtracting fractions. However, those problems were considered only with regard to the use of the least common multiple. Let's state a general rule for adding any two rational numbers.

If $\frac{a}{b}$ and $\frac{c}{d}$ are rational numbers, then

$$\frac{a}{b} + \frac{c}{d} = \frac{ad + bc}{bd}$$

Therefore, for the examples $\frac{1}{5} + \frac{2}{3}$, we have

$$\frac{1}{5} + \frac{2}{3} = \frac{1 \times 3 + 5 \times 2}{5 \times 3} = \frac{3 + 10}{15} = \frac{13}{15}$$

Recall that in order to add two fractions, we rewrite them so that they have the same denominator, and then we add the numerators. The above procedure is just another way of doing this. Because the two given fractions do not have the same denominator, we could have found a common denominator by finding the least common multiple of 5 and 3, which is 15. This would result in the same answer, as shown previously.

$$\frac{1}{5} + \frac{2}{3} = \frac{3}{15} + \frac{10}{15} = \frac{13}{15}$$

EXAMPLE 3 Add $\frac{1}{3} + \frac{2}{5}$.

SOLUTION Using the rule for addition, we have

$$\frac{1}{3} + \frac{2}{5} = \frac{1 \times 5 + 3 \times 2}{3 \times 5} = \frac{5 + 6}{15} = \frac{11}{15}$$

EXAMPLE 4 Add $\frac{2}{6} + \frac{3}{9}$.

SOLUTION $$\frac{2}{6} + \frac{3}{9} = \frac{2 \times 9 + 6 \times 3}{6 \times 9} = \frac{18 + 18}{54} = \frac{36}{54} = \frac{\not{2} \times 2 \times \not{3} \times \not{3}}{\not{2} \times 3 \times \not{3} \times \not{3}} = \frac{2}{3}$$

How do we subtract two rational numbers? Consider the problem $\frac{2}{3} - \frac{1}{5}$. The expression $-\frac{1}{5}$ is equivalent to $\frac{1}{-5}$ and also to $\frac{-1}{5}$. Hence the problem $\frac{2}{3} - \frac{1}{5}$ is the equivalent to $\frac{2}{3} + \frac{-1}{5}$, which turns out to be an addition problem similar to those that we have been considering. Therefore

$$\frac{2}{3} - \frac{1}{5} = \frac{2}{3} + \frac{-1}{5} = \frac{2 \times 5 + 3 \times (-1)}{3 \times 5} = \frac{10 + (-3)}{15} = \frac{7}{15}$$

EXAMPLE 5 Subtract $\frac{4}{5} - \frac{1}{3}$.

SOLUTION $$\frac{4}{5} - \frac{1}{3} = \frac{4}{5} + \frac{-1}{3} = \frac{4 \times 3 + 5 \times (-1)}{5 \times 3} = \frac{12 + (-5)}{15} = \frac{7}{15}$$

Now that we have examined the operations of addition and subtraction for rational numbers, we next examine multiplication. Most students feel that multiplication is the easiest operation to perform with fractions. In order to multiply two fractions, we simply multiply numerator times numerator and denominator times denominator.

If $\frac{a}{b}$ and $\frac{c}{d}$ are rational numbers, then

$$\frac{a}{b} \times \frac{c}{d} = \frac{a \times c}{b \times d}$$

If we want to find the product of $\frac{3}{5}$ and $\frac{2}{7}$, we simply multiply the numerators together to find the numerator of the product and multiply the denominators together to find the denominator of the product. Therefore

$$\frac{3}{5} \times \frac{2}{7} = \frac{3 \times 2}{5 \times 7} = \frac{6}{35}$$

Consider the problem $\frac{5}{18} \times \frac{6}{25}$. We can do this problem in the same manner as the previous example; that is,

$$\frac{5}{18} \times \frac{6}{25} = \frac{5 \times 6}{18 \times 25} = \frac{30}{450} = \frac{\cancel{2} \times \cancel{3} \times \cancel{5}}{\cancel{2} \times 3 \times \cancel{3} \times \cancel{5} \times 5} = \frac{1}{15}$$

Or we can make the problem a little easier by simplifying it before performing the actual multiplication.

$$\frac{5}{18} \times \frac{6}{25} = \frac{5}{2 \times 3 \times 3} \times \frac{2 \times 3}{5 \times 5} = \frac{\cancel{5} \times \cancel{2} \times \cancel{3}}{\cancel{2} \times 3 \times \cancel{3} \times \cancel{5} \times 5} = \frac{1}{15}$$

Note that we eliminated the factors that are common to both the numerator and denominator.

EXAMPLE 6 Multiply $\frac{4}{9} \times \frac{2}{5}$.

SOLUTION $$\frac{4}{9} \times \frac{2}{5} = \frac{4 \times 2}{9 \times 5} = \frac{8}{45}$$

EXAMPLE 7 Multiply $\frac{7}{16} \times \frac{40}{42}$.

SOLUTION $$\frac{7}{16} \times \frac{40}{42} = \frac{7}{2 \times 2 \times 2 \times 2} \times \frac{2 \times 2 \times 2 \times 5}{2 \times 3 \times 7}$$

$$= \frac{\cancel{7} \times \cancel{2} \times \cancel{2} \times \cancel{2} \times 5}{\cancel{2} \times \cancel{2} \times \cancel{2} \times 2 \times 2 \times 3 \times \cancel{7}} = \frac{5}{2 \times 2 \times 3} = \frac{5}{12}$$

Division of fractions can be defined in terms of multiplication.

$$\frac{a}{b} \div \frac{c}{d} = \frac{a}{b} \times \frac{d}{c}$$

You may recall a rule that you learned previously: "In order to divide two fractions, invert the divisor and multiply." Why does this work? Consider the problem $\frac{2}{3} \div \frac{1}{2}$. According to the rule,

$$\frac{2}{3} \div \frac{1}{2} = \frac{2}{3} \times \frac{2}{1} = \frac{2 \times 2}{3 \times 1} = \frac{4}{3}$$

Another way to look at this problem is

$$\frac{\frac{2}{3}}{\frac{1}{2}}$$

This is a **complex fraction,** because the numerator or (inclusive *or*) denominator of the fraction is also a fraction. The complex fraction would no longer be complex if the denominator were 1. In order to convert this denominator to 1, we must multiply $\frac{1}{2}$ by its *reciprocal*, $\frac{2}{1}$. But, if we multiply the denominator by $\frac{2}{1}$, we must multiply the numerator by $\frac{2}{1}$. Therefore,

$$\frac{\frac{2}{3}}{\frac{1}{2}} = \frac{\frac{2}{3} \times \frac{2}{1}}{\frac{1}{2} \times \frac{2}{1}} = \frac{\frac{2 \times 2}{3 \times 1}}{\frac{1 \times 2}{2 \times 1}} = \frac{\frac{4}{3}}{\frac{2}{2}} = \frac{\frac{4}{3}}{1} = \frac{4}{3}$$

In general terms, we have

$$\frac{\frac{a}{b}}{\frac{c}{d}} = \frac{\frac{a}{b} \times \frac{d}{c}}{\frac{c}{d} \times \frac{d}{c}} = \frac{\frac{a}{b} \times \frac{d}{c}}{1} = \frac{a}{b} \times \frac{d}{c}$$

From the illustrative example, we can see that in order to divide two rational numbers, we multiply the first rational number (the *dividend*) by the multiplicative inverse of the second rational number (the *divisor*).

For any rational numbers $\frac{a}{b}$ and $\frac{c}{d}$,

$$\frac{a}{b} \div \frac{c}{d} = \frac{a}{b} \times \frac{d}{c} \qquad b \neq 0, c \neq 0$$

EXAMPLE 8 Divide $\frac{9}{11} \div \frac{5}{4}$.

SOLUTION $\frac{9}{11} \div \frac{5}{4} = \frac{9}{11} \times \frac{4}{5} = \frac{9 \times 4}{11 \times 5} = \frac{36}{55}$

EXAMPLE 9 Divide $\frac{6}{7} \div \frac{9}{14}$.

SOLUTION $\frac{6}{7} \div \frac{9}{14} = \frac{6}{7} \times \frac{14}{9} = \frac{2 \times 3}{7} \times \frac{2 \times 7}{3 \times 3} = \frac{2 \times \not{3} \times 2 \times \not{7}}{\not{7} \times 3 \times \not{3}} = \frac{4}{3}$

ALTERNATE SOLUTION $\frac{6}{7} \div \frac{9}{14} = \frac{6}{7} \times \frac{14}{9} = \frac{84}{63} = \frac{2 \times 2 \times \not{3} \times \not{7}}{3 \times \not{3} \times \not{7}} = \frac{4}{3}$

Note that the only difference between the two solutions in Example 9 is that in the first solution, the prime factors are determined first and those common to both the numerator and denominator are eliminated before the answer is determined.

EXAMPLE 10 Evaluate the following.

a. $\dfrac{\frac{2}{3}}{\frac{5}{7}}$ b. $\dfrac{4}{\frac{2}{3}}$ c. $\dfrac{\frac{3}{5}}{7}$

SOLUTION All of these expressions are complex fractions. The major fraction line (division line) is made longer to avoid confusion about the numerator and denominator of the complex fraction. We can evaluate these by performing the indicated division.

a. $\dfrac{\frac{2}{3}}{\frac{5}{7}} = \frac{2}{3} \div \frac{5}{7} = \frac{2}{3} \times \frac{7}{5} = \frac{14}{15}$

b. $\dfrac{4}{\frac{2}{3}} = 4 \div \frac{2}{3} = \frac{4}{1} \times \frac{3}{2} = \frac{12}{2} = 6$

c. $\dfrac{\frac{3}{5}}{7} = \frac{3}{5} \div 7 = \frac{3}{5} \times \frac{1}{7} = \frac{3}{35}$

Sometimes a complex fraction may contain more than one fraction or whole number in the numerator or denominator (or both). For example, consider

$$\frac{1 - \frac{2}{3}}{\frac{1}{2} + \frac{1}{3}}$$

To simplify this fraction we must first perform the indicated operations in the numerator and denominator, and then simplify the resulting complex fraction by performing the indicated division. Therefore,

$$\frac{1 - \frac{2}{3}}{\frac{1}{2} + \frac{1}{3}} = \frac{\frac{1}{3}}{\frac{5}{6}} = \frac{1}{3} \times \frac{6}{5} = \frac{6}{15} = \frac{2}{5}$$

EXERCISES FOR SECTION 8.5

1. Reduce each fraction to lowest terms.
   a. $\frac{6}{16}$ b. $\frac{8}{72}$ c. $\frac{81}{129}$ d. $\frac{54}{448}$

2. Reduce each fraction to lowest terms.
   a. $\frac{24}{72}$ b. $\frac{4}{9}$ c. $\frac{484}{576}$ d. $\frac{775}{1{,}325}$

3. Perform the indicated operations.
   a. $\frac{4}{5} + \frac{1}{7}$ b. $\frac{2}{3} + \frac{1}{4}$ c. $\frac{1}{9} + \frac{1}{8}$
   d. $\frac{9}{11} - \frac{2}{3}$ e. $\frac{13}{16} - \frac{4}{5}$ f. $\frac{8}{9} - \frac{1}{3}$

4. Perform the indicated operations.
   a. $\frac{4}{11} + \frac{5}{13}$ b. $\frac{6}{11} + \frac{2}{15}$ c. $\frac{8}{33} + \frac{3}{16}$
   d. $\frac{9}{11} - \frac{1}{3}$ e. $\frac{13}{16} - \frac{3}{5}$ f. $\frac{12}{33} - \frac{4}{11}$

5. Perform the indicated operations.
   a. $\frac{4}{5} \times \frac{2}{7}$ b. $\frac{3}{11} \times \frac{4}{5}$ c. $\frac{8}{13} \times \frac{4}{7}$
   d. $\frac{4}{5} \div \frac{2}{7}$ e. $\frac{3}{11} \div \frac{4}{9}$ f. $\frac{8}{13} \div \frac{4}{7}$

6. Perform the indicated operations.
   a. $\frac{6}{11} \times \frac{2}{9}$ b. $\frac{8}{9} \times \frac{3}{4}$ c. $\frac{7}{14} \times \frac{22}{14}$
   d. $\frac{8}{33} \div \frac{4}{11}$ e. $\frac{8}{9} \div \frac{2}{3}$ f. $\frac{12}{33} \div \frac{4}{11}$

7. Simplify each of the following.
   a. $\frac{\frac{3}{11}}{\frac{4}{9}}$ b. $\frac{\frac{2}{13}}{\frac{5}{7}}$ c. $\frac{\frac{2}{5}}{3}$
   d. $\frac{1 + \frac{1}{2}}{2 - \frac{1}{3}}$ e. $\frac{2 + \frac{1}{4}}{3 - \frac{1}{2}}$ f. $\frac{\frac{1}{2} + \frac{1}{3}}{\frac{1}{4} + \frac{4}{5}}$

8. Simplify each of the following.
   a. $\frac{\frac{4}{7}}{5}$ b. $\frac{6}{\frac{1}{2}}$ c. $\frac{\frac{3}{4}}{\frac{6}{7}}$
   d. $\frac{2 - \frac{1}{2}}{3 + \frac{2}{5}}$ e. $\frac{1 - \frac{2}{3}}{2 + \frac{3}{4}}$ f. $\frac{\frac{3}{5} + \frac{4}{7}}{\frac{5}{8} - \frac{1}{2}}$

9. Simplify each of the following.

a. $\dfrac{\frac{3}{4}}{\frac{2}{5}}$ b. $\dfrac{8}{\frac{3}{5}}$ c. $\dfrac{\frac{4}{5}}{3}$

d. $\dfrac{\frac{3}{8}}{\frac{5}{4}}$ e. $\dfrac{2\frac{1}{5}}{3\frac{3}{4}}$ f. $\dfrac{1\frac{1}{2}}{2\frac{3}{4}}$

10. Simplify each of the following.

a. $\dfrac{1+\frac{1}{5}}{2-\frac{1}{2}}$ b. $\dfrac{1+\frac{3}{4}}{2-\frac{2}{3}}$ c. $\dfrac{\frac{2}{3}+\frac{3}{4}}{\frac{5}{8}-\frac{1}{4}}$

d. $\dfrac{3\frac{1}{2}}{2-\frac{1}{4}}$ e. $\dfrac{2\frac{5}{6}}{3+\frac{1}{3}}$ f. $\dfrac{3-\frac{3}{4}}{2+\frac{1}{2}}$

11. Simplify each of the following.

a. $\left(\frac{3}{4}+\frac{2}{3}\right)\times\left(\frac{7}{8}-\frac{1}{2}\right)$ b. $\frac{1}{2}\times\left(\frac{5}{6}+2\right)$

c. $\left(1-\frac{2}{3}\right)\times\left(\frac{3}{8}+\frac{1}{2}\right)$ d. $\left(\frac{3}{4}\times\frac{5}{6}+\frac{1}{2}\right)\times 2$

12. Simplify each of the following.

a. $\left(\frac{1}{2}+\frac{3}{4}\right)\div\left(\frac{7}{8}-\frac{1}{2}\right)$ b. $3\div\left(\frac{1}{4}+\frac{2}{3}-\frac{1}{5}\right)$

c. $\left(\frac{1}{4}+\frac{1}{3}\right)\times\left(\frac{3}{4}-\frac{1}{3}\right)$ d. $\frac{3}{8}\times\left(\frac{4}{5}-\frac{1}{2}+\frac{1}{3}\right)$

13. Determine whether each statement is true or false. (*Hint:* Convert the fractions under consideration to fractions with the same denominator.)

a. $\frac{4}{7}>\frac{2}{3}$ b. $\frac{3}{4}<\frac{7}{8}$ c. $\frac{4}{11}>\frac{3}{7}$

d. $\frac{4}{9}<\frac{16}{36}$ e. $\frac{5}{11}>\frac{11}{5}$ f. $\frac{6}{7}<\frac{8}{8}$

14. Determine whether each statement is true or false (see Exercise 13).

a. $\frac{4}{3}>\frac{5}{4}$ b. $\frac{6}{4}<\frac{8}{9}$ c. $\frac{8}{9}>\frac{6}{5}$

d. $\frac{8}{33}<\frac{4}{11}$ e. $\frac{8}{9}<\frac{2}{3}$ f. $\frac{12}{33}=\frac{4}{11}$

15. Determine whether each statement is true or false.

a. Every rational number is an integer.
b. Every integer is a rational number.
c. Every rational number is a natural number.
d. Every natural number is a rational number.
e. Every rational number is a whole number.

16. Determine whether each statement is true or false.

a. Every whole number is a rational number.
b. Every whole number is an integer.
c. Every integer is a whole number.
d. The rationals are a subset of the integers.
e. The integers are a subset of the rationals.

JUST FOR FUN If you double $\frac{1}{4}$ of a certain fraction and multiply it by that fraction, the answer is $\frac{1}{8}$. What is the fraction?

8.6 RATIONAL NUMBERS AND DECIMALS

Before we continue our discussion regarding what other numbers belong on the number line, let's review the topic of *decimals*. **Decimals** are fractions that have a power of 10, such as 10, 100,

Courtesy Texas Instruments

1,000, or 10,000, for their denominator. The word *decimal* comes from the Latin word *decem,* which means *ten.* The fraction $\frac{3}{10}$ is represented by the decimal 0.3.

The *decimal point* is a period (.) that appears just to the left of the tenths place in the decimal. Some other examples of fractions expressed as decimals are

$$\frac{4}{10} = 0.4, \quad \frac{31}{100} = 0.31, \quad \frac{471}{1000} = 0.471$$

The decimal 0.4 is read as "4 tenths," 0.31 is read as "31 hundredths," and 0.471 is read as "471 thousandths." The following example indicates the names of the places for decimals.

thousands	hundreds	tens	units		tenths	hundredths	thousandths	ten-thousandths
4	3	4	1	•	2	1	3	4
Integers					Decimals			

In order to change any fraction in the form a/b to a decimal, we divide the denominator into the numerator.

$$\begin{array}{r} 0.2 \\ 5\overline{)1.0} \\ \underline{1\,0} \\ 0 \end{array} \quad \frac{1}{5} = 0.2 \qquad\qquad \begin{array}{r} 0.375 \\ 8\overline{)3.000} \\ \underline{2\,4} \\ 60 \\ \underline{56} \\ 40 \\ \underline{40} \\ 0 \end{array} \quad \frac{3}{8} = 0.375$$

Decimals such as 0.2 and 0.375 are called **terminating decimals** because at some point in the division a remainder of zero is obtained; that is, when we change the fraction to a decimal, the division terminates.

Not all fractions can be expressed as terminating decimals. For example, consider the rational number $\frac{1}{3}$. Converting $\frac{1}{3}$ to a decimal, we have

$$\begin{array}{r} 0.3333\ldots \\ 3\overline{)1.0000} \\ \underline{9} \\ 10 \\ \underline{9} \\ 10 \\ \underline{9} \\ 10 \\ \underline{9} \\ 1 \end{array}$$

The division does not terminate: we will never obtain a remainder of zero, but the remainder of 1 will keep reappearing at regular intervals.

Instead of writing the decimal expression for $\frac{1}{3}$ as $0.3333\ldots$, we can express it in a more convenient and efficient way by placing a bar over the 3, which indicates that the 3 repeats endlessly, that is, $0.333\ldots = 0.\overline{3}$. In the same manner, $0.121212\ldots = 0.\overline{12}$. In this case the digits 12 repeat endlessly, so we place a bar over both the 1 and the 2. Decimals such as $0.\overline{3}$ and $0.\overline{12}$ are called **repeating nonterminating decimals.**

EXAMPLE 1 Express $\frac{5}{8}$ as a decimal.

SOLUTION

$$\begin{array}{r} 0.625 \\ 8\overline{)5.000} \\ \underline{4\,8} \\ 20 \\ \underline{16} \\ 40 \\ \underline{40} \\ 0 \end{array} \qquad \frac{5}{8} = 0.625$$

EXAMPLE 2 Express $\frac{4}{9}$ as a decimal.

SOLUTION

$$\begin{array}{r} 0.44\ldots \\ 9\overline{)4.00} \\ \underline{3\,6} \\ 40 \\ \underline{36} \\ 4 \end{array} \qquad \frac{4}{9} = 0.\overline{4}$$

EXAMPLE 3 Express $\frac{3}{7}$ as a decimal.

SOLUTION

$$
\begin{array}{r}
0.4285714 \\
7\overline{)3.0000000} \\
\underline{2\,8} \\
20 \\
\underline{14} \\
60 \\
\underline{56} \\
40 \\
\underline{35} \\
50 \\
\underline{49} \\
10 \\
\underline{7} \\
30 \\
\underline{28} \\
2
\end{array}
\qquad \frac{3}{7} = 0.\overline{428571}
$$

Note that we place a bar over the six digits $0.\overline{428571}$. The last remainder of 2 is a repeat of a remainder that we had previously. Therefore the pattern of division will repeat, and the digits 428571 will repeat endlessly.

In examining the set of rational numbers, we learned how to express a fraction as a decimal. For example, $\frac{5}{8} = 0.625$, $\frac{4}{9} = 0.\overline{4}$, and $\frac{3}{7} = 0.\overline{428571}$. The next thing to consider is whether, given the decimal expression, we can find its equivalent fraction. We already know that $0.\overline{3} = \frac{1}{3}$ and $0.25 = \frac{1}{4}$, but what about other examples we might encounter?

To convert a terminating decimal to a fraction, we simply omit the decimal point and supply the proper denominator. For example,

$$0.3 = \frac{3}{10}, \quad 0.25 = \frac{25}{100} = \frac{1}{4}, \quad 0.125 = \frac{125}{1000} = \frac{1}{8}$$

This technique does not work for repeating decimals, so we will need to develop another method for repeating decimals. Consider the decimal $0.3\overline{3}$. Let $x = 0.3\overline{3}$. Then $10x = 3.3\overline{3}$. (Multiplying a number by 10 moves the decimal point one place to the right, multiplying by 100 moves the decimal point two places to the right, and so on.) Thus far we have

$$\begin{aligned} 10x &= 3.3\overline{3} \\ x &= 0.3\overline{3} \end{aligned}$$

The decimal points are lined up in the same position.

Now subtract x from $10x$ and $0.3\overline{3}$ from $3.3\overline{3}$.

$$\begin{aligned} 10x &= 3.3\overline{3} \\ x &= 0.3\overline{3} \\ \hline 9x &= 3 \end{aligned}$$

All of the repeating 3s are subtracted from repeating 3s.

Next divide both sides of the resulting equation by 9.

$$\frac{9x}{9} = \frac{3}{9}$$

$$x = \frac{3}{9} = \frac{1}{3}$$

Therefore

$$0.3\overline{3} = \frac{1}{3}$$

Let's try another example. Suppose we wish to convert the repeating decimal $0.\overline{13}$ to a fraction. Let $x = 0.\overline{13}$; because the digits repeat in cycles of two, we multiply both sides of the equation by 100. If $x = 0.\overline{13}$, then $100x = 13.\overline{13}$. We subtract, which gives

$$\begin{aligned} 100x &= 13.\overline{13} \\ x &= 0.\overline{13} \\ \hline 99x &= 13 \end{aligned}$$

Dividing both sides of the equation by 99, we have

$$\frac{99x}{99} = \frac{13}{99}$$

$$x = \frac{13}{99}$$

Therefore

$$0.\overline{13} = \frac{13}{99}$$

EXAMPLE 4 Express $0.\overline{25}$ as a quotient of integers (that is, convert $0.\overline{25}$ to a fraction).

SOLUTION Let $x = 0.\overline{25}$ and multiply both sides of the equation by 100 (two digits repeating). Subtracting, we have

$$\begin{array}{r} 100x = 25.\overline{25} \\ x = \ \ 0.\overline{25} \\ \hline 99x = 25 \end{array}$$

Dividing both sides by 99,

$$\frac{99x}{99} = \frac{25}{99}$$

$$x = \frac{25}{99}$$

Therefore

$$0.\overline{25} = \frac{25}{99}$$

EXAMPLE 5 Express $3.\overline{162}$ as a quotient of integers.

SOLUTION Let $x = 3.\overline{162}$. Multiplying both sides of the equation by 1,000 (three digits repeating) and subtracting, we have

$$\begin{array}{r} 1{,}000x = 3162.\overline{162} \\ x = \ \ \ \ 3.\overline{162} \\ \hline 999x = 3159 \end{array}$$

Dividing both sides by 999,

$$\frac{999x}{999} = \frac{3{,}159}{999}$$

$$x = \frac{3{,}159}{999}$$

$$= \frac{117}{37}$$

Therefore

$$3.\overline{162} = \frac{117}{37}$$

EXAMPLE 6 Express $2.14\overline{27}$ as a quotient of integers.

SOLUTION Let $x = 2.14\overline{27}$, and multiply both sides of the equation by 100 (two digits repeating). We place the decimal points in the same position and subtract.

$$\begin{aligned} 100x &= 214.27\overline{27} \\ x &= 2.14\overline{27} \\ \hline 99x &= 212.13 \end{aligned}$$

Dividing both sides by 99,

$$\frac{99x}{99} = \frac{212.13}{99}$$

But we are not finished. We were supposed to express $2.14\overline{27}$ as a quotient of two integers, and 212.13 is not an integer—it is a decimal.

In our earlier discussion of rational numbers (Sec. 8.5), we noted that

$$\frac{a}{b} = \frac{a \times k}{b \times k}$$

Using this idea, we multiply $\frac{212.13}{99}$ by $\frac{100}{100}$ (we use 100, as we have two decimal places). Therefore

$$\begin{aligned} x &= \frac{212.13}{99} \times \frac{100}{100} = \frac{21{,}213}{9{,}900} \\ &= \frac{2{,}357 \times \not{3} \times \not{3}}{1{,}100 \times \not{3} \times \not{3}} = \frac{2{,}357}{1{,}100} \end{aligned}$$

Therefore

$$2.14\overline{27} = \frac{2{,}357}{1{,}100}$$

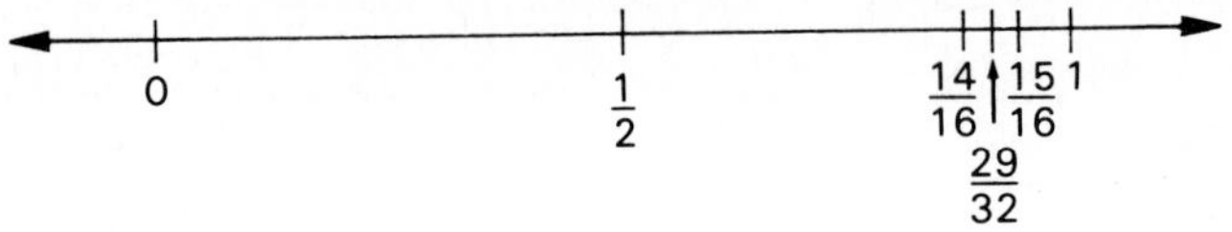

FIGURE 12

Thus far we have examined the set of natural numbers, the set of whole numbers, the set of integers, and the set of rational numbers. All of these can be shown on a number line. For example, Figure 12 is a number line showing the integers zero and 1 and the rational numbers $\frac{1}{2}$, $\frac{14}{16}$, and $\frac{15}{16}$.

If two rational numbers are indicated on a number line, can we find other numbers that fit between them? The answer is yes. As an example, we will find a number that lies between $\frac{14}{16}$ and $\frac{15}{16}$ on the line in Figure 12. To do this, we convert $\frac{14}{16}$ to $\frac{28}{32}$ and $\frac{15}{16}$ to $\frac{30}{32}$. Now we can see that the number $\frac{29}{32}$ is between the two given fractions. This process can be continued indefinitely: it is possible to find a rational number between any two given rational numbers. This process may seem to "fill up" the number line, but this is not the case, because there is always room for one more number. In fact, between any two rational numbers, there is always another rational number. This particular property is called the **density property of rational numbers.** We can also say that the rational numbers are **dense.** Note that the density property does not hold for all kinds of numbers. The natural numbers are not dense, because there is no natural number between the natural numbers 3 and 4.

We have seen one example of how to find a rational number between two given rational numbers, but how do we do it for *any* two rational numbers? One way to do this is to find the arithmetic mean of the two given rational numbers. In general terms, if we let x equal the number we are seeking, and we let a/b and c/d equal the given rational numbers, then

$$x = \frac{1}{2} \times \left(\frac{a}{b} + \frac{c}{d}\right)$$

Using $\frac{14}{16}$ and $\frac{15}{16}$ from our previous discussion, we have

$$x = \frac{1}{2} \times \left(\frac{14}{16} + \frac{15}{16}\right)$$

$$x = \frac{1}{2} \times \left(\frac{29}{16}\right)$$

$$x = \frac{29}{32}$$

EXAMPLE 7 Find a rational number between $\frac{5}{7}$ and $\frac{6}{7}$.

SOLUTION Finding the arithmetic mean of the given numbers, we have

$$x = \frac{1}{2} \times \left(\frac{5}{7} + \frac{6}{7}\right)$$

$$x = \frac{1}{2} \times \left(\frac{11}{7}\right)$$

$$x = \frac{11}{14}$$

Check:

$$\frac{5}{7} = \frac{10}{14} \quad \text{and} \quad \frac{6}{7} = \frac{12}{14}$$

$$\frac{10}{14} < \frac{11}{14} \quad \text{and} \quad \frac{11}{14} < \frac{12}{14}$$

EXAMPLE 8 Find a rational number between $\frac{1}{5}$ and $\frac{1}{9}$.

SOLUTION

$$x = \frac{1}{2} \times \left(\frac{1}{5} + \frac{1}{9}\right)$$

$$x = \frac{1}{2} \times \left(\frac{9+5}{45}\right) = \frac{1}{\cancel{2}} \times \left(\frac{\overset{7}{\cancel{14}}}{45}\right)$$

$$x = \frac{7}{45}$$

Check:

$$\frac{1}{9} = \frac{5}{45} \quad \text{and} \quad \frac{1}{5} = \frac{9}{45}$$

$$\frac{5}{45} < \frac{7}{45} \quad \text{and} \quad \frac{7}{45} < \frac{9}{45}$$

Note that there are other rational numbers that are also between $\frac{1}{5}$ and $\frac{1}{9}$. However, using the formula $x = \frac{1}{2} \times [(a/b) + (c/d)]$, we found the arithmetic mean, which is exactly halfway between the two given numbers.

In summary, whenever we are given two rational numbers a/b and c/d, we can find another rational number (call it x) such that $a/b < x$ and $x < c/d$, or $a/b > x$ and $x > c/d$. This particular property is called the *density property of rational numbers.* The rational numbers are dense, but the integers are not. For example, there is no integer between the integers 1 and 2. The density of rational numbers can also be stated as

Between every two rational numbers there is another rational number.

EXERCISES FOR SECTION 8.6

1. Express each fraction as a decimal.
   a. $\frac{3}{8}$ b. $\frac{5}{16}$ c. $\frac{2}{3}$
   d. $\frac{7}{33}$ e. $\frac{1}{11}$ f. $\frac{15}{37}$

2. Express each fraction as a decimal.
   a. $\frac{5}{8}$ b. $\frac{15}{16}$ c. $\frac{2}{33}$
   d. $\frac{7}{11}$ e. $\frac{19}{37}$ f. $\frac{1}{7}$

3. Express each fraction as a decimal.
   a. $\frac{7}{8}$ b. $\frac{4}{9}$ c. $\frac{9}{11}$
   d. $\frac{1}{16}$ e. $\frac{5}{7}$ f. $\frac{8}{17}$

4. Express each decimal as a quotient of integers, in simplest form.
   a. 0.45 b. 0.035 c. $0.\overline{6}$
   d. $0.\overline{12}$ e. $0.\overline{134}$ f. $2.1\overline{78}$

5. Express each decimal as a quotient of integers, in simplest form.
   a. 0.125 b. 0.0025 c. $0.\overline{7}$
   d. $0.\overline{34}$ e. $6.2\overline{81}$ f. $0.\overline{9}$

6. Express each decimal as a quotient of integers, in simplest form.
   a. 0.012 b. 0.0005 c. 0.55
   d. $1.\overline{3}$ e. $0.\overline{45}$ f. $3.\overline{12}$

7. Express each decimal as a quotient of integers, in simplest form.
   a. 0.75 b. 0.875 c. $0.\overline{8}$
   d. $0.\overline{45}$ e. $0.\overline{235}$ f. $4.59\overline{6}$

8. Express each decimal as a quotient of integers, in simplest form.
   a. 0.43 b. 0.375 c. $0.\overline{4}$
   d. $2.\overline{147}$ e. $3.1\overline{45}$ f. $2.54\overline{9}$

9. Find a rational number between each pair of rational numbers.
   a. $\frac{1}{2}, \frac{1}{3}$ b. $\frac{1}{3}, \frac{1}{4}$ c. $\frac{1}{4}, \frac{1}{5}$
   d. $\frac{2}{3}, \frac{7}{8}$ e. $\frac{3}{4}, \frac{9}{11}$ f. $\frac{7}{11}, \frac{15}{16}$

10. Find a rational number between each pair of rational numbers.
    a. $\frac{2}{9}, \frac{3}{9}$ b. $\frac{4}{7}, \frac{5}{7}$ c. $\frac{3}{7}, \frac{4}{11}$
    d. $\frac{3}{5}, \frac{7}{9}$ e. $\frac{4}{9}, \frac{11}{12}$ f. $\frac{7}{13}, \frac{9}{17}$

11. Find a rational number between each pair of rational numbers.
    a. $\frac{2}{8}, \frac{3}{8}$ b. $\frac{4}{9}, \frac{5}{9}$ c. $\frac{1}{8}$, 0
    d. $\frac{9}{10}$, 1 e. $\frac{5}{7}, \frac{3}{8}$ ⋆f. 1.999, 2

12. Find a positive number that is between 0 and 0.1.

13. Find a positive number that is between 0 and 0.01.

14. Find a positive number that is between 0 and 0.001.

JUST FOR FUN Write down four odd digits that will add up to 19.

8.7 IRRATIONAL NUMBERS AND THE SET OF REAL NUMBERS

We have seen that any rational number can be expressed as a decimal, and that this decimal will be a terminating decimal or a repeating nonterminating decimal. For example, $\frac{5}{8}$ and $\frac{1}{11}$ are rational numbers and, expressing each as a decimal, we have

$$\frac{5}{8} = 0.625 \qquad \text{(a terminating decimal)}$$

$$\frac{7}{11} = 0.\overline{09} \qquad \text{(a repeating nonterminating decimal)}$$

Are all decimals either terminating or repeating nonterminating decimals? The answer is no. We can construct a decimal that does not terminate, yet does not repeat, as follows: choose any digit and write it after the decimal point, then write a zero; repeat the digit, then write two zeros; repeat the digit, then write three zeros; repeat the digit, then write four zeros; repeat the digit, and so on. Using this idea we construct decimals like

$$0.101001000100001000001\ldots$$
$$0.202002000200002000002\ldots$$
$$0.909009000900009000009\ldots$$

For each of these decimals, we have no repeating cycle as we did for $\frac{1}{3}$ and $\frac{3}{7}$. Regardless of how far we extend these decimals, there will be no repeating set of digits. These decimals have a pattern, but no repeating cycle. Decimals that are *nonterminating* and *nonrepeating* are called **irrational numbers.**

Probably the most famous irrational number is *pi* (π). The value of π for the first 50 decimal places is

$$\pi = 3.14159265358979323846264338327950288419716939937510$$

You will note that there is no repeating sequence of digits for this decimal.

The formula for finding the circumference of a circle says that to find the circumference of a circle, we multiply its diameter times *pi*; that is, $C = \pi d$. Therefore π is the ratio of the circumference of a circle to its diameter. In other words, to find

HISTORICAL NOTE

Rather than describe a number as the square root of another number, the ancient Greeks referred to it as a side of a square number. For example, 2 is the side of a square number 4, 3 is the side of a square number 9, and so on. From this the Arabs developed the idea of a root of a number, that is, 2 is the root of 4, 3 is the root of 9, and so forth. Later, European mathematicians used the Latin word *radix* to indicate the root of a number. When abbreviations began to be used in Algebra (about 100 A.D.), the word radix was abbreviated as R_x. Therefore, $R_x4 = 2$, $R_x9 = 3$. This notation was used for some time, until a small *r* began to be used in place of R_x. R_x4 became *r*4. Most historians believe that the radical sign $\sqrt{\ }$, as we know it today, is a representation of the small letter *r* as it was written by mathematicians at the time (because everything had to be copied by hand).

the value of π, you must divide the circumference of a circle by its diameter. You will obtain a nonrepeating nonterminating decimal. Computers have been used to find the value of π well beyond the 50 decimal places shown, but no one has ever reached the end, or ever will.

There are many other examples of irrational numbers—numbers that, when they are expressed as decimals, are nonterminating and nonrepeating. Some other examples are:

$$2.718281824\ldots$$
$$0.12122122212222122222\ldots$$
$$\sqrt{2},\ \sqrt{3},\ \sqrt{5},\ \sqrt{6},\ \sqrt{7}$$

A **perfect square** is a number that is the product of an integer times itself. For example, 4 is a perfect square because $4 = 2 \cdot 2$. Some other examples of perfect squares are 1, 9, 16, 25, and 36. The square root of any positive integer that is *not* a perfect square is an irrational number.

For example, $\sqrt{2}$ is an irrational number. If we were to express $\sqrt{2}$ as a decimal, it would be a nonterminating, nonrepeating decimal. Some other examples of irrational numbers are $\sqrt{3}$, $\sqrt{5}$, $\sqrt{6}$, $\sqrt{7}$, and so on. These are all square roots of nonnegative numbers that are not perfect squares.

EXAMPLE 1 Classify each number as rational or irrational.

a. $0.\overline{13}$ b. $0.131131113\ldots$
c. $\sqrt{11}$ d. $\sqrt{49}$

SOLUTION a. $0.\overline{13}$ is a rational number. It is a nonterminating decimal, but it is repeating.
b. $0.131131113\ldots$ is an irrational number. It is a decimal that is nonterminating and nonrepeating.
c. $\sqrt{11}$ is an irrational number. The square root of a nonnegative number that is not a perfect square is an irrational number.
d. $\sqrt{49}$ is a rational number because 49 is a perfect square. In fact, $\sqrt{49} = 7$.

It is interesting to note that the set of rational numbers and the set of irrational numbers are disjoint sets; that is, their intersection is empty. If we take the union of these two sets, then we have all of the numbers on the number line. The union of the set of rational numbers and the set of irrational numbers yields a set of numbers that is called the set of **real numbers.** Therefore, any rational or irrational number is a real number. Note that a rational number may be expressed as either a terminating decimal or a repeating nonterminating decimal, and an irrational number may be expressed as a nonrepeating nonterminating decimal. Hence, every decimal is a real number and every real number may be expressed as a decimal.

Recall that we began our discussion of the classification of numbers with the set of natural numbers. Including zero with the set of natural numbers produced the set of whole numbers. The set of integers was composed of the set of whole numbers and their additive inverses. The set of integers and the set of fractions yielded the set of rational numbers. Finally came the set of irrational numbers, and the union of the set of rationals and the set of irrationals yielded the set of real numbers. This process is illustrated by the following diagram.

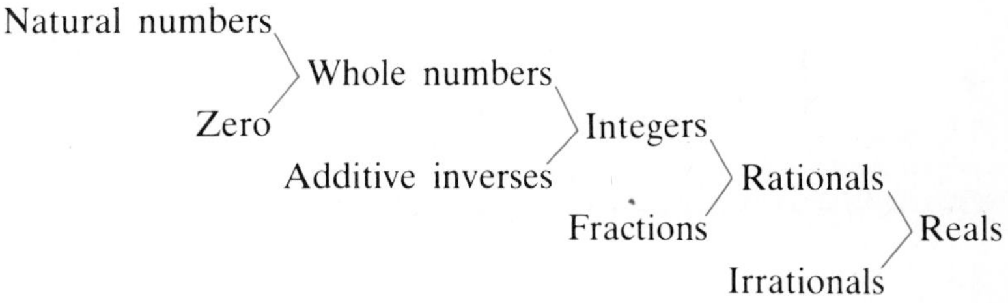

Figure 13 illustrates the relationship of these sets of numbers to the set of real numbers.

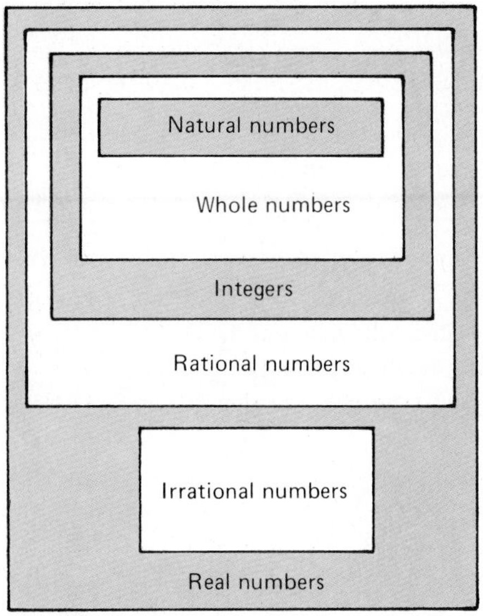

FIGURE 13

The real numbers.

HISTORICAL NOTE

Irrational numbers were studied as early as 500 B.C. by Pythagoras, a famous Greek mathematician. He discovered that there is no rational number for the square root of 2; that is, there is no rational number whose square is 2.

Pythagoras and his colleagues also found that there is no common unit of length for the length of a side of a square and its diagonal at the same time. That is, no unit of length will ever measure exactly both the diagonal of a square and its side.

These mathematicians were supposedly so upset with this discovery that they vowed not to reveal it, and threatened to punish any member of the group who divulged this "secret." They were afraid that this information would cause people to doubt their ability as mathematicians.

Consider the Pythagorean theorem, which in essence states that the square of the hypotenuse of a right triangle is equal to the sum of the squares of the other two sides, or given right triangle ABC, $c^2 = a^2 + b^2$.

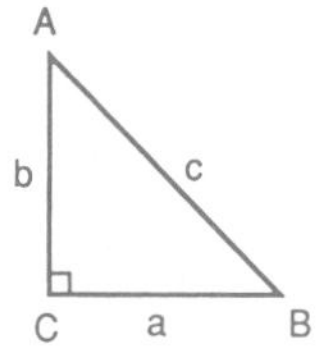

Hence, the length of the hypotenuse of a right triangle that has sides of length 1 is $\sqrt{2}$. This implies that we can easily construct a line of a certain length. However, it is not measurable, and this is what concerned Pythagoras.

EXAMPLE 2 Let R = {real numbers}, I = {integers}, Y = {rational numbers}, Z = {irrational numbers}
Find

a. $R \cap I$ b. $I \cap Y$ c. $Y \cap Z$ d. $Y \cup Z$

SOLUTION We can find these sets by examining Figure 13.

a. $R \cap I = I$ (The set of integers is common to both sets.)
b. $I \cap Y = I$ (The set of integers is common to both sets.)
c. $Y \cap Z = \varnothing$ (The set of rational numbers and the set of irrational numbers have no elements in common.)
d. $Y \cup Z = R$ (The set of rational numbers together with the set of irrational numbers forms the set of real numbers.)

EXERCISES FOR SECTION 8.7

1. Classify each numbers as rational or irrational.
   a. $\frac{1}{3}$ b. -2 c. $0.\overline{3}$
   d. $\sqrt{2}$ e. $\sqrt{3}$ f. $\sqrt{4}$
2. Classify each number as rational or irrational.
   a. 0 b. π
   c. $\sqrt{9}$ d. 2.343343334...
   e. 1.010010001... f. 3.14
3. Classify each number as rational or irrational.
   a. $\frac{3}{4}$ b. $\frac{2}{3}$ c. $\sqrt{16}$
   d. 3.14159 e. $0.\overline{21}$ f. $\sqrt{99}$
4. Determine whether each of the following can be represented by a terminating decimal, a repeating decimal, or a nonterminating nonrepeating decimal.
   a. $\frac{3}{4}$ b. $\frac{2}{3}$ c. $\sqrt{2}$
   d. $\sqrt{25}$ e. $\frac{1}{7}$ f. π
5. Determine whether each of the following can be represented by a terminating decimal, a repeating decimal, or a nonterminating nonrepeating decimal.
   a. $\frac{1}{8}$ b. $\frac{3}{7}$ c. $\sqrt{3}$
   d. $\sqrt{\frac{9}{16}}$ e. $\sqrt{99}$ f. $\frac{1}{11}$
6. Determine whether each of the following can be represented by a terminating decimal, a repeating decimal, or a nonterminating nonrepeating decimal.
   a. $\sqrt{5}$ b. $\sqrt{25}$ c. $\frac{3}{8}$
   d. $\frac{5}{7}$ e. $\frac{22}{7}$ f. $\sqrt{225}$
7. Determine whether each sentence is true or false.
   a. A real number is either a rational or irrational number.
   b. A real number is positive, negative, or zero.
   c. A repeating nonterminating decimal is a rational number.
   d. A nonrepeating nonterminating decimal is a rational number.
   e. A terminating decimal is an irrational number.

8. Determine whether each sentence is true or false.
   a. The intersection of the set of rational numbers and the set of irrational numbers is not empty.
   b. The union of the set of rational numbers and the set of irrational numbers is the set of real numbers.
   c. All rational numbers are real numbers.
   d. All real numbers are rational numbers.
   e. Every real number can be expressed as a terminating decimal or a repeating decimal.

9. Let $R = \{\text{real numbers}\}$, $I = \{\text{integers}\}$, $Y = \{\text{rational numbers}\}$, and $Z = \{\text{irrational numbers}\}$. Determine whether each of the following is true or false.
   a. $R \cap Y = I$
   b. $I \cup Y = Z$
   c. $Y \subset R$
   d. $R \subset Z$
   e. $Y \cap Z = R$
   f. $Y \cup Z = R$

10. Answer yes or no to tell whether each of the following numbers is (I) a natural number, (II) a whole number, (III) an integer, (IV) a rational number, (V) an irrational number, and (VI) a real number.
   a. 7
   b. −3
   c. 3.14
   d. $\sqrt{2}$
   e. 0
   f. $2.\overline{123}$

JUST FOR FUN

If *pi* is an irrational number (a number that is a nonrepeating nonterminating decimal), then why is it that we are usually *told* that pi (π) is $\frac{22}{7}$, when $\frac{22}{7} = 3.\overline{142857}$?

8.8 SCIENTIFIC NOTATION (OPTIONAL)

It is estimated that the average human breath contains as many as ten sextillion atoms. This extremely large number, when expressed in *standard form,* appears as

$$10{,}000{,}000{,}000{,}000{,}000{,}000{,}000.$$

To write this number in *scientific notation* we imagine a decimal point after the first nonzero digit, and count the number of decimal places from there to the actual decimal point.

Standard Form	*Scientific Notation*
$10{,}000{,}000{,}000{,}000{,}000{,}000{,}000 =$	1.0×10^{22}
22 places	Write the number as a product of 1.0 and 10^{22}.

Note: Multiplying by 10^{22} moves the decimal point 22 places to the *right.*

To write a number in scientific notation, write it as a product so that

1. The first factor is greater than or equal to 1 and less than 10.
2. The second factor is a power of 10.

EXAMPLE 1 Write each of the following in scientific notation.

a. 190,000,000 b. 67,200 c. 150

SOLUTION

a.	b.	c.
190,000,000	67,200	150
8 places	4 places	2 places
1.9×10^8	6.72×10^4	1.5×10^2

One millimicron is approximately 0.00000394 inch. Scientific notation can also be used to write very small numbers more compactly. Remember that scientific notation consists of writing a number in the general form $N \times 10^m$ where N is greater than or equal to 1 and less than 10, and m is an integer. Imagine a decimal point after the first nonzero digit. Count the number of places from there to the actual decimal point. Note that we use a negative exponent with 10 when the original number is less than 1.

Standard Form		*Scientific Notation*
0.00000394 (6 places)	=	3.94×10^{-6}

Note: Multiplying by 10^{-6} moves the decimal point 6 places to the *left.*

EXAMPLE 2 Write each of the following in scientific notation.

a. 0.00245 b. 0.00074 c. 0.0271

SOLUTION

a.	b.	c.
0.00245	0.00074	0.0271
3 places	4 places	2 places
2.45×10^{-3}	7.4×10^{-4}	2.71×10^{-2}

EXAMPLE 3 Express each of the following in standard form.

a. 5.21×10^5 b. 3.9×10^{-3} c. 1.05×10^{-6}

SOLUTION a. The exponent of 10 is positive. Hence, move the decimal point five places to the right.

$$5.21 \times 10^5 = 521{,}000$$

b. The exponent of 10 is negative. Therefore, move the decimal point three places to the left.

$$3.9 \times 10^{-3} = 0.0039$$

c. Move the decimal point six places to the left.

$$1.05 \times 10^{-6} = 0.00000105$$

We can use scientific notation in computing answers to problems that contain very large or very small numbers. Since scientific notation makes use of integral exponents, we can apply the rules for exponents. If b is any real number and m and n are natural numbers, then $b^m \times b^n = b^{m+n}$. For example, $10^3 \times 10^2 = 10^{3+2} = 10^5$ and $10^{-2} \times 10^{-4} = 10^{-2+-4} = 10^{-6}$.

Consider this multiplication problem: $12{,}000{,}000 \times 130{,}000{,}000$. Many of the pocket calculators that are so popular today would not be able to handle this problem because of the number of digits involved. An electronic calculator that can handle this multiplication most likely uses scientific notation in performing the necessary computation.

To perform the indicated multiplication, we must first write the given numbers in scientific notation.

$$12{,}000{,}000 \times 130{,}000{,}000 = (1.2 \times 10^7)(1.3 \times 10^8)$$

Next, we make use of the commutative and associative properties of multiplication, and write an equivalent expression.

$$(1.2 \times 10^7)(1.3 \times 10^8) = (1.2 \times 1.3)(10^7 \times 10^8)$$

Performing these multiplications, we obtain

$$(1.2 \times 1.3)(10^7 \times 10^8) = 1.56 \times 10^{15}$$

Therefore,

$$\begin{aligned} 12{,}000{,}000 \times 130{,}000{,}000 &= 1.56 \times 10^{15} \\ &= 1{,}560{,}000{,}000{,}000{,}000 \end{aligned}$$

Note that we performed the operations separately. That is, we performed the multiplication of the numbers between 1 and 10, and separately we performed the multiplication of the powers of 10.

A similar procedure can be used for the operation of division. But, instead of multiplying powers of 10, we divide powers of 10.

If b is any real number ($b \neq 0$) and m and n are natural numbers, then

$$\frac{b^m}{b^n} = b^{m-n}$$

For example,

$$\frac{10^6}{10^2} = 10^{6-2} = 10^4, \qquad \frac{10^2}{10^4} = 10^{2-4} = 10^{-2}, \qquad \text{and}$$

$$\frac{10^3}{10^{-2}} = 10^{3-(-2)} = 10^{3+2} = 10^5$$

Consider the division problem 144,000,000 ÷ 7,200,000. We write this as $\frac{144{,}000{,}000}{7{,}200{,}000}$. Now we write the given numbers in scientific notation.

$$\frac{144{,}000{,}000}{7{,}200{,}000} = \frac{1.44 \times 10^8}{7.2 \times 10^6} = \frac{1.44}{7.2} \times \frac{10^8}{10^6}$$

Performing the indicated operations separately, we obtain:

$$\frac{1.44}{7.2} \times \frac{10^8}{10^6} \qquad = 0.2 \times 10^{8-6} = 0.2 \times 10^2 = 20$$

EXAMPLE 4 Use scientific notation to compute the following.

a. $0.0035 \times 2{,}400{,}000$ b. $0.000143 \div 0.0013$

SOLUTION We first express all numbers in scientific notation and then separately perform the operations with the numbers between 1 and 10, and the powers of 10.

$$\begin{aligned}\text{a. } 0.0035 \times 2{,}400{,}000 &= (3.5 \times 10^{-3})(2.4 \times 10^6)\\ &= (3.5 \times 2.4)(10^{-3} \times 10^6)\\ &= 8.4 \times 10^{-3+6}\\ &= 8.4 \times 10^3 = 8400\end{aligned}$$

$$\begin{aligned}\text{b. } 0.000143 \div 0.0013 &= \frac{0.000143}{0.0013}\\ &= \frac{1.43 \times 10^{-4}}{1.3 \times 10^{-3}} = \frac{1.43}{1.3} \times \frac{10^{-4}}{10^{-3}}\\ &= 1.1 \times 10^{-4-(-3)} = 1.1 \times 10^{-4+3}\\ &= 1.1 \times 10^{-1}\\ &= 0.11\end{aligned}$$

In computations with scientific notation, it is best to perform separately the multiplication or division of the numbers between 1 and 10, and then separately perform the multiplication or division of the powers of 10.

EXERCISES FOR SECTION 8.8

For Exercises 1–16, write each number in scientific notation.

1. 160
2. 4100
3. 235,000
4. 168,200
5. 3,400,000
6. 86,100,000
7. 937.4
8. 234.21
9. 0.123
10. 0.0086
11. 0.0002345
12. 0.0000008
13. 0.0012
14. 0.00301
15. 0.00000009
16. 0.000209

17. The Earth is approximately 4,000,000,000 years old. Write this number in scientific notation.

18. An electron is approximately 0.0000000000001 centimeter in diameter. Write this number in scientific notation.

19. An angstrom is a unit of length and it is 0.0000001 millimeter long. Write this number in scientific notation.

20. The sun is believed to be 4,500,000,000 years old. Write this number in scientific notation.

For Exercises 21–28, write the given expressions in standard form.

21. 4.1×10^{-3}
22. 9.12×10^{4}
23. 3.142×10^{5}
24. 3.17×10^{-5}
25. 4.859×10^{7}
26. 5.4321×10^{-6}
27. 4.8×10^{-5}
28. 9.8×10^{-7}

For Exercises 29–49, use scientific notation to evaluate each of the given expressions. Answers may be left in scientific notation.

29. $(3.4 \times 10^{4}) \times (2.6 \times 10^{3})$
30. $(1.5 \times 10^{-2}) \times (4.7 \times 10^{-5})$
31. 1800×180
32. 0.000301×0.003
33. $60{,}500 \times 0.0034$
34. $75{,}000 \times 1300$
35. 0.0058×750
36. $76{,}000 \times 120{,}000$
37. $\dfrac{4.2 \times 10^{6}}{1.4 \times 10^{3}}$
38. $\dfrac{6.25 \times 10^{7}}{2.5 \times 10^{2}}$
39. $\dfrac{450{,}000}{9000}$
40. $\dfrac{105{,}000{,}000}{21{,}000}$
41. $\dfrac{255{,}000{,}000}{170{,}000}$
42. $\dfrac{0.000143}{0.0013}$
43. $\dfrac{0.0204}{0.00012}$
44. $\dfrac{10^{5} \times 10^{-6} \times 10^{2}}{10^{-3} \times 10^{-5}}$
45. $\dfrac{10^{1} \times 10^{4} \times 10^{-5}}{10^{3} \times 10^{-2} \times 10^{-3} \times 10^{4}}$

★46. $\dfrac{0.0018 \times 12000}{0.000027 \times 0.002}$
★47. $\dfrac{24{,}000 \times 140{,}000}{210{,}000 \times 0.0016}$
★48. $\dfrac{320{,}000 \times 18{,}000}{0.0012 \times 0.000048}$
★49. $\dfrac{0.00036 \times 0.024}{0.00012 \times 0.0015}$

50. Loose snow weighs approximately 0.125 gram per cubic centimeter. Approximately how much does 1,000,000,000 cubic centimeters of loose snow weigh?

JUST FOR FUN

Four men, in different colored overcoats, sat on two facing bench seats in a train. Two sat next to the window and two next to the aisle. The Englishman sat on Mr. Bell's left. Alto wore a tan-colored coat. The man in olive was on the German's right. Mr. Cardinal was the only cigar smoker. Mr. Dunn was across from the American. The Russian was in khaki, and the Englishman stared out the window on his left. Who was the man in the rust-colored coat?

Summary

Numbers, like many other things, can be classified in a variety of ways. The first set of numbers that we discussed in this chapter was the set of natural numbers, that is, the set of numbers {1, 2, 3, 4, ...}. Every natural number is divisible by itself and 1. Some natural numbers are called *prime numbers*. A *prime number* is any natural number greater than 1 that is divisible only by itself and 1. The numbers 2, 3, 5, and 7 are examples of prime numbers. If a natural number is not prime, then it is called a *composite number.* Disregarding order, every composite natural number can be expressed as a product of prime factors in one and only one way.

The *greatest common divisor* (GCD) of a pair of natural numbers is the greatest natural number that divides both of the given numbers. Two numbers whose GCD is 1 are said to be relatively prime. The *least common multiple* (LCM) of two natural numbers is the least natural number that is a multiple of each of the given numbers.

Expanding the set of natural numbers to include zero gives us the set of numbers {0, 1, 2, 3, 4, ...}. This is called the set of *whole numbers.* The set of numbers consisting of the set of whole numbers and their additive inverses is called the set of *integers;* that is, the set of integers is the set {...,−3, −2, −1, 0, 1, 2, 3, ...}. It should be noted that the set of natural numbers and the set of whole numbers are subsets of the integers. We can use the number line to determine the order of the integers. For example, $-1 > -3$ because -1 is to the right of -3 on the number line.

In combining integers, we should remember the following:

1. If we add two positive integers, the sum is a positive integer.
2. If we add two negative integers, the sum is a negative integer.
3. If we add a positive integer and a negative integer, the sum may be positive or negative, or zero.

The product of two positive integers is positive, as is the product of two negative integers. The product of a positive integer and a negative integer is negative. An integer is *even* if it can be expressed in the form $2 \times n$, where n represents an integer. An integer is *odd* if it can be expressed in the form $2 \times n + 1$.

A *rational number* is a number that can be expressed as a quotient of integers, that is, in the form a/b, where a and b are integers and $b \neq 0$. Rational numbers can also be classified as either terminating decimals or repeating nonterminating decimals. Decimals that are nonterminating and nonrepeating are called *irrational numbers.* Some examples of irrational numbers are the following: π, $\sqrt{2}$, $\sqrt{3}$, and 0.1010010001 ... If we take the union of the set of rational numbers and the set of irrational numbers, then we have all of the numbers that belong on the number line. This set of numbers is called the set of *real numbers.*

Scientific notation can be used to write very large or small numbers more compactly. It is the process of writing a number that is greater than or equal to one and less than 10, times an integral power of 10. Scientific notation, therefore, consists of $N \times 10^m$ where $1 \leq N < 10$, and m is an integer.

Vocabulary Check

natural number
rational number
density property
irrational number
Fundamental Theorem of Arithmetic
whole number
complex fraction
greatest common factor
real number
integer
Sieve of Eratosthenes
least common multiple

Review Exercises for Chapter 8

1. Determine whether each number is prime or composite.
   a. 99 b. 97 c. 83
   d. 431 e. 657 f. 10,101

2. Determine the prime factors of each natural number.
   a. 78 b. 111 c. 475
   d. 147 e. 903 f. 1,111

3. Find the greatest common divisor for each pair of numbers.
   a. 30 and 48 b. 42 and 55
   c. 48 and 72 d. 66 and 90
   e. 111 and 231 f. 342 and 612

4. Find the least common multiple for each pair of numbers.
   a. 30 and 48 b. 42 and 55
   c. 48 and 72 d. 66 and 90
   ★e. 111 and 231 ★f. 342 and 612

5. Evaluate each of the following:
   a. $3 + (-7)$ b. $4 + (-5)$
   c. $-8 + 7$ d. $-7 - (-3)$
   e. $13 - (-3)$ f. $-8 - 5$

6. Evaluate each of the following:
   a. $(-3) \times (-2)$
   b. $(-4) \times 4$
   c. $(-2) \times (-3) \times (-4)$
   d. $(-2 + 5) \times (-3)$
   e. $(1 - 5) \times 2$
   f. $(-2 - 6) \times (-2 - 3)$

7. Evaluate each of the following.
   a. $3(4 - 3) + 5$
   b. $(7 \times 6 - 2) \div 10$
   c. $4 \times 3 - 5 + 8 \div 4$
   d. $(8 \times 2 - 4) \div 6 \times 2$
   e. $48 \div [(10 - 6) \times 6]$
   f. $16 \div 4 \times 8 + 10 - 6$

8. Perform the indicated operation.
   a. $\frac{1}{3} + \frac{2}{5}$ b. $\frac{3}{7} + \frac{4}{9}$ c. $\frac{2}{11} + \frac{1}{13}$
   d. $\frac{4}{7} - \frac{1}{3}$ e. $\frac{3}{5} - \frac{1}{2}$ f. $\frac{7}{8} - \frac{2}{5}$

9. Perform the indicated operation.
   a. $\frac{1}{3} \times \frac{2}{5}$ b. $\frac{3}{7} \times \frac{4}{9}$ c. $\frac{2}{11} \times \frac{3}{5}$
   d. $\frac{2}{3} \div \frac{4}{7}$ e. $\frac{3}{5} \div \frac{1}{2}$ f. $\frac{6}{11} \div \frac{3}{5}$

10. Simplify each of the following.
   a. $\dfrac{\frac{3}{8}}{\frac{4}{9}}$ b. $\dfrac{\frac{3}{4}}{5}$ c. $\dfrac{5}{\frac{3}{4}}$
   d. $\dfrac{3 - \frac{1}{2}}{2 + \frac{4}{5}}$ e. $\dfrac{\frac{1}{2} + \frac{3}{4}}{\frac{2}{3} + \frac{1}{6}}$ f. $\dfrac{\frac{3}{8} - \frac{1}{4}}{\frac{2}{5} + \frac{1}{10}}$

11. Express each of the following as a decimal.
   a. $\frac{7}{8}$ b. $\frac{7}{16}$ c. $\frac{2}{11}$
   d. $\frac{13}{37}$ e. $\frac{5}{13}$ f. $\frac{3}{7}$

12. Express each of the following as a quotient of integers.
   a. 0.75 b. 0.213 c. 3.14
   d. $0.\overline{46}$ e. $2.\overline{49}$ f. $4.\overline{123}$

13. Which of the following numbers are rational and which are irrational?
   a. 2.1 b. $2.\overline{1}$
   c. $\sqrt{5}$ d. 3.141141114 . . .
   e. 2.121121112 . . . f. 3.1415926

14. Determine whether each statement is true or false.
   a. Every real number is an irrational number.
   b. Every real number is a rational number.
   c. Every real number is either a rational number or an irrational number.
   d. Every rational number is a real number.
   e. Every irrational number is a real number.
   f. The union of the sets of rational and irrational numbers is the set of real numbers.

15. Answer yes or no to tell whether each of the following numbers is (I) a natural number, (II) a whole number, (III) an integer, (IV) a rational number, (V) an irrational number, (VI) a real number.
   a. 6 b. -1 c. 0
   d. 2.89 e. $1.\overline{34}$ f. π
   g. $\sqrt{16}$ h. $\sqrt{5}$ i. $\frac{2}{3}$

16. Use scientific notation to evaluate each of the given expressions. Answers should be left in scientific notation.
   a. $355{,}000 \times 8{,}000$ b. 0.000065×0.00084
   c. $\dfrac{24{,}000 \times 140{,}000}{210{,}000 \times 0.0016}$ d. $\dfrac{0.024 \times 0.00036}{0.0015 \times 0.00012}$

JUST FOR FUN

The value of π has been computed to hundreds of decimal places by modern computers. The value of π to seven decimal places is

$$\pi = 3.1415926 \ldots$$

The following diagram shows a trick for remembering these digits:

3	1	4	1	5	9	2	6
May	I	have	a	large	container	of	coffee?

How does this trick work?

Chapter Quiz

Determine whether the following statements are true or false.

1. "Janet finished third" is an example of the use of a cardinal number.
2. An example of a prime number is 111.
3. The smallest prime number is 1.
4. An example of a composite number is 97.
5. The greatest common divisor of two natural numbers is the largest natural number that is a multiple of each of the two given numbers.
6. Two numbers whose greatest common divisor (or greatest common factor) is 1 are said to be relatively prime.
7. A prime number cannot be written as the product of natural numbers that are less than itself.
8. The least common multiple of two natural numbers is the smallest natural number that divides a given pair of natural numbers with remainders of zero.
9. Whenever we multiply two negative integers, the product is always positive.
10. If n and k are even integers then $2 \times (n + k + 1)$ is an even integer.
11. A rational number is a number that can be expressed in the form a/b, where a and b are integers and $b \neq 0$.
12. $\dfrac{1 + \frac{1}{2}}{2 - \frac{1}{3}} = \dfrac{10}{9}$
13. $\left(\frac{3}{4} \times \frac{5}{6} - \frac{1}{2}\right) \times 2 = \frac{1}{4}$
14. $0.\overline{142857} = \frac{1}{7}$
15. The set of rational numbers between $\frac{5}{11}$ and $\frac{6}{11}$ is empty.
16. Decimals that are nonterminating are called irrational numbers.
17. Decimals that are nonrepeating are called irrational numbers.
18. The union of the set of rational numbers and the set of irrational numbers is called the set of real numbers.
19. Every irrational number is a real number.
20. Every real number is a rational number.
21. $\dfrac{\frac{1}{2} + \frac{3}{5}}{\frac{7}{8} - \frac{1}{2}} = \dfrac{15}{44}$
22. $-4 - (-4) = -8$
23. The least common multiple of 48 and 72 is 288.

★24. $0.00000065 = 6.5 \times 10^7$

★25. $\dfrac{630{,}000}{0.09} = 7{,}000$

CHAPTER 9

AN INTRODUCTION TO ALGEBRA

AFTER STUDYING THIS CHAPTER YOU WILL BE ABLE TO DO THE FOLLOWING:

1. Find the solution sets of some simple **open sentences.**
2. **Graph** the solution sets of inequalities in one variable on the number line.
3. Solve equations in one variable.
4. Translate word problems into mathematical equations, and solve verbal problems involving one variable.
5. Find at least three solutions for a **linear equation** in two unknowns, and graph a linear equation in two unknowns on the **Cartesian plane.**
6. Graph a **parabola** of the form $y = ax^2 + bx + c$ on the Cartesian plane.
7. Graph an inequality in two variables on the Cartesian plane.
8. Solve a **linear-programming problem** by translating the word problem into mathematical inequalities and equations, and use graphing and corner-point evaluation to determine the values of the variables that will provide the desired maximum or minimum value for a given expression.

★9. Solve any quadratic equation by means of the *quadratic formula.* If $ax^2 + bx + c = 0$, then

$$x = \frac{-b \pm \sqrt{b^2 - 4ac}}{2a}$$

9.1 INTRODUCTION

What is algebra? Basically, algebra is the area of mathematics that generalizes the facts of arithmetic. In arithmetic we encounter expressions such as $3 + 2 = 5$, whereas in algebra we encounter an expression such as $x + 2 = y$. In algebra, letters are used to denote numbers or a certain set of numbers. An arithmetic expression is one such as $4 + 2$, whereas an algebraic expression is one such as $4x + 2$, or $3x + 2y$. The operations in algebra are similar to those in arithmetic; that is, they are addition, subtraction, multiplication, and division.

Letters used in equations are called **variables.** They are symbols that represent an unknown member of a set. The symbols x and y are variables in the expression $x + y = y + x$. In this chapter, we shall examine the basic properties of algebra and its relationship to some of the topics covered in previous chapters.

9.2 OPEN SENTENCES AND THEIR GRAPHS

You may recall that in Chapter 2 we encountered statements like the following.

February has 30 days.

$$3 + 2 = 1$$

Ronald Reagan was President of the United States.

$$5 + 4 + 3 + 2 + 1 = 14$$

Each of these statements is either true or false (but not both). We can tell whether one of these statements is true or false upon reading the statement.

Consider the sentence

$$x + 1 = 3$$

Is this sentence true of false? We cannot answer that question yet. The sentence is neither true nor false until we replace x with a number. Sentences like $x + 1 = 3$, which cannot be classified as true or false until we replace x by a number, are called **open sentences.** Note that any real number can be substituted for x in the open sentence $x + 1 = 3$. The sentence will be true or false depending on the value we substitute for x. Because x represents an unknown number, we can call x a variable.

If we replace x by 4 in the open sentence $x + 1 = 3$, we have $4 + 1 = 3$, which is a false statement. If we replace x by 2, we have $2 + 1 = 3$, which is a true statement. Therefore, a solution to the open sentence $x + 1 = 3$. Because it can be shown that 2 is the only solution to $x + 1 = 3$, we call 2 the *solution set* for the open sentence. The **solution set** of an open sentence is the set of numbers that make the sentence true when they are substituted for x.

If we replace the equal sign in the open sentence $x + 1 = 3$ with the symbol $>$, we have

$$x + 1 > 3$$

an open sentence of *inequality*. The sentence $x + 1 > 3$ is read as "$x + 1$ is greater than 3." Note that if we replace x by 2 in this statement, we have $2 + 1 > 3$, which is not a true statement. Suppose we restrict our replacements for x to the set of integers. What will the solution set be? The integers 3, 4, 5, and so on will make the sentence true. Therefore the solution set is $\{3, 4, 5, \ldots\}$.

Some other examples of open sentences of inequality are

$x > 2$	x is greater than 2
$x < 2$	x is less than 2
$x \geq 2$	x is greater than or equal to 2
$x \leq 2$	x is less than or equal to 2

It is important to realize than an open sentence may have different solution sets, depending on the restrictions placed on the set of permissible replacements, the **replacement set.** For instance, the equation $x + 1 = 0$ has no solution set (the solution is the empty set, $\varnothing$) if we restrict the set of permissible replacements to the set of natural numbers. But if the replacement set is the set of integers, then the solution set for $x + 1 = 0$ is $\{-1\}$.

Consider the following sentence.

$$x + 1 < 5$$

If the replacement set is the set of natural numbers, then the solution set is $\{1, 2, 3\}$. If the replacement set is the set of whole numbers, then the solution set is $\{0, 1, 2, 3\}$. If the replacement set is the set of integers, then the replacement set is $\{\ldots, -2, -1, 0, 1, 2, 3\}$. If the replacement set is the set of real numbers, then the solution set is {all real numbers less than 4}.

HISTORICAL NOTE

The study of algebra began in early Egypt. The first documented occurrence of algebra-type problems appear in a papyrus copied by the Egyptian priest Ahmes (approximately 1600 B.C.). His papyrus was a copy of a book written even earlier, approximately 2200 B.C. It is significant because it illustrates the first recorded use of a "symbol" to represent an unknown quantity. The Egyptians used the word *hau* to represent any unknown quantity in a mathematical problem.

Diophantas (approximately 200 A.D.), a Greek mathematician, was one of the first mathematicians to use symbols to represent unknown quantities. He supposedly used a combination of the first two letters of the Greek word *arithmos*, which means "number."

One of the most famous Arabian mathematicians was Mohammed ibn Musa al-Khowarizmi (approximately 800 A.D.), who wrote a book titled *Al-Jabr wal-Muzabalah*, which contributed greatly to the adoption of Hindu-Arabic numerals by the scholars of that time. It is believed that the title word *Al-Jabr* gave rise to the word *algebra* as we know it today.

EXAMPLE 1 Find the solution set for $x + 5 = 9$, where x is any real number.

SOLUTION There is only one real number that will make this sentence true. Because $4 + 5 = 9$, the solution set is $\{4\}$.

EXAMPLE 2 Find the solution set for $x + 2 \geq 7$, where x is an integer.

SOLUTION The sentence is read as "$x + 2$ is greater than or equal to 7," and the integers that satisfy this sentence are 5, 6, 7, and so on. Hence, the solution set is $\{5, 6, 7, \ldots\}$. Note that 5 is a member of this set because the sentence reads "greater than *or* equal to."

EXAMPLE 3 Find the solution set for $x + 2 < 3$, where x is a natural number.

SOLUTION We want the sum $(x + 2)$ of two natural numbers to be less than 3. There are no natural numbers that we may substitute for x to make this statement true; hence the solution set is the empty set, $\varnothing$.

EXAMPLE 4 Find the solution set of $x + 2 < 3$, where x is an integer.

SOLUTION Now we are considering a different replacement set than in Example 3. There are integers that can be added to 2 to yield a sum less than 3, namely, $0, -1 -2$, and so on. Therefore, the solution set is $\{\ldots, -2, -1, 0\}$.

Once we find the solution set for an open sentence, we can make a "picture" of the solution set. That is, we can **graph** the solution set. In Chapter 8 we introduced the concept of a *number line.* First, we drew a horizontal line, picked a point on the line, and labeled it zero. Next, we marked off equal units to the right and left of zero. We labeled the end points of the intervals to the right of zero with the positive integers, and those to the left of zero with the negative integers. The number line can be extended indefinitely in either direction (see Fig. 1).

Recall that each integer corresponds to a particular point on the number line. Also, each point on the number line corresponds to a real number. The set of real numbers is composed of the set

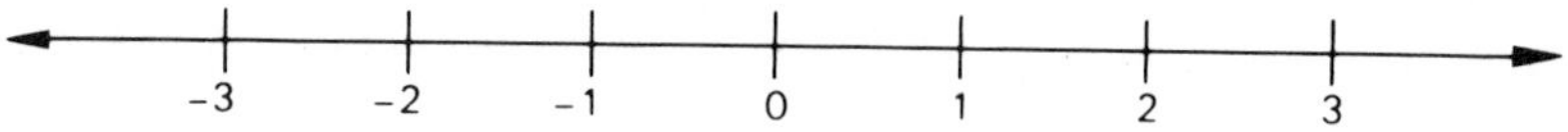

FIGURE 1

of rational numbers and the set of irrational numbers. It is not physically possible to label every point on the number line, but we do know that there exists a point $\frac{3}{4}$ unit from 0, a point -1.5 units from 0, and points $\sqrt{3}$ units and π units from 0. There is a point on the number line that corresponds to every rational or irrational number, and there is a rational or irrational number that corresponds to every point on the number line. Hence we can use the number line to represent the set of real numbers.

How can we picture these numbers on the real number line? We can represent individual numbers on the number line by marking a heavy solid dot on the line at the point that corresponds to that number. For example, Figure 2 is a graph of the number 1. Figure 3 shows the graph of the set of integers $\{-1, 0, 1\}$.

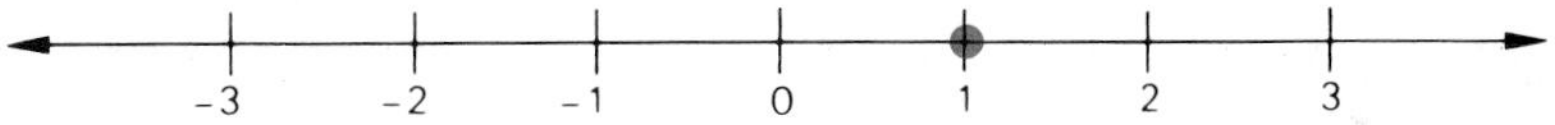

FIGURE 2

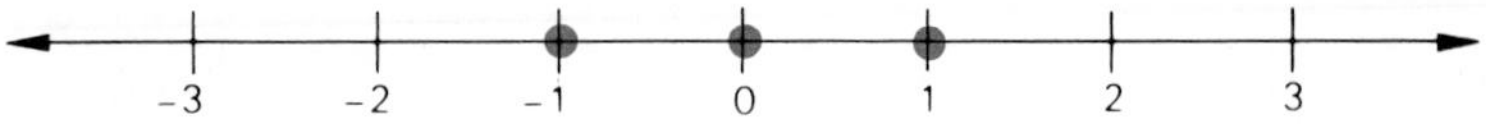

FIGURE 3

Note that we can also call Figure 3 the graph of the integers between -2 and 2. Figure 3 is the correct graph because "between -2 and 2" means that we do not include the integers -2 and 2.

Suppose we want a graph that represents all of the real numbers between -2 and 2. This graph would have to include all of the integers, rational numbers, and irrational numbers between -2 and 2. We indicate this by using hollow dots at -2 and 2 and drawing a solid bar between the hollow dots, as shown in Figure 4.

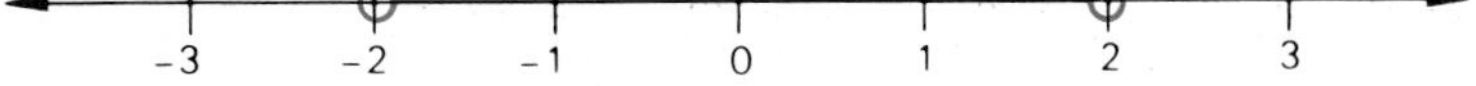

FIGURE 4

The fact that the dots at points -2 and 2 are not solid indicates that these points are not included in our solution set. Speaking in terms of algebra and the set of real numbers, Figure 4 is a graph of all real numbers x such that $-2 < x$ and $x < 2$. We can shorten this expression to $-2 < x < 2$.

Now consider the algebraic sentence $-2 \leq x \leq 2$, read "-2 is less than or equal to x, and x is less than or equal to 2." If we are to represent the solution set of this sentence on the number line,

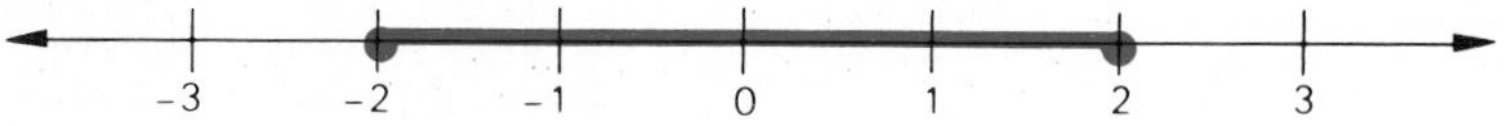

FIGURE 5

where x is any real number, then we must include the end points. Consequently, they would be colored in, as shown in Figure 5.

Next, let's consider the open sentence $x + 2 < 3$, where x is a real number. This statement is true for any real number that is less than 1, that is, $x < 1$. We can represent this on the number line as shown in Figure 6.

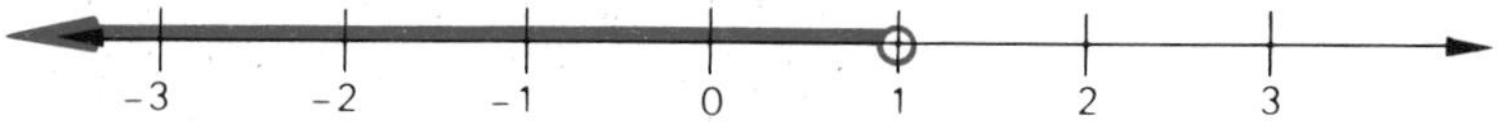

FIGURE 6

The solid arrow indicates that the solution set extends to the left indefinitely. Note that we have $x < 1$ (x is less than 1) and therefore we do not include 1 in the solution. If we had the statement $x \leq 1$ (x is less than or equal to 1), then we would have included 1 in the solution, and 1 would have been marked with a solid dot.

The graph in Figure 6 is sometimes called a **half-line.** If the point at 1 were included, then the graph would be called a **ray.** Figure 4 depicts an **open line segment,** whereas Figure 5 depicts a **line segment.**

EXAMPLE 5 Graph the solution of $x - 2 > 1$, where x is a real number.

SOLUTION In order for the open sentence $x - 2 > 1$ to be true, x must be a number greater than 3, that is, $x > 3$. The solution is the half-line shown in Figure 7.

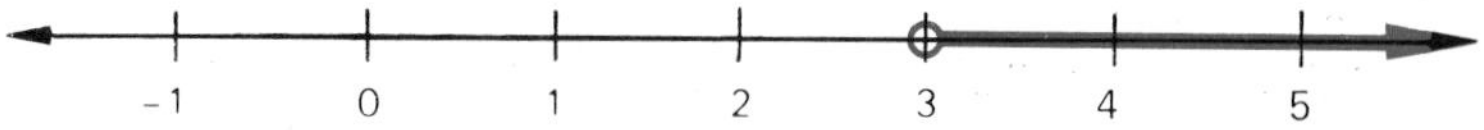

FIGURE 7

EXAMPLE 6 Graph the solution of $x + 2 \leq 1$, where x is a real number.

SOLUTION In order for the open sentence $x + 2 \leq 1$ to be true, x must be a number less than or equal to negative 1, that is, $x \leq -1$. The solution is the ray shown in Figure 8.

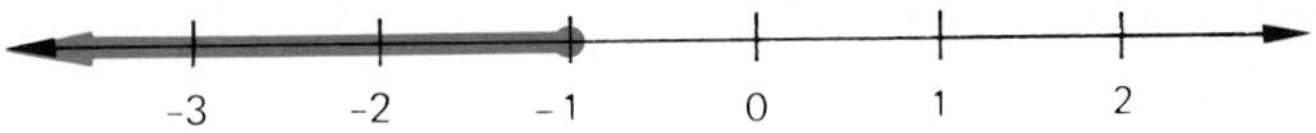

FIGURE 8

EXAMPLE 7 Graph the solution of $x + 2 > x$, where x is a real number.

SOLUTION Note that there are no restrictions on x in order for the open sentence $x + 2 > x$ to be true. Regardless of the number with which we replace x, we have a true statement. For example, $1 + 2 > 1$, $0 + 2 > 0$, and $-2 + 2 > -2$. Therefore, the solution set is the set of all real numbers. The graph of the set of all real numbers is a **line**, as shown in Figure 9.

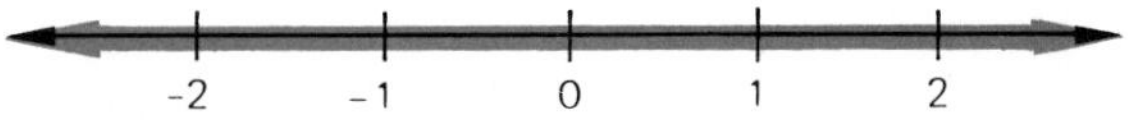

FIGURE 9

EXAMPLE 8 Graph the solution of the open sentence $-1 < x < 3$, where x is a real number.

SOLUTION The sentence $-1 < x < 3$ is read as "-1 is less than x and x is less than 3." This means that a member of the solution set must satisfy both conditions at the same time. That is, if 0 is a member of the solution set, then -1 must be less than it and at the same time it must be less than 3.

The solution set for this sentence is the set of real numbers between -1 and 3. The graph of the solution set is the open line segment shown in Figure 10.

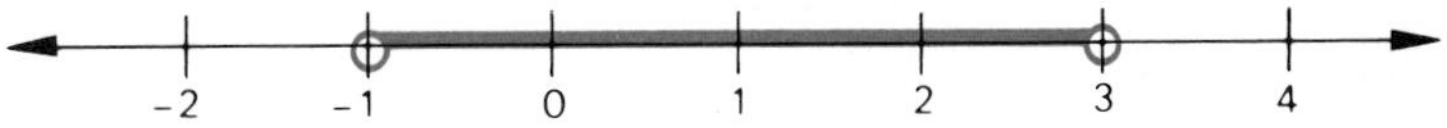

FIGURE 10

In this section we discussed open sentences, that is, sentences that are neither true nor false. A solution of an open sentence is a member of the solution set for the open sentence, and a "picture" of the solution set on the number line is called a graph of the solution set.

EXERCISES FOR SECTION 9.2

For Exercises 1–20, find the solution set for each open sentence. Unless otherwise noted, the replacement set is the set of real numbers.

1. $x + 2 = 5$
2. $x + 1 = 5$
3. $0 = 1 + x$
4. $5 = x + 4$
5. $5 - 2 = 4 + x$ (x is a whole number)
6. $x - 3 = 4 + 1$ (x is a whole number)
7. $x - 17 = 23$
8. $19 = x - 5$
9. $13 = x + 5$
10. $16 = 4 + x$
11. $x + 2 < 7$ (x is a natural number)
12. $x + 3 \leq 5$ (x is a natural number)
13. $x + 4 > 5$
14. $x - 3 > -5$
15. $0 < x < 3$ (x is an integer)
16. $-1 \leq x \leq 3$ (x is an integer)
17. $0 \leq x < 5$ (x is an integer)
18. $0 < x \leq 5$ (x is an integer)
19. $-2 < x \leq 3$ (x is an integer)
20. $-2 \leq x \leq 2$ (x is a whole number)

For Exercises 21–36, graph the solution set for each open sentence on the number line. In each case, the replacement set is the set of real numbers.

21. $x + 3 = 4$
22. $x - 1 = 0$
23. $4 = x - 5$
24. $6 = 7 - x$
25. $5 - x = 2$
26. $3 = 7 - x$
27. $x < 4$
28. $x \geq 2$
29. $x + 2 \geq 0$
30. $x - 2 > 1$
31. $2 < x < 4$
32. $-1 \leq x \leq 3$
33. $-2 < x \leq 1$
34. $-1 \leq x < 1$

★35. $x + 1 > x$

★36. $x + 1 < x$

JUST FOR FUN Can you name 100 different words that do not contain the letters *A*, *B*, *C*, or *Q*? (*Hint:* What is one of the first things you do with numbers?) Time limit: 3 minutes.

9.3 ALGEBRAIC NOTATION

In the preceding section, we examined open sentences such as $x + 3 = 4$ and $x - 2 > 1$. The open sentences that we have encountered up to now have contained expressions involving only x, not $2x$, $3x$, and so on. You should be aware that if we wish to express the product of a number such as 2 and the variable y, we can do this by writing any of the following.

$$(2)(y), \quad (2)y, \quad 2(y), \quad 2 \cdot y, \quad \text{or} \quad 2y$$

Note that we avoid the use of $2 \times y$, because the symbol "×" is easily confused with x, which is reserved for use as a variable. In algebra, the most common method of expressing the product of 2 and y is to write $2y$. Similarly, $5x$ is a method of expressing "5 times x," and $7z$ means "7 times z." But the product of 2 and 3 must *not* be written as 23. We use the above method for expressing a product only when we are working with variables.

If we have an expression such as $2y + 3y$, we can simplify this expression by writing $2y + 3y = 5y$. This can be illustrated in the following manner.

$$2y + 3y = (2 + 3)y = 5y$$

In the second step, the y has been factored out of $2y + 3y$ using the distributive property; we then add $(2 + 3)$ and obtain 5. We can also think of this as combining two like quantities: if we have 2 yaks plus 3 yaks, we will obtain 5 yaks; that is, 2 yaks + 3 yaks = 5 yaks.

An expression such as $5z - 2z$ can also be simplified.

$$5z - 2z = (5 - 2)z = 3z$$

We can also think of this as 5 zebras minus 2 zebras, resulting in 3 zebras; that is, 5 zebras − 2 zebras = 3 zebras.

It should be noted that when we wish to write $1x$, we usually do this by writing x by itself. Similarly, $1y$ is written as y. Therefore, an expression such as

$$3x + x$$

is the indicated sum of $3x$ and $1x$, so that $3x + x = 4x$. Also, $4y - y = 3y$ and $x + x = 2x$.

If we have $3x$ and we wish to double the amount, we can multiply $3x$ by 2, that is,

$$2(3x) = 3x + 3x = 6x$$

We may use the distributive property to perform indicated multiplications, such as $5(2x + 6)$. Basically, the distributive property says that for any numbers a, b, and c, $a(b + c) = ab + ac$. Hence,

$$5(2x + 6) = 5(2x) + 5(6)$$
$$5(2x + 6) = 10x + 30$$

The following are some other examples of indicated multiplications and the resulting products.

$$3(4x + 2) = 12x + 6, \qquad 2(5x - 3) = 10x - 6$$
$$3(4x) = 12x, \qquad 2(6y) = 12y, \qquad 4(2z) = 8z$$
$$9(z) = 9(1z) = 9z, \qquad \tfrac{1}{2}(4y) = 2y$$

The last multiplication example also illustrates another operation. If we multiply a quantity by $\frac{1}{2}$, then this is the same as dividing the quantity by 2. In each of the examples above, we are only working with the numbers, not the variables. For example, $2x + 3x = 5x$, $3y - y = 2y$, and $3(2z) = 6z$. The same holds true for the operation of division. If we have $9x$ and divide that quantity by a number such as 3, the quotient is $3x$.

$$9x \div 3 = \frac{9x}{3} = 3x$$

The following are some other examples of indicated divisions and the resulting quotients:

$$\frac{5y}{5} = y, \qquad \frac{10z}{5} = 2z, \qquad \frac{6x}{2} = 3x$$
$$\frac{3y}{1} = 3y, \qquad \frac{8x}{2} = 4x$$

EXERCISES FOR SECTION 9.3

Simplify each of the following algebraic expressions.

1. $2x + 3x$
2. $4y + 5y$
3. $2x + x$
4. $4x - 2x$
5. $3y - y$
6. $5z - 3z$
7. $2(4x)$
8. $3(2z)$
9. $2(7y)$
10. $8z \div 4$
11. $10x \div 2$
12. $2y \div 1$
13. $\dfrac{4x}{2}$
14. $\dfrac{16z}{4}$
15. $\dfrac{15x}{3}$
16. $x + 3x + 4$
17. $2y + 2 + 3y$
18. $4z - 2z + 3$
19. $8y + 2y + 4$
20. $8x - x + 3$
21. $2z - z + 2$
22. $2x + 3 - x - 2$
23. $4y - 2 - y + 3$
24. $6z + 2 - z - 3$
25. $4(3x) + 2$
26. $6(2x + 3)$
27. $3(x - 2)$
28. $2(4y - 3) + 2y$
29. $5(2z) + 3 - z$
30. $y + 3(4 - y)$
31. $0(3x - 4) + 1$
32. $3(4y) - 4(2y)$
33. $4x + 3x + 5 - 3$
34. $2y - y + 7 + 2$
35. $3(5z) + 2$
36. $(16x + 4) \div 4$

37. $(3x - x) \div 2$

38. $4(2y) + 3(2x)$

39. $4z + 3z - 7 + 5$

40. $2x - x + 7 + 4x$

41. $2(4x) + 3(5x)$

42. $8y + 3y + 2(2y + 4)$

43. $4x - 2 - x + 3(x + 1)$

44. $4(x - 2) + 3$

45. $(3y + 6) \div (2 + 1)$

46. $(8y - 2) \div (5 - 3)$

47. $y + 4(5 - 2y)$

48. $(3x - x)(5 + 3)$

JUST FOR FUN How many triangles are there in the figure?

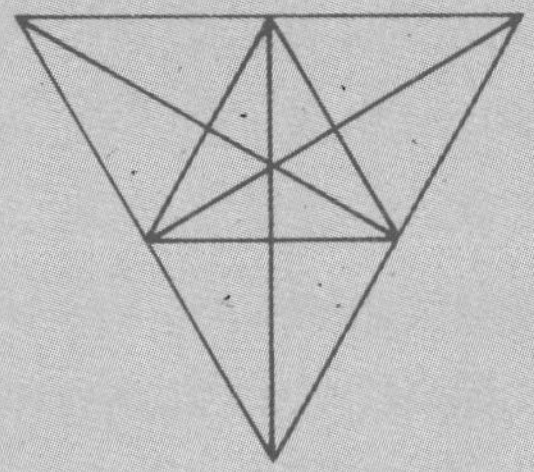

9.4 MORE OPEN SENTENCES

Thus far in our discussion of open sentences, we have only considered sentences such as $x - 2 > 1$ and $x + 2 = 3$. It is not too difficult to find replacements for x such that the given sentences will be true, as we can usually do this by observation or by trial and error. That is, we can replace x by a number and see if the resulting statement is true. But how about a sentence such as the following?

$$4x - 2 = x + 7$$

This open sentence is an equation because it is a statement of equality. We should probably find the solution to the equation $4x - 2 = x + 7$ by trial and error, but there does exist a more efficient method. By using certain techniques, we can systematically solve many equations in one unknown.

Before we generate specific techniques for solving equations, let's make sure that we understand the following axioms, or rules, that we will use in solving equations.

AXIOM 1 If the same quantity or two equal quantities are added to two equal quantities, then the sums are equal.

As an example of axiom 1, we have

$$\begin{array}{rl} x - 1 = 2 & \text{given} \\ +\ 1 = +\ 1 & \text{adding 1 to both sides} \\ \hline x - 1 + 1 = 2 + 1 & \text{sums are equal} \\ x = 3 & \end{array}$$

AXIOM 2 If the same quantity or two equal quantities are subtracted from two equal quantities, then the remainders (differences) are equal.

The following example illustrates axiom 2.

$$\begin{array}{rl} x + 4 = 5 & \text{given} \\ -\ 4 = -\ 4 & \text{subtracting 4 from both sides} \\ \hline x + 4 - 4 = 5 - 4 & \text{remainders are equal} \\ x = 1 & \end{array}$$

AXIOM 3 If equal quantities are multiplied by equal quantities, then the products are equal.

As an example of axiom 3, we have

$$\begin{array}{rl} \frac{x}{2} = 4 & \text{given} \\ \times\ 2 = 2 & \text{multiplying both sides by 2} \\ \hline \frac{x}{2} \cdot 2 = 4 \cdot 2 & \text{products are equal} \\ x = 8 & \end{array}$$

AXIOM 4 If equal quantities are divided by equal quantities (other than zero), then the quotients are equal.

To illustrate axiom 4, we have

$$\begin{array}{rl} 3x = 9 & \text{given} \\ \div\ 3 = 3 & \text{dividing both sides by 3} \\ \hline \frac{3x}{3} = \frac{9}{3} & \text{quotients are equal} \\ x = 3 & \end{array}$$

One thing to keep in mind in solving any equation is that an equation is like a balance scale, and in order to keep the equation in balance, whatever we do to one side of the equation, we must do to the other side of the equation. Our ultimate goal is to get x

by itself on one side of the equal sign. It does not matter whether we wind up with x on the left side or the right side of the equal sign, as long as it is by itself.

Now let's solve the equation $4x - 2 = x + 7$ using the axioms whenever necessary. We must get x by itself on one side of the equation. We will first eliminate the -2 from the left side. We can do this by adding 2 to the left side; but if we add 2 to the left side, then, by axiom 1, we must also add 2 to the right side. Hence we have

$$4x - 2 + 2 = x + 7 + 2$$

which simplifies to

$$4x = x + 9$$

Remember that we must get all of the x's on one side of the equal sign. We can eliminate x from the right side, $x + 9$, by subtracting x from it. But if we subtract x from the right side, we must subtract x from the left side in order to maintain the balance (axiom 2). Therefore we have

$$4x - x = x - x + 9$$

or

$$3x = 9$$

Because we want to solve for x, we must divide the expression $3x$ (3 times x) by 3. Therefore we divide both sides by 3 (axiom 4).

$$\frac{3x}{3} = \frac{9}{3}$$
$$x = 3$$

We can check our answer by replacing x by 3 in the original equation to see if the left side of the equation equals the right side. Replacing x by 3, we have

$$4(3) - 2 \stackrel{?}{=} 3 + 7$$
$$12 - 2 \stackrel{?}{=} 10$$
$$10 = 10$$

The solution checks.

Below is the solution of the equation $4x - 2 = x + 7$ without the discussion. We list the reasons for each step of the solution.

$4x - 2 = x + 7$	given
$4x - 2 + 2 = x + 7 + 2$	axiom 1: adding 2 to both sides
$4x = x + 9$	combining like terms
$4x - x = x - x + 9$	axiom 2: subtracting x from both sides
$3x = 9$	combining like terms
$\frac{3x}{3} = \frac{9}{3}$	axiom 4: dividing both sides by 3
$x = 3$	simplifying

EXAMPLE 1 Solve $4x - 7 = x + 5$ for x.

SOLUTION

$4x - 7 = x + 5$	given
$4x - 7 + 7 = x + 5 + 7$	axiom 1: adding 7 to both sides
$4x = x + 12$	combining like terms
$4x - x = x - x + 12$	axiom 2: subtracting x from both sides
$3x = 12$	combining like terms
$\frac{3x}{3} = \frac{12}{3}$	axiom 4: dividing both sides by 3
$x = 4$	simplifying

EXAMPLE 2 Solve $2y + 6 = 6y - 10$ for y.

SOLUTION

$2y + 6 = 6y - 10$	given
$2y + 6 + 10 = 6y - 10 + 10$	axiom 1: adding 10 to both sides
$2y + 16 = 6y$	combining like terms
$2y - 2y + 16 = 6y - 2y$	axiom 2: subtracting $2y$ from both sides
$16 = 4y$	combining like terms
$\frac{16}{4} = \frac{4y}{4}$	axiom 4: dividing both sides by 4
$4 = y$	simplifying

EXAMPLE 3 Solve $\frac{3z}{4} + 2 = 8$ for z.

SOLUTION

$\frac{3z}{4} + 2 = 8$	given
$\frac{3z}{4} + 2 - 2 = 8 - 2$	axiom 2: subtracting 2 from both sides
$\frac{3z}{4} = 6$	combining like terms
$\frac{3z}{4} \cdot 4 = 6 \cdot 4$	axiom 3: multiplying both sides by 4
$3z = 24$	simplifying
$\frac{3z}{3} = \frac{24}{3}$	axiom 4: dividing both sides by 3
$z = 8$	simplifying

EXAMPLE 4 Solve $2(3x + 2) = 10$ for x.

SOLUTION

$2(3x + 2) = 10$	given
$2 \cdot 3x + 2 \cdot 2 = 10$	distributive property
$6x + 4 = 10$	simplifying
$6x + 4 - 4 = 10 - 4$	axiom 2: subtracting 4 from both sides
$6x = 6$	combining like terms
$\frac{6x}{6} = \frac{6}{6}$	axiom 4: dividing both sides by 6
$x = 1$	simplifying

EXERCISES FOR SECTION 9.4

Solve each of the following equations in the system of real numbers.

1. $x + 3 = 5$
2. $x - 4 = 2$
3. $6 = x - 2$
4. $4 = x + 1$
5. $2y = 12$
6. $3z = 12$
7. $2y - 2 = 10$
8. $3x + 4 = 16$
9. $2x = x + 5$
10. $z - 4 = 2z$
11. $3y = 2y + 4$
12. $3y - 4 = 4y$
13. $2x + 2 = x + 3$
14. $3y - 3 = 2y + 3$
15. $4x + 3 = 2x + 9$
16. $5x - 3 = 2x + 3$
17. $3y - 3 = y + 3$
18. $2y + 6 = 5y - 3$

19. $\frac{x}{2} = 5$
20. $\frac{y}{3} = 12$
21. $\frac{x}{4} + 1 = 15$
22. $\frac{z}{2} - 1 = 5$
23. $\frac{x}{2} = x - 1$
24. $\frac{y}{3} + y = 12$
25. $\frac{z}{4} + z = 5$
26. $\frac{x}{3} = 2 - x$
27. $\frac{x}{2} - 1 = x - 3$
28. $\frac{z}{5} + 2 = 1 - z$
29. $\frac{x}{2} + 3 = x + 2$
30. $\frac{y}{3} - 1 = 2y + 1$
31. $2(3x + 1) = 8$
32. $3(x + 2) = 12$
33. $4(x + 2) = 24$
34. $2(x - 1) = 4$
35. $6(2x - 3) = 6$
36. $7(3x + 1) = 7$
37. $x + 2(3x + 1) = 9$
38. $2x + 3(x - 2) = 4$
39. $12 = 3(2x + 2)$
40. $2(x + 1) = 3(x - 1)$
41. $4(x - 1) = 2(x + 4)$
42. $2(3x - 3) = 3(4x - 2)$

JUST FOR FUN

Ask a friend to perform the steps in the left column while you do the steps in the right column.

1. Pick a number.	1. Let x = the number
2. Subtract 2.	2. $x - 2$
3. Multiply by 4.	3. $4(x - 2) = 4x - 8$
4. Add 10.	4. $4x - 8 + 10 = 4x + 2$
5. Divide by 2.	5. $(4x + 2) \div 2 = 2x + 1$
6. Add 9.	6. $2x + 1 + 9 = 2x + 10$
7. Divide by 2.	7. $(2x + 10) \div 2 = x + 5$
8. Give the result.	8. Give the original number. The result will always be 5 more than the starting number.

9.5 PROBLEM SOLVING

One of the oldest applications of algebra is solving word problems. As early as 2000 B.C., the Egyptians worked on word problems. You can probably solve some word problems by means of simple arithmetic, but most word problems require the use of algebra in order to find the solution in a systematic manner, as opposed to trial and error.

Consider the following sentence.

What number when decreased by 5 equals 15?

This sentence is an example of a word problem. What is the number? We can find the number if we can translate the sentence

HISTORICAL NOTE

Diophantus of Alexandria was an ancient mathematician who contributed greatly to the development of algebra. He produced three different known works in the area of mathematics. The titles of these books are: *Arithmetica*, *Porisms*, and *Polygonal Numbers*. It is thought that all these works dealt with the properties of rational or integral numbers. His most famous work, *Arithmetica*, deals with algebraic equations and certain problems that require positive rational numbers as solutions. In his presentation, Diophantus used special signs and abbreviations, such as the first letter of a word. Therefore a great deal of his work had standardized notation that made it easier to understand. This was one of the first step toward a formalization of algebra and the creation of mathematical notation.

Not much is known about Diophantus other than his mathematical contributions. Historians disagree as to when he lived: perhaps 75 A.D. or 250 A.D. They even disagree on the spelling of his name: Diophantas or Diophantus. Nevertheless he was immortalized because of an epitaph that appeared in a Greek anthology. It gives the following information.

> Diophantas spent one-sixth of his life as a child, one-twelfth as a youth, one-seventh more as a bachelor. In the fifth year of his marriage he had a son. The son died 4 years prior to the death of Diophantas, which made the life span of the son half that of Diophantas. This information gives us the equation
>
> $$\frac{x}{6} + \frac{x}{12} + \frac{x}{7} + 5 + \frac{x}{2} + 4 = x$$
>
> where x equals the age of Diophantas.

English Phrase	Mathematical Phrase
Addition	
4 more than a number	$x + 4$
a number increased by 5	$y + 5$
the sum of x and y	$x + y$
a number added to 3	$3 + n$
x plus y	$x + y$
Subtraction	
a number decreased by 5	$y - 5$
5 less than a number	$n - 5$
the difference between x and y	$x - y$
x minus y	$x - y$
Multiplication	
the product of a and b	$a \cdot b$
x multiplied by a	$a \cdot x$
twice a number	$2 \cdot n$
$\frac{1}{2}$ of y	$\frac{1}{2} \cdot y$
32 percent of z	$(0.32)(z)$
x times y	$x \cdot y$
Division	
the quotient of x and y	$\frac{x}{y}$
the quotient of y and x	$\frac{y}{x}$
the ratio of x and y	$\frac{x}{y}$
x divided by y	$\frac{x}{y}$

into an equation. In translating word problems into algebra, we must pay strict attention to what the words say. For instance, in the given sentence the phrase "what number" is the key phrase; it tells us that we are looking for a certain number. Therefore we let x equal the number. (We could have just as easily let some other variable such as n or y equal the number.) Next we have the phrase "decreased by 5." What does this mean? Exactly what it says; that is, "decreased by 5" is the same as "minus 5," or "−5." The last part of the sentence, "equals 15," can be expressed as "=15." Therefore, the equation for the sentence

What number when decreased by 5 equals 15?

is

$$x - 5 = 15$$

Now we solve the equation.

$$x - 5 = 15$$
$$x - 5 + 5 = 15 + 5$$
$$x = 20$$

As a check, observe that $20 - 5 = 15$.

Whenever we attempt to solve a word problem, it is necessary to translate the given sentence into an equation. The first thing we must do is let the unknown be represented by a variable such as x, y, z, m, or n. Next we must translate the given relationship with the unknown into an equation. The next step is to solve the equation in order to find the value of the unknown.

Mathematical translations of common English phrases are listed in the table on page 457. The underlined phrase is represented in the corresponding mathematical phrase. Note that we can choose a variety of letters to represent the unknown. Normally x is used, but it does not have to be the chosen letter.

The word "is" or some form of the verb "to be" is often used to indicate equality (=). For example, the English sentence, "The sum of a number and 10 is 32," is translated into the mathematical sentence, $x + 10 = 32$. In the following table are some examples of English sentences translated into mathematical sentences.

English Sentence	Mathematical Sentence
A number increased by 2 is 6.	$x + 2 = 6$
The product of a number and 3 is 18.	$3y = 18$
The difference between a number and 3 is 10.	$n - 3 = 10$
Two less than four times a number is 34.	$4z - 2 = 34$
One-half of a number and 6 more is 14.	$\frac{1}{2}n + 6 = 14$
Three more than twice a number is 9.	$2x + 3 = 9$
If a number is increased by 4, twice the sum is 50.	$2(y + 4) = 50$

EXAMPLE 1 Write an equation to illustrate "What number increased by 2 is 5?"

SOLUTION Let x equal the unknown. The phrase "increased by 2" means "plus 2," or "+2." Another word for "is" is "equals." Therefore we have

$$x + 2 = 5$$

EXAMPLE 2 Write an equation to illustrate "The sum of what number and twice that number equals 9?"

SOLUTION Let n equal the unknown. Twice a number can be expressed as $2n$. Therefore we have

$$n + 2n = 9$$

EXAMPLE 3 Write an equation to illustrate "Two less than four times what number is 34?"

SOLUTION Let x equal the unknown. The phrase "two less" indicates that we subtract 2, but from what? We subtract 2 from "four times what number," that is, $4x$. Therefore we have

$$4x - 2 = 34$$

EXAMPLE 4 Write an equation to illustrate "One-half of what number and 5 more is 12?"

SOLUTION Let n equal the unknown. One-half of a number can be expressed as $\frac{1}{2}n$ or $n/2$. The phrase "and 5 more" indicates that we add 5 to $n/2$. Therefore we have

$$\frac{n}{2} + 5 = 12$$

EXAMPLE 5 A rectangular swimming pool is 6 feet longer than twice its width w. If the perimeter of the pool is 120 feet, write an equation to find its dimensions.

SOLUTION This is stated in a little more formal language, but we treat it the same as the other examples. We are already given that the width is w, so the length l can be expressed as $2w + 6$. Using the fact that the perimeter of a rectangle is $2w + 2l$, we express the perimeter as

$$2w + 2(2w + 6) = 120$$

EXAMPLE 6 The sum of two consecutive odd integers is 40. Write an equation for this problem.

SOLUTION Let x equal the first odd integer. The next (consecutive) odd integer is $x + 2$. (It cannot be $x + 1$, because if x is odd, then

$x + 1$ would be even.) Therefore the sum of the two consecutive odd integers is $x + (x + 2)$. Hence we have

$$x + (x + 2) = 40$$

Following is a list of suggestions that will aid us in solving word or verbal problems.

1. Read the problem carefully.
2. Reread the problem carefully.
3. If possible, draw a diagram to assist in interpreting the given information.
4. Translate the English phrases into mathematical phrases and choose a variable for the unknown quantity.
5. Write the equation using all of the above information.
6. Solve the equation.
7. Check the solution to determine whether it satisfies the original problem (not the equation).

Now that we have had some practice in translating verbal problems into algebraic expressions, let us complete the process and solve the problem. Consider the following problem.

The sum of two consecutive integers is 99. Find the value of the smaller integer.

The first thing we must do is represent the unknown in terms of a variable. Let x equal the first integer; then $x + 1$ will equal the next consecutive integer. Translating the given relationship with the unknown into an equation, we have

$$x + (x + 1) = 99$$

Next we solve the equation.

$$2x + 1 = 99 \qquad \text{combining like terms}$$

$$2x + 1 - 1 = 99 - 1 \qquad \text{subtracting 1 from both sides}$$

$$2x = 98$$

$$\frac{2x}{2} = \frac{98}{2} \qquad \text{dividing both sides by 2}$$

$$x = 49 \qquad \text{(the solution)}$$

We can check our solution to see if it is correct. If the first integer is 49, then the next consecutive integer is $x + 1$, that is, $49 + 1$, or 50. Is the sum of 49 and 50 equal to 99? Yes, $49 + 50$ does equal 99. Our answer checks, so the solution is correct.

EXAMPLE 7 Scott is 2 years older than Joe and the sum of their ages is 18. What are the ages of Scott and Joe?

SOLUTION Let x = Joe's age; then $x + 2$ = Scott's age. The sum of their ages is 18. Therefore we have the equation

$$x + x + 2 = 18$$

Solving the equation, we have

$2x + 2 = 18$	combining like terms
$2x + 2 - 2 = 18 - 2$	subtracting 2 from both sides
$2x = 16$	
$\frac{2x}{2} = \frac{16}{2}$	dividing both sides by 2
$x = 8$	Joe's age
$x + 2 = 8 + 2 = 10$	Scott's age

EXAMPLE 8 Two less than four times what number is 34?

SOLUTION Let x = the number. Two less than four times x is 34 gives us the equation

$$4x - 2 = 34$$

Next, we solve the equation.

$4x - 2 + 2 = 34 + 2$	adding 2 to both sides
$4x = 36$	
$\frac{4x}{4} = \frac{36}{4}$	dividing both sides by 4
$x = 9$	(the solution)

EXAMPLE 9 The width of a rectangle is 2 feet less than its length. If the perimeter is 32 feet, find the dimensions of the rectangle.

SOLUTION Let x = the length of the rectangle; then $x - 2$ = the width of the rectangle. Because the perimeter is equal to the sum of the lengths of all the sides, we have the equation

$$x + x - 2 + x + x - 2 = 32$$

Solving the equation,

$4x - 4 = 32$	combining like terms
$4x - 4 + 4 = 32 + 4$	adding 4 to both sides
$4x = 36$	
$\frac{4x}{4} = \frac{36}{4}$	dividing both sides by 4
$x = 9$	length of the rectangle
$x - 2 = 9 - 2 = 7$	width of the rectangle

EXAMPLE 10 The sum of two angles of a triangle is 90 degrees. If one angle is 10 degrees more than three times the smaller angle, find the angles.

SOLUTION Let y = the smaller angle; then $3y + 10$ = the larger angle. Because the sum of the two angles is 90 degrees, we have the equation

$$y + 3y + 10 = 90$$

Solving the equation,

$4y + 10 = 90$	combining like terms
$4y + 10 - 10 = 90 - 10$	subtracting 10 from both sides
$4y = 80$	
$\frac{4y}{4} = \frac{80}{4}$	dividing both sides by 4
$y = 20$	the smaller angle
$3y + 10 = 3 \cdot 20 + 10 = 60 + 10 = 70$	the larger angle

EXAMPLE 11 Pam has 85 cents in her change purse. If there are only nickels and dimes in her change purse, and she has 13 coins altogether, how many nickels and dimes does she have?

SOLUTION Let x = number of dimes; then $13 - x$ = the number of nickels. Now if Pam has x dimes in her purse, then the value of these dimes in cents is $10x$. For instance, if she had three dimes, then the value is $10 \cdot 3$, or 30 cents. Similarly, the value of the nickels in Pam's purse is $5(13 - x)$. The total value of the coins in her purse is 85 cents, and hence we have the equation

$$10x + 5(13 - x) = 85$$

Solving the equation,

$10x + 65 - 5x = 85$	distributive property
$5x + 65 = 85$	combining like terms
$5x + 65 - 65 = 85 - 65$	subtracting 65 from both sides
$5x = 20$	
$\frac{5x}{5} = \frac{20}{5}$	dividing both sides by 5
$x = 4$	number of dimes
$13 - x = 13 - 4 = 9$	number of nickels

Check: $x = 4$, and $13 - x = 9$, so it checks that Pam has 13 coins. But we must also check the value of the coins; that is, $4 \cdot 10 = 40$ (the value of the dimes), $9 \cdot 5 = 45$ (the value of the nickels), and $40 + 45 = 85$ cents (the total value of the coins). Our solution is verified.

EXAMPLE 12 The sum of three consecutive even integers is 30. Find the integers.

SOLUTION Let z = the first even integer; then the next consecutive even integer is $z + 2$. The third consecutive even integer is 2 more than the second, that is, $z + 2 + 2$, or $z + 4$. Hence, we have

$$z = \text{first even integer}$$
$$z + 2 = \text{second consecutive even integer}$$
$$z + 4 = \text{third consecutive even integer}$$

As the sum of these integers is 30, we have the equation

$$z + (z + 2) + (z + 4) = 30$$

Solving the equation,

$3z + 6 = 30$	combining like terms
$3z + 6 - 6 = 30 - 6$	subtracting 6 from both sides
$3z = 24$	
$\frac{3z}{3} = \frac{24}{3}$	dividing both sides by 3
$z = 8$	the first even integer

HISTORICAL NOTE

We mentioned earlier that the first algebra-type problems occurred in a papyrus copied by Ahmes. This document also contained one of the oldest verbal problems in the world. It was stated as "Hau, its whole, its seventh, it makes 19," which roughly translates to "Find a number such that if the whole number is added to one-seventh of it, then the result is 19." This problem was considered quite difficult by the Egyptians and its solution was complicated. If we translate the English phrases into mathematical phrases, we can obtain an equation and determine the solution to the problem. The desired equation is

$$y + \frac{y}{7} = 19$$

$z + 2 = 10$ the second consecutive even integer

$z + 4 = 12$ the third consecutive even integer

Check: The three consecutive even integers are 8, 10, and 12. Their sum is 8 + 10 + 12, or 30. This satisfies the original problem and shows that the solution is correct.

EXERCISES FOR SECTION 9.5

Solve each of the following problems. Only an algebraic solution will be accepted.

1. Three more than a certain number is 10. Find the number.
2. Four less than a certain number is 10. Find the number.
3. The sum of a certain number and 5 is 12. Find the number.
4. Two more than twice a certain number is 6. Find the number.
5. The sum of two consecutive integers is 15. Find the numbers.
6. Mary is 5 years older than Tom and the sum of their ages is 51. How old is each?
7. Julia is 7 years younger than Lewis and the sum of their ages is 53. How old is each?
8. Seven less than four times a number is 41. Find the number.
9. The sum of five times a number and 8 is 63. Find the number.
10. The sum of one-half of a number and 5 is 12. Find the number.
11. The width of a rectangle is 3 feet less than its length. If the perimeter of the rectangle is 50 feet, find the dimensions of the rectangle.
12. The length of a rectangle is 4 meters longer than its width. If the perimeter of the rectangle is 60 meters, find the dimensions of the rectangle.
13. The length of a rectangle is 2 meters longer than twice its width. If the perimeter of the rectangle is 100 meters, find the dimensions of the rectangle.
14. The sum of two consecutive integers is 101. Find the numbers.
15. The sum of two consecutive odd integers is 40. Find the numbers.
16. The sum of two consecutive even integers is 38. Find the numbers.
17. The sum of three consecutive integers is 93. Find the numbers.
18. The sum of three consecutive odd integers is 123. Find the numbers.
19. David has \$2.25 in dimes and quarters in his pocket. If he has twice as many dimes as quarters, how many of each type of coin does he have?
20. Daniel, a newspaper carrier, has \$2.90 in nickels, dimes, and quarters. If he has three more nickels than dimes and twice as many dimes as quarters, how many of each type of coin does he have?
21. In a collection of 60 coins the number of quarters is one-third the number of dimes and the number of nickels is 10 less than twice the number of dimes. How many of each type of coin are in the collection?
22. A rectangular garden is enclosed by 460 feet of fencing. If the length of the garden is 10 feet less than three times the width, find the dimensions of the garden.
23. Joe emptied his bank, which contained only nickels, dimes, and quarters. Joe discovered that he had the same number of quarters and dimes and

five more nickels than quarters. If the total value of the coins was \$4.25, how many of each type of coin was in the bank?

24. The width of a rectangle is 3 inches more than one-half of its length. If the perimeter is 60 inches, find the length and width of the rectangle.

25. Two angles of a triangle are equal and the third angle is 20 degrees less than twice one of the equal angles. Find the number of degrees in each angle of the triangle. (*Hint:* The sum of the angles of a triangle is 180 degrees.)

26. One number is 20 more than another. If the greater number is increased by 4, the result is five times the smaller. Find the two numbers.

27. A collection of 30 quarters and dimes amounts to \$5.55. How many of each kind of coin are there?

28. The perimeter of a triangle is 17 centimeters. One side is 3 centimeters longer than the shortest side, whereas the third side is 2 centimeters shorter than twice the shortest side. Find the lengths of the three sides.

29. A square and an equilateral triangle have the same perimeter. Each side of the triangle is 8 meters. Find the length of each side of the square.

30. Seventy-seven mathematics students are separated into two groups. The first group is 4 less than twice the second group. How many students are in each group?

JUST FOR FUN

Riddles and puzzles have also helped to develop the study of word problems. Many ancient texts or manuscripts contained sets of various types of riddles. The following is a very old children's rhyme from England.

As I was going to St. Ives,
I met a man with seven wives,
Every wife had seven sacks,
Every sack had seven cats,
Every cat had seven kits,
Kits, cats, sacks, and wives—
How many were going to St. Ives?

9.6 LINEAR EQUATIONS IN TWO VARIABLES

Thus far in our discussion of open sentences and equations, we have dealt with sentences that contained only one unknown, or variable. But these are not the only type of open sentences that exist. Consider the equation

$$x + 2y = 10$$

This is an equation in two unknowns, or two variables. If we replace x by 2, that is,

$$2 + 2y = 10$$

we still have an open sentence, but now we can solve the equation $2 + 2y = 10$ for y. When $x = 2$ and $y = 4$, the open sentence $x + 2y = 10$ is true. This means that $x = 2$ and $y = 4$ is a *solution* to the equation $x + 2y = 10$.

Rather than write $x = 2$ and $y = 4$, we can denote this in another manner, namely, by means of the ordered pair (2, 4). We say *ordered pair* because the order in which the numbers appear is important. By convention, the first number in an ordered pair is always a value for x, and the second number is always a value for y. We know that the ordered pair (2, 4) is a solution to the equation $x + 2y = 10$: when we replace x by 2 and y by 4, the resulting statement is true; that is, $2 + 2(4) = 10$, or $2 + 8 = 10$. Note that the ordered pair (4, 2) is not a solution to the equation $x + 2y = 10$; that is, $4 + 2(2) \neq 10$.

An ordered pair is always of the form (x, y); that is, the x value is listed first and the y value is listed second. The x value is formally called the **abscissa** and the y value is called the **ordinate,** but in our discussion we shall refer to them as the x and y values.

It should be noted that the ordered pair (2, 4) is not the only ordered pair that will make the open sentence $x + 2y = 10$ a true statement; it is not the only solution for the given equation. In fact, there are infinitely many ordered pairs in the solution set of the equation $x + 2y = 10$.

Any equation of the form

$$Ax + By = C$$

where A, B, and C are real numbers, is called a **linear equation.** Thus $x + 2y = 10$ is a linear equation. It is called a linear equation because when we graph such an equation in the Cartesian plane, we get a straight line.

We mentioned before that there are infinitely many ordered pairs in the solution set of the equation $x + 2y = 10$. We already have the ordered pair (2, 4); now let's obtain some more ordered pairs that are in the solution set. One way to do this is to let x be any value we choose and then solve for y. Suppose $x = 0$. Then we have

$$x + 2y = 10$$

$$\text{let } x = 0\text{:} \quad 0 + 2y = 10$$

$$\frac{2y}{2} = \frac{10}{2} \qquad \text{dividing both sides by 2}$$

$$y = 5$$

Hence when $x = 0$ and $y = 5$, we have another solution to the equation $x + 2y = 10$. We can therefore say that the ordered pair (0, 5) is a solution.

To find another solution for the equation $x + 2y = 10$, let $x = 4$. Then we have

$$x + 2y = 10$$

let $x = 4$: $$4 + 2y = 10$$

$$4 + 2y - 4 = 10 - 4 \quad \text{subtracting 4 from both sides}$$

$$2y = 6$$

$$\frac{2y}{2} = \frac{6}{2} \quad \text{dividing both sides by 2}$$

$$y = 3$$

Hence (4, 3) is a solution.

Thus far, we have three ordered pairs that satisfy the linear equation $x + 2y = 10$, namely, (2, 4), (0, 5), and (4, 3).

In order to find the solutions for any linear equation, select a value for x and substitute it for x in the equation; then solve the resulting equation for y. We can also find a solution for a linear equation by selecting a value for y, substituting it for y in the equation, and then solving the resulting equation for x.

EXAMPLE 1 Find three solutions for $3x + 4y = 15$.

SOLUTION a. We let $x = 1$. Then we have

$$3x + 4y = 15$$

let $x = 1$: $$3(1) + 4y = 15$$

$$3 + 4y = 15$$

$$3 + 4y - 3 = 15 - 3 \quad \text{subtracting 3 from both sides}$$

$$4y = 12$$

$$\frac{4y}{4} = \frac{12}{4} \quad \text{dividing both sides by 4}$$

$$y = 3$$

Therefore (1, 3) is a solution.

b. Let $x = 5$; then

$$3(5) + 4y = 15$$

$$15 + 4y = 15$$

$$15 + 4y - 15 = 15 - 15 \qquad \text{subtracting 15 from both sides}$$
$$4y = 0$$
$$\frac{4y}{4} = \frac{0}{4} \qquad \text{dividing both sides by 4}$$
$$y = 0$$

Therefore (5, 0) is a solution.

c. Let $x = 0$; then

$$3(0) + 4y = 15$$
$$0 + 4y = 15$$
$$4y = 15$$
$$\frac{4y}{4} = \frac{15}{4} \qquad \text{dividing both sides by 4}$$
$$y = \frac{15}{4}$$

Therefore $(0, \frac{15}{4})$ is a solution.

The third solution obtained in Example 1, $(0, \frac{15}{4})$, contains a fraction. However, as we shall see in the next section, it is advantageous to obtain integral solutions, that is, solutions that contain only integers. We can avoid obtaining a fractional solution if we are careful about the values that we choose for x. For instance, in example 1, if $x = -3$, than $y = 6$, and $(-3, 6)$ is a solution for $3x + 4y = 15$.

EXAMPLE 2 Find three solutions for $2x - 3y = 12$.

SOLUTION a. Selecting a value for x, we let $x = 0$. Hence we have

$$2x - 3y = 12$$
$$\text{let } x = 0: \quad 2(0) - 3y = 12$$
$$0 - 3y = 12$$
$$-3y = 12$$
$$\frac{-3y}{-3} = \frac{12}{-3} \qquad \text{dividing both sides by } -3$$
$$y = -4$$

Therefore $(0, -4)$ is a solution.

b. Let $x = 3$; then

$$\begin{aligned} 2(3) - 3y &= 12 \\ 6 - 3y &= 12 \\ 6 - 3y - 6 &= 12 - 6 && \text{subtracting 6 from both sides} \\ -3y &= 6 \\ \frac{-3y}{-3} &= \frac{6}{-3} && \text{dividing both sides by } -3 \\ y &= -2 \end{aligned}$$

Therefore $(3, -2)$ is a solution.

c. Let us try a different approach, selecting a value for y. Let $y = 0$; then

$$\begin{aligned} 2x - 3(0) &= 12 \\ 2x - 0 &= 12 \\ 2x &= 12 \\ \frac{2x}{2} &= \frac{12}{2} && \text{dividing both sides by 2} \\ x &= 6 \end{aligned}$$

Therefore $(6, 0)$ is a solution. Note that even though we have selected a value for y first, we list the x value first and the y value second when we write the solution as an ordered pair.

Instead of listing the solutions to the equation $2x - 3y = 12$ in Example 2 as the ordered pairs $(0, -4)$, $(3, -2)$, and $(6, 0)$, we can list the solutions as a **table of values,** as shown below.

$2x - 3y = 12$

x	y
0	−4
3	−2
6	0

EXERCISES FOR SECTION 9.6

Find three solutions for each of the following equations.

1. $x + y = 5$
2. $x + y = 8$
3. $x - y = 3$
4. $x - y = 1$
5. $2x + y = -6$
6. $2x - y = 3$
7. $3x - y = -10$
8. $2x + 3y = -6$
9. $3x - 2y = 8$
10. $5x + 3y = 15$

11. $3x + 5y = -15$
12. $x + y = 0$
13. $x - y = 0$
14. $x - 2y = 0$
15. $x + 2y = 0$
16. $x + 3y = 13$
17. $2x - 3y = -11$
18. $3x - 5y = -9$
19. $-2x - 3y = 7$
20. $-3x - 2y = -3$
21. $-5x - 4y = -12$

JUST FOR FUN In the "old, old days" students had to know the following kinds of measure: the tierce, hogshead, pipe, butt, and tun. These are similar kinds of measure. What do they measure?

9.7 GRAPHING EQUATIONS

Now that we are able to find solution sets for linear equations, our next goal is to illustrate these solution sets by means of a graph. We can do this on a grid called the **Cartesian plane,** named after the French mathematician-philosopher René Descartes.

In order to develop an understanding of the Cartesian plane and how to graph ordered pairs on it, consider the map of Anytown in Figure 11. Assume that we are at the intersection of Main Street (east–west) and Euclid Avenue (north–south). In order to get to the town hall, we must go three blocks east and then two blocks north; to get to the school, we must go two blocks east and then three blocks south. Similarly, to get to the hospital, we must go three blocks west and then three blocks north, and to get to the library we must go one block west and then three blocks south.

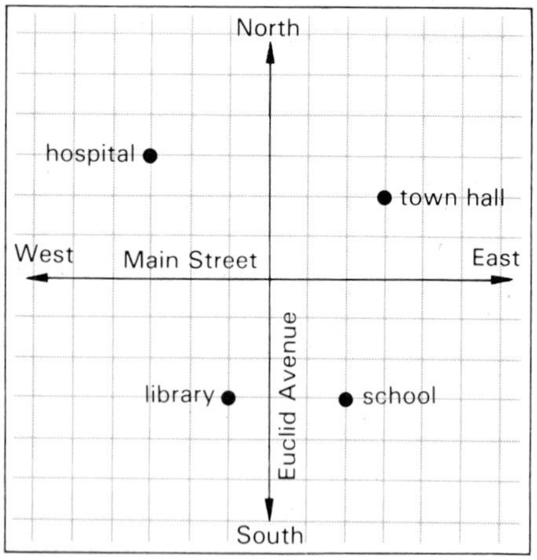

FIGURE 11

We can use this same idea to graph, or **plot,** ordered pairs. In Chapter 8, we used a horizontal number line like the line representing the east–west Main Street in Figure 11. Recall that zero was in the middle of the line, the positive numbers were to the right (east) of zero, and the negative numbers were to the left (west) of zero. We shall use such a horizontal line and call it the

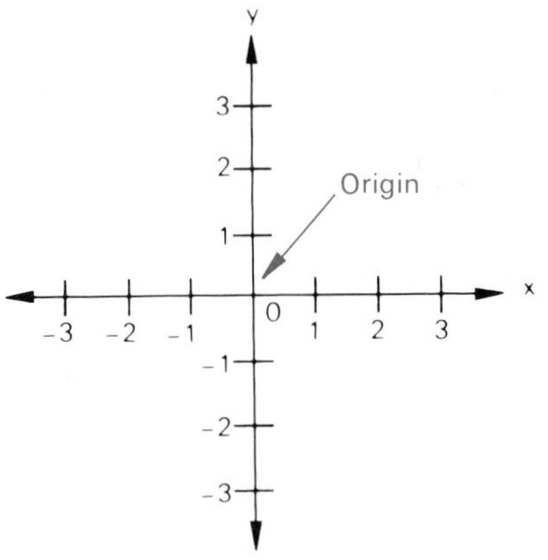

FIGURE 12

x-number line, or the **x-axis.** Now we construct another number line perpendicular to the x-axis (like the north–south street Euclid Avenue in Fig. 11) and passing through zero. We mark off the numbers on this new line as we did for the x-number line. The numbers above the x-axis (north) are positive, and those below the x-axis (south) are negative. This vertical number line is called the **y-number line,** or **y-axis.** The Cartesian plane is based on the x- and y-axes. They intersect at the common point where $x = 0$ and $y = 0$. This particular point is represented by the ordered pair $(0, 0)$ and is called the **origin** (see Fig. 12).

Instead of giving directions to the town hall in Figure 11 by saying "go three blocks east and then two blocks north," we can now say "go 3 units in the positive x-direction, and then 2 units in the positive y-direction." We can shorten this even more by writing the ordered pair $(3, 2)$. If we are asked to plot the ordered pair $(3, 2)$ on the Cartesian plane, we start at the origin and move 3 units in the positive x-direction, and then 2 units in the positive y-direction (see Fig. 13).

Using this idea, we can graph the ordered pairs $(2, -3)$, $(-3, 3)$, and $(-1, -3)$ on the Cartesian plane shown in Figure 13. In order to graph an ordered pair, we always start at the origin $(0, 0)$ and proceed from there. The point represented by $(2, -3)$ is obtained by moving 2 units in a positive x-direction and then 3 units in a negative y-direction. The point represented by $(-3, 3)$ is obtained by starting at the origin and moving 3 units in a negative x-direction and then 3 units in a positive y-direction. To plot the point represented by $(-1, -3)$, we start at the origin and move 1 unit in a negative x-direction and then 3 units in a negative y-direction.

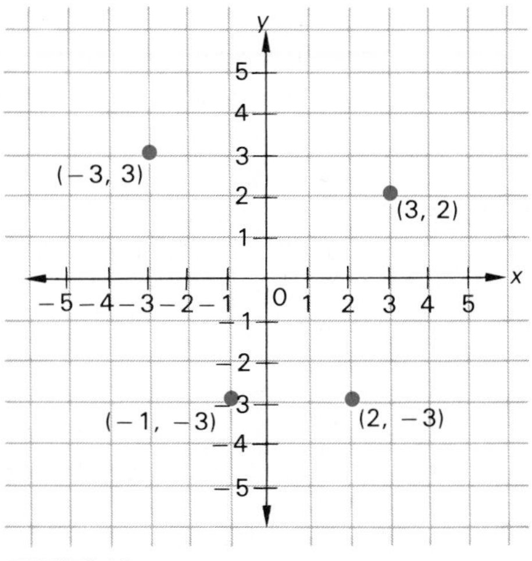

FIGURE 13

In order to obtain any point on the plane, we start at the origin and then move in a positive or negative x-direction (east–west, right–left); next we move in a positive or negative y-direction (north–south, up–down). The first number in an ordered pair is the x-value and the second number is the y-value. These two numbers are called the **coordinates** of the point; hence the Cartesian plane is also called the **Cartesian coordinate system.**

EXAMPLE 1 Locate the points corresponding to the ordered pairs $A(1, 2)$ and $B(2, -1)$.

SOLUTION The given ordered pairs are labeled A and B to aid us in our discussion. Point A, corresponding to the ordered pair $(1, 2)$, is found by moving 1 unit in a positive x-direction from the origin and then 2 units in a positive y-direction. Point B, corresponding to $(2, -1)$, is found by moving 2 units in a positive x-direction from the origin and then 1 unit in a negative y-direction. Points A and B are shown in Figure 14.

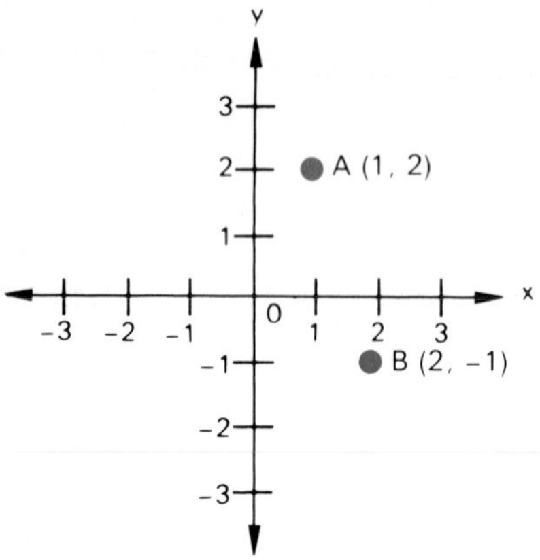

FIGURE 14

EXAMPLE 2 Locate the points corresponding to the ordered pairs $C(-3, 2)$ and $D(-2, -3)$.

SOLUTION Point C, $(-3, 2)$, is found by moving 3 units in a negative x-direction from the origin and then 2 units in a positive y-direction. Point D, $(-2, -3)$, is found by moving 2 units in a negative x-direction from the origin and then 3 units in a negative y-direction (see Fig. 15).

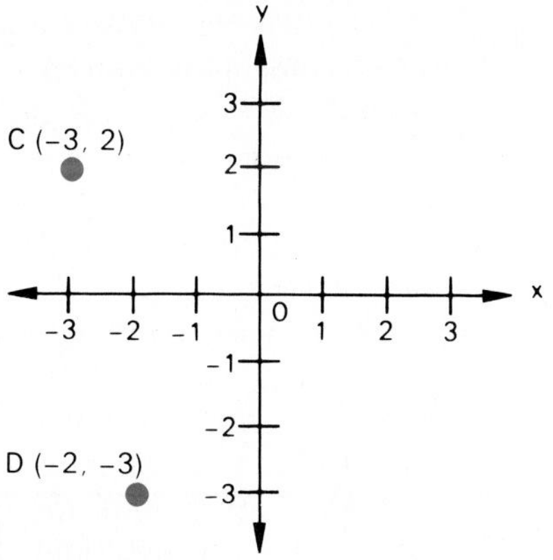

FIGURE 15

EXAMPLE 3 Locate the points corresponding to the ordered pairs $A(0, 2)$, $B(3, 0)$, $C(0, -1)$, $D(-2, 0)$, and $E(0, 0)$.

SOLUTION Each of the given ordered pairs contains 0 as one of its values. Point $A(0, 2)$ is found by moving 0 units in the x-direction from the origin and then 2 units in a positive y-direction. Point $B(3, 0)$ is found by moving 3 units in a positive x-direction from the origin and 0 units in the y-direction.

Point $C(0, -1)$ is found by moving 0 units in the x-direction from the origin and then 1 unit in the negative y-direction. Point $D(-2, 0)$ is found by moving 2 units in a negative x-direction from the origin and then 0 units in the y-direction. Point $E(0, 0)$ is the origin.

Points A, B, C, D, and E are shown in Figure 16.

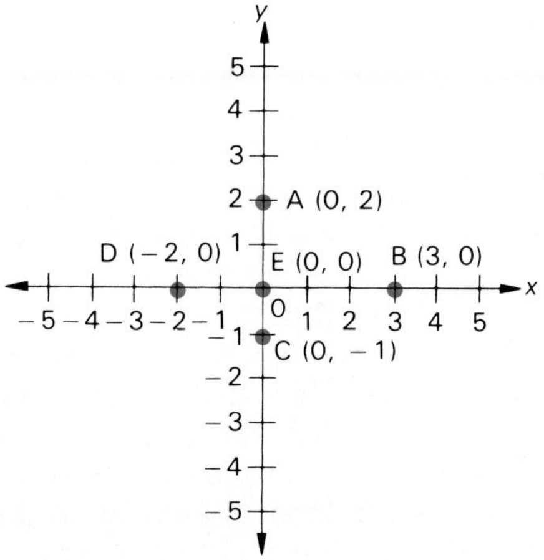

FIGURE 16

Because each ordered pair represents a point on the plane, let us now locate some ordered pairs that are solutions to the linear equations that we discussed in the previous section.

Consider the equation

$$x + 2y = 10$$

Three ordered pairs that satisfy this equation are (4, 3), (2, 4), and (0, 5). We can plot all of these ordered pairs in the Cartesian plane, as shown in Figure 17. Note that these points tend to form a pattern; that is, they appear to lie on the same path. If we connect these points, we see that all of these points do in fact lie on the same straight line. This line, shown in Figure 18, is a graph of the equation $x + 2y = 10$.

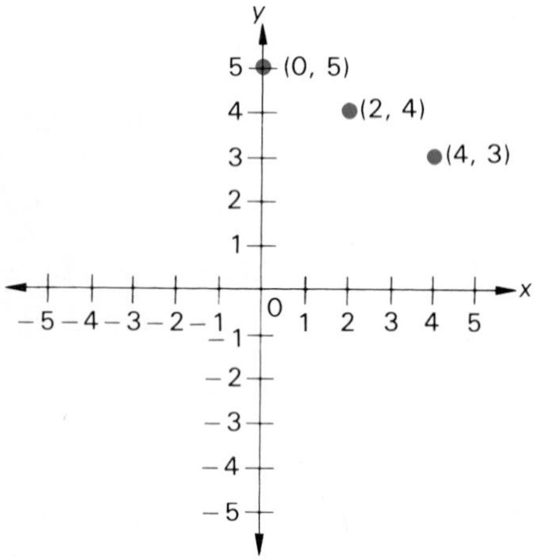

FIGURE 17

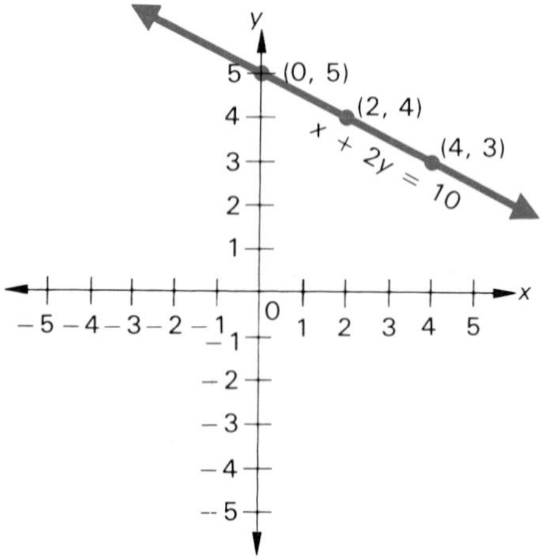

FIGURE 18

The fact that the graph of this type of equation—that is, an equation of the form $Ax + By = C$, where A, B, and C are real numbers—is a straight line is the reason that such equations are called **linear equations.**

If the graph of a linear equation is always a straight line, then why did we use three points to determine the graph of $x + 2y = 10$? It is true that two points determine a line, that is, through two given points one and only one straight line can be drawn. We use three points because the third point is a check. We locate the third point to check that all three points lie on the same straight line. If one of the points is not on the line, then we must go back and check for an error in determining the solutions to the linear equation.

EXAMPLE 4 Graph $x + y = 5$.

SOLUTION In order to graph $x + y = 5$, we must first find three ordered pairs that are solutions to the equation.

a. Selecting a value for x, we let $x = 1$. Hence, we have

$$\begin{aligned} & x + y = 5 \\ \text{let } x = 1: \quad & 1 + y = 5 \\ & y = 4 \end{aligned}$$

and (1, 4) is a solution.

b. Let $x = 4$; then

$$\begin{aligned} 4 + y &= 5 \\ y &= 1 \end{aligned}$$

and (4, 1) is a solution.

c. Let $x = 2$; then

$$\begin{aligned} 2 + y &= 5 \\ y &= 3 \end{aligned}$$

and (2, 3) is a solution.

Now that we have three different ordered pairs, (1, 4), (4, 1), and (2, 3), that are solutions to the equation $x + y = 5$, we plot these points and draw the line that contains them, as shown in Figure 19.

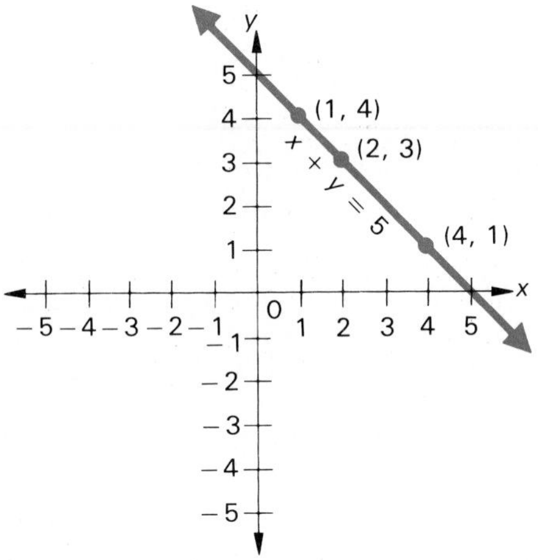

FIGURE 19

When we graph a linear equation, it is often convenient to find the points where the graph crosses each axis. Note that $y = 0$ at the point where a line crosses the x-axis, and $x = 0$ at the point where a line crosses the y-axis. The point at which a line crosses the x-axis is called the **x-intercept,** and similarly, the point at which it crosses the y-axis is called the **y-intercept.** In order to find the intercepts of the equation in example 4, $x + y = 5$, we proceed as follows.

$$\begin{aligned} & x + y = 5 \\ \text{let } x = 0: \quad & 0 + y = 5 \\ & y = 5 \qquad (0, 5) \text{ is a solution} \end{aligned}$$

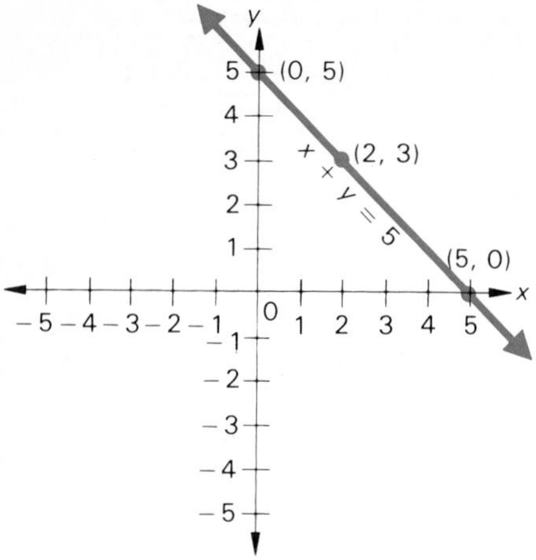

FIGURE 20

The y-intercept is $(0,5)$. Now let $y = 0$; then

$$x + 0 = 5$$
$$x = 5 \qquad (5,0) \text{ is a solution}$$

The x-intercept is (5, 0).

If we want to graph $x + y = 5$ by using the intercepts, we should find a third point as a check.

$$\text{let } x = 2: \quad 2 + y = 5$$
$$y = 3 \qquad (2,3) \text{ is a solution}$$

Using this point and the intercepts, we graph $x + y = 5$, as in Figure 20. Note that we get the same graph as in Figure 19, although we used different ordered pairs.

Remember that there is an infinite number of solutions to a linear equation and that each solution (ordered pair) is a point on the line. All the solutions of the equation are points on the line, and all the points on the line are solutions of the equation. In graphing an equation it is sometimes convenient to find the intercepts because they contain zeros, which simplifies computation.

EXAMPLE 5 Graph $2x - y = 4$.

SOLUTION First we shall find the intercepts. Therefore the value we select for x is 0.

$$2x - y = 4$$
$$\text{let } x = 0: \quad 2(0) - y = 4$$
$$0 - y = 4$$
$$-y = 4$$
$$-1(-y) = -1(4) \qquad \text{multiplying both sides by } -1 \text{ to make the } y\text{-term positive}$$
$$y = -4$$

The ordered pair $(0, -4)$ is a solution; $(0, -4)$ is also the y-intercept. Next, let $y = 0$; then

$$2x - 0 = 4$$
$$2x = 4$$
$$x = 2$$

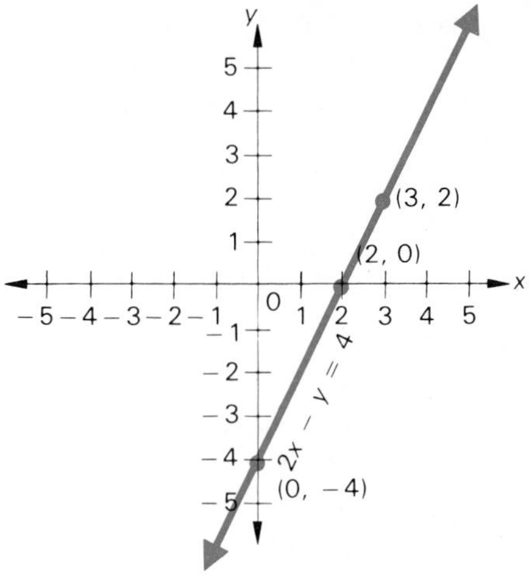

FIGURE 21

The ordered pair $(2, 0)$ is a solution; $(2, 0)$ is also the x-intercept.

Selecting a third point as a check, we let $x = 3$; then

$$\begin{aligned} 2(3) - y &= 4 \\ 6 - y &= 4 \\ -y &= -2 \\ -1(-y) &= -1(-2) \quad \text{multiplying both sides by } -1 \text{ to make the } y\text{-term positive} \\ y &= 2 \end{aligned}$$

The pair $(3, 2)$ is a solution.

Using these points, we graph $2x - y = 4$ in Figure 21.

EXAMPLE 6 Graph $2x + 3y = 12$.

SOLUTION We find the intercepts first.

$$\begin{aligned} 2x + 3y &= 12 \\ \text{let } x = 0: \quad 2(0) + 3y &= 12 \\ 0 + 3y &= 12 \\ 3y &= 12 \\ y &= 4 \end{aligned}$$

The y-intercept is $(0, 4)$.

$$\begin{aligned} \text{let } y = 0: \quad 2x + 3(0) &= 12 \\ 2x + 0 &= 12 \\ 2x &= 12 \\ x &= 6 \end{aligned}$$

The x-intercept is $(6, 0)$.

We then select a third point.

$$\begin{aligned} \text{let } x = 3: \quad 2(3) + 3y &= 12 \\ 6 + 3y &= 12 \\ 3y &= 6 \\ y &= 2 \end{aligned}$$

The point $(3, 2)$ is a solution.

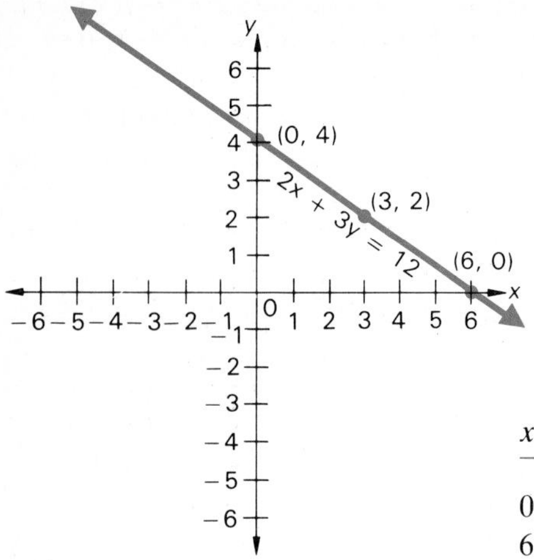

FIGURE 22

x	y
0	4
6	0
3	2

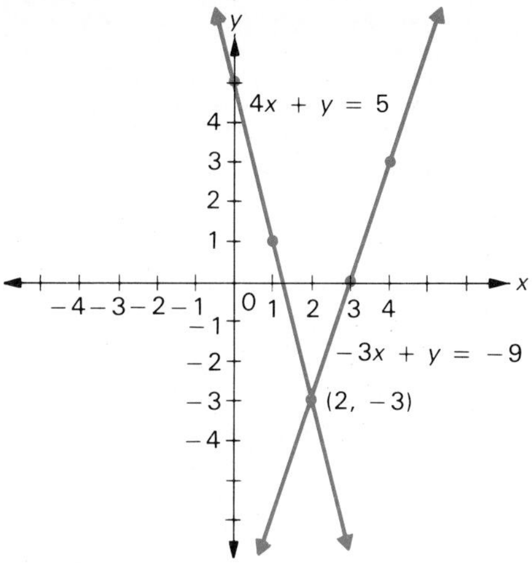

FIGURE 23

Using these points, we graph $2x + 3y = 12$ in Figure 22. Note that we can also list the solutions to the equation $2x + 3y = 12$ in a table of values such as the one shown next to Figure 22.

EXAMPLE 7 Graph $-3x + y = -9$ and $4x + y = 5$ on the same set of axes.

SOLUTION We determine a table of values for each equation as we did in previous examples. Then we graph each equation (see Fig. 23).

$-3x + y = -9$

x	y
3	0
4	3
2	−3

$4x + y = 5$

x	y
0	5
1	1
2	−3

Note that in Example 7, the graphs of the two lines intersect in a common point, $(2, -3)$. This point is common to both lines and the coordinates $x = 2$, $y = -3$ satisfy both equations, $-3x + y = 9$ and $4x + y = 5$. Therefore we can say that $(2, -3)$ is a solution to the system of equations: $-3x + y = 9$ and $4x + y = 5$.

To solve a system of equations by graphing, we graph each equation on the same set of axes. Next, we read the coordinates of the point of intersection from the graph. It should be noted that two lines may have no points in common (if they are parallel) and therefore never intersect. When this occurs, we say that the *given* system of equations is **inconsistent.**

For example, if we graph $x + y = 5$ and $x + y = 2$ on the same set of axes, the graphs are parallel lines and therefore have no common point of intersection. The equations are said to be inconsistent. Another possibility that can exist when determining the solution for a system of linear equations is that the two graphs may be the same line. When this occurs, any solution of one equation is also a solution of the other. In this case, the equations are called **dependent.**

When we graph a linear equation, it is usually convenient to find the points where the graph crosses each axis, that is, the x-intercept and the y-intercept. But if letting x or y be zero produces an intercept with a fractional value, then it is better to find solutions to the equation that contain only integers. For example, consider the equation $3x + 4y = 15$. If $x = 0$, then $y = \frac{15}{4}$, and $(0, \frac{15}{4})$ is the y-intercept. It will be more difficult to locate this point accurately on the Cartesian plane than a pair of integers, such as $(-3, 6)$. Therefore it is better to use $(-3, 6)$ than $(0, \frac{15}{4})$ as one of the ordered pairs used to graph $3x + 4y = 15$.

Reprinted by permission: Tribune Media Services.

EXERCISES FOR SECTION 9.7

1. Locate the points corresponding to the given ordered pairs on a Cartesian plane.
   a. $(4, 3)$ b. $(2, 5)$ c. $(3, -1)$
   d. $(-2, 4)$ e. $(0, 0)$ f. $(-2, -3)$
   g. $(3, -2)$ h. $(-1, 5)$ i. $(-4, 0)$

For Exercises 2–22, graph the given equation on a Cartesian plane.

2. $x + y = 5$
3. $x + y = 8$
4. $x - y = 3$
5. $x - y = 1$
6. $2x + y = 6$
7. $2x - y = 4$
8. $3x - y = -12$
9. $2x + 3y = -6$
10. $3x - 2y = 12$
11. $5x + 3y = 15$
12. $3x - 5y = -15$
13. $x + y = 0$
14. $x - y = 0$
15. $x - 2y = 0$
16. $x + 2y = 0$
17. $2x + 3y = 5$
18. $3x - 2y = 7$
19. $x - 2y = 3$

20. $-2x - 3y = 6$

21. $-x - 2y = -4$

22. $-2x + y = -8$

For Exercises 23–30, solve each system of equations graphically.

23. $2x + y = -3$
$-x + 2y = 4$

24. $2x + 3y = -2$
$-x + 4y = 1$

25. $2x + y = 0$
$3x + 2y = 1$

26. $-x + y = -6$
$x + 2y = 3$

27. $-3x + 6y = 0$
$x - 2y = 6$

28. $2x + 3y = 0$
$x - y = 5$

29. $2x + 3y = 7$
$-3x - y = -7$

30. $2x - 3y = -6$
$-4x + 6y = 18$

JUST FOR FUN A bicycle dealer was asked how many bikes he had in stock. He answered, "If one-half, one-third, and one-quarter of the number of bikes were added together, they would make 13." How many bikes did he have in stock?

9.8 GRAPHING $y = ax^2 + bx + c$

In the preceding section, we graphed linear equations in two variables. The graphs of these equations were straight lines. In this section, we shall examine graphs of equations of the form $y = ax^2 + bx + c$. An equation of this form is called a **quadratic equation** because it contains a term, ax^2, whose exponent is 2—that is, a **second-degree** term.

A **relation** is defined to be a set of ordered pairs. For example, $\{(2, 3), (4, 5)\}$ is a relation, as is $\{(0, 1), (2, 3), (5, 7)\}$. A **function** is a relation in which no two ordered pairs have the same first coordinate and different second coordinates. Consider, for example, the equation $y = x^2$. Some of the ordered pairs that satisfy this equation are $(0, 0)$, $(1, 1)$, $(-1, 1)$, $(2, 4)$, and $(-2, 4)$. Note that no two of these ordered pairs have the same first coordinate

© George Gardner/The Image Works

and different second coordinates. Thus the set of ordered pairs (x, y) that satisfy $y = x^2$ is a function.

Now consider $x = y^2$. Some of the ordered pairs that satisfy this equation are $(0, 0)$, $(1, 1)$, $(4, 2)$, and $(4, -2)$. Some of these ordered pairs do have the same first coordinate and different second coordinates. Hence, the set of ordered pairs (x, y) that satisfy $x = y^2$ is not a function. In this section, we shall consider only *quadratic functions*. **Quadratic functions** are sets of ordered pairs (x, y) that satisfy equations of the general form $y = ax^2 + bx + c$.

Consider $y = x^2$. This is a quadratic equation of the form $y = ax^2 + bx + c$: in this case, $a = 1$, $b = 0$, and $c = 0$. To graph $y = x^2$, we first find some ordered pairs that satisfy the given equation. Recall that the ordered pairs that satisfy $y = x^2$ are a function. Therefore, for every value of x, there is a unique value of y for which $y = x^2$. For example, if $x = 0$, then $y = (0)^2 = 0$. Similarly, if $x = 1$, then $y = (1)^2 = 1$, and if $x = -1$, then $y = (-1)^2 = 1$. If $x = 2$, $y = 4$ and if $x = -2$, $y = 4$. These points can be listed in a table of values as follows.

x	-2	-1	0	1	2
y	4	1	0	1	4

If we plot these points and connect them by means of a smooth curve, we obtain the curve shown in Figure 24. Note that the graph cannot be a straight line, because $y = x^2$ is not a linear equation.

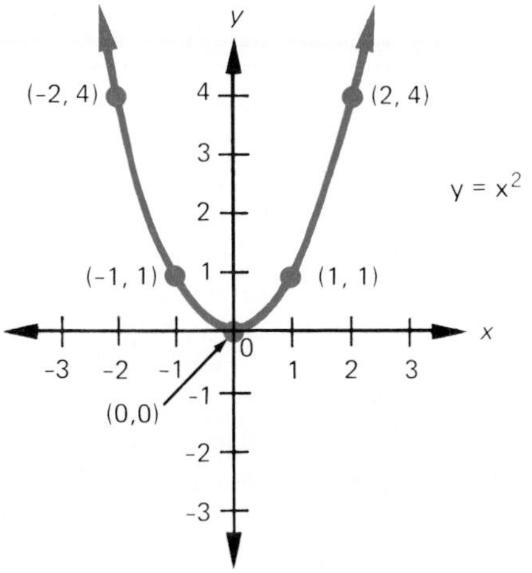

FIGURE 24

The smooth curve in Figure 24 is called a **parabola.** The path of most projectiles is shaped like a parabola and the cables on suspension bridges hang in the shape of a parabola.

In Figure 24, note that the point (0, 0) is a turning point of the parabola $y = x^2$. That is, as we proceed from left to right along the x-axis, the values of y decrease until we reach (0, 0); then the values of y begin to increase. The turning point of a parabola is called the **vertex** of the parabola.

For any parabola with equation of the form $y = ax^2 + bx + c$, the x-value of the vertex is

$$\frac{-b}{2a}$$

We can use this information to help graph a given parabola. For example, $y = x^2$ is of the form $y = ax^2 + bx + c$, with $a = 1$ and $b = 0$. Hence the x-value of the vertex is

$$\frac{-b}{2a} = \frac{-0}{2 \cdot 1} = -\frac{0}{2} = 0$$

The y-value is found by substituting zero for x in the original equation. If $x = 0$, $y = 0^2 = 0$. Therefore the vertex of $y = x^2$ is (0, 0).

© *George Gardner/The Image Works*

These freeways form a pattern of intersecting parabolas and straight lines.

It should be noted that the value of a in $y = x^2$ is $+1$. If a is positive in any parabola with an equation of the form $y = ax^2 + bx + c$, then the parabola opens upward. If a is negative, then the parabola opens downward. These two facts are useful when graphing parabolas.

EXAMPLE 1 Graph $y = x^2 - 4$.

SOLUTION The general form is $y = ax^2 + bx + c$. In this case, $a = 1, b = 0$, and $c = -4$. Since $a = 1$ is positive, the parabola opens upward. The x-value of the vertex is

$$\frac{-b}{2a} = \frac{-0}{2 \cdot 1} = \frac{0}{2} = 0$$

If $x = 0$, $y = 0^2 - 4 = -4$. Therefore the vertex is $(0, -4)$.

To obtain a better idea of the shape of the parabola, we choose two or three values for x on each side of the vertex and find the corresponding values of y. To the left of the vertex, we choose $x = -1$, $x = -2$, and $x = -3$, and to the right of the vertex, we choose $x = 1$, $x = 2$, and $x = 3$. The corresponding values of y are listed in the following table.

x	-3	-2	-1	0	1	2	3
y	5	0	-3	-4	-3	0	5

Using these points, we graph $y = x^2 - 4$ in Figure 25.

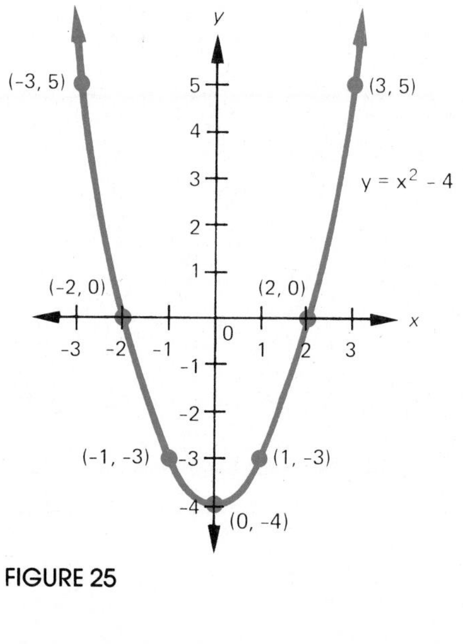

FIGURE 25

EXAMPLE 2 Graph $y = -x^2 + 4x$.

SOLUTION In this case, $a = -1$, $b = 4$, and $c = 0$. Because $a = -1$ is negative, the parabola opens downward. The x-value of the vertex is

$$\frac{-b}{2a} = \frac{-(4)}{2(-1)} = \frac{-4}{-2} = 2$$

If $x = 2$, $y = -2^2 + 4 \cdot 2 = -4 + 8 = 4$. Therefore the vertex is $(2, 4)$. To graph the parabola, we choose $x = 1$, $x = 0$, and $x = -1$ to the left of the vertex, and $x = 3$, $x = 4$, and $x = 5$ to the right of the vertex. The corresponding values of y are listed in the following table.

x	-1	0	1	2	3	4	5
y	-5	0	3	4	3	0	-5

Using these points, we graph $y = -x^2 + 4x$ in Figure 26.

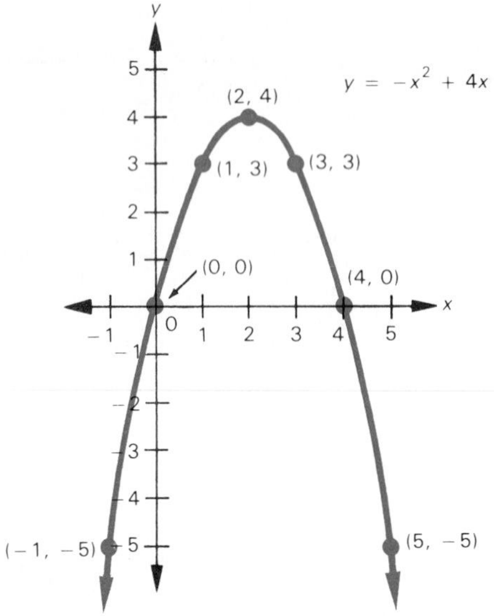

FIGURE 26

EXAMPLE 3 Graph $y = x^2 - 8x + 12$.

SOLUTION In this case, $a = 1$, $b = -8$, and $c = 12$. Since $a = 1$ is positive, the parabola opens upward. The x-value of the vertex is

$$\frac{-b}{2a} = \frac{-(-8)}{2 \cdot 1} = \frac{8}{2} = 4$$

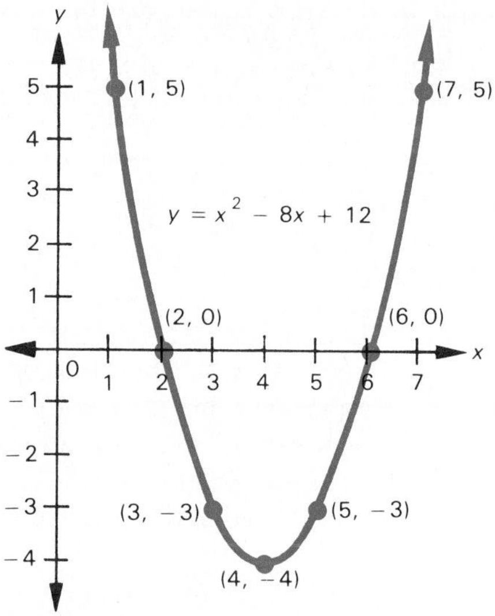

FIGURE 27

If $x = 4$, $y = 4^2 - 8 \cdot 4 + 12 = 16 - 32 + 12 = -4$. Therefore the vertex is $(4, -4)$. To graph the parabola, we choose $x = 3$, $x = 2$, and $x = 1$ to the left of the vertex, and $x = 5$, $x = 6$, and $x = 7$ to the right of the vertex. The corresponding values of y are listed in the following table.

x	1	2	3	4	5	6	7
y	5	0	−3	−4	−3	0	5

Using these points, we graph $y = x^2 - 8x + 12$ in Figure 27.

EXERCISES FOR SECTION 9.8

For Exercises 1–20, graph the given equation on a Cartesian plane.

1. $y = x^2 - 1$
2. $y = -x^2 + 1$
3. $y = -x^2 + 2$
4. $y = -x^2 + 5$
5. $y = -x^2 + 6x$
6. $y = -x^2 - 4x$
7. $y = x^2 + 2x + 1$
8. $y = x^2 + 2x - 3$
9. $y = x^2 - 4x + 3$
10. $y = -2x^2 - 4x - 2$
11. $y = -x^2 + 6x - 9$
12. $y = x^2 + 2x - 8$
13. $y = x^2 - 6x + 5$
14. $y = x^2 - 4x + 2$
15. $y = -x^2 + 4$
16. $y = x^2 - 2x + 1$
17. $y = x^2 + 2x$
18. $y = 2x^2 + 4x - 6$
19. $y = x^2 + 3x + 2$
20. $y = x^2 + 5x + 6$

JUST FOR FUN Which is worth more, a box full of \$10 gold pieces, or an identical box half full of \$20 gold pieces?

9.9 INEQUALITIES IN TWO VARIABLES

In Section 9.2, we graphed the solution sets of open sentences in one variable. The open sentences that we considered were either equations or inequalities, such as $x + 3 = 4$ and $x > 4$. In Sections 9.7 and 9.8, we graphed the solution sets of equations in two unknowns, such as $2x - 2y = 10$ and $y = x^2 + x + 1$. In this section we shall graph the solution sets of inequalities involving two variables.

When the equation $x + y = 3$ is graphed on the Cartesian plane, we note that the points (3, 0) and (0, 3) are solutions of the given equation. In fact, there is an infinite number of solutions to the linear equation $x + y = 3$. Each solution (ordered pair) is a point on the line. All the solutions of the equation are points on the line, and all the points on the line are solutions of the equation. But what about the points (ordered pairs) that are not on the line? What about points such as (0, 0) and (4, 0)? Because they are not on the line, they are not solutions of the equation $x + y = 3$. But how do they compare with the solutions? Let's find out.

We can evaluate the equation $x + y = 3$ for the point (0, 0). Let $x = 0$ and $y = 0$; then we have

$$0 + 0 = 3$$
$$0 = 3 \quad \text{(not true)}$$

Substituting 0 for x and 0 for y in the equation $x + y = 3$ gives us a false statement of equality. The value 0 on the left side of the equal sign is *less than* the value 3 on the right side.

$$0 < 3 \quad \text{(true)}$$

Let's try the other point, (4, 0). Let $x = 4$ and $y = 0$; then we have

$$4 + 0 = 3$$
$$4 = 3 \quad \text{(not true)}$$

Again we have a false statement of equality, but note that this time the value 4 on the left side of the equal sign is *greater than* the value 3 on the right side.

$$4 > 3 \qquad \text{(true)}$$

Notice that $(0, 0)$ is a solution of the inequality $x + y < 3$ and $(4, 0)$ is a solution of $x + y > 3$. We have found three different types of points: those like $(3, 0)$ and $(0, 3)$, for which $x + y = 3$; those like $(0, 0)$, for which $x + y < 3$; and those like $(4, 0)$, for which $x + y > 3$.

In fact, the equation $x + y = 3$ separates the Cartesian plane into three sets. One of these sets is the graph of $x + y = 3$. Because this graph is a line, we call it "the line $x + y = 3$." The other two sets are the **half-planes** that lie above and below this line. One of these half-planes contains $(0, 0)$ and is the graph of the inequality $x + y < 3$; the other contains $(4, 0)$ and is the graph of the inequality $x + y > 3$. These two half-planes are called "the half-plane $x + y < 3$" and "the half-plane $x + y > 3$," respectively.

How do we go about graphing an inequality such as $x + y < 3$? Recall that the line $x + y = 3$ divides the plane into two half-planes. One half-plane contains those points (x, y) such that $x + y < 3$, whereas the other half-plane contains those points (x, y) such that $x + y > 3$. The set of points that satisfy the equation $x + y = 3$ are those points that lie on the line. In order to graph $x + y < 3$, we locate the boundary of this half-plane, which is the line $x + y = 3$ (see Fig. 28). Note that we draw the graph of $x + y = 3$ as a dashed line because the points on the line do not satisfy the inequality $x + y < 3$.

Which half-plane is the correct half-plane, the one whose points satisfy the inequality $x + y < 3$? Let's test a point on either side of the line to see which one satisfies the given inequality. One convenient point to try is the origin, $(0, 0)$. Let $x = 0$ and $y = 0$; then we have

$$0 + 0 < 3$$
$$0 < 3 \qquad \text{(true)}$$

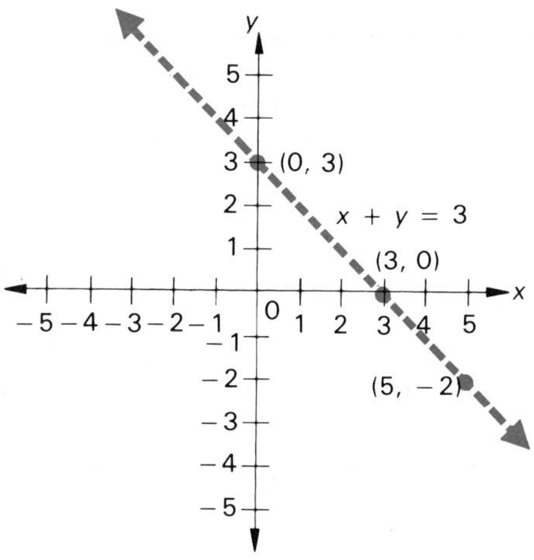

FIGURE 28

Now, if a point satisfies the given inequality, then all of the points in the half-plane must also satisfy the same inequality. It should also be noted that if a point does not satisfy the given inequality, then none of the points in that half-plane satisfies the given inequality and the desired half-plane must be on the other side of the line.

Because the point $(0, 0)$ does satisfy the given inequality, $x + y < 3$, we shade the half-plane that contains this point. In

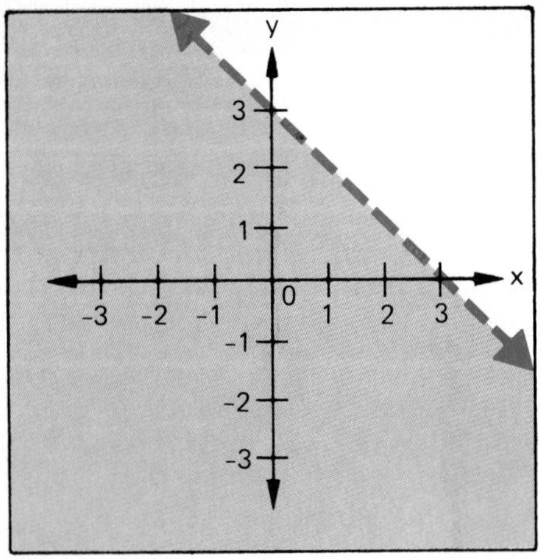

FIGURE 29
$x + y < 3$

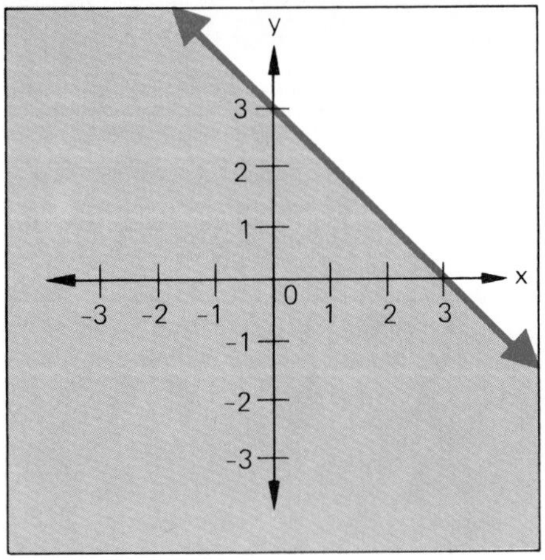

FIGURE 30
$x + y \leq 3$

this case we shade the half-plane that is below the line $x + y = 3$ to indicate the solution (see Fig. 29).

If we were asked to graph $x + y \leq 3$ instead of $x + y < 3$, then we would have the same picture with one exception. The exception would be to draw the line $x + y = 3$ as a solid line instead of a dotted line. The reason for this is that $x + y \leq 3$ means that we want the set of points that satisfy the inequality $x + y < 3$ *or* the equation $x + y = 3$. The sentence $x + y \leq 3$ is read as "$x + y$ is *less than or equal to* 3." Recall that *or* is the same as set union, and therefore we unite the half-plane $x + y < 3$ and the line $x + y = 3$. Figure 30 is the graph of $x + y \leq 3$.

In both Figures 29 and 30, the line $x + y = 3$ is the boundary for the solution set, a half-plane. In Figure 29 the boundary is not included in the solution, whereas in Figure 30 the boundary is included in the solution.

In order to graph inequalities in two variables, we must do the following.

1. Find the boundary of the half-plane. We do this by graphing the equation derived from the inequality.
2. The boundary should be a dashed line for *greater than* (>) or *less than* (<). For ≥ or ≤, the boundary should be a solid line.
3. Indicate the half-plane that is the solution by shading. We do this by testing a point. If a point satisfies the given inequality, then all of the points in that half-plane must also satisfy the same inequality.

EXAMPLE 1 Graph $x + 2y > 6$.

SOLUTION First we determine the boundary by graphing the equation $x + 2y = 6$. Note that this boundary will be a dashed line because we have a strict inequality in $x + 2y > 6$. Next we test a point to see if it satisfies the given inequality. Let's try the origin, $(0, 0)$. Let $x = 0$ and $y = 0$; then we have

$$\begin{aligned} 0 + 2(0) &> 6 \\ 0 + 0 &> 6 \\ 0 &> 6 \qquad \text{(not true)} \end{aligned}$$

The point $(0, 0)$ does not satisfy the statement $x + 2y > 6$. Thus we can try a point on the other side of the line $x + 2y = 6$, $(3, 3)$. Let $x = 3$ and $y = 3$; then we have

$$\begin{aligned} 3 + 2(3) &> 6 \\ 3 + 6 &> 6 \\ 9 &> 6 \qquad \text{(true)} \end{aligned}$$

If a point satisfies a given inequality, then all of the points in that half-plane must also satisfy the same inequality. Therefore we shade the half-plane that contains the point $(3, 3)$. In this case, we shade the half-plane that is above the line $x + 2y = 6$ to indicate the solution (see Fig. 31).

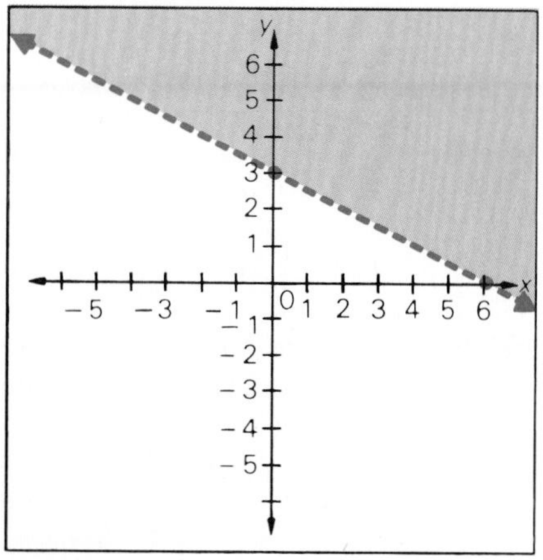

FIGURE 31
$x + 2y > 6$

EXAMPLE 2 Graph $2x - y \le 6$.

SOLUTION First we locate the boundary by graphing $2x - y = 6$. Note that this boundary will be a solid line because of the relationship $\le$. Next we test a point to see if it satisfies the given inequality. Let's try the origin $(0, 0)$. Let $x = 0$ and $y = 0$; then we have

$$2(0) - 0 \le 6$$
$$0 - 0 \le 6$$
$$0 \le 6 \quad \text{(true)}$$

If a point satisfies a given inequality, then all of the points in that half-plane must also satisfy the same inequality. Therefore, we shade the half-plane that contains the point $(0, 0)$, that is, the half-plane above the line $2x - y = 6$ (see Fig. 32). Note that the boundary is a solid line.

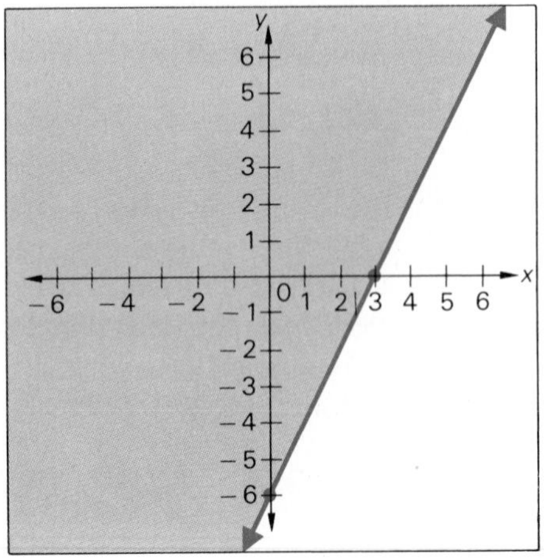

FIGURE 32
$2x - y \le 6$

EXAMPLE 3 Graph $-3x + 4y < 12$.

SOLUTION First we locate the boundary by graphing the equation $-3x + 4y = 12$. This boundary will be a dashed line because we have a strict inequality. Next we test a point to see if it satisfies the given inequality. Testing the origin, we let $x = 0$ and $y = 0$; then we have

$$-3(0) + 4(0) < 12$$
$$0 + 0 < 12$$
$$0 < 12 \quad \text{(true)}$$

Since $(0,0)$ satisfies the given inequality, all of the points in that half-plane will also satisfy the given inequality. Therefore we shade the half-plane that contains the point $(0,0)$, which is the half-plane below the line $-3x + 4y = 12$ (see Fig. 33). Note that the boundary is a dotted line.

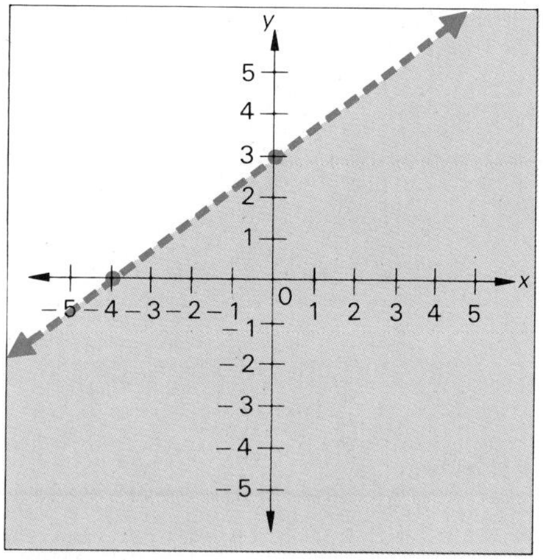

FIGURE 33
$-3x + 4y < 12$

In the preceding section we graphed equations, and in this section we have graphed inequalities. We can also combine equations or inequalities to get compound sentences, for example, the conjunction of $x - y \leq 2$ and $x + y < 1$. The graph of the conjunction of $x - y \leq 2$ and $x + y < 1$ is the intersection of the individual graphs of $x - y \leq 2$ and $x + y < 1$. In order to obtain this intersection, we draw both graphs on the same Cartesian plane. The solution is the region where the two half-planes intersect. The coordinates of all the points in the intersection will satisfy both of the given sentences and hence form the solution set of their conjunction. The graphs of $x - y \leq 2$ and $x + y < 1$ are shown separately in Figures 34a and 34b. Figure 34c is the graph of $x - y \leq 2$ and $x + y < 1$. Note that Figure 34c is the intersection of the graphs in Figures 34a and 34b.

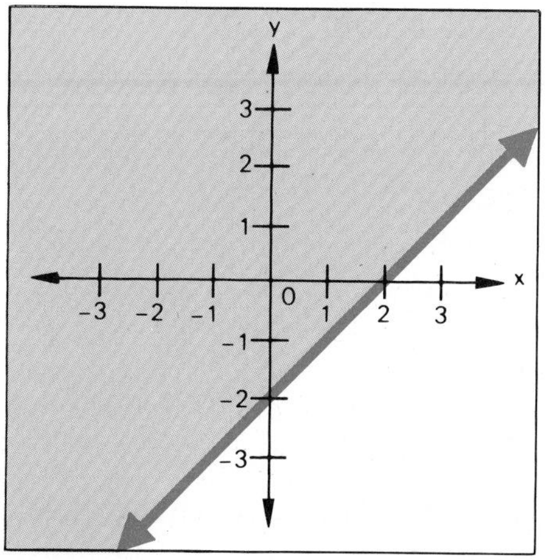

FIGURE 34 a
$x - y \leq 2$

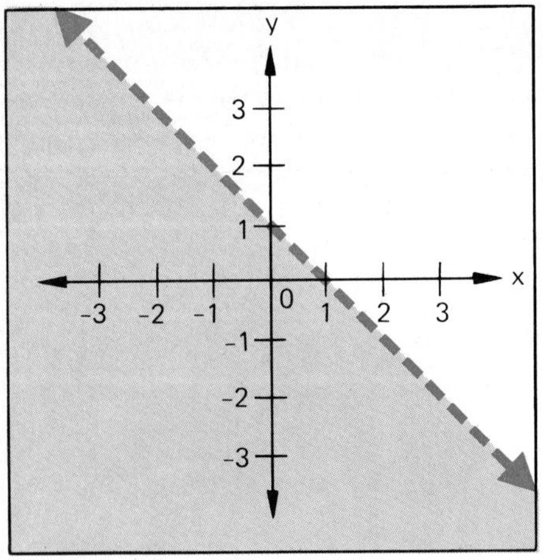

FIGURE 34 b
$x + y < 1$

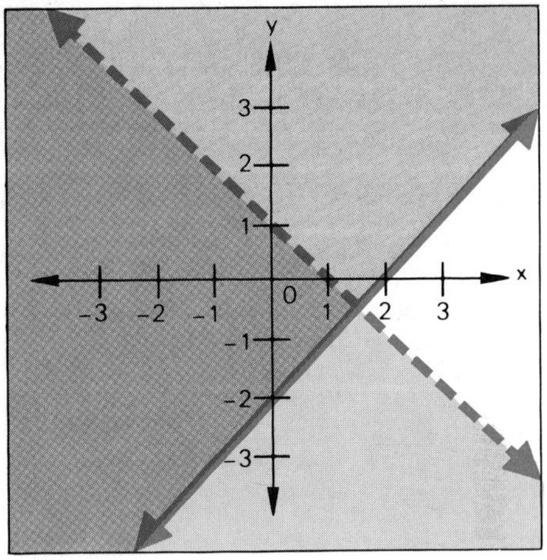

FIGURE 34 c
$x - y \leq 2$ and $x + y < 1$

EXERCISES FOR SECTION 9.9

Graph each of the following inequalities on the Cartesian plane.

1. $x + y > 5$
2. $x + y < 8$
3. $x - y < 3$
4. $x - y > 4$
5. $x - y \leq 1$
6. $2x - y \geq 4$
7. $-2x + y \geq 4$
8. $x - y \geq 0$
9. $y - x < 0$
10. $x - 2y < 0$
11. $-2x - 2y > 0$
12. $3x - 5y \leq 15$
13. $-3x + 5y \leq 15$
14. $-3x - 5y < -15$
15. $3x + 5y > 15$
16. $-2x - 3y > -6$
17. $-3x - 2y < -6$
18. $x + y \geq 2$
19. $x + y \geq 2$ and $x - y < 2$
20. $x + y \leq 3$ and $x - y > 3$
21. $y \geq x$ and $y \geq -x$

JUST FOR FUN If one peacock lays three eggs in one day, then how many eggs will 33 peacocks lay in 11 days? (Be careful!)

9.10 LINEAR PROGRAMMING

In the preceding section, we graphed linear inequalities. Such graphs can be used to solve practical problems. As an example, consider a problem encountered by the E-Z Furniture Company, which manufactures tables and chairs for family rooms and patios: all of the E-Z Company's furniture undergoes some construction steps in building I and others in building II. Each table requires 3 hours' work in building I and 2 hours in building II to produce it. Each chair requires 2 hours in building I and 4 hours in building II. The profit from each table is \$6, and the profit from each chair is \$5. Due to union regulations, the two buildings can only operate for at most 8 hours a day. Given these restrictions, how many tables and how many chairs should the company produce each day in order to maximize its profits?

This problem is an example of a **linear programming problem.** The theory of linear programming was developed to solve problems like these, where the goal is to maximize profits subject to a few simple restrictions. The theory has been expanded to solve similar problems of a much more complex nature. Many linear programming problems involve many more unknowns than our example and require the use of a computer to solve them. However, simple problems like that of the E-Z Furniture Company can be solved by graphing linear inequalities, as we shall see.

To solve a linear programming problem, we must first translate the word problem into mathematical statements. We shall now proceed to do this for the E-Z Furniture Company example. We begin by letting

x = the number of tables to be produced

y = the number of chairs to be produced

We also list all of the given information in a table.

	Time Needed (hours)		
	Building I	**Building II**	**Profit for Each**
Table	3	2	\$6
Chair	2	4	\$5
Time limit	8	8	

From this table, we can construct the mathematical statements needed to solve the problem. For example, if a table requires 3 hours in building I, then 2 tables would require $2 \cdot 3 = 6$ hours in building I. Therefore x tables would require $x \cdot 3 = 3x$ hours in building I. Similarly, y chairs would require $2y$ hours in building I. What do we know about the time spent in building I? It must be less than or equal to 8 hours, due to union rules. Therefore we obtain the open sentence

$$3x + 2y \leq 8$$

Similarly, x tables require $2x$ hours in building II, and y chairs require $4y$ hours in building II. Building II can also only be used for a maximum of 8 hours. Therefore, we obtain the open sentence

$$2x + 4y \leq 8$$

The same type of reasoning can be used to express the profit P in terms of x and y. A profit of \$6 is made on each table. Therefore two tables would yield a profit of $6 \cdot 2 = 12$ dollars. Similarly, x tables would yield a profit of $6x$ dollars. A profit of \$5 is made on each chair. Hence y chairs would yield a profit of $5y$ dollars. The total profit from both tables and chairs is $6x + 5y$. Therefore we say that

$$P = 6x + 5y$$

We now have three mathematical statements from the information in the table.

Building I:	$3x + 2y \leq 8$
Building II:	$2x + 4y \leq 8$
Profit:	$P = 6x + 5y$

The first two statements are linear inequalities; they describe the restrictions on the numbers of chairs and tables that can be manufactured each day. These inequalities are called the **constraints** of the problem. The third mathematical statement describes the relationship between the number of chairs and tables produced and the profit that can be made by selling them. This expression is called the **objective function.** Because the E-Z Furniture Company would like its profits to be as high as possible, we want to maximize the value of the objective function, $P = 6x + 5y$, subject to the constraints $3x + 2y \leq 8$ and $2x + 4y \leq 8$. In other words, we will solve the E-Z Company's problem if we can find numbers x and y such that

(1) $3x + 2y \leq 8$
(2) $2x + 4y \leq 8$
(3) $P = 6x + 5y$ is as large as possible

Any numbers x and y that solve the problem must satisfy inequalities (1) and (2). Graphing both of these inequalities on the Cartesian plane gives us a region in which the coordinates of

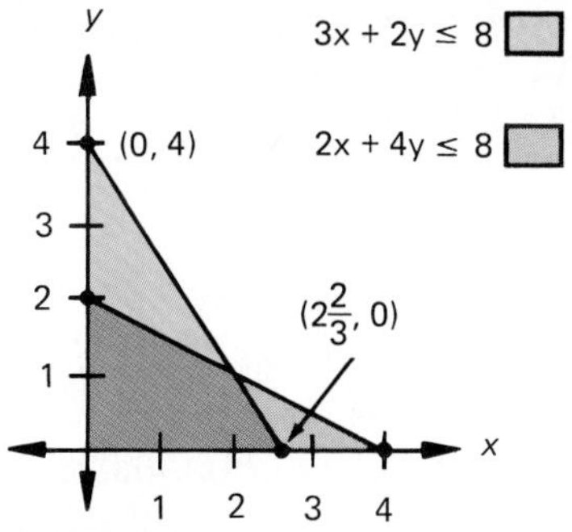

FIGURE 35
$3x + 2y \le 8$ and $2x + 4y \le 8$

each point satisfy inequalities (1) and (2). Therefore the coordinates of the points in this region will be the values of x and y that we shall consider when we look for solutions to the problem.

We will now graph inequalities (1) and (2) on the same set of axes. However, we shall graph them only in the region where $x \ge 0$ and $y \ge 0$. We do this because x and y represent the numbers of tables and chairs to be manufactured, respectively, and therefore x and y cannot be negative quantities. Hence $x \ge 0$ and $y \ge 0$.

The graphs of the two inequalities are shown in Figure 35. We are concerned only with the portion of this figure where the two half-planes intersect. This is because the coordinates of each point in the intersection satisfy both of the inequalities, and therefore the intersection forms the solution set of the conjunction of the two inequalities.

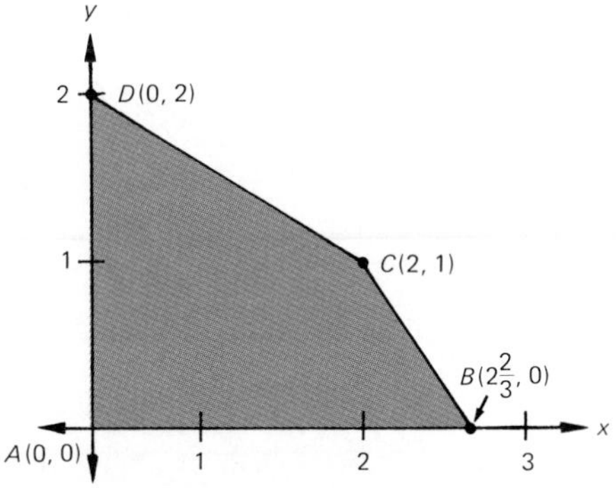

FIGURE 36

Let us examine this region more closely: Figure 36 shows this region with its four corners labeled. The corner points are A (0, 0), B $(2\frac{2}{3}, 0)$, C (2, 1), and D (0, 2).

It can be proved, using more advanced mathematics, that in a linear programming problem, any maximum or minimum values of the objective function always occur at a vertex. Hence our profit expression P will have its maximum or minimum value at a corner of the region in Figure 36. We can test these values by substituting them in the profit expression, $P = 6x + 5y$.

At A (0, 0), we have $P = 6 \cdot 0 + 5 \cdot 0 = 0 + 0 = 0$.
At B $(2\frac{2}{3}, 0)$, we have $P = 6 \cdot \frac{8}{3} + 5 \cdot 0 = \frac{48}{3} + 0 = 16$.
At C (2, 1), we have $P = 6 \cdot 2 + 5 \cdot 1 = 12 + 5 = 17$.
At D (0, 2), we have $P = 6 \cdot 0 + 5 \cdot 2 = 0 + 10 = 10$.

The maximum profit occurs when $x = 2$ and $y = 1$, and it is \$17. The E-Z Furniture Company should produce two tables and one chair to obtain a maximum profit. Note that the minimum profit, \$0, occurs at A (0, 0).

Keep in mind that this is an illustrative example of a linear programming problem and one method for solving such problems. More complex problems involve more variables and must be solved in a different manner.

In solving this example, we evaluated the corner points of the graph of the constraint inequalities. As frequently happens, one of these corner points—B $(2\frac{2}{3}, 0)$—does not have integer coordinates. What if the profit expression P had its maximum value at B $(2\frac{2}{3}, 0)$ instead of at C (2, 1)? (This could easily happen if the problem were slightly different.) In terms of the original problem, this would mean that the E-Z Furniture Company should manufacture $2\frac{2}{3}$ tables and 0 chairs. But it does not make much sense to manufacture $\frac{2}{3}$ of a table. In such a case, linear programming yields only an approximate answer to the problem. A more sophisticated mathematical technique, *integer programming,* must be used to find exact answers to problems that require whole numbers as answers.

EXAMPLE 1 The Blivit Electronic Company manufactures bleeps and peeps. Manufacturing a bleep requires 2 hours on machine A and 1 hour on machine B. Manufacturing a peep requires 1 hour on machine A and 1 hour on machine B. Machine A cannot be used more than 7 hours a day, and machine B cannot be used more than 5 hours a day. If the profit from a bleep is \$5 and that from a peep is \$4, how many of each should be produced to maximize profit?

SOLUTION We begin by letting

x = the number of bleeps to be manufactured
y = the number of peeps to be manufactured

Next we list all of the given information in a table.

	Time Needed (hours)		
	Machine A	Machine B	Profit for Each
Bleep	2	1	\$5
Peep	1	1	\$4
Time limit	7	5	

x bleeps require $2 \cdot x$, or $2x$, hours on machine A and y peeps require $1y$, or y, hours on machine A. The time limit on machine A is 7 hours. Therefore we have

$$2x + y \leq 7$$

x bleeps require $1x$, or x, hours on machine B and y peeps require $1y$, or y, hours on machine B. The time limit on machine B is 5 hours. Therefore we have

$$x + y \leq 5$$

The profit for x bleeps is $5x$ and the profit for y peeps is $4y$. The total profit, P, is $5x + 4y$. That is,

$$P = 5x + 4y$$

We want to maximize P.

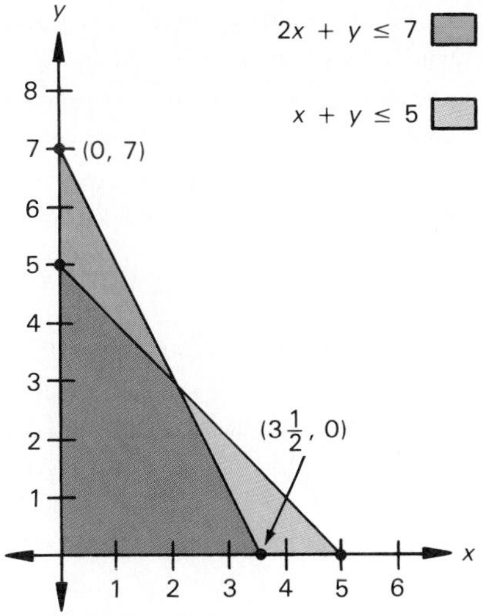

FIGURE 37
$2x + y \le 7$ and $x + y \le 5$

We now have three mathematical statements from the given information in the table.

Machine A:	$2x + y \le 7$
Machine B:	$x + y \le 5$
Profit:	$P = 5x + 4y$

To find the desired region, we graph the inequalities for machines A and B on the same set of axes. We do this only where $x \ge 0$ and $y \ge 0$, as shown in Figure 37.

The feasible region whose vertices will provide a maximum or minimum is that region where the half-planes intersect. The region and its vertices are shown in Figure 38.

We now test the profit expression $P = 5x + 4y$ at each of these vertices.

At A $(0,0)$, $P = 5 \cdot 0 + 4 \cdot 0 = 0 + 0 = 0$.
At B $(3\frac{1}{2},0)$, $P = 5 \cdot \frac{7}{2} + 4 \cdot 0 = \frac{35}{2} + 0 = 17.5$.
At C $(2,3)$, $P = 5 \cdot 2 + 4 \cdot 3 = 10 + 12 = 22$.
At D $(0,5)$, $P = 5 \cdot 0 + 4 \cdot 5 = 0 + 20 = 20$.

The maximum profit occurs when $x = 2$ and $y = 3$, and it is \$22. The Blivit Electronic Company should produce two bleeps and three peeps to obtain a maximum profit.

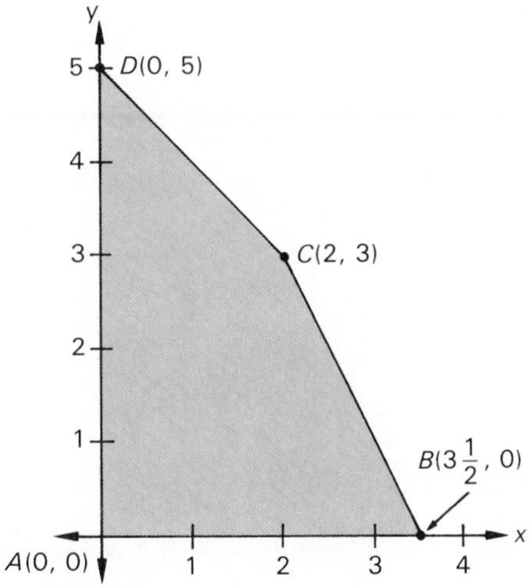

FIGURE 38

When linear programming problems involve only two variables, the technique used in Example 1 (and in the E-Z Furniture Company problem) is convenient, and it is the only technique we will consider here. For problems containing many variables, more complicated techniques, usually executed by a computer, must be used. The most common of these techniques is called the *simplex method*.

EXERCISES FOR SECTION 9.10

1. Find the maximum value of $P = 3x + 2y$ under the following conditions.

$$x \geq 0, \quad y \geq 0, \quad x + y \leq 5, \quad x - y \leq 1$$

2. Find the maximum value of $P = 5x + 8y$ under the following conditions.

$$x \geq 0, \quad y \geq 0, \quad x + y \leq 7, \quad 3x + y \leq 15$$

3. Find the maximum value of $P = 4x + 3y$ under the following conditions.

$$x \geq 0, \quad y \geq 0, \quad 2x + y \leq 12, \quad x + y \leq 7$$

4. Find the maximum value of $P = x + 4y$ under the following conditions.

$$x \geq 0, \quad y \geq 0, \quad x + y \leq 10, \quad 3x - y \leq 6$$

5. A manufacturer makes bikes and wagons. To produce a bike requires 2 hours on machine A and 4 hours on machine B. To produce a wagon requires 3 hours on machine A and 2 hours on machine B. Machine A can operate at most 12 hours per day and machine B can operate at most 16 hours per day. If the manufacturer makes a profit of $12 on a bike and $10 on a wagon, how many of each should be produced in order to maximize profit?

6. A manufacturer makes lawn mowers and snow blowers. To produce a lawn mower requires 3 hours on machine A and 2 hours on machine B. To produce a snow blower requires 2 hours on machine A and 4 hours on machine B. Machine A can operate at most 18 hours per day and machine B can operate at most 20 hours per day. If the manufacturer makes a profit of $20 on a lawn mower and $30 on a snow blower, how many of each should be produced in order to maximize profit?

7. Frank Sloane raises pheasants and partridges and has room for at most 100 birds. It costs him $2 to raise a pheasant and $3 to raise a partridge, and he has $240 to cover these costs. If he can make a profit of $7 on each pheasant and $8 on each partridge, how many of each bird should be raise in order to maximize his profit?

8. Betty Juarez has a 100-acre farm where she raises two crops, potatoes and cauliflower. It costs her $20 to raise an acre of potatoes and $40 to raise an acre of cauliflower, and she has $2,600 to cover the costs. If she can make a profit of $35 on each acre of potatoes and $60 on each acre of cauliflower, how many acres of each crop should she plant in order to maximize her profit?

9. The Long Island Shellfish Company processes (cleans, sorts, opens, and freezes) oysters and clams. In a given week, the company can process 600 bushels of shellfish, of which 100 bushels of oysters and 200 bushels of clams are required by regular customers (restaurants). The profit on a bushel of oysters is $8 and on a bushel of clams is

$10. How many bushels of oysters and of clams should the company process in order to maximize its profit?

10. A service station manager stocks two brands of motor oil, x and y. The manager has room for no more than 60 cans of oil in the station. He also knows that at least twice as much Brand x is sold as Brand y. If he makes a profit of 10 cents a can on Brand x and 12 cents a can on Brand y, how many cans of each kind should he stock in his station to make the maximum profit?

JUST FOR FUN

Two people have the same parents. They were born on the same day, and at the same place, but they are not twins. How are they related?

9.11 QUADRATIC EQUATIONS (OPTIONAL)

In previous sections (9.4 and 9.5) we solved equations of the form $ax + b = 0,\ a \neq 0$. Equations of this form are called **linear equations.** An equation of the form

$$ax^2 + bx + c = 0, \quad a \neq 0$$

is called a **quadratic equation.** If a quadratic equation is of the form $ax^2 + bx + c = 0$, it is said to be in **standard form.** Quadratic equations that are in standard form can be solved by means of the quadratic formula. That is, if the equation $ax^2 + bx + c = 0$ has solutions in the real number system, then they are

$$x = \frac{-b \pm \sqrt{b^2 - 4ac}}{2a}$$

This form of the solution is commonly known as the "quadratic formula"; since it is the solution for $ax^2 + bx + c = 0$, the general quadratic equation, we can use it to solve any quadratic equation. You should learn it and be able to state it as easily as your phone number.

The $\pm$ sign represents the two solutions of the equation. To find each of the solutions, first use $+$ and then use $-$. This will be demonstrated in the examples that follow. The development of the quadratic formula is based upon a method called completing the square. The derivation of this formula can be found in most algebra texts.

We shall now examine some examples to see how we can apply the formula. Consider again the equation

$$x^2 + 6x + 8 = 0$$

We shall use the quadratic formula to find the roots of the equation. Since the equation is already in standard form, our next task is to determine the values of a, b, and c. Note that $x^2 + 6x + 8 = 0$ is the same as $1x^2 + 6x + 8 = 0$, and the standard form is $ax^2 + bx + c = 0$. Therefore we have

$$a = 1, \qquad b = 6, \qquad c = 8$$

Next, we substitute these values in the quadratic formula

$$x = \frac{-b \pm \sqrt{b^2 - 4ac}}{2a}$$

$$x = \frac{-6 \pm \sqrt{6^2 - 4 \cdot 1 \cdot 8}}{2 \cdot 1}$$

Simplifying, we have

$$x = \frac{-6 \pm \sqrt{36 - 32}}{2}$$

$$x = \frac{-6 \pm \sqrt{4}}{2}$$

$$x = \frac{-6 \pm 2}{2}$$

$$x = \frac{-6 + 2}{2} \quad \text{or} \quad x = \frac{-6 - 2}{2}$$

$$x = \frac{-4}{2} \qquad\qquad x = \frac{-8}{2}$$

$$x = -2 \qquad\qquad x = -4$$

The solution set is $\{-2, -4\}$.

Check: If $x = -2$, $\quad$ if $x = -4$,

$$(-2)^2 + 6(-2) + 8 = 0 \qquad (-4)^2 + 6(-4) + 8 = 0$$

$$4 - 12 + 8 = 0 \qquad 16 - 24 + 8 = 0$$

$$0 = 0 \qquad 0 = 0$$

The solution checks.

To solve a quadratic equation by means of the quadratic formula, we must be sure that the equation is in standard form before we determine the values for a, b, and c. Also, in determining these values, we must be sure to include the correct signs for a, b, and c.

EXAMPLE 1 Solve for x by means of the quadratic formula.

$$2x^2 - 3x + 1 = 0$$

SOLUTION The equation is in standard form. Therefore, $a = 2$, $b = -3$, and $c = 1$.

$$x = \frac{-b \pm \sqrt{b^2 - 4ac}}{2a}$$

$$x = \frac{-(-3) \pm \sqrt{(-3)^2 - 4(2)(1)}}{2(2)}$$

$$x = \frac{3 \pm \sqrt{9 - 8}}{4}$$

$$x = \frac{3 \pm \sqrt{1}}{4}$$

$$x = \frac{3 \pm 1}{4}$$

$$x = \frac{3 + 1}{4} \quad \text{or} \quad x = \frac{3 - 1}{4}$$

$$x = \frac{4}{4} \qquad x = \frac{2}{4}$$

$$x = 1 \qquad x = \frac{1}{2}$$

The solution set is $\left\{1, \frac{1}{2}\right\}$.

Check: If $x = 1$ if $x = \frac{1}{2}$

$$2 \cdot 1^2 - 3 \cdot 1 + 1 = 0 \qquad 2 \cdot \left(\frac{1}{2}\right)^2 - 3 \cdot \frac{1}{2} + 1 = 0$$

$$2 \cdot 1 - 3 + 1 = 0 \qquad 2 \cdot \frac{1}{4} - \frac{3}{2} + 1 = 0$$

$$2 - 3 + 1 = 0 \qquad \frac{1}{2} - \frac{3}{2} + 1 = 0$$

$$0 = 0 \qquad 0 = 0$$

The solution checks.

EXAMPLE 2 Solve for x by means of the quadratic formula.

$$6x^2 + 7x = 3$$

SOLUTION First write the equation in standard form

$$6x^2 + 7x - 3 = 0$$

Now that the equation is in standard form, we have $a = 6$, $b = 7$, $c = -3$.

$$x = \frac{-b \pm \sqrt{b^2 - 4ac}}{2a}$$

$$x = \frac{-7 \pm \sqrt{7^2 - 4(6)(-3)}}{2 \cdot 6}$$

$$x = \frac{-7 \pm \sqrt{49 + 72}}{12}$$

$$x = \frac{-7 \pm \sqrt{121}}{12}$$

$$x = \frac{-7 + 11}{12} \quad \text{or} \quad x = \frac{-7 - 11}{12}$$

$$x = \frac{4}{12} \qquad\qquad x = \frac{-18}{12}$$

$$x = \frac{1}{3} \qquad\qquad x = -\frac{3}{2}$$

The solution set is $\left\{\frac{1}{3}, -\frac{3}{2}\right\}$.

Check: If $x = \frac{1}{3}$ if $x = -\frac{3}{2}$

$$6\left(\frac{1}{3}\right)^2 + 7\left(\frac{1}{3}\right) = 3 \qquad 6\left(-\frac{3}{2}\right)^2 + 7\left(-\frac{3}{2}\right) = 3$$

$$6 \cdot \frac{1}{9} + \frac{7}{3} = 3 \qquad 6 \cdot \frac{9}{4} - \frac{21}{2} = 3$$

$$\frac{2}{3} + \frac{7}{3} = 3 \qquad \frac{27}{2} - \frac{21}{2} = 3$$

$$\frac{9}{3} = 3 \qquad\qquad \frac{6}{2} = 3$$

$$3 = 3 \qquad\qquad 3 = 3$$

The solution checks.

EXAMPLE 3 Solve for x by means of the quadratic formula.

$$x^2 + 2x - 24 = 0$$

SOLUTION The equation is in standard form. Therefore, $a = 1$, $b = 2$, $c = -24$. Note that the value of c is -24, not 24. Next we substitute these values in the quadratic formula to get

$$x = \frac{-2 \pm \sqrt{2^2 - 4(1)(-24)}}{2(1)}$$

Simplifying the expression, we obtain

$$x = \frac{-2 \pm \sqrt{4 + 96}}{2}$$

$$x = \frac{-2 \pm \sqrt{100}}{2}$$

$$x = \frac{-2 \pm 10}{2}$$

$$x = \frac{-2 + 10}{2} \quad \text{or} \quad x = \frac{-2 - 10}{2}$$

$$x = \frac{8}{2} \qquad\qquad x = \frac{-12}{2}$$

$$x = 4 \qquad\qquad x = -6$$

The solution set is $\{4, -6\}$.

Check: If $x = 4$, if $x = -6$,

$$4^2 + 2(4) - 24 = 0 \qquad (-6)^2 + 2(-6) - 24 = 0$$

$$16 + 8 - 24 = 0 \qquad 36 - 12 - 24 = 0$$

$$24 - 24 = 0 \qquad 36 - 36 = 0$$

$$0 = 0 \qquad 0 = 0$$

The solution checks.

In this section we have developed a formula that can be used to solve any quadratic equation. The solution set of most quadratic equations is obtained by means of this formula. Therefore, for the sake of convenience, this formula should be memorized just as you would memorize your student identification number.

$$x = \frac{-b \pm \sqrt{b^2 - 4ac}}{2a}$$

EXERCISES FOR SECTION 9.11

Solve each of the following equations by means of the quadratic formula.

1. $2x^2 - 10x + 12 = 0$
2. $2x^2 + 6x + 4 = 0$
3. $2y^2 - 6y - 8 = 0$
4. $2y^2 - 4y - 6 = 0$
5. $6x^2 + 6x - 36 = 0$
6. $x^2 - 8x + 15 = 0$
7. $y^2 - 3y + 2 = 0$
8. $y^2 + y - 6 = 0$
9. $x^2 - 3x - 10 = 0$
10. $2x^2 + 5x + 3 = 0$
11. $6x^2 + 7x + 2 = 0$
12. $15x^2 + 8x + 1 = 0$
13. $6y^2 - 7y - 5 = 0$
14. $15y^2 - y - 2 = 0$
15. $6x^2 = 1 - x$
16. $3y^2 = 4y + 4$
17. $5y^2 + 7y = 6$
18. $6x^2 + 1 = 7x$
19. $2y^2 = y + 1$
20. $8x^2 = 6x - 1$

★21. $x^2 - 2 = 2x$

★22. $x^2 + 3x = 1$

★23. $y^2 - 2y = 5$

★24. $y^2 = 3y + 2$

★25. $4x^2 - 3 = 0$

★26. $16y^2 - 5 = 0$

★27. $3y^2 - 1 = 0$

★28. $2x^2 - 1 = 0$

JUST FOR FUN Where are the following located?
a. Tunnel of Corti b. Island of Reil c. McBurney's Point

Summary

In this chapter, we explored the topic of algebra and some of its basic characteristics. Algebra is the area of mathematics that generalizes the facts of arithmetic. In algebra, letters are used to denote numbers or a certain set of numbers. The expression $4 + 2$ is an arithmetic expression, whereas $3x + 2y$ is an algebraic expression. The letters are called *variables,* because they are used to represent an unknown member of the set. Sentences such as $x + 1 = 3$, which cannot be classified as true or false until we replace x by a number, are called *open sentences*. There are open sentences of *equality* and open sentences of *inequality* (for example, $x + 1 > 3$). We can find the solution sets for these sentences, and we can graph these solution sets on the number line.

In solving most equations, we make use of the following axioms.

1. If the same quantity or two equal quantities are added to two equal quantities, then the sums are equal.
2. If the same quantity or two equal quantities are subtracted from two equal quantities, then the remainders (differences) are equal.
3. If equal quantities are multiplied by equal quantities, then the products are equal.
4. If equal quantities are divided by equal quantities (except zero), then the quotients are equal.

Once we have found a solution to an equation, we can check our answer by replacing the variable by the solution in the original equation to see if the left side of the equation equals the right side of the equation.

One of the oldest applications of algebra is in solving word problems. Most word problems require the use of algebra in order to find the solution in a systematic manner, as opposed to trial and error.

Linear equations in two unknowns are equations of the form $Ax + By = C$, where A, B, and C are real numbers. They are called linear equations because the graph of such an equation is a straight line. Because a linear equation such as $x + y = 1$ contains two variables, x and y, each of its solutions is an ordered pair (x, y). There are infinitely many ordered pairs that will satisfy a given linear equation. In order to find solutions for a linear equation, select a value for x and substitute it for x in the equation; then solve the resulting equation for y. In listing the solution as an ordered pair, the x value is always the first value, and the y value is listed second: (x, y).

Since each ordered pair represents a point on the Cartesian plane, we can locate several solutions for a linear equation on the plane and connect these points to obtain the graph of the equation. When we graph a linear equation, it is usually convenient to find the points where the line crosses each axis, that is, the *x-intercept* and the *y-intercept.* We should also find a third solution to the equation to check that all three points do lie on the same straight line. If one of the points is not on the line, then we must go back and check for an error in determining the solutions to the given linear equation.

A solution for a pair of linear equations can be obtained by graphing both equations on the same set of axes. If the graphs of the two equations intersect at a point, the equations are *consistent,* and the coordinates of this point represent the solution for the given system. The coordinates of the point of intersection must satisfy both equations simultaneously. Two linear equations whose solution set is the empty set are called *inconsistent. Dependent* equations are those where the solution of one equation is also a solution of the other equation.

The graph of an equation of the form $y = ax^2 + bx + c$ is a parabola. If $a > 0$, then the parabola opens upward, whereas if $a < 0$, then the graph opens downward. The turning point of a parabola is called the *vertex* of the parabola. For any parabola of the form $y = ax^2 + bx + c$, the x-value of the vertex is $-b/2a$. The y-value is found by substituting $-b/2a$ for x in the original equation. After determining the vertex of the parabola and which way it opens, the next step in graphing the parabola is to make a table of values. This is done by choosing two or three values of x on each side of the vertex and finding the corresponding values of y. Then the points are plotted and connected by a smooth curve to obtain the graph of the parabola.

In order to graph an inequality in two variables, we do the following.

1. Find the boundary of the half-plane. We do this by graphing the equation derived from the inequality.
2. Make the boundary a dashed line for a strict inequality such as $>$ or $<$. But for $\geq$ or $\leq$, make the boundary a solid line.
3. Use shading to indicate the half-plane that is the solution. We do this by testing a point. If a point satisfies the given inequality, then all of the points in the half-plane must also satisfy the same inequality.

Graphs of linear inequalities can be used to solve some linear programming problems. To solve a linear programming problem, first translate the problem into mathematical statements. These mathematical statements can be obtained by placing all of the given information in a table. Next, we express the constraints in terms of linear inequalities in x and y, and graph this system of inequalities in the region of the Cartesian plane where $x \geq 0$ and $y \geq 0$. The solution set for the system of inequalities is a polygonal region, one of whose vertices will provide the desired maximum or minimum value for an expression of profit, cost, or some other quantity.

You should be aware that this chapter is intended only to provide an introduction to algebra. There are many topics in algebra that are beyond the scope of this text.

The two solutions of the general quadratic equation $ax^2 + bx + c = 0$ are

$$x = \frac{-b \pm \sqrt{b^2 - 4ac}}{2a}$$

This form of the solution is commonly known as the quadratic formula, and we can use it to solve any quadratic equation. To solve a quadratic equation by means of the formula, we must make sure that the equation is in standard form before we determine the values for a, b, and c. Also, when determining these values, we must be sure to include the correct sign for a, b, and c.

Vocabulary Check

abscissa	standard form	ordinate	inconsistent	origin	consistent
linear equation	objective function	coordinate	relation	dependent	function
Cartesian plane	linear programming	x-intercept	parabola	y-intercept	vertex
quadratic equation					

JUST FOR FUN

A train 1 mile long travels through a tunnel 1 mile long at a rate of 1 mile per hour. How long will it take the train to pass completely through the tunnel?

Review Exercises for Chapter 9

1. Find the solution set for each of the following open sentences. Unless otherwise noted, the replacement set is the set of real numbers.
   a. $x - 2 = 0$
   b. $x + 4 = -2$
   c. $x - 2 = 6 - 2$
   d. $x + 4 \leq 7$ (x is a natural number)
   e. $x - 2 < 4$ (x is a whole number)
   f. $-3 < x < 3$ (x is an integer)
   g. $-4 \leq x \leq 3$ (x is a natural number)

2. Graph the solution set (on a number line) for each of the following open sentences. In each case, the replacement set is the set of real numbers.
   a. $x < 3$ b. $x - 1 \geq 1$
   c. $-1 < x < 2$ d. $-2 \leq x \leq 1$
   e. $0 < x \leq 3$ f. $x - 2 < x$

3. Solve each of the following equations. (The replacement set is the set of real numbers.)
   a. $2x - 2 = 10$ b. $2y + 4 = y + 6$
   c. $5z - 3 = 2z + 3$ d. $2y + 6 = 5y - 3$
   e. $\frac{x}{2} - 2 = 3$ f. $\frac{z}{3} = 2 - z$

4. The sum of two consecutive odd integers is 28. Find the numbers.

5. Ike is 2 years older than Bill and the sum of their ages is 50. How old is each?

6. Five less than three times a number is 16. Find the number.

7. The width of a rectangle is 5 meters less than its length. If the perimeter of the rectangle is 30 meters, find the dimensions of the rectangle.

8. Frank has \$3.00 in dimes and quarters in his pocket. If he has twice as many quarters as dimes, how many of each type of coin does he have?

9. Ten less than four times a number is equal to twice that number. Find the number.

10. A piece of wire 64 meters long is bent into the shape of a rectangle whose length is three times its width. Find the dimensions of the rectangle.

11. In a collection of 50 coins, the number of quarters is five more than twice the number of nickels and the number of dimes is five less than twice the number of nickels. How many of each type of coin are in the collection?

12. Tickets for a particular concert cost \$5 each if purchased at the advance sale and \$7 each if bought at the box office on the day of the concert. For this particular concert, 1,200 tickets were sold and the receipts were \$6,700. How many tickets were bought at the box office on the day of the concert?

13. Find three solutions for each of the following equations, and then use the solutions to graph the equation on a Cartesian plane.
   a. $x - y = 2$ b. $2x + y = 4$
   c. $2x - y = 4$ d. $-2x + y = 6$
   e. $3x - 5y = 15$ f. $y = x$

14. Solve each system of equations graphically.
   a. $x + y = 1$, $2x + y = 3$ b. $x - y = 1$, $x + y = -5$ c. $2x - y = -3$, $-x + 2y = 0$

15. Graph the given equations on a Cartesian plane.
   a. $y = x^2$ b. $y = x^2 - 4$
   c. $y = -x^2 + 2$ d. $y = x^2 - 6x$
   e. $y = x^2 + 2x - 3$ f. $y = -x^2 + 6x - 5$

16. Graph each of the following inequalities on the Cartesian plane.
   a. $x + y > 3$ b. $x - y \leq 2$
   c. $-x + y > 2$ d. $3x + 5y > -15$
   e. $x - 2y \leq 0$ f. $-3x - 2y < -6$

17. Find the maximum value of $P = 2x + 3y$ under the following conditions.

$$x \geq 0,\ y \geq 0,\ 2x + y \leq 10,\ x + 3y \leq 15$$

18. A manufacturer makes couches and recliners. To produce a couch requires 3 hours in the frame shop and 2 hours in the upholstery shop. To produce a recliner requires 4 hours in the frame shop and 4 hours in the upholstery shop. The frame shop can operate at most 24 hours per day, and the upholstery shop can operate at most 20 hours per day. If the manufacturer makes a profit

of \$30 on a couch and \$45 on a recliner, how many of each should be produced in order to maximize profit?

19. The Safety-First Corporation manufactures two types of smoke alarms, a standard model and a deluxe model. To produce a standard smoke alarm requires 1 hour on machine A and 1 hour on machine B. To produce a deluxe smoke alarm requires 1 hour on machine A and 2 hours on machine B. Due to costs and safety regulations, machine A can operate at most 35 hours per week and machine B can operate at most 40 hours per week. If the manufacturer makes a profit of \$7 on each standard model and \$10 profit on each deluxe model, how many of each should the company produce in order to maximize profits?

Solve the following equations by means of the quadratic formula.

⋆20. $3y^2 - 4y - 4 = 0$

⋆21. $8x^2 = 6x - 1$

⋆22. $y^2 - y = 2$

⋆23. $x^2 + 2x = 2$

⋆24. $4x^2 - 4 = 0$

⋆25. $2y^2 - y - 1 = 0$

⋆26. $x^2 - 4x = 0$

⋆27. $2z^2 + 6z - 20 = 0$

Chapter Quiz

Determine whether the following statements are true or false.

1. The graph of $x < 1$ is a ray.
2. The graph of $-2 \le x \le 3$ is a line segment.
3. If the replacement set is the set of whole numbers, then the equation $x + 4 = 5 - 2$ has no solution.
4. $(4x + 2) \div 2 = 2x + 2$.
5. $4(5x + 3 - 2x + 1 - 3x) = 16$.
6. If $2x + 10 = 6x - 6$, then $x = 3$.
7. If $\frac{y}{3} + \frac{y}{2} = 5$, then $y = 6$.
8. If Mary is 5 years older than Margaret and the sum of their ages is 27, then Margaret is 16 and Mary is 11.
9. The width of a rectangle is 3 feet less than its length. If the perimeter is 30 feet, then its width is 6 feet and its length is 9 feet.
10. If $3x + 4y = 15$, then $(1, 3)$, $(-3, 6)$ and $(0, 5)$ are all solutions of the equation.
11. The graph of the equation $2x - 3y = 12$ passes through the point $(-2, 3)$.
12. The origin is located at the point where the x- and y-axes interest on the Cartesian plane.
13. The point at which a line crosses the x-axis is called the y-intercept.
14. The solution to the system of equations: $4x + y = 5$ and $-3x + y = 9$ is $(2, -3)$.
15. The system of equations $-2x + y = 1$ and $2x - y = 5$ is inconsistent.
16. The graph of the parabola $y = -x^2 + 6x$ opens downward and its vertex is located at the origin.
17. The boundary of the graph of $2x - y \le 6$ is a solid line.
18. The graph of $y - x \ge 0$ is located entirely to the right of the origin on the Cartesian plane.
19. The objective function in a linear programming problem describes the relationship between the items produced and the profit that can be realized from selling them.
20. Any maximum or minimum values of the objective function always occur at a vertex.
21. The system of equations $x + y = 5$ and $x + y = 2$ is dependent.
22. The set of ordered pairs $\{(1, 3), (2, 5), (3, 6), (1, 0)\}$ is a function.
23. The graph of $y = x - 2$ is a vertical straight line.

⋆24. The solution of $5y^2 = 6y - 1$ is $\{\frac{1}{5}, 1\}$.

⋆25. The quadratic equation $4x^2 - 4 = 0$ is in standard form.

CHAPTER 10

AN INTRODUCTION TO GEOMETRY

AFTER STUDYING THIS CHAPTER YOU WILL BE ABLE TO DO THE FOLLOWING:

1. Identify **points, lines, half-lines, planes, half-planes, rays,** and **line segments.**
2. Find the **intersection** and **union** of lines, rays, line segments and half-lines, including lines that are parallel or perpendicular, and horizontal or vertical.
3. Identify and find the measure of **acute angles, obtuse angles, right angles, straight angles, vertical angles, complementary angles, supplementary angles,** and **angles formed by two parallel lines cut by a transversal.**
4. Use the **Pythagorean theorem** to find the length of a side of a right triangle when given the measure of the other two sides.
5. Identify, from a figure or stated properties, and find the values of certain parts of **equilateral triangles, isosceles triangles, right triangles, acute triangles, obtuse triangles, equiangular triangles, similar triangles, congruent triangles.**
6. Identify **trapezoids, parallelograms, rhombuses, rectangles, squares,** and **pentagons,** from a figure or stated properties.
7. Find the **perimeter** of any given **polygon,** nonregular or regular.
8. Find the **circumference** of any given **circle.**
9. Find the **area** of **rectangles, squares, parallelograms, triangles, trapezoids,** and **circles.**
10. Find the **surface area** of certain **solids,** such as rectangular solids, right circular cylinders, and spheres.
11. Find the **volume** of rectangular **solids,** right circular cylinders, circular cones, and spheres.
12. Determine whether a **network** is **traversable** by identifying the number of even and odd vertices in the network.

SYMBOLS FREQUENTLY USED IN THIS CHAPTER

$\cdot C$	point C	$\overleftrightarrow{MN} \perp \overleftrightarrow{BT}$	line MN is perpendicular to line BT
$\overleftrightarrow{AB}$	line AB		
$\overrightarrow{AB}$	ray AB		
$\overline{AB}$	line segment AB	$\overleftrightarrow{MN} \parallel \overleftrightarrow{BT}$	line MN is parallel to line BT
$\overset{\circ\!\!\longrightarrow}{AB}$	half-line AB		
$\angle RST$	angle RST	$\triangle ABC$	triangle ABC
$m\angle RST$	measure of angle RST	∟	right angle
$\overset{\circ\!-\!\circ}{AB}$	open line segment AB	$\sim$	(is) similar to
$m(\overline{AB})$	measure of line segment AB	$\cong$	(is) congruent to
		$\square ABCD$	parallelogram $ABCD$

10.1 INTRODUCTION

What is geometry? To begin with, the word *geometry* is derived from two Greek words. The first part of the word *geometry* is taken from the Greek word *ge,* which means *earth,* and the second part is taken from the Greek word *metron,* which means measure. Therefore we can safely assume that early geometry concerned itself with the measure of the earth, or earth measurement.

The Egyptians were one of the first people to use geometry. One of the ways that the Egyptians used geometry was to survey land, or "measure earth." The Egyptians had to pay taxes for their land. The kings sent workers to measure the people's land, and taxes were levied accordingly. The Egyptians used this form of taxation as early as 1300 B.C.

Reproduced by courtesy of the Trustees of the British Museum

From this beginning, the study of geometry was carried on by the Greeks. Thales was one of the first Greeks to concern himself with proving that mathematical statements were true. Pythagoras was the next major contributor to the study of geometry (approximately 550 B.C.). You may be familiar with the Pythagorean theorem, which states that the sum of the squares of the legs of a right triangle is equal to the square of the hypotenuse.

The next great Greek scholar was Plato. He founded a school in the city of Athens. Plato considered geometry so important that he had the slogan "Let no one ignorant of geometry enter my doors" posted at the entrance to his school.

Euclid, another Greek, is known as the "father of geometry." In approximately 300 B.C., Euclid gathered all of the mathematical works known to exist at that time and produced a book called *The Elements*.

10.2 POINTS AND LINES

The most basic terms of geometry are point, line, and plane. Each of these terms presents a problem for us since we cannot define it. Granted, you can find descriptions for these words in the dictionary—but can you define them? For instance, what is a **point?** You might want to describe a point as a dot on a piece of paper, but what size dot? Other concepts of a point are of a point on the number line or a point on the Cartesian plane. The important thing to remember is that a geometric point has no dimension; that is, it has no length, breadth, or thickness. The one thing that a geometric point does possess is position. We can represent a point by a dot, just as we used the symbol "2" to represent the number two. Points are labeled or named by using capital letters.

Therefore, if we wish to represent a point C, we make a dot and label it with a capital C as shown.

$$\cdot C$$

We now have an intuitive idea of what a point is and how we shall represent it. We shall not define a point, just as we did not define a set in Chapter 1. (Recall that one of the ingredients of an axiomatic system is a collection of undefined terms that are used to define other terms.)

A **line** is formed by the intersection of two flat surfaces. For example, the intersection of two walls in one corner of a room can be thought of as a line. A point in geometry has only position, whereas a line in geometry has only one dimension, length. A geometric line has no width. Note that when we refer to a *line* we are talking about a *straight line,* unless otherwise noted.

We can think of a line as a set of points. The points are *on* the line and the line passes *through* the points. It is interesting to note that a line may be extended infinitely far in either direction. What this means is that a line has no end points. Normally we work with only pieces or parts of lines; these are described later in this section.

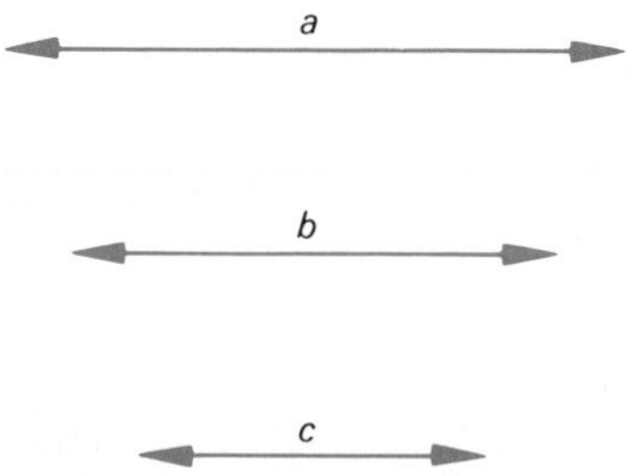

FIGURE 1

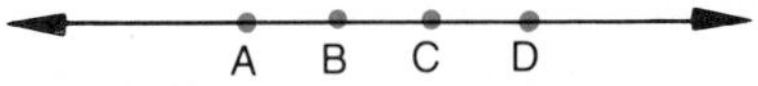

FIGURE 2

We can name a line in different ways. We can, for example, name a line by writing a lowercase letter such as a, b, c, or d near the line. Figure 1 shows three lines, line a, line b, and line c. It is more common to name a line by using two points on the line. The line shown in Figure 2 can be denoted by $\overleftrightarrow{AB}$, $\overleftrightarrow{AC}$, $\overleftrightarrow{BC}$, $\overleftrightarrow{CD}$, and so on. Since all the points A, B, C, and D lie on the same line, we say that they are *collinear.* **Collinear points** are points that lie on the same line.

We can name a line by using only two letters, such as $\overleftrightarrow{AB}$, because one and only one straight line can be drawn through any two points. This is another way of saying that any two different points determine a unique line; or we could say that there is exactly one line containing any two different points.

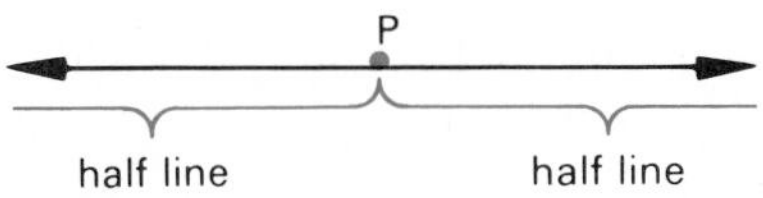

FIGURE 3

A point on a line separates one part of the line from another part. In fact, when we place a point on a line, as in Figure 3, we separate the line into three sets: the given point and two *half-lines.* The point P is not a point on either of the half-lines. A **half-line** is a set of points; if we include point P with the set of the points that constitute a half-line, we get what is known as a *ray.* A **ray** has only one end point and may be extended indefinitely in only one direction from that end point. Figure 4 illustrates ray AB. The notation used to denote ray AB is $\overrightarrow{AB}$.

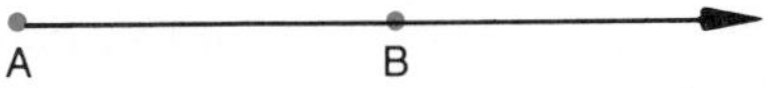

FIGURE 4

FIGURE 5

A **line segment** has two end points. The end points are used to name the segment, as in Figure 5. Line segment AB is denoted by

$\overline{AB}$. Remember that a line may be extended indefinitely in either direction, a ray can be extended indefinitely in only one direction, and a line segment cannot be extended at all.

Thus far in our discussion, we have covered the concepts of point, line, half-line, ray, and line segment. These terms and their corresponding notations are

Description	Diagram	Notation
point P	•	P
line PQ	P Q	$\overleftrightarrow{PQ}$
half-line PQ	P Q	$\overset{\circ\!\!\rightarrow}{PQ}$
ray PQ	P Q	$\overrightarrow{PQ}$
ray QP	P Q	$\overrightarrow{QP}$
line segment PQ	P Q	$\overline{PQ}$

Ray PQ is denoted by $\overrightarrow{PQ}$, which means that the end point of the ray is P and the ray is directed toward the point Q. Ray QP is denoted by $\overrightarrow{QP}$, which means that the end point of the ray is Q and the ray is directed toward the point P. (Refer to the chart.) Ray PQ ($\overrightarrow{PQ}$) and ray QP ($\overrightarrow{QP}$) are distinct; they involve different sets of points.

By now you have probably discovered that lines, rays, half-lines, and line segments that pass through a given pair of points have some points not in common. We can illustrate this by means of the two set operations, intersection and union. Consider line PQ in Figure 6. Using this line, what is $\overrightarrow{PQ} \cap \overrightarrow{QP}$?

FIGURE 6

$\overrightarrow{PQ} \cap \overrightarrow{QP}$ is the intersection of ray PQ and ray QP. Ray PQ consists of the set of points that has the end point P and is directed toward the right through Q. Ray QP consists of the set of points that has the end point Q and is directed toward the left through P. Their intersection is the set of points common to both rays, that is, line segment PQ.

$$\overrightarrow{PQ} \cap \overrightarrow{QP} = \overline{PQ}$$

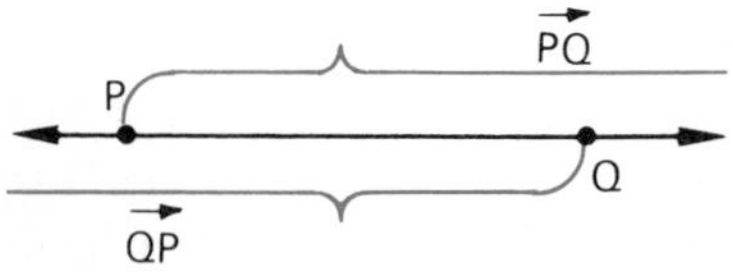

FIGURE 7

See Figure 7.

What is $\overrightarrow{PQ} \cup \overrightarrow{QP}$? We again refer to Figures 6 and 7, but instead of the intersection we want the union of the two sets of points. That gives us all of the points on the line PQ. Therefore, we have

$$\overrightarrow{PQ} \cup \overrightarrow{QP} = \overleftrightarrow{PQ}$$

EXAMPLE 1 Use the line in Figure 8 with the indicated points to find each of the following.

a. $\overrightarrow{AB} \cap \overrightarrow{CA}$ b. $\overrightarrow{AB} \cap \overrightarrow{BC}$
c. $\overrightarrow{BA} \cup \overrightarrow{BC}$ d. $\overline{AB} \cap \overline{CD}$

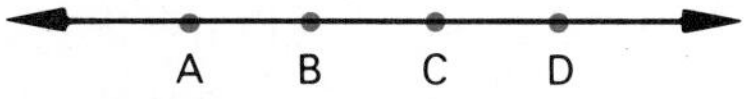

FIGURE 8

SOLUTION a. $\overrightarrow{AB} \cap \overrightarrow{CA}$ is the intersection of ray AB and ray CA. The set of points common to both of these rays is line segment AC. Therefore $\overrightarrow{AB} \cap \overrightarrow{CA} = \overline{AC}$.

b. $\overrightarrow{AB} \cap \overrightarrow{BC}$ is the intersection of ray AB and ray BC. These two rays are both directed to the right and have all of the points in ray BC in common. Therefore $\overrightarrow{AB} \cap \overrightarrow{BC} = \overrightarrow{BC}$.

c. $\overrightarrow{BA} \cup \overrightarrow{BC}$ is the union of the two rays, and, because they are directed in opposite directions, their union will result in all of the points on the line. Therefore $\overrightarrow{BA} \cup \overrightarrow{BC} = \overleftrightarrow{AC}$. Note that it would also be correct to denote the answer as $\overleftrightarrow{AB}$, $\overleftrightarrow{BC}$, $\overleftrightarrow{CD}$, $\overleftrightarrow{BD}$, and so on.

d. $\overline{AB} \cap \overline{CD}$ is intersection of line segment AB and line segment CD. Examining the diagram, we see that there are no points common to these two line segments. Their intersection is empty. Therefore, $\overline{AB} \cap \overline{CD} = \varnothing$.

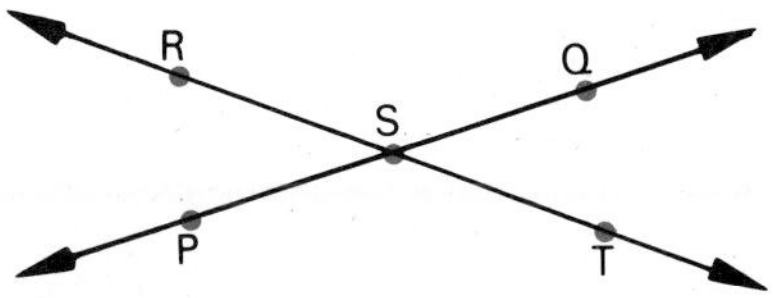

FIGURE 9

When we have two different lines that contain the same point, these lines are said to **intersect** at that point. In Figure 9, lines PQ and RT intersect in point S; that is, $\overleftrightarrow{PQ} \cap \overleftrightarrow{RT} = \{S\}$.

The set of points formed by two intersecting lines has many interesting subsets. For example, consider $\overrightarrow{SQ} \cup \overrightarrow{ST}$. What is the result when we unite the sets of points in ray SQ and ray ST? The result is a geometric figure formed by two rays drawn from the same point, as shown in Figure 10. This figure is called an angle. An **angle** ($\angle$) is the union of two rays that have a common end point. The rays are called the **sides** of the angle and the common end point is called the **vertex** of the angle. Therefore

$$\overrightarrow{SQ} \cup \overrightarrow{ST} = \angle QST$$

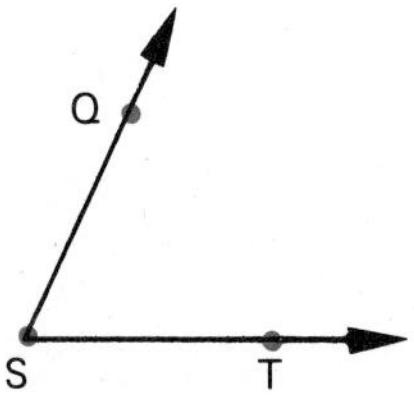

FIGURE 10
$\overrightarrow{SQ} \cup \overrightarrow{ST} = \angle QST$

We use capital letters to label an angle, and the name of the angle is written with the vertex letter in the middle. The first and third letters are used to designate the sides of the angle. In Figure 9, we can see that

$$\overrightarrow{SR} \cup \overrightarrow{SP} = \angle RSP$$
$$\overrightarrow{SR} \cup \overrightarrow{SQ} = \angle RSQ$$
$$\overrightarrow{SP} \cup \overrightarrow{ST} = \angle PST$$

Angles RSP and RSQ are **adjacent angles** because they have the same vertex and a common side between them. Angles RSQ and QST are also adjacent angles, but angles RSP and QST are *not* adjacent angles. They do have the same vertex, but they do not have a common side. Angles RSP and QST are angles where the sides of one angle extend through the vertex and form the sides of the other. Angles of this type are called **vertical angles.** In Figure 9, angles RSQ and PST are also vertical angles.

Consider the union of rays SR and ST in Figure 9. We see that $\overrightarrow{SR} \cup \overrightarrow{ST} = \overleftrightarrow{RT}$. But because an angle is the union of two rays that have a common end point, we can also say that $\overrightarrow{SR} \cup \overrightarrow{ST} = \angle RST$. Angle RST is a special kind of angle because its sides form a straight line. Angle RST is referred to as a **straight angle.** Therefore a line such as line PR in Figure 11 can also be thought of as $\angle PQR$. (Remember that the vertex letter of an angle is always written in the middle when we label the angle.)

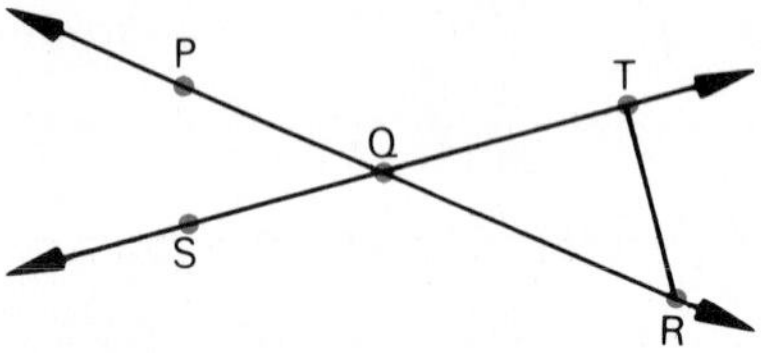

FIGURE 11

EXAMPLE 2 Use Figure 11 with the indicated points to find each of the following:

a. $\overleftrightarrow{PR} \cap \overleftrightarrow{ST}$ b. $\overrightarrow{QP} \cup \overrightarrow{QT}$
c. $\overrightarrow{PR} \cap \overrightarrow{RP}$ d. $\overrightarrow{QP} \cup \overrightarrow{QR}$

SOLUTION a. $\overleftrightarrow{PR} \cap \overleftrightarrow{ST}$ is the intersection of lines PR and ST, and the two lines intersect at point Q. Therefore, $\overleftrightarrow{PR} \cap \overleftrightarrow{ST} = \{Q\}$.

b. $\overrightarrow{QP} \cup \overrightarrow{QT}$ is the union of rays QP and QT. The union of two rays that have a common end point is an angle. Therefore we see that $\overrightarrow{QP} \cup \overrightarrow{QT} = \angle PQT$.

c. $\overrightarrow{PR} \cap \overrightarrow{RP}$ is the intersection of two rays having opposite direction. The set of points that they have in common is line segment PR. Therefore $\overrightarrow{PR} \cap \overrightarrow{RP} = \overline{PR}$.

d. $\overrightarrow{QP} \cup \overrightarrow{QR}$ is the union of two rays of opposite direction, so their union will result in all of the points on the line. Therefore $\overrightarrow{QP} \cup \overrightarrow{QR} = \overleftrightarrow{PR}$. (*Note:* Recall that we can also say that $\overrightarrow{QP} \cup \overrightarrow{QR} = \angle PQR$.)

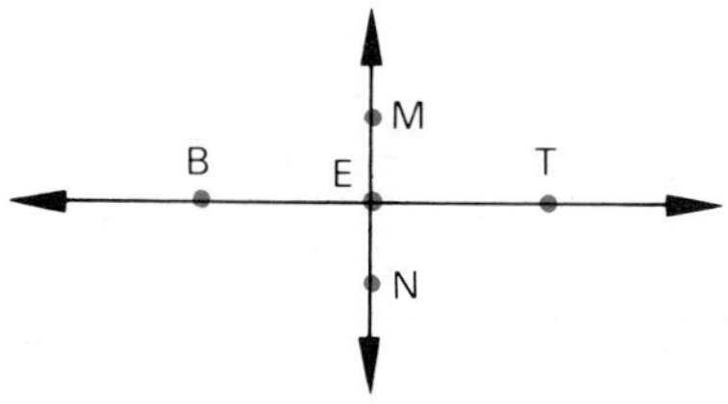

FIGURE 12

There are many other kinds of angles that can be defined. However, before we can discuss these angles, we must define *perpendicular lines.* Two lines that intersect so as to form a pair of congruent adjacent angles are called **perpendicular lines.** Each line is said to be perpendicular to the other. In Figure 12, $\overleftrightarrow{MN}$ is perpendicular to $\overleftrightarrow{BT}$; we denote this by $\overleftrightarrow{MN} \perp \overleftrightarrow{BT}$. Note also that $\overleftrightarrow{EM} \perp \overleftrightarrow{BT}$.

A right angle is an angle whose sides are perpendicular. In Figure 12, $\angle MET$ is a **right angle,** as are $\angle BEM$, $\angle BEN$, and $\angle NET$. An angle that is wider than a right angle but narrower than a straight angle is called an **obtuse angle.** An angle that is narrower than a right angle is called an **acute angle.** In Figure 13, $\angle FID$ is an obtuse angle, and $\angle DIG$ is an acute angle.

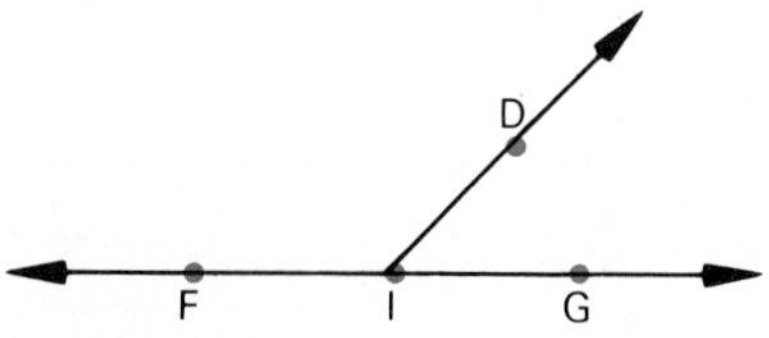

FIGURE 13

EXAMPLE 3 Use Figure 13 with the indicated points to find each of the following:

a. $\overrightarrow{FG} \cap \overrightarrow{ID}$ b. $\overrightarrow{IF} \cup \overrightarrow{ID}$
c. $\overline{FI} \cap \overline{IG}$ d. $\angle FID \cap \angle DIG$

SOLUTION a. $\overrightarrow{FG} \cap \overrightarrow{ID}$ is the intersection of two rays; the only thing they have in common is point I. Therefore $\overrightarrow{FG} \cap \overrightarrow{ID} = \{I\}$.

b. $\overrightarrow{IF} \cup \overrightarrow{ID}$ is the union of two rays with a common end point. Therefore $\overrightarrow{IF} \cup \overrightarrow{ID} = \angle FID$.

c. $\overline{FI} \cap \overline{IG}$ is the intersection of two line segments with point I in common. Therefore, $\overline{FI} \cap \overline{IG} = \{I\}$.

d. $\angle FID \cap \angle DIG$ is the intersection of two angles. They are adjacent angles because they have the same vertex and a common side. The intersection of these two angles is the common side, namely, $\overrightarrow{ID}$. Therefore $\angle FID \cap \angle DIG = \overrightarrow{ID}$.

Remember that any dot we place on a piece of paper will have some measurement, such as height, width, and even thickness. The same can be said for a line segment that we may draw on paper; that is, any line that we draw will have some width and thickness in addition to length. But in geometry points and lines do not have such characteristics. The points and lines that we draw are diagrams that represent the points and lines of geometry.

EXERCISES FOR SECTION 10.2

For Exercises 1–14, use Figure 14 to find each of the following:

1. $\overline{QI} \cap \overline{IK}$
2. $\overline{QI} \cap \overline{CK}$
3. $\overline{QI} \cup \overline{IK}$
4. $\overline{UI} \cap \overline{CK}$
5. $\overrightarrow{QI} \cap \overline{QU}$
6. $\overrightarrow{QI} \cap \overrightarrow{CI}$
7. $\overrightarrow{IU} \cup \overrightarrow{IK}$
8. $\overleftrightarrow{UI} \cap \overline{CK}$
9. $\overrightarrow{IK} \cap \overline{IU}$
10. $\overrightarrow{IK} \cup \overline{IU}$
11. $\overrightarrow{IK} \cap \overrightarrow{IQ}$
12. $\overrightarrow{IK} \cap \overrightarrow{IC}$
13. $\overrightarrow{CI} \cap \overline{QU}$
14. $\overline{QU} \cap \overrightarrow{IC}$

FIGURE 14

For Exercises 15–28, use Figure 15 to find each of the following:

15. $\overleftrightarrow{BT} \cap \overleftrightarrow{SW}$
16. $\overrightarrow{BT} \cap \overrightarrow{SW}$
17. $\overrightarrow{TB} \cap \overrightarrow{WS}$
18. $\overline{BE} \cup \overline{ET}$
19. $\overline{SE} \cap \overline{ET}$
20. $\overline{BE} \cap \overline{ET}$
21. $\overrightarrow{ET} \cup \overrightarrow{EW}$
22. $\overrightarrow{EB} \cup \overrightarrow{ES}$
23. $\overrightarrow{ET} \cup \overrightarrow{EB}$
24. $\overrightarrow{BE} \cap \overrightarrow{ET}$
25. $\overrightarrow{BE} \cup \overrightarrow{ET}$
26. $\overrightarrow{EW} \cup \overrightarrow{EB}$
27. $\angle BEW \cap \angle WET$
28. $\angle SET \cap \angle WET$

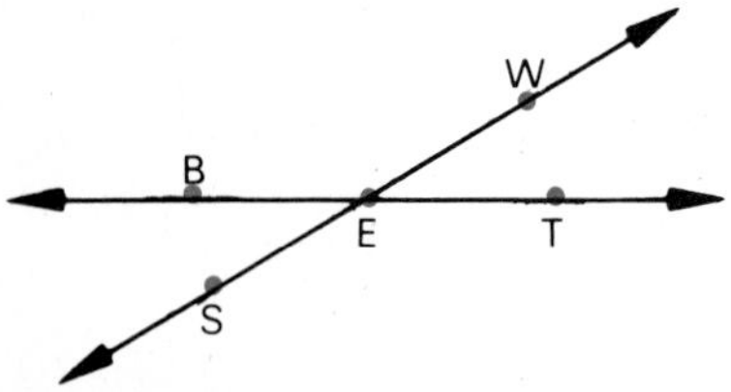

FIGURE 15

For Exercises 29–41, use Figure 16 to find each of the following:

29. $\overleftrightarrow{WM} \cap \overleftrightarrow{ZP}$
30. $\overrightarrow{ZA} \cap \overrightarrow{PA}$
31. $\overrightarrow{AP} \cup \overrightarrow{AZ}$
32. $\overrightarrow{AP} \cap \overrightarrow{AZ}$
33. $\overline{ZW} \cap \overline{MP}$
34. $\overrightarrow{AP} \cup \overrightarrow{AM}$
35. $\overrightarrow{AZ} \cup \overrightarrow{AW}$
36. $\overrightarrow{PA} \cap \overline{ZW}$
37. $\angle PAM \cap \angle ZAM$
38. $\angle PAW \cap \angle ZAW$
39. $\angle PAM \cap \angle ZAW$
40. $\angle ZAP \cap \angle WAM$

★41. $\overline{ZA} \cup \overline{AW} \cup \overline{ZW}$

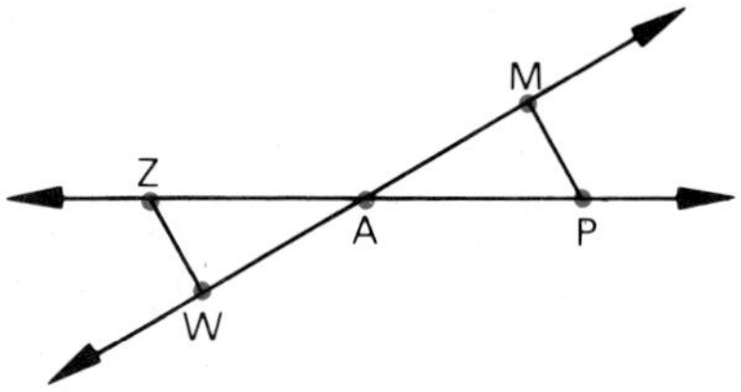

FIGURE 16

JUST FOR FUN

In this section we discovered that one and only one straight line can be drawn through any two given points. We can also say that two points determine a unique line. How many points determine a unique circle? What is the minimum number of points required?

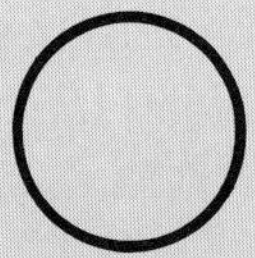

10.3 PLANES

A **plane** can be thought of as a flat surface such as the floor of a room, or a table top, or a desk top. A plane or surface divides, or separates, one portion of space from another. A floor separates the space above the floor from the space below the floor. A wall is a plane that separates the space in one room from the space in the adjoining room. The wall of a building is a plane that separates the space inside from the space outside. In Figure 17, each of the faces of the pyramid is a flat, or plane, surface.

A plane has two dimensions, length and width. In geometry, a plane does not have any thickness. **Note, a flat surface suggests a plane, and a true plane is endless or infinite.**

Just as *point* and *line* are undefined terms in geometry, so is *plane*. Although our concept of a plane is intuitive, we can still discuss some of the properties of a plane. For example, we can think of a line as a set of points, and we can do the same for a plane. A plane is a set of points. The points are on the plane, and the plane contains the points. **Coplanar** points are points that are on the same plane, just as collinear points are points that are on the same line.

A unique plane is determined by any three noncollinear points. In other words, if we are given three distinct points that are not all on the same line, then there is one and only one plane that can contain all three points. For example, in Figure 17 points *F, S,* and *T* are not on the same line, and they determine a unique plane, namely, plane *FST*. Note that none of the other planes in Figure 17 contains all three of these points. The other planes do contain two of the points; by using a different third point, a different plane is determined.

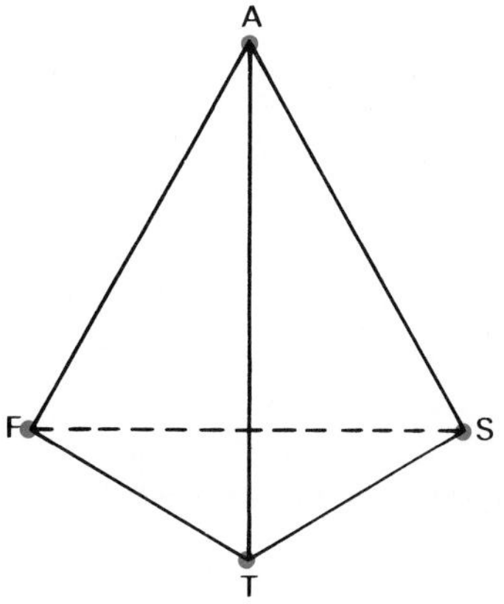

FIGURE 17

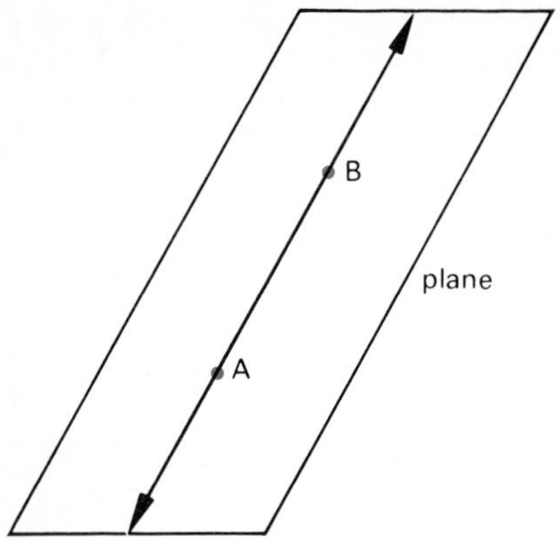

FIGURE 18

Have you ever noticed that things such as easels, telescopes, cameras, and Christmas trees are mounted on stands with three legs? These stands, or tripods, have only three legs because three legs will always rest on some plane. A stand, table, or chair with four legs tends to "wobble" unless all four legs are exactly the same length; this is not the case for an object with three legs.

Since lines and planes are both composed of points, we can make the following observation: If two different points of a line are on a plane, then all the points on the line are also on the plane. In other words, if two points of a line are on a plane, then the line must also be on the plane, since two points determine a line. For an example, see line *AB* in Figure 18.

Figure 18 illustrates another important concept regarding the relationship between lines and planes. Any line on a given plane divides that plane into two half-planes. Note that in Figure 18 $\overleftrightarrow{AB}$ separates the points on the plane. There are those points to the right of $\overleftrightarrow{AB}$, and those points to the left of $\overleftrightarrow{AB}$. The points that are on $\overleftrightarrow{AB}$ are not points on either half-plane. This concept is similar to that of a point dividing a line into two half-lines. A line on a plane divides the plane into two half-planes, but the result is three sets of points: the points on the line, the points on one half-plane, and the points on the other half-plane.

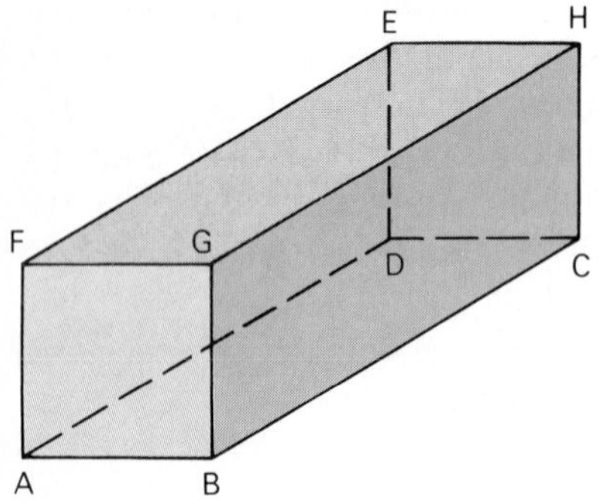

FIGURE 19

Either two different planes intersect or they do not. That is, two distinct planes must meet in a line or they must be parallel. The plane of a wall intersects the plane of the floor; that is, they meet in a line. But the floor and ceiling planes will never meet in a line, no matter how far they are extended. These two planes are **parallel planes.** In Figure 19, planes *ABCD* and *EFGH* are parallel planes, but planes *ABCD* and *BGHC* intersect in a line.

It should also be noted that lines *BC* and *AD* in Figure 19 are parallel lines. We can denote this by $\overleftrightarrow{BC} \parallel \overleftrightarrow{AD}$. Two lines that are on the same plane and do not intersect, however far they are extended, are called **parallel lines.** In Figure 19, lines *AB* and *GH* will not intersect, no matter how far we extend them, but they are not parallel lines. Why not? They are not parallel lines because they are not in the same plane. If two lines do not lie in the same plane, they are *skew lines*.

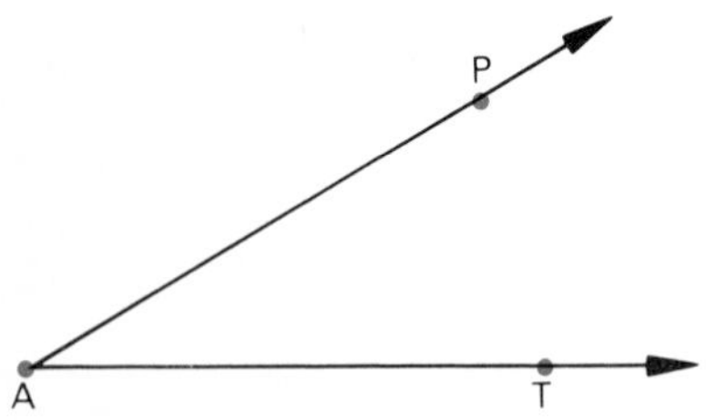

FIGURE 20

In the preceding section, we noted that an angle is the union of two rays that have a common end point, like angle *PAT* in Figure 20. An angle on a plane divides that plane into two half-planes, and that angle produces three sets of points: the points on the angle, the points on one half-plane, and the points on the other half-plane. Consider this page as a plane. It is a set of points, and angle *PAT* divides those points into three parts. There are those points on angle *PAT*, such as point *P*. There are also points inside angle *PAT;* that is, they are in the interior of angle *PAT*. Finally, there are points outside angle *PAT;* that is, in the

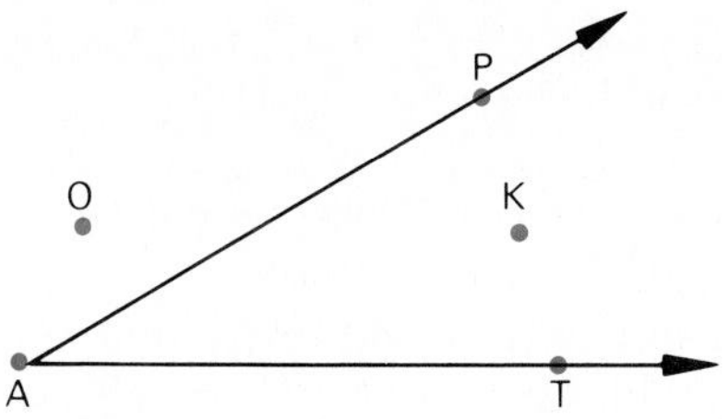

FIGURE 21

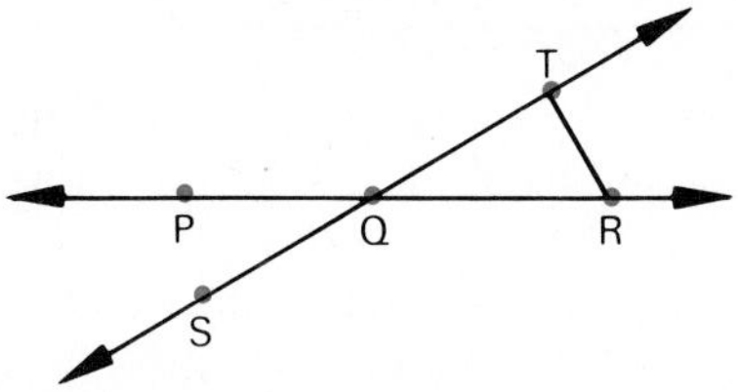

FIGURE 22

exterior of angle PAT. In Figure 21, point O is an exterior point, and point K is an interior point. Note that points P, A, and T are neither interior nor exterior points, because they are on the angle.

Let's combine these new concepts of interior and exterior points with set intersection. Consider Figure 22 with the indicated points. What is the intersection of angle TQR and $\overline{TR}$? That is, what is $\angle TQR \cap \overline{TR}$? The given figure is on a plane, namely, this page. An angle is the union of two rays with a common end point. When we are discussing $\angle TQR$, we are in effect discussing those points on the angle. Therefore, the points that $\angle TQR$ and $\overline{TR}$ have in common are points T and R; that is $\angle TQR \cap \overline{TR} = \{T, R\}$.

Now consider the intersection of $\overline{TR}$ and the interior of $\angle TQR$, that is, $\overline{TR} \cap$ (interior $\angle TQR$). The points that these two sets of points have in common are those points that are in the interior of $\angle TQR$ and are also on line segment TR. Is the answer $\overline{TR}$? No. We cannot list our answer as line segment TR because that would mean that points T and R are also members of the solution, and from our previous discussion we know that points T and R are on $\angle TQR$; therefore, they cannot be inside $\angle TQR$. Our solution consists of all those points on $\overline{TR}$ except T and R. This is an open line segment, and we denote it by $\overset{\circ\!-\!\circ}{TR}$. Therefore, $\overline{TR} \cap$ (interior $\angle TQR$) $= \overset{\circ\!-\!\circ}{TR}$.

What is the intersection of ray QP and the exterior of angle TQR? That is, what is $\overrightarrow{QP} \cap$ (exterior $\angle TQR$)? The exterior of $\angle TQR$ consists of all those points that are not on $\angle TQR$ and not inside $\angle TQR$. Now all of the points on ray QP lie outside of $\angle TQR$, except one point, namely Q. Therefore, the solution is the half-line QP (excluding Q), and we denote this by $\overset{\circ\!\rightarrow}{QP}$. Therefore, $\overrightarrow{QP} \cap$ (exterior $\angle TQR$) $= \overset{\circ\!\rightarrow}{QP}$.

EXAMPLE 1 Consider Figure 23 with the indicated points. Find each of the following:

a. $\overrightarrow{AR} \cup \overrightarrow{AN}$
b. $\overrightarrow{AF} \cap$ (exterior $\angle RAN$)

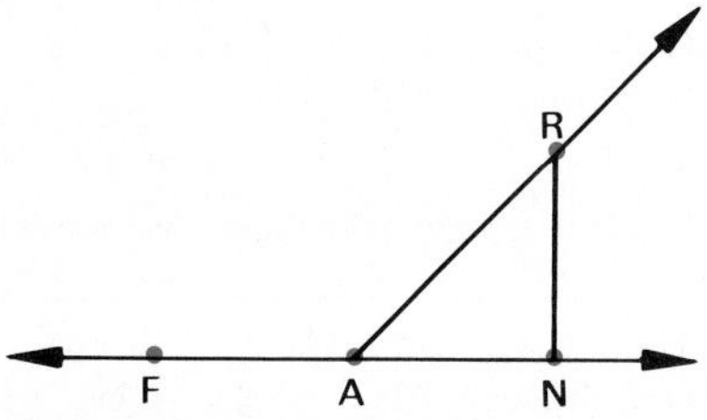

FIGURE 23

c. $\overline{RN} \cap (\text{interior } \angle RAN)$
d. $(\text{interior } \angle RAN) \cap (\text{exterior } \angle RAN)$

SOLUTION

a. $\overrightarrow{AR} \cup \overrightarrow{AN}$ is the union of two rays with a common end point, that is, an angle. Therefore, $\overrightarrow{AR} \cup \overrightarrow{AN} = \angle RAN$.

b. $\overrightarrow{AF} \cap (\text{exterior } \angle RAN)$ is the intersection of those points on $\overrightarrow{AF}$ with those points that are in the exterior of $\angle RAN$. This is all the points on $\overrightarrow{AF}$ except A. Therefore, $\overrightarrow{AF} \cap (\text{exterior } \angle RAN) = \overset{\circ\!\!\longrightarrow}{AF}$.

c. $\overline{RN} \cap (\text{interior } \angle RAN)$ is the intersection of those points that are on $\overline{RN}$ and also in the interior of $\angle RAN$. This is all the points on $\overline{RN}$ except R and N. Therefore, $\overline{RN} \cap (\text{interior } \angle RAN) = \overset{\circ\!\!-\!\!\circ}{RN}$.

d. The set of points that are in the interior of $\angle RAN$ and the points that are in the exterior of $\angle RAN$ do not intersect. They are separated by those points that are on $\angle RAN$. Therefore, $(\text{interior } \angle RAN) \cap (\text{exterior } \angle RAN) = \varnothing$.

American Petroleum Institute/Courtesy of Texaco, Inc.

This surveyor's instrument is mounted on a three-legged stand for stability on any kind of terrain.

Remember that a plane is a flat surface, and that it separates one part of space from an adjoining part of space. It is also a set of points. A line on a plane divides that plane into half-planes, but there are other possible relationships between lines and planes. Line AB in Figure 24 is an example of a line on a plane; since two points of $\overleftrightarrow{AB}$ are on the plane, all of the points on $\overleftrightarrow{AB}$ are on the plane. Line CD intersects the plane at only one point, while line EF does not intersect the plane at all. Line EF is parallel to the

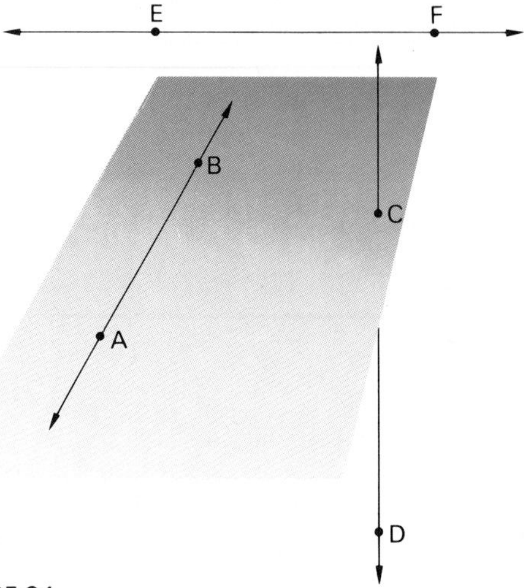

FIGURE 24

given plane and skew to line AB. A line parallel to the horizon is called a horizontal line, while a line that is perpendicular to a horizontal line is called a vertical line.

EXERCISES FOR SECTION 10.3

For Exercises 1–8, consider Figure 25 with the indicated points. Find each of the following.

1. $\overline{AR} \cap \overrightarrow{RS}$
2. $\overrightarrow{RA} \cap \overrightarrow{RS}$
3. $\overline{AS} \cap (\text{interior } \angle ARS)$
4. $\overline{AS} \cap (\text{exterior } \angle ARS)$
5. $\overrightarrow{RA} \cup \overrightarrow{RD}$
6. $\overrightarrow{RS} \cap (\text{interior } \angle DRA)$
7. $\overrightarrow{RS} \cap (\text{exterior } \angle DRA)$
8. $(\text{interior } \angle DRA) \cap (\text{exterior } \angle ARS)$

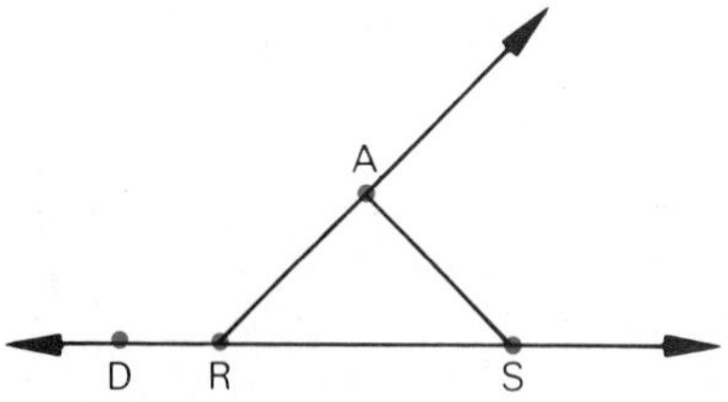

FIGURE 25

For Exercises 9–18, consider Figure 26 with the indicated points. Find each of the following.

9. $\overrightarrow{OT} \cup \overrightarrow{ON}$
10. $\overrightarrow{OT} \cap \overrightarrow{ON}$
11. $\overrightarrow{OE} \cap (\text{interior } \angle TON)$
12. $(\text{interior } \angle TOE) \cap (\text{exterior } \angle EON)$
13. $(\text{exterior } \angle TOL) \cap (\text{interior } \angle EON)$

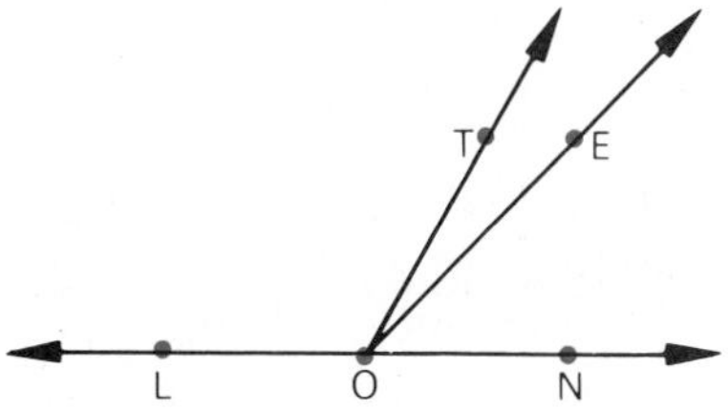

FIGURE 26

14. $(\text{interior } \angle EON) \cup (\text{interior } \angle TOE)$
15. $(\text{interior } \angle TOE) \cap (\text{interior } \angle EON)$
16. $(\text{exterior } \angle EON) \cap \overrightarrow{OL}$
17. $(\text{exterior } \angle TOE) \cap (\text{interior } \angle EON)$
18. $(\text{interior } \angle TOE) \cap (\text{exterior } \angle LOT)$

For Exercises 19–30, consider Figure 27 with the indicated points. Find each of the following.

19. $\overrightarrow{OT} \cup \overrightarrow{OC}$
20. $\overrightarrow{OC} \cup \overrightarrow{OS}$
21. $\overrightarrow{OJ} \cup \overrightarrow{OS}$
22. $\overrightarrow{OS} \cap \overrightarrow{OT}$
23. $(\text{interior } \angle SOC) \cap (\text{interior } \angle SOT)$
24. $(\text{exterior } \angle COT) \cap (\text{interior } \angle JOS)$
25. $\overrightarrow{OT} \cap (\text{exterior } \angle JOS)$
26. $\angle COT \cap \angle SOC$
27. $\angle JOS \cap \angle COT$
28. $(\text{exterior } \angle COT) \cap (\text{exterior } \angle COS)$
29. $(\text{exterior } \angle SOJ) \cap (\text{exterior } \angle SOC)$
30. $(\text{interior } \angle JOS) \cap (\text{interior } \angle JOC)$

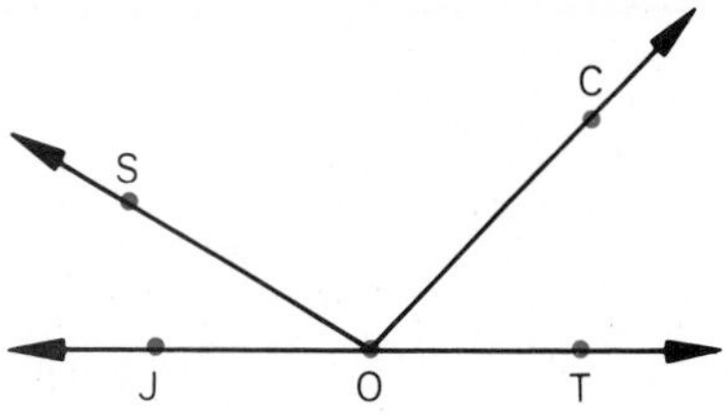

FIGURE 27

For Exercises 31–38, indicate whether each of the following is true or false.

31. Two parallel planes will never intersect.
32. The intersection of two planes that are not parallel is a line.

33. The intersection of a line and a plane is never a point.

34. Two lines on the same plane must intersect at some point.

35. Skew lines intersect at some point.

36. Parallel lines intersect at some point.

37. An angle is formed by the union of two rays with a common end point.

38. The interior of an angle is the intersection of two half-planes.

39. Three intersecting planes can have (**a**) no common intersection, (**b**) intersection in a common line, or (**c**) intersection in one common point. Conditions (**a**) and (**b**) are shown in Figure 28. Can you draw three planes so that their intersection is one common point?

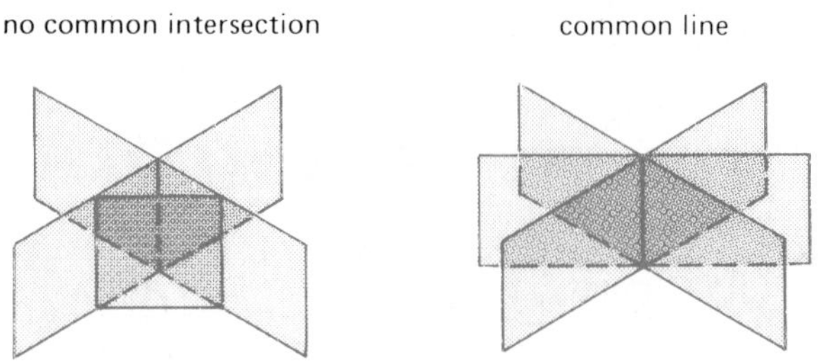

FIGURE 28

JUST FOR FUN

Irene and Tom were each hired to work at a camp for 20 days. They had two choices regarding their salary: (A) they could receive $25 per day, or (B) receive 1 cent the first day, 2 cents the second day, 4 cents the third day, and so on. That is, the first day's pay was to be $0.01 and each succeeding day's pay would be twice the previous day's pay.

Irene promptly chose plan B, whereas Tom thought awhile and then chose plan A. The person in charge then announced that Irene would be chief counselor and Tom assistant counselor. Why?

10.4 ANGLES

In the previous sections we have discussed angles to some extent. Thus far, you should know that an **angle** is the union of two rays that have a common end point. The rays are the *sides* of the angle and the common end point is the *vertex* of the angle. When we label an angle, we use three capital letters, and put the vertex letter in the middle.

Thus far, we have discussed three different kinds of angles: **adjacent angles, vertical angles,** and **straight angles.** Adjacent angles are angles that have the same vertex and a common side between them. Vertical angles are two angles such that the sides of one angle are extended through the vertex to form the sides of

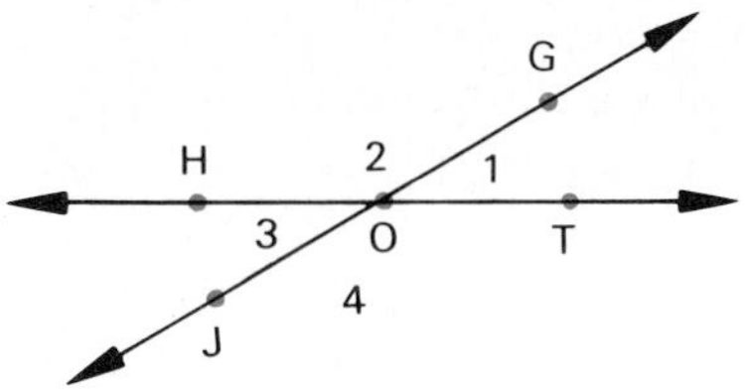

FIGURE 29

the other angle. A straight angle is an angle whose sides lie on the same straight line and extend in opposite directions from the vertex.

We can use Figure 29 to illustrate these various angles. Angle *GOT* and angle *TOJ* are adjacent angles: they have the same vertex and a common side between them. Angle *HOJ* and angle *GOT* are vertical angles: the sides of one angle are extended through the vertex and form the sides of the other angle. Angle *JOG* is an example of a straight angle: its sides lie on the same straight line and extend in opposite directions from the vertex. Note that angle *GOT* can also be named $\angle 1$ and similarly that angle $GOH = \angle 2$, $\angle HOJ = \angle 3$, $\angle JOT = \angle 4$.

EXAMPLE 1 Name six different angles in Figure 30. Line *AS* is a straight line.

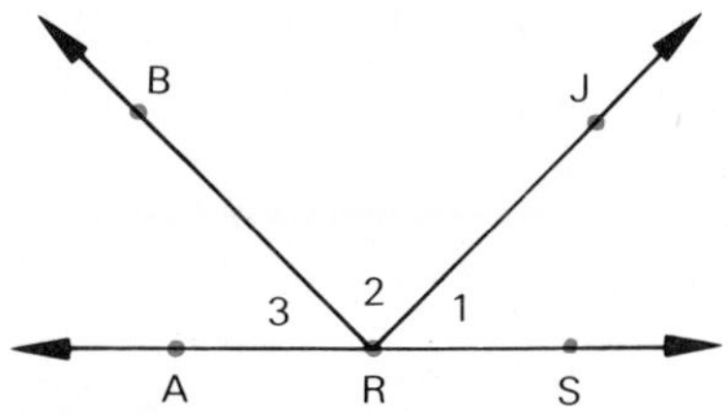

FIGURE 30

SOLUTION At first glance, there may appear to be only three angles in the given figure. But remember that an angle is formed by the union of two rays that have a common end point. We shall number the angles as we list them, but this does not mean that they must be listed in this order.

1. $\angle JRS$ or $\angle 1$	3. $\angle BRA$ or $\angle 3$	5. $\angle JRA$
2. $\angle JRB$ or $\angle 2$	4. $\angle SRB$	6. $\angle ARS$

There do exist other types of angles, and we shall examine them in this section.

A **right angle** is an angle whose sides are perpendicular. In Figure 31, $\overleftrightarrow{MN}$ is perpendicular to $\overleftrightarrow{BT}$; we denote this by $\overleftrightarrow{MN} \perp \overleftrightarrow{BT}$. Note also that $\overleftrightarrow{EM} \perp \overleftrightarrow{BT}$. In Figure 31, $\angle MET$ is a right angle, as are $\angle BEM$, $\angle BEN$, and $\angle NET$.

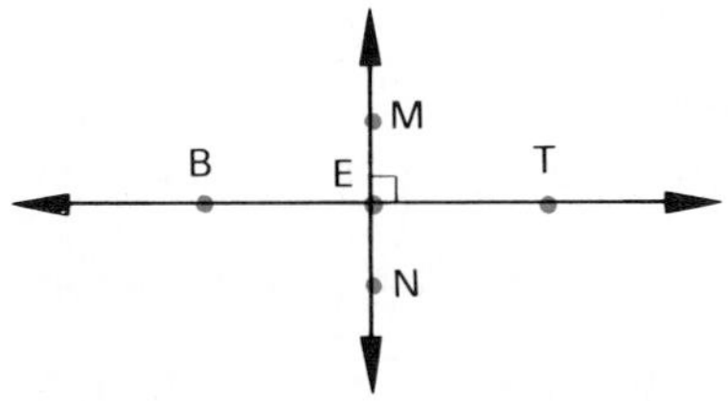

FIGURE 31

Angles are usually measured in **degrees.** If the measure of an angle is 90 degrees, this is denoted by 90°. Just as feet can be divided into inches, and meters can be divided into centimeters, we can divide a degree into smaller units. This usually is done when better accuracy in measuring an angle is desired. Each

HISTORICAL NOTE

Born in Germany, and the daughter of a mathematician, Amalie Emmy Noether (1882–1935) grew up in a world of mathematics. She received a degree from the University of Erlangen, and in 1916 Noether began further studies at the University of Göttingen, where she worked with two well-known and respected mathematicians, Hilbert and Klein. Together they developed the mathematical aspects of the theory of relativity.

At that time women were not allowed to hold faculty rank at universities; however, at the insistence of the mathematics faculty she was given a non-faculty teaching position. She taught classes of enthusiastic students at Göttingen until 1933 when the Nazis came to power and people of Jewish descent were denied the right to teach. Emmy Noether then came to the United States and taught at Bryn Mawr College until her death. Her greatest contributions were in the area of abstract algebra, particularly with the theory of rings.

degree is divided into 60 smaller units called **minutes.** A minute is denoted by $'$. Therefore, $1° = 60'$. Similarly, a minute is divided into 60 smaller units called **seconds.** A second is denoted by $''$. Therefore, $1' = 60''$. The fact that a degree is divided into 60 equal minutes and a minute is divided into 60 equal seconds is interesting, especially if we recall from Chapter 7 that the Babylonians used a sexagesimal system of numeration; that is, their system was based on 60.

The measure of a straight angle is 180°, and the measure of a right angle is 90°. Given the measure of an angle, we can find the measure of its **supplement** and its **complement.** Two angles whose measures sum to that of a straight angle (180°) are called *supplementary angles,* and two angles whose measures sum to that of a right angle (90°) are called *complementary angles.* Therefore, if we want to find the measure of the supplement of an angle which measures 60°, we must subtract 60° from 180°.

$$180° - 60° = 120°$$

In order to find the measure of the complement of an angle measuring 60°, we must subtract 60° from 90°.

$$90° - 60° = 30°$$

Suppose we want to find the measure of the complement of an angle measuring 42°45′. Since we can't subtract 42°45′ from 90° we write 90° as 89°60′, because a degree is divided into 60 minutes. Therefore, we have

$$\begin{array}{r} 89°60' \\ -42°45' \\ \hline 47°15' \end{array}$$

EXAMPLE 2 Find the measure of the complement of angles with each of the following measures.

a. 62° b. 45° c. 22°15′

SOLUTION Two angles whose measures sum to that of a right angle are called complementary angles. Therefore, in order to find the measure of the complement of a given angle, we must subtract the measure of that angle from 90°, since the measure of a right angle is 90°.

a. $90° - 62° = 28°$. Therefore, the measure of the complement of an angle measuring 62° is 28°.

b. $90° - 45° = 45°$. Therefore, the measure of the complement of an angle measuring 45° is 45°.

c. In order to find the measure of the complement of an angle measuring 22°15′, we rewrite 90° as 89°60′, and then subtract 22°15′ from it:

$$\begin{array}{r} 89°60' \\ -22°15' \\ \hline 67°45' \end{array}$$

Therefore, the measure of the complement of an angle measuring 22°15′ is 67°45′.

EXAMPLE 3 Find the measure of the supplement of angles with each of the following measures.

a. 62° b. 22°15′ c. 128°42′16″

SOLUTION Two angles whose measures sum to that of a straight angle are called supplementary angles. Therefore, in order to find the measure of the supplement of a given angle, we must subtract the measure of that angle from 180°, since a straight angle contains 180°.

a. $180° - 62° = 118°$. Therefore, the measure of the supplement of an angle measuring 62° is 118°.

b. In order to find the measure of the supplement of an angle measuring 22°15′, we must rewrite 180° as 179°60′, and then subtract 22°15′ from it.

$$\begin{array}{r} 179°60' \\ -\ 22°15' \\ \hline 157°45' \end{array}$$

Therefore, the measure of the supplement of an angle measuring 22°15′ is 157°45′.

c. In order to find the measure of the supplement of an angle measuring 128°42′16″, we must rewrite 180° as 179°59′60″, and then subtract 128°42′16″ from it.

$$\begin{array}{r} 179°59'60'' \\ -128°42'16'' \\ \hline 51°17'44'' \end{array}$$

Therefore, the measure of the supplement of an angle measuring 128°42′16″ is 51°17′44″.

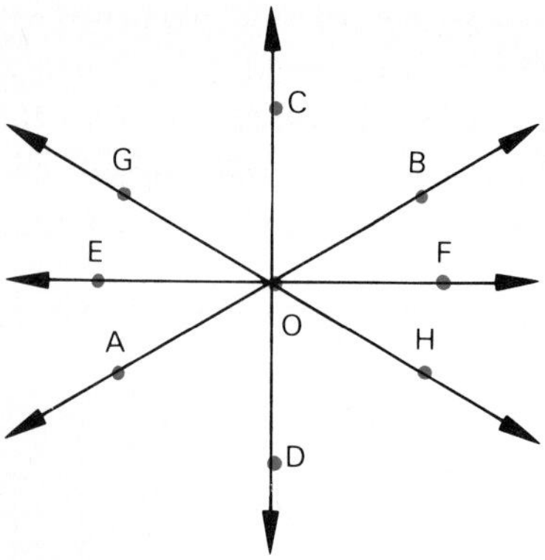

FIGURE 32

In Figure 32, we have four lines intersecting at point O, and $\overleftrightarrow{CD} \perp \overleftrightarrow{EF}$. An **acute angle** is an angle whose measure is greater than 0° and less than 90°. In Figure 32, $\angle COF$ is a right angle and $\angle BOF$ is an acute angle. An **obtuse angle** is an angle whose measure is greater than 90° and less than 180°. In Figure 32, $\angle EOF$ is a straight angle and $\angle GOF$ is an obtuse angle.

Two angles whose measures sum to that of a right angle are called complementary angles. In Figure 32, $\angle COB$ and $\angle BOF$ are complementary angles. $\angle COB$ is the complement of $\angle BOF$, and $\angle BOF$ is the complement of $\angle COB$. Note that there exist other complementary angles in Figure 32, such as $\angle EOG$ and $\angle GOC$.

Two angles whose measures sum to that of a straight angle are called supplementary angles. In Figure 32, $\angle EOC$ and $\angle COF$ are supplementary angles, as are $\angle EOG$ and $\angle GOF$. $\angle EOG$ is the supplement of $\angle GOF$, and $\angle GOF$ is the supplement of $\angle EOG$. Note that $\angle HOF$ is also the supplement of $\angle GOF$. Therefore, $m\angle EOG = m\angle HOF$, where m means "the measure of." (Supplements of the same angle must be equal in measure.) Therefore, since $\angle HOF$ and $\angle EOG$ are vertical angles, we may conclude that the measures of vertical angles are equal.

EXAMPLE 4 In Figure 33, straight lines $\overleftrightarrow{AS}$ and $\overleftrightarrow{HR}$ intersect at point O and $\overleftrightarrow{ON} \perp \overleftrightarrow{HR}$. Using this information, find each of the following.

a. an acute angle
b. an obtuse angle
c. a right angle
d. a straight angle
e. adjacent angles
f. vertical angles
g. complementary angles
h. supplementary angles

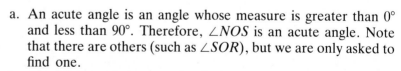

SOLUTION $\overleftrightarrow{AS}$ and $\overleftrightarrow{HR}$ are straight lines and therefore form straight angles. $\overleftrightarrow{ON} \perp \overleftrightarrow{HR}$ and therefore $\angle NOR$ and $\angle NOH$ are right angles.

a. An acute angle is an angle whose measure is greater than 0° and less than 90°. Therefore, $\angle NOS$ is an acute angle. Note that there are others (such as $\angle SOR$), but we are only asked to find one.
b. An obtuse angle is an angle whose measure is greater than 90° and less than 180°. Therefore, $\angle ROL$ is an obtuse angle.
c. Angle NOR is a right angle.
d. A straight angle is an angle whose sides lie on the same straight line and extend in opposite directions from the vertex. Therefore, $\angle AOS$ is a straight angle.
e. Adjacent angles are angles that have the same vertex and a common side between them. Therefore, $\angle NOS$ and $\angle SOR$ are adjacent angles.

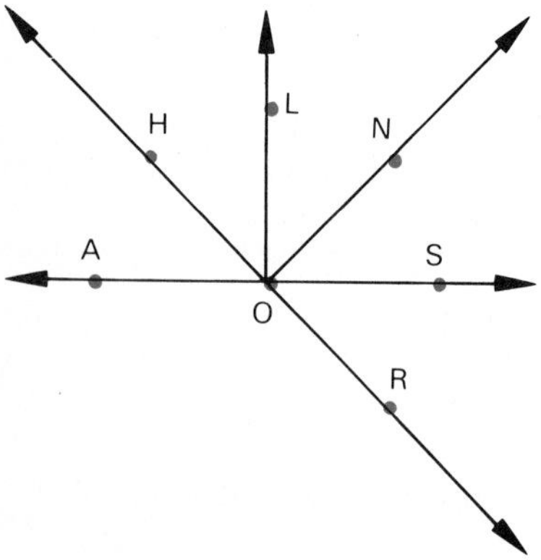

FIGURE 33

f. Vertical angles are two angles where the sides of one angle extend through the vertex to form the sides of the other angle. Therefore, $\angle HOA$ and $\angle SOR$ are vertical angles.
g. Two angles whose measures sum to that of a right angle are complementary angles. Therefore, $\angle NOS$ and $\angle SOR$ are complementary angles.
h. Two angles whose measures sum to that of a straight angle are supplementary angles. Therefore, $\angle HOL$ and $\angle LOR$ are supplementary angles.

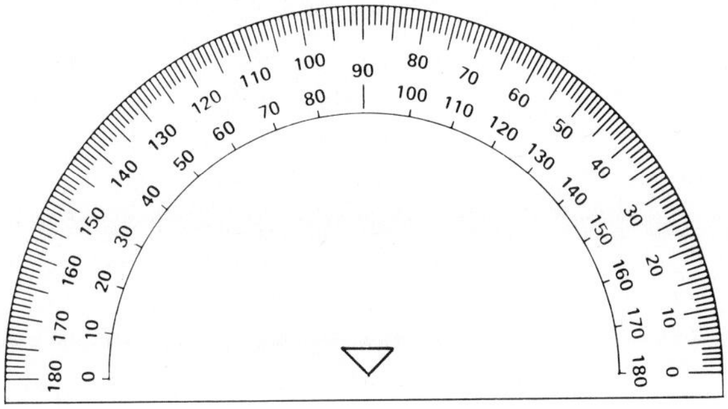

FIGURE 34

A **protractor** is the instrument used to measure the number of degrees contained in an angle. One type of protractor is shown in Figure 34. A protractor can also be used to draw an angle of any size. Note that 0° is marked on the right side of the protractor and 180° is marked on the left side. This is because an angle is *generated* in a counterclockwise direction. An angle is generated in the following manner: We are given two rays, $\overrightarrow{AB}$ and $\overrightarrow{AC}$, that lie in the same straight line and extend in the same direction, as shown in Figure 35a. Now we rotate $\overrightarrow{AC}$ in a counterclockwise direction for a certain number of degrees and then stop as shown in Figure 35b.

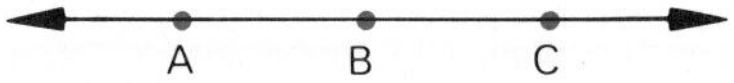

FIGURE 35a

In generating $\angle CAB$, $\overrightarrow{AB}$ is called the **initial side,** since it is the ray where the angle begins, and $\overrightarrow{AC}$ is called the **terminal side,** since it is the ray where the angle ends, or terminates.

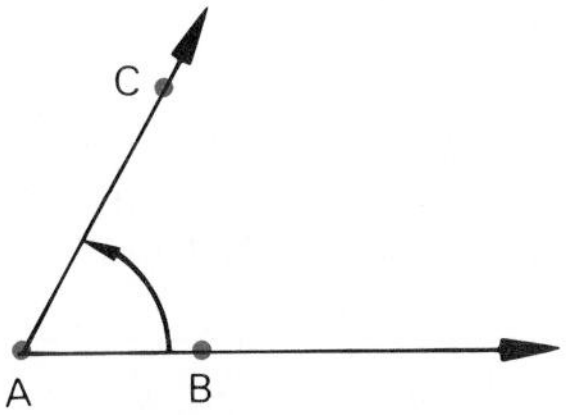

FIGURE 35b

In order to measure the number of degrees in an angle, we place the marked point of the protractor at the vertex of the angle, and the 0° line along the initial side of the angle. The number of degrees in the angle is found by reading the position of the terminal side of the angle on the edge of the protractor. Using this technique, as shown in Figure 36, we see that angle CAB contains 60°. We denote this by writing $m\angle CAB = 60°$. For example, $m\angle BOE = 88°$ means that the measure of $\angle BOE$ is 88°.

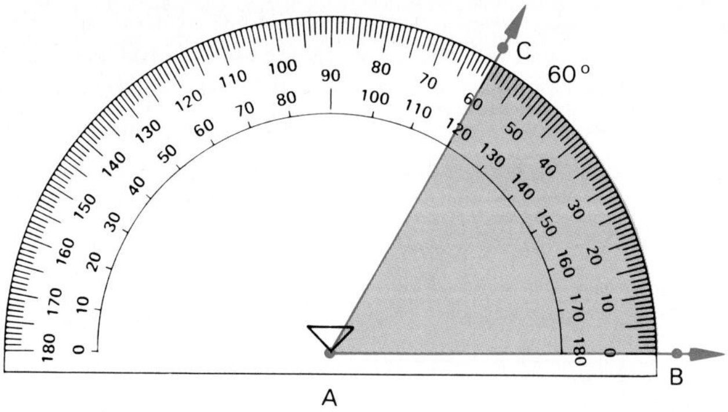

FIGURE 36

Notice that a protractor can be used to measure any angle up to and including 180°. But there are angles of greater magnitude. If we move the hands of a clock back one hour, as we do when we change from daylight saving time to standard time, then the minute hand is rotated one complete revolution, or 360°. The hands of the clock have generated an angle greater than 180°. When the terminal side of an angle has generated an angle of 360°, the angle is called a **round angle** or **perigon.** When the terminal side of an angle has generated an angle greater than 180°, but less than 360°, the angle is called a **reflex angle.** Two angles whose measures sum to that of a round angle are called **conjugate angles.**

Angles are normally generated so that the terminal side of the angle has been moved in a counterclockwise direction. The measure of these angles is considered to be positive. But angles can be generated so that the terminal side of the angle has been moved in a clockwise direction. The measure of angles generated in a clockwise direction is considered to be negative. If the measure of $\angle BAC$ in Figure 37 is 120°, we could also say that its measure is $-240°$.

We can also use algebra to find the number of degrees in an angle. Consider $\angle ARS$ in Figure 38. Suppose we are given that $\angle ARS$ is a right angle, and that $m\angle BRS$ is 30° larger than $m\angle ARB$. How many degrees are in each angle? Since we are

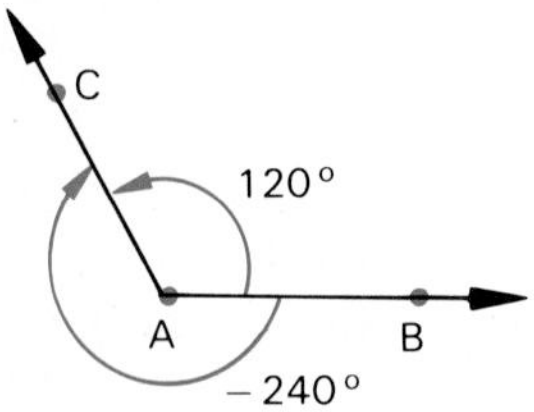

FIGURE 37

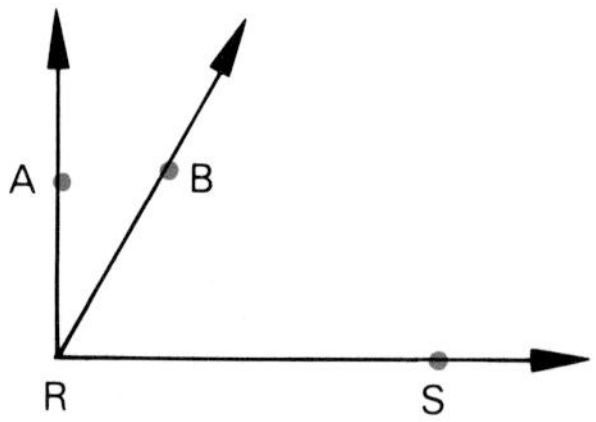

FIGURE 38

given that $\angle ARS$ is a right angle, we know that its measure is 90°. Also, $\angle ARB$ and $\angle BRS$ are complementary angles because the sum of their measure is 90°. Let $x = m\angle ARB$; then $m\angle BRS = x + 30$ because $m\angle BRS$ is 30° larger than $m\angle ARB$. We can now write the equation $x + x + 30 = 90$. Next, we solve the equation for x and find the number of degrees in each angle.

$$x + x + 30 = 90$$
$$2x + 30 = 90$$
$$2x = 60$$
$$x = 30° \qquad m\angle ARB$$
$$x + 30 = 60° \qquad m\angle BRS$$

EXAMPLE 5 Two angles are complementary and the measure of one angle is 20° larger than the measure of the other. How many degrees are there in each angle?

SOLUTION We can draw a diagram (Fig. 39) to aid us in solving this problem. The two angles are complementary, so we know their measures sum to that of a right angle, and one angle is 20° larger than the other. The resulting equation is

$$x + x + 20 = 90$$

Solving the equation for x,

$$2x + 20 = 90$$
$$2x = 70$$
$$x = 35° \qquad \text{first angle}$$
$$x + 20 = 55° \qquad \text{second angle}$$

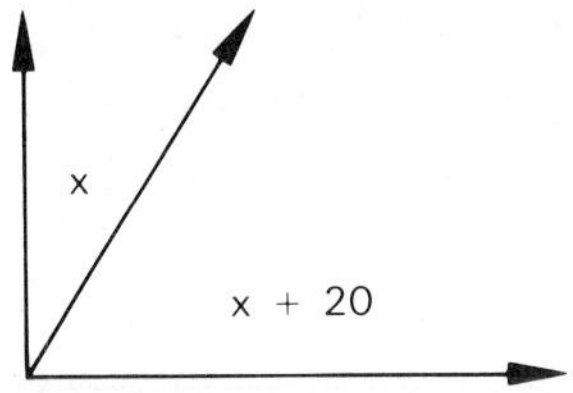

FIGURE 39

EXAMPLE 6 Two angles are supplementary and the measure of one angle is 60° less than the measure of the other. How many degrees are there in each angle?

SOLUTION Drawing a diagram to aid us in solving this problem, we have Figure 40.

The two angles are supplementary, therefore their measures sum to that of a straight angle. Also, one angle is 60° less than the other, so if one angle is x, then the other is $x - 60°$. The resulting equation is

$$x + x - 60 = 180$$

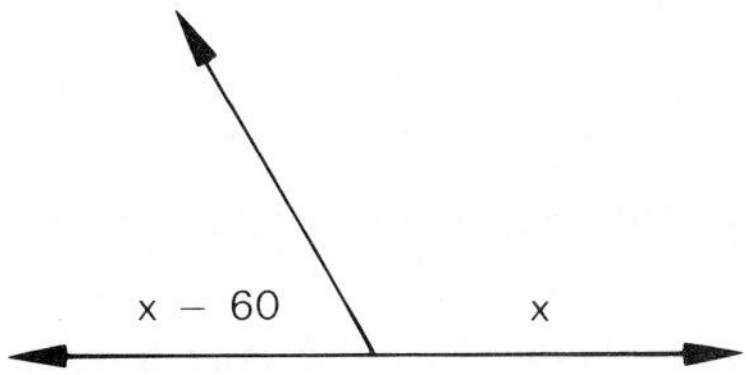

FIGURE 40

Solving the equation for x,

$$\begin{aligned} 2x - 60 &= 180 \\ 2x &= 240 \\ x &= 120^\circ && \text{first angle} \\ x - 60 &= 60^\circ && \text{second angle} \end{aligned}$$

EXAMPLE 7 In Figure 41, ST is a straight line. What is the value of x?

SOLUTION If ST is a straight line, then $\angle SKT$ is a straight angle and its measure is 180°. Therefore, the sum of $m\angle SKE$ and $m\angle EKT$ is 180°, and the resulting equation is

$$3x + 2x + 10 = 180$$

Solving the equation for x,

$$\begin{aligned} 5x + 10 &= 180 \\ 5x &= 170 \\ x &= 34 \end{aligned}$$

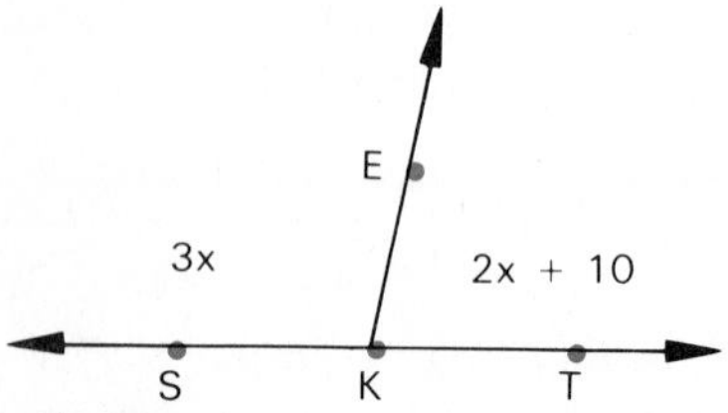

FIGURE 41

Note that we are not asked to find the measure of the two angles, but only to find the value of x.

EXAMPLE 8 In Figure 42, AS is a straight line. What is the value of x?

SOLUTION If AS is a straight line, then $\angle ARS$ must be a straight angle and its measure is 180°. Therefore, the sum of $m\angle ARB$, $m\angle BRJ$, and $m\angle JRS$ is 180°, and the resulting equation is

$$3x + 20 + 2x + x + 10 = 180$$

Solving the equation for x,

$$\begin{aligned} 6x + 30 &= 180 \\ 6x &= 150 \\ x &= 25 \end{aligned}$$

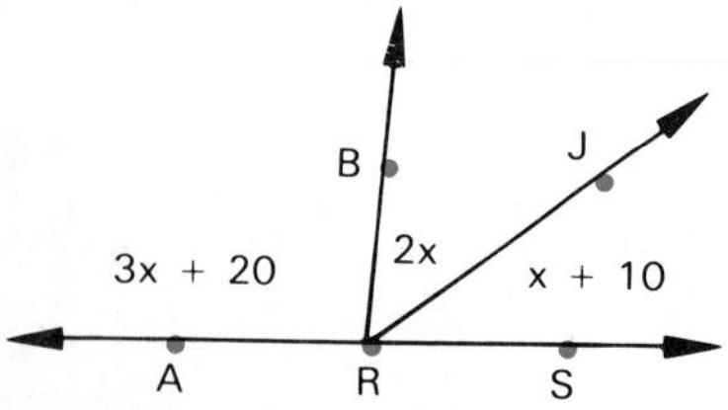

FIGURE 42

Note that the size of an angle is a function of how many degrees it contains, not of how big we draw it. For example, $\angle ARS$ and $\angle WMS$ in Figure 43 each contain 30° and are therefore equal in size: $m\angle ARS = m\angle WMS$.

In Figure 44, lines AB and CD are parallel and they are cut by a transversal EF. A **transversal** is a line that intersects (cuts) two or more lines. (Note: To mark two given parallel lines, use

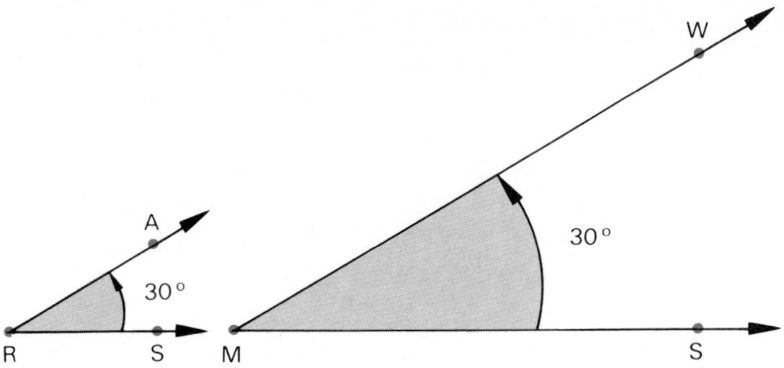

FIGURE 43

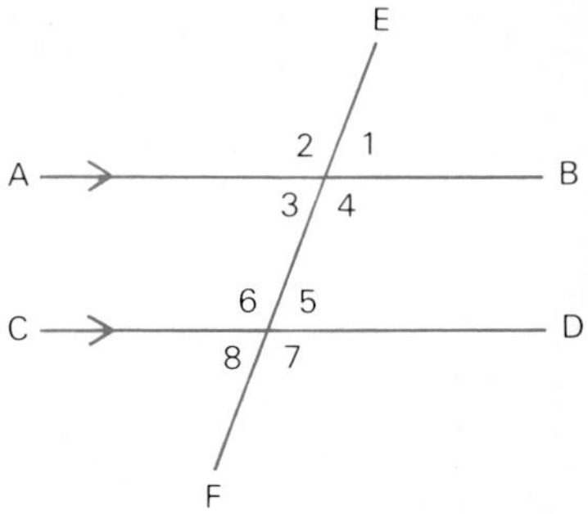

FIGURE 44

arrowheads as shown in Figure 44.) In addition to vertical angles, there are other types of angles in the given figure. *Interior angles* are angles formed by a transversal and a line, and inside the lines cut by a transversal. Hence, angles 3, 6, 4, 5 are interior angles. *Alternate interior angles* are a pair of nonadjacent interior angles on opposite sides of the transversal. Therefore in the given figure, angles 3 and 5 are alternate interior angles, as are angles 4 and 6. *Corresponding angles* have the same position with respect to their lines and the transversal. In Figure 44, $\angle 1$ and $\angle 5$, or $\angle 4$ and $\angle 7$, or $\angle 2$ and $\angle 6$, or $\angle 3$ and $\angle 8$, are a pair of corresponding angles.

It can be shown that if two parallel lines are cut by a transversal, then

1. alternate interior angles are **congruent.**
2. corresponding angles are **congruent.**
3. interior angles on the same side of the transversal are **supplementary.**

EXAMPLE 9 In Figure 45, $\overline{AB} \parallel \overline{CD}$ cut by a transversal EF.

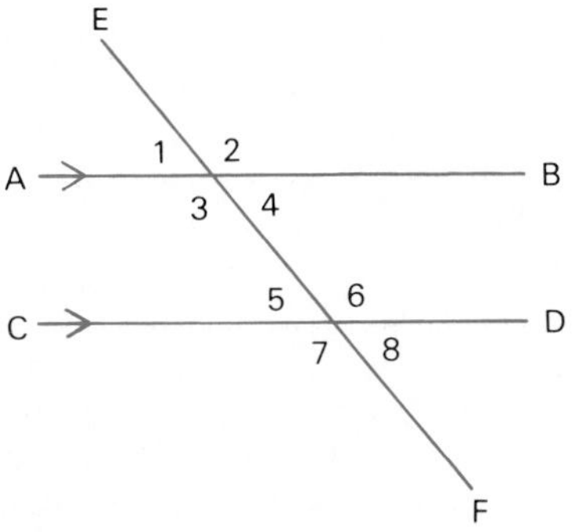

FIGURE 45

If $m\angle 1 = 50°$, find the number of degrees in each of the other seven angles.

SOLUTION
a. $m\angle 2 = 130°$, supplementary angles, $\angle 1$ and $\angle 2$.
b. $m\angle 3 = 130°$, vertical angles, $\angle 2$ and $\angle 3$.
c. $m\angle 4 = 50°$, vertical angles, or supplementary angles.
d. $m\angle 5 = 50°$, alternate interior angles, $\angle 5$ and $\angle 4$.
e. $m\angle 6 = 130°$, interior angles on the same side of the transversal are supplementary, $\angle 4$ and $\angle 6$.
f. $m\angle 7 = 130°$, corresponding angles, $\angle 3$ and $\angle 7$.
g. $m\angle 8 = 50°$, supplementary angles, $\angle 7$ and $\angle 8$.

EXERCISES FOR SECTION 10.4

1. In Figure 46, how many different angles are there? Name them.

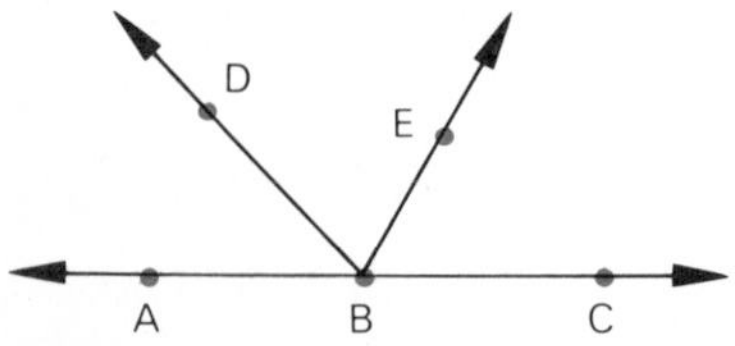

FIGURE 46

2. In Figure 47, how many different angles are there? Name them.

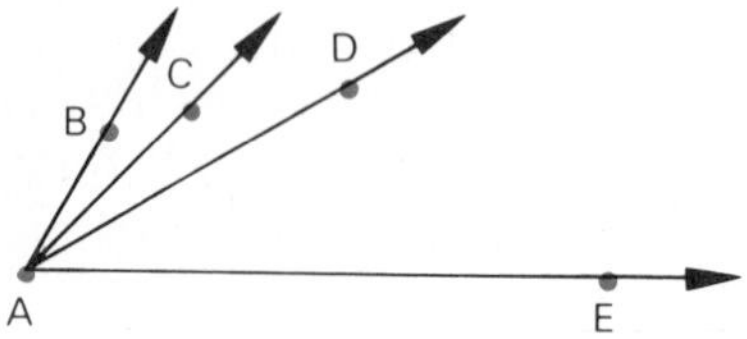

FIGURE 47

3. In Figure 48, straight lines $\overleftrightarrow{NS}$, $\overleftrightarrow{WE}$, $\overleftrightarrow{LR}$, and $\overleftrightarrow{BF}$ intersect at O. Also, $\overleftrightarrow{NS} \perp \overleftrightarrow{WE}$ and $\overleftrightarrow{LR} \perp \overleftrightarrow{BF}$. Using this information, find each of the following.

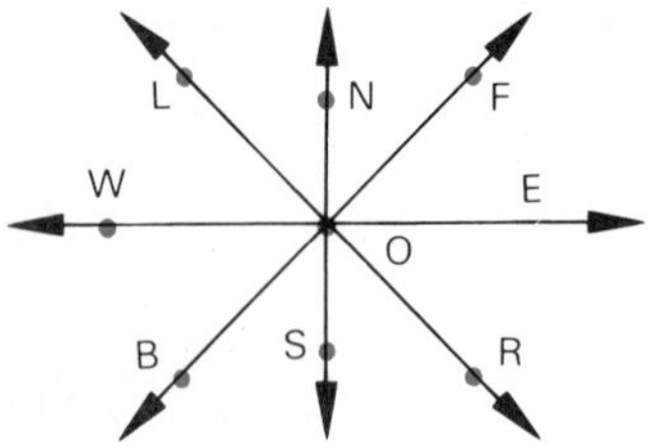

FIGURE 48

a. two acute angles
b. two obtuse angles
c. two right angles
d. two straight angles
e. two pairs of adjacent angles
f. two pairs of vertical angles
g. two pairs of complementary angles
h. two pairs of supplementary angles

4. Find the measure of the complement of the angle with each of the following measures.
a. 30°
b. 60°
c. 72°
d. 28°28′
e. 30°48′
f. 42°35′20″

5. Find the measure of the supplement of the angle with each of the following measures.
a. 90°
b. 110°
c. 60°
d. 120°20′
e. 38°30′
f. 100°50′25″

6. Two angles are complementary and one angle measures 10° less than the other. How many degrees are there in each angle?

7. Two angles are supplementary and one angle measures 60° more than the other. How many degrees are there in each angle?

8. Two angles are supplementary and one angle measures 30° more than two times the other. How many degrees are there in each angle?

9. Two angles are complementary and one angle measures 30° less than three times the other. How many degrees are there in each angle?

10. In Figure 49, $\overleftrightarrow{AB}$ is a straight line. How many degrees are in $\angle AOC$ and $\angle COB$?

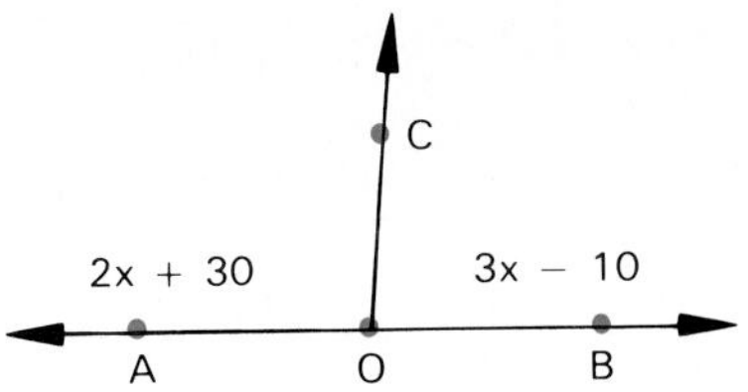

FIGURE 49

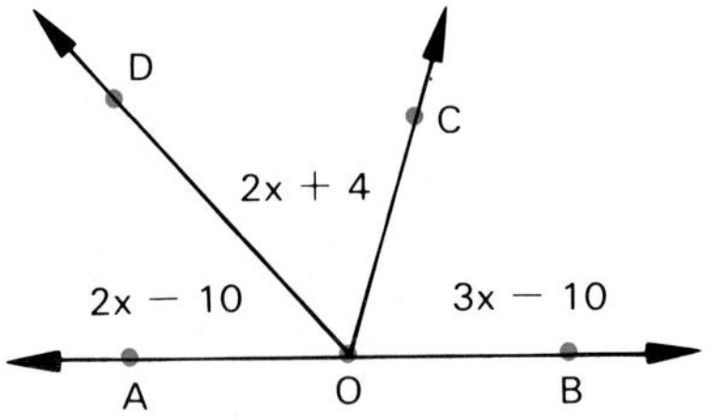

FIGURE 50

11. In Figure 50, what value of x will make AB a straight line?

12. The sum of the measures of the complement and the supplement of a certain angle is 100°. Find the number of degrees in the angle.

13. The number of degrees in angle ACB is equal to one-third the number of degrees in its supplement. Find the number of degrees in angle ACB.

For Exercises 14–21, indicate whether each is true or false.

14. The sides of a right angle are perpendicular to each other.

15. The complement of an acute angle is acute.

16. The supplement of an acute angle is acute.

17. If an acute angle is doubled, then the result must be an acute angle.

18. If an obtuse angle is doubled, then the result must be a reflex angle.

19. If the sum of two angles is a straight angle, then one of the angles must be acute.

20. $m\angle BAC + m\angle ABC + m\angle ACB = 180°$. Therefore, these angles are supplementary angles.

21. $m\angle BAC + m\angle ABC = 90°$. Therefore, these angles are complementary angles.

22. Find the measure of all the angles in Figure 51 if $\overline{AB} \parallel \overline{CD}$ and $m\angle 1 = 40°$.

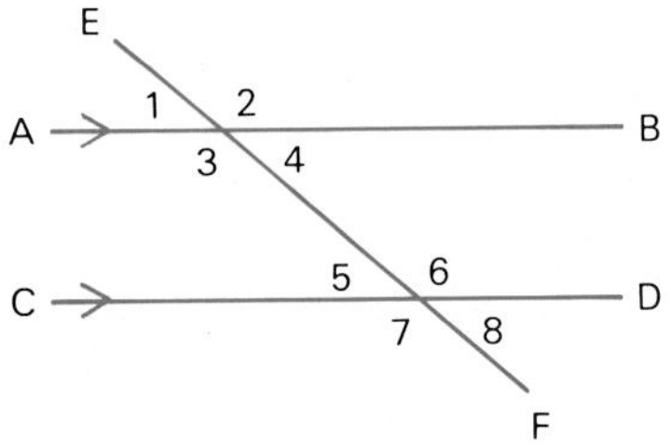

FIGURE 51

23. Find the measure of all the angles in Figure 52 if $\overline{CD} \parallel \overline{EF}$ and $m\angle 3 = 50°$.

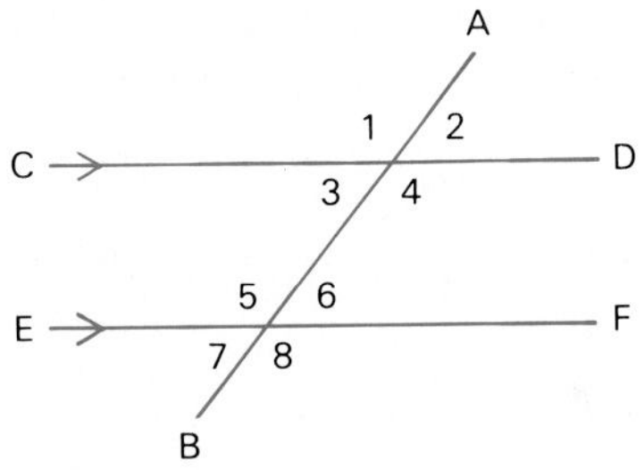

FIGURE 52

24. Find the measure of all the angles in Figure 53 if $\overline{MN} \parallel \overline{OP}$ and $m\angle 7 = 135°$.

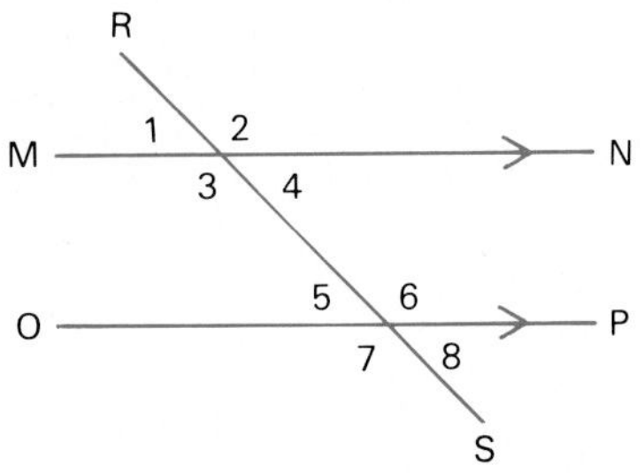

FIGURE 53

25. Find the measure of all the angles in Figure 54 if $\overline{AB} \parallel \overline{CD}$ and $m\angle 5 = 155°$.

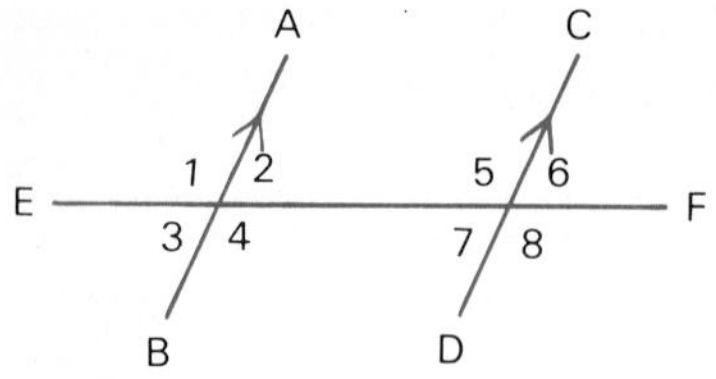

FIGURE 54

For Exercises 26–30, refer to Figure 55 in which $\overline{AB} \parallel \overline{CD}$.

26. Find $m\angle 1$ and $m\angle 5$, if $m\angle 1 = (130 - x)^\circ$ and $m\angle 5 = 9x^\circ$.

27. Find $m\angle 3$ and $m\angle 5$, if $m\angle 3 = (x + 20)^\circ$ and $m\angle 5 = x^\circ$.

28. Find $m\angle 4$ and $m\angle 5$, if $m\angle 4 = 2x^\circ$ and $m\angle 5 = (x + 40)^\circ$.

29. Find $m\angle 4$ and $m\angle 8$, if $m\angle 4 = 2x^\circ$ and $m\angle 8 = (x + 60)^\circ$.

30. Find $m\angle 5$ and $m\angle 7$, if $m\angle 1 = (x + 20)^\circ$ and $m\angle 6 = (x + 60)^\circ$.

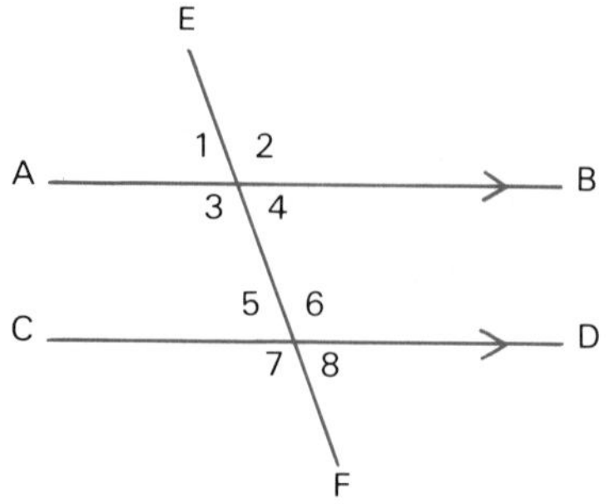

FIGURE 55

JUST FOR FUN What has four equal sides, but is drawn with six lines?

10.5 POLYGONS

A **broken line** is a set of connected line segments, that is, a set of line segments that have been placed end to end. The four figures in Figure 56 are examples of broken lines.

Now let us take each of the broken lines in Figure 56 and "close" it so that it appears as shown in Figure 57. A **closed broken line** begins and ends at the same point. You will note that each of the figures in Figure 57 is a closed broken line. However, in addition each is also a *simple closed broken line.* A **simple**

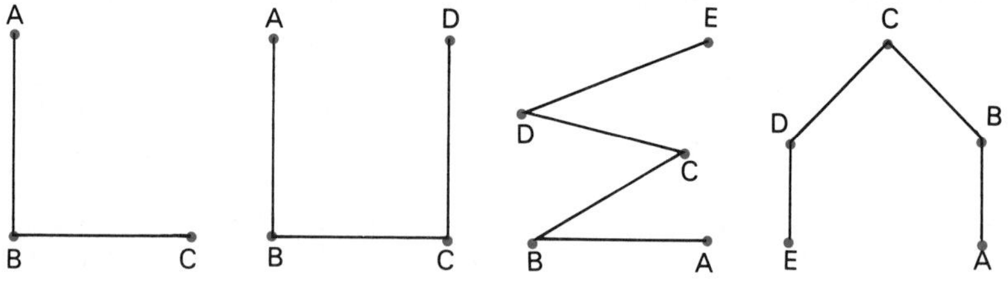

FIGURE 56

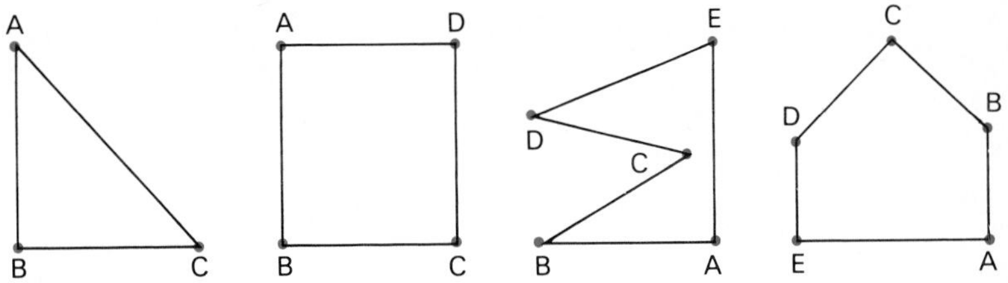

FIGURE 57

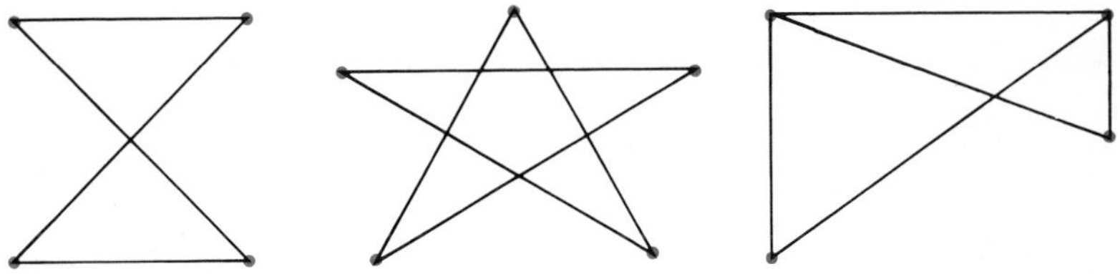

FIGURE 58

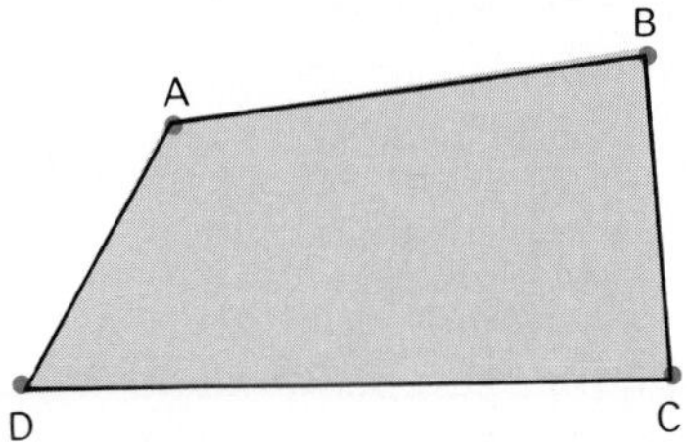

FIGURE 59

closed broken line is one that does not intersect itself. None of the closed broken line figures shown in Figure 58 are simple because in each case the broken line intersects itself.

A simple closed broken line is called a **polygon.** The connected line segments are the **sides** of the polygon, and the points at which the line segments are connected are the **vertices** of the polygon. Note that any two consecutive sides of a polygon form an angle of the polygon. In Figure 59, polygon $ABCD$ is a simple closed broken line. Its sides are $\overline{AB}$, $\overline{BC}$, $\overline{CD}$, and $\overline{DA}$, its vertices are A, B, C, and D, and its angles are $\angle DAB$, $\angle ABC$, $\angle BCD$, and $\angle CDA$.

Polygons are classified according to the number of sides they have. Below is a partial list of some types of polygons and the number of sides of each.

Polygon	Number of Sides	Polygon	Number of Sides
Triangle	3	Octagon	8
Quadrilateral	4	Nonagon	9
Pentagon	5	Decagon	10
Hexagon	6	Dodecagon	12
Heptagon	7	Icosagon	20

In order to form a simple closed broken line, we need at least three line segments. If we form such a closed broken line, then we have a polygon with three sides, that is, a **triangle.** There are many different kinds of triangles, and they can be classified

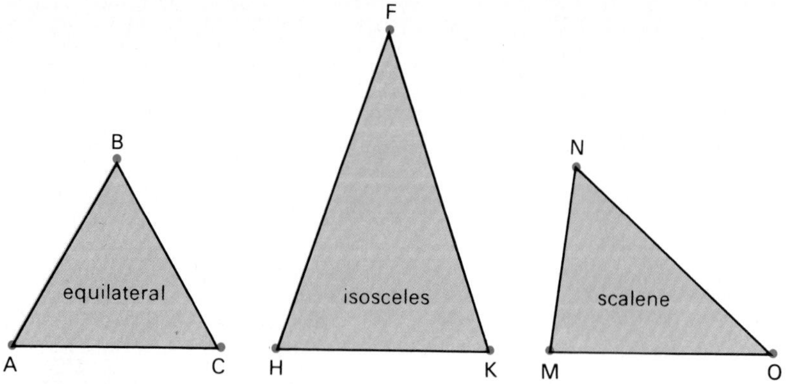

FIGURE 60

according to the characteristics of their sides or their angles. A triangle in which the measures of all three sides are equal is called an **equilateral triangle.** A triangle in which two sides are of equal length is called an **isosceles triangle.** A triangle in which no two sides are of equal length is called a **scalene triangle.** An example of each type of triangle is shown in Figure 60.

Can we form a triangle when given any three line segments? The answer is no. There is a certain requirement that must be fulfilled in order to construct a triangle from three segments. For example, suppose we are given three lines segments that measure 10 centimeters, 5 centimeters, and 2.5 centimeters. Can we form a triangle with these three line segments?

10 centimeters

5 centimeters

2.5 centimeters

If we attempt to construct a triangle with these segments, we have the situation illustrated in Figure 61.

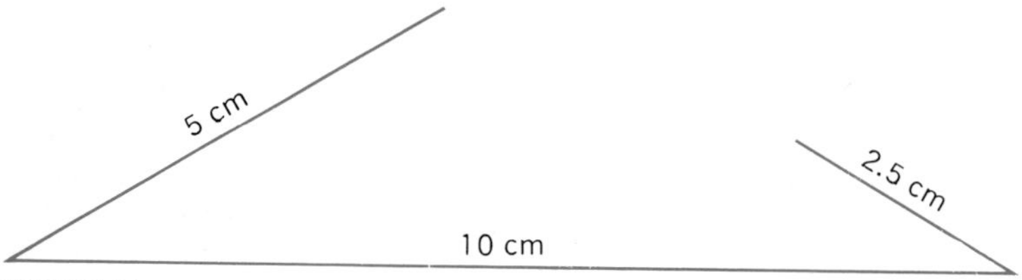

FIGURE 61

Mark Antman/The Image Works

As you can see, we are not able to form a simple *closed* broken line using these three line segments. In fact, in order to construct any triangle, the sum of the measures of any two of the line segments must be greater than the measure of the third line segment. If we already have a triangle, then we know that the sum of the lengths of any two of the sides is greater than the length of the third side.

Is it possible to construct a triangle whose sides measure 4, 4, and 8? The answer is no, since $4 + 4$ is not greater than 8. Is it possible to construct a triangle whose sides measure 4, 5, 6? Yes, because $4 + 5 > 6$, $4 + 6 > 5$, and $5 + 6 > 4$.

We can also classify triangles according to the types of angles that are in the triangle. If all of the angles in the triangle are acute angles, then the triangle is called an **acute triangle.** If a triangle has an obtuse angle in it, then it is called an **obtuse triangle.** Can a triangle have more than one obtuse angle in it? We cannot form such a triangle because the figure formed is not a closed broken line, as shown in Figure 62.

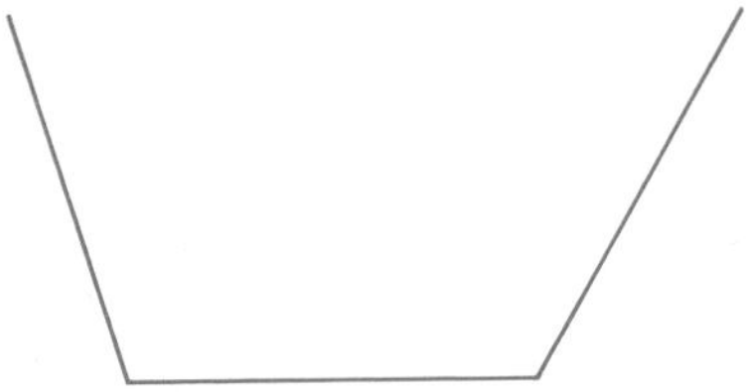

FIGURE 62

If the measures of all of the angles in a triangle are equal, then the triangle is called an **equiangular triangle.** It can be shown that the sum of the measures of the interior angles of a triangle is equal to 180 degrees. Therefore, if a triangle is equiangular, each angle will measure exactly 60 degrees and it is also an acute triangle.

If a triangle contains a right angle, then it is called a **right triangle.**

Examples of an acute triangle, an obtuse triangle, a right triangle, and an equiangular triangle are shown in Figure 63.

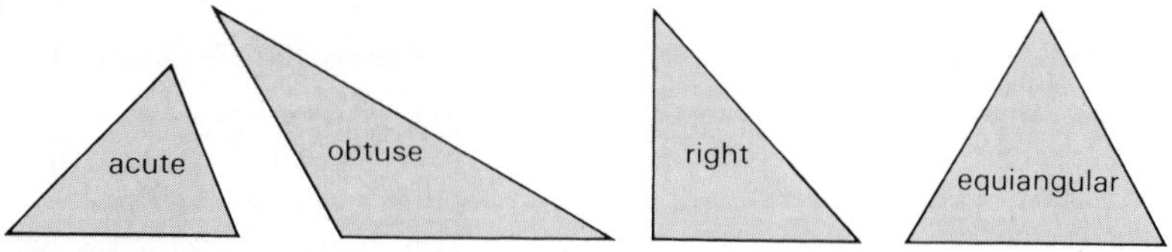

FIGURE 63

Consider right triangle ABC in Figure 64. The right angle is $\angle BCA$. The sides of the triangle that form the right angle are called the **legs** of the right triangle (a and b in Fig. 64) and the side opposite the right angle is called the **hypotenuse** (c in Fig. 64).

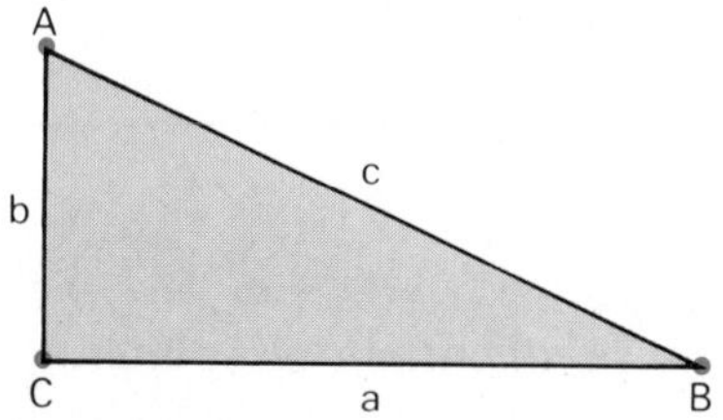

FIGURE 64

Probably one of the most famous theorems in geometry deals with the right triangle. It is called the **Pythagorean theorem** because its proof was supposedly discovered by Pythagoras. The Pythagorean theorem states that

The square of the hypotenuse of a right triangle is equal to the sum of the squares of the legs.

Using right triangle ABC in Figure 64 as a reference, we can also state the Pythagorean theorem as

$$c^2 = a^2 + b^2$$

Suppose in Figure 64 we are given that $m(\overline{AC}) = 6$ and $m(\overline{BC}) = 8$, where $m(\overline{AC})$ is the measure of $\overline{AC}$ and $m(\overline{BC})$ is the measure of $\overline{BC}$. What is the length of $\overline{AB}$?

We can use the Pythagorean theorem to find the answer. Using the formula $c^2 = a^2 + b^2$, we are given $a = 8, b = 6$. Therefore, we can substitute the appropriate values in the formula and solve it for c.

HISTORICAL NOTE

Unfortunately, much of the documented history of Greek geometry was either lost or destroyed through the centuries. However, scholarly research has created an approximate history of early Greek mathematics.

It is believed that Pythagoras was born in approximately 570 B.C. on the Aegean island called Samos. After becoming an educated young man, Pythagoras spent several years traveling to different lands to learn all that he could about mathematics. After this, he settled in the Greek colonial seaport of Crotona, which is located in southern Italy. It was there that he founded the famous Pythagorean academy (in approximately 540 B.C.) In addition to mathematics, he taught his students (disciples) to worship numbers, to believe in reincarnation, and to sign the name of the Pythagorean brotherhood to any writing or discovery. As a result, it is now difficult to know which mathematical findings should be credited to Pythagoras himself, and which to other members of the brotherhood.

The best known of the Pythagorean teachings is the theorem that states: *In any right triangle, the square of the hypotenuse is equal to the sum of the squares of the two other sides.* The Babylonians discovered this property much earlier; however, the Pythagorean school is credited with being the first to prove it.

An interesting theory Pythagoras promoted was that the entire universe could be expressed by whole number relationships. He even classified numbers as "amicable" and "perfect." He labeled the even whole numbers as feminine and the odd whole numbers as masculine, except 1, which he considered to be the generator of all the numbers. The number 5 represented marriage, the sum of the first feminine number (2) and the first masculine number (3).

$$c^2 = a^2 + b^2$$

substituting $a = 8$, $b = 6$ $\quad c^2 = 8^2 + 6^2$

$$c^2 = 64 + 36$$
$$c^2 = 100$$
$$c = \sqrt{100}$$
$$c = 10 \quad \text{length of } \overline{AB}$$

It should be noted that Pythagoras and his followers were studying geometry long before the development of algebraic notation. Therefore, they expressed the theorem in terms of its geometric representation. They probably stated the theorem in a manner similar to the following.

The sum of the areas of the squares on the legs of any right triangle is equal to the area of the square on the hypotenuse.

Figure 65 illustrates this statement for right triangle ABC whose sides are 3 and 4 and whose hypotenuse is 5.

Area of square I = $3 \times 3 = 9$

Area of square II = $4 \times 4 = 16$

Area of square III = $5 \times 5 = 25$

9 + 16 = 25

Area of square I + area of square II = area of square III

This can be shown for all right triangles.

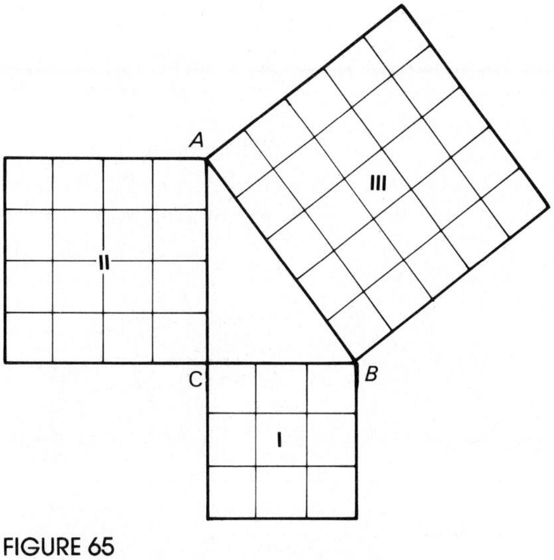

FIGURE 65

EXAMPLE 1 Find the length of hypotenuse $\overline{AB}$ in right triangle ABC, given that $m(\overline{AC}) = 12$ and $m(\overline{BC}) = 5$. See Figure 66.

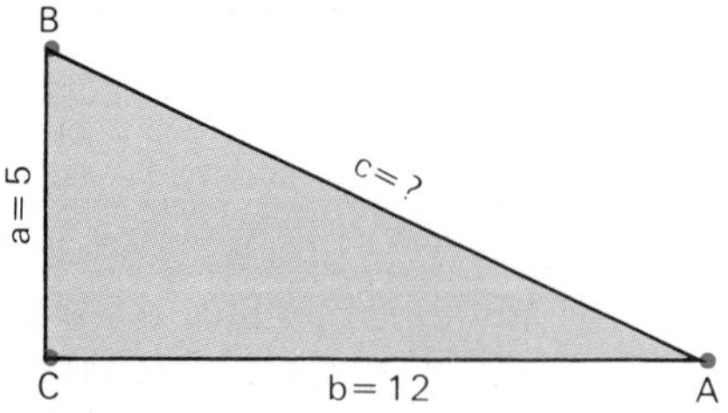

FIGURE 66

SOLUTION Using the formula $c^2 = a^2 + b^2$, we have $a = 5$ and $b = 12$. Therefore,

$$\begin{aligned} c^2 &= a^2 + b^2 \\ c^2 &= 5^2 + 12^2 \\ c^2 &= 25 + 144 \\ c^2 &= 169 \\ c &= \sqrt{169} \\ c &= 13 \qquad \text{length of } \overline{AB} \end{aligned}$$

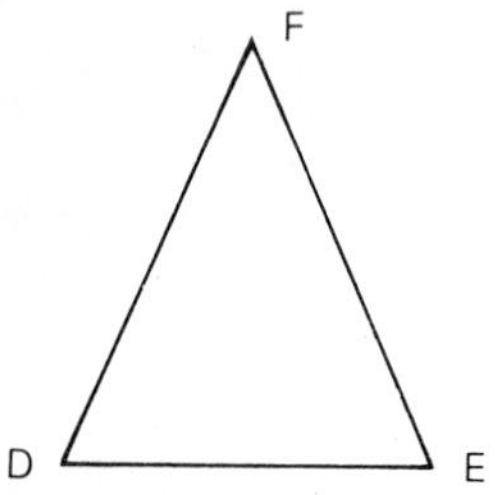

FIGURE 67

If a triangle is isosceles, then it has two sides of equal length. It can also be shown that the measures of the two angles opposite the sides of equal length are equal. For example, given isosceles triangle DEF in Figure 67, with $m(\overline{DF}) = m(\overline{FE})$, then the angles opposite these sides are also equal. That is, $m\angle FDE = m\angle FED$. The measure of these two angles depends on the size of the third angle.

If we form a simple closed broken line using four line segments, such as $ABCD$ in Figure 68, we have a polygon with four sides; this polygon is called a **quadrilateral.**

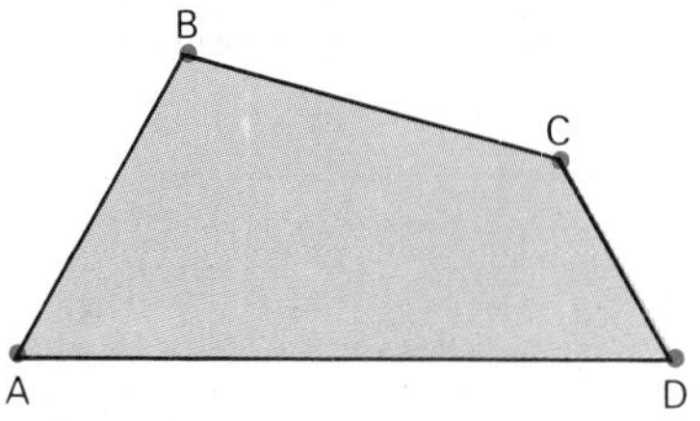

FIGURE 68
Quadrilateral *ABCD*

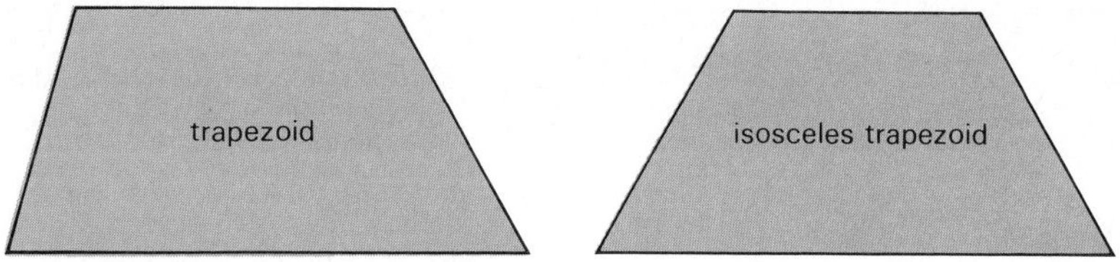

FIGURE 69

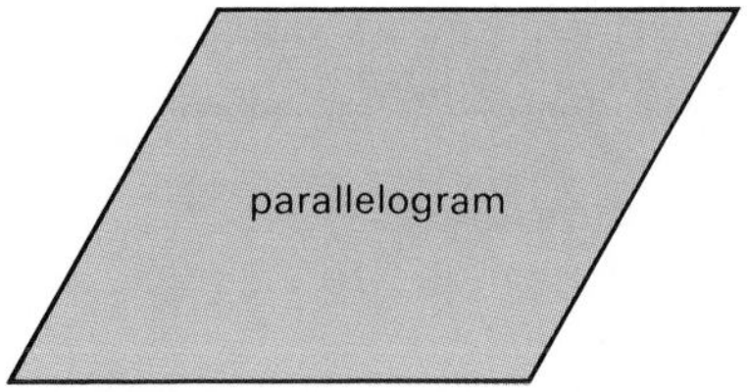

FIGURE 70

Any four-sided polygon is a quadrilateral, but a quadrilateral that has only two parallel sides is called a **trapezoid.** The two parallel sides are called the **bases** of the trapezoid, and if the two non-parallel sides are equal in length, then the trapezoid is an **isosceles trapezoid.** A trapezoid and an isosceles trapezoid are shown in Figure 69.

If a quadrilateral has both pairs of opposite sides parallel, then it is called a **parallelogram.** Note that we did not say anything about the angles of a parallelogram; they can be any type, as long as both pairs of opposite sides are parallel. Figure 70 shows a parallelogram.

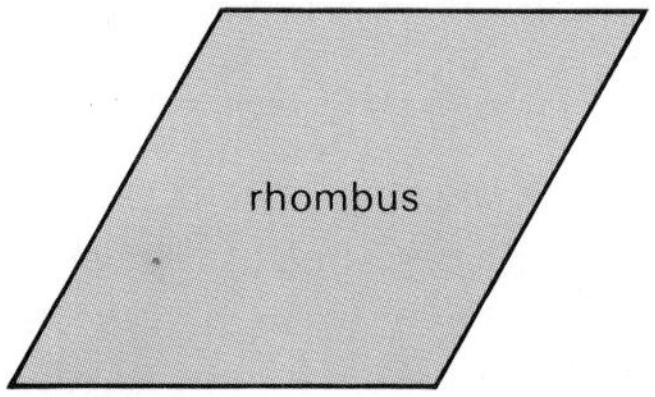

FIGURE 71

A parallelogram that has adjacent sides of equal length is called a **rhombus.** Note that if the adjacent sides of a parallelogram are of equal length, then all of its sides are of equal length. Therefore, we can also describe a rhombus as a parallelogram with four equal sides (see Fig. 71).

A parallelogram that contains a right angle is called a **rectangle.** If a parallelogram contains one right angle, then it must contain four right angles. Also, any quadrilateral with four right angles is a rectangle, as shown in Figure 72.

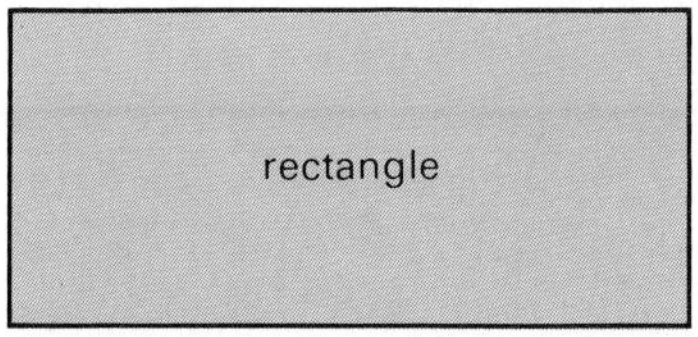

FIGURE 72

A **square** can be described in a number of ways. For example, a square is a rectangle with two adjacent sides of equal length. We can also say that a square is a quadrilateral with four sides of equal length and four angles of equal measure. Note that the square can be considered to be a special case of the rhombus, because a rhombus is a parallelogram with four sides of equal length (see Fig. 73).

Following is a diagram that illustrates the relationships among the various quadrilaterals.

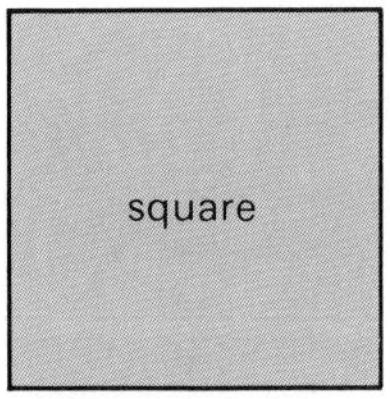

FIGURE 73

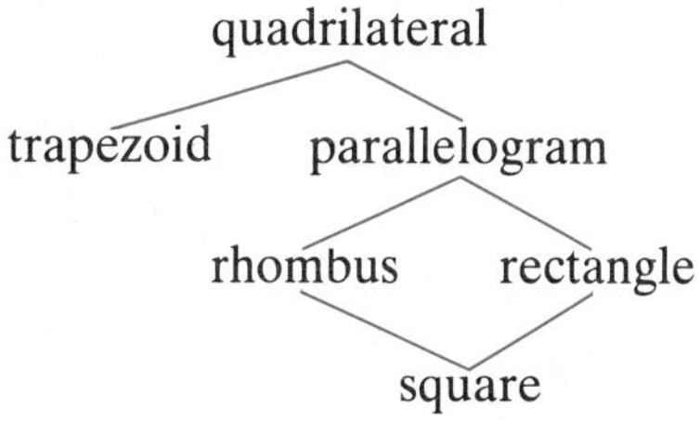

EXAMPLE 2 Determine whether each sentence is true or false.

a. Every square is a rectangle.
b. Every rectangle is a square.
c. Every square is a rhombus.
d. Every rhombus is a square.
e. Every parallelogram is a rectangle.
f. Every trapezoid is a parallelogram.

SOLUTION a. True, a square is a rectangle with two adjacent sides equal.
b. False, a rectangle is a quadrilateral with four right angles; the adjacent sides are not necessarily equal.
c. True, a square is a special case of the rhombus.
d. False, a rhombus is a parallelogram with four equal sides; the four angles are not necessarily equal.
e. False, a parallelogram is a quadrilateral with both pairs of opposite sides parallel. The angles of a parallelogram are not necessarily right angles.
f. False, a trapezoid is a quadrilateral that has only two sides parallel.

Now that we have discussed the different types of quadrilaterals, we shall examine some of them more closely regarding some of their unique properties. For example, it can be shown that the diagonals of an isosceles trapezoid are **congruent.** That is, they coincide exactly when superimposed. This is also true for a square and a rectangle. In other words, the diagonals of a square, rectangle and isosceles trapezoid are congruent to each other.

A **regular polygon** is one that is equilateral and equiangular. That is, its sides are equal in measure, and its angles are equal in measure. Hence, a square is a **regular quadrilateral.** In addition to its diagonals being congruent, they are perpendicular to each other and also bisect each other.

Two other quadrilaterals whose diagonals bisect each other are the parallelogram and the rhombus. It should be noted that the diagonals of a rhombus are also perpendicular to each other. But, in neither case are the diagonals congruent. Hence, if we are given a quadrilateral whose diagonals bisect each other, we cannot conclude what the quadrilateral is. This is a property of parallelograms, rhombuses, rectangles, and squares. We need more information to determine the nature of the quadrilateral.

EXAMPLE 3 For what kind(s) of quadrilateral is each of the following true?

a. The diagonals bisect each other.
b. The opposite sides are congruent and parallel.

c. Two sides only are parallel.
d. Two sides are parallel and the other two are congruent, but not parallel.

SOLUTION a. The diagonals of a parallelogram, rhombus, rectangle, and square bisect each other.
b. The opposite sides are congruent and parallel for a parallelogram, rhombus, rectangle, and square.
c. A trapezoid has only two sides parallel.
d. An isosceles trapezoid has two sides parallel and the other two congruent, but not parallel.

EXERCISES FOR SECTION 10.5

1. Which of the following are broken lines?

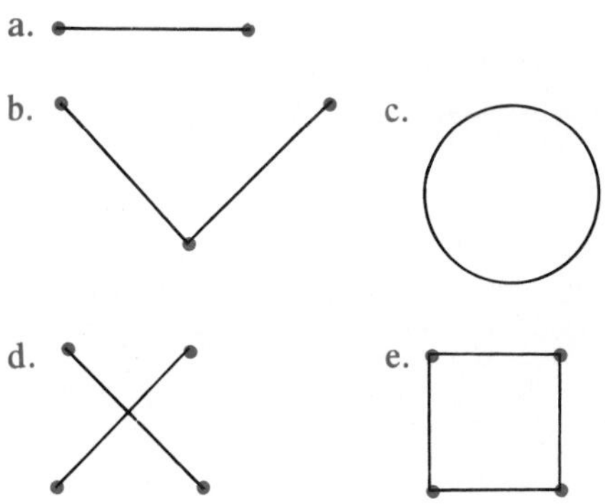

2. Which of the following are simple closed broken lines?

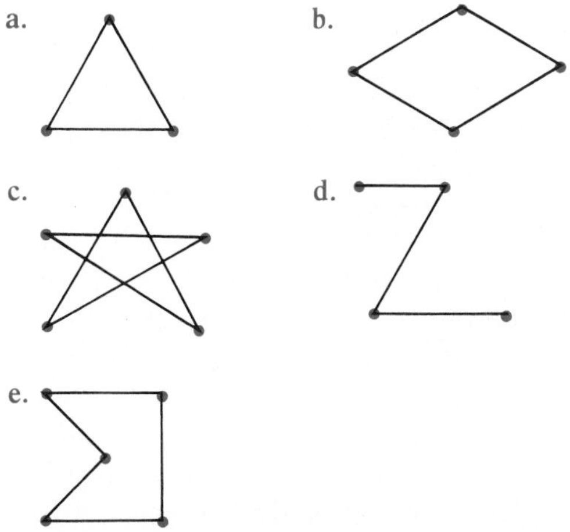

3. Which of the following are polygons?

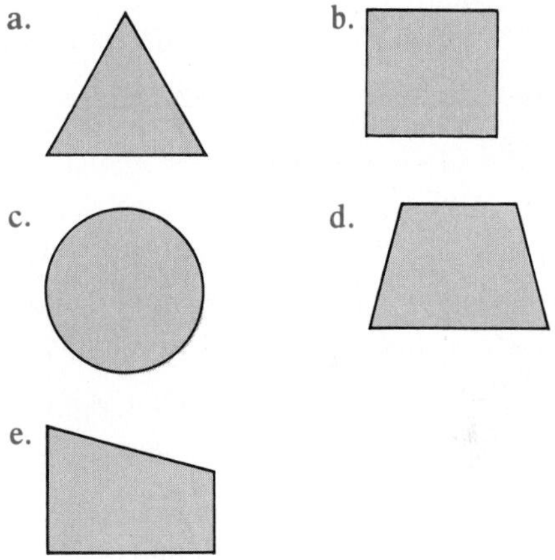

4. Identify each of the following polygons.

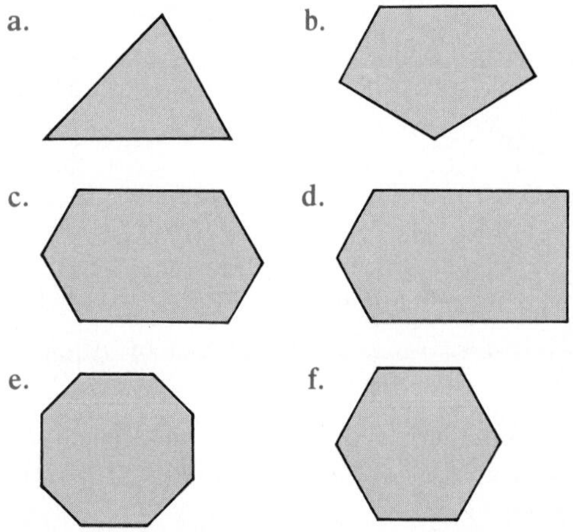

g. 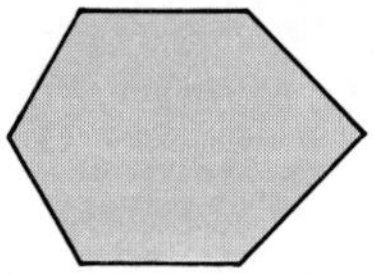

h. 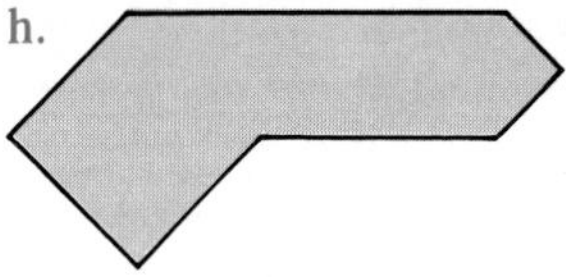

5. Tell whether each of the following is true or false.
   a. A scalene triangle has no sides equal.
   b. An isosceles triangle has all three sides equal.
   c. An equiangular triangle is an acute triangle.
   d. A triangle may contain two obtuse angles.
   e. In a right triangle, the side opposite the right angle is called the hypotenuse.

6. Tell whether each of the following is true or false.
   a. The set of numbers {2, 7, 10} may represent the lengths of the sides of a triangle.
   b. The lengths of two sides of a triangle are 5 and 6. The third side may have length 11.
   c. The set of numbers {2, 3, 4} may represent the lengths of the sides of a right triangle.
   d. The set of numbers {3, 4, 5} may represent the lengths of the sides of a right triangle.
   e. In any triangle, the square of one side is equal to the sum of the squares of the other two sides.

7. Find the length of hypotenuse $\overline{AB}$ in right triangle ABC, given that $m(\overline{AC}) = 9$ and $m(\overline{BC}) = 12$.

8. Find the length of hypotenuse $\overline{AB}$ in right triangle ABC, given that $m(\overline{AC}) = 15$ and $m(\overline{BC}) = 8$.

9. Find the length of hypotenuse $\overline{AB}$ in right triangle ABC, given that $m(\overline{AC}) = 10$ and $m(\overline{BC}) = 24$.

10. A rectangular field is 75 meters wide and 100 meters long. What is the length of a diagonal path connecting two opposite corners?

11. Mary rode her moped 3 miles south and then 4 miles east. How far was she from her starting point?

12. A 13-foot ladder is leaning against the side of a building. The bottom of the ladder is 5 feet from the base of the building. At what height does it touch the building?

13. A 25-foot ramp covers 24 feet of ground. How high does it rise?

14. Find the length of leg $\overline{AC}$ in right triangle ABC, given that $m(\overline{BC}) = 12$, $m(\overline{AB}) = 13$, and $\overline{AB}$ is the hypotenuse.

15. Find the length of leg $\overline{BC}$ in right triangle ABC, given that $m(\overline{AC}) = 4$, $m(\overline{AB}) = 5$, and $\overline{AB}$ is the hypotenuse.

16. Identify each of the following quadrilaterals:

a. 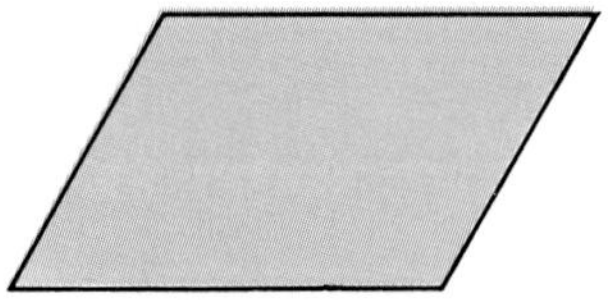

b. 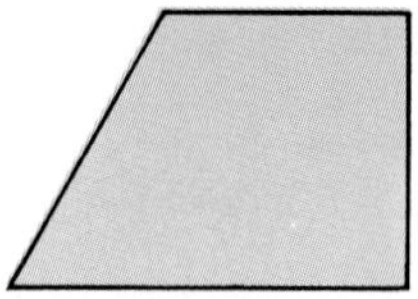

c. 

d.  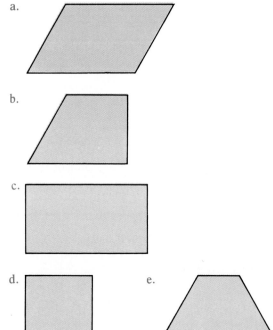

e. 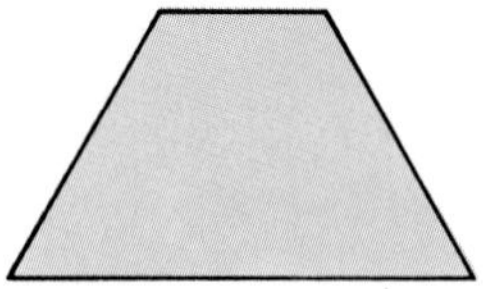

f. 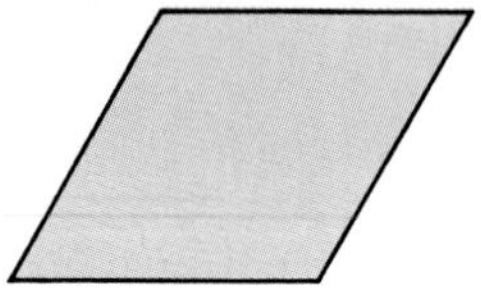

17. Tell whether each of the following is true or false.
   a. A square is a rhombus.
   b. A parallelogram is a rectangle with a right angle.
   c. A parallelogram is a polygon whose opposite sides are parallel.
   d. A rectangle is a parallelogram with four right angles.
   e. A trapezoid is a quadrilateral whose opposite sides are parallel.

18. Tell whether each of the following is true or false.
   a. An isosceles triangle is a triangle that has exactly two sides that are equal in length.

b. An acute triangle has only one acute angle.
c. A triangle can be both isosceles and obtuse.
d. An equilateral triangle can be an obtuse triangle.
e. A square is a rectangle.

19. Arrange the following terms in the order in which the definitions of each should be given: triangle, hypotenuse, polygon, right triangle.

20. Arrange the following terms in the order in which the definitions of each should be given: square, quadrilateral, parallelogram, polygon, rectangle.

21. For what kind(s) of parallelogram is each statement true?
a. The diagonals are equal.
b. The diagonals are perpendicular to each other.

22. Which figure does *not* always have congruent diagonals?
a. rectangle b. square
c. rhombus d. isosceles trapezoid

23. In quadrilateral $ABCD$ if $m\angle A = 40°$, $m\angle B = 140°$, quadrilateral $ABCD$ must (choose one)
a. be a rhombus.
b. have at least one pair of sides parallel.
c. be an isosceles trapezoid.
d. have at least one right angle.

24. In parallelogram $ABCD$, $m(\overline{AB}) = 5x - 4$ and $m(\overline{CD}) = 2x + 14$. Find the value of x.

25. In parallelogram $ABCD$, $m\angle A = 3x$ and $m\angle B = x + 40$. What is the value of x?

26. Which statement about the diagonals of an isosceles trapezoid is *always* true?
a. They bisect each other.
b. They are congruent.
c. They are perpendicular to each other.
d. They divide the trapezoid into four congruent triangles.

27. In isosceles trapezoid $DEFG$, $m\angle D$ is three times $m\angle F$. Find $m\angle F$.

28. A rectangle has a diagonal of length 10 and one side of length 6. What is the length of the other diagonal?

29. Which statement is true?
a. All parallelograms are quadrilaterals.
b. All parallelograms are rectangles.
c. All quadrilaterals are trapezoids.
d. All trapezoids are parallelograms.

30. In rhombus $ABCD$, $\overline{AB} = 4x - 2$ and $\overline{BC} = 3x + 3$. Find the value of x.

For Exercises 31 and 32 choose the best answer.

31. The opposite angles of a parallelogram are
a. complementary b. congruent
c. supplementary d. right

32. Two opposite angles of an isosceles trapezoid are
a. equal b. complementary
c. supplementary

B.C. by permission of Johnny Hart and Creators Syndicate, Inc.

JUST FOR FUN

In Chapter 9, we located natural numbers, whole numbers, integers, and rational numbers on the number line. We can also locate irrational numbers on the number line.

For example, suppose we want to locate the point that corresponds to $\sqrt{2}$. We can do this by drawing the hypotenuse of a right triangle whose legs are 1 unit in length along the number line with one vertex at 0 (see the figure).

Recall that the Pythagorean theorem tells us that the square of the hypotenuse of a right triangle equals the sum of the squares of the legs; that is, if a and b are the lengths of the legs, and c is the length of the hypotenuse, then $c^2 = a^2 + b^2$.

If $a = 1$ and $b = 1$, then

$$c^2 = 1^2 + 1^2 = 1 + 1 = 2$$

and therefore

$$c = \sqrt{2}$$

The length corresponds to a point on the number line in the figure.

Using the fact that $1^2 + 2^2 = 5$, can you graph $\sqrt{5}$ on the number line?

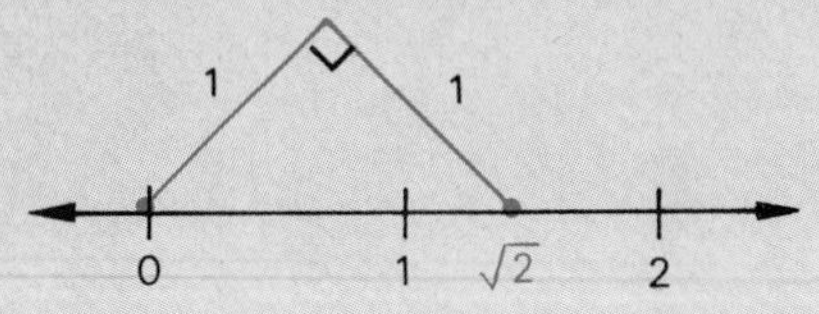

10.6 PERIMETER AND AREA

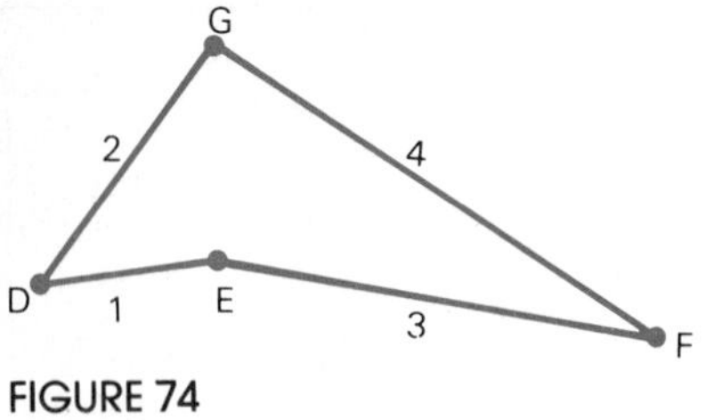

FIGURE 74
Perimeter = 10

Geometry in its beginnings dealt with the measure of surfaces, and this has remained one of its most practical applications. We can measure a polygon in different ways. The **perimeter** of a polygon is the sum of the lengths of its sides. The perimeter of quadrilateral $DEFG$ in Figure 74 is determined by finding the sum of the lengths of its sides. Therefore, the perimeter of quadrilateral $DEFG$ is $1 + 3 + 4 + 2 = 10$.

Formulas can be used to find the perimeters of certain polygons. For example, the perimeter of a square whose side is s

units in length is $4s$, that is, $P = 4s$. The perimeter of a rectangle with length l and width w is twice the length plus twice the width, that is $P = 2l + 2w$. The perimeter of a parallelogram with sides a and b is $P = 2a + 2b$. These formulas are shown in Figure 75.

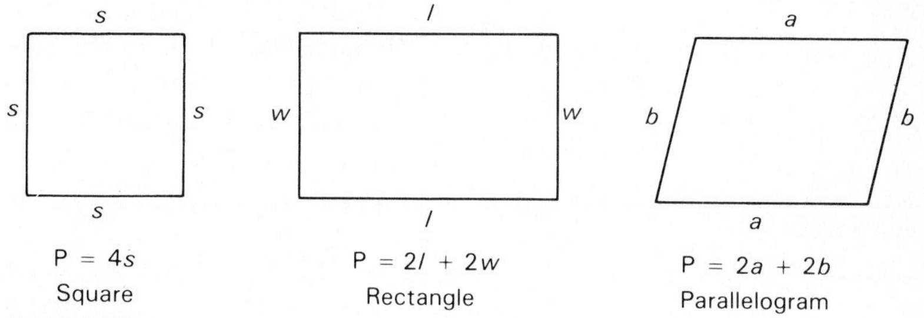

FIGURE 75

EXAMPLE 1 Find the perimeter of a square whose side measures 1.2 centimeters.

SOLUTION
$P = 4s$

$P = 4(1.2)$

$P = 4.8$ cm

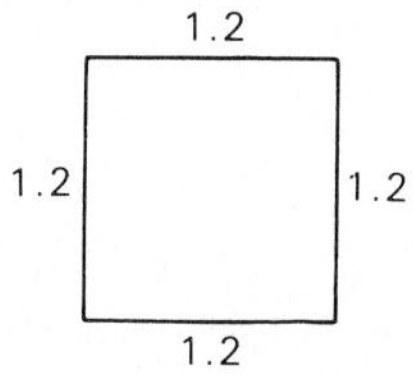

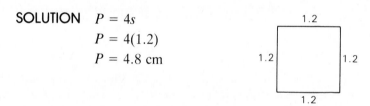

EXAMPLE 2 Find the perimeter of a flag that measures 0.8 by 1.2 meters.

SOLUTION The longer side, 1.2, is the length and the shorter side, 0.8, is the width. The shape of the flag is a rectangle. Hence,

$P = 2l + 2w$

$P = 2(1.2) + 2(0.8)$

$P = 2.4 + 1.6$

$P = 4.0$ m

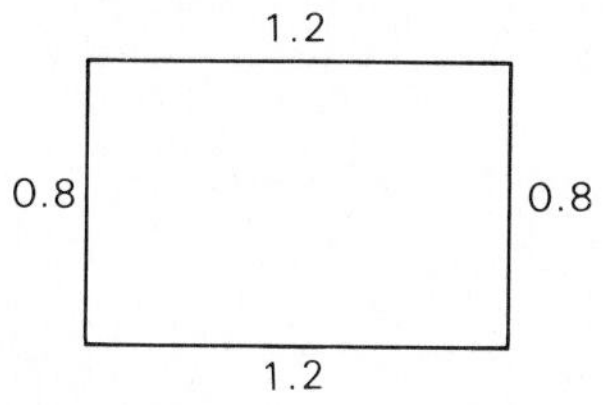

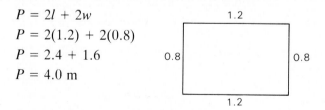

Another measure of a polygon is *area*. The **area** of a polygon tells us how many square units (inches, feet, yards, meters, centimeters, and so on) that polygon contains.

Suppose we want to put carpet tiles on the floor of a room. Suppose that the room is of a fairly standard size, 10 feet by 8 feet, and the carpet tiles are 1 foot by 1 foot. How many carpet tiles do

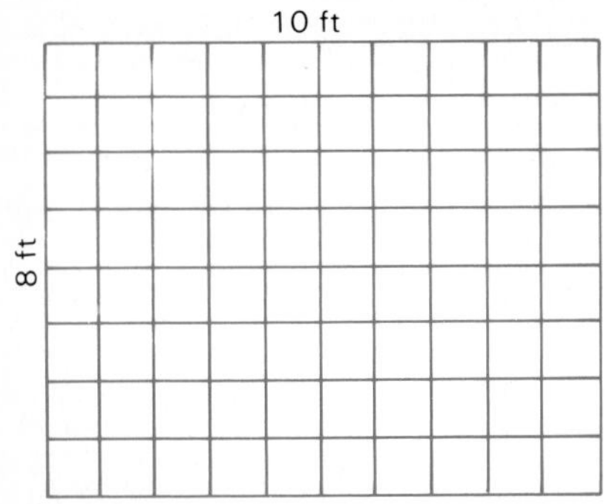

FIGURE 76

we need to cover the floor of the room? One way to do this would be to start laying the tiles down on the floor, placing one next to the other until the entire floor is covered. The result would look like Figure 76. Because the carpet tiles measure 1 foot by 1 foot, they are squares, and we can say that their measure is 1 square foot. Counting the number of carpet tiles in Figure 76 we see that there are 80 of them, that is, there are 80 square feet of carpet tiles on the floor. We can also say that the *area* of the floor is 80 square feet. Recall that the area of a polygon tells how many square units of a certain kind that polygon contains. Note that we could have also obtained the answer 80 square feet by multiplying 10 times 8; that is, $10 \times 8 = 80$.

Figure 76 is a rectangle whose length is 10 feet and width is 8 feet, and the rectangle contains 80 square feet. Therefore, we can conclude that

The area of a rectangle is equal to the product of its length and width; that is, area of a rectangle = length × width

Symbolically, we have

$$A = l \times w \qquad \text{or} \qquad A = l \cdot w$$

We can also say that the area of a rectangle is equal to the product of its base and its height. The term **base** is another name for the length of a rectangle, and the term **height** is another name for the width of a rectangle. Therefore,

$$A = b \times h$$

EXAMPLE 3 Find the area of a rectangle whose base is 7 centimeters and whose height is 4 centimeters.

SOLUTION The area of a rectangle is equal to the product of its base and height. Therefore, the area of the given rectangle is $7 \times 4 = 28$ square centimeters. Remember that area is measured in terms of square units.

EXAMPLE 4 The area of a rectangle is 40 square meters, and the length is 8 meters. What is the width?

SOLUTION Using the formula $A = lw$, we have $A = 40$ and $l = 8$. Substituting these values in the formula, we have

$$40 = 8w$$

Now we solve for w.

$$\frac{40}{8} = \frac{8w}{8} \quad \text{dividing both sides by 8}$$
$$5 = w$$

The width is 5 meters.

EXAMPLE 5 Find the area of a rectangle, given that its length is 15 inches and the length of its diagonal is 17 inches.

SOLUTION This problem is a little more involved than the previous examples. In order to find the area, we must first find the width of the rectangle. We are given its length and diagonal. A diagonal of a polygon is a line segment that connects two vertices that are not consecutive. Therefore, the given rectangle could appear as in Figure 77.

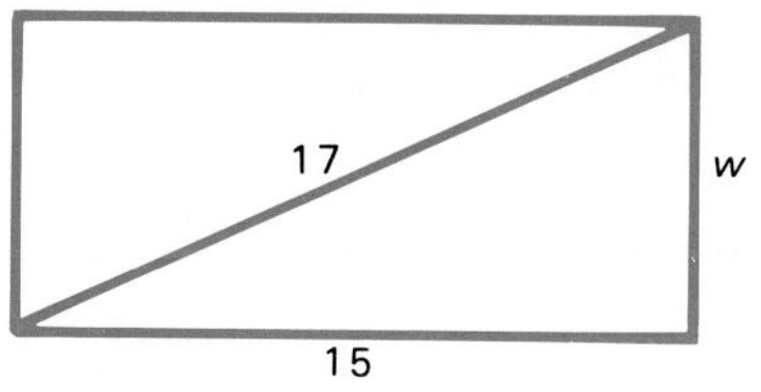

FIGURE 77

Note that the triangle in the rectangle is a right triangle. We can find w by means of the Pythagorean theorem, which states that the square of the hypotenuse of a right triangle is equal to the sum of the squares of the legs. Therefore we have

$$17^2 = 15^2 + w^2$$
$$289 = 225 + w^2$$
$$64 = w^2$$
$$\sqrt{64} = w$$
$$8 = w$$

The width is 8 inches.

Now, we can find the area using the formula.

$$A = lw$$
$$A = 15 \times 8$$
$$A = 120 \text{ square inches}$$

EXAMPLE 6 Find the area of a square whose diagonal is 10 inches in length.

SOLUTION Recall that a square is a rectangle with two adjacent sides equal. Therefore, the area of a square is also equal to the product of its length and width. But the length and width of a square are equal; hence, the area of a square is $s \times s$, or s^2, where s is the length of any side. Therefore, once we find the side of a square we can find its area. A square whose diagonal is 10 inches could appear as in Figure 78.

Note that we again have a right triangle (as in Example 5), and we can find s by means of the Pythagorean theorem. Therefore, we have

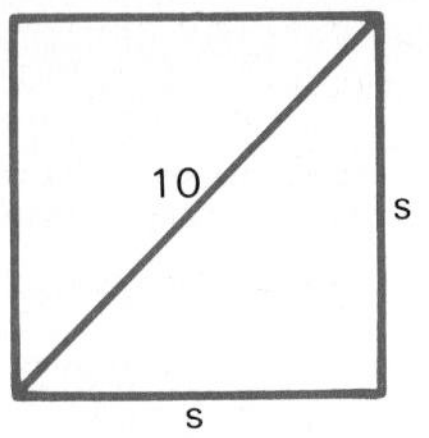

FIGURE 78

$$10^2 = s^2 + s^2$$
$$100 = 2s^2$$
$$50 = s^2$$

This is the answer; we do not need to do any more computation. Recall that the area of a square is equal to the product of two of its sides, that is, $s \times s$, or s^2, which is what we have. Therefore, the area of a square whose diagonal is 10 inches is 50 square inches.

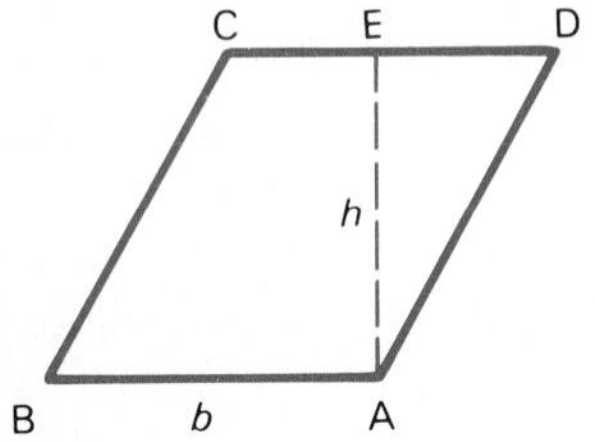

FIGURE 79a

Now that we are able to find the area of a rectangle and a square, let's consider the area of a parallelogram. We can state formally that:

The area of a parallelogram is equal to the product of its base and height; that is, $A = bh$.

Let's see why this formula works. Consider parallelogram $ABCD$ in Figure 79a, with the base AB denoted by b and its height AE denoted by h.

The height of a parallelogram is a line segment drawn perpendicular to the side from a point in the opposite side. We have chosen point A in order to form triangle ADE. Now, if we were to cut off triangle ADE from parallelogram $ABCD$, we would have Figure 79b.

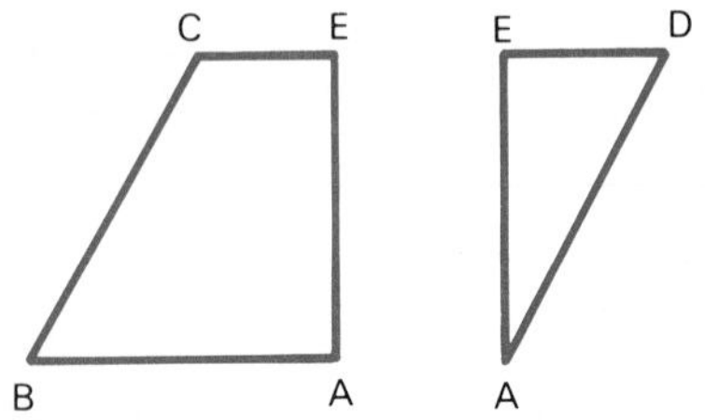

FIGURE 79 b

Now let's move the triangle to the left side of $ABCE$ and attach it to $\overline{CB}$ so that $\overline{DA}$ and $\overline{CB}$ coincide. Note that $\overline{DA}$ and $\overline{CB}$ are equal, since the opposite sides of a parallelogram are equal. Therefore, we have Figure 79c.

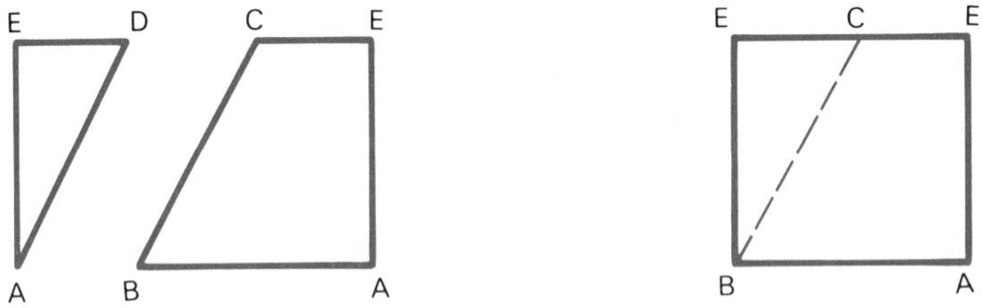

FIGURE 79 c

The resulting figure is a rectangle, and the area of a rectangle is equal to the product of its base and its height. Recall that the term *base* is another name for the length of a rectangle, and the term *height* is another name for the width of a rectangle. The area of a parallelogram is equal to the product of its base and height, because the area of a rectangle is equal to the product of its base and height, and Figures 79a–c show that the area of a parallelogram is the same as the area of a rectangle. That is, $A = bh$.

Note that the discussion concerning parallelogram $ABCD$ is not a formal proof. It is merely an attempt to show why the area of a parallelogram can be found by multiplying its base times its height.

EXAMPLE 7 Find the area of a parallelogram whose base is 12 centimeters and whose height is 6 centimeters.

SOLUTION The area of a parallelogram is equal to the product of its base and height. Therefore, the area of the given parallelogram is $12 \times 6 = 72$ square centimeters.

EXAMPLE 8 The area of a parallelogram is 140 square feet and the base is 20 feet. What is its height?

SOLUTION Using the formula $A = bh$, we have $A = 140$ and $b = 20$. Substituting these values in the formula, we have

$$140 = 20h$$

Solving for h,

$$\frac{140}{20} = \frac{20h}{20} \qquad \text{dividing both sides by 20}$$

$$7 = h$$

The height is 7 feet.

Now that we are able to find the area of a parallelogram, we can determine a method for finding the area of a triangle. Consider parallelogram $ABCD$ in Figure 80, with diagonal $\overline{DB}$ and height $\overline{DE}$.

We know that the area of parallelogram $ABCD$ is equal to the product of its base and height, that is, $A = bh$. What is the area of triangle ABD?

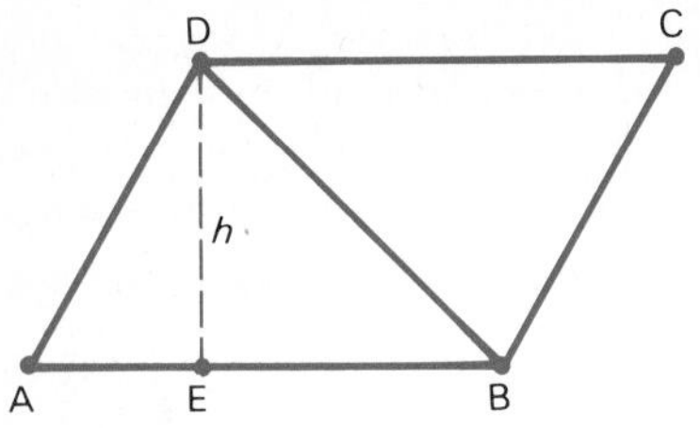

FIGURE 80

If we were to trace triangle ABD on a piece of paper and then take the paper and turn it so that the tracing of triangle ABD fitted over triangle DBC, we would see that the two triangles are exactly the same, that is, they are *congruent.* Two figures that can be made to coincide are said to be *congruent.* They match exactly in shape and size. The two triangles are congruent; together they form a parallelogram and we already know how to determine the area of a parallelogram. Because the parallelogram is formed by two congruent triangles, the area of one triangle (ABD) is equal to one-half of the area of the parallelogram. Therefore, the area of triangle ABD is $\frac{1}{2}bh$. We can state formally that

The area of a triangle is equal to one-half the product of its base and height; that is, $A = \frac{1}{2}bh$.

EXAMPLE 9 Find the area of a triangle whose base is 12 inches and whose height is 6 inches.

SOLUTION The area of a triangle is equal to one-half the product of its base and height. Therefore, the area of the given triangle is $\frac{1}{2} \times 12 \times 6 = 36$ square inches.

EXAMPLE 10 The area of a triangle is 70 square centimeters and the base is 20 centimeters. What is its height?

SOLUTION Using the formula $A = \frac{1}{2}bh$, we have $A = 70$ and $b = 20$. Substituting the values in the formula, we have

$$70 = \frac{1}{2}(20)h$$

Solving for h,

$$70 = 10h$$

$$\frac{70}{10} = \frac{10h}{10} \qquad \text{dividing both sides by 10}$$

$$7 = h$$

The height is 7 centimeters.

EXAMPLE 11 Find the area of right triangle ABC in Figure 81 given that $m(\overline{AC}) = 10$, $m(\overline{BC}) = 6$, and $\angle BCA$ is the right angle.

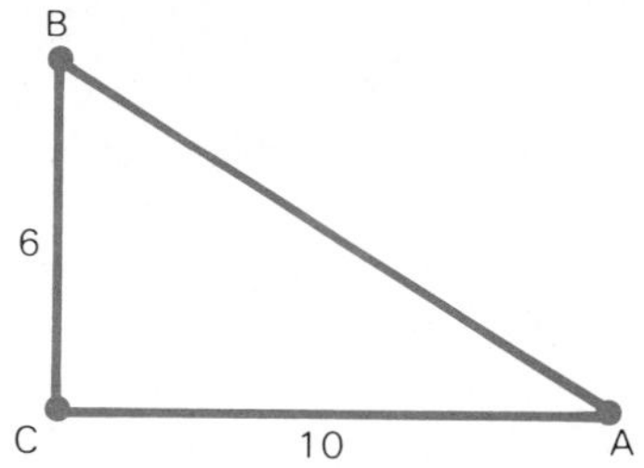

FIGURE 81

SOLUTION The base and height are not given directly in this problem, but the triangle is a right triangle, and therefore the legs of the triangle are its base and height, respectively. The base is $\overline{AC}$, whose length is 10, and the height is $\overline{BC}$, whose length is 6. Using the formula $A = \frac{1}{2}bh$, we have

$$A = \frac{1}{2}(10)(6)$$

$$A = 30 \text{ square units}$$

Note that the lengths of the sides were not given in terms of a specific measurement. Therefore, the area is described in terms of *square units.*

The area of a trapezoid can be determined if we think of the trapezoid as half of a parallelogram (see Fig. 82).

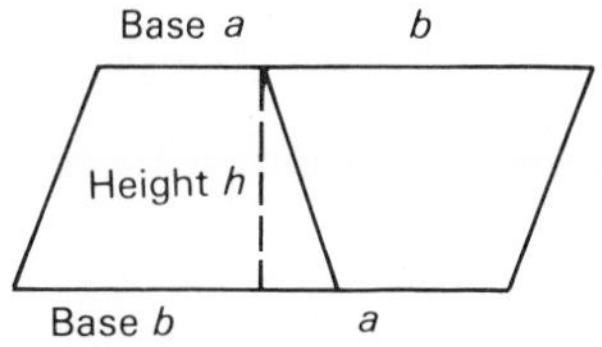

FIGURE 82

The base of this parallelogram is $a + b$. The area of the parallelogram is $(a + b)h$. Hence the area of the trapezoid is $\frac{1}{2}(a + b)h$.

The area of a trapezoid is equal to one-half the height times the sum of the bases; that is,

$$A = \tfrac{1}{2}h(a + b)$$

EXAMPLE 12 Find the area of a trapezoid whose bases are 8 centimeters and 22 centimeters and whose height is 10 centimeters.

SOLUTION The area of a trapezoid is equal to one-half the height times the sum of the bases. Therefore,

$$A = \frac{1}{2}h(a + b)$$

$$A = \frac{1}{2}(10)(8 + 22)$$

$$A = 5(30)$$

$$A = 150 \text{ square centimeters}$$

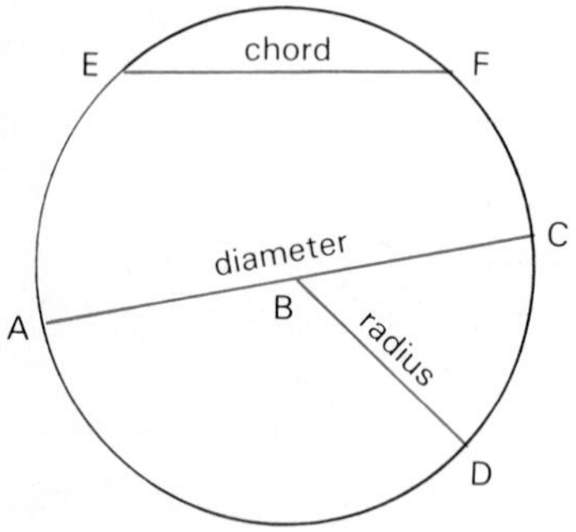

FIGURE 83

Thus far we have discussed the perimeter and area of certain polygons. It is also important to be able to do the same for another common geometric figure, the circle. The distance around a circle is called the **circumference** of a circle. Figure 83 is a circle whose center is at point B. A **chord** is a line segment that joins two points on a circle. $\overline{EF}$ is a chord. A **diameter** is a chord that passes through the center of the circle. $\overline{AC}$ is a diameter. A **radius** is a line segment that has the center and a point on the circle as end points. $\overline{BA}$, $\overline{BC}$, and $\overline{BD}$ are radii.

The constant ratio of the circumference of a circle to its diameter is represented by the Greek letter π (pi). That is, $c/d = \pi$.

The value of π, correct to the nearest ten thousandth, is 3.1416. The fraction $\frac{22}{7}$ gives a value of π correct to within 0.04 of one percent. Pi is a decimal that never ends and has no repeating pattern.

$$\pi = 3.141592653589793238 \ldots$$

It has been shown that the value of π to 10 decimal places is sufficiently accurate to give the circumference of a circle as large as the earth's equator correct to a small fraction of an inch.

For any circle, the circumference is equal to π times the diameter. That is,

$$C = \pi d$$

Since a diameter is composed of two radii, we can also say: for any circle, the circumference is equal to two times π times the radius. That is,

$$C = 2\pi r$$

EXAMPLE 13 Find the circumference of a circular swimming pool if its diameter is 10 meters. Use 3.14 for π.

SOLUTION The circumference of a circle is equal to π times its diameter. Therefore,

$$C = \pi d$$
$$C = (3.14)(10)$$
$$C = 31.4 \text{ m}$$

EXAMPLE 14 Find the circumference of a circle that has a radius of 21 cm. Use $\pi = \frac{22}{7}$.

SOLUTION The circumference of a circle is equal to two times π times its radius. Hence,

$$C = 2\pi r$$
$$C = 2 \cdot \frac{22}{7} \cdot 21$$
$$C = \frac{924}{7} = 132 \text{ cm}$$

The area of a circle is defined as the area enclosed by the circle.

For any circle, the area is equal to π times the radius squared. That is,

$$A = \pi r^2$$

If the radius of a given circle is 10 cm and we wish to find its area, then we have

$$A = \pi r^2$$
$$\text{let } \pi = 3.14: \quad A = (3.14)(10^2)$$
$$A = 3.14(100)$$
$$A = 314 \text{ cm}^2$$

Please note: The notation cm^2 represents the label "square centimeters" and is a common notation for expressing square units. Similarly, 9 m^2 means 9 square meters and 32 mm^2 means 32 square millimeters.

EXAMPLE 15 The radius of a circular rug is 3 m; find its area. Use $\pi = 3.14$.

SOLUTION The area of a circle is equal to π times the radius squared. Therefore,

$$A = \pi r^2$$
$$A = (3.14)3^2$$
$$A = (3.14)9$$
$$A = 28.26 \text{ m}^2$$

EXAMPLE 16 What is the area of a circle whose diameter measures 18 dm? Use $\pi = 3.14$.

SOLUTION We must still use the formula $A = \pi r^2$, so first recall that the diameter of a circle is composed of two radii. Hence for this problem, since $d = 18$, r must equal 9. Therefore,

$$A = \pi r^2$$
$$A = (3.14)9^2$$
$$A = (3.14)81$$
$$A = 254.34 \text{ dm}^2$$

EXERCISES FOR SECTION 10.6

1. Find the perimeter of each polygon.

a.

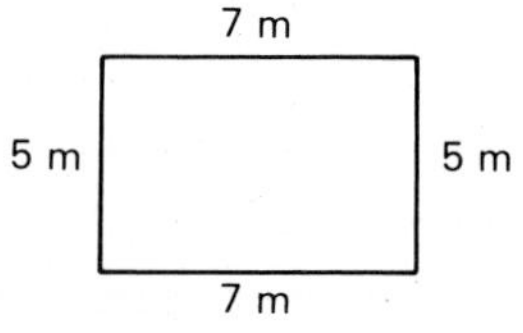

b.

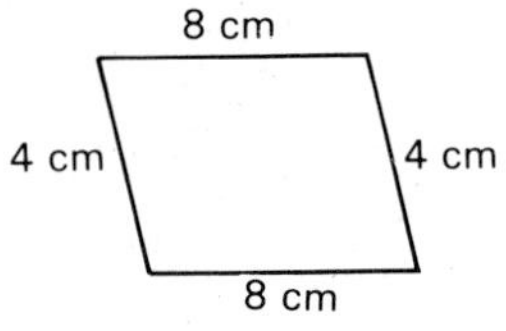

c.

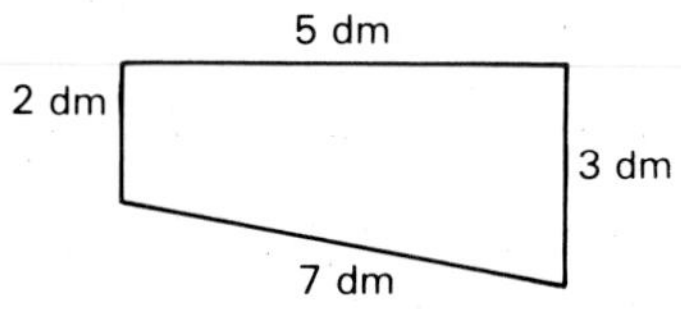

2. Find the perimeter of each polygon.

a.

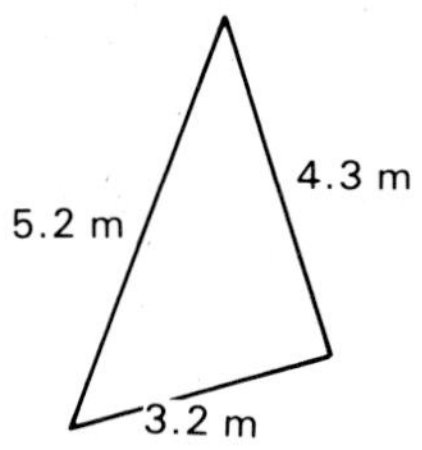

b.

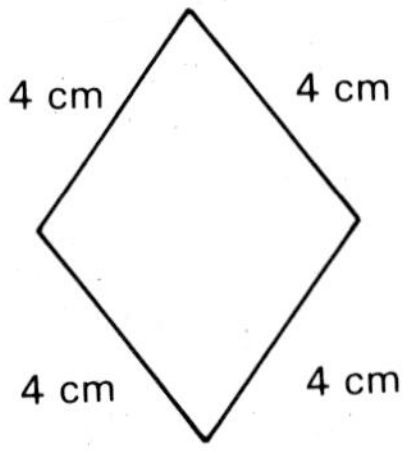

c.

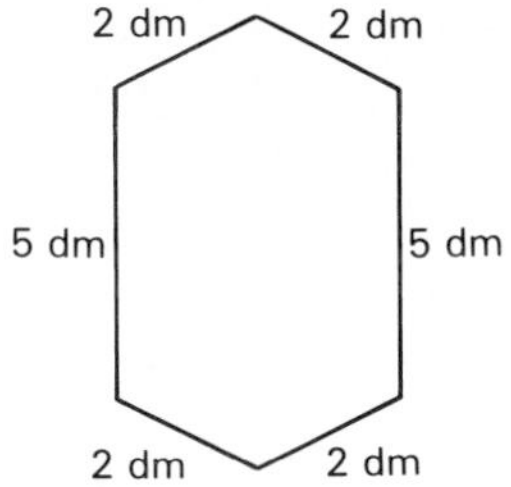

3. Find the perimeter of a rectangle whose length is 10 meters and whose width is 5 meters.

4. Find the perimeter of a square whose side is 6.5 centimeters in length.

5. Find the perimeter of a rhombus whose side is 1.4 meters in length.

6. Find the perimeter of a parallelogram whose sides are 8 inches and 14 inches.

7. Find the perimeter of a parallelogram whose sides are 42 centimeters and 65 centimeters.

8. Find the perimeter of a rectangle, given that its width is 3 and the length of its diagonal is 5.

9. Find the perimeter of a rectangle, given that its length is 12 and the length of its diagonal is 13.

10. Find the perimeter of a rectangle, given that its length is 15 and the length of its diagonal is 17.

11. The area and perimeter of a certain square are the same number. What is the length of a side of the square?

12. Find the area of a rectangle whose length is 10 meters and whose width is 5 meters.

13. The area of a rectangle is 30 square inches and the length is 6 inches. What is the width?

14. The area of a rectangle is 78 square centimeters and the width is 6 centimeters. What is the length?

15. Find the area of a rectangle, given that its length is 12 and the length of its diagonal is 13.

16. Find the area of a rectangle, given that its width is 3 and the length of its diagonal is 5.

17. Find the area of a square whose diagonal measures 5 inches.

18. The length of a diagonal of a square is 8 centimeters. Find the number of square centimeters in the area of the square.

19. If the length of a diagonal of a square is 4 inches, find the number of square inches in the area of the square.

20. Find the area of a parallelogram whose base is 14 inches and whose height is 8 inches.

21. The area of a parallelogram is 300 square meters and the base is 30 meters. What is its height?

22. The area of a parallelogram is 30 square feet and the height is 4 feet. What is the length of the base?

23. Find the area of a triangle whose base is 13 and whose height is 7.

24. The area of a triangle is 62.5 square centimeters and the base is 10 centimeters. Find the height.

25. The area of a triangle is 37 square feet and the height is 10 feet. Find the base.

26. Find the area of a right triangle whose legs measure 5 inches and 6 inches, respectively.

27. Find the area of a right triangle whose legs measure 5 and 7, respectively.

28. Find the area of a trapezoid whose bases are 8 and 12, and height is 5.

29. Find the area of a trapezoid whose bases are 10 and 20, and height is 6.

30. The area of a trapezoid is 200, the bases are 15 and 25, respectively. Find the height.

31. The bases of a trapezoid are 14 centimeters and 6 centimeters. Both nonparallel sides are 5 centimeters. Find the area.

32. The bases of an isosceles trapezoid are 9 centimeters and 21 centimeters. The nonparallel sides are each 10 centimeters. Find the area.

★33. Find the area of an isosceles right triangle whose hypotenuse is 10 meters in length.

★34. The area of a square is 16 square units. Find the length of a diagonal.

35. Find the area and perimeter of each of the following figures. Let $\pi = 3.14$.

a.

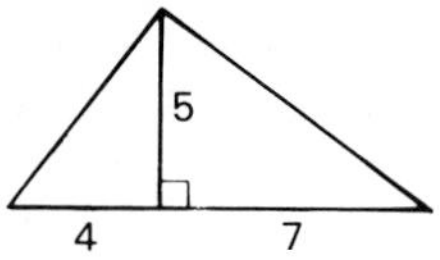

b.

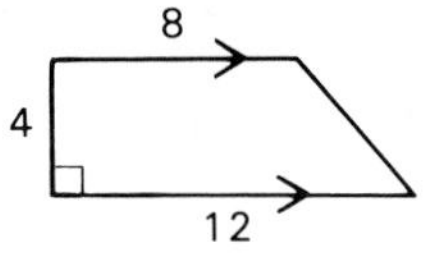

c.

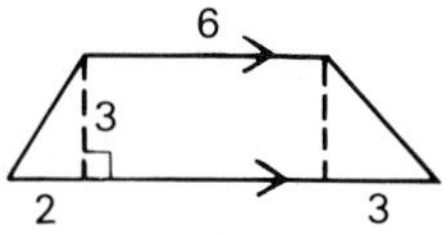

d.

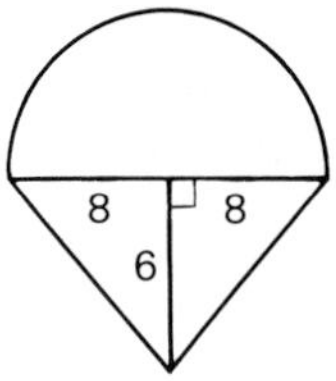

e.

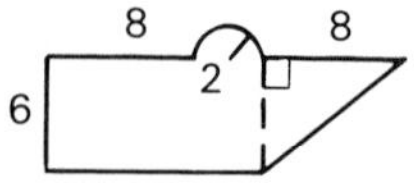

In Exercises 36–50, use 3.14 for π.

36. The diameter of a tree trunk is 20 cm. Find its circumference.

37. The diameter of a wheel is 30 cm. Find its circumference.

38. The radius of a circle is 3.5 mm. Find the circumference.

39. The minute hand of a clock is 15 cm. What is the distance that the tip of the minute hand moves in one hour?

40. The minute hand of a wrist watch is 14 mm. What is the distance that the tip of the minute hand moves in two hours?

41. How much fencing would be needed to enclose a circular garden that has a diameter of 10 m?

42. A circular rug has a diameter of 3 m. Find the area.

43. The floor of a circus tent is circular with a radius of 30 m. What is the area of the floor?

44. Find the area of a circle whose diameter is 20 dm.

45. Find the area of a circle whose radius is 5 mm.

46. What is the area of a circular garden that has a diameter of 8 m?

47. Find the area of a circle if its circumference is 314 m.

★48. If the radius of a circle is increased by 1 cm, by how many cm is its circumference increased?

★49. A circle and a square each have a perimeter of 140. Which has the greater area and how much?

★50. Find the radius of a circle whose area is numerically equal to its circumference.

JUST FOR FUN Are you good at counting squares? How many squares are there in this figure?

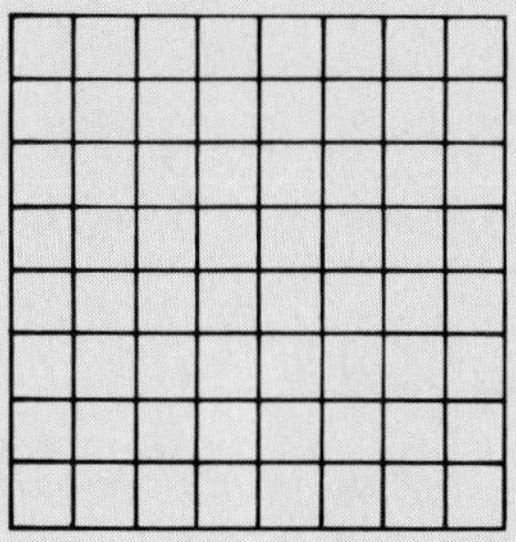

10.7 SOLIDS

In previous sections we discussed lines and planes. Two distinct points A and B determine a unique line segment AB, whereas any three distinct points A, B, and C that are not on the same line determine a unique plane. What happens if we have any four distinct points A, B, C, and D that are in a space, but not on the same plane? Figure 84 shows the outcome for this situation. The object shown in Figure 84 is called a **tetrahedron.** The four points A, B, C, D are the **vertices** of the tetrahedron. The six line segments $\overline{AB}$, $\overline{BC}$, $\overline{CD}$, $\overline{DB}$, $\overline{AD}$, $\overline{AC}$ are the **edges** of the tetrahedron. The four triangles, ABD, ABC, BCD, ACD, are called the **faces** of the tetrahedron. The solid in Figure 86 is called a tetrahedron because it has *four* faces. A **polyhedron** is a space figure or solid bounded by polygonal regions. For example, Figure 85 shows a different polyhedron. This polyhedron is called a square pyramid.

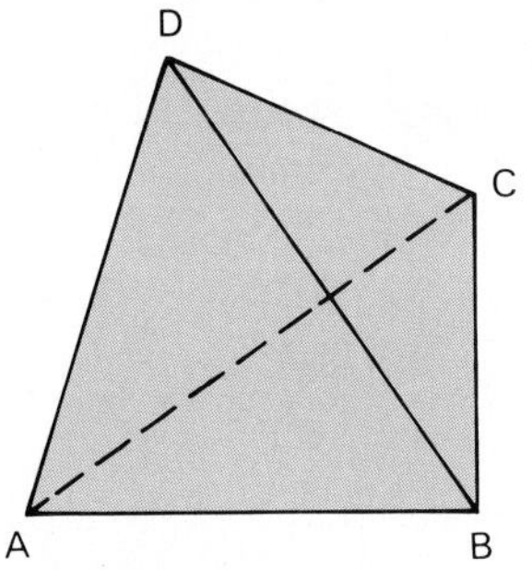

FIGURE 84

EXAMPLE 1 Figure 85 contains the square pyramid $ABCDE$. Identify (name) its

a. vertices b. edges c. faces

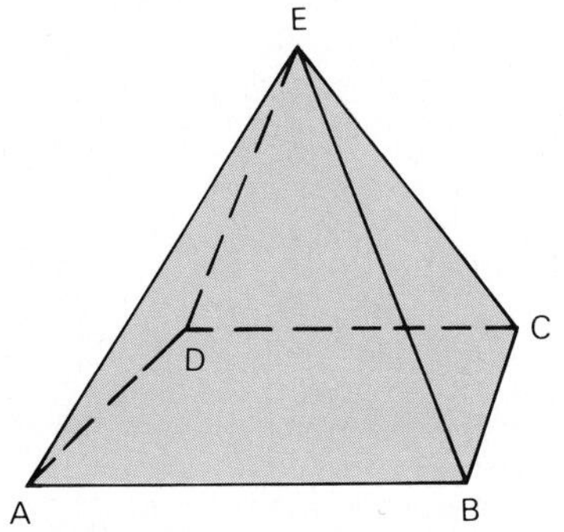

FIGURE 85

SOLUTION
a. The vertices are the end points of the line segments. That is, A, B, C, D, E.
b. The edges are the line segments $\overline{AB}$, $\overline{BC}$, $\overline{CD}$, $\overline{DA}$, $\overline{AE}$, $\overline{BE}$, $\overline{CE}$, $\overline{DE}$.
c. The faces are the polygonal regions, namely, triangles ABE, BCE, CDE, ADE and square $ABCD$.

HISTORICAL NOTE

The French mathematician Pierre de Fermat (1601–1665), is famous for his contributions to number theory. He was the son of a leather merchant and received his early education at home. Later in life he was a counselor for the local parliament at Toulouse. He devoted most of his leisure time to the study of mathematics.

There are a number of mathematical problems that are still mysteries to mathematicians. *Fermat's last theorem* is one of them. He wrote, "If n is a number greater than 2, there are no whole numbers a, b, c such that $a^n + b^n = c^n$." It can be shown that when $n = 2$, $x^2 + y^2 = z^2$, for example, (3, 4, 5) (5, 12, 13) and (8, 15, 17). Yet, no one has ever been able to find positive whole number such that $x^3 + y^3 = z^3$ or $x^4 + y^4 = z^4$. Fermat said that he could prove that no such numbers could be found.

Although many prominent mathematicians have tried, no one has solved the problem—that is, no correct proof has been found. Fermat's last theorem has the distinction of being the mathematical problem for which the greatest number of incorrect proofs have been published.

As recently as March 1988, Yoichi Miyaoka, a Japanese working at the Max Planck Institute for Mathematics in Bonn, West Germany, claimed to have finally proved Fermat's last theorem. But eminent mathematicians claim that there is a flaw in Miyaoka's proof and it is not an actual proof.

The famous Euler formula for polyhedrons states that

For any single polyhedron

$$V - E + F = 2$$

where V is the number of vertices, E is the number of edges, and F the number of faces.

Let's verify this formula for the square pyramid in Figure 85.

$$V = 5, \qquad E = 8, \qquad F = 5 \qquad \text{(see Example 1)}$$

Hence,

$$\begin{aligned} V - E + F &= 2 \\ 5 - 8 + 5 &= 2 \qquad \text{(substituting)} \\ 10 - 8 &= 2 \\ 2 &= 2 \end{aligned}$$

Thus far in our treatment of polyhedrons we have discussed the square pyramid where all the faces are triangles and the base is a square. A rectangular pyramid has a rectangle for a base and its other faces are triangular regions. Similarly, a pentagonal pyramid has a pentagon as its base and its other faces are triangular regions.

A **prism** is a polyhedron with two parallel and congruent bases. Its other faces are formed by parallelograms. (Note: these

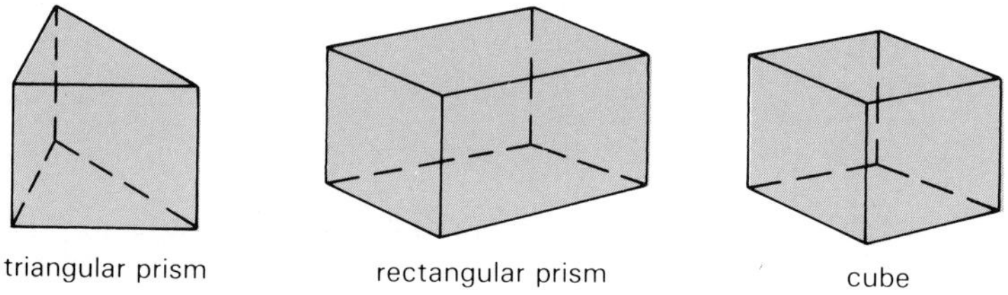

FIGURE 86

parallelograms are usually rectangles or squares.) **If the edges of the prism are perpendicular to the base,** as in Figure 86, **the prism is called a right prism.**

The total **surface area** of a polyhedron is the sum of the areas of all the faces.

EXAMPLE 2 Find the surface area of the prism in Figure 87.

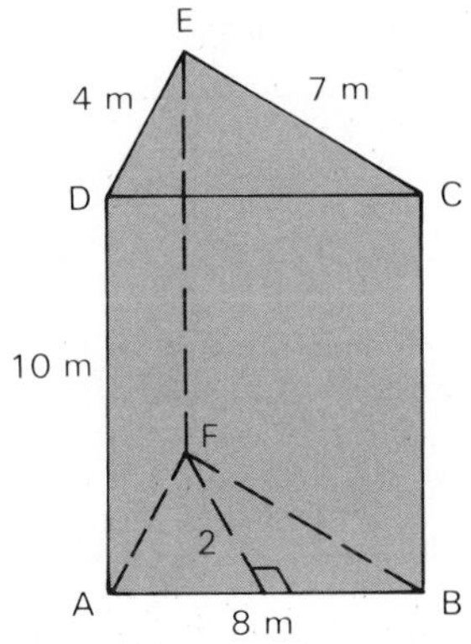

FIGURE 87

SOLUTION We find the area of each face, and then find the total.

$$\text{Area } ABF = \frac{1}{2}(8)(2) = 8$$

$$\text{Area } DCE = \frac{1}{2}(8)(2) = 8$$

$$\text{Area } ABCD = (8)(10) = 80$$
$$\text{Area } ADEF = (4)(10) = 40$$
$$\text{Area } BCEF = (7)(10) = 70$$
$$\text{Total surface area} = 206 \text{ m}^2$$

A **cylinder** (can) has two congruent circular bases. The total surface area is the sum of the areas of the bases and the area of the curved surface. The area of a circular base is πr^2. Hence, the total area of the two circular bases is $2\pi r^2$. If the cylinder is opened up, the curved surface flattens out to form a rectangle as shown in Figure 88. The length of the rectangle is $2\pi r^2$ because that is the circumference of the circular base; it still has the same length

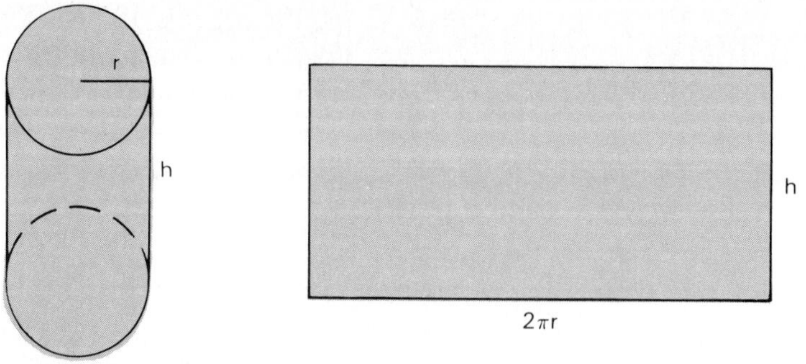

FIGURE 88

when it is "opened" up. The area of the rectangle (curved surface) is $2\pi rh$. Therefore,

The total surface area of any cylinder is equal to two times π times the radius squared plus two times π times the radius times the height. That is,

$$\text{surface area of a cylinder} = 2\pi r^2 + 2\pi rh$$

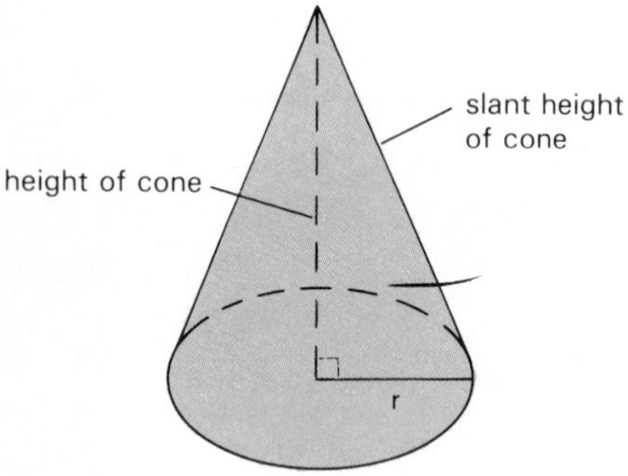

FIGURE 89

A **cone** has one circular base and a slant height s. See Figure 89.

The total surface area for any cone is equal to π times the radius squared (area of the base) plus π times the radius times the slant height. That is

$$\text{surface area of a cone} = \pi r^2 + \pi rs$$

EXAMPLE 3 Find the total surface area of a cylinder with radius of 3 m and height of 10 m. Use $\pi = 3.14$.

SOLUTION The formula for the total surface area of a cylinder is

$$A = 2\pi r^2 + 2\pi rh$$

Substituting,

$$A = 2(3.14)(3^2) + 2(3.14)(3)(10)$$
$$A = (6.28)9 + (6.28)(30)$$
$$A = 56.52 + 188.40$$
$$A = 244.92 \text{ m}^2$$

EXAMPLE 4 Find the total surface area of a cone with diameter 8 m and slant height of 10 m. Use $\pi = 3.14$.

SOLUTION The formula for the total surface area of a cone is

$$A = \pi r^2 + \pi rs$$

Note that $d = 8$; therefore $r = 4$.

$$A = (3.14)(4^2) + (3.14)(4)(10)$$
$$A = (3.14)(16) + (3.14)(40)$$
$$A = 50.24 + 125.60$$
$$A = 175.84 \text{ m}^2$$

The surface area of a **sphere** is four times the area of a circle with the same radius. The area of a circle is πr^2. Hence, the surface area of a sphere is equal to $4\pi r^2$. That is,

$$\text{surface area of a sphere} = 4\pi r^2$$

EXAMPLE 5 Find the surface area of a baseball, given that its radius is 3.3 cm. Use $\pi = 3.14$. Express your answer to the nearest tenth.

SOLUTION The formula for the surface area of a sphere is

$$A = 4\pi r^2$$

Therefore,

$$A = 4(3.14)(3.3)^2$$
$$A = (12.56)(10.89)$$
$$A = 136.7784$$
$$A = 136.8 \text{ cm}^2$$

The amount of space that is enclosed by a space figure is called **volume.** Volume is measured in cubic units. Hence, when we find the volume of a space figure or solid, we are finding the number of cubic units enclosed by the given space figure.

Recall that the formula for finding the volume of a cube or rectangular prism is length times width times height. That is,

$$V = lwh \qquad \text{(see Sec. 5.4)}$$

For any prism, the volume is equal to the area of the base times the height. That is,

$$V = Bh, \qquad \text{where } B = \text{area of the base}$$

EXAMPLE 6 Find the volume of the prism shown in Figure 90.

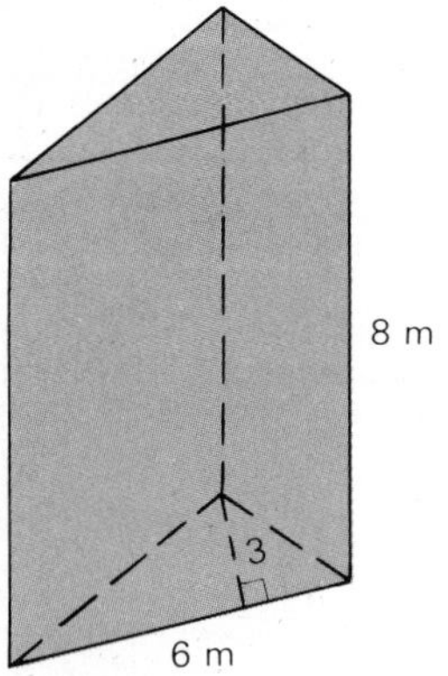

FIGURE 90

SOLUTION The base of the prism is a triangle. The formula for the area of a triangle is

$$A = \frac{1}{2}bh$$

Therefore,

$$A = \frac{1}{2}(6)(3)$$
$$A = (3)(3)$$
$$A = 9$$

Note, this is the area of the base. Now,

$$V = Bh$$
$$V = (9)(8)$$
$$V = 72 \text{ m}^3$$

For any pyramid, the volume is equal to one third the area of the base times the height. That is,

$$V = \frac{1}{3}Bh$$

This formula is applicable whether the pyramid has a triangular or rectangular base. For example, the Great Pyramid of Egypt has a square base of 227 m on a side and its height is 144 m. To find its volume we use the formula

$$V = \frac{1}{3}Bh$$

The area of its base is (227)(227) = 51,529; hence, we now have

$$V = \frac{1}{3}(51{,}529)(144)$$

$$V = \frac{1}{3}(7{,}420{,}176)$$

$$V = 2{,}473{,}392 \text{ m}^3$$

The volume of a cylinder is found in the same fashion as a prism. For any prism, its volume is equal to the area of the base times its height. But the base of a cylinder is a circle, and its area is πr^2. Hence,

For any cylinder, the volume is equal to π times the radius squared times the height. That is,

$$V = \pi r^2 h \qquad \text{(volume of a cylinder)}$$

A typical coffee can has a diameter of 10 cm and a height of 13 cm. To find its volume using the formula $V = \pi r^2 h$, we must first determine r. Since $d = 10$, r must equal 5, because $2r = d$. Therefore, substituting in the formula, we have

$$V = (3.14)(5^2)(13)$$
$$V = (3.14)(25)(13)$$
$$V = 1020.5 \text{ cm}^3$$

The volume of a cone is found in a similar manner to that of a pyramid. Recall that the formula for the volume of a pyramid is $V = \frac{1}{3}Bh$, where B is the area of the base.

The volume of a cone is equal to one third times π times the radius squared times the height. That is,

$$V = \frac{1}{3}\pi r^2 h \qquad \text{(volume of a cone)}$$

Note that πr^2 is the area of a circle, which is the base of a cone. The volume of a cone is one third the volume of a cylinder with base and height the same size as in the cone.

EXAMPLE 7 Find the volume of a cone whose height is 10 cm and the radius of whose base is 4 cm. Use $\pi = 3.14$. Express your answer to the nearest tenth.

SOLUTION The formula for the volume of a cone is

$$V = \frac{1}{3}\pi r^2 h$$

Substituting,

$$V = \frac{1}{3}(3.14)(4^2)(10)$$

$$V = \frac{1}{3}(3.14)(160)$$

$$V = \frac{1}{3}(502.4)$$

$$V = 167.46\overline{6}$$
$$V = 167.5 \text{ cm}^3$$

The volume of a sphere is two-thirds the volume of a cylinder with the same radius and a height equal to twice the radius. Recall that the volume of a cylinder is equal to $\pi r^2 h$. Now if h is equal to $2r$, we have

$$V = \pi r^2(2r) \quad \text{(substituting for } h\text{)}$$
$$V = 2\pi r^3 \quad \text{(volume of a cylinder)}$$

The volume of a sphere is equal to two-thirds of this. That is,

$$\frac{2}{3} \text{ of } 2\pi r^3 = \frac{2}{3}(2\pi r^3) = \frac{4}{3}\pi r^3$$

$$V = \frac{4}{3}\pi r^3 \quad \text{(volume of a sphere)}$$

EXAMPLE 8 Find the volume of a baseball, given that its radius is 3.3 cm. Use $\pi = 3.14$. Express your answer to the nearest tenth.

SOLUTION The formula for the volume of a sphere is

$$V = \frac{4}{3}\pi r^3$$

Substituting,

$$V = \frac{4}{3}(3.14)(3.3)^3$$

$$V = \frac{4}{3}(3.14)(35.937)$$

$$V = \frac{4}{3}(112.84218) = \frac{451.36872}{3}$$

$$V = 150.45624$$

$$V = 150.5 \text{ cm}^3$$

In this section we have discussed a variety of geometric solids and how to find their surface area and volume. The total surface area of a polyhedron is the sum of the areas of all the faces. Following is a list of the formulas that we used in this section.

Name	Surface Area	Volume
Cylinder	$A = 2\pi r^2 + 2\pi rh$	$V = \pi r^2 h$
Cone	$A = \pi r^2 + \pi rs$	$V = \frac{1}{3}\pi r^2 h$
Sphere	$A = 4\pi r^2$	$V = \frac{4}{3}\pi r^3$
Prism	sum of the areas of the faces	$V = Bh$
Pyramid	sum of the areas of the faces	$V = \frac{1}{3}Bh$

EXERCISES FOR SECTION 10.7

1. What is the least number of vertices that a polyhedron may have?

2. What is the least number of edges that a polyhedron may have?

3. What is the least number of faces that a polyhedron may have?

4. a. How many faces does a tetrahedron have?
   b. How many vertices does a tetrahedron have?
   c. How many edges does a tetrahedron have?
   d. Verify that $V - E + F = 2$.

5. a. How many faces does a cube have?
   b. How many vertices does a cube have?
   c. How many edges does a cube have?
   d. Verify that $V - E + F = 2$.

6. a. How many faces does a triangular prism have?
   b. How many vertices does a triangular prism have?
   c. How many edges does a triangular prism have?
   d. Verify that $V - E + F = 2$.

7. Figure 91 is called an octahedron.
   a. How many faces does it have?
   b. How many vertices does it have?
   c. How many edges does it have?
   d. Verify that $V - E + F = 2$.

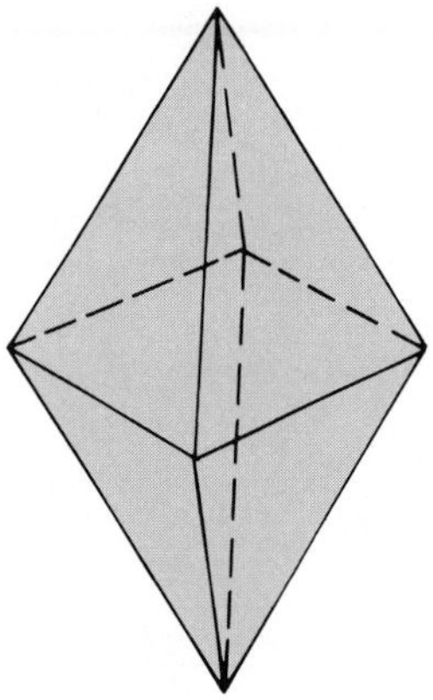

FIGURE 91

For Exercises 8–18, find the total surface area and volume of each polyhedron. Let $\pi = 3.14$. Express decimal answers to the nearest tenth.

8.
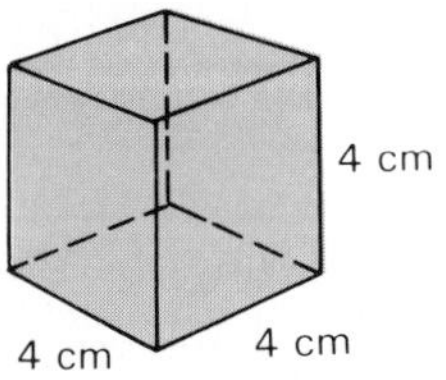

9.
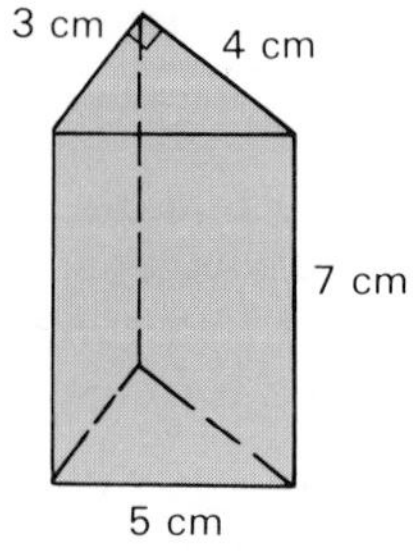

10.
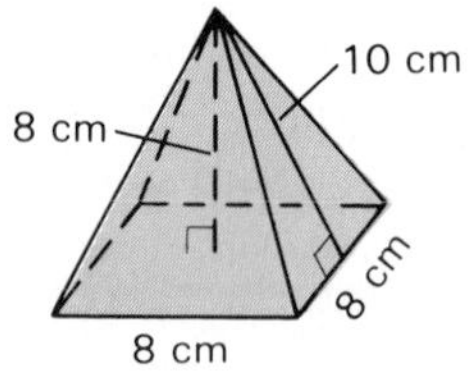

11.
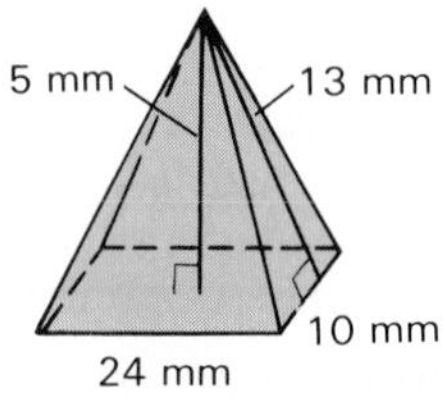

12.
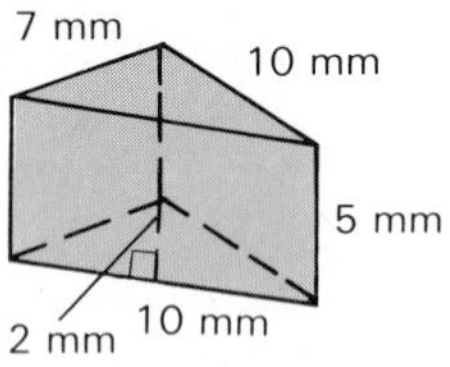

13.
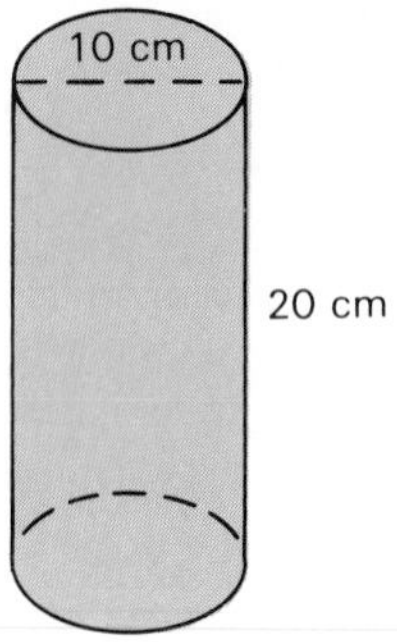

14.
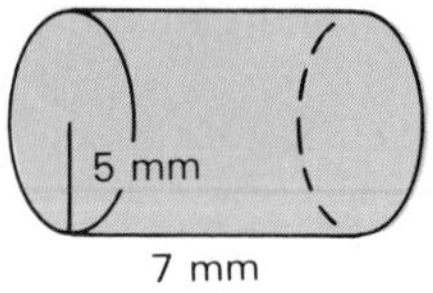

15.
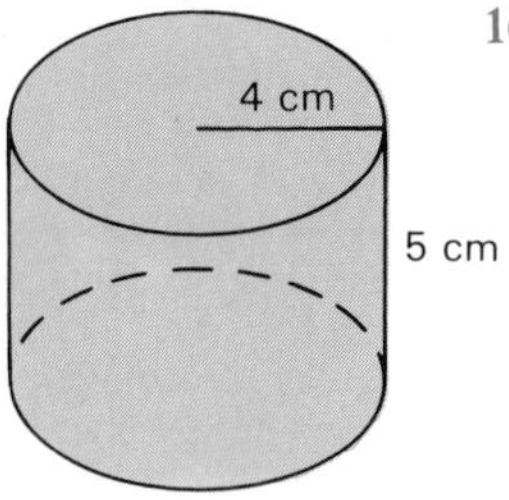

16.
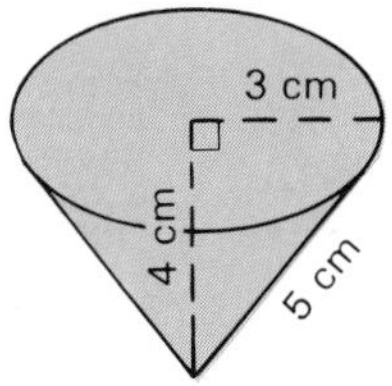

17.
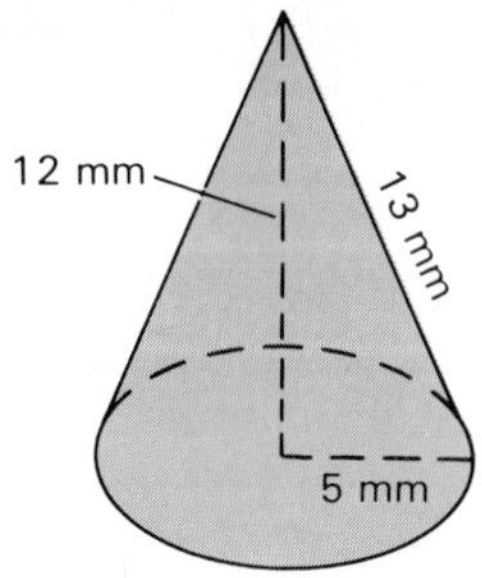

18.
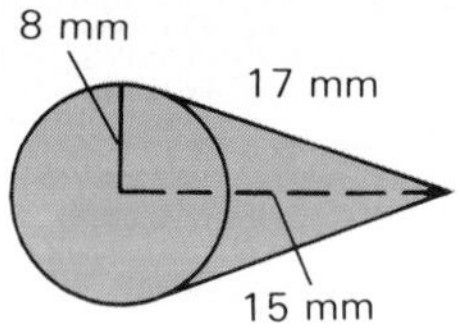

★19. Will the contents of a can whose base has a radius of 5 cm and a height of 20 cm fit into a canister whose base measures 5 cm by 6 cm and whose height is 20 cm?

★20. Will the contents of a can whose base has a radius of 10 cm and a height of 10 cm fit into a canister whose base measures 15 cm by 15 cm and whose height is 15 cm?

21. Find the surface area and volume of a basketball, if its radius is 12 cm. Let $\pi = 3.14$. Express your answer to the nearest tenth.

22. Find the surface area and volume of a marble, if its radius is 0.64 cm. Let $\pi = 3.14$. Express your answer to the nearest tenth.

23. Find the surface area and volume of a ball, if its radius is 9 cm. Let $\pi = 3.14$. Express your answer to the nearest tenth.

24. Find the surface area and volume of a ball bearing if its diameter is 6 mm. Let $\pi = 3.14$.

25. Find the surface area and volume of a spherical container that has a radius of 0.4 m. Let $\pi = 3.14$. Express your answer to the nearest tenth.

★26. The moon has a radius of approximately 1080 miles. What are its surface area and volume? Let $\pi = 3.14$. Express your answer to the nearest tenth.

JUST FOR FUN A farmer who had three and one-quarter piles of potatoes in one row and four and three-eighths piles in another row decided to put them together. How many piles did he then have?

10.8 CONGRUENT AND SIMILAR TRIANGLES

Congruent figures are figures that can be made to coincide. This means that it is possible to place one figure upon the other and have the two exactly match. Hence, two congruent figures have the same shape and the same size. The symbol for congruence is $\cong$ and is read "is congruent to." Therefore, $A \cong B$ is read "A is congruent to B." Sheets of paper in your notebook are congruent to each other.

Figure 92 contains two triangles that are congruent. If made to coincide, which parts would match? If we mentally turn $\triangle DEF$, we can see that $\angle F = \angle C$, $\angle A = \angle D$, and $\angle B = \angle E$. Also $\overline{AC} = \overline{DF}$, $\overline{AB} = \overline{DE}$, and $\overline{CB} = \overline{FE}$. In general, corresponding parts of congruent polygons are **congruent.**

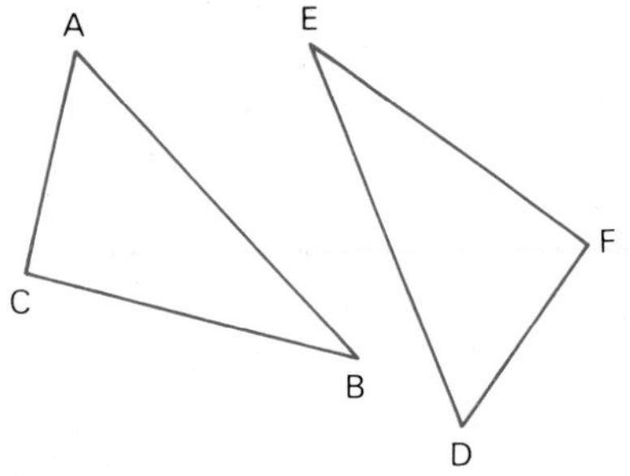

FIGURE 92

If certain parts of one triangle are congruent to parts of another triangle, the two triangles will be congruent. These conditions are summarized in the following postulates.

1. **Two triangles are congruent if measures of two sides and the included angle are equal, respectively, to measures of two sides and the included angle of the other. (Abbreviation: s.a.s.)**

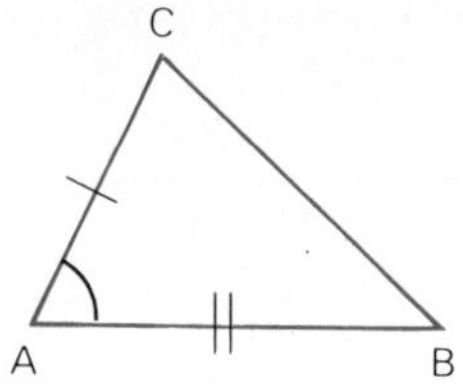

If $m(\overline{AC}) = m(\overline{DF})$

$m(\overline{AB}) = m(\overline{DE})$

$m\angle A = m\angle D$

then

$\triangle ABC \cong \triangle DEF$

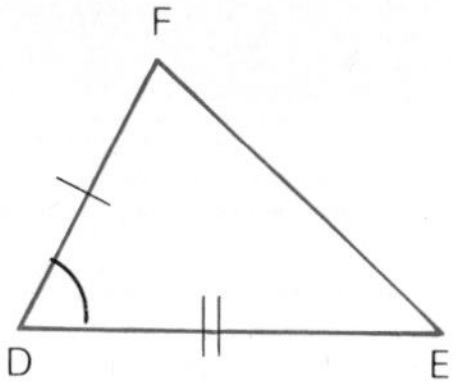

2. **Two triangles are congruent if two angles and the included side of one are congruent, respectively, to two angles and the included side of the other. (Abbreviation: a.s.a.)**

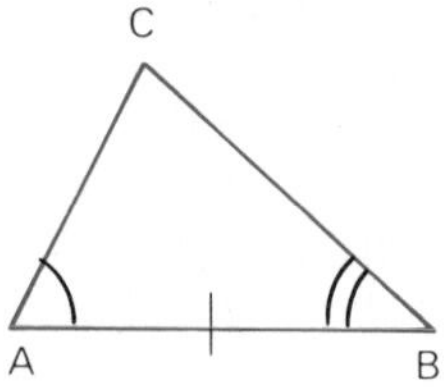

If $\angle A \cong \angle D$

$\angle B \cong \angle E$

$AB \cong DE$

then

$\triangle ABC \cong \triangle DEF$

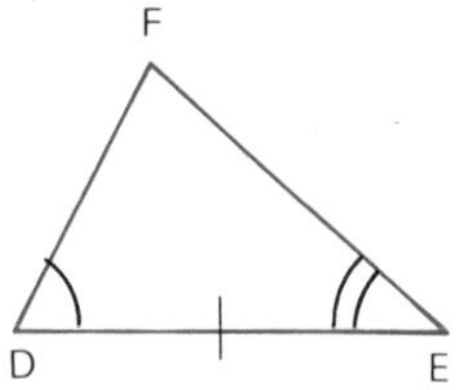

3. **Two triangles are congruent if the three sides of one are congruent, respectively, to the three sides of the other. (Abbreviation: s.s.s.)**

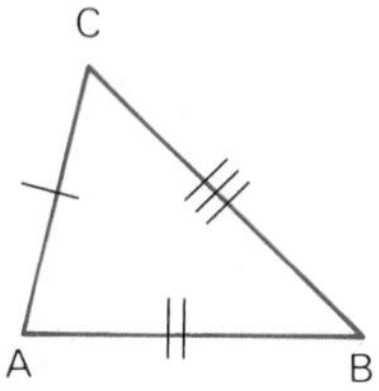

If $AB \cong DE$

$BC \cong EF$

$AC \cong DF$

then

$\triangle ABC \cong \triangle DEF$

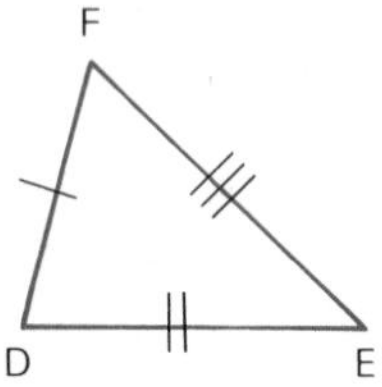

EXAMPLE 1 In triangles ABC and DEF, angles A and B are congruent, respectively, to angles D and E. If side AB is represented by $3a - 6$ and side DE by $a + 8$, for what value of a are the triangles congruent?

SOLUTION These triangles will be congruent if a.s.a. = a.s.a. Hence, in this case AB must equal DE. That is,

$$\begin{aligned} 3a - 6 &= a + 8 \\ 3a &= a + 14 \\ 2a &= 14 \\ a &= 7 \end{aligned}$$

If $a = 7$, then the triangles are congruent.

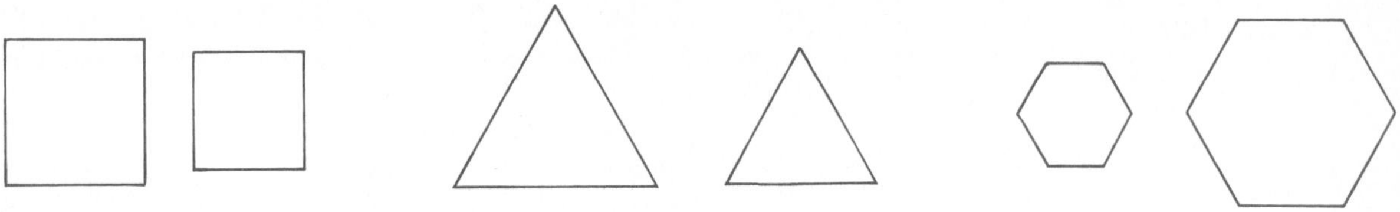

FIGURE 93

Similar polygons are polygons with the same number of sides that have their corresponding angles equal and their corresponding sides in proportion. The symbol ~ represents the word "similar." Some examples of similar polygons are shown in Figure 93.

Note that polygons are similar if and only if their corresponding angles are equal *and* their corresponding sides are in proportion. Both conditions must be met. If we consider a square and a rectangle, they would have corresponding angles equal, but their corresponding sides would not be in proportion.

But when we consider only triangles we have a special case. It can be shown that

If the three angles of one triangle are congruent with the three corresponding angles of another triangle, then the triangles are similar.

It also follows that if the measures of two angles of one triangle are equal to the measures of two angles of a second triangle, then the two triangles are similar. Why? The measures of the third angles must be equal. Hence, the triangles are similar.

Triangle ABC is similar to triangle DEF in Figure 94. We write $\triangle ABC \sim \triangle DEF$. In similar triangles the corresponding angles are congruent or equal and the lengths of the corresponding sides are proportional. That is, we can write

$$\angle A \cong \angle D$$

$$\angle B \cong \angle E \quad \text{and} \quad \frac{AB}{DE} = \frac{BC}{EF} = \frac{CA}{FD}$$

$$\angle C \cong \angle F$$

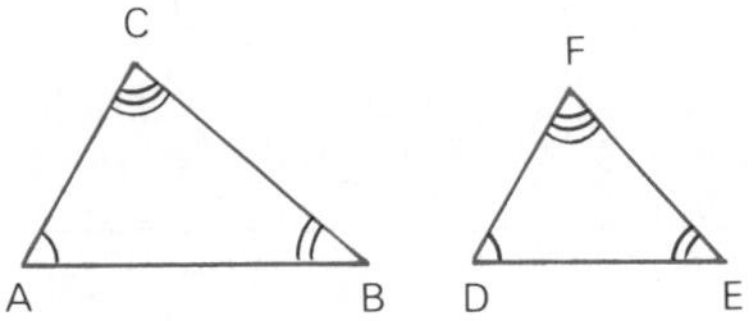

FIGURE 94

In dealing with similar triangles, we will use any two of the three ratios to find a missing side. For a detailed discussion of ratios and proportions, see Section 11.2

We are given that the two triangles in Figure 94 are similar. Suppose the lengths of AC, CB, and AB are 4, 3, and 5, respectively, and the lengths of DF and FE are 16 and 12. We can find the length of DE by using a proportion and solving it. That is,

$$\frac{4}{16} = \frac{5}{x} \quad \text{or} \quad \frac{3}{12} = \frac{5}{x}$$

$$4x = 5(16) \quad \text{or} \quad 3x = 12(5) \quad \text{(cross multiplying)}$$

$$4x = 80 \qquad 3x = 60$$

$$x = 20 \qquad x = 20$$

The length of DE is 20.

EXAMPLE 2 The triangles in Figure 95 are similar. Find the missing length.

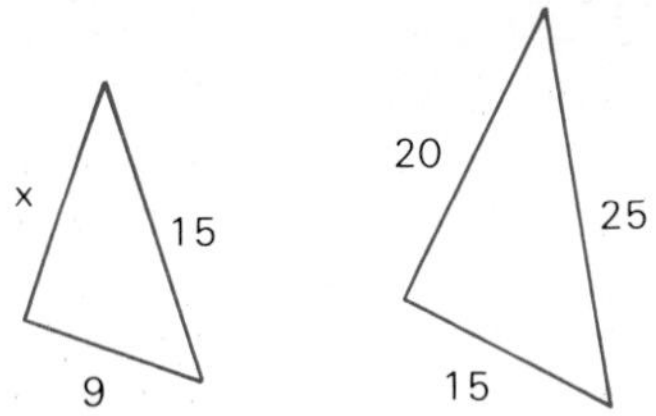

FIGURE 95

SOLUTION

$$\frac{x}{20} = \frac{15}{25}$$

$$25\,x = 15(20) \quad \text{(cross multiplying)}$$

$$25\,x = 300$$

$$\frac{25\,x}{25} = \frac{300}{25} \quad \text{(dividing both sides by 25)}$$

$$x = 12$$

EXAMPLE 3 A flagpole casts a shadow 5 feet long. Sandra, whose height is 5 feet 4 inches, is standing next to the flagpole. If she casts a shadow 16 inches long, what is the height of the flagpole?

SOLUTION Figure 96 represents the situation in terms of similar triangles. The angle of elevation is the same for Sandra and the flagpole,

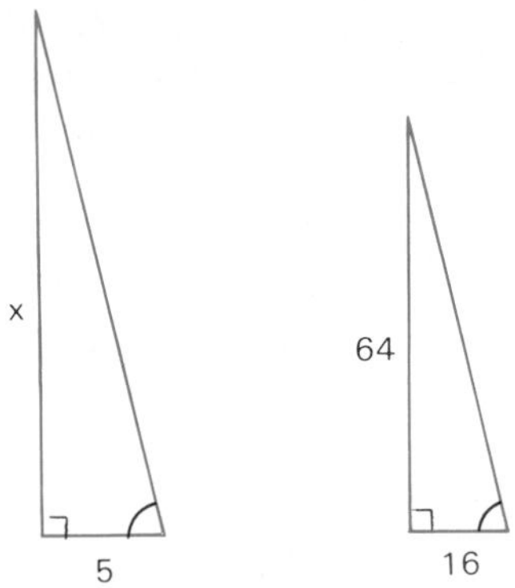

FIGURE 96

and they both are perpendicular to the ground. Two angles of one triangle are equal, respectively, to two angles of another triangle. Hence, the triangles are similar. We can solve this problem using a proportion since the corresponding sides of similar triangles are proportional. One ratio could be $x:5$, and the other is $64:16$.

Note that we set up ratios of like quantities, and the top and bottom of each ratio are in the same units. That is, x and 5 are measured in feet, and Sandra's height is expressed in inches to agree with the measure of her shadow.

The proportion is

$$x:5 = 64:16$$
$$\frac{x}{5} = \frac{64}{16}$$
$$16(x) = 5(64)$$
$$16x = 320$$
$$x = 20 \text{ feet}$$

EXERCISES FOR SECTION 10.8

For Exercises 1–8, determine whether the statements are true or false.

1. The corresponding parts of congruent polygons are equal.
2. Congruent polygons have the same size and shape.
3. S.A.S. = S.A.S. is a way of proving triangles similar.
4. Two triangles are congruent if three angles of one triangle are congruent with three angles of another triangle.
5. If two right triangles have the legs of one congruent with the legs of the other, the triangles are congruent.
6. If two right triangles have a leg and the adjoining acute angle of one congruent with the corresponding parts of the other, the triangles are congruent.

7. If two isosceles triangles have the legs of one congruent with the legs of the other, the triangles are congruent.

8. If two isosceles triangles have a leg and the base of one congruent with the corresponding parts of the other, the triangles are congruent.

9. In triangles *MNO* and *RST*, angles *M* and *N* are congruent, respectively, with angles *R* and *S*. If side *MN* is represented by $6X - 12$ and side *RS* is represented by $2x + 16$, for what value of x are the triangles congruent?

10. In triangles *ABC* and *DEF*, angles *A* and *B* are congruent, respectively, with angles *D* and *E*. If side *AB* is represented by $4x - 3$ and side *DE* by $2x + 9$, for what value of x are the triangles congruent?

11. In triangles *ABC* and *DEF*, angle $A \cong$ angle D and side $AB \cong$ side DE. If *AB* is represented by $3x + 7$, *DE* by $2x + 12$, *AC* by $4x - 1$, and *DF* by $3x + 3$, are the triangles congruent? Why?

12. In Figure 97, $\overline{DBE}$ bisects $\overline{ABC}$ and $\angle A \cong \angle C$. Which postulate could be used to prove $\triangle ABE \cong \triangle CBD$?

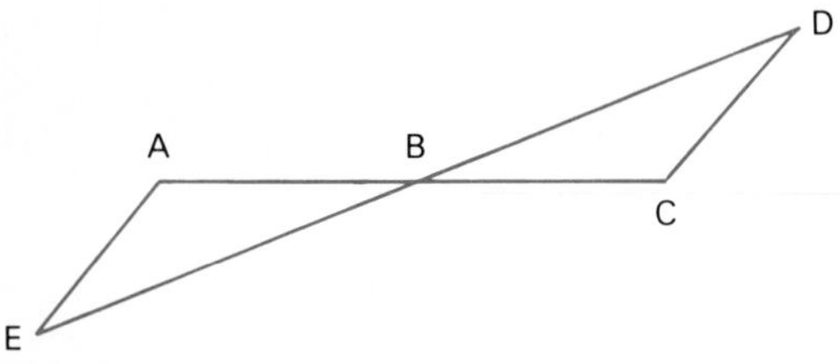

FIGURE 97

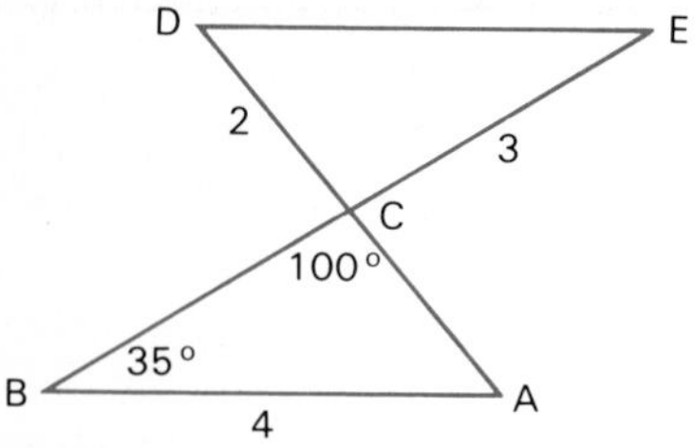

FIGURE 98

Triangles *ABC* and *DEC* in Figure 98 are congruent. Give the measures of the segments and angles in Exercises 13–18.

13. $\overline{DE}$ 14. $\overline{CB}$ 15. $\overline{AC}$

16. $\angle A$ 17. $\angle D$ 18. $\angle ECD$

In Figure 99, triangle $ABC \cong \triangle DEF$. Give the measures of the segments and angles in Exercises 19–24.

19. $\overline{FE}$ 20. $\overline{DE}$ 21. $\overline{FD}$

22. $\angle E$ 23. $\angle B$ 24. $\angle D$

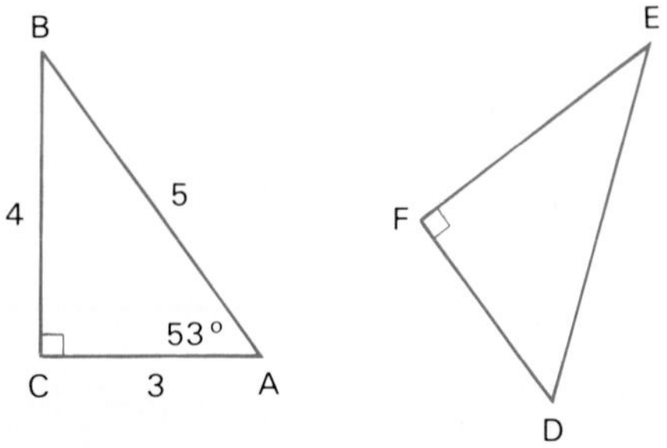

FIGURE 99

In each of Exercises 25–30, the two triangles are similar. Find the missing length.

25.

26.
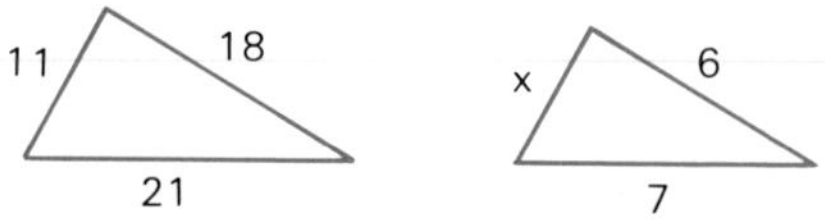

27.
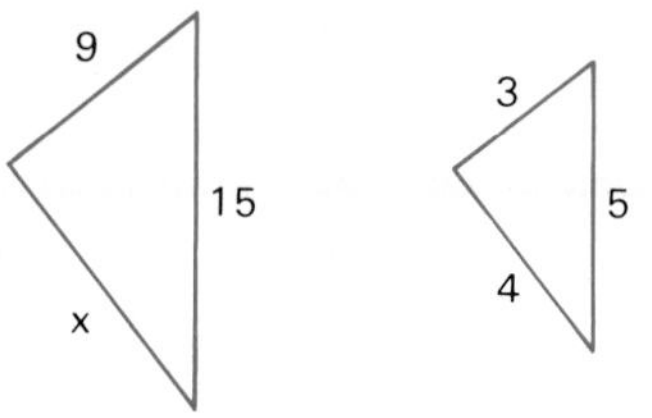

28.
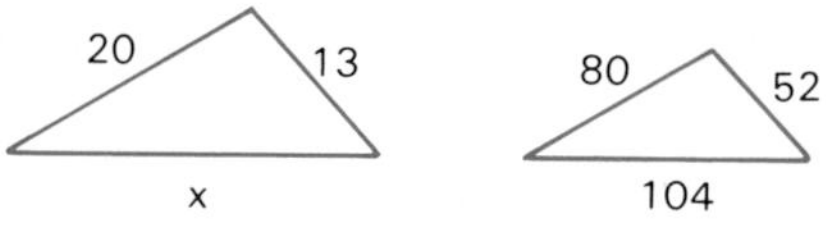

29.

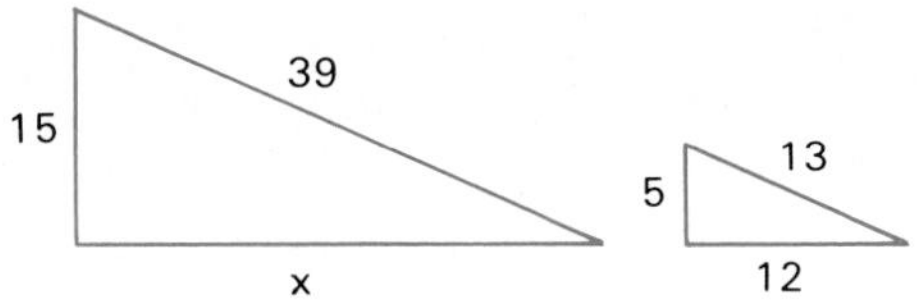

30.

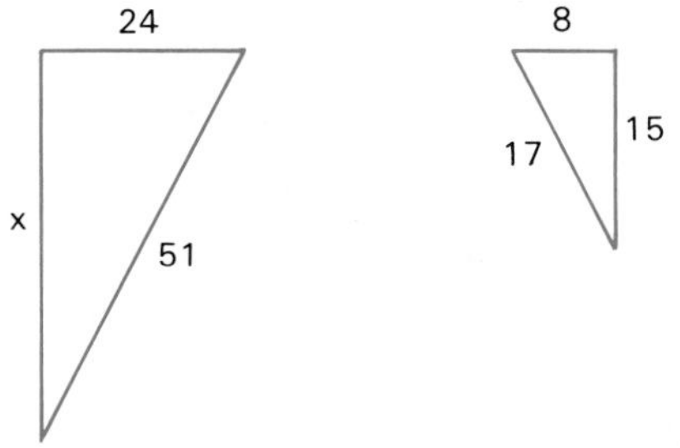

31. A vertical flagpole casts a shadow 12 m long at the same time that a nearby vertical pole 5 m high casts a shadow 3 m long. Find the height of the flagpole.

32. A vertical rod 4 m high casts a shadow 3 m long. At the same time a nearby tree casts a shadow 15 m long. Find the height of the tree.

33. A tree casts a shadow 40 feet long. At the same time, a nearby student 5 feet 6 inches tall casts a shadow 8 feet long. Find the height of the tree.

34. A vertical pole 3 m high casts a shadow 2.5 m long. At the same time, a nearby signal tower casts a shadow 15 m long. Find the height of the signal tower.

⋆35. Two triangles are similar. The sides of one are 6, 8, and 10. If the perimeter of the second triangle is 36, find the lengths of its sides.

⋆36. Two triangles are isosceles. The vertex angle in one measures 50°. An exterior angle at one end of the base of the other measures 110°. Are the triangles similar?

37. The triangles are similar in Figure 100. Find the distance across the river.

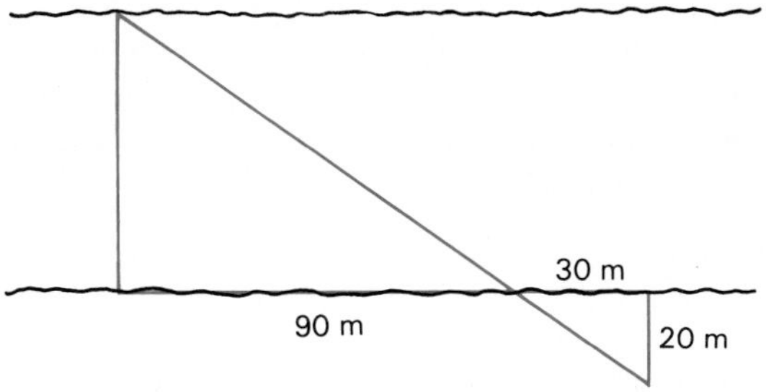

FIGURE 100

38. From the similar triangles in Figure 101, find the length of the lake.

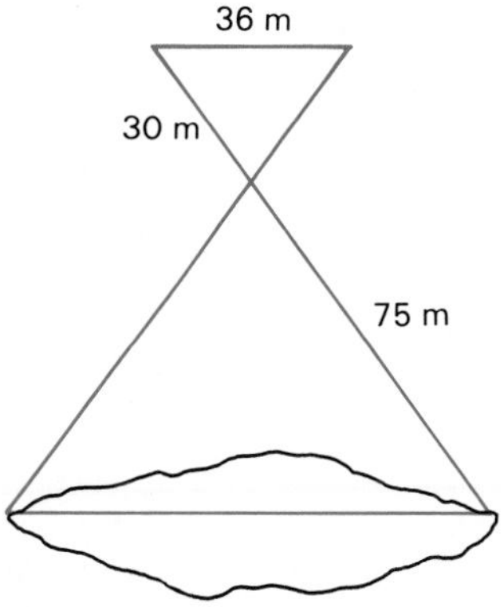

FIGURE 101

JUST FOR FUN What has an inside on the outside?

10.9 NETWORKS

A set of line segments or arcs is called a **graph.** If it is possible to move from any point in the graph to any other point in the graph by moving along the line segments or arcs, then we say the graph is **connected.** A **network** is a connected graph.

A network is **traversable** if it can be drawn by tracing each line segment or arc exactly once without lifting the pencil from the paper. Figure 102 shows some examples of networks that are traversable. The end points of the arcs or line segments are called **vertices.**

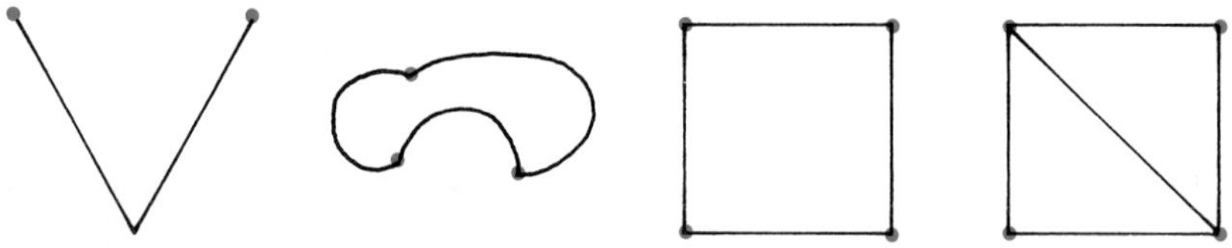

FIGURE 102

A **closed network** divides a plane into two or more regions. A **simple network** is one that does not cross itself. A network that is simple and closed is traversable. Furthermore, you can start at any vertex of a simple closed network to traverse the network. The vertex chosen for the initial point will also be the terminal point.

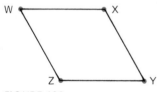

FIGURE 103

Consider the network in Figure 103. It is simple and closed. Therefore we can begin at any vertex and traverse the network.

The network in Figure 104 is closed, but not simple. It can also be traversed, but we must begin at either U or X. If we start at U, then we end at X; if we start at X, then we end at U.

A famous puzzle is largely responsible for beginning the study of network theory. This puzzle, the "Seven Bridges of Königsberg," first attracted attention during the 1700s. Königsberg (now Kaliningrad, U.S.S.R.) was a town in Prussia built on both sides of the Pregel River. Located in the river were two islands, connected to each other and to the city by seven bridges, as shown in Figure 105.

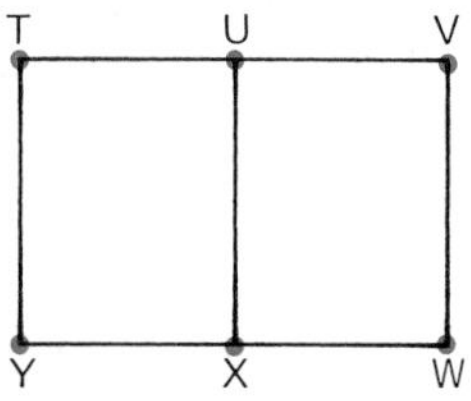

FIGURE 104

The problem associated with these bridges and islands was to determine if a person could start at a given point in the town of Königsberg and follow a path that would cross every bridge once and only once on a continuous walk through the town. The citizens of Königsberg tried many routes, but found that—no matter where they started, or what path they chose—they could not cross each bridge once and only once. However, it was not

HISTORICAL NOTE

Sophie Germain is considered to be one of the truly outstanding women mathematicians. She decided at an early age to study mathematics, but for a different reason than most people would guess.

Germain was born in Paris on April 1, 1776. In 1789, the Bastille fell and a decade of revolution followed. The turmoil that ensued caused Sophie to be confined to her house, and as a result, she spent many hours reading in her father's library. Here she read of the violent death of Archimedes as he studied a mathematical figure in the sand. Sophie was extremely impressed with Archimedes' absorption in the subject, as he had been oblivious to the events around him. She resolved to study mathematics.

She obtained lecture notes from different professors at the Ecole Polytechnique, even though the school did not accept women. Sophie corresponded with many mathematicians, including the famous Lagrange, but she signed herself M. Leblanc. After a period of some time Lagrange learned of her true identity and openly praised her work in number theory and mathematical analysis.

In 1815, Germain was awarded a prize by the Institut de France. She was also recommended for an honorary degree from the University of Göttingen, but died in 1831 before the degree could be conferred.

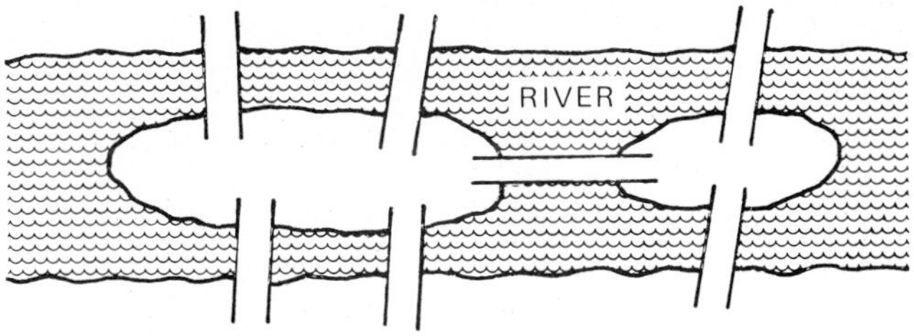

FIGURE 105

until Leonhard Euler (1707–1783), a Swiss mathematician, became interested in the problem that it was proved that each bridge could not be crossed once and only once on a continuous walk through the town. To prove this, Euler analyzed the problem by transforming it into a network similar to that shown in Figure 106.

Euler called the points where the paths of the network came together *vertices*. Furthermore, he classified the vertices of a network as **odd** or **even**, depending on whether an odd or even number of paths passed through the vertex. For example, in Figure 104, vertex T is even because two paths pass through it, and vertex X is odd because three paths pass through it. In Figure 106 all of the vertices are odd.

Euler proved that any network containing only even vertices is traversable by a route beginning at any vertex and ending at the same vertex. He also showed that a network that has exactly two odd vertices is traversable, but the traversing route must start at one of the vertices and end at the other.

Finally, Euler showed that if a network has more than two odd vertices, then it is not traversable. This means that the network in Figure 106 is not traversable. Therefore, Euler proved that each of the bridges of Königsberg could not be crossed once and only once on a continuous walk because the equivalent network in Figure 106 has four odd vertices.

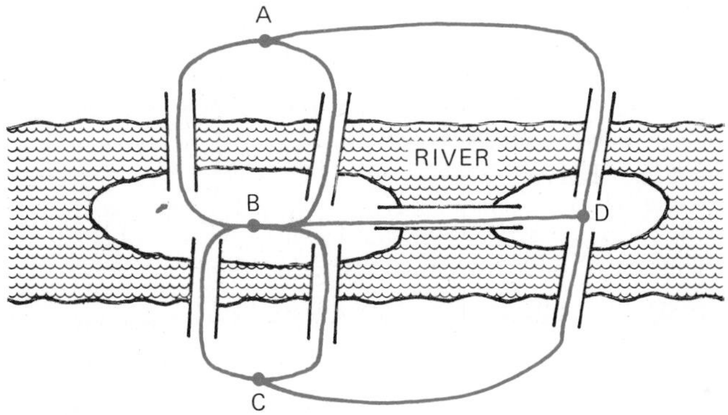

FIGURE 106

EXAMPLE 1 For each network, identify the even and odd vertices.

a. 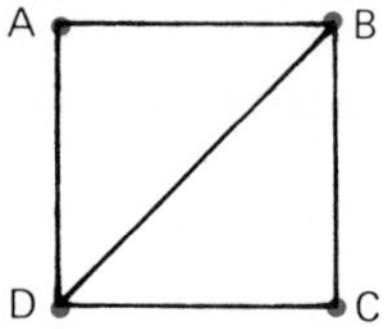

b. 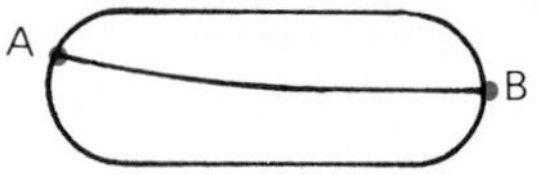

c. 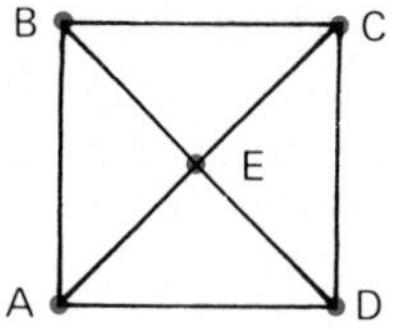

SOLUTION Recall that an even vertex is one that is an end point of an even number of arcs or line segments, and an odd vertex is one that is an end point of an odd number of arcs or line segments.

a. Even vertices: *A* and *C;* odd vertices: *B* and *D*
b. Even vertices: none; odd vertices: *A* and *B*
c. Even vertices: *E;* odd vertices: *A, B, C,* and *D*

EXAMPLE 2 Determine whether the networks in Example 1 are traversable. If the network is traversable, find the possible starting points.

SOLUTION a. The network is traversable; *B* and *D* are the possible starting points.
b. The network is traversable; *A* and *B* are the possible starting points.
c. The network is not traversable because it has more than two odd vertices.

EXERCISES FOR SECTION 10.9

For Exercises 1–10, find (a) the number of even vertices, (b) the number of odd vertices, (c) whether the network is traversable, and (d) the possible starting points if the network is traversable.

1. 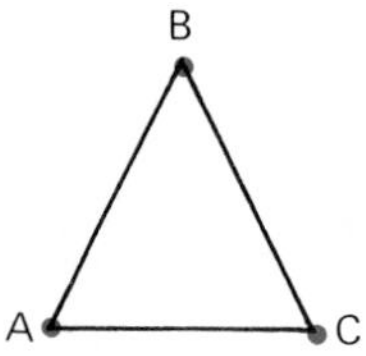

2. 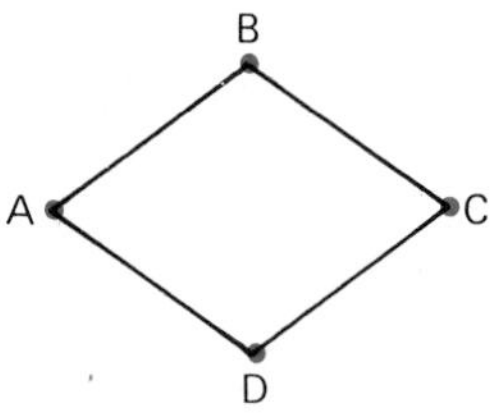

3. 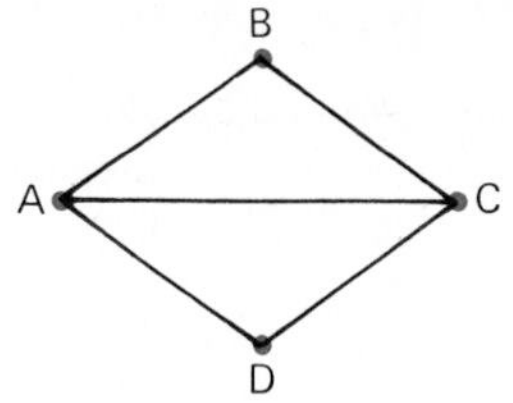

4. 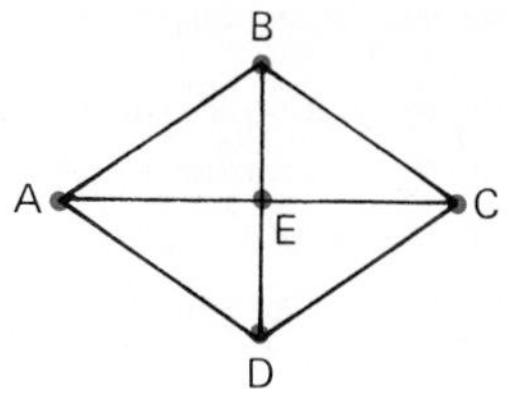

5. 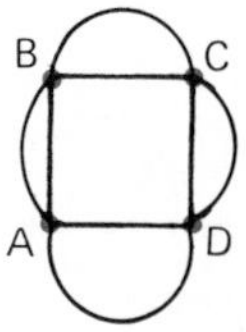

6. 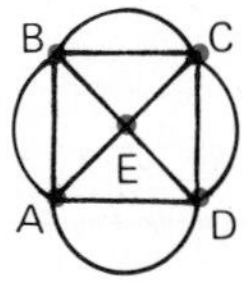

7. 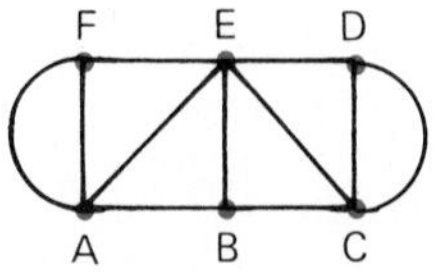

8. 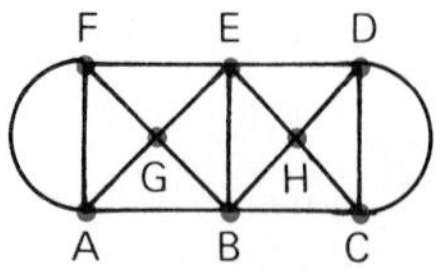

9. 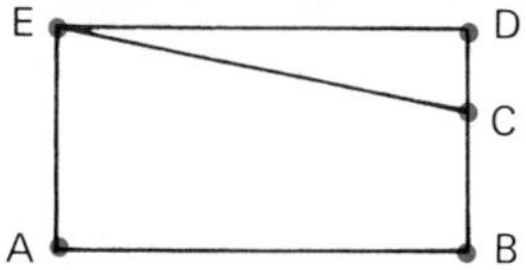

10. 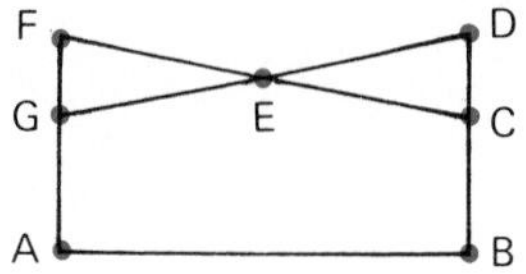

11. Is it possible to walk through a house with the floor plan given in Figure 107 and pass through each doorway exactly once? (*Hint:* Use a network.)

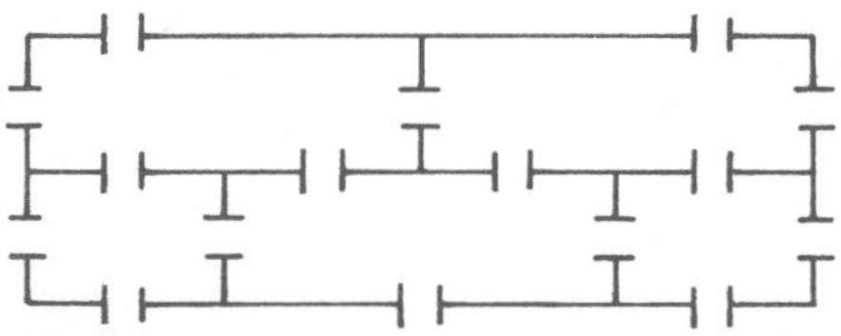

FIGURE 107

12. Is it possible to walk through a house with the floor plan given in Figure 108 and pass through each doorway exactly once? (*Hint:* Use a network.)

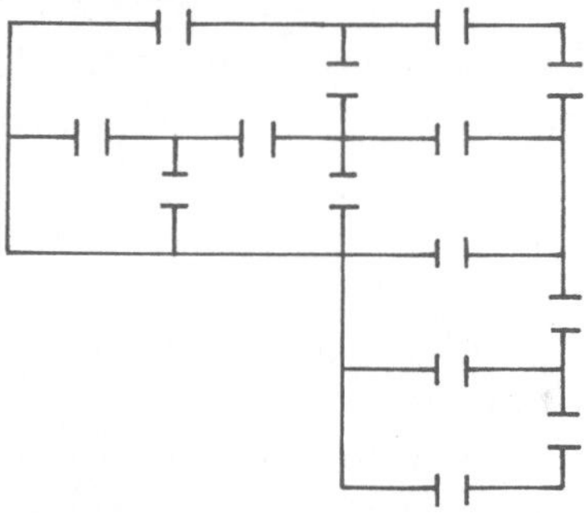

FIGURE 108

B.C. by permission of Johnny Hart and Creators Syndicate, Inc.

JUST FOR FUN

Can you cut a hole in a standard-size piece of paper in such a way that your entire body can pass through it? Take a standard-size piece of paper and cut it as indicated by the broken lines.

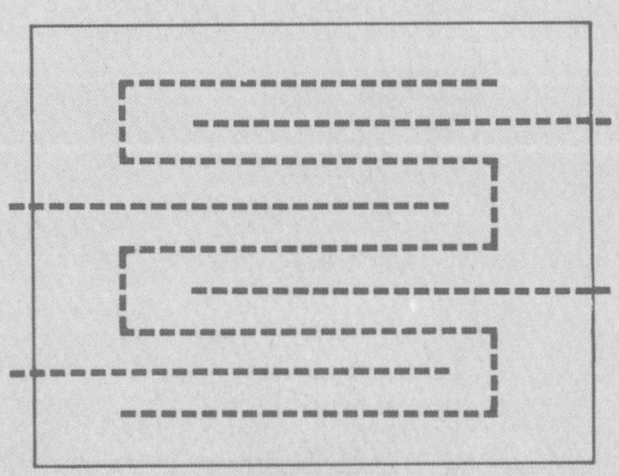

If you have cut correctly, you should be able to pass your body through the hole in the paper. With some practice, you can use a smaller piece of paper and make more cuts.

Summary

Point, line, and *plane* are the most basic terms of geometry. A geometric *point* has no dimension, that is, it has no length, breadth, or thickness. The one thing that a geometric point does possess is position. A geometric *line* is formed by the intersection of two surfaces. A line in geometry has no width, but it does have length. We can think of a line as a set of points. *Collinear points* are points that lie on the same straight line. Some sets of points commonly used in geometry are

Description	Diagram	Notation
point P	•	P
line PQ	←•—•→ P Q	$\overleftrightarrow{PQ}$
half-line PQ	∘—•→ P Q	$\overset{\circ\rightarrow}{PQ}$
ray PQ	•—•→ P Q	$\overrightarrow{PQ}$
ray QP	←•—• P Q	$\overrightarrow{QP}$
line segment PQ	•—• P Q	$\overline{PQ}$
open line segment PQ	∘—∘ P Q	$\overset{\circ-\circ}{PQ}$

An *angle* is the union of two rays that have a common end point.

In this chapter we discussed many different kinds of angles. You should be familiar with the following types of angles: *adjacent angles, vertical angles, straight angles, right angles, acute angles,* and *obtuse angles.*

A *broken line* is a set of connected line segments. A *simple closed broken line* is one that starts and stops at the same point and does not intersect itself. A simple closed broken line is called a *polygon*. The connected line segments are the *sides* of the polygon, and the points at which the line segments are connected are the *vertices* of the polygon. Polygons are classified according to the number of sides that they have. For example, a polygon having five sides is called a *pentagon,* and a polygon having ten sides is called a *decagon*. In this chapter we discussed in detail polygons having three sides and four sides, that is, *triangles* and *quadrilaterals.* You should be familiar with the following: *equilateral triangles, isosceles triangles, scalene triangles, obtuse triangles*, *acute triangles*, and *right triangles*. You should also be familiar with *trapezoids*, *parallelograms*, *rhombuses*, *rectangles*, and *squares*.

The *perimeter* of a polygon is the sum of the lengths of its sides. The perimeter of a square can be determined by the formula $P = 4s$, whereas the perimeter of a rectangle is found by using $P = 2l + 2w$, and the perimeter of a parallelogram is found by using $P = 2a + 2b$, where a and b are two adjacent sides. The *area* of a polygon tells us how many square units that polygon contains. The area of a rectangle is equal to the product of its length and width; that is, $A = lw$. The area of a square is equal to the product of two sides; that is $A = s \times s$, $A = s^2$.

For any circle, the circumference is equal to π times the diameter, and its area is equal to π times the radius squared.

The area of a parallelogram is equal to the product of its base and height; that is, $A = bh$. The area of a triangle is equal to one-half the product of its base and height; that is, $A = \frac{1}{2}bh$. The area of a trapezoid is equal to one-half the height times the sum of its bases; that is, $A = \frac{1}{2}h(a + b)$.

A *polyhedron* is bounded by polygonal regions. A prism is a polyhedron with two parallel and congruent bases. See Section 10.7 for a list of formulas used to find the surface areas and volumes of certain geometric solids.

A set of line segments or arcs is called a *network*. A network is *traversable* if it can be drawn by tracing each line segment or arc exactly once without lifting the pencil point from the paper. Any network that has only even vertices is traversable. If a network has exactly two odd vertices, it is traversable providing we begin at one of the odd vertices. A network that has more than two odd vertices is not traversable.

Vocabulary Check

vertical angles	circumference	tetrahedron	square
similar triangles	perpendicular	prism	polygon
supplementary angles	acute angle	complementary angles	pi (π)
right angle	obtuse angle	parallel	area
trapezoid	straight angle	Pythagorean theorem	volume
rhombus	parallelogram	isosceles triangle	network
perimeter	rectangle	congruent triangles	

Review Exercises for Chapter 10

For Exercises 1–10, use Figure 109 with the indicated points to find each of the following.

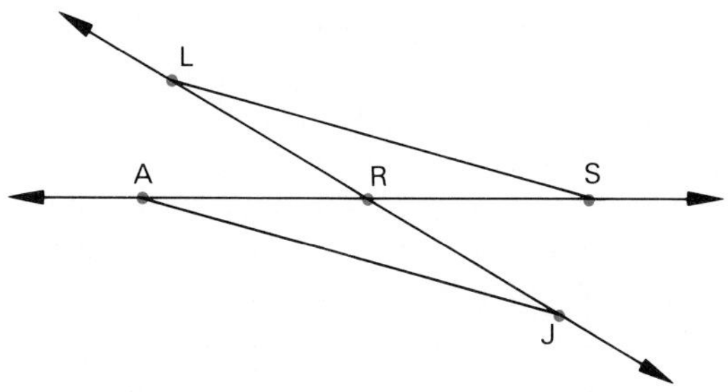

FIGURE 109

1. $\overleftrightarrow{AR} \cap \overleftrightarrow{LJ}$
2. $\overrightarrow{AR} \cap \overrightarrow{SR}$
3. $\overrightarrow{RS} \cap \overrightarrow{RA}$
4. $\overrightarrow{RS} \cup \overrightarrow{RA}$
5. $\overline{LS} \cap \overline{AJ}$
6. $\overrightarrow{RA} \cap \overline{LS}$
7. $\overrightarrow{RA} \cup \overrightarrow{RL}$
8. $\overrightarrow{RS} \cup \overrightarrow{RJ}$
9. $\angle SRJ \cap \angle ARJ$
10. $\angle SRJ \cap \angle LRA$

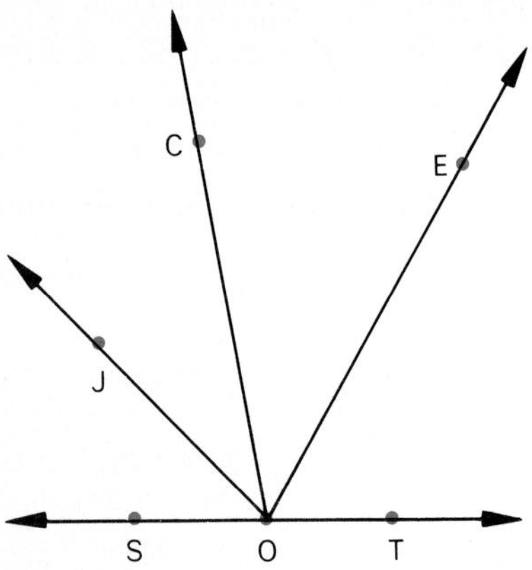

FIGURE 110

For Exercises 11–20, use Figure 110 with the indicated points to find each of the following.

11. $\overrightarrow{OS} \cup \overrightarrow{OJ}$ 12. $\overrightarrow{OT} \cup \overrightarrow{OE}$

13. $\overrightarrow{OS} \cap \overrightarrow{OJ}$ 14. $\overline{OT} \cap \overline{ST}$

15. (interior $\angle COT$) $\cap$ (interior $\angle EOT$)

16. (exterior $\angle COT$) $\cap$ (interior $\angle JOS$)

17. $\angle JOS \cap \angle EOT$ 18. $\angle COT \cap \angle COS$

19. (exterior $\angle$EOT) $\cap$ (exterior $\angle JOS$)

20. (interior $\angle JOE$) $\cap$ (interior $\angle COE$)

21. Find the measures of the complement and supplement of angles with each of the following measures.
   a. 42° b. 37° c. 79°
   d. 82°30′ e. 3°45′ f. 33°33′33″

22. Two angles are complementary, and one angle measures 42° less than the other. How many degrees are there in each angle?

23. Two angles are supplementary, and one angle measures 30° more than twice the other. How many degrees are there in each angle?

24. The lengths of two sides of a triangle are 4 and 7. The third side may be (choose one)
   a. 1 b. 2 c. 3 d. 4

25. Which set of numbers may represent the lengths of the sides of a right triangle?
   a. {2, 3, 4} b. {4, 5, 6}
   c. {7, 8, 9} d. {5, 12, 13}

26. Find the length of hypotenuse $\overline{AB}$ in right triangle ABC, given that $m(\overline{AC}) = 8$ and $m(\overline{BC}) = 15$.

27. Given isosceles triangle ABC with $m(\overline{AB}) = m(\overline{AC})$ and $m\angle BAC = 48°$, what are the measures of $\angle ABC$ and $\angle ACB$?

28. The area of a rectangle is 72 cm^2 and the width is 5 cm. What is the length?

29. Find the area of a square whose diagonal measures 10 inches.

30. Find the area of a parallelogram whose base is 13 m and whose height is 7 m.

31. The area of a parallelogram is 200 m^2 and the base is 25 m. What is the height?

32. Find the area of a right triangle whose legs measure 5 cm and 12 cm, respectively.

33. Find the circumference of a circle if its diameter is 10 cm. Use 3.14 for π.

34. Find the area of a circle whose diameter measures 18 m. Use 3.14 for π.

35. The diameter of a wheel is 30 cm. Find its circumference and area. Use 3.14 for π.

Find (a) the total surface area and (b) the volume of each polygon in Exercise 36–38. Let $\pi = 3.14$.

36.

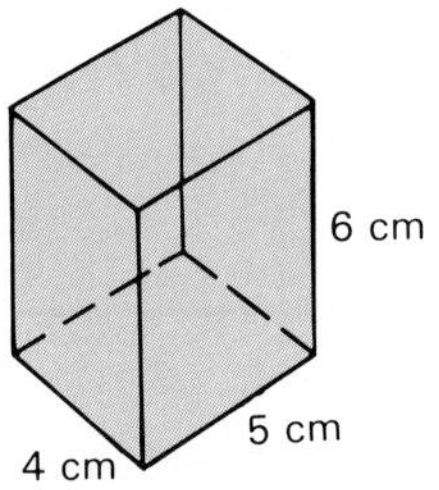

37.

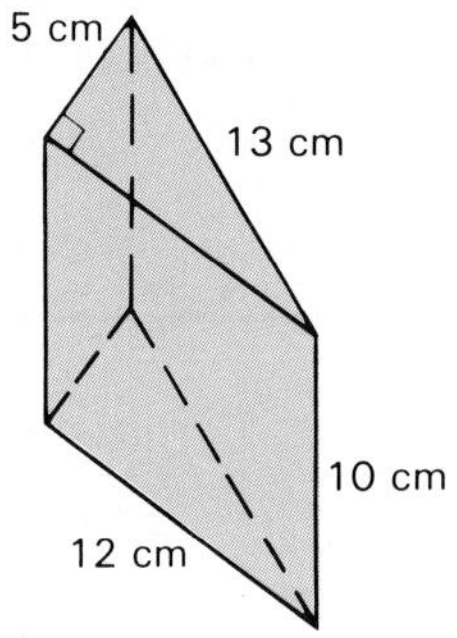

38.

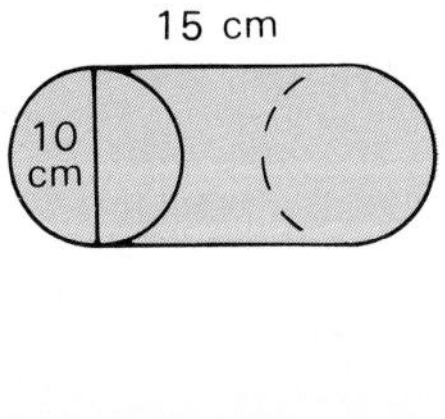

39. Find the surface area and volume of a cone if its radius is 5 mm, its height is 12 mm, and its slant height is 13 mm. Let $\pi = 3.14$.

40. Find the surface area and volume of a ball if its radius is 12 cm. Let $\pi = 3.14$. Express your answer to the nearest tenth.

41. In triangles ABC and DEF, angles A and B are equal, respectively, to angles D and E. If side AB is represented by $4x - 6$ and DE by $2x + 18$, for what values of x are the triangles congruent?

42. A vertical pole 8 m high casts a shadow 6 m long. At the same time a nearby tree casts a shadow 30 m long. Find the height of the tree.

For Exercises 43–46 find (a) the number of even vertices, (b) the number of odd vertices, (c) whether the network is traversable, and (d) the possible starting points if the network is traversable.

43.
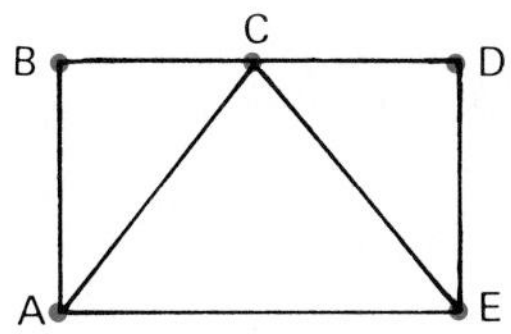

44.
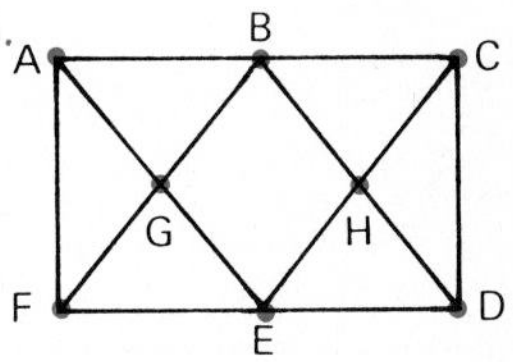

45.
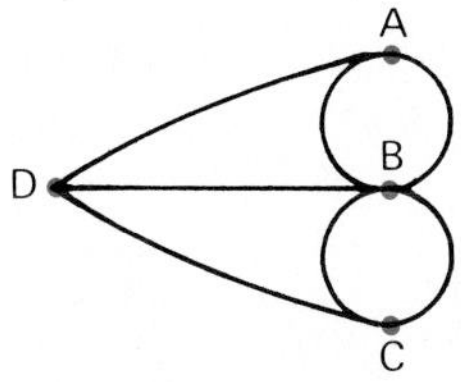

46.
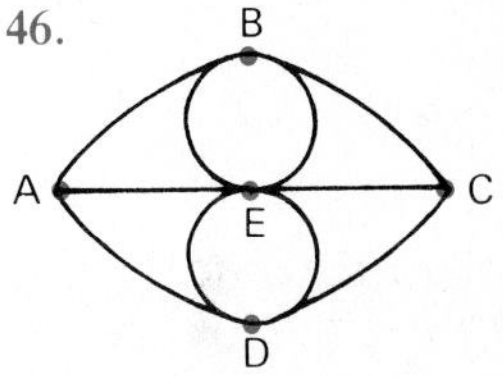

Chapter Quiz

Determine whether the following statements are true or false.

1. The sum of the squares of two sides of a triangle is equal to the square of the third side.

2. Two rectangles can have equal areas and unequal perimeters.

3. Two rectangles can have equal perimeters and unequal areas.

4. A parallelogram is a rhombus if the diagonals of the parallelogram are equal.

5. If the number of degrees in a pair of supplementary angles is represented by $2x - 40$ and $3x + 10$ then $x = 41°$.

6. Two adjacent angles are angles that have a common vertex.

7. A parallelogram is a polygon whose opposite sides are equal in length.

8. A square is a rectangle with adjacent sides equal in length.

9. A rectangle is a parallelogram with four right angles.

10. Parallel lines are lines that lie in the same plane and do not intersect however far they are extended.

11. An acute triangle is a triangle in which one angle is acute.

12. If the opposite sides of a quadrilateral are not parallel, then the quadrilateral is a trapezoid.

13. If two angles of a triangle are equal in measure, the sides opposite these angles are of equal length.

14. An isosceles trapezoid has both bases equal.

15. The area of a triangle is determined by the product of its base and height.

16. If two sides of a triangle are 5 and 12, respectively, then the third side must have a length of 13.

17. Pythagoras is known as the "father of geometry."

18. An angle is the union of two rays that have a common end point.

19. A right triangle cannot have two congruent sides.

20. A solid cone is an example of a polyhedron.

21. A chord is a line segment that joins two points on a circle.

22. A prism is a polygon with two parallel and congruent bases.

23. For any prism, the volume is equal to the area of the base times the height.

24. A scalene triangle is a triangle in which two sides are of equal length.

25. If two lines do not lie in the same plane, they are skew lines.

JUST FOR FUN

Cut a strip from a standard-size piece of paper: the strip of paper should measure approximately $\frac{1}{2}$ by 11 inches. Give the paper a half-twist and then tape or glue the two ends together. You have constructed a Möbius strip. It should resemble (a).

The Möbius strip has some interesting properties. For example, it has only one surface. You can demonstrate this by drawing a continuous line, or shading one side, all the way around without lifting your pencil. You will note that this will mark or shade the entire surface.

Next, cut the strip in the middle, as you normally would to obtain two loops, indicated by the broken line in (b). If you have done everything correctly, you should get one bigger loop.

(a) (b)

CHAPTER 11

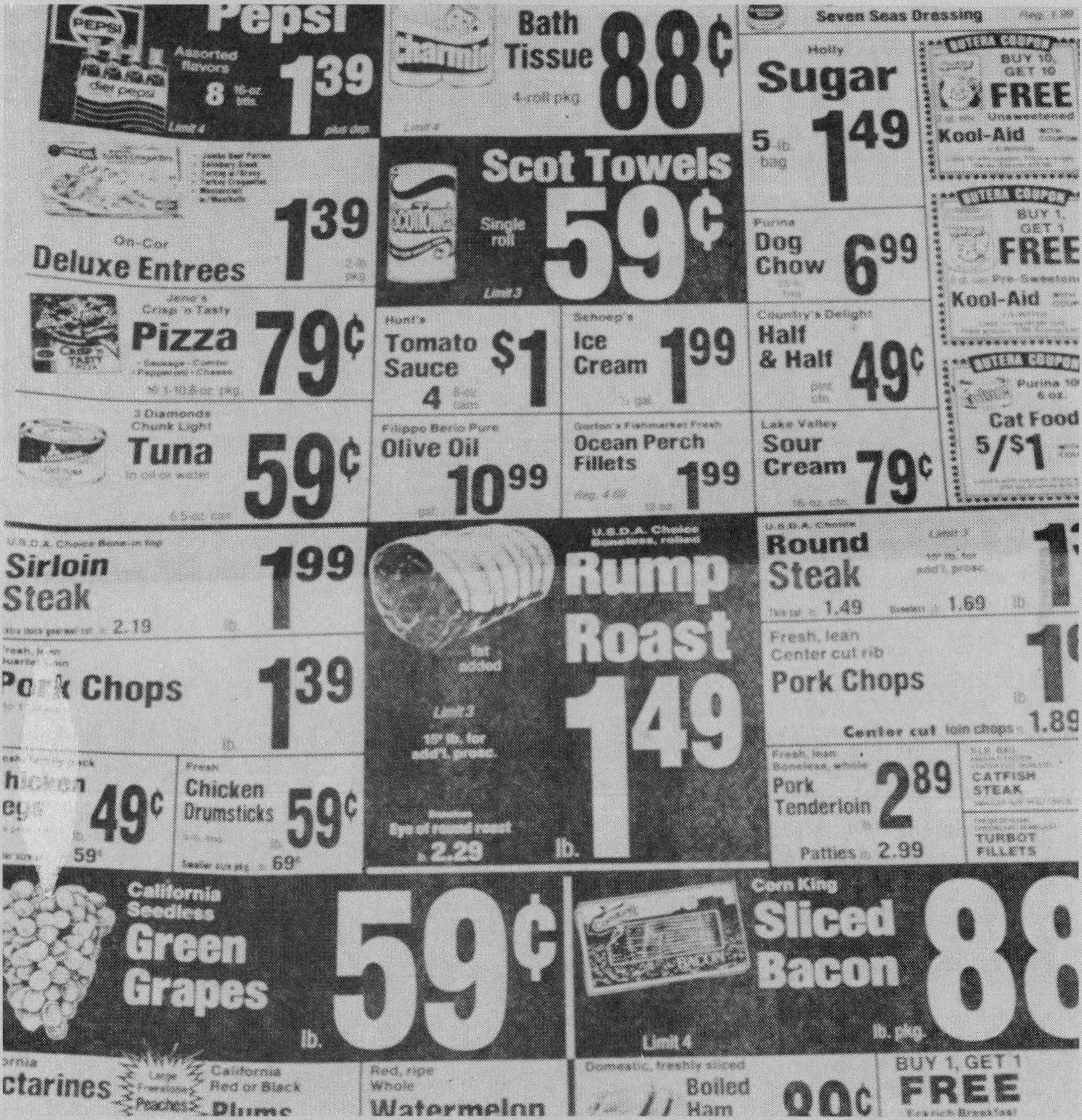

Pepsi
8
1.39
Bath Tissue
88¢
4-roll pkg.
Seven Seas Dressing
Holly
Sugar
5 lb. bag
1.49
BUY 10, GET 10 FREE
Kool-Aid
Scot Towels
Single roll
59¢
Limit 3
On-Cor
Deluxe Entrees
1.39
Purina
Dog Chow
6.99
BUY 1, GET 1 FREE
Kool-Aid
Jeno's
Crisp 'n Tasty
Pizza
79¢
Hunt's
Tomato Sauce
4
$1
Schoep's
Ice Cream
1.99
Country's Delight
Half & Half
49¢
Purina
Cat Food
5/$1
3 Diamonds
Chunk Light
Tuna
in oil or water
59¢
Filippo Berio Pure
Olive Oil
10.99
Gorton's Fishmarket Fresh
Ocean Perch Fillets
1.99
Lake Valley
Sour Cream
79¢
U.S.D.A. Choice Bone-in top
Sirloin Steak
1.99
2.19
U.S.D.A. Choice
Boneless, rolled
Rump Roast
Limit 3
1.49
lb.
Eye of round roast
2.29
U.S.D.A. Choice
Round Steak
1.49
1.69
Pork Chops
1.39
Fresh, lean
Center cut rib
Pork Chops
Center cut loin chops 1.89
49¢
59¢
Fresh
Chicken Drumsticks
59¢
69¢
Fresh, lean
Boneless, whole
Pork Tenderloin
2.89
Patties 2.99
CATFISH STEAK
TURBOT FILLETS
California Seedless
Green Grapes
lb.
59¢
Corn King
Sliced Bacon
Limit 4
lb. pkg.
ctarines
California
Red or Black
Plums
Red, ripe
Whole
Watermelon
Domestic, freshly sliced
Boiled Ham
BUY 1, GET 1 FREE

CONSUMER MATHEMATICS

AFTER STUDYING THIS CHAPTER YOU WILL BE ABLE TO DO THE FOLLOWING:

1. Express the relationship between two quantities as a **ratio,** and solve a **proportion** for the missing term.
2. Convert a **percent** to a **decimal** or **fraction,** and convert decimals or fractions to percents.
3. Find the **markup** on an item when given the cost or selling price and the percent of markup.
4. Find the amount of **markdown** and **sale price** when given the original retail price and the percent of markdown.
5. Use the formula $I = Prt$ to compute **simple interest.**
6. Compute **compound interest** and **compound amounts.**
7. Find the **effective annual interest rate** when money is compounded annually, semiannually, or quarterly.
8. Determine the annual **premium** for the following types of life insurance policies: **term** (5 year), **straight life, limited-payment life** (20 year), and **endowment** (20 year).
9. Find the **true annual interest rate** when an item is purchased on the installment plan.
10. Find the **amount of interest paid** when an item is purchased on the installment plan.
11. Determine the monthly payments for **principal** and **interest** for mortgages of various lengths at various interest rates.

11.1 INTRODUCTION

Most of us have probably heard the phrase *caveat emptor* at one time or another. It means "let the buyer beware," which can also be interpreted as "the buyer buys at his own risk." Unfortunately, many businesses maintain this attitude when dealing with customers. As consumers, we must be wise and discerning shoppers. That is, we must look for the best buy when we purchase an item. When buying items on credit, purchasing insurance, or taking out a loan, consumers must have an understanding of *decimals, percents, simple interest, compound interest,* and *effective rate of interest.* When faced with several alternatives—for example, buying on the installment plan or paying cash—we must make an intelligent decision based on our own particular situation. The topics in this chapter are designed to give you the information necessary to be a more intelligent shopper and a wiser consumer.

11.2 RATIO AND PROPORTION

The **ratio** of two quantities a and b is the quotient or indicated quotient obtained by dividing a by b. The ratio of a to b is written as

$$a \div b \quad \text{or} \quad \frac{a}{b} \quad \text{or} \quad a:b$$

For example, to indicate the ratio of 3 to 5, we could use

$$3 \div 5 \quad \text{or} \quad \frac{3}{5} \quad \text{or} \quad 3:5$$

A ratio provides us with a way of comparing two numbers by means of division. The ratio of one number to another number is the quotient of the first number divided by the second number. Therefore, the ratio of 12 to 3 is $12 \div 3$, or 4 to 1. That is,

$$12 \text{ to } 3 = \frac{12}{3} = \frac{4}{1} = 4 \text{ to } 1$$

EXAMPLE 1 Express each ratio in simplest form.

a. 12 to 36 b. 49 : 14 c. $\frac{14}{12}$ d. $15 \div 75$

SOLUTION a. $12 \text{ to } 36 = \frac{12}{36} = \frac{1}{3}$

b. $49:14 = \frac{49}{14} = \frac{7}{2}$

c. $\frac{14}{12} = \frac{7}{6}$

d. $15 \div 75 = \frac{15}{75} = \frac{1}{5}$

EXAMPLE 2 Find the ratio of 210 minutes to 3 hours.

SOLUTION The quantities compared by ratio must represent objects measured in the same units. In this case, we have minutes compared to hours. However, to form a ratio, either both quantities should be in terms of minutes, or both should be in terms of hours. We choose to convert both to minutes. Since there are 60 minutes in an hour, we have

$$\frac{210}{180} = \frac{21}{18} = \frac{7}{6}$$

EXAMPLE 3 Express the ratio of $1\frac{1}{2}$ to $2\frac{1}{4}$ in simplest form.

SOLUTION The ratio of one number to another number is the quotient of the first number divided by the second number. Therefore, the ratio of $1\frac{1}{2}$ to $2\frac{1}{4}$ is the same as

$$1\frac{1}{2} \div 2\frac{1}{4} \qquad \text{or} \qquad \frac{3}{2} \div \frac{9}{4}$$

This problem involves division of rational numbers. (See Sec. 8.5, Examples 8 and 9.) Hence

$$1\frac{1}{2} : 2\frac{1}{4} = 1\frac{1}{2} \div 2\frac{1}{4} = \frac{3}{2} \div \frac{9}{4} = \frac{3}{2} \cdot \frac{4}{9} = \frac{12}{18} = \frac{2}{3}$$

EXAMPLE 4 A baseball team played 20 games and won 15 of them.

a. What is the ratio of the number of games won to the number of games played?
b. What is the ratio of the number of games lost to the number of games played?
c. What is the ratio of the number of games won to the number of games lost?

SOLUTION a. The ratio of games won to games played is 15 : 20.

$$15:20 = \frac{15}{20} = \frac{3}{4}$$

b. Games lost = games played − games won
Games lost = 20 − 15 = 5
The ratio of games lost to games played is 5 : 20.

$$5:20 = \frac{5}{20} = \frac{1}{4}$$

c. The ratio of games won to games lost is 15 : 5.

$$15:5 = \frac{15}{5} = \frac{3}{1}$$

A **proportion** is the equality of two ratios. That is, a proportion is a statement that says two ratios are equal. For example,

$$\frac{1}{2} = \frac{4}{8}$$

is a proportion. It states that the two ratios $\frac{1}{2}$ and $\frac{4}{8}$ are equal. If the ratios $a:b$ and $c:d$ are equal, then we can form a proportion by writing

$$a:b = c:d \quad \text{or} \quad \frac{a}{b} = \frac{c}{d}$$

We can read this proportion by saying that "a is to b as c is to d." In this proportion, b and c are called the **means**, whereas a and d are called the **extremes.**

In any proportion, the product of the means equals the product of the extremes.

$$\text{If } \frac{a}{b} = \frac{c}{d}, \text{ then } ad = bc.$$

For example, in the proportion $\frac{1}{2} = \frac{4}{8}$, 2 and 4 are the means, whereas 1 and 8 are the extremes.

$$\text{If } \frac{1}{2} = \frac{4}{8}, \text{ then } \underbrace{1 \cdot 8}_{\text{extremes}} = \underbrace{2 \cdot 4}_{\text{means}}, \text{ or } 8 = 8.$$

Many times we know three terms of a proportion and need to find the fourth term. For example, suppose that in a certain mathematics class, the ratio of the number of men to the number of women is 4 : 3. If there are 12 women in the class, how many men are in the class?

To solve this problem, note that the ratio 4: 3 is the same as the ratio of the number of men to the number of women (12). We can therefore set up the proportion

$$\frac{x}{12} = \frac{4}{3}$$

where x represents the number of men in the class. By applying the rule that the product of the means equals the product of the extremes, we have

$$\frac{x}{12} = \frac{4}{3}$$

$$3x = 12 \cdot 4 \quad \text{cross multiplying}$$

$$3x = 48$$

$$\frac{3x}{3} = \frac{48}{3}$$

$$x = 16 \text{ men}$$

EXAMPLE 5 A car travels 400 miles on 20 gallons of gasoline. At this rate, how many gallons of gasoline will be consumed on a trip of 900 miles?

SOLUTION Let x = the number of gallons used. We can then write the proportion

© *Brent Jones*

$$\frac{20}{x} = \frac{400}{900}$$

Solving this proportion,

$$400 \cdot x = 20 \cdot 900 \quad \text{cross multiplying}$$

$$400x = 18{,}000$$

$$\frac{400x}{400} = \frac{18{,}000}{400}$$

$$x = 45 \text{ gallons}$$

EXAMPLE 6 Find the value of x in each of the following proportions.

a. $4:8 = 3:x$ b. $3:7 = x:28$ c. $x:6 = 10:12$

SOLUTION To find the values of x in each proportion, we use the property that, for any proportion, the product of the means equals the product of the extremes.

a. $4:8 = 3:x$

$$\frac{4}{8} = \frac{3}{x}$$

$$4 \cdot x = 8 \cdot 3$$

$$4x = 24$$

$$x = 6$$

b. $3:7 = x:28$

$$\frac{3}{7} = \frac{x}{28}$$

$$3 \cdot 28 = 7 \cdot x$$

$$84 = 7x$$

$$12 = x$$

c. $x:6 = 10:12$

$$\frac{x}{6} = \frac{10}{12}$$

$$12x = 60$$

$$x = 5$$

EXAMPLE 7 A flagpole casts a shadow 5 feet long. Sandra, whose height is 5 feet 4 inches, is standing next to the flagpole. If she casts a shadow 16 inches long, what is the height of the flagpole?

SOLUTION We can solve this problem using a proportion. Let x represent the height of the pole. Therefore, one ratio could be the height of the pole to its shadow, $x:5$. Similarly, the ratio of Sandra's height to her shadow is $64:16$. Note that we set up ratios of like quantities and that the top and bottom of each ratio are in the same units. The proportion is

$$x:5 = 64:16$$

$$\frac{x}{5} = \frac{64}{16}$$

$$16 \cdot x = 5 \cdot 64$$

$$16x = 320$$

$$x = 20 \text{ ft}$$

EXERCISES FOR SECTION 11.2

1. Express each ratio as a fraction.
   a. 3 to 2 b. 4 to 7 c. 8:5 d. 5:6

2. Express each ratio as a fraction.
   a. 4 to 11 b. 7 to 9 c. 9:7 d. 11:7

3. Express each ratio in simplest form.
   a. 12 to 4 b. 14:28 c. $\frac{18}{12}$
   d. 36:9 e. $3\frac{1}{2}$ to $4\frac{2}{3}$ f. $1\frac{1}{2}:2\frac{1}{4}$

4. Express each ratio in simplest form.
   a. 14 to 56 b. $\frac{14}{12}$ c. 2:18
   d. 14:49 e. $3\frac{1}{3}:2\frac{2}{3}$ f. $2\frac{1}{3}:2\frac{5}{6}$

5. A mathematics class has 30 students in it. There are 18 women and 12 men in the class.
   a. What is the ratio of men to women?
   b. What is the ratio of women to men?
   c. What is the ratio of the number of men to the number of students in the class?

6. The perimeter of a rectangle is 36 meters and the width is 8 meters. Find the ratio of the length of the rectangle to its width.

7. The perimeter of a rectangle is 44 meters and the length is 19 meters. Find the ratio of the width of the rectangle to its length.

8. A final examination has 40 true–false questions and 60 multiple-choice questions.
   a. What is the ratio of true–false questions to multiple-choice questions?
   b. What is the ratio of multiple-choice questions to true–false questions?
   c. What is the ratio of true–false questions to the total number of questions on the examination?

9. Find the value of x in each of the following proportions.
   a. $x:7 = 6:21$ b. $3:x = 14:28$
   c. $5:1 = x:6$ d. $1:7 = 5:x$

10. Find the value of x in each of the following proportions.
   a. $1:2 = 4:x$ b. $5:7 = x:21$
   c. $2:x = 8:12$ d. $x:6 = 4:8$

11. Find the value of x in each of the following proportions.
   a. $3:8 = 12:x$ b. $6:7 = x:21$
   c. $2:x = 18:12$ d. $x:6 = 5:8$

12. Find the value of x in each of the following proportions.
   a. $x:7 = 35:49$ b. $2:x = 8:44$
   c. $7:9 = x:18$ d. $5:13 = 35:x$

13. A tree casts a shadow 6 feet long. Gerry, whose height is 5 feet 6 inches, is standing next to the tree. If she casts a shadow 16 inches long, what is the height of the tree?

14. Marlene can type at a rate of 55 words per minute. At this rate, how long will it take her to type a report that contains approximately 1100 words?

15. A certain athletic squad has 40 members. There are 18 seniors, 6 juniors, 12 sophomores, and 4 freshmen on the team.
   a. What is the ratio of juniors to seniors?
   b. What is the ratio of seniors to freshmen?
   c. What is the ratio of sophomores to seniors?
   d. What is the ratio of freshmen to the total number of team members?

16. When decomposed by an electric current, 36 grams of water yields 4 grams of hydrogen. How much hydrogen can be obtained from 50 grams of water?

17. An alloy is composed of two parts tin and one part lead. How much of each is needed to make 60 kilograms of this alloy?

18. On a map, a line segment 3 inches long represents a distance of 18 miles. Using the same scale, how many miles long is a road that measures 2 inches on the map?

19. Addie takes 3 minutes to read an article of 350 words. At the same rate, how many minutes will it take her to read another article of 875 words?

20. If one-half of a number is 30, then how much is two-thirds of the same number?

21. A student who is 6 feet tall stands next to a flagpole that is 50 feet tall. If the shadow cast by the student is 9 feet long, how long is the shadow cast by the flagpole?

22. What is the gear ratio of a bicycle that has 20 teeth on the rear sprocket and 46 teeth on the front chainwheel?

23. Sally takes 35 minutes to do a certain math assignment. Hugh takes 1 hour to do the same job. What is the ratio of their job times?

24. Use a tape measure to find the circumference of any circle. (Use a can, jar, clock, etc.) Next, measure its diameter. Find the ratio of the circumference to its diameter. Express your answer as a decimal. You should obtain approximately 3.14, which is the value of π (pi) to two decimal places.

25. Find the value of x in each of the following proportions.
    a. $3:4 = 12:x$ b. $7:10 = x:5$
    c. $6:(5 - x) = 7:3$ d. $(x - 6):4 = 4:5$

26. Find the value of x in each of the following proportions.
    a. $x:4 = 10:16$ b. $6:x = 16:7$
    c. $5:0.3 = 20:x$ d. $6:9 = (x - 2):12$

27. If for every three males in a mathematics class there are two females, how many males are there in the class if there are eight females?

JUST FOR FUN **How long will it take to cut a 12-foot log into 1-foot lengths, allowing 2 minutes for each cut?**

11.3 PERCENTS, DECIMALS, AND FRACTIONS

The concept of percent is one that occurs daily in our lives. For example, when we read the newspaper, listen to the radio, or watch television, we might encounter such statements as

> "There is a 30 percent chance of rain tomorrow."
> "The sales tax is 7 percent."
> "Sibley's Department Store's fall sale on women's fashions will feature savings of 20 percent and more on all items in stock."
> "The consumer price index rose one-tenth of 1 percent last month."
> "First Federal Savings Bank offers loans at 14 percent."

We can describe a percent as a ratio with a denominator of 100. That is, a percent is the ratio of a number to 100. The word *percent* means "per one hundred." The symbol for percent is %. Hence, "20 percent" can be written as "20%." Because a percent is the ratio of any number to 100, 20% means the ratio of 20 to 100; that is, $20\% = \frac{20}{100}$. Similarly,

$$18\% \quad \text{means} \quad 18:100, \quad \text{or} \quad \frac{18}{100}$$

$$6\% \quad \text{means} \quad 6:100, \quad \text{or} \quad \frac{6}{100}$$

$$\frac{1}{2}\% \quad \text{means} \quad \frac{1}{2}:100, \quad \text{or} \quad \frac{\frac{1}{2}}{100}$$

EXAMPLE 1 Express each percent as a fraction in simplest form.

a. 20% b. $33\frac{1}{3}\%$ c. $\frac{1}{2}\%$

SOLUTION To change a percent to a fraction, we drop the percent sign, place the number over 100, and reduce the resulting fraction.

a. $20\% = \frac{20}{100} = \frac{1}{5}$

b. $33\frac{1}{3}\% = \frac{33\frac{1}{3}}{100} = \frac{\frac{100}{3}}{100} = \frac{100}{3} \div \frac{100}{1} = \frac{100}{3} \cdot \frac{1}{100} = \frac{1}{3}$

c. $\frac{1}{2}\% = \frac{\frac{1}{2}}{100} = \frac{1}{2} \div \frac{100}{1} = \frac{1}{2} \cdot \frac{1}{100} = \frac{1}{200}$

It is sometimes necessary to express a percent as a decimal, such as in finding the sales tax on a purchase of \$15 if the sales tax is 7%. To convert a percent to a decimal, we use a technique similar to that used for converting a percent to a fraction. Seven percent means $\frac{7}{100}$, and this is the same as 0.07. Therefore, 7% = 0.07. The 7% sales tax on \$15 is found by multiplying 15 by 0.07.

$$\begin{array}{r} 15 \\ \times\ 0.07 \\ \hline 1.05 \end{array}$$

The sales tax is \$1.05.

EXAMPLE 2 Express 15% as a decimal.

SOLUTION $15\% = \frac{15}{100} = 0.15$

To convert a percent to a decimal, we can drop the percent sign and move the decimal two places to the left. We can do this because a percent is the ratio of a number to 100, and to divide a number by 100 is the same as moving the decimal point two places to the left. For example, $125 \div 100 = 1.25$. Therefore we can convert 15% to a decimal directly.

$$15\% = 0.15$$

Note that we drop the percent sign and move the decimal point two places to the left. When the number in the percent is a whole number, the decimal point is understood to be to the right of the last digit.

EXAMPLE 3 Express each percent as a decimal.

a. 3% b. 18% c. $\frac{1}{2}\%$

SOLUTION a. $3\% = 0.03$

b. $18\% = 0.18$

c. $\frac{1}{2}\% = 0.5\% = 0.005$

Note: To express $\frac{1}{2}\%$ as a decimal, we first had to convert $\frac{1}{2}$ to a decimal: $\frac{1}{2} = 0.5$. For a review of converting fractions to decimals, see Section 8.6.

Thus far we have converted percents to fractions and to decimals. Next we want to consider changing fractions and decimals to percents.

When expressing a decimal as a percent, it is important to remember that one decimal place to the right of the decimal point represents tenths, two decimal places represent hundredths, three decimal places represent thousandths, and so on. For example,

$$0.5 = \frac{5}{10}, \quad 0.12 = \frac{12}{100}, \quad \text{and} \quad 0.125 = \frac{125}{1{,}000}$$

A percent is the ratio of a number to 100. Therefore, to convert a decimal to a percent, we must obtain an equivalent expression with a denominator of 100. To express 0.15 as a percent, we first rewrite 0.15 as $\frac{15}{100}$. Next we drop the denominator (100) and add a percent sign. Therefore $0.15 = \frac{15}{100} = 15\%$.

To express 0.5 as a percent, we first express it as $\frac{5}{10}$. However, $\frac{5}{10}$ does not have a denominator of 100. Hence we use a proportion to find an equivalent ratio. That is,

$$\frac{5}{10} = \frac{x}{100}$$

$$10 \cdot x = 5 \cdot 100$$

$$10x = 500$$

$$x = 50$$

Therefore, $0.5 = \frac{5}{10} = \frac{50}{100}$, or 50%.

After examining these examples, you may have discovered another way to express a decimal as a percent.

To write a decimal as a percent, move the decimal point two places to the right and add a percent sign.

For example, 0.15 = 15% and 0.5 = 50%. In each case, the decimal point has been moved two places to the right and a percent sign has been added.

EXAMPLE 4 Express each decimal as a percent.

a. 0.07 b. 0.1 c. 3.2 d. 0.003

SOLUTION To convert a decimal to a percent, move the decimal point two places to the right and add a percent sign.

a. 0.07 = 7% b. 0.1 = 10%
c. 3.2 = 320% d. 0.003 = 0.3%

To change a fraction such as $\frac{1}{4}$ to a percent, we can make use of the process for converting a decimal to a percent because we can express $\frac{1}{4}$ as a decimal. To change a fraction to a decimal, we divide the numerator by the denominator. That is,

$$\frac{1}{4} = 1 \div 4 = 0.25$$

Now that we have a decimal, we can express it as a percent by moving the decimal point two places to the right and adding a percent sign. Hence,

$$\frac{1}{4} = 1 \div 4 = 0.25 = 25\%$$

EXAMPLE 5 Express each fraction as a percent.

a. $\frac{1}{2}$ b. $\frac{1}{8}$ c. $1\frac{3}{4}$

SOLUTION To change a fraction to a percent, first change the fraction to a decimal, and then change the decimal to a percent.

a. $\frac{1}{2} = 0.5 = 50\%$

b. $\frac{1}{8} = 0.125 = 12.5\%$

c. First we rewrite $1\frac{3}{4} = \frac{7}{4}$ and then proceed as before.

$$1\frac{3}{4} = \frac{7}{4} = 1.75 = 175\%$$

EXAMPLE 6 If there are 30 questions on an exam and a student answered 80% of them correctly, how many did she answer correctly?

SOLUTION To calculate the solution, we express 80% as a decimal, 80% = .80, and multiply 30 by 0.80. That is,

$$30 \times 0.80 = 24$$

EXERCISES FOR SECTION 11.3

In Exercises 1–6, express each percent as a fraction in simplest terms.

1. a. 15% b. 25% c. 75%
2. a. 10% b. 30% c. 48%
3. a. $4\frac{1}{2}\%$ b. $2\frac{1}{3}\%$ c. $6\frac{1}{4}\%$
4. a. $7\frac{3}{4}\%$ b. $8\frac{1}{2}\%$ c. $20\frac{1}{3}\%$
5. a. 6.5% b. 2.3% c. 150%
6. a. 13.7% b. 3.6% c. 250%

In Exercises 7–9, express each percent as a decimal.

7. a. 17% b. 3% c. $4\frac{1}{2}\%$
8. a. 12% b. 4% c. $5\frac{1}{2}\%$
9. a. 6.5% b. 300% c. 0.25%

In Exercises 10–13, express each decimal as a percent.

10. a. 0.05 b. 0.32 c. 0.5
11. a. 0.09 b. 2.14 c. 0.9
12. a. 0.005 b. 0.314 c. 5.12

13. a. 1.125 b. 0.010 c. 3.01

In Exercises 14–19, express each fraction as a percent.

14. a. $\frac{1}{4}$ b. $\frac{2}{5}$ c. $\frac{3}{8}$

15. a. $\frac{3}{4}$ b. $\frac{4}{5}$ c. $\frac{5}{8}$

16. a. $1\frac{1}{2}$ b. $\frac{1}{25}$ c. $\frac{3}{25}$

17. a. $2\frac{3}{5}$ b. $1\frac{1}{8}$ c. $\frac{7}{8}$

18. a. $\frac{1}{3}$ b. $1\frac{4}{5}$ c. $\frac{3}{16}$

19. a. $\frac{2}{3}$ b. $2\frac{3}{4}$ c. $\frac{1}{16}$

20. It rained on 40% of the days of November. On how many days did it rain in November?

21. Forty percent of a certain mathematics class are females. If there are 50 students in this class, how many are males?

JUST FOR FUN

Three sisters met for dinner at a restaurant. Their bill was $30. They divided the amount equally, and each paid $10. The cashier discovered that an error had been made in tabulating their bill. It should have been $25 instead of $30. The cashier informed the waiter of this error and gave him $5 to return to the sisters. On the way back to the table, the waiter decided to keep $2 and return $1 to each sister.

If each sister received $1 back, then each paid $10 − $1, or $9 for dinner, which is a total of $27! The waiter kept $2, which yields a total of $29! Where is the missing dollar?

11.4 MARKUPS AND MARKDOWNS

The price that we pay for an item when we buy it from a retailer is the retail price, or **selling price.** The amount that a retailer pays for goods is called the **cost** of the item. The difference between the *selling price* of an item and the *cost* of that item is the retailer's **profit margin.** For example, if a color television has a selling price of $400 and it cost the dealer $300, then the profit margin is

$$\$400 - \$300 = \$100$$

The $100 represents a profit margin, but it is not all profit. Out of this $100, the dealer has to meet such expenses as utility costs (heat, light, phone), employees' wages, insurance premiums, taxes, rent or mortgage payments, and so on. Another term commonly used to describe this profit margin is **markup.** Markup is the difference between the selling price and the cost of an item. That is,

$$\text{Markup} = \text{selling price} - \text{cost}$$

Using this equation, we can derive two other equations:

$$\text{Selling price} = \text{markup} + \text{cost}$$
$$\text{Cost} = \text{selling price} - \text{markup}$$

If the selling price of a ring is \$90 and the cost is \$50, then the markup is \$90 − \$50 = \$40. The amount of markup on an item tells you how much of the price you pay goes to the retailer to cover overhead (the cost of running the business) and profit, and how much goes to the manufacturer or wholesaler. Markup can also be given as a percentage of the cost, or as a percentage of the selling price. A **percent markup on cost** tells you the amount by which the cost of an item was increased to obtain the price you pay, the selling price. A **percent markup on selling price** indicates the amount of the selling price that the retailer retains for overhead and profits.

Most retailers work with percent markups because they deal with large lots of merchandise. A markup of 50% of cost can be applied to a whole group of items whose individual selling prices might vary widely. Then the total markup for the group is easily figured as a percentage of the total cost, without any need to count the items in the group or figure individual markup costs or selling prices.

EXAMPLE 1 The pro shop at the National Golf Club sells a certain brand of golf clubs at prices based on a markup of 40% of the cost. If the cost of a set of these golf clubs is \$150, what is the selling price?

SOLUTION Because the markup is determined by the cost, we find 40% of \$150.

$$40\% \text{ of } \$150 = 0.40 \times \$150 = \$60.00 = \text{markup}$$
$$\text{Selling price} = \text{markup} + \text{cost}$$
$$\text{Selling price} = \$60 + \$150 = \$210$$

EXAMPLE 2 Another pro shop, at Shinnecock Hills Golf Club, sells a different brand of golf clubs for \$300 per set. If the cost of a set of these golf clubs is \$200, what is the percent markup on the cost?

SOLUTION Markup = selling price − cost. Therefore,

$$\text{Markup} = \$300 - \$200 = \$100$$

Now we must find what the percent markup is, based on the cost. That is, we must find what percentage \$100 (the markup) is of

\$200 (the cost). To do this, we form a ratio, markup : cost. Therefore,

$$\frac{\text{Markup}}{\text{Cost}} = \frac{\$100}{\$200} = \frac{1}{2} = 0.5 = 50\%$$

The markup (\$100) is 50% of the cost (\$200).
Check: 50% of \$200 = 0.50 × \$200 = \$100 = markup.

To find what percentage one number is of another, we form a ratio between the two numbers and convert the ratio (fraction) to a percentage. For example, suppose you took a quiz and had 12 out of 15 questions correct. To find the percentage of correct answers, we form a ratio, 12 : 15, and convert it to a percentage.

$$\frac{12}{15} = \frac{4}{5} = 0.8 = 80\%$$

Recall that we can also do this by means of a proportion. That is,

$$\frac{12}{15} = \frac{x}{100}$$

$$15 \cdot x = 12 \cdot 100$$

$$15x = 1{,}200$$

$$x = 80, \quad \text{and} \quad \frac{80}{100} = 80\%$$

EXAMPLE 3 a. What percentage of 108 is 27?
b. What percentage of 60 is 48?
c. What percentage of 60 is 90?

SOLUTION a.

$$\frac{27}{108} = \frac{x}{100}$$

$$108 \cdot x = 27 \cdot 100$$

$$108x = 2{,}700$$

$$x = 25, \quad \text{and} \quad \frac{25}{100} = 25\%$$

b.

$$\frac{48}{60} = \frac{x}{100}$$

$$60 \cdot x = 48 \cdot 100$$

$$60x = 4{,}800$$

$$x = 80, \quad \text{and} \quad \frac{80}{100} = 80\%$$

c. $$\frac{90}{60} = \frac{x}{100}$$

$$60 \cdot x = 90 \cdot 100$$

$$60x = 9{,}000$$

$$x = 150, \quad \text{and} \quad \frac{150}{100} = 150\%$$

ALTERNATE SOLUTION a. Express $\frac{27}{108}$ as a decimal and convert the resulting decimal to a percent:

$$\frac{27}{108} = \frac{1}{4} = 0.25 = 25\%$$

b. $$\frac{48}{60} = \frac{4}{5} = 0.8 = 80\%$$

c. $$\frac{90}{60} = \frac{3}{2} = 1.5 = 150\%$$

EXAMPLE 4 Al's Appliance Outlet sells a particular black-and-white television for \$110. If the set costs Al \$80, what is the percent markup on the cost?

SOLUTION Markup = selling price − cost. Therefore,

$$\text{Markup} = \$110 - \$80 = \$30$$

Since the percent markup on cost is equal to the ratio of markup to cost, we have

$$\frac{\$30}{\$80} = \frac{3}{8} = 0.375 = 37.5\%$$

The markup (\$30) is 37.5% of the cost (\$80).
Check: 37.5% of \$80 = 0.375 × \$80 = \$30 = markup.

EXAMPLE 5 A stereo sells for \$320. The markup is 60% of the cost. What is the cost of the stereo system?

SOLUTION Selling price = markup + cost. Since the markup is 60% of the cost, we can also state that

$$\text{Selling price} = (60\% \text{ of cost}) + \text{cost}$$

This means that the selling price is 60% of the cost plus the full cost. But the full cost is 100% of the cost. Therefore,

$$\text{Selling price} = (60\% \text{ of cost}) + (100\% \text{ of cost})$$

Thus, the selling price is 160% of the cost, so \$320 represents 160% of the cost. To find the cost, we divide \$320 by 160%, or 1.60.

$$\frac{\$320}{160\%} = \frac{\$320}{1.60} = \$200 = \text{cost}$$

Check:

$$\begin{aligned} 60\% \text{ of } \$200 &= 0.60 \times \$200 = \$120 = \text{markup} \\ \text{Selling price} &= \text{markup} + \text{cost} \\ \$320 &= \$120 + \$200 \\ \$320 &= \$320 \end{aligned}$$

Thus far we have discussed markup in terms of percentage of cost. However, there are many businesses that figure their markup on the selling price. Suppose a coat costs a retailer \$50 and the markup is 20% of the selling price. What is the selling price for this particular coat? Recall that

$$\text{Cost} = \text{selling price} - \text{markup}$$

Then

$$\text{Cost} = \text{selling price} - (20\% \text{ of selling price})$$

Note that the selling price is 100% of the selling price. Hence,

$$\text{Cost} = (100\% \text{ of selling price}) - (20\% \text{ of selling price})$$

This means that

$$\text{Cost} = 80\% \text{ of selling price}$$

To find the selling price, we divide \$50 by 80%, or 0.80.

$$\frac{\$50}{80\%} = \frac{\$50}{0.80} = \$62.50 = \text{selling price}$$

We can check our work. The selling price is \$62.50 and the cost is \$50. Therefore the markup is \$62.50 − \$50.00 = \$12.50. We now find 20% of \$62.50 and check to see if it is \$12.50.

$$20\% \text{ of } \$62.50 = 0.20 \times \$62.50 = \$12.50 = \text{markup}$$

The markup is \$12.50, which is 20% of the selling price, \$62.50.

EXAMPLE 6 A 10-speed bike costs a retailer \$90. The markup is 25% of the selling price. Find the selling price.

SOLUTION

$$\begin{aligned}\text{Selling price} &= \text{markup} + \text{cost}\\ &= (25\% \text{ of selling price}) + \text{cost}\end{aligned}$$

This means that the cost is 100% − 25% = 75% of the selling price. That is, \$90 is 75% of the selling price. Therefore, to find the selling price, we divide \$90 by 75%, or 0.75.

$$\frac{\$90}{75\%} = \frac{\$90}{0.75} = \$120 = \text{selling price}$$

Check: The markup is \$120 − \$90 = \$30. Hence, 25% of \$120 should be \$30.

$$25\% \text{ of } \$120 = 0.25 \times \$120 = \$30 = \text{markup}$$

The solution checks.

EXAMPLE 7 A color television costs a retailer \$300. The markup is 30% of the selling price. Find the selling price.

SOLUTION

$$\begin{aligned}\text{Selling price} &= \text{markup} + \text{cost}\\ &= (30\% \text{ of selling price}) + \text{cost}\end{aligned}$$

Therefore, the cost is 100% − 30% = 70% of the selling price. That is, \$300 is 70% of the selling price. To find the selling price, we divide \$300 by 70%, or 0.70.

$$\frac{\$300}{70\%} = \frac{\$300}{0.70} = \$428.57 = \text{selling price, to the nearest cent}$$

Check: The markup is

$$\$428.57 - \$300.00 = \$128.57$$

Therefore, 30% of \$428.57 should be \$128.57.

$$\begin{aligned} 30\% \text{ of } \$428.57 &= 0.30 \times \$428.57 \\ &= \$128.57 = \text{markup, to the nearest cent} \end{aligned}$$

The solution checks.

Many times, retailers cannot sell everything at the retail selling price. This happens with defective merchandise, overstocked items, discontinued models, unpopular styles, and so on. A retailer still wants to sell the merchandise in stock, but, since it cannot be sold at the original price, it must be reduced in price. Therefore, the merchandise is sold at a new, lower price called the **sale price.** The change, or difference, between the original price and the sale price is called the **markdown.**

$$\text{Markdown} = \text{original price} - \text{sale price}$$

Markdown can be expressed in terms of a dollar amount, and it can also be expressed in terms of percent reduction.

Percent markdowns are usually a better indicator of savings than the dollar amount of the markdown. For example, a \$10 markdown on an item that ordinarily sells for \$100 is a saving of 10%, but a \$10 markdown on a \$20 item is a saving of 50%.

If a retailer is selling coats for \$30 that were originally priced at \$40, we can find the amount of markdown by subtracting \$30 from \$40. That is,

$$\text{Markdown} = \text{original price} - \text{sale price}$$

In this case, the markdown is \$40 − \$30 = \$10. But what is the percent reduction? To find the percent reduction, we must find what percent \$10 is of \$40. Therefore we divide \$10 by \$40.

$$\frac{\$10}{\$40} = \frac{1}{4} = 0.25 = 25\%$$

The original price was reduced by 25%.

If we want to find the percent markdown based on the sale price, we divide the markdown (\$10) by the sale price (\$30).

$$\frac{\$10}{\$30} = \frac{1}{3} = 33\tfrac{1}{3}\%$$

The markdown of \$10 is $33\frac{1}{3}\%$ of the sale price.

EXAMPLE 8 Al's Appliance Outlet had a clearance sale on last year's color television sets. A certain set originally selling for \$480 was advertised at a reduction of 20%.

a. What was the dollar markdown?
b. What was the sale price?

SOLUTION a. We find 20% of the original price, \$480.

$$20\% \text{ of } \$480 = 0.20 \times \$480 = \$96 = \text{markdown}$$

b. Sale price = original price − markdown
Sale price = \$480 − \$96 = \$384

EXAMPLE 9 During a clearance sale, a pair of boots that originally sold at \$50 was reduced to \$40.

a. What was the percent markdown based on the original price?
b. What was the percent markdown based on the sale price?

SOLUTION

$$\begin{aligned} \text{Markdown} &= \text{original price} - \text{sale price} \\ &= \$50 - \$40 = \$10 \end{aligned}$$

a. To find the percent markdown based on the original price, we divide \$10 by \$50.

$$\frac{\$10}{\$50} = \frac{1}{5} = 0.20 = 20\%$$

b. To find the percent markdown based on the sale price, we divide \$10 by \$40.

$$\frac{\$10}{\$40} = \frac{1}{4} = 0.25 = 25\%.$$

EXERCISES FOR SECTION 11.4

For Exercises 1 and 2, express your answers to the nearest tenth of a percent.

1. a. What percent of 48 is 24?
   b. What percent of 48 is 12?
   c. What percent of 30 is 40?
   d. What percent of 24 is 16?
   e. What percent of 216 is 54?
   f. What percent of 60 is 96?
2. a. What percent of 76 is 38?
   b. What percent of 56 is 7?

c. What percent of 114 is 38?
d. What percent of 7 is 2?
e. What percent of 55 is 95?
f. What percent of 54 is 170?

3. Find the dollar amount of markup, to the nearest cent.

	Cost	Percent Markup on Cost
a.	$ 50.00	25%
b.	$300.00	$33\frac{1}{3}$%
c.	$ 25.00	30%
d.	$175.00	$12\frac{1}{2}$%

4. Find the dollar amount of markup, to the nearest cent.

	Cost	Percent Markup on Cost
a.	$ 35.00	7%
b.	$210.00	$12\frac{1}{2}$%
c.	$ 29.95	20%
d.	$ 49.95	$18\frac{1}{2}$%

5. Find the markup and the percent markup on the cost. Express each percentage to the nearest tenth.

	Selling Price	Cost
a.	$ 50.00	$20.00
b.	$100.00	$75.00
c.	$ 13.50	$10.50
d.	$ 19.95	$15.95

6. Find the markup and the percent markup on the cost. Express each percentage to the nearest tenth.

	Selling Price	Cost
a.	$220.00	$110.00
b.	$ 8.00	$ 6.00
c.	$ 14.75	$ 10.00
d.	$ 99.99	$ 75.49

7. A coat sells for $125. The markup is 40% of the cost. What is the cost of the coat, to the nearest cent?

8. A pair of boots retails for $60. The markup is $33\frac{1}{3}$% of the cost. What is the cost of the boots?

9. A television sells for $358. The markup is 30% of the cost. What is the cost of the television, to the nearest cent?

10. A lawnmower costs a retailer $100. The markup is 25% of the selling price. Find the selling price, to the nearest cent.

11. A snowblower costs a retailer $310. The markup is 40% of the selling price. Find the selling price, to the nearest cent.

12. A sporting-goods dealer pays $18.50 for each basketball that she buys. If she wants a markup of 35% on the selling price, what should she set as the selling price?

13. Find the markup and the cost, to the nearest cent.

	Selling Price	Percent Markup on Selling Price
a.	$400.00	40%
b.	$ 10.98	20%
c.	$ 49.95	$12\frac{1}{2}$%

14. Find the markup and the cost, to the nearest cent.

	Selling Price	Percent Markup on Selling Price
a.	$ 60.00	$33\frac{1}{3}$%
b.	$ 99.99	25%
c.	$299.95	$12\frac{1}{2}$%

15. Al's Appliance Outlet had a clearance sale on refrigerators. A certain model, originally priced at $395, was advertised at 20% off. What was the dollar markdown? What was the sale price?

16. A bike shop advertised one of its 10-speed bikes for $150. After some time had passed, the bike had not sold, so the dealer lowered the price to $120. What was the percentage of markdown on the original price?

17. Dinah's Donut Shop sells fresh doughnuts for \$1.50 per dozen. Day-old doughnuts are sold for \$1.00 per dozen. What percentage (to the nearest tenth) is the markdown on the original price?

18. A department store sells men's suits for \$199.95. During a clearance sale, the price is reduced 25%. What are the dollar markdown and the sale price, to the nearest cent?

19. During an inventory clearance sale, a used-car dealer reduced the price of all the cars on his lot by \$300. If a certain car was originally priced at \$1195, what was the sale price? What percent (to the nearest tenth) is the markdown on the sale price?

20. A coat that was originally priced to sell for \$72.50 was reduced to \$60.00. What percent (to the nearest tenth) was the markdown on the original price? What percent is the markdown (to the nearest tenth) on the sale price?

21. Monro Hardware paid \$150 each for a shipment of lawnmowers. Each mower was marked up by 50% of the cost. At the end of the season, the mowers that had not been sold were reduced 30%. Find the regular selling price of a mower, and the sale price at the end of the season.

22. Scott's Service Station paid \$40 each for a shipment of snow tires. Each tire was marked up by 50% of the selling price. At the end of the winter, the tires that had not been sold were reduced 40%. Find the regular selling price of a snow tire, and the sale price at the end of the season.

23. For each of the items in exercises 15, 17, 19, and 21 find the amount of sales tax, and the total amount due, if a sales tax of 7% is charged on each item.

24. For each of the items in exercises 16, 18, 20, and 22 find the amount of sales tax, and the total amount due, if a sales tax of $7\frac{1}{2}$% is charged on each item.

JUST FOR FUN The Bills are in first place and the Dolphins are in fifth, and the Patriots are midway between them. If the Jets are ahead of the Dolphins and the Colts are immediately behind the Patriots, then who is in second place?

11.5 SIMPLE INTEREST

Borrowing money is a necessary transaction for most people and businesses. The borrower may use the money to do something that otherwise would be impossible, whereas the lender charges *interest* and therefore makes a profit on the transaction. Whenever merchandise is bought on credit, the customer pays **interest** on the unpaid balance. Most department stores and credit card companies charge 1.75% to 2% **interest per month** on the unpaid balance, after 30 days. Banks also pay interest to people who deposit their money in savings accounts.

If a person borrows $1,000 from a bank, then the $1,000 is called the *principal.* Similarly, if a person has $1,000 in a savings account, this is also called the principal. The amount of money on which interest is paid is always called the **principal.**

The amount of interest depends on the principal; that is, the interest is a certain percentage of the principal. This percent is called the *rate of interest,* or **rate.** Unless otherwise noted, the rate of interest is an annual one.

When money is borrowed, the borrower agrees to pay back the principal and the interest within a specified period of time. For example, an auto loan may be given for a period of 3 years. At the end of 3 years, both the principal and the interest have been paid, so the interest for such a loan is computed for a period of 3 years. This period is called the **time** of the loan. The interest due depends on three things: the principal, the rate of interest, and the time.

Interest is calculated by multiplying the principal times the rate of interest times the time (in years). That is,

$$\textbf{Interest} = \textbf{principal} \times \textbf{rate} \times \textbf{time}$$

or

$$I = Prt$$

This formula determines the *simple interest.* **Simple interest** is the cost of borrowing money computed on the original principal only. This formula can also be used to find the simple interest earned on an investment. If a person borrows $500 for a period of 1 year at 12% simple interest, we can find the amount of interest by means of the formula $I = Prt$. In this case, we have $P = \$500$, $r = 0.12$ (because $12\% = 0.12$), and $t = 1$ year.

$$\begin{aligned} I &= Prt \\ &= \$500 \times 0.12 \times 1 \\ &= \$60 \end{aligned}$$

Therefore, the borrower would have to pay $60 interest on the loan. The total amount that must be repaid is $560 (the principal plus the interest). The amount repaid is equal to the principal plus the interest.

$$A = P + I$$

EXAMPLE 1 Irene borrows \$3,000 for 3 years at a simple interest rate of 13%. How much interest does she pay? What is the total amount that must be repaid?

SOLUTION We use the formula $I = Prt$ with $P = \$3{,}000$, $r = 0.13$, and $t = 3$.

$$\begin{aligned} I &= Prt \\ &= \$3{,}000 \times 0.13 \times 3 \\ &= \$1{,}170 \end{aligned}$$

The total amount that must be repaid is

$$\begin{aligned} A &= P + I \\ &= \$3{,}000 + \$1{,}170 \\ &= \$4{,}170 \end{aligned}$$

EXAMPLE 2 Carl invested \$12,000 for 6 months at a simple interest rate of 10%. How much interest did he earn?

SOLUTION This problem is similar to Example 1, but note that the investment was for a period of time less than 1 year. To use the simple-interest formula, we must express 6 months as a fraction of a year.

> No savings bank or savings and loan can give higher rates on a 6-month deposit of \$10,000 or more.
>
> 6-MONTH C.D. RATE
>
> Rates effective July 6–12, 1988
>
> **8.35%**
>
> PER YEAR
>
> **8.71%**
>
> EFFECTIVE ANNUAL YIELD

$$6 \text{ months} = \frac{6}{12}, \quad \text{or } \frac{1}{2} \text{ year}$$

Now we have $P = \$12{,}000$, $r = 0.10$, and $t = \frac{1}{2}$.

$$\begin{aligned} I &= Prt \\ &= \$12{,}000 \times 0.10 \times \frac{1}{2} \\ &= \$600 \end{aligned}$$

EXAMPLE 3 How much will \$5,000 earn in 2 years at $9\frac{1}{2}\%$ simple interest?

SOLUTION We use the formula for simple interest with $P = \$5{,}000$, $r = 9\frac{1}{2}\% = 9.5\% = 0.095$, and $t = 2$.

$$\begin{aligned} I &= Prt \\ &= \$5{,}000 \times 0.095 \times 2 \\ &= \$950 \end{aligned}$$

EXERCISES FOR SECTION 11.5

1. Find the simple interest on a $2,000 loan for 3 years at 12%.
2. Find the simple interest on a $3,500 loan for 1 year at 15%.
3. Find the simple interest on a $2,000 loan for 6 months at 12%.
4. Find the simple interest on a $3,000 loan for 9 months at 15%.
5. Find the simple interest on a $5,000 loan for 6 months at $11\frac{1}{2}$%.
6. A merchant borrowed $2,000, agreeing to repay the principal and 12% simple interest at the end of 6 months. Find the amount of interest and the total sum that must be paid.
7. Luke borrowed $3000 at $10\frac{1}{2}$% simple interest for a period of 3 years. How much interest will he pay? What is the total amount that he must pay?
8. To help pay her tuition bill, Sally borrowed $1,100 at $9\frac{1}{2}$% simple interest for 18 months. How much interest will she pay? What is the total amount she will pay?
9. How much simple interest will $5,000 earn in 3 years at a rate of $8\frac{1}{2}$%? How much interest will it earn in 3 months?
10. How much simple interest will $6,000 earn in 4 years if the rate of interest is 9%? How much will it earn in 4 months?
11. Connie has a balance due of $240 on her charge account. The rate is $1\frac{1}{2}$% simple interest *per month* on the unpaid balance. If Connie decides not to pay anything toward her balance this month, and she does not charge anything to her account next month, what will be her new balance due next month?
12. Sam has a balance due of $300 on his charge account. The rate is $1\frac{1}{2}$% simple interest per month on the unpaid balance. If Sam pays $180 toward his balance this month, and does not charge anything else next month, what will be the balance due on his charge account next month?
13. Benny deposited $10,000 in a 26-week certificate of deposit that paid 14.9% simple interest. How much interest will he receive at the end of 26 weeks?
14. If a certificate of deposit pays 15.2% simple interest, how much interest will $20,000 earn at the end of 26 weeks?
15. Find the interest earned on $1850 invested for 30 months in an account that pays 12% simple interest per year.
16. Bob borrowed $12,000 from the Friendly Finance Company for 18 months. If the finance company charges 18% simple interest per year, find the total amount that Bob will have to pay back.
17. Find the simple interest on $20,000 loan for 3 years at $8\frac{1}{2}$%. How much interest will it earn in 3 months?
18. How long must $10,000 be kept at 5% simple interest to become $12,000?
19. Find the time required for $35,000 to earn $1,750 at 10% simple interest.
20. What principal will yield $9,280 (principal and interest) in 2 years at 8% simple interest?

★21. Mr. Gilligan has $10,000 and invests $3,000 for 1 year at 8% simple interest. At what rate of simple interest should the remainder be invested, if the total income from the $10,000 is to be $660?

★22. An investment firm invested $20,000 for 1 year at simple interest. One portion was invested at 6% and the remainder at 8%. How much money was invested at each rate of interest, if the total interest was the same as it would have been if all the money had been invested at 7%?

JUST FOR FUN

How much is a billion dollars? To give you some idea, suppose that you could spend \$1,000 a day every day. How long would it take you to spend \$1,000,000,000? Assume that each year contains 365 days.

11.6 COMPOUND INTEREST

If a person borrows \$1,000 at 8% simple interest for a period of 1 year, then at the end of that year \$1,080 must be repaid.

$$\begin{aligned} I &= Prt & A &= P + I \\ &= \$1{,}000 \times 0.08 \times 1 & &= \$1{,}000 + \$80 \\ &= \$80 & &= \$1{,}080 \end{aligned}$$

If the borrower did not pay back any of the loan or the interest by the end of the first year, and if he or she wanted to continue the loan for another year at the same rate, then he or she would owe \$1,080 plus the interest on \$1,080, which is \$86.40. That is, \$1,166.40 would have to be repaid.

$$\begin{aligned} I &= Prt & A &= P + I \\ &= \$1{,}080 \times 0.08 \times 1 & &= \$1{,}080 + \$86.40 \\ &= \$86.40 & &= \$1{,}166.40 \end{aligned}$$

This is an example of *compound interest.* For this example, we would say that \$1,000 was loaned for a period of 2 years, with interest *compounded annually*. Banks pay compound interest on their savings accounts. Most banks pay interest that is compounded quarterly, and some banks pay interest compounded daily.

When the interest due at the end of a certain period is added to the principal and that sum earns interest for the next period, the interest paid is called **compound interest.** The interest for each succeeding period is greater than the previous one, because the principal keeps increasing.

EXAMPLE 1

a. Find the amount of simple interest when \$1,000 is invested for 3 years at 9%.
b. Find the amount of *compound* interest when \$1,000 is invested for 3 years at 9% compounded annually.

SOLUTION a. $I = Prt$
$= \$1{,}000 \times 0.09 \times 3$
$= \$270$

b. Because this is compound interest, we must find the interest at the end of each year and add it to the principal before computing the interest for the next year.

$$\text{1st year:} \quad I = Prt = \$1{,}000 \times 0.09 \times 1 = \$90$$

Amount at end of 1 year = \$1,000 + \$90 = \$1,090.

$$\text{2nd year:} \quad I = Prt = \$1{,}090 \times 0.09 \times 1 = \$98.10$$

Amount at end of 2 years = \$1,090 + \$98.10 = \$1,188.10.

$$\text{3rd year:} \quad I = Prt = \$1{,}188.10 \times 0.09 \times 1 = \$106.93$$

Amount at end of 3 years = \$1,188.10 + 106.93 = \$1,295.03. The total interest is equal to the total amount due at the end of 3 years minus the principal.

$$\text{Interest} = \text{amount} - \text{principal} = \$1{,}295.03 - \$1{,}000 = \$295.03$$

Note that in Example 1 the interest earned in 3 years on \$1,000 at 9% simple interest was \$270, whereas the interest earned on the same amount over the same time period at 9% *compounded annually* was \$295.03—a difference of \$25.03. This points up the advantage of compound interest over simple interest when money is invested at a given rate over a number of interest periods. Imagine what the difference would be for a large investment such as \$100,000 with an interest rate in the neighborhood of 20%! The compound interest is greater because it is computed more often, and each time on a larger principal.

The computation in part b of Example 1 was somewhat tedious. Banks use computers and accountants use calculators to determine compound interest. But they also use books of tables

NOTE OF INTEREST

The highest denomination of currency in circulation is the U.S. Federal Reserve Bank note for \$10,000. It bears the picture of Salmon P. Chase, former Secretary of the Treasury. It is believed that there are some 400 of these bills in circulation. No bills higher than \$100 have been issued since 1969.

to quote different interest amounts, loan payments, and so on. Table 1 can be used to compute compound interest. It shows the amounts that must be paid at the end of different interest periods if $1 is invested at compound interest at one of the given interest rates.

TABLE 1
COMPOUNDED AMOUNT OF $1

	Interest Rate per Period								
Period	2%	3%	4%	6%	8%	10%	12%	14%	16%
1	1.020	1.030	1.040	1.060	1.080	1.100	1.120	1.140	1.160
2	1.040	1.061	1.082	1.124	1.166	1.210	1.254	1.300	1.346
4	1.082	1.126	1.170	1.262	1.360	1.464	1.574	1.689	1.811
6	1.126	1.194	1.265	1.419	1.587	1.772	1.974	2.195	2.436
8	1.172	1.267	1.369	1.594	1.851	2.144	2.476	2.853	3.278
10	1.219	1.344	1.480	1.791	2.159	2.594	3.106	3.707	4.411
12	1.268	1.426	1.601	2.012	2.518	3.138	3.896	4.818	5.936
14	1.319	1.513	1.732	2.261	2.937	3.797	4.887	6.261	7.988
16	1.373	1.605	1.873	2.540	3.426	4.595	6.130	8.137	10.748
20	1.486	1.806	2.191	3.207	4.661	6.728	9.646	13.743	19.461
24	1.608	2.033	2.563	4.049	6.341	9.850	15.179	23.212	35.236
28	1.741	2.288	2.999	5.112	8.627	14.421	23.884	39.204	63.800

Table 1 is an example of a compound-interest table. It will enable us to do the problems in this section. It should be noted that there are tables that are much more detailed. Interest in these tables may be carried out to four or five decimal places, and values are given for many more interest rates and numbers of periods. Table 1 shows the amount of interest that $1 will accumulate when interest is paid at the indicated rate and compounded for the indicated number of interest periods. For example, the entry in the third row under the 8% column of Table 1 is 1.360. This indicates that if $1 is placed in an account that pays 8% and is compounded for each of four interest periods, then the total accumulation in the account will be $1.36. That is, $1 will grow to $1.36 in four interest periods at 8% per period. The amount accumulated in this way is called the **compound amount.** To find the compound amount on other amounts of money, we multiply the principal by the compound amount for $1. For example, to find the compound amount if $1,000 is deposited at 6% compounded annually for 10 years, we first find the compound amount for $1 using these figures for time and interest rate. Using the table, we obtain $1.791. The compound amount for $1,000 is therefore

$$\$1,000 \times 1.791 = \$1,791$$

EXAMPLE 2 Find the compound amount on deposit when $1,200 is deposited for 6 years at 8% compounded annually.

SOLUTION From Table 1, the compound amount for $1 at 8% for six interest periods is $1.587. Therefore, to find the compound amount for $1,200 at 8% for six periods, we multiply $1,200 times 1.587.

$$\text{Compound amount} = \$1{,}200 \times 1.587 = \$1{,}904.40$$

We can find the total compound interest by subtracting the principal from the compound amount. That is,

$$\text{Compound interest} = \text{compound amount} - \text{principal}$$

In Example 2, the compound amount is $1,904.40 and the principal is $1,200. Hence,

$$\text{Compound interest} = \$1{,}904.40 - \$1{,}200 = \$704.40$$

We noted earlier that banks usually compound interest more often than once a year. There are some banks that compound interest semiannually, others that compound interest quarterly, and others that compound interest daily. If a bank offers 6% interest compounded semiannually, then there are two interest periods every year, and the bank pays 3% interest every 6 months.

To find the compound amount when $1,000 is deposited for 4 years at 6% compounded semiannually, we must first determine the number of interest periods and the interest rate per period. Because the principal is deposited for 4 years compounded semiannually, there are eight interest periods. The rate of 6% per year is the same as 3% per half-year. Using Table 1, we find 8 under the period column and go across to the 3% column. The entry is 1.267. Now we multiply 1.267 by $1,000.

$$\text{Compound amount} = \$1{,}000 \times 1.267 = \$1{,}267$$

EXAMPLE 3 Find the compound amount on deposit when $500 is deposited for 5 years at 8% compounded quarterly.

SOLUTION Interest compounded quarterly means that there are four interest periods per year, so for 5 years there are 20 interest periods. Eight percent interest per year is the same as 2% every quarter-year. The compound amount for $1 at 2% for 20 interest periods is $1.486. Therefore,

$$\text{Compound amount} = \$500 \times 1.486 = \$743$$

EXAMPLE 4 Find the compound amount and the compound interest on \$2,000 invested at 12% compounded quarterly for 6 years.

SOLUTION

$$\text{Number of interest periods per year} = 4$$
$$\text{Total number of interest periods} = 6 \times 4 = 24$$
$$\text{Interest rate each interest period} = \frac{1}{4} \text{ of } 12\% = 3\%$$

The compound amount for \$1 at 3% for 24 interest periods is \$2.033. Therefore,

$$\text{Compound amount} = \$2{,}000 \times 2.033 = \$4{,}066$$
$$\text{Compound interest} = \$4{,}066 - \$2{,}000 = \$2{,}066$$

EXAMPLE 5 If \$1,000 were invested at 12% compounded semiannually, how long would it take for the investment to double itself?

SOLUTION The rate per interest period is $\frac{1}{2}$ of 12%, or 6%. Examining the column headed 6% in Table 1, we see that \$1 will double itself in 12 periods; that is, the compound amount for \$1 at 6% for 12 periods is \$2.012. The interest is compounded semiannually, so 12 periods is 12 ÷ 2 = 6 years. Therefore, \$1,000 will double itself in 6 years when invested at 12% compounded semiannually. The compounded amount would be \$1,000 × 2.012 = \$2,012. (*Note:* This compound amount is a little more than twice \$1,000, but the answer to the original question is correct to the nearest year.)

EXERCISES FOR SECTION 11.6

For Exercises 1–6, use Table 1 to find the compound amount for each investment if interest is compounded annually.

1. \$700 at 6% for 4 years
2. \$900 at 8% for 6 years
3. \$2,500 at 8% for 10 years
4. \$3,300 at 10% for 2 years
5. \$10,000 at 14% for 2 years
6. \$12,000 at 16% for 6 years

For Exercises 7–12, use Table 1 to find the compound amount for each investment if interest is compounded semiannually.

7. \$500 at 8% for 10 years
8. \$800 at 6% for 6 years
9. \$1,000 at 4% for 3 years
10. \$2,500 at 12% for 3 years
11. \$5,000 at 12% for 5 years
12. \$10,000 at 16% for 2 years

For Exercises 13–18, use Table 1 to find the compound amount for each investment if interest is compounded quarterly.

13. \$500 at 8% for 2 years

14. \$750 at 12% for 3 years

15. \$1,000 at 16% for 5 years

16. \$2,500 at 8% for 6 years

17. \$10,000 at 12% for 6 years

18. \$20,000 at 16% for 7 years

19. Find the compound amount and the compound interest on \$3,000 invested for 5 years at 12% compounded quarterly.

20. Find the compound amount and the compound interest on \$2,500 invested for 10 years at 16% compounded semiannually.

21. If \$1,000 were invested at 6% compounded semiannually, how long would it take for the investment to double itself?

22. If \$1,000 were invested at 24% compounded quarterly, how long would it take for the investment to *quadruple* itself?

23. In 1988, the price of a medium-sized automobile was \$9,000. What can we expect the price of this same type of car to be in the year 2000, if we assume that the annual rate of inflation is 10%? (*Hint:* Use the compound-interest table.)

24. A video-cassette recorder (VCR) cost approximately \$400 in 1988. If we assume the annual rate of inflation is 12%, what can we expect the price of this same type of VCR to be in the year 2000?

25. A typical weekly magazine had a cover price of \$2.50 in 1988. If we assume the annual rate of inflation is 8%, what can we expect the price of this magazine to be in the year 2000?

26. In 1988, a certain newspaper sold for \$.50. What can we expect the price of this same newspaper to be in the year 2000, if we assume that the annual rate of inflation is 8%?

27. The sum of \$10,000 is deposited in each of four banks, each paying 8% interest per year. The first bank compounds interest annually, the second bank semiannually, and the third bank quarterly. The fourth bank pays simple interest. If no further deposits or withdrawals are made, how much will be in each account at the end of 4 years?

★28. A father wishes to give his daughter \$20,000 at the age of 25. If the daughter is now 13 years old, how much money must her father invest at 8% interest compounded semiannually in order to have \$20,000 when his daughter is 25?

29. A grandmother deposits \$1,000 in an account that pays 12% compounded annually when her granddaughter is born. What will be the value of the account when the granddaughter reaches her twenty-fourth birthday, assuming that no other deposits or withdrawals are made during this time?

★30. A grandfather wishes to have \$30,000 available for his grandson's college education. If the grandson is now 4 years old, how much money must the grandfather invest at 6% interest compounded semiannually in order to have \$30,000 when his grandson is 18?

JUST FOR FUN There are 11 denominations of Federal Reserve notes now in circulation. Two of these are the one-dollar bill and the two-dollar bill. Can you name the others?

11.7 EFFECTIVE RATE OF INTEREST

8 $\frac{25}{}$%
Effective Annual Yield
on **7** $\frac{95}{}$%
Annual Interest Rate

In our previous discussions pertaining to interest, we have seen that interest rates are usually stated on a yearly basis, even though the period of payment is not a year. For example, an interest rate of 8% compounded quarterly means that we use an interest rate of 2% for each interest period. Similarly, an interest rate of 6% compounded semiannually means that we use an interest rate of 3% for each period.

To find the compound amount on deposit when \$1 is deposited for 1 year at 8% compounded quarterly, we can use Table 1 from Section 11.6 to find that the amount is \$1.082. That is, \$1 will compound to \$1.082 in 1 year with interest compounded quarterly. The compound amount on deposit when \$1 is deposited for 1 year at 6% compounded semiannually is \$1.061. That is, \$1 will yield \$1.061 in one year with interest compounded semiannually.

We have chosen these two examples to help illustrate the *effective annual interest rate,* also referred to as **effective rate.** For example, an interest rate of 8% compounded quarterly is equivalent to an effective rate of 8.2% compounded annually. This is because, as noted in the preceding paragraph, \$1 compounds to \$1.082 in 1 year at 8% compounded quarterly. Therefore, the total interest earned in 1 year is

$$\$1.082 - \$1.00 = \$0.082$$

This is an effective rate of 8.2% since

$$8.2\% \text{ of } \$1.00 = \$0.082$$

Thus the interest earned at 8% compounded quarterly is the same as the interest earned at an annual interest rate of 8.2%.

Similarly, a 6% interest rate compounded semiannually is equivalent to an effective rate of 6.1% compounded annually. (Note that \$1 compounds to \$1.061 in 1 year at 6% compounded semiannually.)

If interest is paid more than once a year, the effective annual rate is greater than the stated annual rate. If interest is paid annually, then the effective rate is the same as the stated annual rate. In the preceding examples, the effective rate was greater than the stated annual rate.

Effective rate is often used to compare interest rates, which are compounded at different intervals. For example, if Bank A offers 5% interest on its deposits compounded quarterly and Bank B offers $5\frac{1}{2}$% interest on its deposits compounded semiannually, we can find the effective annual rate for each bank to determine which offers the best investment. We can also compare effective rates when borrowing money. In this case, we would select the lowest effective rate. Many banks publish pamphlets listing the various types of accounts they offer, together with the stated interest rate and the effective annual yield. This is also a common practice in newspaper advertisements.

We can determine the effective annual rate by means of the formula

$$E = (1 + r)^n - 1$$

where E = effective rate
n = number of payment periods per year
r = interest rate per period

We shall now use this formula to find the effective annual rate for 6% compounded semiannually.

$$E = (1 + r)^n - 1$$

The interest rate is compounded semiannually, so there are $n = 2$ payment periods per year. Therefore, $r = 6\% \div 2 = 3\% = 0.03$. (Note that r is expressed without a percent sign.) Substituting these values in the formula, we have

$$\square\ E = (1 + 0.03)^2 - 1$$
$$E = (1.03)^2 - 1$$
$$E = 1.0609 - 1$$
$$E = 0.0609, \text{ or } 6.1\%$$

Note: The symbol ▯ identifies a problem that is more readily done with a calculator.

EXAMPLE 1 What is the effective rate, if money is invested at 8% compounded quarterly?

SOLUTION The number of payment periods is four, because interest is compounded quarterly. The interest rate per period is $8\% \div 4 = 2\%$. Therefore $r = 0.02$. Substituting these values in the formula, we have

$$\square\ E = (1 + r)^n - 1$$
$$E = (1 + 0.02)^4 - 1$$
$$E = (1.02)^4 - 1$$
$$E = 1.0824 - 1$$
$$E = 0.0824, \text{ or } 8.2\% \quad \text{(to the nearest tenth)}$$

This answer checks with the one we obtained earlier using the compound-interest table.

In Example 1 we rounded our answer to the nearest tenth so that it could be compared with the first illustrative example in this section. However, all other problems in this section will be rounded to the nearest hundredth of a percent. Also, it is recommended that any calculation involving the use of this formula be done on a pocket calculator.

EXAMPLE 2 What is the effective rate, if money is invested at 7% compounded semiannually?

SOLUTION $n = 2$, $r = 7\% \div 2 = 3.5\% = 0.035$.

$$\begin{aligned} E &= (1 + r)^n - 1 \\ E &= (1 + 0.035)^2 - 1 \\ E &= (1.035)^2 - 1 \\ E &= 1.071225 - 1 = 0.071225 \\ E &= 7.12\% \end{aligned}$$

EXAMPLE 3 Bank A offers its depositors an interest rate of 5% compounded quarterly and Bank B offers its depositors a rate of $5\frac{1}{2}\%$ compounded semiannually. Which bank makes the better offer?

SOLUTION We must compare the effective annual rate for each bank.

Bank *A*: $n = 4$, $r = 5\% \div 4 = 1.25\% = 0.0125$

$$\begin{aligned} E &= (1 + r)^n - 1 \\ E &= (1 + 0.0125)^4 - 1 \\ E &= (1.0125)^4 - 1 \\ E &= 1.0509 - 1 = 0.0509 \\ E &= 5.09\% \end{aligned}$$

Bank *B*: $n = 2$, $r = 5\frac{1}{2}\% \div 2 = 2.75\% = 0.0275$

$$\begin{aligned} E &= (1 + r)^n - 1 \\ E &= (1 + 0.0275)^2 - 1 \\ E &= (1.0275)^2 - 1 \\ E &= 1.0558 - 1 = 0.0558 \\ E &= 5.58\% \end{aligned}$$

The effective rate of Bank B is greater than that of Bank A by 5.58% − 5.09% = 0.49%. Therefore Bank B offers a better interest rate to its depositors.

EXERCISES FOR SECTION 11.7

Answers involving fractional parts of a percentage should be rounded to the nearest hundredth of a percent.

1. What is effective annual interest rate?

□ 2. What is the effective rate if money is invested at 8% compounded semiannually?

□ 3. What is the effective rate if money is invested at 12% compounded semiannually?

□ 4. What is the effective rate if money is invested at 10% compounded semiannually?

□ 5. What is the effective rate if money is invested at 10% compounded quarterly?

□ 6. What is the effective rate if money is invested at 12% compounded quarterly?

□ 7. What is the effective rate if money is invested at 6% compounded quarterly?

□ 8. What is the effective rate if money is invested at 7.5% compounded quarterly?

□ 9. What is the effective rate if money is invested at 7.75% compounded quarterly?

□ 10. What is the effective rate if money is invested at 5% compounded (a) annually? (b) semiannually? (c) quarterly?

□ 11. What is the effective rate if money is invested at 6% compounded (a) annually? (b) semiannually? (c) quarterly?

□ 12. What is the effective rate if money is invested at 9% compounded (a) annually? (b) semiannually? (c) quarterly?

□ 13. Which is the higher interest rate: 5% compounded quarterly or 5.5% compounded semiannually?

□ 14. Which is the higher interest rate: 6.5% compounded quarterly or 6.8% compounded semiannually?

□ 15. Marlene invested her money at 6.4% compounded quarterly, whereas Pamela invested her money at 6.6% compounded semiannually. Who receives the better interest rate?

□ 16. Bob invested his money at 5.5% compounded quarterly, whereas Larry invested his money at 5.75% compounded semiannually. Who receives the better interest rate?

□ 17. Ruth invested her money at 6.75% compounded quarterly, whereas Julia invested her money at 6.9% compounded semiannually. Who receives the better interest rate?

★□ 18. There are banks that compound interest daily. (a) Compare the effective rate for 6% compounded quarterly with the effective rate for 6% compounded *daily*. Assume a 360-day year. (*Note:* You will need a calculator with an exponent key.) (b) What is the difference in interest paid in 1 year on $1,000,000 between the two methods in part (a)?

JUST FOR FUN Mario and Roy arranged an unusual race. They agreed that the man whose car crossed the finish line first would be the loser. The man whose car crossed the finish line second would be the winner. How should Mario and Roy drive to have such a race?

11.8 LIFE INSURANCE

Almost all consumers purchase insurance at some point in their lifetimes. The types of insurance that a person may obtain include car insurance, health insurance, fire insurance, and life insurance. Many people obtain all of these types of insurance to protect themselves, their homes, and their families.

The basic concept of insurance is the sharing of risks or losses. That is, a person who buys an insurance policy is agreeing to share the risks involved with other people buying a similar policy. A policy is a contract between the insured person, or **policyholder**, and the insurer, or **underwriter.** The fee that an insured person pays to the insurer is called a **premium.** Premiums can be paid monthly, quarterly, semiannually, or annually. In this section we shall consider life insurance. There are three basic types of life insurance policies: (1) *ordinary life* insurance, which is also known as *whole life* or *straight life;* (2) *term;* and (3) *endowment.* An **ordinary life** insurance policy provides protection for the life of the insured person. That is, it offers permanent protection. The insurer will pay the face value of the policy to a designated person, called the **beneficiary**, when the insured person dies, regardless of when this occurs. Premiums for a **whole life** policy can be paid in two different ways. A **straight life** policy is one in which the policyholder pays premiums until death. A **limited-payment life** policy is one in which the policyholder pays premiums for a certain number of years. Whether the policy is a straight life policy or a limited-payment policy, it is in effect for the insured person's lifetime, and the beneficiary receives the face value of the policy whenever the insured person dies. Annual premiums for a limited-payment life policy are higher than those for a

NOTE OF INTEREST

The highest life insurance payout is believed to be over $18,000,000. This was for a man in Oklahoma, who paid more than $300,000 in premiums the year before his death.

straight-life policy. A **term insurance policy** is one that provides protection for a specified term, or number of years. The insurer will pay the face value of the policy to a beneficiary if the insured person dies during the term or period of time stated in the policy. At the end of the term, the policy is no longer in effect and is worthless. A common term is 5 years. For example, if a 20-year-old college student purchased a 5-year term life insurance policy, then the insurance company is liable for payment of the face value of the policy if the insured dies within the 5-year period. Once the term of the policy expires, the insurance company is no longer liable. Usually term insurance can be renewed for another term, but at a higher rate.

An **endowment insurance policy** is similar to a term policy in that it insures the life of the insured for a specified term or number of years. But if the insured is living at the end of the term of the policy, then the face value of the policy will be paid to the policyholder. Hence the use of the word *endowment.* A common term for an endowment policy is 20 years.

Premiums vary for the different types of life insurance policies. Term insurance is the cheapest, followed by straight life, then limited payment. Endowment is the most expensive. Different rates are also applied in each individual case. Typical annual premium rates per $1,000 of life insurance are given in Table 2. A person who is going to buy life insurance should investigate plans and prices thoroughly because different companies offer different prices and options. It should be noted that premiums are usually lower for females than males because women live longer than men. (For the sake of convenience in making computations, we shall assume that the rates are the same for both males and females of the same age.)

TABLE 2
ANNUAL PREMIUM RATE PER $1,000 OF LIFE INSURANCE

Age	5-Year Term	Straight Life	Limited Payment (20 year)	Endowment (20 year)
20	$ 6.99	$15.48	$25.96	$49.32
25	7.57	17.50	28.83	49.73
30	8.60	20.15	32.10	50.17
35	9.98	23.84	35.92	51.13
40	12.44	28.37	40.74	52.69
45	15.59	33.90	46.57	55.33
50	19.89	41.23	53.01	59.00
55	27.09	50.88	59.73	64.55
60	—	63.95	70.00	—

Insurance rates are determined by using data about large samples of the population. The most important pieces of information used are the average death rates for different age groups. It stands to reason that a 20-year-old male should not be charged the same premium as a 65-year-old male, as the 65-year-old will pay fewer premiums. The average death rate of 20-year-old males is about 2 per 1,000, and they can expect to live for 50 years more. The average death rate of 65-year-old males is about 32 per 1,000, and they can expect to live for 13 years more. This kind of information can be found in tables called **mortality tables.** People who help construct these tables, which determine what premiums an insurance company should charge, are called **actuaries.**

EXAMPLE 1 Tom Jones is 20 years old and wishes to purchase $10,000 worth of life insurance. Determine the annual premium of a straight life policy.

SOLUTION Using Table 2, we see that the premium for a 20-year-old person is $15.48 per $1,000 of straight life insurance. The policy is worth $10,000; hence we have

$$10 \times \$15.48 = \$154.80 \quad \text{annual premium}$$

EXAMPLE 2 John Jackson is 40 years old and wishes to purchase $20,000 of life insurance. Determine the annual cost of a straight life policy and of a limited-payment life policy for 20 years.

SOLUTION From the table, the straight life premium per $1,000 for a 40-year-old is $28.37. Therefore, the annual cost on a $20,000 policy is

$$20 \times \$28.37 = \$567.40$$

The limited-payment life (also called 20-payment life) premium per $1,000 for a 40-year-old is $40.74. Therefore, the annual cost on a $20,000 policy is 20 × $40.74 = $814.80.

EXAMPLE 3 If John Jackson (see Example 2) lives for 25 years after he purchases his insurance policy, which policy (straight or limited payment) will cost more? How much more?

SOLUTION If John Jackson lives for 25 years after purchasing a straight life policy, he would pay 25 annual premiums. Hence, the cost is 25 times $567.40, the annual straight life premium.

$$25 \times \$567.40 = \$14,185.00$$

On a 20-year, limited-payment life policy, John Jackson would pay 20 annual premiums. Therefore, the cost is 20 times \$814.80, the annual limited-payment life premium.

$$20 \times \$814.80 = \$16{,}296.00$$

The limited-payment life policy would cost more. It would cost

$$\$16{,}296 - \$14{,}185 = \$2{,}111 \text{ more}$$

Thus far we have discussed only annual premiums. But premiums may also be semiannual, quarterly, or monthly. These rates are a percentage of the annual rate and may be found in Table 3.

TABLE 3
PREMIUM RATES

Payment Period	Percentage of Annual Premium
Semiannual	51%
Quarterly	26%
Monthly	9%

EXAMPLE 4 Susan Simons is 30 years old and wants to purchase a \$15,000 straight life policy. Determine the annual cost if she pays (a) annually, (b) semiannually, (c) quarterly, or (d) monthly.

SOLUTION
a. Using Table 2, we see that the premium for a 30-year-old person is \$20.15 per \$1,000 of straight life insurance. Therefore, the annual premium is 15 × \$20.15 = \$302.25.
b. The semiannual premium is 51% of the annual premium (see Table 3), or 0.051 × \$302.25 = \$154.15. The annual cost is 2 × \$154.15 = \$308.30.
c. The quarterly premium is 26% of the annual premium (see Table 3), or 0.26 × \$302.25 = \$78.59. The annual cost is 4 × \$78.59 = \$314.36.
d. The monthly premium is 9% of the annual premium (see Table 3), or 0.09 × \$302.25 = \$27.20. The annual cost is 12 × \$27.20 = \$326.40.

From the previous example, we can see that the annual cost of a policy increases as the number of payments increases. But many people would prefer to pay a smaller amount more often as opposed to paying a large amount once a year.

EXERCISES FOR SECTION 11.8

Use Tables 2 and 3 for these exercises.

1. Find the annual premium for each of the given insurance policies.

	Face Value of Policy	Age at Issue	Type of Policy
a.	$25,000	25	straight life
b.	20,000	20	20-year limited-payment life
c.	15,000	30	5-year term
d.	50,000	35	20-year endowment
e.	36,000	40	straight life

2. Find the annual premium for each of the given insurance policies.

	Face Value of Policy	Age at Issue	Type of Policy
a.	$40,000	35	5-year term
b.	15,000	25	20-year limited-payment life
c.	50,000	45	straight life
d.	25,000	30	20-year endowment
e.	22,000	55	5-year term

3. Laurie Adams is 25 years old and wishes to purchase a $20,000 5-year term life insurance policy. Determine the annual cost if she pays (a) annually, (b) semiannually, (c) quarterly, or (d) monthly.

4. Carlos Lopez is 30 years old and wishes to purchase a $25,000 limited-payment (20-year) life insurance policy. Determine the annual cost if he pays (a) annually, (b) semiannually, (c) quarterly, or (d) monthly.

5. Irene Gilligan is 35 years old and wishes to purchase a $30,000 straight life insurance policy. Determine the annual cost if she pays (a) annually, (b) semiannually, (c) quarterly, or (d) monthly.

6. Joan Zambriski is 20 years old and wishes to purchase $20,000 worth of life insurance. She decides to purchase a 20-year endowment policy. Determine the annual cost if she pays (a) annually, (b) semiannually, (c) quarterly, or (d) monthly.

7. Paul Thomas is 35 years old and wishes to purchase a $50,000 endowment (20-year) life insurance policy. How much will Paul save over the 20-year period if he pays his premium annually as opposed to monthly?

8. Stan Simon is 20 years old and wishes to purchase $10,000 worth of life insurance. Determine the annual cost of a 5-year term policy. If Stan renews his policy when it expires, what is the difference in total cost between the first and second policies?

9. Jan Peters is 25 years old and wishes to purchase $12,000 worth of life insurance. Determine the annual cost of a 5-year term policy. If Jan renews her policy when it expires, what is the difference in total cost between the first and second policies?

10. Jane Lark is 30 years old and wishes to purchase $20,000 worth of life insurance. Determine the annual cost of a straight life policy and also of a limited-payment life policy for 20 years. If Jane lives for 30 years after she purchases her policy, which policy would cost more? How much more?

11. Carl Ronold is 35 years old and wishes to purchase $25,000 worth of life insurance. Determine the annual cost of a straight life policy and also of a limited-payment life policy for 20 years. If Carl lives for 20 years after he purchases his policy, which policy would cost more? How much more?

12. Rich Nelson obtained a $20,000 limited-payment life policy for 20 years when he was 40 years old. Determine the annual cost of this policy. What is the total cost if Rich Nelson lives to be 65 years old? If he had obtained the same policy when he was 25 years old, what total would he have paid for the insurance?

13. Lois Carson obtained a $20,000 straight life insurance policy at age 40. Determine the annual cost for this policy. If she died 18 years later, how much more did her beneficiary collect than Lois Carson paid in premiums?

14. Tom Tillard is 25 years old and just purchased a $30,000 straight life policy. If he lives to age 65, how much could he have saved by purchasing a 20-year limited-payment life policy with the same face value?

JUST FOR FUN An Amtrak train leaves Miami for Boston traveling at the rate of 80 miles per hour. Two hours later, another train leaves Boston for Miami traveling at the rate of 60 miles per hour. When the two trains meet, which train is closer to Miami?

11.9 INSTALLMENT PLANS AND MORTGAGES

Many consumers purchase items on an installment plan. That is, they obtain possession of an item immediately and agree to pay the purchase price plus an additional charge in a series of regular payments, usually monthly.

If a stereo set that sells for $1,000 can be purchased on an installment plan for $200 down and 12 monthly payments of $70 each, we find the total cost of the set when it is purchased on credit by multiplying the monthly payment times the number of payments to obtain the total amount paid in monthly payments, and then we add the down payment.

$$\$70 \times 12 = \$840 \quad \text{total amount of monthly payments}$$
$$\$840 + \$200 = \$1040 \quad \text{total cost}$$

To find the amount of interest paid (also called *service charge* or *finance charge*), we subtract the cash price from the total installment cost.

$$\$1040 - \$1000 = \$40 \quad \text{interest}$$

This amount does not reflect the true interest rate. But because the actual dollar amount involved is relatively small, consumers are more than willing to pay it.

Consider the case of Sam Larson, who borrows $1,200 from the Friendly Finance Company. The finance company advertised an interest rate of 10% simple interest. Sam wanted to borrow the money for 6 months and repay the loan in six monthly payments.

Hence we have

$$I = Prt$$
$$I = \$1{,}200 \times 0.10 \times \frac{6}{12}$$
$$= \$60$$

The total amount that Sam has to repay is \$1,200 + \$60, or \$1,260. If he is going to repay the loan in six monthly payments, he will pay \$210 each month.

It should be noted that the Friendly Finance Company computed the interest as though Sam Larson owed the \$1,200 for the entire 6 months. But he did not! Sam owed \$1,200 for only one month. At the end of the first month, he paid the finance company \$210. Of this amount, \$200 was applied toward the principal (the other \$10 is interest). Because each payment includes \$200 of the principal, the amount that Sam actually owes decreases by \$200 each month. The following list shows what Sam owes each month:

\$1,200		(original amount, owed for first month only)
\$1,200 − \$200 =	\$1,000	(amount owed for second month only)
\$1,000 − \$200 =	\$800	(amount owed for third month only)
\$800 − \$200 =	\$600	(amount owed for fourth month only)
\$600 − \$200 =	\$400	(amount owed for fifth month only)
\$400 − \$200 =	\$200	(amount owed for sixth month only)
Total	\$4,200	

Thus, Sam's debt to the finance company is equivalent to owing \$4,200 for one month only. The interest that Sam paid was \$60. Using the formula $I = Prt$, we can find the value of r.

$$\$60 = \$4{,}200 \times r \times \frac{1}{12}$$
$$60 = 350 \times r$$

Solving this equation for r (correct to the nearest thousandth),

$$r = 0.171$$
$$= 17.1\% \quad \text{(to the nearest tenth of a percent)}$$

Therefore, Sam paid about 17.1% interest on his loan, not 10% as advertised.

In 1969, Congress passed a Truth-in-Lending Act that requires all sellers to reveal the **true annual interest rate** that they charge and the total finance charge, which includes such additional fees as service charges. The true annual interest rate is also known as the *annual percentage rate* and is often quoted as the **APR.**

The Truth-in-Lending Act does not establish any maximums on interest rates or finance charges. It does not regulate interest charges, but it does make consumers aware of the cost of credit so that the consumer can compare terms.

There is a formula that can be used to estimate the true annual interest rate on installment loans. It is

$$i = \frac{2nr}{n + 1}$$

where i = the true interest
n = number of payments
r = the stated rate of interest

Applying this formula to the previous example, where $n = 6$ and $r = 10\% = 0.10$, we have

$$\begin{aligned} i &= \frac{2nr}{n + 1} \\ &= \frac{2 \times 6 \times (0.10)}{6 + 1} = \frac{12(0.10)}{7} \\ &= \frac{1.2}{7} \\ &= 0.171 \quad \text{(to the nearest thousandth)} \\ &= 17.1\% \quad \text{(to the nearest tenth of a percent)} \end{aligned}$$

EXAMPLE 1 Stella Frisco purchased a stereo set advertised for \$1,000. She bought the set on the installment plan, paying \$200 down and agreeing to pay the balance in 12 monthly payments. The finance charge was 10% simple interest on the balance. What was the true annual interest rate?

SOLUTION Using the formula $i = 2nr/(n + 1)$, we have $n = 12$, $r = 10\% = 0.10$.

Therefore,

$$i = \frac{2nr}{n+1}$$

$$= \frac{2 \times 12 \times 0.10}{12 + 1} = \frac{24(0.10)}{13}$$

$$= \frac{2.4}{13} = 0.185 \quad \text{(to the nearest thousandth)}$$

$$= 18.5\% \quad \text{(to the nearest tenth of a percent)}$$

Most charge accounts and credit cards (MasterCard, Visa, etc.) use an interest rate of 1.5% or 1.75% per month. Note that 1.5% per month is 18% per year and 1.75% per month is 21% per year!

Many bank credit cards and store charge accounts charge no interest if the balance is paid by the due date (usually 25 to 30 days). There exist a variety of methods for computing interest charges on amounts owed past the due date. One of the methods used is to charge an interest rate of 1.5% per month on the unpaid balance. For example, if a person purchases an item that costs \$100 (total amount due) and charges it, then at the end of the month (or billing period) he has the choice of paying the full \$100 or a portion of the \$100. If he chooses to pay a portion of the \$100 (say \$40), then the next monthly statement would be for \$60 plus 1.5% of \$60, the unpaid balance, or \$0.90. Thus, the next monthly statement would be for \$60.90 (assuming no other charges). If another payment is made (say \$25), then the following monthly statement would be for \$60.90 − \$25.00 = \$35.90, plus 1.5% of \$35.90 or \$0.54. Thus, the statement would be for \$36.44, and a payment for this amount would clear the account. (*Note:* A store will typically add a minimum finance charge of \$0.50 if the interest due is less than \$0.50, but for our purposes we shall use the exact amount due.)

EXAMPLE 2 Fred Worth purchased a chair for \$300 and charged it. The store uses an interest rate of 1.75% per month on the unpaid balance. If Fred decides to make payments of \$100 per month, how much interest does he pay?

SOLUTION The first bill will be for \$300 (assuming no other charges).

1. Fred makes a payment of \$100. The balance is \$200. The interest is 1.75% of \$200, or \$3.50. Hence the next bill is for \$203.50.

2. Fred makes another payment of $100. The balance is $103.50. The interest is 1.75% of $103.50, or $1.81. Therefore the next bill is for $105.31
3. Fred makes another payment of $100. The balance is $5.31. The interest is 1.75% of $5.31, or $0.09. Hence the next bill is for $5.40.
4. Fred makes a payment of $5.40. Fred has paid a total of $305.40, so he has paid $305.40 − $300 = $5.40 in interest.

The largest purchase that a person is likely to make in a lifetime is buying a house. Typically, most people purchase houses by making a specified down payment and borrowing the balance from a bank or other lending institution. The borrower agrees to make regular payments on the principal and interest until the loan is paid off. This process is called **amortizing** the loan. Mortgages on new homes are usually for a period of 20, 25, or 30 years. Mortgages on older homes sometimes run for shorter periods of time.

The monthly payments necessary to amortize a loan (pay off the principal and interest) are compiled in tables called **amortization tables.** Table 4 lists typical monthly payments per $1,000 of a mortgage. The amount that must be paid monthly for principal and interest is given for selected interest rates and different periods of time. (*Note:* It is predicted that the average price of a new home will increase to $110,000 by 1992.)

HISTORICAL NOTE

According to historical records, the first bank was founded in 1148. It was the Bank of San Giorgio in Genoa, Italy. Before this time, the lending of money was a personal matter, but the Romans did make some attempt to control money lending by passing certain regulations.

The first bank in the United States was founded in 1781 in Philadelphia. It was the Bank of North America, and it stayed in business only until 1784. Later, in 1791, Congress issued a charter for another bank in Philadelphia called the First Bank of the United States. After that, state governments began to issue charters to banks. The National Banking System was then established by the federal government, and it attempted to guide and regulate banking in the United States.

On December 23, 1913, Congress established the Federal Reserve System, the central banking system of the United States, to improve supervision of banking. Today, the primary function of the system is to promote a flow of credit and money that will determine orderly economic growth, a stable dollar, and a balance of international payments.

TABLE 4

MONTHLY MORTGAGE PAYMENT PER $1,000

	Length of Mortgage		
Rate (%)	10 Years	20 Years	30 Years
8	$12.13	$ 8.36	$ 7.34
9	12.67	9.00	8.05
10	13.22	9.65	8.78
$10\frac{1}{2}$	13.49	9.98	9.15
12	14.35	11.01	10.29
$12\frac{1}{2}$	14.64	11.36	10.67
14	15.53	12.44	11.85
15	16.13	13.17	12.64
16	16.75	13.91	13.45
18	18.02	15.43	15.07
20	19.33	16.99	16.71

Note: This table lists payments per $1,000 for "traditional" fixed-rate mortgages. Some lending institutions are now offering another type of mortgage: variable-rate mortgages, where the interest rate may be raised or lowered depending on the economic climate.

If a person assumes a \$30,000 mortgage for 20 years at $12\frac{1}{2}\%$, then according to Table 4, the monthly payment for each \$1,000 is \$11.36. The mortgage is for \$30,000, so we multiply \$11.36 by 30 to find the total monthly payment. Hence, the monthly payment on this mortgage is 30 × \$11.36 or \$350.10. It should be noted that the payments are higher than shown because the mortgagee usually has to make monthly payments on property taxes and insurance. For example, if property taxes were \$1,200 a year, then an additional \$100 (\$1,200 ÷ 12) would have to be paid each month. Also, life and fire insurance might cost \$20 per month. Therefore, the total monthly payment would be \$350.10 + \$100 + \$20 = \$470.10.

EXAMPLE 3 The Carsons assumed a \$40,000 mortgage for 30 years at 12%. What is their monthly payment for principal and interest?

SOLUTION The monthly payment per \$1,000 at 12% for a 30-year mortgage is \$10.29. The mortgage is for \$40,000. Hence, we multiply \$10.29 by 40.

$$40 \times \$10.29 = \$411.60 \quad \text{(monthly payment)}$$

EXAMPLE 4 How much total interest will the Carsons (Example 3) pay on their mortgage?

SOLUTION Their monthly payment is \$411.60 and the mortgage is for 30 years. Hence there will be 30 × 12, or 360, payments.

$$360 \times \$411.60 = \$148{,}176 \quad \text{(total payment)}$$
$$\$148{,}176 - \$40{,}000 = \$108{,}176 \quad \text{(total interest)}$$

EXAMPLE 5 How much interest would the Carsons save if they assume the same \$40,000 mortgage at 12% for 20 years instead of 30 years?

SOLUTION The monthly payment per \$1,000 at 12% for a 20-year mortgage is \$11.01. The mortgage is for \$40,000. Therefore,

$$\text{Monthly payment} = 40 \times \$11.01 = \$440.40$$

The mortgage is for 20 years. Hence, there will be 12 × 20, or 240 payments

$$20 \times \$440.40 = \$105{,}696 \quad \text{(total payment)}$$
$$\$105{,}696 - \$40{,}000 = \$\ 65{,}696 \quad \text{(total interest)}$$
$$\text{Interest on 30-year mortgage} = \$108{,}176$$
$$\text{Interest on 20-year mortgage} = \underline{\$\ 65{,}696} \quad \text{(subtracting)}$$
$$\$\ 42{,}480 \quad \text{(interest saved)}$$

Note that in both cases in Example 5, the interest is more than the face value of the loan. This is the reason banks lend money—to make money! Homebuyers should be aware of the costs involved in assuming a mortgage. A disadvantage of a mortgage is the total amount of interest to be paid on such a loan. But people are willing to do this for a variety of reasons. Advantages to consider include: (1) the satisfaction of owning a home; (2) the value of a home is sure to increase, usually faster than the rate of inflation (a house that sells for $60,000 this year would sell for $66,000 next year with a 10% annual inflation rate); (3) the loan will be repaid with cheaper dollars because of inflation; and (4) owning a home provides protection against inflation. Another thing to consider is that the annual interest paid on a mortgage loan may be used as an itemized deduction on an income tax return for that year, which results in an income tax saving.

EXERCISES FOR SECTION 11.9

1. Carl Thomas purchased a television set advertised for $500. He bought the television on the installment plan, paying $100 down and agreeing to pay the balance in 12 monthly payments. The store charged a finance charge of 12% simple interest on the balance. (a) What was the amount of each payment? (b) What was the true annual interest rate (correct to the nearest tenth of a percent)?

2. A used car is advertised for $3,000. It may be purchased on the installment plan by paying $400 down and agreeing to pay the balance plus 12% simple interest on the balance in 24 monthly payments. (a) What was the amount of each payment? (b) What was the true annual interest rate (correct to the nearest tenth of a percent)?

3. An easy chair is advertised for $450. It may be purchased on the installment plan by paying $100 down and agreeing to pay the balance plus 18% simple interest in six monthly payments. (a) What is the amount of each payment? (b) What is the true annual interest rate (correct to the nearest tenth of a percent)?

4. The Jaxsons purchased a new couch for $500. They put no money down and agreed to pay for it in 24 monthly payments. The store stated a finance charge of 12% simple interest. (a) What was the amount of each payment? (b) What was the true annual interest rate (correct to the nearest tenth of a percent)? (c) How much more will the Jaxsons pay for the couch by buying it on the installment plan?

5. Sandra West purchased a coat for $100 and charged it. The store charges an interest rate of $1\frac{1}{2}$% per month on the unpaid balance. If Sandra decides to make payments of $30 per month, how much interest will she pay by the time the coat is paid for? (Assume that there are no additional purchases and that the first payment is made before the end of the month, so no interest is charged on the first payment.)

6. Bill North received his statement from Stone's Department Store. The amount due is $220. If Bill decides to make payments of $50 per month and the interest rate is $1\frac{1}{2}$% per month on the unpaid balance, then how many months will it be before the account is paid in full? What is the total amount to be paid in interest? (Assume that there are no additional purchases and that the first payment is made before the end of the month, so no interest is charged on the first payment.)

7. Julia Swanson used her credit card to charge purchases totaling $200 at the Super Discount House. If Julia decides to make payments of $50 per month and the interest rate is 1.75% per

month on the unpaid balance, then how many months will it be before the account is paid in full? What is the total amount to be paid in interest? (Assume that there are no additional purchases and that the first payment is made before the end of the month, so no interest is charged on the first payment.)

8. Nancy Anderson purchased a television set for $400 and charged it. The store charges an interest rate of 1.75% per month on the unpaid balance. If Nancy decides to make payments of $100 per month, then how many months will it be before the account is paid in full? What is the total amount to be paid in interest? (Assume that there are no additional purchases and that the first payment is made before the end of the month, so no interest is charged on the first payment.)

For Exercises 9–14, find the monthly payment for principal and interest for each mortgage.

	Amount of Mortgage	Interest Rate (%)	Term of Mortgage (years)
9.	$20,000	9	10
10.	30,000	$10\frac{1}{2}$	30
11.	25,000	12	20
12.	45,000	10	20
13.	55,000	$12\frac{1}{2}$	30
14.	22,000	15	10

15. The Smiths assumed a $40,000 mortgage for 20 years at $10\frac{1}{2}\%$. What is their monthly payment? How much total interest will the Smiths pay on their mortgage?

16. The Garcias assumed a $30,000 mortgage for 30 years at 12%. What is their monthly payment? How much total interest will the Garcias pay on their mortgage?

17. The Donovans need to borrow $30,000 to buy a house. Bank A will give them a 30-year mortgage at $10\frac{1}{2}\%$. Bank B will give them a 20-year mortgage at 12%. Which mortgage should the Donovans assume so that they will pay the smallest amount of interest? How much will they save?

18. The Sullivans need to borrow $40,000 to buy a house. Bank A will give them a 30-year mortgage at $12\frac{1}{2}\%$. Bank B will give them a 20-year mortgage at 15%. Which mortgage should the Sullivans assume so that they will pay the smallest amount of interest? How much will they save?

19. The Seteks purchased a condominium for $100,000. They made a down payment of 20% and borrowed the remainder. If they assumed a 12% mortgage for 30 years, what is their monthly payment? How much total interest will the Seteks pay on their mortgage?

20. The Nennos purchased a house for $150,000. They made a down payment of 25% and borrowed the remainder. If they assumed a 14% mortgage for 20 years, what is their monthly payment? How much total interest will the Nennos pay on their mortgage?

JUST FOR FUN Whose picture is on the front of a $10 bill? What is pictured on the back of it?

Summary

The *ratio* of two quantities a and b is the quotient or indicated quotient obtained by dividing a by b. The ratio of a to b is written as $a \div b$, a/b, or $a:b$. A *proportion* is defined to be an equality of two ratios. In any proportion, the product of the means equals the product of the extremes. That is,

$$\text{if } \frac{a}{b} = \frac{c}{d}, \quad \text{then} \quad ad = bc.$$

A *percent* is a ratio with a denominator of 100. We can also say that a percent is the ratio of any number to 100. The symbol for percent is %. For example, 6% means 6 : 100, or $\frac{6}{100}$. We can also express a percent as a decimal; that is,

$$6\% = \frac{6}{100} = 0.06$$

To convert a percent to a decimal, drop the percent sign and move the decimal point two places to the left. To change a decimal to a percent, move the decimal point two places to the right and add a percent sign. To change a fraction to a percent, express the fraction as a decimal, and then change the decimal to a percent.

Markup is the difference between the selling price and the cost of an item. *Markdown* is the difference between the original selling price and the sale price. Both of these terms can be expressed in terms of a dollar amount or in terms of a percentage.

Simple interest is found by means of the formula

$$I = Prt$$

where I = interest, P = principal, r = rate of interest, and t = time in years. *Compound interest* occurs when interest due at the end of a certain period of time is added to the principal, and both the principal and the interest from the first period earn interest for the next period. Table 1 in Section 11.6 shows the amount that $1 will accumulate when interest is paid at the indicated rate for the indicated number of interest periods.

Effective annual interest rate is the annual interest rate that gives the same yield as a nominal interest rate compounded several times a year. We can determine the effective annual rate by means of the formula

$$E = (1 + r)^n - 1$$

where E = effective rate, n = number of payment periods per year, and r = interest rate per period.

There are three basic kinds of life insurance policies: *whole life, term,* and *endowment.* A *whole life insurance policy* is one in which the insurance company will pay the face value of the policy to the beneficiary when the insured person dies, regardless of when this happens. Premiums for a whole life policy can be paid in two different ways. A *straight life* policy is one in which the policyholder pays premiums until death. A *limited-payment life* policy is one in which the policyholder pays premiums for a certain number of years. A *term insurance policy* is one where the insurance company will pay the face value of the policy to a beneficiary when the insured person dies, providing this occurs during the period of time (the *term*) stated in the policy. An *endowment policy* is similar to a term policy in that it insures the life of the insured for a specific term. However, if the insured is living at the end of the term, then the face value of the policy will be paid to the policyholder. Typical annual premium rates per $1,000 of life insurance for various policies and different age groups are listed in Table 2, Section 11.8.

Installment buying allows a consumer to obtain possession of an item immediately by agreeing to pay the purchase price plus an additional charge in a series of regular payments. When a person buys an article on the installment plan, he or she should know the *true annual interest rate* being charged. This is also known as the *annual percentage rate,* or APR. A formula that can be used to determine the true annual interest rate is

$$i = \frac{2nr}{n + 1}$$

where i = the true interest rate, n = number of payments, and r = the stated interest rate.

A *home mortgage* is probably the largest loan that most people assume. The monthly payments necessary to amortize a loan are compiled in tables called *amortization tables.* Table 4, Section 11.9, lists typical monthly payments per $1,000 of mortgage for various interest rates and periods of time.

Vocabulary Check

ratio	compound amount	proportion	effective (annual) rate	percent	mortgage
markup	premium	markdown	APR	simple interest	underwriter
decimal		fraction		compound interest	principal

Review Exercises for Chapter 11

1. Express each ratio in simplest form.
   a. $48:120$ b. $\frac{54}{36}$ c. 10 to 25 d. $1\frac{1}{2}:2\frac{1}{4}$

2. Do the ratios form a proportion? Write yes or no.
   a. $\frac{8}{6}:\frac{3}{4}$ b. $\frac{16}{28}:\frac{4}{7}$ c. $\frac{13}{17}:\frac{78}{108}$ d. $\frac{0.3}{0.9}:\frac{0.9}{2.7}$

3. Find the value of x in each proportion.
   a. $2:x = 5:10$ b. $x:6 = 5:10$
   c. $4:3 = x:6$ d. $2:8 = 6:x$

4. Express each percent as a fraction in simplest form.
   a. 38% b. $12\frac{1}{2}\%$ c. 2.3% d. 125%

5. Express each percent as a decimal.
   a. 48% b. $10\frac{1}{2}\%$ c. 4.7% d. 0.25%

6. Express each decimal as a percent.
   a. 0.82 b. 1.3 c. 0.78 d. 2.134

7. Express each fraction as a percent.
   a. $\frac{3}{5}$ b. $\frac{1}{8}$ c. $2\frac{1}{5}$ d. $\frac{3}{16}$

8. During a certain television program, pollsters found out that 80 out of 200 people surveyed were watching the program. Predict the number of viewers out of 50,000 people.

9. Twenty-three percent of a certain group of mathematics students had previously studied algebra. If 46 students in this group had studied algebra, how many students are in the group?

10. A suit sells for $125. The markup is 30% of the cost. What is the cost of the suit, correct to the nearest cent?

11. A portable television costs a retailer $60. If the markup is 25% of the selling price, find the selling price, correct to the nearest cent.

12. Find the markup and the cost, correct to the nearest cent, of a new couch that retails for $400, if the percent markup on the selling price is 40%.

13. A clothing store sells topcoats for $124.99. During a clearance sale the price is reduced by 25%. What is the dollar markdown and what is the sale price, correct to the nearest cent?

14. Find the amount of simple interest on a $5,000 loan for 3 years at 11%.

15. Find the amount of simple interest on a $10,000 loan for 9 months at 12%.

16. How much interest will $10,000 earn in 4 years at 10% simple interest? How much in 4 months?

17. Use Table 1 to find the compound amount and the compound interest on $4,000 invested at 16% compounded semiannually for 10 years.

18. Use Table 1 to find the compound amount and the compound interest on $5,000 invested at 12% compounded quarterly for 6 years.

19. What is the effective rate if money is invested at 5% compounded (a) annually, (b) semiannually, (c) quarterly?

20. Use Table 2 to find the annual premium for each type of insurance policy listed.

	Face Value of Policy	Age at Issue	Type of Policy
a.	$20,000	30	5-year term
b.	$15,000	25	straight life
c.	$30,000	35	20-year limited-payment life
d.	$40,000	40	20-year endowment

21. Lewis Scott is 25 years old and wants to purchase a $30,000 straight-life policy. How much will Lewis save if he pays his premium annually instead of monthly? (Use Tables 2 and 3).

22. What is the APR?

23. Find the monthly payment for the principal and interest for each of the following mortgages (use Table 4).

	Amount of Mortgage	Interest Rate (%)	Term of Mortgage (years)
a.	$30,000	9	10
b.	45,000	$10\frac{1}{2}$	30
c.	60,000	12	30
d.	55,000	15	20

24. Julie Thomas purchased a tape recorder for $125 and charged it. The store charges an interest rate of 1.75% per month on any unpaid balance. If Julie decides to make payments of $40 per month, what will she pay for the tape recorder? (Assume that there are no other purchases and that the first payment is made before the end of the month, so no interest is charged on the first payment.)

25. A video-cassette recorder is advertised for $600. It may be purchased on the installment plan by paying $50 down and agreeing to pay the balance plus 12% simple interest on the balance in 24 monthly payments. (a) What is the finance charge? (b) What is the amount of each payment? (c) What is the total cost of the video-cassette recorder? (d) What is the true annual interest rate (correct to the nearest tenth of a percent)?

26. The Stones assumed a $50,000 mortgage for 30 years at $12\frac{1}{2}$%. Use Table 4 to find their monthly payment. How much total interest will the Stones pay on their mortgage?

JUST FOR FUN True or false: Arabic numerals were invented by the Arabs.

Chapter Quiz

Determine whether the following statements are true or false.

1. Given the following proportion, $4:x = 6:18$, then $x = 12$.

2. If stereo tapes are on sale, three for $12.99, then two stereo tapes should cost $8.68.

3. Given that 1 centimeter = 0.3937 inches, the 9 is in the tenths position.

4. To write a decimal as a percent, move the decimal point two places to the right and add a percent sign.

5. If a jacket that normally sells for $49.95 is marked down 20%, and all transactions are subject to a 7% sales tax, then the final amount due on the jacket is $43.46.

6. If a television sells for $320 and the markup is 60% of the cost, then the cost is $200.

7. The simple interest on a $10,000 loan for 6 months at 12% is $60.00.

8. The compound interest on $5,000 invested at 12% compounded quarterly for 6 years is $5,165.

9. If money is invested at 12% compounded semiannually, then it will double itself in approximately 6 years.

10. An effective annual interest rate is the annual interest rate that gives the same yield as a nominal interest rate compounded several times a year.

11. If money is invested at 5% compounded quarterly, then the effective rate is 5.5%.

12. Straight-life insurance provides coverage for the life of the insured person and he or she must pay premiums for life.

13. James Sparks is 25 years old and wishes to purchase $10,000 worth of life insurance. The annual cost of a 5-year term policy is $175.

14. The true annual interest rate is the same as the effective annual interest rate.

15. If Carlos purchases a car and the dealer states that he can pay for it in 36 monthly payments with an interest rate of 12% simple interest, then the true annual interest rate is 12%.

16. Connie Dawkins purchased a coat for $100 and charged it. The store uses an interest rate of 1.75% per month on the unpaid balance. If Connie decides to make payments of $20 per month, then she must make five payments.

17. The Dohlins assumed a $40,000 mortgage for 20 years at $12\frac{1}{2}$%. Their monthly payment is $454.40.

18. If you assume a mortgage at $12\frac{1}{2}$% per year for 30 years, you will pay less interest than if you assume a mortgage at 15% per year for 20 years.

19. If you assume a $50,000 mortgage at 14% for 30 years, then you will pay approximately $100,000 in interest during the life of the loan.

20. If money is invested at 10% compounded annually, then the effective rate of interest is 10.5%.

21. The compound interest on $10,000 invested at 16% compounded quarterly for 7 years is $20,000.

22. If money is invested at 7% compounded semiannually, then the effective rate is 7.21%.

23. Scott is 30 years old and purchases a $10,000 straight life policy. He decides to pay his premium monthly. Hence, his annual cost is $167.16.

24. Mary purchased a camera for $200. She paid $50 down and agreed to pay the balance plus 12% simple interest on the balance in 24 monthly payments. Her monthly payments were $7.75.

25. If a certain soda cost $.75 per can in 1988, and we assume an annual inflation rate of 8%, the can of soda will cost $1.89 in the year 2000.

CHAPTER 12

AN INTRODUCTION TO COMPUTERS

AFTER STUDYING THIS CHAPTER YOU WILL BE ABLE TO DO THE FOLLOWING:

1. Identify those inventions and people that played a significant role in the development of the computer.
2. Multiply numbers using the technique of **Napier's rods.**
3. Identify the distinguishing characteristics of computers that belong to a particular **generation.**
4. Identify the main **components** of a computer system, and describe the function of each.
5. Name at least four different **programming languages** and the areas of study to which they are most adaptable.
6. Construct a **flowchart** for a particular problem.
7. Express common arithmetic and algebraic expressions in **BASIC notation.**
8. Evaluate given BASIC **expressions.**
9. Write a simple BASIC **program.**
10. Write a BASIC program that contains a **loop.**

12.1 INTRODUCTION

What is a computer? A computer is an electronic machine that is capable of solving problems by accepting data (called **input**), performing prescribed operations on the data (called **processing**), and supplying the results of these operations (called **output**). The processing of data is governed by a **computer program,** that is, a set of instructions that tells the computer how to process the data or what to do with it. We say that a computer **executes** or **runs** the program when the computer carries out these instructions. Computer programs are written using a special **computer language.** One such language is BASIC, and we will study it later in this chapter.

Essentially, computers have only three primary functions. They can perform arithmetic calculations, compare data values (test the relationship between data), and store and retrieve data. But it is the speed at which computers can perform these functions that, in part, make them so powerful. Today, computers can perform millions of calculations in one second. In fact some computers can perform more operations in one hour than we could perform in a lifetime using pencil and paper.

Some other features of computers that make them both useful and powerful are: versatility, accuracy, and memory. Computers can be programmed to execute many different sets of instructions and therefore are capable of performing a variety of tasks. Computers are also capable of storing millions of characters in their memories. This in itself provides a capability that would be humanly impossible.

Computers produce accurate results. But this does not imply that what comes out of a computer is correct! It is possible for computers to produce incorrect results. This can occur in any one of three ways. First, a computer is controlled by a program and it will only do what the program instructs it to do. Hence, if a program has been written in a manner that will yield incorrect results, the computer executing the program will simply process the data and produce these incorrect results. The computer did not make the mistake, the person who wrote the program did.

Bad input may also produce incorrect results. That is, the data given to the computer may be incorrect initially. Hence, the computer cannot be expected to process it correctly. In computerese, this is commonly referred to as GIGO (garbage in, garbage out).

Finally, a third way in which faulty results may be obtained is via some type of electrical disturbance, such as a power surge, or changes in temperature or humidity. A computer is an electrical device that can malfunction and produce errors as a result of these electrical disturbances.

NOTE OF INTEREST

90000

9 780024 092014

Computers are used for inventory and price control. The Universal Product Code was adopted by the supermarket industry in 1973. It is a system that identifies each item sold in the stores with a unique 12-digit code. A scanner reads the bar code and the computer registers the price and notes that the item has been purchased. The code for a book such as this gives country and publisher (the first six digits) and the identifying number for the particular title.

Computers and related technology have had (and continue to have) a tangible effect on many aspects of our daily lives. For example, in most major grocery and chain stores, the price of an item is no longer "rung up" at the checkout counter. Instead, the product's product code is entered into a computerized cash register, usually by means of a scanning device or a keyboard. The computerized cash register is called a point of sale (POS) terminal. In the banking industry, customers now have 24-hour access to their accounts as they use automated teller machines (ATM). Bank customers can make cash withdrawals from savings, checking, or credit card accounts. They may also make payments on various types of bank loans. Certain retail businesses now use special ATM/POS terminals that provide an immediate electronic transfer of funds from a customer's bank account to a merchant's bank account when a sale occurs. Thus, there is no need for a hand-to-hand exchange of funds.

Today there are dozens of communication satellites in orbit handling data, voice, and video communications. For example, cable television uses a number of communication satellites to bring distant and special programs into local areas. Telecommunications, the combining of a communications facility (such as a telephone) with computers, is providing us with a host of electronic information services. By using the proper equipment, it is possible to shop, bank, or even work from home. A modem is needed to transmit data from a computer through a telephone line. The modem converts the computer code into a suitable form for transmission and, after it is received at the destination point, converts it back for the receiving computer.

Following are some other areas where computers are used and have an impact in our daily lives. Computers can act as teaching machines for *computer-assisted instruction* in all subjects and at any grade level. In industry computers are used in a variety of ways: quality control, inventory, simulation, and assembly line production. They are also used to control robots. Robots can perform repetitive, precise, and dangerous jobs. One example where robots are used is the automotive industry. Computers are also used in health-related fields. They are commonly used for medical record keeping and to furnish a tentative diagnosis based on a patient's test results. Computers are used in microscopic surgery, ultrasound analysis, and nuclear medicine. The *computed axial tomography* (CAT) scan is a recent development in medicine that has helped save many lives. The CAT scan allows doctors to obtain information about internal organs that previously could only be obtained through surgery.

It should also be noted that with the advent of computer networks and on-line information utilities, many people now receive news reports, entertainment information, and mail via a computer. Also, experimentation with home banking, electronic funds transfer, and shopping services is taking place in various parts of the country.

Computers certainly have become an integral part of our society, and consequently crimes are being committed using computers. Computer-related crimes usually involve large sums of money. For example, the average amount stolen in each computer bank fraud is estimated at $620,000, whereas the conventional "bank robber" averages $9,000 per theft. Basically, there are three types of computer-related crime: embezzlement, the use of computers for personal gain, and destruction of computer equipment. All of these mean that lawyers, lawmakers, and courts are being faced with new and different challenges. These examples should provide an idea of the tremendous impact computers and computer technology have had (and are having) on our daily lives and on all of society.

As computers become more a part of our everyday life, it is our responsibility to understand them so that we can think and speak intelligently about them and form realistic opinions and attitudes regarding the role that they play in our daily living.

12.2 HISTORY OF COMPUTERS

Ever since humans learned to count, they have been searching for ways to perform calculations more easily. Prior to the development of writing materials, such as papyrus and paper, people used pointed sticks to make calculations on tables covered with sand or dust. Later, lines were etched on the tables, and loose counters were put on the lines to record numbers. This technique was refined to the extent that grooves were cut in a board and small disks representing numbers were moved along the grooves. Counting devices similar to this were commonly used in Europe until the early 1600s. The earliest recorded mechanical counting device is the *abacus*. The Chinese developed the abacus sometime during the twelfth century. The abacus consists of movable beads on rods; skilled operators can add, subtract, multiply, and divide on them quickly and efficiently. But the abacus requires a great deal of skill to operate quickly. For example, when adding 9 + 8 on the abacus, 7 is obtained in the 1s column, but the 1 must be carried to the 10s column by hand, that is, by the person doing the computation. Thus, many people saw the need for, and tried to develop, more sophisticated calculating devices.

Trustees of the Science Museum, London
An abacus

In the early 1600s John Napier, a Scottish mathematician (1550–1617), developed a calculating device called **Napier's rods**

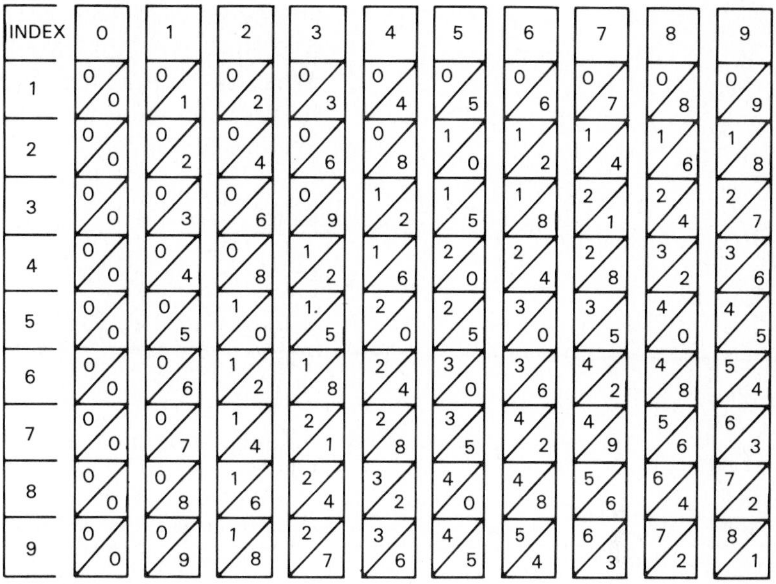

FIGURE 1
Napier's bones

or *Napier's bones,* which were used to perform multiplication. This device was supposedly made of lengths of bones. The numerals 0 through 9 were inscribed at the top of each "bone" (rod) and multiples of the numerals were listed below it in the manner shown in Figure 1. To multiply 34 by 5 in a manner similar to how the "bones" work, first note that

$$\begin{array}{r} 34 \\ \times\ \ 5 \\ \hline 20 \\ 15\ \ \\ \hline 170 \end{array}$$

Now consider the row of numerals alongside the "index" 5 and below the 3 and 4 columns. Hence we have

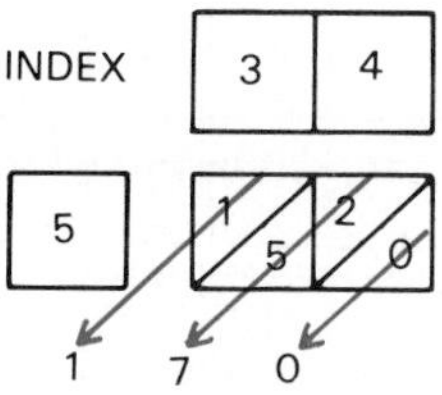

NOTE OF INTEREST

In 1614, John Napier invented logarithms. The word *logarithm* means "ratio number." It was derived from the Greek words *logos,* which means "ratio," and *arithmos,* which means "number." Napier first published a table of logarithms for trigonometric functions. Later, Henry Briggs (1561–1631), an English mathematician, visited Napier to discuss the logarithms. As a result of this visit, Briggs introduced what are now known as common logarithms. The combined work of these two men led to the invention of the slide rule in England by Edmund Gunter in 1620. A slide rule can be used to find products, quotients, powers, and roots using the basic laws of logarithms. Before the advent of today's pocket calculators, a slide rule was a common calculating device for thousands of college students studying mathematics, science, and engineering.

Napier announced his discovery of logarithms in a book titled *Mirifici Logarithmorum Canonis Description (A Description of the Admirable Table of Logarithms).* In a prefatory paragraph, he wrote

> Seeing there is nothing, right well-beloved students of mathematics, that is so troublesome to mathematical practice, nor doth more molest and hinder calculators, than multiplication, division, square and cubical extractions of great numbers, which besides the tedious expense of time are for the most part subject to many slippery errors, I began therefore to consider in my mind by what certain and ready art I might remove those hinderances.

If we add along the paths or diagonals, we obtain 170. Therefore, 5 × 34 = 170 using Napier's rods. Let's try another example. What is 9 × 76? First, we find 9 on the "index" and examine the numerals in the row alongside it and below the 7 and 6 columns. We find

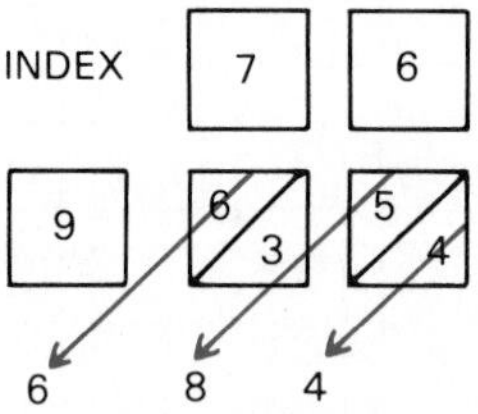

Adding along the diagonals, we obtain 684. Therefore, 9 × 76 = 684 by means of Napier's rods.

EXAMPLE 1 By means of Napier's rods, find the product of 34 × 578.

SOLUTION We are only able to find one-digit products from Napier's rods. Hence, we must perform the calculations in the following manner:

$$\begin{array}{r} 578 \\ \underline{\times \ \ 34} \end{array} \quad \text{is the same as} \quad 4 \times 578 + 30 \times 578$$

For 4 × 578 we have

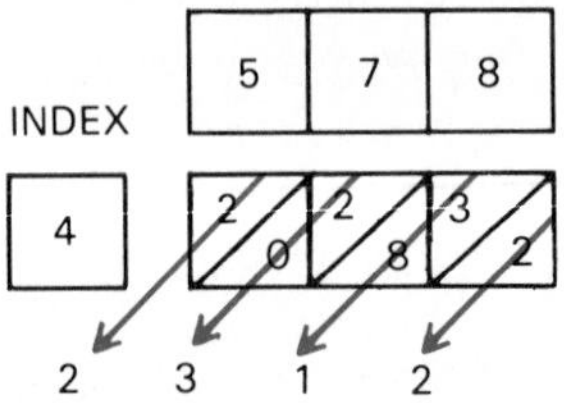

which equals 2,312.

For 30 × 578 we have

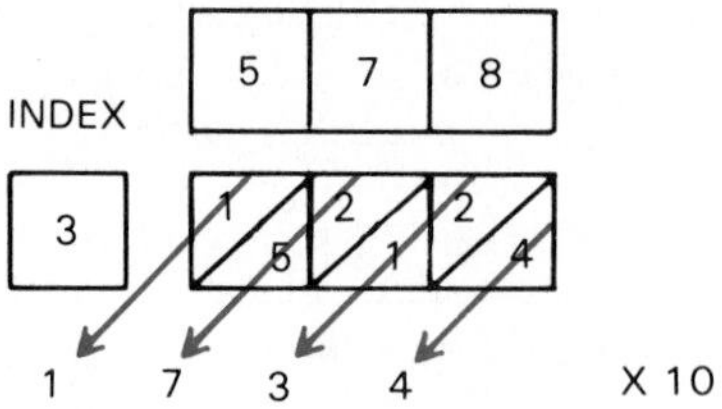

which equals 17,340.

Hence

$$\begin{array}{r} 4 \times 578 = \ \ 2{,}312 \\ \underline{30 \times 578} = \underline{17{,}340} \\ 34 \times 578 = 19{,}652 \end{array}$$

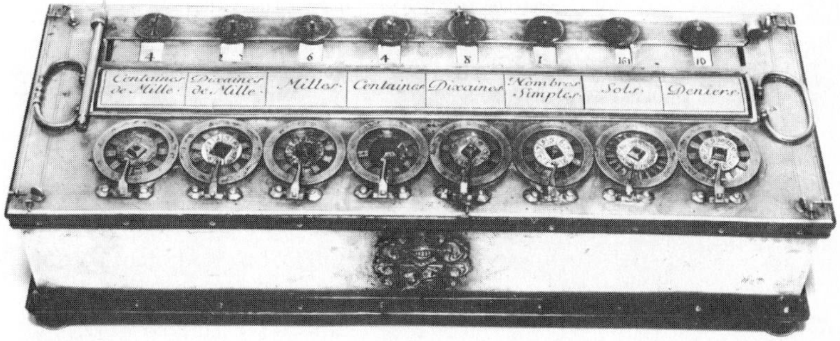

Pascal's calculator *Courtesy International Business Machines Corporation*

In 1642, Blaise Pascal (1623–1662), invented the first mechanical adding machine. His main reason for doing so was that he was tired of tallying figures for his tax-collector father. The machine was a box-like contrivance (see photo) composed of cog wheels, each of which contained the digits 0 through 9. It could carry numbers automatically. The person using this machine would enter each number on a dial by hand and then pull a handle to register the number. This technique is still used by some desktop adding machines.

In 1671, a German mathematician, Gottfried Wilhelm von Leibniz (1646–1716), produced a more advanced calculating machine than Pascal's. Pascal's machine could add and subtract, but Leibniz's calculator could add, subtract, multiply, divide, and find square roots.

Courtesy International Business Machines Corporation

Gottfried Wilhelm von Leibniz

Leibniz's calculator *Courtesy International Business Machines Corporation*

Courtesy International Business Machines Corporation

Charles Babbage

An Englishman, Charles Babbage (1792–1871), made two important contributions to the development of computers. He first attempted to make a *difference engine* (see photo), a device capable of calculating the powers of a given number. While working on this project, Babbage had another idea for a machine that would calculate, store data, and print the solutions to problems. About 1830, Babbage attempted to build this *Analytical Engine*. But, largely because of the technological inadequacies of the early 19th century, neither project was completed. It should be noted that the ideas and logic that Babbage used for the Analytical Engine were sound. In fact, a working model of this machine was eventually built, proving that Babbage's theories were indeed correct.

The United States census was largely responsible for the next major development in computers. Of course, the results of a census are tabulated, and it took the U.S. Bureau of the Census almost 9 years to tabulate the results of the 1880 census. The 1890 census was approached with trepidation, both because the time taken on the 1880 census was considerable and because the U.S. population was believed to have increased by almost 25%! Therefore, it was thought that it might take nearly 20 years to tabulate the 1890 census.

A few years earlier, Herman Hollerith (1860–1929) had learned that a weaver, Joseph Jacquard, used punched cards to control complicated weaving designs on a loom. In essence, Jacquard had developed an automatic loom. Hollerith was sure that such cards could be used to tabulate data. Hollerith and James Sperry together devised ways of entering data on punched cards, and also invented machines that could sort and compile this information. Hence the very beginning of the phrase, "Do not fold, spindle, or mutilate." These machines were the forerunners of automatic data processing machines and were used by the U.S. Bureau of the Census for greater efficiency. As a result, the 1890 census was completed in approximately 3 years.

It is interesting to note that Hollerith later founded the Tabulating Machine Company, which later became the International Business Machines Corporation, commonly known today as IBM. Sperry also formed a data equipment company, which became the Sperry Rand Corporation (now part of Unisys).

Courtesy International Business Machines Corporation

Babbage's difference engine

The next significant advance in computer development took place during the 1930s and 1940s. The Federal Insurance Contributions Act of 1935 (the Social Security Act) created a tremendous demand for the processing of data quickly and efficiently. World War II also created a need for a quick, efficient method of processing large amounts of information. For example, hundreds of thousands of soldiers' personnel files had to be processed. Simu-

Courtesy International Business Machines Corporation

Jacquard's loom

lated systems of warfare had to be examined. An inventory of supplies and equipment had to be controlled. Probably one of the most important uses of a computer at this time was in calculating artillery trajectories. When United States forces landed in Africa in November 1942, the Army discovered that the change in environment required recalculation of thousands of firing tables for artillery guns. At that time no machine existed that could work efficiently and quickly enough.

In 1944, Howard Aiken and IBM completed work on the Mark I computer. It was an automatic sequence-controlled computer that used electromagnetic relays rather than mechanical gears to perform calculations. It was the first machine to possess all of the characteristics of a computer. In 1946, the Remington Rand Corporation completed work on a new computer, the ENIAC (Electronic Numerical Integrator and Computer). It contained 19,000 electron vacuum tubes instead of electrical relays and could perform 5,000 additions or subtractions per second. Unfortunately, ENIAC was set up so that only a certain sequence of calculations could be performed; therefore, it sometimes took days to solve complex problems. Also, the large number of vacuum tubes gave off great amounts of heat. This in turn caused many other components to fail. Hence ENIAC was often shut

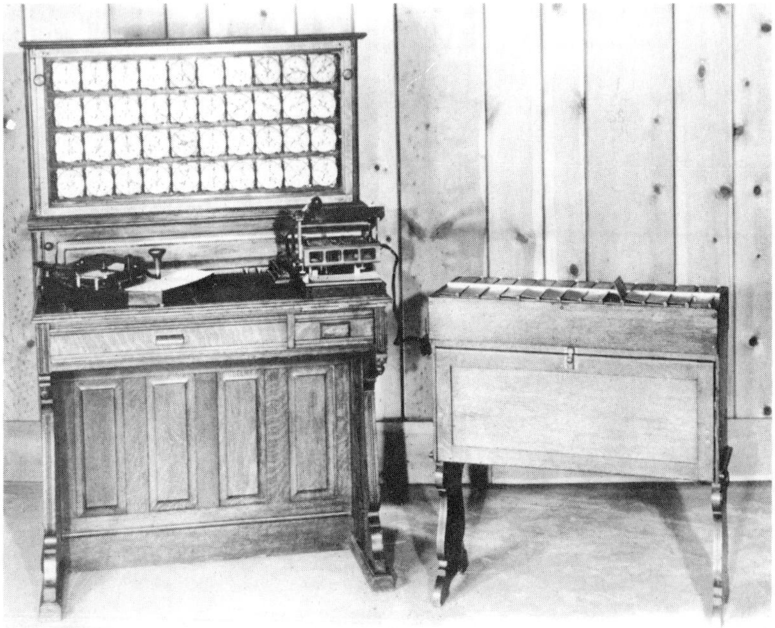

Hollerith's tabulating machine

Courtesy International Business Machines Corporation

NOTE OF INTEREST

In less than an hour a computer can perform more calculations than you could handle in a lifetime using pencil and paper. According to the *Guinness Book of World Records* (1988), the fastest computer is the liquid-cooled CRAY-2, named after Seymour R. Cray of Cray Research, Inc., Minneapolis. Its memory has a capacity of 256 million 64-bit words, and it attains speeds of 250 million operations per second. In 1985 the cost of such system was quoted at $17 million.

down for repairs. It is interesting to note that ENIAC occupied 1,800 square feet of floor space. This is equivalent to the amount of floor space in a one-story house that measures 40 by 45 feet. This is a larger-sized house than most people live in! ENIAC also weighed over 60,000 pounds, or more than 30 tons!

In 1951, J. Presper Eckert and John W. Mauchly completed work on UNIVAC I (Universal Automatic Computer), which was designed for the Bureau of the Census. It was the world's first commercial computer. Eventually about 40 UNIVAC I models were sold to businesses for more than $1 million apiece.

In 1952, John Von Neumann completed work on the EDVAC computer (Electronic Discrete Variable Automatic Computer).

TABLE 1
A TIME LINE OF THE DEVELOPMENT OF COMPUTERS AND COMPUTATION

Date	Event
9th centurv B.C.	Place-value concepts developed.
12th century A.D.	Chinese use an abacus made from beads and rods.
1614	John Napier invented logarithms.
1617	John Napier developed "bones."
1620	Edmund Gunter built the first slide rule.
1624	First comprehensive table of logarithms is published.
1642	Pascal's first calculating machine.
1671	Leibniz invents calculating machine that also multiplies, divides, and finds square roots.
1812	Babbage develops his "difference machine."
1820	First successful commercial calculating machine introduced in France.
1850	Runner added to slide rule.
1850	First key-driven adding machine is patented in the United States by D. D. Parmalee.
1872	First tape adding machine is developed in the United States by E. D. Barbour.
1889	Herman Hollerith (U.S.) develops the punched card for automatic tabulation.
1919	First electronic "flip-flop" circuit is developed.
1944	MARK I
1946	ENIAC

TABLE 1 (*continued*)

Date	Event
1951–1955 (first-generation computers)	Electronic circuits used vacuum tubes. Magnetic drum as primary internal storage unit. Limited main storage capacity. Slow input and output, punched cards. Problems with heat and maintenance. Uses: payroll, record keeping. Example: UNIVAC I, EDVAC
1956–1963 (second-generation computers)	Transistors replaced vacuum tubes. Magnetic core as primary internal storage unit. Increased speed of input and output, tape. Sophisticated programming languages such as COBOL and FORTRAN. Reduced size and heat generation. Increased speed and reliability. Uses: payroll, billing, updating inventory. Examples: IBM 1401, Honeywell 200
1964–1970 (third-generation computers)	Integrated circuits replaced transistors. Solid state storage. Flexible input and output, disk orientation. Smaller size, greater reliability, increased processing speed. Development of minicomputers. Applications: Credit card billing, reservations, financial analysis. Examples: IBM System 360, NCR 395
1971 to present (fourth-generation computers)	Microprocessor chip, large-scale integration. Increased speed and storage capacity. Increased versatility of input and output. Development of microcomputer. Applications: home computers, computer-assisted instruction, simulation. Examples: IBM 3033, Apple II, IBM PC Manufacturers of microcomputers include Radio Shack (a division of Tandy Corporation), Apple Computer Corporation, Commodore Computers, and IBM.

The EDVAC was the first computer developed in the United States that could store both program instructions and data. It should be noted that a student of Von Neumann's, Maurice Wilkes, completed a similar type of machine, the EDSAC (Electronic Delay Storage Automatic Computer) in 1949 in England.

In the late 1950s transistors began to replace vacuum tubes. (In 1956, J. Bardeen, W. Brattain, and W. Shockley of Bell Telephone Laboratories were awarded the Nobel Prize in Physics for their work in inventing the transistor in 1948.) Some of the advantages of transistors are that they are extremely small compared to vacuum tubes; they allow calculations to be performed much more quickly; unlike vacuum tubes, they produce very little heat; and they last almost forever. Those computers that used vacuum tubes are classified as **first-generation computers; second-generation computers** with transistors, were faster and more efficient.

Further improvements took place in the 1960s when transistors were replaced by integrated circuits. This technology (called solid state technology) combined transistors, diodes, and resistors into a single chip of silicon. Solid state technology gave birth to yet a third generation of computers. These circuits not only increased the processing speed of the computer, but also made

TABLE 2
TYPES OF COMPUTERS

Supercomputers	Extremely powerful, fast and expensive; used for scientific problems that require large amounts of computations. Examples: weather predictions, cryptography
Maxicomputers (mainframes)	Large scale; capable of meeting all or most data-processing requirements of a large organization.
Supermini computers	Capabilities similar to maxicomputers.
Minicomputers	Not built around a microprocessor; between maxicomputers and microcomputers in terms of size, cost, and processing capabilities.
Microcomputers	Built around microprocessor; relatively low-cost desktop machines; usually less processing capability than minicomputers.

possible the construction of smaller, less expensive computers called **minicomputers.** Since the arrival of the minicomputer, the term **mainframe computer** assigned to the larger, more expensive higher-powered computers that preceded the minicomputer. This is sometimes now called a **maxicomputer.**

In the mid-1970s, Intel Corporation discovered how to put thousands of transistors onto a single chip of silicon. This became known as a large-scale integration (LSI) chip, or microprocessor chip. It is approximately 0.2 inch square, and can control the operation of a computer. In 1975, the Altair 8800 became the first home computer kit, based on the Intel 8080 microprocessor. It was the first **microcomputer** and thus ushered in a **fourth generation** of computers.

The growth of microcomputers has an interesting history, thanks in part to people like Steve Wozniak and Steve Jobs, the creators of Apple Computer Inc. Full-scale computing power is now available in desktop sizes and at affordable prices. Future computer generations will most likely employ very large-scale integration (VLSI) chips, which will provide still more power and capability.

EXERCISES FOR SECTION 12.2

1. Name the earliest recorded mechanical counting device. What was a disadvantage of using it?
2. Name four people who helped to develop the computer prior to 1900, and state their contributions.
3. Name two people who helped to develop the computer after 1900, and state their contributions.
4. Name two distinct improvements today's computers have, as compared to earlier models.
5. List the distinguishing characteristics of a first-generation computer, second-generation computer, third-generation computer, and fourth-generation computer.
6. How was Jacquard's automated loom related to computers?
7. What prompted the development of techniques for entering data on punched cards?

Each diagram in Exercises 8–13 illustrates a use of Napier's rods. Determine the indicated product and find the product.

8.
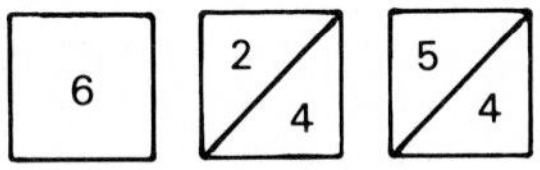

9.
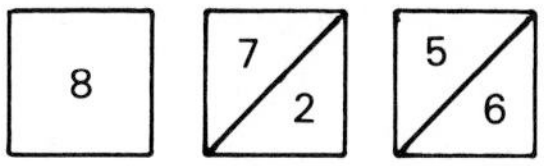

10.
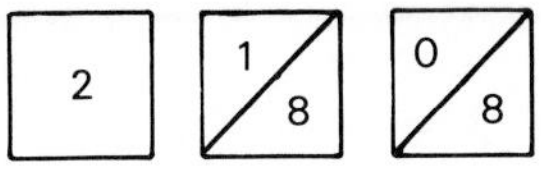

11.
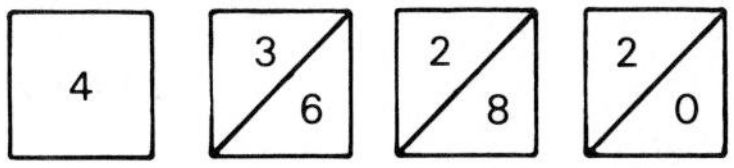

12.
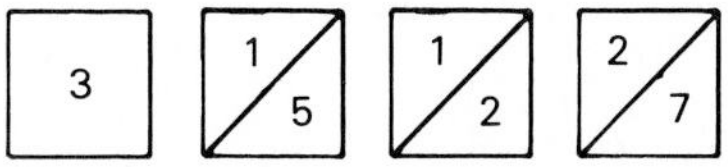

13.
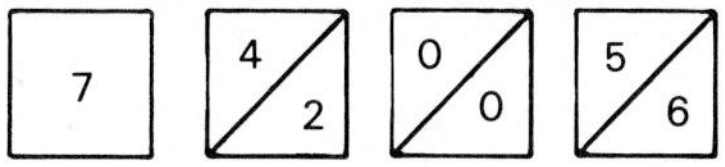

Construct a set of Napier's rods and use them to find each of the following products.

14.   382
      × 4

15.   437
      × 5

16.   497
      × 23

17.   2847
      × 345

JUST FOR FUN

Given the set of digits $A = \{0, 1, 2, 3, 4, 5, 6, 7, 8, 9\}$, construct the following addition problem; that is, fill in each box with a digit. Each digit must be used once and only once.

$$\begin{array}{r} \Box\,\Box\,\Box \\ +\ \Box\,\Box\,\Box \\ \hline \Box\,\Box\,\Box\,\Box \end{array}$$

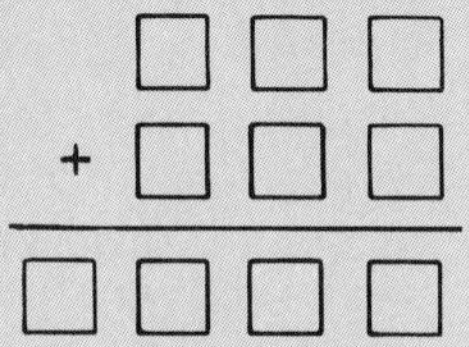

12.3   HOW A COMPUTER SYSTEM WORKS

Basically, there are five parts of a computer system. These parts consist of a **central processing unit** (CPU), two forms of **memory** (**main** and **auxiliary**), an **input unit,** and an **output unit.** These parts represent the actual physical components of a computer system and are referred to as the hardware of the system. See Figure 2.

At the heart of any computer system is the **CPU.** The CPU performs the fundamental arithmetic and logic functions and oversees the operation of the entire system. Contained within the CPU are a control unit, which controls all operations and activities of the rest of the system; an arithmetic/logic unit (ALU), which takes care of all necessary calculations/comparisons; and temporary storage units, where temporary results of an operation are stored.

Main memory is an extension of the CPU. It is directly accessible by the CPU and is regarded as the computer's primary storage. **Auxiliary memory** is a secondary storage medium used to store data permanently. Prior to program execution, program instructions or data must first be placed in main memory. This can be accomplished by an **input unit.** A typical input unit is a keyboard. Once processing has been completed, the results may be given to the user via the **output unit** or placed in auxiliary

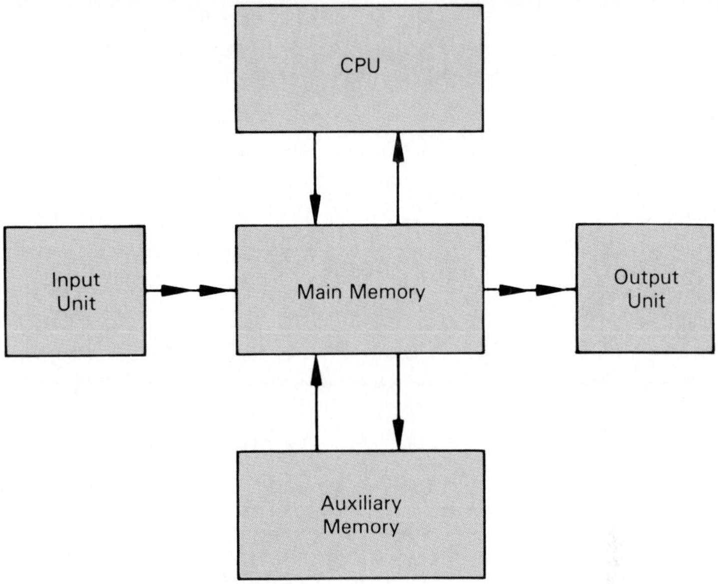

FIGURE 2

storage. Examples of output units are printers or visual display screens.

When a computer system is operating, the components of the system are constantly interacting with each other, often resulting in the execution of millions of instructions per second by the CPU. Typically this interaction is as follows.

1. The input unit places data to be processed into main memory.
2. Once data are in the main memory, the CPU fetches these data and processes them according to the instructions of the program.
3. The processed data are then usually returned to main memory where the output unit in turn makes the results available to the users of the system, or stores the results.

The instructions executed by the CPU must be in a form that is suitable for machine execution. Since it is undesirable to write computer programs in machine language, we use "high-level" languages such as assembly language (a symbolic form of machine language) or problem-solving-oriented languages (which consist of English-like statements). Programs written in a high-level language must be converted to machine language to be executed. This translation can be done by an assembler (for assembly language programs) or by a compiler or interpreter for the problem-solving-oriented languages. Hence assemblers, compilers, and interpreters all perform the same function: they transfer program instructions written in a high-level language to machine-executable code.

To program the solution to a problem means to organize statements into a logical sequence that the computer will follow to solve the given problem. The instructions that people give to the computers are written in special programming *languages*. Different languages are used in different areas. Some of these languages and the areas in which they are used are

ALGOL	ALGOrithmic Language is used extensively in Europe for scientific purposes.
BASIC	Beginner's All-purpose Symbolic Instruction Code was designed to be a simple language that could be learned quickly.
COBOL	COmmon Business Oriented Language is used for solving business problems.
FORTRAN	FORmula TRANslation is used to solve mathematical and scientific problems.
Pascal	Pascal is named in honor of the French mathematician Blaise Pascal, who designed and built one of the world's first mechanical calculators. The Pascal language is one of the newer languages in use today. It was originally developed to teach programming; however, now it is becoming one of the standard languages being implemented with the latest generation of microcomputers.
Ada	A relatively new programming language used by the Department of Defense.
Modula-2	A recently created general-purpose programming language for writing modern, well-structured systems using nontrivial data structures. It reduces complexity in programming and enhances program reliability.
C	A programming language designed by Dennis Ritchie of Bell Laboratories and used there on a PDP-11. It was designed to be the systems language for the UNIX[1] operating system. C was invented to overcome the limitations of a language named B.

To write a program that the computer can use to solve a problem, certain steps must be followed. First we must be sure of what the problem is; that is, we must define the problem. Then we must find a method of solution: how can the problem be solved?

NOTE OF INTEREST

Ada was the daughter of the poet Lord Byron. She was the Countess of Lovelace, known as Lady Lovelace. Ada is credited with writing the world's first computer program because she wrote a demonstration program for Babbage's Analytical Machine. She also was Babbage's public advocate and confidante.

[1]UNIX is a registered trademark of Bell Laboratories.

To find a solution requires systematic and logical steps, which can be illustrated by means of a **flowchart.** A flowchart is a diagram that shows the logical sequence of steps that must be followed in order to solve a problem. In a flowchart there can be as many as 12 standard symbols: each one has a special meaning. We shall be using some of these symbols in our discussion.

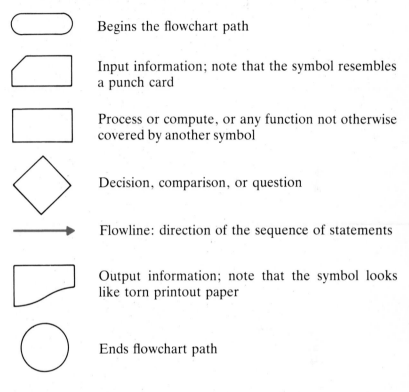

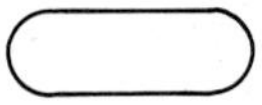
Begins the flowchart path

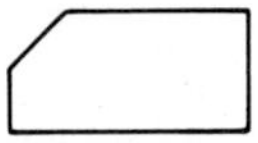
Input information; note that the symbol resembles a punch card

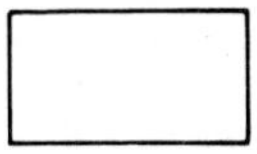
Process or compute, or any function not otherwise covered by another symbol

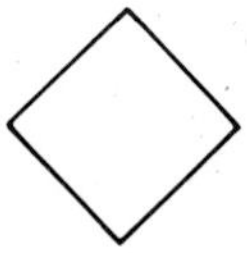
Decision, comparison, or question

Flowline: direction of the sequence of statements

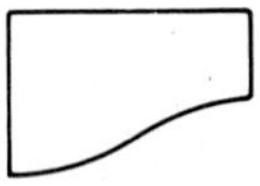
Output information; note that the symbol looks like torn printout paper

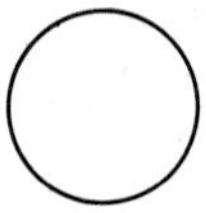
Ends flowchart path

Two examples of simple flowcharts are shown in Figure 3.

Flowcharts can be useful tools when programming problems. Besides showing the logic and process for solving problems, they can also serve as ready references for the solutions of problems. Regardless of what programming language is used, the flowchart format remains basically the same. There are three methods of structuring program actions.

1. *Simple sequence* statements are executed one after the other.
2. *Decision* or *selection* requires a decision; only one of two or more paths will be taken.
3. *Iteration* (or repetition): a set of instructions is executed more than once.

The first two methods of program structure are illustrated in Figure 3; the third will be shown in Section 12.5.

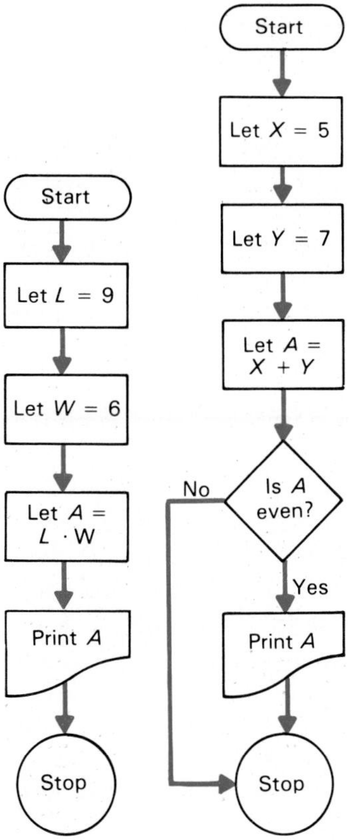

FIGURE 3

EXAMPLE 1 Use the flowchart shown and the input numbers listed below to obtain a result (output).

Let X equal
a. 1
b. 4
c. 5

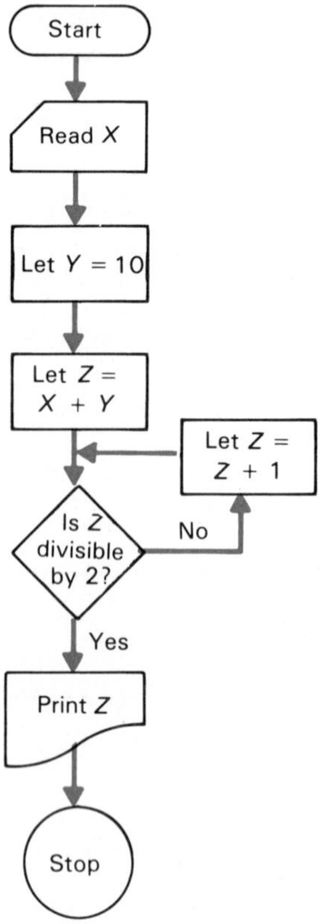

SOLUTION a. 12 b. 14 c. 16

The decision symbol (the fifth step in Example 1) is often called a *branch* because it allows the program to branch off to a different operation when such a decision is made. The branch in turn may form a *loop* within the program. Branches and loops allow long, repetitive calculations to be done by the computer. Because a computer can perform such tasks with lightning speed and never gets tired or bored, humans have been freed from many dull, time-consuming tasks.

EXERCISES FOR SECTION 12.3

1. Name the basic units of a computer.
2. What is the function of the central processing unit?
3. What is the function of the compiler?
4. What does the control unit do?
5. Name a typical input unit.
6. Name a typical output unit.
7. How are the assembler, compiler, and interpreter similar?
8. Name at least three programming languages that are in use today.
9. What does BASIC stand for?
10. What does COBOL stand for?
11. What does FORTRAN stand for?
12. What does Pascal stand for?

For Exercises 13–16, use the flowcharts shown to obtain a result. In each case, let X = 3.

13.

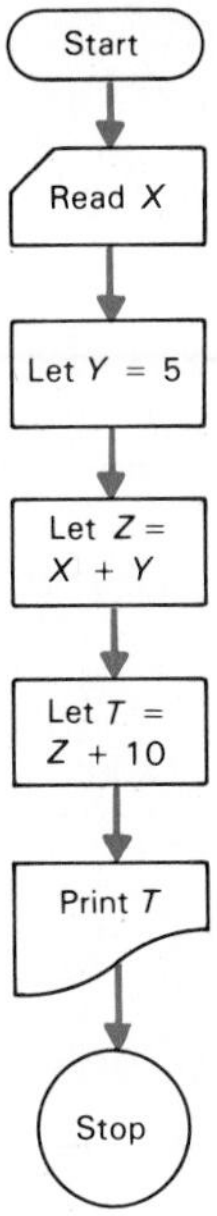

14.

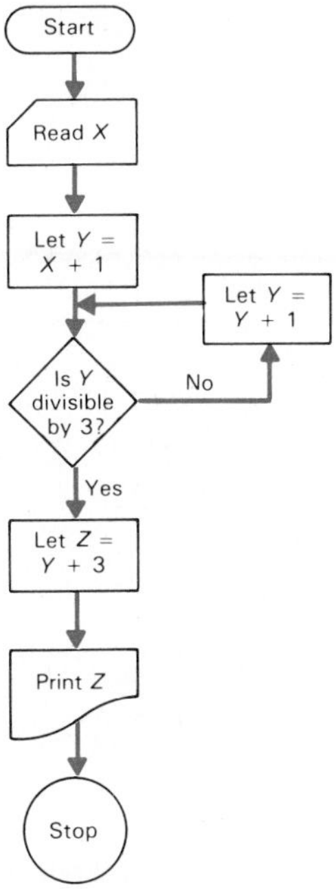

15.

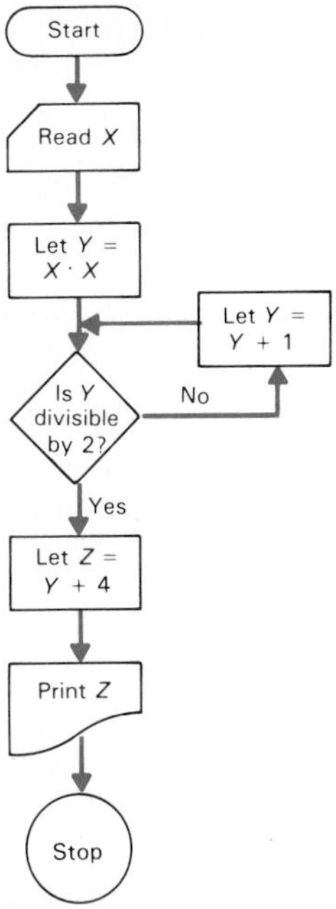

16.

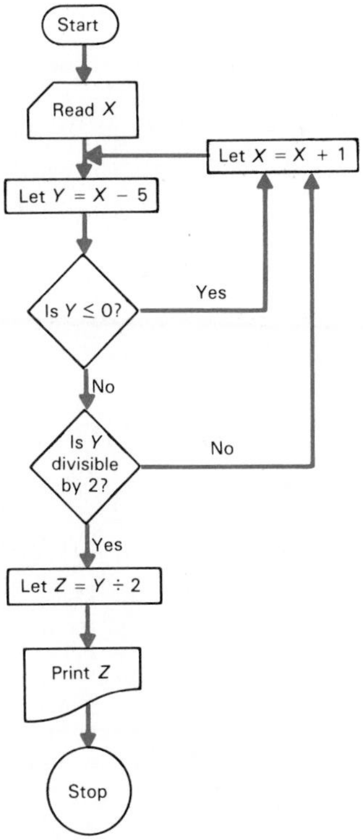

17. Construct a flowchart for finding the perimeter of a rectangle where $l = 4$ and $w = 3$.

18. Construct a flowchart for finding the area of a triangle where $b = 4$ and $h = 6$.

19. Construct a flowchart for finding the area of a square whose side is 8.

20. Construct a flowchart for finding the simple interest on a $10,000 loan for 2 years at 12%.

If a dog makes 27 leaps while a rabbit makes 25 of the same length, in how many leaps will the dog overtake the rabbit, which has a head start of 50 leaps?

12.4 USING BASIC

In the preceding section we mentioned some of the different programming languages in use today. In this section we shall examine one of these languages, BASIC, and use it to write programs. BASIC has certain rules and symbols that must be used, as do other programming languages. Some of these symbols and their meanings are listed in the table that follows.

Symbol	Meaning	Example	Answer
+	addition	5 + 7	12
−	subtraction	7 − 5	2
*	multiplication	7 * 5	35
/	division	8/2	4
=	equals	12/2 = 3 * 2	6 = 6
↑	exponent	2 ↑ 3	8
∧		2 ∧ 3	8
**		2 ** 3	8
( )	parentheses	(1 + 2) ↑ 2	9

Other symbols (called **relational operations**) used in BASIC to assist in developing a program are

Symbol	Meaning	Common Algebraic Expression
>	greater than	$x > 2$
<	less than	$x < 3$
> =	greater than or equal to	$x \geq 2$
< =	less than or equal to	$x \leq 3$
< >	not equal to	$x \neq 5$

Many other symbols and operations are used in BASIC, but we have listed only those that are most commonly used. In addition, certain rules are followed when evaluating expressions in BASIC, as follows.

1. Powers are always evaluated first, in a left to right direction.
2. Multiplications and divisions are performed next, in a left to right direction.
3. Additions and subtractions are performed next, in a left to right direction.

The order of operations is superseded by parentheses. That is, for $(1 + 2) \uparrow 3$, the computer would first do $1 + 2$ inside the parentheses and then raise that sum to the indicated power.

(*Note:* Be sure to check for the actual implementation of the operations or symbols on the computer that you are using. Even though the programming language you are using is BASIC, one computer may vary somewhat from another. For example, with the expression $2 \uparrow 3$ in BASIC notation, which means 2^3 in ordinary arithmetic notation, one machine may use $2 \wedge 3$, whereas another machine may use $2 * * 3$. Check your keyboard and user's manual.)

For example, to evaluate the BASIC statement

$$8 \uparrow 2 - 4 * 3/2 + 4$$

we proceed as follows.

$$\begin{aligned} 8 \uparrow 2 - 4 * 3/2 + 4 &= 64 - 4 * 3/2 + 4 \\ &= 64 - 12/2 + 4 \\ &= 64 - 6 + 4 \\ &= 68 - 6 \\ &= 62 \end{aligned}$$

EXAMPLE 1 Evaluate the following.

a. $3 * 4 \uparrow 2$
b. $18/3 \uparrow 2$
c. $4 \uparrow 2 + 8 * 3 - 1$
d. $2 \uparrow 4 + 4 * 4/2 - 3$

SOLUTION

a. $3 * 4 \uparrow 2 = 3 * 16 = 48$
b. $18/3 \uparrow 2 = 18/9 = 2$
c. $4 \uparrow 2 + 8 * 3 - 1 = 16 + 8 * 3 - 1 = 16 + 24 - 1$
$= 40 - 1 = 39$
d. $2 \uparrow 4 + 4 * 4/2 - 3 = 16 + 4 * 4/2 - 3 = 16 + 16/2 - 3$
$= 16 + 8 - 3$
$= 24 - 3$
$= 21$

Parentheses are used whenever we want to perform an operation on an expression. For example, $(1 + 2)^3$ is written as $(1 + 2) \uparrow 3$ and $\frac{5 + 4}{4 - 1}$ is written as $(5 + 4)/(4 - 1)$.

EXAMPLE 2 Express each of the following in BASIC notation.

a. 4^3 b. $3^2 \times 4$ c. $2^3 \cdot 4 + 3^2$ d. $\frac{3 + 5^2}{1 + 3}$

SOLUTION a. $4^3 = 4 \uparrow 3$ b. $3^2 \times 4 = 3 \uparrow 2 * 4$

c. $2^3 \cdot 4 + 3^2 = 2 \uparrow 3 * 4 + 3 \uparrow 2$

d. $\frac{3 + 5^2}{1 + 3} = (3 + 5 \uparrow 2)/(1 + 3)$

EXAMPLE 3 Write each BASIC expression in ordinary algebraic notation.

a. $4 * X \uparrow 2 - 2 * X + 3$
b. $8 * X \uparrow 3 + 4 * X \uparrow 2 - 3 * X + 14/5$
c. $4 * X \uparrow 3 + (15/2) * X \uparrow 2 - (7/4)$

SOLUTION a. $4 * X \uparrow 2 - 2 * X + 3 = 4x^2 - 2x + 3$

b. $8 * X \uparrow 3 + 4 * X \uparrow 2 - 3 * X + 14/5$

$= 8x^3 + 4x^2 - 3x + \frac{14}{5}$

c. $4 * X \uparrow 3 + (15/2) * X \uparrow 2 - (7/4) = 4x^3 + \frac{15x^2}{2} - \frac{7}{4}$

EXAMPLE 4 Express each of the following in BASIC notation.

a. $2x^3 + 5$ b. $4x^2 + 3x - \frac{1}{2}$

c. $\frac{8x^2 + 3}{2}$ d. $[(x - 2)(x + 3)]^2$

SOLUTION a. $2x^3 + 5 = 2 * X \uparrow 3 + 5$

b. $4x^2 + 3x - \frac{1}{2} = 4 * X \uparrow 2 + 3 * X - (1/2)$

c. $\frac{8x^2 + 3}{2} = (8 * X \uparrow 2 + 3)/2$

d. $[(x - 2)(x + 3)]^2 = ((X - 2) * (X + 3)) \uparrow 2$

In addition to symbols, certain instructions are used in executing, manipulating, or providing information about programs. These instructions are called **commands.** A command is a direct instruction to the computer to take a particular action. Examples of commands are

RUN	Starts execution of the program. The computer begins with the lowest-numbered instructions and carries out the program.
LIST	Lists the program that is in memory. The computer will list, in sequential order, each step of the program.
SAVE	Stores programs in external storage (frequently on disk) for permanent storage.
NEW	Erases any program that may be in memory, in preparation for the acceptance of a new program.
LOAD	Takes a program from external storage and puts it in the computer's memory.
CATALOG	Displays names of all programs stored in external storage.

Instructions that are part of the program are known as **statements.** They have line numbers and consist of **keywords,** each of which gives a specific instruction to the computer. When the computer encounters these instructions, it will automatically follow them. Some of the commonly used statements are given in the following list. (*Note:* We will examine more statements in Section 12.5.)

DATA	Provides data to the variables listed in READ.
READ	Causes the computer to read the information in the DATA statement in sequential order.
INPUT	Sets a variable equal to the data that you type in on the keyboard.
LET	Assigns a value to a variable.
PRINT	Causes the computer to print a series of letters or the value of a variable, or to leave a blank line.
REM	Allows the programmer to make a comment (or REMark) in the program. The computer takes no action because of a REM statement. Its purpose is to help the person writing the program, not to make the computer perform.
END	Terminates a program.

Each statement in a program has to be labeled with a line number, which is usually a multiple of 5 or 10. For example, the statements might be labeled 5, 1Ø, 15, and so on, or 1Ø, 2Ø, 3Ø, and so on. Therefore, new statements can be inserted into the program (between existing statements) without having to renumber all of the existing statements. It should be noted that computers sometimes use Ø for "zero" and O for "oh" in order to avoid or prevent confusion.

The following program is one used for finding the area of a rectangle.

```
 5    PRINT "AREA"
10    PRINT
15    LET L = 9
20    LET W = 6
25    LET A = L * W
30    PRINT A
35    END
RUN
AREA

54
```

Note that the word AREA (line 5) is in quotation marks after the PRINT keyword; we are using this as a label for the answer to the program. Other names can also be used. The computer will type (print) exactly what is within the quotation marks. Line 1Ø contains another PRINT statement, but it is by itself. When we use a PRINT statement in this fashion, the computer will leave a blank line on the printout. In other words, PRINT is used for *formatting output*.

Lines 15 and 2Ø tell the computer to assign the values of L and W. The letters L and W are called "variables." BASIC allows variables to consist of a single letter or a single letter followed by a single digit. For example, L3, B, W5 are permissible, but BY, 3L, and B15 are not.

Line 25 directs the computer to multiply L times W and to call the resulting product A. Line 3Ø tells the computer to print the value of A (the answer), and line 35 tells the computer that the program is completed.

Once the program is entered into the computer, we can instruct the computer to carry out the program by typing RUN. It will then print AREA, and when it reaches line 3Ø it will print the value of A, which is 54.

It should be noted that this previous example is intended to illustrate how a program is set up and how certain statements are used (in a correct manner) to make the computer work for us.

NOTE OF INTEREST

Debugging is the term used to describe the process of checking a program for, and correcting, errors. According to folklore, it was adopted by Grace Hopper, the designer of COBOL. At one time, one of her programs was not working, and after analysis and research, it was discovered that one of the computer components was not operating properly. The reason why the computer component was not working properly was that a real BUG had crawled into it. Since then, a problem in a program is known as a bug, and debugging is the way in which these problems are solved.

Remember that we must talk to the computer in terms of its language, and then it will do exactly what we want. If we make a mistake, then the computer will not perform, or it will tell you that you have made a mistake. Such mistakes are frequently language errors. The most common error message is a SYNTAX error. A SYNTAX error usually means that you have left something out or pressed the wrong key. Some computers will identify exactly where the error occurred, whereas others will simply identify the line number where the error occurred.

Different people can write different programs to solve the same problems. There does not have to be one unique set of instructions for a program. The following is another example of an AREA program with slight modifications; however, note that the results are the same.

```
10    REM - AREA
20    LET L = 9
30    LET W = 6
40    LET A = L * W
50    PRINT "THE AREA IS ";A
60    END
RUN
THE AREA IS 54
```

The REM statement (line 1∅) gives the program a descriptive title so that it can be easily identified. Statement 1∅ is a remark. The computer disregards REM statements when it executes programs. It's useful to put remarks into programs so that you can remind yourself what the program does when you look at it sometime later.

Line 5∅ has the computer print THE AREA IS (note the quotation marks), calculate A (the area), and record the area directly after the statement in quotation marks. If we had used a comma instead of a semicolon, the area (54) would have been moved further to the right of the statement in quotation marks.

The following is an example of another program that employs some of the other statements not used in previous examples. The new statements that appear in this program are READ and DATA. The READ statement instructs the computer to read the numbers from the DATA statement (5∅), which are the values for L and W. The computer automatically assigned L the first data value, 9 and W the second data value, 6. Whenever the computer encounters a READ statement, it seeks the lowest-numbered

DATA statement. It then takes the values in the DATA statement and assigns these values to the variables in the READ statement.

```
10  REM - AREA
20  READ L, W
30  LET A = L * W
40  PRINT "THE AREA IS ";A
50  DATA 9,6
60  END
RUN
THE AREA IS 54
```

EXAMPLE 5 Write two BASIC programs for finding the area of a triangle where the base is 4 and height is 6.

SOLUTION There are a variety of ways to write a program for this problem. Two such methods are

a.
```
10  REM - AREA
20  LET B = 4
30  LET H = 6
40  LET A = B * H/2
50  PRINT A
60  END
```

b.
```
10  REM - AREA
20  READ B,H
30  LET A = B * H/2
40  PRINT A
50  DATA 4,6
60  END
```

The INPUT statement can be used to input data into the computer. Using the INPUT statement we can write another program for the problem in Example 5. For example,

```
10  REM - AREA
20  INPUT B,H
30  LET A = B * H/2
40  PRINT A
50  END
RUN
? 4,6
12
```

Note that the values for B and H are not contained in the program. When we run this program, the computer notes the INPUT statement and responds with a question mark. We then input the data 4 and 6, and the computer prints the answer.

One of the major advantages of a program that uses an INPUT statement is that we can continue to use the program repeatedly to determine the answer to a series of similar problems, not just a particular one. In this case we can use the program to find the area of any triangle as long as we know the values of B and H. We simply type in the values when the computer asks for them. (Of course, if this program were to be used by someone other than its author, we would include a PRINT statement before the INPUT statement to advise the user which data to enter and in what order.)

EXERCISES FOR SECTION 12.4

For Exercises 1–15, evaluate the given statement.

1. $4 * 3$
2. $8/2$
3. $5 + 3 * 2$
4. $2 \uparrow 3$
5. $3 \uparrow 3$
6. $3 \uparrow 2 - 3$
7. $32/2 \uparrow 5$
8. $4 * 3 \uparrow 2$
9. $3 \uparrow 4 + 4 * 2 - 1$
10. $3 * 2 + 4 \uparrow 2/8$
11. $4 \uparrow 2 + 3 * 4/2$
12. $16/4 \uparrow 2 - 1$
13. $9 * 4/3 + 8 - 3$
14. $9 \uparrow (1/2) + 16 \uparrow (1/2)$
15. $((3 + 4) \uparrow 2)/(3 \uparrow 2 - 2)$

For Exercises 16–30, write the given statement in BASIC notation.

16. $3 + 5 - 1$
17. $3 \times 5 + 4$
18. $8 \div 2 + 3$
19. $2^3 - 3$
20. $(4 + 3)^2$
21. $3(4 - 2)$
22. $4^2 \times 3$
23. $3^4 - 2^2$
24. $\dfrac{3 + 4}{1 + 2}$
25. $7(3 + 4)^2$
26. $3 \times 2 + 3 - 2 + 5^2$
27. $(8 \cdot 3) \div 2 + 2^3$
28. $(4 \times 3) + 2 \div 3$
29. $\dfrac{2^2 + 3^2}{2^3}$
30. $5^2(3 + 4)^3$

For Exercises 31–36, write each BASIC expression in ordinary algebraic notation.

31. $3 * X \uparrow 2 - 2 * X + 3$
32. $8 * X \uparrow 3 - 2 * X \uparrow 2 + 5 * X$
33. $(X + 2) * (X + 3)$
34. $(5 * X - 1) * (3 * X + 2)$
35. $(2 * X - 1) * (3 * X + 1)/(2 * X + 1)$
36. $(4 * X \uparrow 2 - 1)/(2 * X + 1)$

For Exercises 37–42, write the given statement in BASIC notation.

37. $3x^2 + 2x + 1$
38. $4x^3 + 2x^2 - x + 7$
39. $\dfrac{x^2 - 8x + 16}{x + 4}$
40. $\dfrac{(3x + 2)(x - 7)}{(2x - 1)}$
41. $[(x + 3)(2x + 1)]^2$
★42. $\sqrt{b^2 - 4ac}$

For Exercises 43–49, determine the printed output for each program.

43.
```
10  LET L = 4
20  LET W = 3
30  LET P = 2 * L + 2 * W
40  PRINT P
50  END
```

44.
```
10  LET A = 1
20  LET B = 2
30  LET C = A + B
40  LET D = 5 ↑ C
50  PRINT D
60  END
```

45.
```
5   READ X, Y
10  LET Z = X ↑ 3 + Y
15  PRINT Z
20  DATA 2,4
25  END
```

46.
```
5   READ X
10  LET Y = X + 1
15  LET Z = Y ↑ 2
20  LET M = Z/2
25  PRINT M
30  DATA 12
35  END
```

47.
```
5   REM - PERIMETER
10  LET L = 9
15  LET W = 7
20  LET P = 2 * L + 2 * W
25  PRINT P
30  END
```

48.
```
10  REM - INTEREST
20  INPUT P,R,T
30  LET I = P * R * T
40  PRINT I
50  END
```

49.
```
10  REM - PYTHAGORAS
20  READ A,B
30  LET C = (A ↑ 2 + B ↑ 2)
      ↑ (1/2)
40  PRINT "THE VALUE OF C
      IS";C
50  DATA 5,12
60  END
```

50. Write a BASIC program for finding the area of a square whose side is 5 (use $A = s^2$).

51. Write a BASIC program for finding the simple interest on a \$1,000 loan for 2 years at $8\frac{1}{2}\%$ (use $I = Prt$).

52. Write a BASIC program for finding the area of a parallelogram whose base is 8 and whose height is 5 (use $A = bh$).

53. Write a BASIC program for finding the area of a triangle whose base is 10 and whose height is 6 (use $A = \frac{1}{2}bh$).

54. Write a BASIC program for finding the average of the following exam scores: 85, 90, 84, 76, 70.

55. Write a BASIC program for finding the average of a set of any five exam scores.

56. Given that $A = 9$, $B = 6$, $C = 5$, and $D = 3$, write a BASIC program that will evaluate $(A + B)/(C - D)$.

57. Given that $A = 2$, $B = 3$, and $C = 4$, write a BASIC program that will evaluate $A^3(B + C)$.

58. Write a BASIC program for finding the square root of 5. (*Hint:* the square root of x, $\sqrt{x}$, can also be written as $x^{1/2}$.)

59. Write a BASIC program for finding the square root of any number that is given.

60. Write a BASIC program for finding the simple interest on any amount, at any rate, for any number of years.

JUST FOR FUN

Following is an algebraic expression for an indicated product.

$$(x - a)(x - b)(x - c) \cdots (x - z) = ?$$

Can you determine the product? Time limit for this question is 10 seconds.

12.5 MORE BASIC STATEMENTS

It is often the case that a computer is used to perform certain calculations over and over for different sets of data. In the previous section the only way this could be done was by means of the INPUT statement. But in order to use the INPUT statement, we must stay at the terminal to respond to the computer and input the data whenever it asks for it. That is, after we enter RUN, it responds with a question mark.

Another BASIC instruction that we can use is the GOTO statement. The GOTO statement sends the computer to a specific step in the program and thus can be used to repeat a series of steps. Consider the following program and its result.

```
10    REM - SQUARES
15    READ X
20    LET Y = X ↑ 2
25    PRINT X, Y
30    GOTO 15
35    DATA 1,2,3,4,5
40    END
RUN
1          1
2          4
3          9
4         16
5         25
OUT OF DATA LINE 15
```

The GOTO statement sends the computer back to line 15, where it reads the next value for X and repeats lines 2∅ and 25 for the program. At the end of the program we have the values for X and Y. When a computer does this, we say that it is in a *loop*. It will continue to find the squares of the numbers in line 35 until it is OUT OF DATA.

The previous illustrative example contains a **loop,** but we still had to provide the computer with the necessary information by typing in the data. Recall that the data were sequential, that is, 1, 2, 3, 4, 5. We can control the looping process by means of two new statements; FOR and NEXT.

```
10  REM - SQUARES
15  FOR X = 1 TO 5
20  LET Y = X ↑ 2
25  PRINT  X, Y
30  NEXT X
35  END
```

This program will provide the same results as the previous example without typing in the data.

X is called a control variable. This program will cycle five times. On each iteration, line 3∅ adds 1 to X. If X is less than 5 or equal to 5, the computer goes back to line 2∅. Lines 2∅ and 25 form the body of the loop and will be executed five times.

Another instruction that we can use to form a loop in a program is the IF-THEN statement. IF and THEN are used together to test a condition. If the condition is true, the computer will follow the instructions after THEN. If the condition is not true, the computer will go to the next line in the program. The following is another SQUARES program that uses the IF-THEN statement.

```
10  REM - SQUARES
15  LET X = 1
20  LET Y = X ↑ 2
25  PRINT X, Y
30  LET X = X + 1
35  IF X < 6 THEN 20
40  END
```

This program will provide the same results as the previous examples. In line 35, the computer checks to see if $X < 6$. If $X = 6$ or $X > 6$, then the computer goes to line 4Ø. It should be noted that some machines require a GOTO after THEN.

Note that the IF-THEN statement allows us to test for a condition that will result in leaving the loop. This useful statement can now be used to end the execution of a loop and thereby avoid an OUT OF DATA error (as occurred in the first program in this section). We can thus test for an **end-of-data signal** (some value, placed after the last item in the DATA statement, that will not be used as data; for example, a negative number could be used in a list of positive numbers). Example 1 uses this method to end execution of the loop.

EXAMPLE 1 Determine the printout for the following program.

```
10  REM - CUBES
20  READ X
30  IF X < 0 THEN 80
40  LET Y = X ↑ 3
50  PRINT X, Y
60  GOTO 20
70  DATA 1, 2, 3, 4, 5, -1
80  END
```

SOLUTION This program contains a GOTO statement that sends the computer back to line 2Ø, where it reads the next piece of data and repeats lines 3Ø and 4Ø of the program. The resulting printout is

```
1       1
2       8
3      27
4      64
5     125
```

EXAMPLE 2 Determine the printout for the following program.

```
 5  REM - NUMBERS
10  FOR X = 1 TO 10
15  PRINT X
20  NEXT X
25  END
```

SOLUTION This program contains the FOR and NEXT statements. The program will cycle 10 times. On each cycle line 20 adds 1 to the current value of X. This program will print the counting numbers 1 through 10 in the following manner.

```
 1
 2
 3
 4
 5
 6
 7
 8
 9
10
```

EXAMPLE 3 Determine the printout for the following program.

```
 5  REM - MULTIPLES
10  LET X = 1
15  LET Y = 8 * X
20  PRINT Y ;"  " ;
25  LET X = X + 1
30  IF X < 13 THEN 15
35  END
```

Reprinted by permission: Tribune Media Services

SOLUTION This program will print the multiples of 8 through 96. The IF statement in line 3Ø will compare the current value of X with 13. IF X is less than 13, the program will branch to line 15. This will continue until the current value of X is not less than 13 (that is, greater than or equal to). When this occurs, the program will then execute line 35, which will cause the program to terminate. Note that we placed a semicolon after the Y and spaces in line 2Ø. This will cause the computer to type the responses on the same line. Hence, we have

```
8  16  24  32  40  48  56  64  72  80  88  96
```

EXAMPLE 4 Write a BASIC program for finding the fifth powers of the natural numbers 1 through 10. Print the number and its fifth power.

SOLUTION As we stated earlier, different people write programs differently. Hence, your program may differ from those listed below. To be sure that your program works, try it at the computer terminal available to you. We list three possible solutions. Each contains one of the statements discussed in this section.

a.
```
 5  REM - POWERS
10  READ X
15  IF X < 0 THEN 40
20  LET Y = X ↑ 5
25  PRINT X, Y
30  GOTO 10
35  DATA 1,2,3,4,5,6,7,8,9,10,-1
40  END
```

b.
```
 5  REM - POWERS
10  FOR X = 1 TO 10
15  LET Y = X ↑ 5
20  PRINT X, Y
25  NEXT X
30  END
```

c.
```
 5  REM - POWERS
10  LET X = 1
15  LET Y = X ↑ 5
20  PRINT X, Y
25  LET X = X + 1
30  IF X < 11 THEN 15
35  END
```

There are many different ways to write programs to solve problems. In this section we encountered the GOTO, FOR-

NEXT, and IF-THEN statements. All of these are useful for having the computer loop or cycle. We want to do this when it is necessary to perform certain calculations repeatedly for different sets of data.

Regardless of the area of interest, we can now use computers to perform many different functions with speed and accuracy. Now many new discoveries and advancements, not possible before, are being made with the help of computers.

A GOTO statement is an unconditional transfer of control statement. A GOTO statement creates a loop. As a result it is important that an exit condition be associated with the loop. Failure to do so could lead to an infinite loop, which (in theory) will cause the program to never end.

The IF-THEN statements form a *conditional branch.* Additionally the FOR-NEXT statements constitute a *conditional branch* or loop.

A loop may be created in any one of three ways: using a GOTO statement, using IF-THEN statements, and using FOR-NEXT statements.

EXERCISES FOR SECTION 12.5

1. What does it mean when we say that a program contains a loop?
2. Name three BASIC statements that allow us to put a loop in a program.
3. When does an OUT OF DATA message occur?
4. What are three procedures that will terminate a loop?
5. **a.** What is the printout for the following program?

```
10  REM - HELLO
15  PRINT "HI"
20  PRINT
25  PRINT "MY NAME IS"
30  PRINT
35  PRINT "KETES THE COMPUTER"
40  PRINT
45  GOTO 15
50  END
```

   b. Is there a problem with this program? If so, what?

For Exercises 6–13, determine the printout for the given program.

6.

```
 5  REM - FORMULA
10  READ X
15  IF X < 0 THEN 45
20  LET Y = X ↑ 3
25  LET Z = Y/4
30  PRINT Z
35  GOTO 5
40  DATA 2, 4, 6, -1
45  END
```

7.

```
 5  REM - MIDWAY
10  READ X, Y
15  IF X < 0 THEN 40
20  LET Z = (X + Y)/2
25  PRINT Z
30  GOTO 10
35  DATA 22,35,41,37,80,92,-1,-1
40  END
```

8.
```
5  REM - MIDWAY
10 READ X, Y
15 IF X < 0 THEN 45
20 IF X = 0 THEN 40
25 LET Z = (X + Y)/2
30 PRINT Z
35 GOTO 10
40 DATA 44,56,68,72,0,4,-1,-1
45 END
```

9.
```
10 REM - MULTIPLES
20 LET X = 1
30 LET Y = 2 * X
40 PRINT Y
50 LET X = X + 1
60 IF X < 51 THEN 30
70 END
```

10.
```
10 REM - NUMBERS
20 FOR X = 1 TO 100
30 PRINT X
40 NEXT X
50 END
```

11.
```
10 REM - PAIRS
20 FOR X = 1 TO 10
30 LET Y = X ↑ 2 - 4
40 PRINT X, Y
50 NEXT X
60 END
```

12.
```
5  REM - PERIMETER
10 READ L,W
15 IF L < 0 THEN 40
20 LET P = 2 * L + 2 * W
25 PRINT L,W,P
30 GOTO 10
35 DATA 7,4,6,3,5,2,-1,-1
40 END
```

13.
```
5  REM - AREA
10 READ L,W
15 IF L < 0 THEN 40
20 LET A = L * W
25 PRINT "A =";A
30 GOTO 10
35 DATA 7,4,6,3,5,2,-1,-1
40 END
```

14. Using READ and DATA statements, write a BASIC program for finding the third powers of the counting numbers 1 through 10.

15. Using the IF-THEN or FOR-NEXT statements, write a BASIC program for finding the squares of the counting numbers 1 through 100.

16. Write a BASIC program for finding the square roots of the counting numbers 1 through 100. (*Hint:* The square root of x, $\sqrt{x}$, can also be written as $x^{1/2}$.)

17. Write a BASIC program to find the sum of the first 100 counting numbers. That is, write a program to calculate $1 + 2 + 3 + \cdots + 99 + 100$.

18. Write a BASIC program for finding a table of values (x, y) for the equation $y = x^2 - 2x + 1$. Let x assume positive integral values of 1 through 25.

19. Write a BASIC program for finding a table of values (x, y) for the equation $y = \sqrt{x^2 - 1}$. Let x assume positive integral values of 1 through 15.

20. Consider the frequency distribution given below.

x	f
1	5
2	9
3	8
4	8
	30

Write a BASIC program to calculate the mean for this frequency distribution. (*Note:* For more information on how to find the mean, see Chapter 4.)

21. Write a BASIC program to calculate the mean for the following frequency distribution.

x	f
0	2
1	6
2	16
3	10
4	6

(*Note:* For more information on how to find the mean, see Chapter 4.)

22. The slope of a line containing the points (x_2, y_2) and (x_1, y_1) is found by means of the formula

$$m \text{ (slope)} = \frac{y_2 - y_1}{x_2 - x_1}$$

Write a BASIC program for finding the slopes of line AB, which passes through A (4, 3) and B (6, 7), and line CD, which passes through C (1, 0) and D (0, 5).

23. The length of a line segment whose endpoints are (x_1, y_1) and (x_2, y_2) is found by means of the formula

$$d \text{ (distance)} = \sqrt{(x_2 - x_1)^2 + (y_2 - y_1)^2}$$

Write a BASIC program for finding the lengths of line segment $\overline{AB}$, whose endpoints are A (5, 2) and B (7, 1), and line segment $\overline{CD}$, whose endpoints are C (1, 5) and D (11, 0).

24. The sum of the interior angles of a polygon is given by the formula $S = (n - 2)180$, where n equals the number of sides. Write a BASIC program to find the sum of the interior angles of a triangle, quadrilateral, pentagon, hexagon, heptagon, octagon, nonagon, and decagon.

25. The area of an equilateral triangle can be found by means of the formula

$$A = \frac{s^2\sqrt{3}}{4}$$

where s is the length of a side. Write a BASIC program to find the area of the equilateral triangles whose sides are of integral length 1 through 10.

26. Write a BASIC program to calculate the sum of the first 100 even counting numbers. That is, write a program to calculate $2 + 4 + 6 + \cdots + 200$.

JUST FOR FUN

A father made out his will and in it he bequeathed his herd of 17 cows to his children, Peter, Paul, and Mary. He specified that

Peter was to receive one-half of the cows.
Paul was to receive one-third of the cows.
Mary was to receive one-ninth of the cows.

The time came to divide the cows and the children were stumped; they did not know what to do. They sought a solution from their lawyer. She promptly came up with a solution. Here is what she did:

She borrowed a cow from a neighboring farm and placed it with the herd of 17 to form a herd of 18. Next she proceeded according to the will:

Peter received one-half of 18 cows, or 9
Paul received one-third of 18 cows, or 6
Mary received one-ninth of 18 cows, or 2

The total number of cows divided among the three children was 17. The lawyer returned the neighbor's cow and announced that everything had been settled according to the will. What do you think?

Summary

Many different people and events are responsible for the development of the computer as we know it today. The time line presented in Table 1 lists some of the dates, developments, and people that played an important role in the evolution of the *first generation, second generation, third generation,* and today's computer, the *fourth generation.*

There are five distinct units in a computer system. These parts consist of a *central processing unit* (CPU), two forms of *memory* (*main* and *auxiliary*), an *input unit*, and an *output unit*. The instructions that people give to computers are written in special programming *languages*. In this chapter we studied the Beginners All-purpose Symbolic Instruction Code (BASIC). By means of this language we can instruct the computer to do what we want it to do. There are many different statements in BASIC, and all of them instruct the computer. The GOTO, FOR-NEXT, and IF-THEN statements are three that we used to put loops into our programs. These loops make the computer perform repetitive operations.

A computer assists people in performing certain jobs or solving problems. It works quickly and efficiently without making an error. But a computer cannot think—it follows a set of instructions. People must be able to tell the machine what to do by means of proper instructions or proper programming.

Vocabulary Check

abacus	program	microcomputer	output	RUN	GOTO
minicomputer	input	statement	BASIC	Hollerith	IF-THEN
command	Napier's rods	Pascal	flowchart	FOR-NEXT	

Review Exercises for Chapter 12

1. Name five people who helped to develop the computer prior to 1900 and state their contributions.
2. Name three people who helped to develop the computer after 1900, and state their contributions.
3. What are the distinguishing characteristics of a third-generation computer?

For Exercises 4 and 5, the diagrams illustrate a use of Napier's rods. Determine the indicated product and find the product.

4. 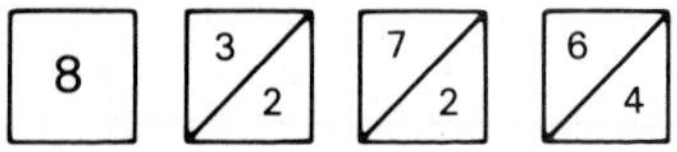

5. 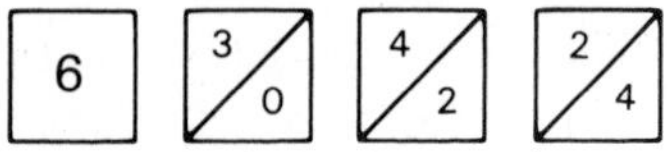

6. Each symbol in a flowchart has a special meaning. Listed below are certain flowchart symbols. Identify each by describing the meaning of each symbol.

a. 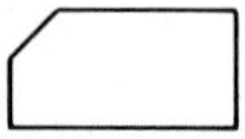

b. 

c. 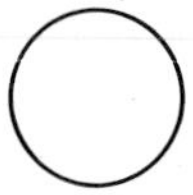

d. 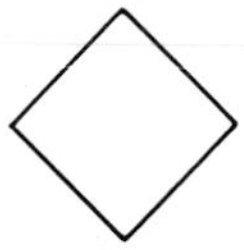

7. What is the function of the CPU?
8. What does BASIC stand for?
9. Evaluate the following.

   a. 3 * 5 ↑ 3 b. 36/3 ↑ 2 c. 2 ↑ 5 − 5 ↑ 2

10. Write each of the following in BASIC notation.

    a. $3x^2 + 2x - 5$ b. $\dfrac{4x^2 - 3x + 2}{x^2 + 2x}$

11. Determine the printed output for the following programs.

a.
```
 10  REM - PAIRS
 20  FOR X = 1 TO 5
 30  LET Y = X ↑ 2 + 53
 40  PRINT X,Y
 50  NEXT X
 60  END
```

b.
```
 10  REM - EVALUATE
 20  READ X
 30  IF X < 0 THEN 100
 40  LET Y = 2 * X - 1
 50  LET Z = Y ↑ 2
 60  LET A = Z/2
 70  PRINT A
 80  GOTO 20
 90  DATA 2,4,6,8,10,-1
100  END
```

12. Write a BASIC program for finding the area of a parallelogram whose base is 8 centimeters and whose height is 5 centimeters.

13. Write a BASIC program for finding the simple interest on a \$10,000 loan for 3 years at 15.5%.

14. Write a BASIC program for finding a table of values (x, y) for the equation $y = 4x^2 - 5x + 1$. Let x assume positive integral values of 1 through 100.

15. Write a BASIC program to find the area of any triangle when its base and height are known. Use the INPUT statement.

16. Some statistics books use the following formula to find the standard deviation (s) for a sample.

$$s = \sqrt{\frac{\Sigma\,(x - \bar{x})^2}{n - 1}}$$

where $\bar{x}$ is the mean and n is the number of pieces of data (Σ represents "the sum of"). Write a BASIC program to find the standard deviation for the following set of data.

{82, 80, 81, 87, 86, 86, 88, 84, 82, 84}

JUST FOR FUN

A card and an envelope together cost 55 cents. The card costs 50 cents more than the envelope. How much does the envelope cost?

Chapter Quiz

Determine whether the following statements are true or false.

1. The earliest recorded mechanical counting device is the abacus.
2. Napier's bones were used to perform division.
3. The first mechanical adding machine was invented by Blaise Pascal.
4. The United States census affected the development of computers.
5. The invention of the EDVAC computer was significant because it was the first computer to use vacuum tubes.
6. The world's first commercial computer was the UNIVAC I.
7. Those computers that used vacuum tubes are classified as second-generation computers.
8. There are five distinct units in a computer system; they are the input unit, control unit, memory unit, arithmetic unit, and processor.

9. COBOL is a programming language that is commonly used to solve mathematical and scientific problems.

10. The flowchart symbol ◇ is used to indicate a decision, comparison, or question.

11. The BASIC symbol <> represents not equal to, that is X <> 6 means $X \neq 6$.

12. If $X = 9 \uparrow 2 - 4 * 3/2 + 5$ in BASIC notation, then $X = 80$.

13. The algebraic expression

$$\frac{5x^2 - 3x + 5}{x - 1}$$

can be expressed in BASIC language as $5 * X \uparrow 2 - 3 * X + 5/X - 1$.

14. The RUN command is used to make a loop in a program.

15. Each instruction in a program has to be labeled with an identification number.

16. Instruction statements are labeled consecutively.

17. BASIC allows variables to consist of a single letter or a single letter followed by a single digit.

18. When the computer encounters an INPUT statement, it responds with a question mark.

19. READ statements always occur in a program with DATA statements.

20. The FOR-NEXT statement is used when building loops into a program.

21. A loop may be created in at least three ways.

22. First generation computers used transistors.

23. A microcomputer is capable of meeting all or most data processing requirements of a large organization.

24. A flowchart is a pictorial representation of the flow of logic of a program.

25. Input, output, and auxilliary units are needed for a computer to process information.

GLOSSARY

Abacus The earliest recorded mechanical counting device, developed by the Chinese during the twelfth century.

Abscissa The x value in the ordered pair (x, y).

Acute angle An angle whose measure is greater than 0° and less than 90°.

Algorithm Some special process of solving a certain type of problem, usually a method that repeats some basic process.

APR Annual percentage rate, also known as the true annual interest rate.

Associative If a system consists of a set of elements $\{a, b, c, \ldots\}$ and an operation $*$, the operation is associative if, for all elements in the system, $(a * b) * c = a * (b * c)$.

Axiom A statement that is accepted as true without proof.

Axiomatic system An axiomatic system consists of four main parts: (1) undefined terms, (2) definitions, (3) axioms, (4) theorems.

Base 5 system A system that groups items by fives.

BASIC A programming language; the name stands for Beginners All-purpose Symbolic Instruction Code.

Biconditional The conjunction of two conditional statements where the antecedent and consequent of the first statement have been switched in the second.

Bimodal A set of data is described as bimodal when the given set of data has two modes.

Binary notation The base 2 system of numeration.

Cardinality The cardinality of a set tells us "how many" elements are in a set.

Cartesian plane The grid or graph divided into four quadrants by the x and y axes, named after the French mathematician-philosopher René Descartes.

Cartesian product The Cartesian product of sets A and B is the set of all possible ordered pairs such that the first element of the ordered pair is an element of A and the second element of the ordered pair is an element of B.

Celsius Anders Celsius (1701–1744), a Swedish astronomer who developed a thermometer scale on which the boiling point of water is 100° and the freezing point of water is 0°.

Centi A prefix which means $\frac{1}{100}$ or 0.01. A centimeter is 0.01 meter.

Central tendency A measure of central tendency describes a set of data by locating the middle region of the set.

Circumference The distance around a circle, the length of a circle.

Closure The characteristic of a system that is closed. A system consisting of a set of elements $\{a, b, c, \ldots\}$ and an operation $*$ is closed if for any two elements a and b in the set, $a * b$ is also a member of the set.

Combination A distinct group of objects without regard to their arrangement.

Command A direct instruction to the computer to take a particular action—for example, RUN, LIST, LOAD.

Commutative In a system consisting of a set of elements $\{a, b, c, \ldots\}$ and an operation $*$, the operation is commutative if for all elements a and b in the system, $a * b = b * a$.

Complement The complement of a set A is the set of all the elements in the given universal set, U, that are not in the set A; it is denoted by A'.

Complementary angles Two angles whose measures sum to that of a right angle.

Complex fraction A fraction that has a fraction in the numerator or denominator or both.

Compound amount The total amount of principal and compound interest combined.

Compound interest The interest due at the end of a period is added to the principal, and thereafter earns interest.

Conclusion The statement which follows as a consequence of the hypothesis or premises.

Conditional The connective "If . . . , then . . ." is used in compound statements referred to as conditionals.

Congruent triangles Triangles that can be made to coincide. They have the identical shape and size.

Conjunction Two simple statements that are connected using the word *and.*

Consistent (1) Two linear equations are consistent if they intersect. (2) Statements that can be true together.

Contradiction A statement that is always false.

Contrapositive The conditional which results from replacing the antecedent by the negation of the consequent, and the consequent by the negation of the antecedent.

Converse The conditional which results from replacing the antecedent by the consequent and the consequent by the antecedent.

Coordinate A point on the Cartesian plane whose position is determined by the values of its abscissa and its ordinate.

Deci A prefix which means $\frac{1}{10}$ of 0.1. A decimeter is 0.1 meter.

Decimal Any number written in decimal notation.

Decimal notation Using the decimal system to write a number in standard notation.

Decimal system A system of notation for real numbers that uses the base 10.

Defined terms Definitions that may use undefined terms or terms that have been previously defined. Definitions should be concise, consistent, and not circular.

De Morgan's law (Augustus De Morgan, 1806–1871) Rules that enable us to change a disjunction, conjunction, or the negation of one of these to an equivalent statement.

Density property Between every two rational numbers there is another rational number.

Dependent Two equations whose graphs are the same line. The solution of one equation is also a solution to the other.

Dependent events Two events are dependent if the occurrence of one affects the occurrence of the other.

Deka A prefix which means 10. A dekameter is 10 meters.

Disjoint sets Two sets whose intersection is the empty set. Disjoint sets have no elements in common.

Disjunction Two simple statements that are connected using the word *or.*

Divisible An integer m is divisible by an integer n if there is an integer q such that $m = nq$.

Duodecimal system The base 12 system of numeration.

Effective annual rate The annual interest rate which gives the same yield as a nominal interest rate compounded several times a year.

Element Any member of the set.

Equally likely events Each outcome of the experiment has the same (or equal) chance of occurring.

Equal sets The sets that contain exactly the same elements.

Equivalent sets Two sets that contain exactly the same number of elements.

Expanded notation Writing a number in powers of the base system. For example, $978 = (9 \times 10^2) + (7 \times 10^1) + (8 \times 10^0)$.

Expectation Expectation tells us the expected value or "fair price" to play a game. It is found by multiplying the probability that an event will occur times the amount that will be won if that event occurs.

Factorial For a positive integer, n, the product of all the positive integers less than or equal to n. For example, $3! = 1 \cdot 2 \cdot 3$ and $n! = 1 \cdot 2 \cdot 3 \cdot \cdots \cdot n$.

Finite (re: finite sets) A finite set contains a limited number of elements. It is a set that has just n members for some integer n.

Flowchart A diagram that shows the logical sequence of steps that must be followed in order to solve a problem.

Foot A unit of linear measure, equal to 12 inches.

FOR-NEXT A programming statement in BASIC that enables the computer to perform certain calculations repeatedly for different sets of data.

Fraction An indicated quotient of two quantities.

Frequency distribution A table that shows the number of times a datum occurs.

Frequency polygon A line graph that results when the upper ends of the middles of the intervals of a histogram are connected by line segments.

Function A relation in which no two ordered pairs have the same first element and different second elements.

Fundamental theorem of arithmetic Every natural number greater than 1 is either a prime or can be expressed as a product of prime factors. Except for the order of the factors, this can be done in one and only one way.

Fundamental counting principle If one experiment has m different outcomes and a second experiment has n different outcomes, then the first and second events performed together have $m \times n$ different outcomes. This may be extended if there are other experiments to follow: we would have $m \times n \times r \times \cdots \times t$ outcomes.

Going metric The process of converting a system of weights and measures to the metric system.

GOTO A programming statement in BASIC that sends the computer to a specific step in the program and thus can be used to repeat a series of steps.

Gram A unit of weight in the metric system.

Greatest common factor The largest natural number that divides a given pair of natural numbers with a remainder of zero.

Great gross One great gross = 12 gross or 12 × 144 or 1,728 units.

Gross One gross = twelve 12's or 12 × 12 or 144 units.

Group A set of elements and an operation that satisfy the closure property, identity property, inverse property, and associative property.

Hecto A prefix that means 100. A hectometer is 100 meters.

Histogram A histogram consists of series of bars that are drawn all with the same size widths on the horizontal axis, and uniform units of the vertical axis that illustrates a frequency distribution.

Hollerith Herman Hollerith (1860–1929) helped devise ways of entering data on punched cards and invented machines that could sort and compile this information.

Identity element A system consisting of a set of elements $\{a, b, c, \ldots\}$ and an operation $*$ has an identity element (e), if for every element a in the system, $a * e = a$ and $e * a = a$.

Iff Abbreviation for the phrase "if and only if" (biconditional)

IF-THEN A programming statement in BASIC that can be used to conditionally transfer control to another location in a program.

Implication An equivalent statement in logic. $P \rightarrow Q \equiv \sim P \vee Q$.

Inclusive or One or the other, or both.

Inconsistent (1) Two lines that have no points in common and therefore never intersect. (2) Statements that cannot be true together.

Independent events Two events are independent if the occurrence of one does not affect the occurrence of the other.

Infinite Becoming large beyond any fixed bound.

INPUT A programming statement in BASIC that sets a variable equal to the data that you type in on the keyboard.

Integer An element of the set consisting of the set of whole numbers and their additive inverses $\{\ldots, -2, -1, 0, 1, 2, \ldots\}$.

Intersection The intersection of sets A and B is a set of elements that are members of both A and B.

Invalid An argument in which all the premises are true, and the conclusion may be false.

Inverse The conditional which results from negating the antecedent and negating the consequent.

Inverse element Each element in a system consisting of a set of elements $\{a, b, c, \ldots\}$ and an operation $*$ has an inverse if for every element a in the system there exists an element b such that $a * b = e$ and $b * a = e$, where e is the identity element of the system.

Irrational numbers Numbers that when expressed as a decimal are nonterminating and nonrepeating.

Isosceles triangle A triangle in which two sides are of equal length.

Kilo A prefix that means 1000. A kilometer is 1000 meters.

Kilogram A kilogram is equal to 1000 grams.

Least common multiple The least natural number that is a multiple of each of two given numbers.

Linear In a straight line, having only one dimension.

Linear equation An equation of the form $Ax + By = C$ where $A, B,$ and C are real numbers. The graph of such an equation is a straight line.

Linear programming A system for solving a system of linear inequalities where the goal is to maximize a value subject to certain restrictions.

Liter A unit of volume in the metric system.

Logically equivalent Two statements that have exactly the same truth tables.

Markdown The change, or difference, between the original price and the sale price of an item.

Markup The difference between the selling price and the cost of an item.

Mathematical system A set of elements together with one or more operations (rules) for combining elements of the set.

Mean The mean for a set of data is found by determining the sum of the data and dividing this sum by the total number of elements in the set.

Median The median can be described as the middle value of a set of data when the data are listed in order.

Meter A unit of length in the metric system.

Metric system The system of measurement in which the meter is the basic unit of length and the gram is the basic unit of weight.

Metric ton A metric ton is 1000 kilograms and equivalent to 2,200 pounds.

Microcomputer Name commonly given to desktop home computers that make use of silicon chips.

Midrange The midrange is found by adding the least value in the given set of data to the greatest value and dividing the sum by 2.

Milli A prefix that means $\frac{1}{1000}$ or 0.001. A millimeter is 0.001 meter.

Minicomputer A third-generation computer that first made use of integrated circuits and silicon chips.

Mode The mode for a given set of data is that number, item, or value that occurs most frequently.

Model A diagram (figure) for a proof that satisfies all of the given axioms.

Mortgage A conditional pledge of property as a security for money lent.

Modulo A system that repeats itself in a cycle. The 12-hour clock is called a modulo 12 system.

Multiplicative grouping system A system that uses certain symbols for numbers in a basic group, together with a second symbol or notation to represent numbers that are multiples of the basic group.

Mutually exclusive Two events that cannot occur together at the same time.

Napier's rods A calculating device developed by John Napier (1550–1617) that was used to perform multiplication.

Natural number Counting number, or an element of the set $\{1, 2, 3, \ldots\}$.

Negation A statement that is formed by prefixing "It is false that."

Normal distribution A normal distribution occurs when the mean, median, and mode all have the same value, and all occur exactly at the center of the distribution.

Number A member of the set of positive integers.

Numeral Any symbol for a number. Symbols like V or 5 are numerals that represent the number five.

Objective function The function in a linear programming problem that describes the relationship between the variables.

Obtuse angle An angle whose measure is greater than 90° and less than 180°.

Ordinate The y value in the ordered pair (x, y).

Origin The point where the x and y axes intersect, whose value is $(0, 0)$.

Output The results given to the user once processing of a program has been completed.

Parallel Equidistant apart at every point.

Parallelogram A quadrilateral that has both pairs of opposite sides parallel.

Parabola The graph of the equation of the general form $y = ax^2 + bx + c$ or $x = ay^2 + by + c$.

Paradox A contradictory statement.

Pascal A relatively new programming language named in honor of Blaise Pascal, the inventor of the first mechanical adding machine who helped develop the basic mathematical theory of probability. It was originally developed for teaching programming, but is now becoming a standard language.

Percent A ratio with a denominator of 100; per one hundred.

Percentile The value that divides the range of a set of data into two parts such that a given percentage of the data lies below this value.

Perimeter The sum of the lengths of the sides of a polygon; the length of a closed curve.

Permutations An ordered arrangement.

Perpendicular Intersecting at or forming right angles.

Pi The name of the Greek letter π. The symbol π represents the ratio of the circumference to the diameter of a circle, which approximately equals 3.14159.

Place-value system A system in which the position of a symbol matters; the value that any symbol represents depends on the position it occupies within the numeral.

Polygon A simple closed broken line.

Premise One of the first two propositions (statements) of a syllogism.

Premium A fee that an insured person pays to the insurer.

Principal The most important or most significant. In finance, money put at interest, or otherwise invested.

Prism A polyhedron with two parallel and congruent bases.

Probability If an experiment has a total T equally likely possible outcomes, and if exactly S of them are considered successful—then the probability that event A will occur, denoted by $P(A)$ is

$$P(A) = \frac{\text{number of successful outcomes}}{\text{total number of all possible outcomes}} = \frac{S}{T}$$

Program A set of instructions that tells the computer how to process the data or what to do with it.

Proper subset If A is a subset of B, and there is at least one element in B not contained in A, then A is a proper subset of B, denoted by $A \subset B$.

Proportion The equality of two ratios.

Pythagorean theorem The sum of the squares of the legs of any right triangle is equal to the square of the hypotenuse.

Quadratic equation An equation of the form $ax^2 + bx + c = 0$.

Quartile The 25th, 50th, and 75th percentiles are the 1st, 2nd, and 3rd quartiles.

Ratio The ratio of two quantities a and b is the quotient or indicated quotient obtained by dividing a by b. The ratio of a to b is written as $a \div b$ or $a:b$ or $\frac{a}{b}$.

Rational number A number that can be expressed in the form a/b where a and b are integers and $b \neq 0$.

Real number Any rational or irrational number.

Rectangle A parallelogram that contains a right angle.

Relation A set of ordered pairs.

Rhombus A parallelogram that has adjacent sides of equal length.

Right angle An angle whose sides are perpendicular. The measure of a right angle equals 90°.

RUN A command that starts execution of the program.

Sample space The set of all possible outcomes of an experiment.

Second One sixtieth of a minute and one thirty-six hundredth part of a degree.

Set A collection of objects.

Sexagesimal system A system based on powers of 60.

SI An abbreviation for the International System of Units.

Sieve of Eratosthenes A technique, devised by Eratosthenes (276–194 B.C.) used for determining prime numbers by punching holes in a clay tablet.

Similar triangles Triangles whose corresponding angles are equal in measure.

Simple grouping system A system in which the position of a symbol does not affect the number represented, and a different symbol is used to indicate a certain number or group of things.

Simple interest Money paid for the use of money where the interest due at the end of a certain period is computed on the original principal during the entire period.

Square A quadrilateral with four sides of equal length and four angles of equal measure.

Standard deviation The square root of the arithmetic mean of the squares of the deviations from the mean.

Standard form A form that has been universally accepted by mathematicians as such, in the interest of simplicity and uniformity. For example, the standard form of a quadratic equation is $ax^2 + bx + c = 0$.

Statement (1) A declarative sentence that is either true or false, but not both. (2) An instruction that is part of the program. Statements have line numbers and consist of keywords.

Straight angle An angle whose measure is 180°.

Subset Given any two sets A and B, if every element in A is also an element in B, then A is a subset of B, denoted by $A \subseteq B$.

Supplementary angles Two angles whose measures sum to that of a straight angle (180°).

Syllogism A logical argument that contains a major premise, a minor premise, and a conclusion.

Tautology A statement that is always true.

Tetrahedron A four-faced polyhedron.

Theorem Logical deductions that are made from undefined terms, defined terms, and axioms.

Trapezoid A quadrilateral that has only two parallel sides.

Tree diagram A tree diagram consists of a number of "branches" that illustrate the possible outcomes for the experiments.

Undefined term A term used without specific mathematical definition; it satisfies certain axioms, but is not otherwise defined.

Underwriter The company or person that agrees to insure a person or item, that is, the insurer.

Union The union of two sets A and B is the set of elements that are members of A, or members of B, or members of both A and B.

Universal set The set of all objects admissible in a particular problem or discussion.

Valid An argument in which all the premises are true and the conclusion is true. A valid argument is one where the conclusion follows logically from the premises.

Venn diagram A diagram that uses a rectangle to represent the universal set and a circle or circles to represent the set or sets being considered in the discussion.

Vertex (1) The turning point of a parabola. (2) The common endpoint of two intersecting rays of an angle.

Vertical angles Two angles where the sides of one angle extend through the vertex and form the sides of the other.

Whole number An integer that is an element of the set $\{0, 1, 2, 3, \ldots\}$.

***x*-intercept** The point at which a line crosses the x-axis.

***y*-intercept** The point at which a line crosses the y-axis.

Z-score The position of a piece of data in terms of the number of standard deviations it is located from the mean.

APPENDIX

TABLE 1
FACTORIALS

n	$n!$
0	1
1	1
2	2
3	6
4	24
5	120
6	720
7	5,040
8	40,320
9	362,880
10	3,628,800
11	39,916,800
12	479,001,600
13	6,227,020,800
14	87,178,291,200
15	1,307,674,368,000

TABLE 2

SQUARES, SQUARE ROOTS, AND PRIME FACTORS FOR THE NUMBERS 1 THROUGH 100

No.	Square	Square Root	Prime Factors	No.	Square	Square Root	Prime Factors
1	1	1.000		51	2,601	7.141	$3 \cdot 17$
2	4	1.414	2	52	2,704	7.211	$2^2 \cdot 13$
3	9	1.732	3	53	2,809	7.280	53
4	16	2.000	2^2	54	2,916	7.348	$2 \cdot 3^3$
5	25	2.236	5	55	3,025	7.416	$5 \cdot 11$
6	36	2.449	$2 \cdot 3$	56	3,136	7.483	$2^3 \cdot 7$
7	49	2.646	7	57	3,249	7.550	$3 \cdot 19$
8	64	2.828	2^3	58	3,364	7.616	$2 \cdot 29$
9	81	3.000	3^2	59	3,481	7.681	59
10	100	3.162	$2 \cdot 5$	60	3,600	7.746	$2^2 \cdot 3 \cdot 5$
11	121	3.317	11	61	3,721	7.810	61
12	144	3.464	$2^2 \cdot 3$	62	3,844	7.874	$2 \cdot 31$
13	169	3.606	13	63	3,969	7.937	$3^2 \cdot 7$
14	196	3.742	$2 \cdot 7$	64	4,096	8.000	2^6
15	225	3.873	$3 \cdot 5$	65	4,225	8.062	$5 \cdot 13$
16	256	4.000	2^4	66	4,356	8.124	$2 \cdot 3 \cdot 11$
17	289	4.123	17	67	4,489	8.185	67
18	324	4.243	$2 \cdot 3^2$	68	4,624	8.246	$2^2 \cdot 17$
19	361	4.359	19	69	4,761	8.307	$3 \cdot 23$
20	400	4.472	$2^2 \cdot 5$	70	4,900	8.367	$2 \cdot 5 \cdot 7$
21	441	4.583	$3 \cdot 7$	71	5,041	8.426	71
22	484	4.690	$2 \cdot 11$	72	5,184	8.485	$2^3 \cdot 3^2$
23	529	4.796	23	73	5,329	8.544	73
24	576	4.899	$2^3 \cdot 3$	74	5,476	8.602	$2 \cdot 37$
25	625	5.000	5^2	75	5,625	8.660	$3 \cdot 5^2$
26	676	5.099	$2 \cdot 13$	76	5,776	8.718	$2^2 \cdot 19$
27	729	5.196	3^3	77	5,929	8.775	$7 \cdot 11$
28	784	5.292	$2^2 \cdot 7$	78	6,084	8.832	$2 \cdot 3 \cdot 13$
29	841	5.385	29	79	6,241	8.888	79
30	900	5.477	$2 \cdot 3 \cdot 5$	80	6,400	8.944	$2^4 \cdot 5$
31	961	5.568	31	81	6,561	9.000	3^4
32	1,024	5.657	2^5	82	6,724	9.055	$2 \cdot 41$
33	1,089	5.745	$3 \cdot 11$	83	6,889	9.110	83
34	1,156	5.831	$2 \cdot 17$	84	7,056	9.165	$2^2 \cdot 3 \cdot 7$
35	1,225	5.916	$5 \cdot 7$	85	7,225	9.220	$5 \cdot 17$
36	1,296	6.000	$2^2 \cdot 3^2$	86	7,396	9.274	$2 \cdot 43$
37	1,369	6.083	37	87	7,569	9.327	$3 \cdot 29$
38	1,444	6.164	$2 \cdot 19$	88	7,744	9.381	$2^3 \cdot 11$
39	1,521	6.245	$3 \cdot 13$	89	7,921	9.434	89
40	1,600	6.325	$2^3 \cdot 5$	90	8,100	9.487	$2 \cdot 3^2 \cdot 5$
41	1,681	6.403	41	91	8,281	9.539	$7 \cdot 13$
42	1,764	6.481	$2 \cdot 3 \cdot 7$	92	8,464	9.592	$2^2 \cdot 23$
43	1,849	6.557	43	93	8,649	9.644	$3 \cdot 31$
44	1,936	6.633	$2^2 \cdot 11$	94	8,836	9.695	$2 \cdot 47$
45	2,025	6.708	$3^2 \cdot 5$	95	9,025	9.747	$5 \cdot 19$
46	2,116	6.782	$2 \cdot 23$	96	9,216	9.798	$2^5 \cdot 3$
47	2,209	6.856	47	97	9,409	9.849	97
48	2,304	6.928	$2^4 \cdot 3$	98	9,604	9.899	$2 \cdot 7^2$
49	2,401	7.000	7^2	99	9,801	9.950	$3^2 \cdot 11$
50	2,500	7.071	$2 \cdot 5^2$	100	10,000	10.000	$2^2 \cdot 5^2$

ANSWERS TO ODD-NUMBERED EXERCISES, ALL REVIEW EXERCISES, AND CHAPTER QUIZZES

CHAPTER 1

SECTION 1.2

1. a. True b. True c. True
   d. True e. True f. False

3. a. True b. False c. False
   d. True e. False f. False

5. a. {Ontario, Erie, Huron, Superior, Michigan}
   b. {New York, New Jersey, New Mexico, New Hampshire}
   c. {Alabama, Alaska, Arizona, Arkansas}
   d. ∅
   e. {Ohio, New Mexico, Colorado, Idaho}

7. a. {a, b, c, . . ., z}
   b. {Erie, Ontario, Superior, Huron, Michigan}
   c. {Ottawa} d. {2} e. ∅

9. a. $\{x|x$ is a day of the week$\}$
   b. $\{x|x$ is a vowel in the English alphabet$\}$
   c. $\{x|x$ is an odd counting number$\}$
   d. $\{x|x$ is a prime number$\}$

11. Yes, it is well defined. The empty set has no elements and there is no difficulty in determining whether any given element belongs to the set.

Just for Fun

Facetious

SECTION 1.3

1. a. True b. False c. False
   d. True e. False f. False

3. a. False b. False c. False
   d. True e. True f. False

5. a. $\{10, 4\}$, $\{10\}$, $\{4\}$, ∅
   b. $\{m, a, t, h\}$, $\{m, a, t\}$, $\{m, a, h\}$, $\{m, t, h\}$, $\{a, t, h\}$, $\{m, a\}$, $\{m, t\}$, $\{m, h\}$, $\{a, t\}$, $\{a, h\}$, $\{t, h\}$, $\{m\}$, $\{a\}$, $\{t\}$, $\{h\}$, ∅
   c. $\{i, o, u\}$, $\{i, o\}$, $\{i, u\}$, $\{o, u\}$, $\{i\}$, $\{o\}$, $\{u\}$, ∅
   d. ∅

7. a. $\{r, i, c\}$ b. $\{m, e, t\}$
   c. $\{t, r\}$ d. $\{m, t\}$
   e. $\{m, e, t, r, i, c\}$ f. ∅

9. a. 16 b. 90¢

Just for Fun

There aren't any.

SECTION 1.4

1. a. $A \cap B = \{4, 6\}$; $A \cup B = \{1, 2, 3, 4, 6, 7, 8\}$
   b. $A \cap B = \varnothing$; $A \cup B = \{a, b, c, d, e, f\}$
   c. $A \cap B = B$; $A \cup B = A$
   d. $A \cap B = \{t, s\}$; $A \cup B = \{g, i, a, n, t, s, j, e\}$
   e. $A \cap B = B$; $A \cup B = A$
   f. $A \cap B = \varnothing$; $A \cup B = \{1, 2, 3, 4, \ldots\}$

3. a. $\{1, 3, 5\}$ b. $\{2, 4\}$
   c. $\varnothing$ d. $\{1, 2, 3, 4, 5\}$
   e. $\{1, 2, 3, 4, 5\}$ f. $\varnothing$

5. a. $\{1, 9, 10\}$ b. $\{1, 2, 3, 6, 7, 8, 9, 10\}$
   c. $\{1, 2, 3, 6, 7, 8, 9, 10\}$ d. $\{1, 9, 10\}$
   e. $\{4\}$ f. $\{2, 3, 4, 5, 6, 7, 8\}$
   g. $\{1, 2, 3, 4, 6, 7, 8, 9, 10\}$
   h. $\{1, 10\}$

7. a. False b. True c. False
   d. False e. False f. False

Just for Fun

29

SECTION 1.5

1. a. region 2 b. region 4
   c. regions 1, 2, and 3 d. regions 1, 3, and 4
   e. regions 1, 2, and 3 f. region 2

3. a. region V
   b. regions I, II, III, IV, V, VI, and VII
   c. regions II, IV, V, VI, and VII
   d. regions II, IV, and V
   e. regions I, II, IV, V and VI
   f. regions IV, V, VI, VII, and VIII

5. a. $A \cap B$ $(A' \cup B')'$

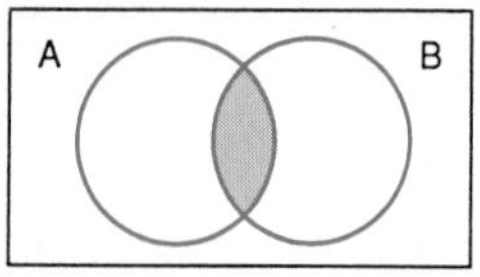

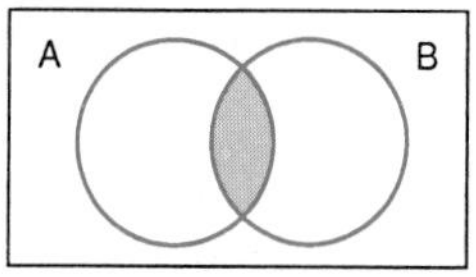

b. $A \cup B$ $(A' \cap B')'$

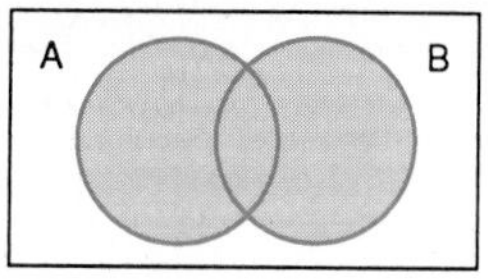

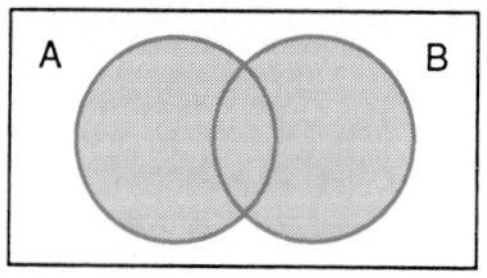

c. $(A \cup B)'$ $A' \cap B'$

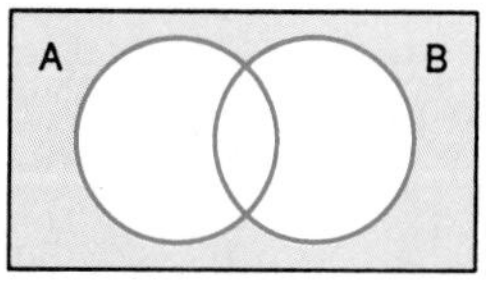

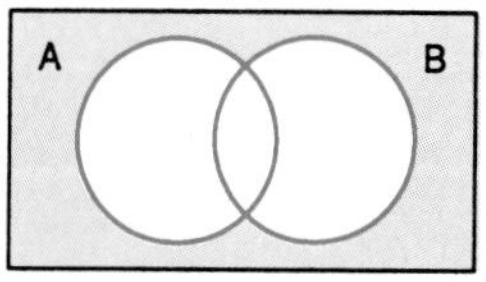

d. $(A \cap B)'$ $A' \cup B'$

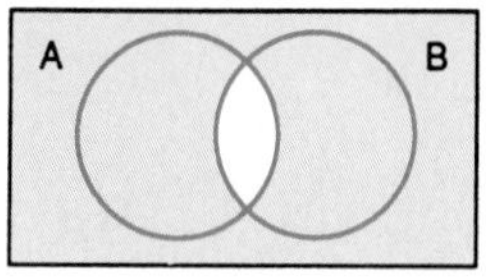

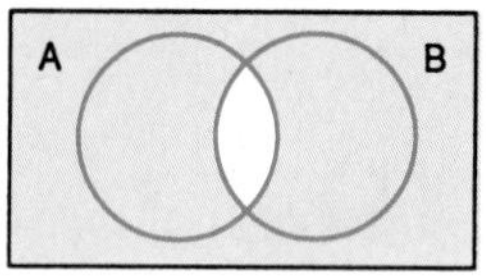

e. $A \cap (B \cup C)$ $(A \cap B) \cup (A \cap C)$

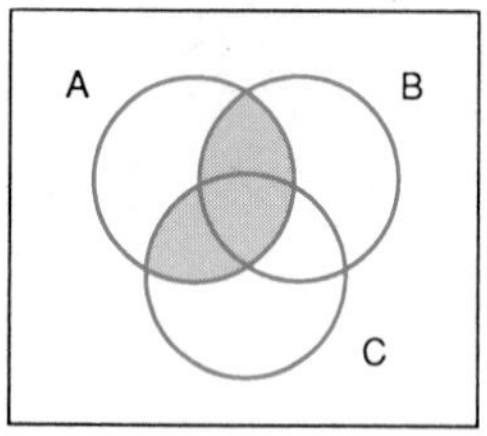

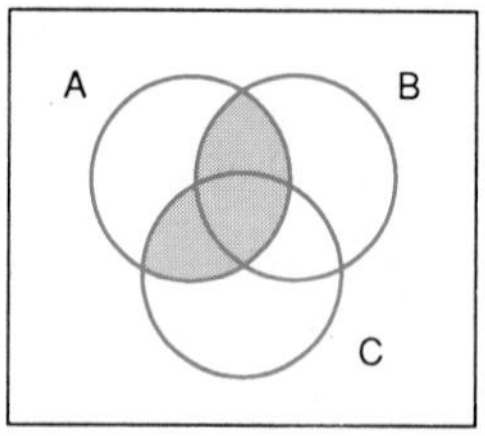

f. $A \cup (B \cap C)$ $(A \cup B) \cap (A \cup C)$

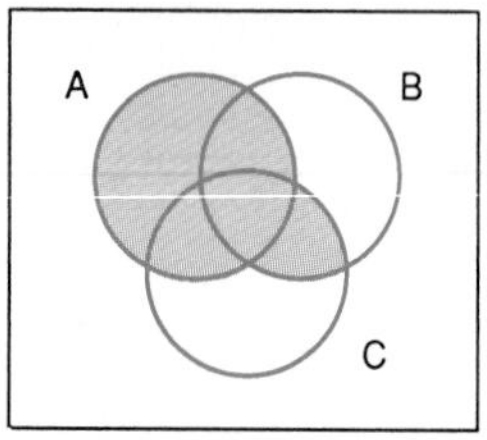

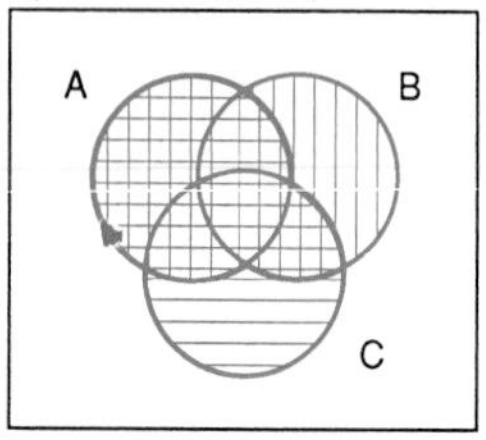

7. a. 22 b. 26 c. 42 d. 106
   e. 56 f. 12 g. 3 h. 63

9. 6; $i \leftrightarrow x$
   $o \leftrightarrow y$
   $u \leftrightarrow z$

11. Zero; the two sets do not have the same cardinality.

Just for Fun

16

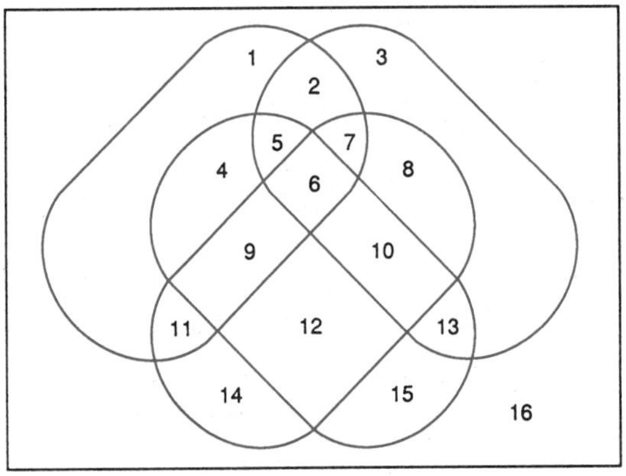

SECTION 1.6

1.	a. 1	b. 19	c. 9	d. 22
3.	a. 18	b. 45	c. 12	d. 43
5.	a. 43	b. 12	c. 100	d. 11
7.	a. 16	b. 54	c. 127	d. 111
9.	a. 82	b. 19	c. 69	d. 49
11.	a. 89	b. 6	c. 8	d. 22

13. Incorrect data: 13 + 6 + 11 = 30 campers, which is more than the 21 stated in the problem.

Just for Fun

Joe—accountant; Sharon—attorney; Jack—architect; Sue—author

SECTION 1.7

1. $A \times B = \{(a,10), (a,20), (b,10), (b,20), (c,10), (c,20)\}$
   $B \times A = \{(10,a), (10,b), (10,c), (20,a), (20,b), (20,c)\}$
   $n(A \times B) = 6$

3. $C \times D = \{(2,1), (2,3), (2,5), (4,1), (4,3), (4,5), (6,1), (6,3), (6,5)\}$
   $D \times C = \{(1,2), (1,4), (1,6), (3,2), (3,4), (3,6), (5,2), (5,4), (5,6)\}$
   $n(C \times D) = 9$

5. $T \times T = \{(t,t), (t,f), (f,t), (f,f)\}$, $n(T \times T) = 4$

7.

B	y	$(4,y)$	$(5,y)$	$(6,y)$
	x	$(4,x)$	$(5,x)$	$(6,x)$
		4	5	6
			A	

9.

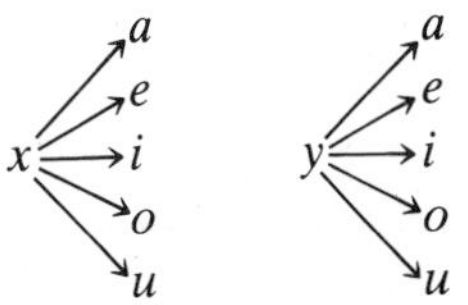

11. a. $\{(a,b), (a,c), (a,d), (b,b), (b,c), (b,d)\}$
    b. $\{(a,c), (a,d), (a,e), (b,c), (b,d), (b,e)\}$
    c. 9
    d. $\{(b,c), (b,d), (b,e)\}$
    e. $\{(c,a), (c,b), (c,c), (c,d), (d,a), (d,b), (d,c), (d,d), (e,a), (e,b), (e,c), (e,d)\}$
    f. $\{(a,c), (a,d), (b,c), (b,d)\}$

13. No, the ordered pairs are different.

Just for Fun

999,999

REVIEW EXERCISES FOR CHAPTER 1

1. A set is any collection of objects.

2. a. {Texas, Tennessee}
   $\{x \mid x$ is a state whose name begins with the letter $t\}$
   b. {Huron, Erie, Michigan, Superior, Ontario}
   $\{x \mid x$ is a Great Lake$\}$
   c. $\{0, 2, 4, 6, \ldots\}$
   $\{x \mid x$ is an even whole number$\}$
   d. $\{1, 2, 3, 4, \ldots\}$
   $\{x \mid x$ is a positive whole number$\}$
   e. $\{1, 2, 3, 4, 5, 6, 7\}$
   $\{x \mid x$ is a counting number less than 8$\}$

3. a. False b. True c. False
   d. True e. False f. True

4. a. True b. True c. True
   d. False e. True f. True

5. a. True b. True c. True
d. True e. False f. False

6. $\{i, o, u\}$, $\{i, o\}$, $\{i, u\}$, $\{o, u\}$, $\{i\}$, $\{o\}$, $\{u\}$, $\varnothing$

7. a. $\{1, 3, 5\}$ b. $\{0, 1, 2, 3, 4, 5\}$
c. $\{2\}$ d. $\{0, 1, 2, 3\}$
e. U f. $\{0, 1, 3, 4, 5\}$
g. U h. $\{0, 1, 2, 3, 4\}$

8. a. regions 1, 2, and 3 b. regions 1, 3, and 4
c. region 4 d. regions II, IV, and V
e. regions I, II, IV, V, and VI
f. regions I, IV, V, VII, and VIII

9. For example: 1—5, 2—10, 3—15; 6

10. a. 11 b. 24 c. 16 d. 5
e. 34 f. 38 g. 47 h. 19

11. a. 6 b. 14 c. 32 d. 68

12. a. For example: m—e, a—a, t—s, h—y
b. $4 \cdot 3 \cdot 2 \cdot 1 = 24$
c. $\{m, a, t, h\}$, $\{m, a, t\}$, $\{m, a, h\}$, $\{m, t, h\}$, $\{a, t, h\}$, $\{m, a\}$, $\{m, t\}$, $\{m, h\}$, $\{a, t\}$, $\{a, h\}$, $\{t, h\}$, $\{m\}$, $\{a\}$, $\{t\}$, $\{h\}$, $\varnothing$
d. $\{(m, e), (m, a), (m, s), (m, y), (a, e), (a, a), (a, s), (a, y), (t, e), (t, a), (t, s), (t, y), (h, e), (h, a), (h, s), (h, y)\}$
e. 16

13. a. when $A \subseteq B$ b. when $A = B$
c. always d. when $B = \varnothing$
e. always

Just for Fun

Step	Amount Left After Each Step: 8 Gallon	5 Gallon	3 Gallon
1. Fill 5 gal from 8 gal.	3	5	0
2. Fill 3 gal from 5 gal.	3	2	3
3. Empty 3 gal into 8 gal.	6	2	0
4. Empty 5 gal into 3 gal.	6	0	2
5. Fill 5 gal from 8 gal.	1	5	2
6. Fill 3 gal from 5 gal.	1	4	3
7. Empty 3 gal into 8 gal.	4	4	0

CHAPTER 1 QUIZ

1. False 2. True 3. True 4. False
5. False 6. False 7. True 8. False
9. False 10. False 11. False 12. True
13. True 14. False 15. False 16. True
17. False 18. True 19. False 20. False
21. True 22. False 23. True 24. False
25. True

CHAPTER 2

SECTION 2.2

1. a. Simple
b. Compound; negation
c. Compound; biconditional
d. Compound; negation
e. Compound; conditional
f. Neither
g. Compound; conjunction
h. Compound; disjunction

3. a. $P \wedge Q$ b. $Q \vee P$ c. $\sim P \vee \sim Q$
d. $P \rightarrow \sim Q$ e. $\sim(\sim P)$ f. $Q \leftrightarrow P$

5. a. $P \vee Q$ b. $\sim Q \wedge P$ c. $\sim(\sim Q)$
d. $Q \leftrightarrow P$ e. $P \rightarrow Q$ f. $\sim P \wedge \sim Q$

7. a. $P \vee M$ b. $G \wedge P$
c. $\sim A \rightarrow D$ d. $B \vee \sim W$
e. $T \leftrightarrow F$ f. $E \rightarrow R$

9. a. I like algebra and I like geometry.
b. If I like algebra, then I do not like geometry.
c. I like algebra or I like geometry.
d. I like algebra or I do not like geometry.
e. I do not like algebra and I do not like geometry.
f. I like algebra iff I like geometry.

Just for Fun

All of them have 28 days.

SECTION 2.3

1. a. $\sim(P \wedge Q \rightarrow R)$ b. none
   c. $\sim P \wedge (Q \rightarrow R)$ d. none
   e. $\sim P \vee (Q \wedge R)$ f. $P \wedge (Q \leftrightarrow R)$

3. a. $Q \wedge P \rightarrow R$ b. $(R \wedge P) \vee Q$
   c. $\sim(Q \wedge P)$ d. $(Q \wedge R) \vee P$
   e. $P \leftrightarrow R \wedge Q$ f. $\sim P \wedge \sim R$

5. a. $E \rightarrow C \wedge G$ b. $S \vee (J \wedge I)$
   c. $C \rightarrow E \wedge Z$ d. $\sim(\sim F)$
   e. $\sim S \wedge \sim L$

7. a. Algebra is difficult, and logic is easy or Latin is interesting.
   b. If algebra is difficult and logic easy, then Latin is interesting.
   c. Algebra is difficult, or logic is easy and Latin is interesting.
   d. Algebra is difficult, and if logic is easy then Latin is interesting.
   e. It is not the case that algebra is difficult and logic is easy.
   f. Algebra is not difficult iff logic is easy and Latin is not interesting.

9.

$\sim$	$(P$	$\vee$	$\sim$	$Q)$
F	T	T	F	
F	T	T	T	
T	F	F	F	
F	F	T	T	

11.

$\sim$	P	$\vee$	$(P$	$\wedge$	$\sim$	$Q)$
F		F	T	F	F	
F		T	T	T	T	
T		T	F	F	F	
T		T	F	F	T	

13.

$\sim$	P	$\underline{\vee}$	Q
F		T	T
F		F	F
T		F	T
T		T	F

15.

$\sim$	$(\sim P$	$\underline{\vee}$	$\sim$	$Q)$
T	F	F	F	
F	F	T	T	
F	T	T	F	
T	T	F	T	

17. a.

$(P$	$\vee$	$Q)$	$\wedge$	$(\sim$	P	$\vee$	$\sim$	$Q)$
T	T	T	F	F		F	F	
T	T	F	T	F		T	T	
F	T	T	T	T		T	F	
F	F	F	F	T		T	T	

b.

$(P$	$\wedge$	$\sim$	$Q)$	$\vee$	$(\sim$	P	$\wedge$	$Q)$
T	F	F		F	F		F	T
T	T	T		T	F		F	F
F	F	F		T	T		T	T
F	F	T		F	T		F	F

Just for Fun

160 ft

SECTION 2.4

1.

$\sim$	$(P$	$\wedge$	$Q)$
F	T	T	T
T	T	F	F
T	F	F	T
T	F	F	F

3.

$\sim$	P	$\wedge$	$\sim$	Q
F		F	F	
F		F	T	
T		F	F	
T		T	T	

5.

P	$\wedge$	$\sim$	P
T	F	F	
F	F	T	

7.

P	$\vee$	$\sim$	Q
T	T	F	
T	T	T	
F	F	F	
F	T	T	

Just for Fun

1	8	6
10	5	0
4	2	9

SECTION 2.5

1.

P	→	*Q*
T	T	T
T	F	F
F	T	T
F	T	F

3.

~	*P*	→	~	*Q*
F		T	F	
F		T	T	
T		F	F	
T		T	T	

5.

~	*P*	→	*Q*
F		T	T
F		T	F
T		T	T
T		F	F

7.

~	*P*	↔	~	*Q*
F		T	F	
F		F	T	
T		F	F	
T		T	T	

9.

P	∨	*Q*	→	~	*Q*
T	T	T	F	F	
T	T	F	T	T	
F	T	T	F	F	
F	F	F	T	T	

11.

(*P*	→	*Q*)	∨	*P*	→	*Q*
T	T	T	T	T	T	T
T	F	F	T	T	F	F
F	T	T	T	F	T	T
F	T	F	T	F	F	F

13.

P	∧	*Q*	↔	*P*	∨	*Q*
T	T	T	T	T	T	T
T	F	F	F	T	T	F
F	F	T	F	F	T	T
F	F	F	T	F	F	F

15.

(*P*	∨	*Q*)	∧	*R*
T	T	T	T	T
T	T	T	F	F
T	T	F	T	T
T	T	F	F	F
F	T	T	T	T
F	T	T	F	F
F	F	F	F	T
F	F	F	F	F

17.

(*P*	∧	*Q*)	∨	(*P*	∧	*R*)
T	T	T	T	T	T	T
T	T	T	T	T	F	F
T	F	F	T	T	T	T
T	F	F	F	T	F	F
F	F	T	F	F	F	T
F	F	T	F	F	F	F
F	F	F	F	F	F	T
F	F	F	F	F	F	F

19.

P	↔	Q	∨	R
T	T	T	T	T
T	T	T	T	F
T	T	F	T	T
T	F	F	F	F
F	F	T	T	T
F	F	T	T	F
F	F	F	T	T
F	T	F	F	F

21.

~	(P	∨	Q)	↔	~	P	∧	~ Q	yes
F	T	T	T	T	F		F	F	
F	T	T	F	T	F		F	T	
F	F	T	T	T	T		F	F	
T	F	F	F	T	T		T	T	

23.

P	∧	~	*Q*	↔	~	(~	*P*	∨	*Q*)	yes
T	F	F		T	F	F		T	T	
T	T	T		T	T	F		F	F	
F	F	F		T	F	T		T	T	
F	F	T		T	F	T		T	F	

25. a. The tide is not out, or we can go clamming.
b. Bill did not drive his van, or he brought the packages.
c. Today is not Wednesday, or tomorrow is not Friday.
d. If two does equal three, then four equals six.
e. If Bob passed the test, then he is unhappy about something else.

Just for Fun

3

SECTION 2.6

1. $P \vee \sim Q$
3. $\sim(\sim P \vee \sim Q)$
5. $\sim(\sim P \vee Q)$
7. $P \wedge Q$
9. $P \wedge Q$
11. John did not go to the party and Janie did not go to the movies.
13. It is false that I passed the test and I did not study too much.
15. Logic is not dull or it is interesting.
17. It is false that the bus is not late and my watch is working correctly.
19. Either x is not greater than zero or x is not negative.
21. It is false that the wind did not come up and we sailed.
23. $A' \cup B'$
25. $(A \cup B)'$
27. $(A' \cup B')'$
29. $A' \cap B$

Just for Fun

They are the same length.

SECTION 2.7

1. Invalid
3. Invalid
5. Valid
7. Valid
9. Valid
11. Invalid
13. Invalid
15. Valid
17. Valid

Just for Fun

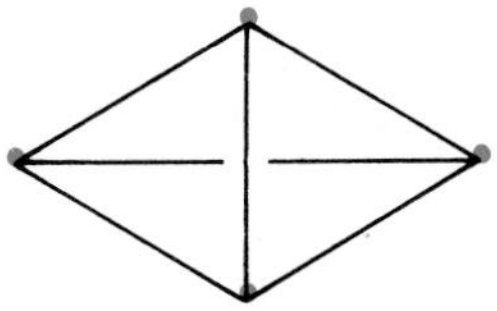

SECTION 2.8

1. Universal negative

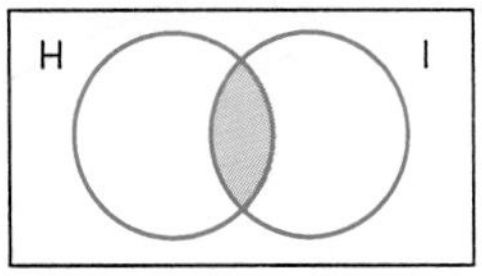

3. Particular negative

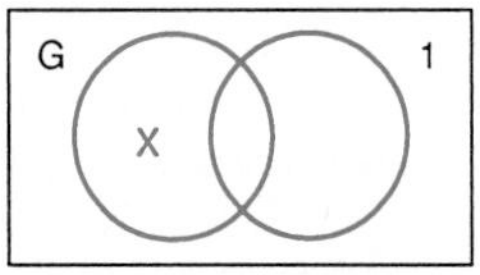

5. Particular affirmative

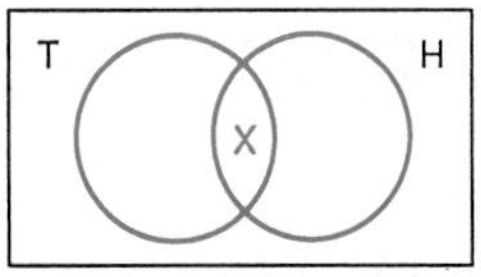

7. Particular negative

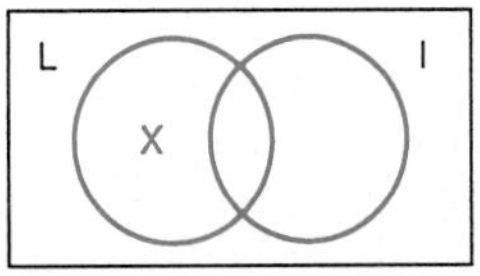

9. Particular affirmative

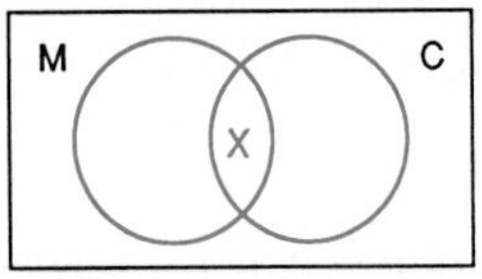

11. Particular negative

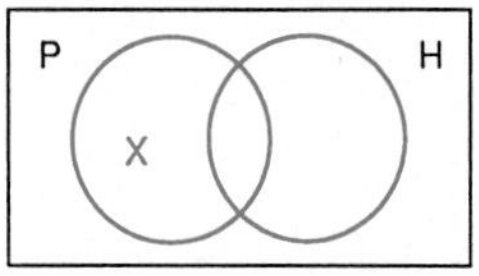

13. Universal negative

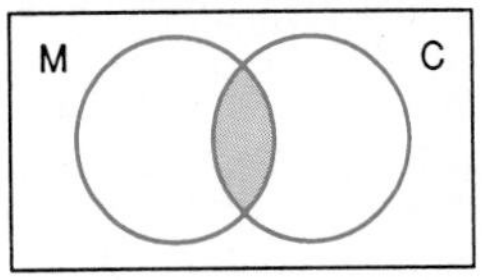

15. Particular negative

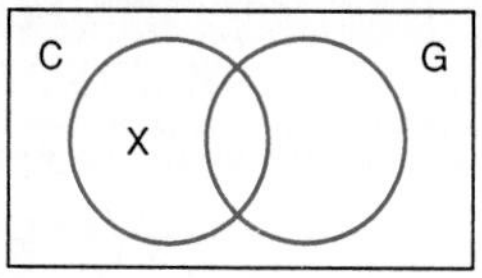

17. Particular negative

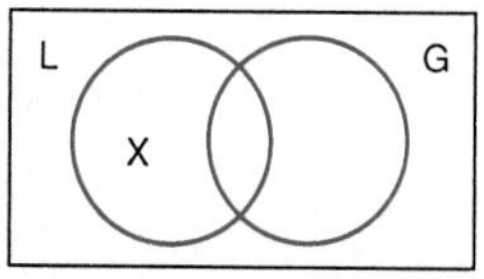

19. Universal negative

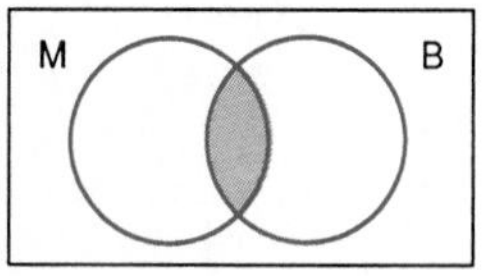

21. Inconsistent
23. Consistent
25. Consistent
27. Inconsistent
29. Consistent
31. Consistent
33. Consistent
35. Consistent

Just for Fun

11.25 seconds

SECTION 2.9

1. Invalid
3. Valid
5. Valid
7. Invalid
9. Valid
11. Invalid
13. Valid
15. Invalid
17. Invalid
19. Valid
21. Valid
23. Invalid
25. Invalid

Just for Fun

Q and Z

SECTION 2.10

1. *Converse:* a. $P \rightarrow S$
   Inverse: $\sim S \rightarrow \sim P$
   Contrapositive: $\sim P \rightarrow \sim S$

   b. $S \rightarrow F$
   $\sim F \rightarrow \sim S$
   $\sim S \rightarrow \sim F$

   c. $F \rightarrow \sim S$
   $S \rightarrow \sim F$
   $\sim F \rightarrow S$

   d. $\sim G \rightarrow L$
   $\sim L \rightarrow G$
   $G \rightarrow \sim L$

   e. $\sim W \rightarrow \sim T$
   $T \rightarrow W$
   $W \rightarrow T$

3. a. $P \rightarrow Q$
   b. $\sim Q \rightarrow \sim P$
   c. $P \rightarrow Q$
   d. $P \rightarrow Q$
   e. $Q \leftrightarrow P$

5. a. $P \rightarrow Q$
   b. $\sim P \rightarrow \sim Q$
   c. $Q \rightarrow P$
   d. $P \leftrightarrow Q$
   e. $Q \rightarrow P$

7. No, the statement is equivalent to "If I pass you, then you will come to class every day."

Just for Fun

1. Special agent
2. Spy
3. Spy

SECTION 2.11

1. $P \vee Q$
3. $P \wedge [(P \wedge Q) \vee R]$
5. $[(P \wedge Q) \vee \text{R}] \wedge P$
7. $[(Q \wedge \sim P) \vee R \vee \sim Q] \wedge P$
9. $(P \wedge \sim Q) \vee [(R \vee P) \wedge (\sim P \vee R)] \vee (Q \wedge P)$
11. 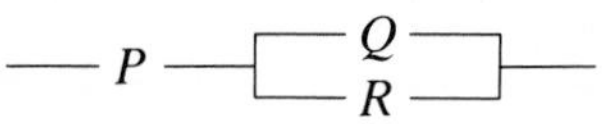

13. 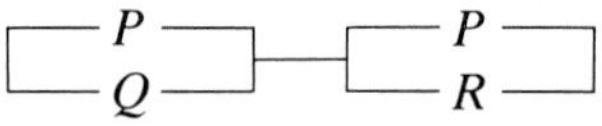

15. 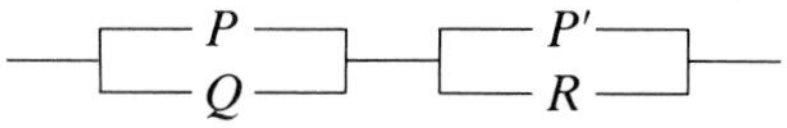

17. 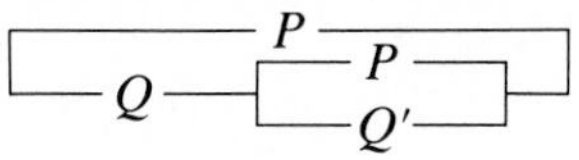

19. $P \rightarrow Q \equiv {\sim}P \vee Q$ 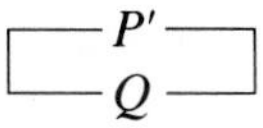

21. a. — P —

b. — P —

c. 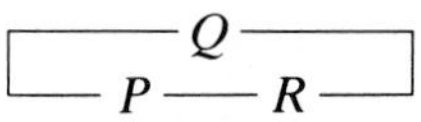

d. 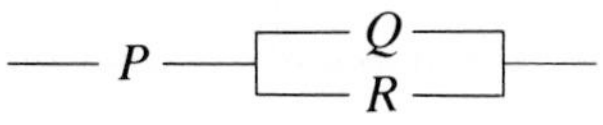

Just for Fun (Meaningless)

REVIEW EXERCISES FOR CHAPTER 2

1. a. Compound; disjunction
   b. Neither
   c. Simple
   d. Compound; conjunction
   e. Compound; negation
   f. Compound; conditional

2. a. $P \wedge Q$ b. ${\sim}Q \rightarrow {\sim}R \vee P$
   c. ${\sim}P \wedge {\sim}Q$ d. $R \rightarrow P \wedge Q$
   e. ${\sim}(P \wedge Q) \vee R$ f. $P \leftrightarrow Q \vee R$

3. a. Sam is sulky, and Tom is tense or Freddy is ready.
   b. Sam is sulky and Tom is tense, or Freddy is ready.
   c. If Sam is sulky or Tom is tense, then Freddy is ready.
   d. Sam is sulky, or if Tom is tense then Freddy is ready.
   e. If Freddy is ready and Tom is tense, then Sam is not sulky.
   f. Sam is not sulky iff Tom is tense and Freddy is not ready.

4. a.

~	P	→	Q
F		T	T
F		T	F
T		T	T
T		F	F

b.

P	∨	~	Q
T	T	F	
T	T	T	
F	F	F	
F	T	T	

c.

P	∨	Q	↔	P
T	T	T	T	T
T	T	F	T	T
F	T	T	F	F
F	F	F	T	F

d.

P	∨	Q	→	R
T	T	T	T	T
T	T	T	F	F
T	T	F	T	T
T	T	F	F	F
F	T	T	T	T
F	T	T	F	F
F	F	F	T	T
F	F	F	T	F

e.

P	∧	~	Q	→	Q	∨	R
T	F	F		T	T	T	T
T	F	F		T	T	T	F
T	T	T		T	F	T	T
T	T	T		F	F	F	F
F	F	F		T	T	T	T
F	F	F		T	T	T	F
F	F	T		T	F	T	T
F	F	T		T	F	F	F

f.

~	(P	∧	~	Q)	→	~	(~	P	∨	Q)
T	T	F	F		F	F	F		T	T
F	T	T	T		T	T	F		F	F
T	F	F	F		F	F	T		T	T
T	F	F	T		F	F	T		T	F

5. a. Today is not Monday or tomorrow is not Sunday.
   b. It is false that Neal is first and David is not second.
   c. It is false that Hugh is not painting and not cutting the grass.

d. Norma did not go to the store and Laurie did not go swimming.
e. It is false that mathematics is difficult and logic is not easy.

6.

~	P	∨	Q	↔	~	(P	∧	~	Q)	yes
F		T	T	T	T	T	F	F		
F		F	F	T	F	T	T	T		
T		T	T	T	T	F	F	F		
T		T	F	T	T	F	F	T		

7. a. Invalid b. Valid

8. a. Inconsistent b. Consistent
c. Consistent d. Inconsistent

9. a. Valid b. Invalid c. Valid
d. Valid e. Invalid

10. a. *Converse:* $G \to P$
Inverse: $\sim P \to \sim G$
Contrapositive: $\sim G \to \sim P$
b. $\sim C \to P$
$\sim P \to C$
$C \to \sim P$
c. $\sim C \to \sim A$
$A \to C$
$C \to A$

11. No, statement is equivalent to "If I marry you, then I will get a job."

12. a. Possible choices are:
Contrapositive: If I will be gullible then I do not study logic.
Implication: I do not study logic or I will not be gullible.
De Morgan: It is false that I study logic and I will be gullible.
b. Use the negation of the given sentence:
It is false that if I study logic then I will not be gullible.

13. a. $P \to Q$ b. $P \to Q$
c. $\sim Q \to \sim P$ d. $\sim Q \to \sim P$
e. $\sim P \vee Q$

14. a. $(P \wedge Q) \vee (\sim P \wedge R)$
b. $(P \vee Q) \vee (R \wedge \sim Q)$
c. $P \wedge ((R \wedge Q) \vee (\sim Q \wedge \sim R))$
d. $P \vee (Q \wedge R) \vee \sim P$

15. a.
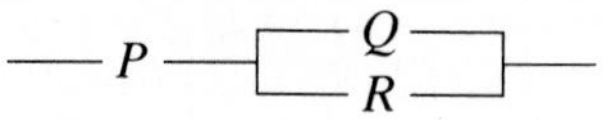

b.
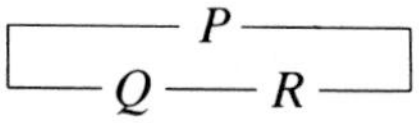

c.
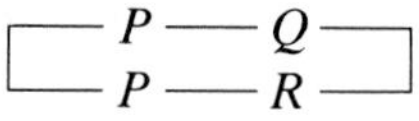

d.
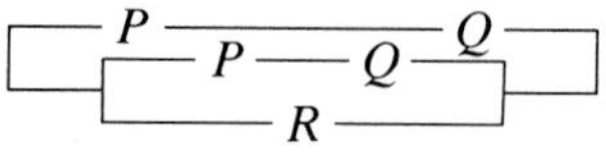

Just for Fun

Barry—painter; Bob—mason; Bart—carpenter

CHAPTER 2 QUIZ

1. False 2. True 3. True 4. True
5. False 6. False 7. True 8. True
9. False 10. True 11. True 12. False
13. False 14. True 15. False 16. True
17. False 18. False 19. False 20. False
21. True 22. False 23. False 24. False
25. False

CHAPTER 3

SECTION 3.2

1. a. $\frac{1}{6}$ b. $\frac{1}{2}$ c. $\frac{1}{2}$ d. $\frac{1}{2}$
e. $\frac{1}{3}$ f. 1

3. a. $\frac{1}{52}$ b. $\frac{1}{52}$ c. $\frac{1}{13}$ d. $\frac{1}{2}$
e. $\frac{1}{4}$ f. $\frac{1}{26}$

5. a. $\frac{3}{13}$ b. $\frac{11}{26}$ c. $\frac{4}{13}$ d. $\frac{1}{52}$
e. $\frac{1}{26}$ f. $\frac{1}{52}$

7. a. $\frac{5}{7}$ b. $\frac{4}{7}$ c. 1 d. $\frac{2}{7}$
e. $\frac{3}{7}$

9. a. $\frac{1}{6}$ b. $\frac{2}{12}$ c. 0 d. $\frac{1}{3}$
e. $\frac{7}{12}$ f. $\frac{11}{12}$

Just for Fun

a. $\frac{0}{1} = 0$ b. $\frac{1}{0}$ is undefined c. $\frac{0}{0}$ is meaningless

SECTION 3.3

1. a. 12
   b. H1 T1
   H2 T2
   H3 T3
   H4 T4
   H5 T5
   H6 T6
   c. $\frac{1}{12}$ d. 0 e. $\frac{1}{2}$ f. $\frac{1}{4}$

3. a. 20
   b.

\$1, \$5	\$5, \$1	\$10, \$1	\$20, \$1	\$50, \$1
\$1, \$10	\$5, \$10	\$10, \$5	\$20, \$5	\$50, \$5
\$1, \$20	\$5, \$20	\$10, \$20	\$20, \$10	\$50, \$10
\$1, \$50	\$5, \$50	\$10, \$50	\$20, \$50	\$50, \$20

   c. $\frac{3}{5}$ d. $\frac{3}{5}$ e. $\frac{3}{10}$ f. $\frac{1}{10}$

5. a. $\frac{1}{4}$ b. 0 c. 1 d. $\frac{1}{4}$
   e. $\frac{3}{4}$ f. $\frac{1}{2}$

7. a. $\frac{1}{6}$ b. $\frac{1}{18}$ c. 0 d. $\frac{1}{2}$
   e. 1 f. $\frac{7}{36}$

9. a. $\frac{2}{5}$ b. $\frac{3}{5}$ c. $\frac{11}{20}$ d. $\frac{9}{20}$
   e. $\frac{7}{20}$ f. $\frac{3}{10}$

11. a. $\frac{1}{10}$ b. $\frac{7}{50}$ c. $\frac{7}{50}$ d. $\frac{1}{25}$
    e. $\frac{2}{25}$ f. $\frac{23}{50}$

Just for Fun

1 + 23 + 4 + 5 + 67 = 100

SECTION 3.4

1.

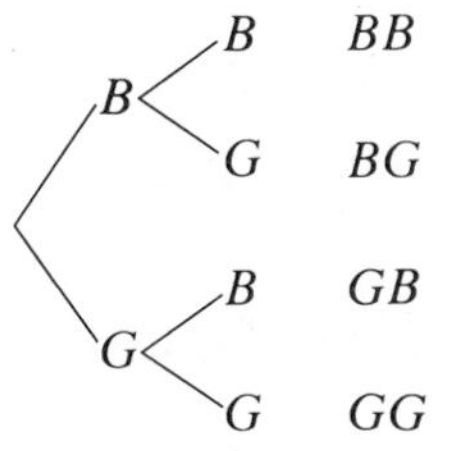

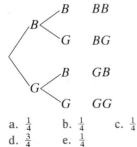

   a. $\frac{1}{4}$ b. $\frac{1}{4}$ c. $\frac{1}{4}$
   d. $\frac{3}{4}$ e. $\frac{1}{4}$

3.

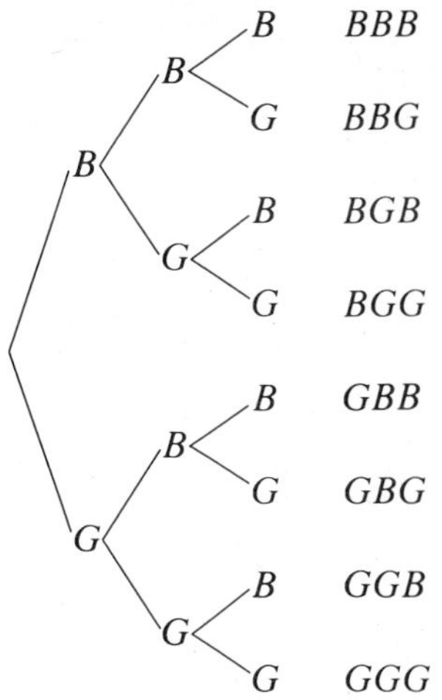

   a. $\frac{1}{8}$ b. $\frac{3}{8}$ c. $\frac{7}{8}$ d. $\frac{1}{8}$ e. $\frac{1}{4}$ f. $\frac{1}{2}$

5.

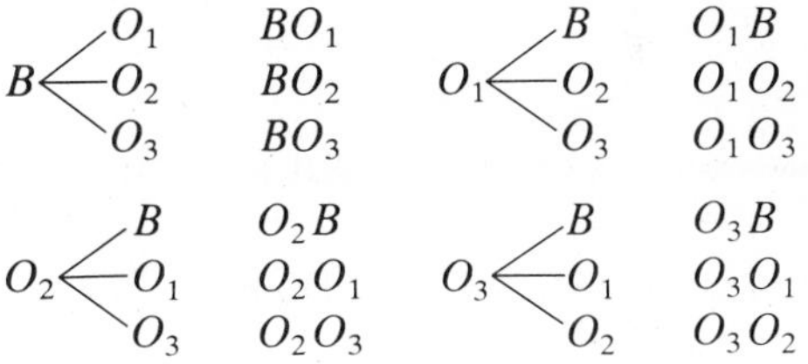

   a. $\frac{1}{2}$ b. 1 c. $\frac{1}{4}$ d. $\frac{1}{4}$ e. $\frac{1}{2}$ f. $\frac{1}{2}$

7. a.

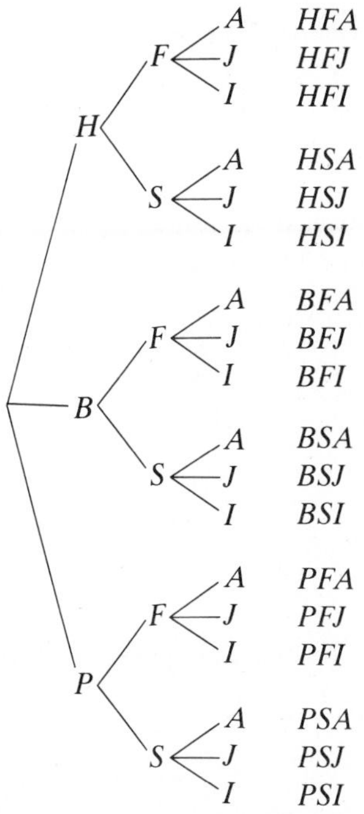

b. $\frac{1}{3}$ c. $\frac{1}{2}$ d. $\frac{1}{18}$ e. $\frac{1}{9}$
f. $\frac{1}{9}$

9. a.

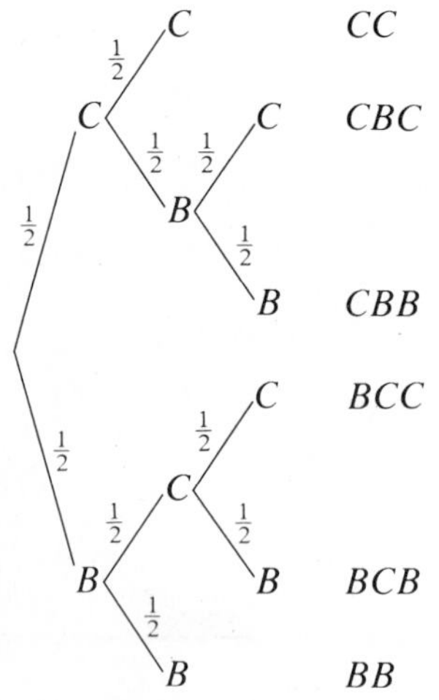

b. $\frac{1}{2}$ c. $\frac{1}{4}$ d. $\frac{3}{8}$
e. $\frac{3}{4}$ f. $\frac{3}{8}$

Just for Fun

One—it is one continuous groove.

SECTION 3.5

1. a. 1 : 5 b. 1 : 17 c. 1 : 35
 d. 35 : 1 e. 31 : 5 f. 11 : 1

3. a. 1 : 12 b. 1 : 3 c. 1 : 1
 d. 12 : 1 e. 3 : 10 f. 15 : 11

5. a. 1 : 3 b. 3 : 1 c. 1 : 1
 d. 3 : 1 e. 3 : 1 f. 1 : 1

7. $\frac{2}{9}$

9. The odds are 1 : 5 in favor of a seven. The odds favor Benny.

11. \$6.14

13. a. $\frac{1}{1000}$ b. 50¢
 c. The cost is too high.

15. \$.14

17. a. \$0.175
 b. The cost is too high as a fair price to pay is less than the price of a stamp.

Just for Fun

Optical illusion; 3 or 5

SECTION 3.6

1. a. $\frac{1}{12}$ b. $\frac{1}{6}$ c. $\frac{1}{18}$
 d. $\frac{1}{4}$ e. $\frac{1}{9}$ f. $\frac{1}{36}$

3. a. $\frac{7}{228}$ b. $\frac{143}{570}$ c. $\frac{7}{95}$
 d. $\frac{11}{76}$ e. $\frac{91}{570}$ f. $\frac{91}{1140}$

5. a. $\frac{1}{4}$ b. $\frac{3}{4}$ c. $\frac{3}{26}$
 d. $\frac{8}{13}$ e. $\frac{3}{52}$ f. $\frac{29}{52}$

7. a. $\frac{11}{221}$ b. $\frac{4}{221}$ c. $\frac{1}{221}$
 d. $\frac{1}{221}$ e. $\frac{1}{17}$ f. $\frac{1}{17}$

9. a. $\frac{1}{55}$ b. $\frac{1}{220}$ c. $\frac{1}{22}$
 d. $\frac{1}{22}$ e. $\frac{1}{22}$ f. $\frac{14}{55}$

11. a. $\frac{1}{6}$ b. $\frac{1}{30}$ c. $\frac{2}{6}$ d. $\frac{3}{4}$
 e. $\frac{1}{360}$

13. a. $\frac{1}{5}$ b. $\frac{91}{13,505}$ c. $\frac{2}{25}$ d. $\frac{2}{5}$
 e. $\frac{3}{37}$ f. $\frac{4}{5}$

15. a. 0.2 b. 0.7 c. 0.6
 d. 0.5 e. 0.2 f. 0.8

Just for Fun

1. $\frac{1}{4}$ 2. $\frac{1}{1,024}$ 3. $\frac{3}{4}$ 4. $\frac{5}{4}$

SECTION 3.7

1. a. 120 b. 216 3. 657,720 5. 12

7. 3,276,000; 2,948,400; 2,835,000

9. 64; 56

11. a. 6 b. 120 c. 1
 d. 210 e. 20 f. 24

13. 90 15. 362,880 17. 132

19. a. 2,520 b. 50,400 c. 5,040
d. 1,260

21. 625

23. 1,000,000,000

Just for Fun

Hawaii

SECTION 3.8

1. a. 10 b. 10 c. 35
d. 35 e. 1 f. 1

3. 210

5. 10; 85¢

7. 3,003

9. 21

11. 1,960

13. 2,970

15. 5,400

17. 108,900

Just for Fun

No, order is important.

SECTION 3.9

1. $\frac{1}{221}$

3. $\frac{11}{850}$

5. a. $\frac{7}{306}$ b. $\frac{7}{102}$ c. $\frac{7}{17}$
d. $\frac{28}{153}$ e. $\frac{95}{102}$

7. a. $\frac{1}{208}$ b. $\frac{3}{104}$ c. $\frac{21}{52}$
d. $\frac{15}{52}$ e. $\frac{15}{208}$ f. $\frac{21}{104}$

9. ${}_{52}C_{13}$; $\frac{{}_{13}C_{13}}{{}_{52}C_{13}}$

11. $\frac{1}{35}$

13. a. $\frac{{}_6C_3}{{}_{13}C_3}$ b. $\frac{{}_4C_3}{{}_{13}C_3}$ c. $\frac{{}_3C_3}{{}_{13}C_3}$
d. $\frac{{}_6C_2 \cdot {}_4C_1}{{}_{13}C_3}$ e. $\frac{{}_4C_2 \cdot {}_3C_1}{{}_{13}C_3}$ f. $\frac{{}_3C_2 \cdot {}_6C_1}{{}_{13}C_3}$

15. $\frac{1287}{2,598,960}$

17. $\frac{1}{108,290}$

19. $\frac{6}{4165}$

Just for Fun

The probability this will occur is greater than $\frac{1}{2}$.

REVIEW EXERCISES FOR CHAPTER 3

1. a. $\frac{4}{13}$ b. $\frac{6}{13}$ c. $\frac{7}{13}$ d. $\frac{9}{13}$
e. $\frac{10}{13}$ f. $\frac{3}{13}$

2. a. $\frac{1}{6}$ b. $\frac{5}{6}$ c. $\frac{1}{6}$ d. $\frac{2}{9}$
e. $\frac{5}{12}$ f. $\frac{1}{2}$

3. 24

4.

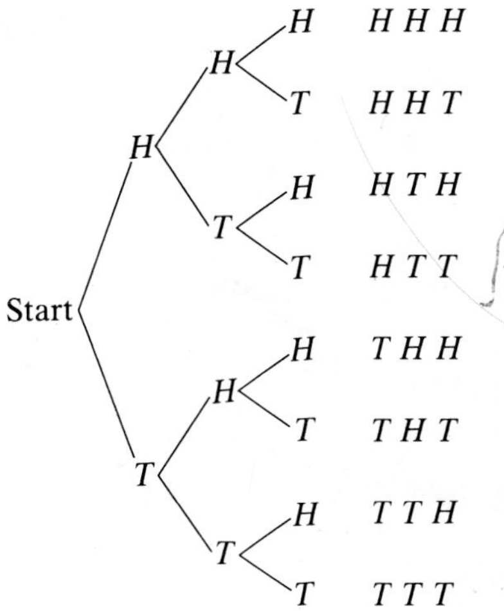

a. $\frac{1}{8}$ b. $\frac{7}{8}$ c. $\frac{1}{8}$

5. a. 1:12 b. 12:1 c. 3:10
d. 4:9 e. 1:51 f. 51:1

6. $\frac{1}{9}$

7. 3:8

8. $2

9. $4, or 80¢ per ticket

10. Independent; it does not matter which ball is chosen first because there is replacement. The occurrence of one event does not affect the occurrence of a second event.

11. Yes; mutually exclusive events cannot happen at the same time. Either a 7 or an 11 may occur, but they cannot occur at the same time.

12. a. 1000 b. $0.50
c. The cost is too high; lose an average of fifty cents for each ticket purchased.

13. a. $\frac{25}{102}$ b. $\frac{1}{221}$ c. $\frac{13}{51}$
d. $\frac{13}{204}$ e. $\frac{40}{221}$ f. $\frac{4}{663}$

14. a. $\frac{1}{12}$ b. $\frac{7}{12}$ c. $\frac{1}{78}$
d. $\frac{3}{13}$ e. $\frac{1}{156}$ f. $\frac{31}{156}$

15. a. 7,776 b. $\frac{1}{1296}$

16. a. 468,000 b. 405,000 c. 302,400

17. 143,640 18. 90 19. 2,520

20. a. 24 b. 1 c. 60 d. 20
e. 30 f. 1

21. a. 10 b. 15 c. 35 d. 35
e. 1 f. 1

22. 196,000

23. 3,600

24. a. $\frac{1}{30}$ b. $\frac{1}{30}$ c. $\frac{1}{30}$
d. $\frac{1}{5}$ e. $\frac{1}{10}$

25. a. $\frac{_{26}C_5}{_{52}C_5}$ b. $\frac{_{26}C_5}{_{52}C_5}$ c. $\frac{_{12}C_5}{_{52}C_5}$

d. $\frac{_{40}C_5}{_{52}C_5}$ e. $\frac{_{12}C_3 \cdot {_{40}C_2}}{_{52}C_5}$

26. a. $\frac{624}{2,598,960}$

b. $\frac{3744}{2,598,960}$

c. $\frac{5148}{2,598,960}$ (counting straight flushes) or
$\frac{5108}{2,598,960}$ (not counting straight flushes)

CHAPTER 3 QUIZ

1. True 2. True 3. False 4. True
5. True 6. False 7. False 8. True
9. True 10. False 11. True 12. False
13. False 14. False 15. True 16. True
17. False 18. False 19. True 20. False
21. True 22. False 23. False 24. False
25. True

CHAPTER 4

SECTION 4.2

1. mean = 5; median = 4; mode = 4; midrange = 6.5
3. mean = 5.5; median = 5.5; no mode; midrange = 5.5
5. mean = 7; median = 7; no mode; midrange = 7
7. mean = 55; median = 55; no mode; midrange = 55
9. mean = 1755.3; median = 1794; no mode; midrange = 1716.5
11. mean = 40.3; median = 40; mode = 39; midrange = 35.5
13. mean = 672.2; median = 672.5; no mode; midrange = 686
15. a. \$15,000 b. \$15,000; \$15,000
c. \$10,000; \$15,000 d. \$17,500; \$15,000
e. mean; median or mode
17. mean = 4; median = 3.5; mode = 3; no
19. 1,001
21. 76.7
23. mean = \$23,125; median = \$19,000; mode = \$16,000; midrange = \$63,000
25. mean = 13.6; median = 14; mode = 10; midrange = 16

Just for Fun

Put one penny in one cup, four in another, and five in the third, then stack the first cup inside the second.

SECTION 4.3

1. $\sigma = 3$ 3. $\sigma = \sqrt{6}$ or 2.4
5. $\sigma = \sqrt{13.7}$ or 3.7 7. $\sigma = \sqrt{26.8}$ or 5.2
9. $\sigma = \sqrt{21.6}$ or 4.6

11. a. 4 b. 3.5 c. 3 d. 5
e. 4.5 f. $\sigma = \sqrt{2.6}$ or 1.6

13. a. 34 b. 32.5 c. $\varnothing$ d. 40
e. 40 f. $\sigma = \sqrt{134.8}$ or 11.6

15. a. 345 b. 370 c. none d. 340
e. 320 f. $\sigma = \sqrt{11055.6}$ or 105.1

17. a. Rudy, Maureen, Jeff, Mark
b. Maureen c. Eric, Maria
d. Maureen e. Maria f. Mark

Just for Fun

35?

SECTION 4.4

1. 94 3. 93 5. 24

7. 8th 9. 15th

11. a. 3rd b. 75 c. 83 d. 75

13. Larry

15. It is false; it is impossible to score at the 100th percentile. Ricci is one of the class and that would exclude Ricci.

17. 0

Just for Fun

qoph, qiviut

SECTION 4.5

1. Bus = 180°
Car = 120°
Walk = 60°

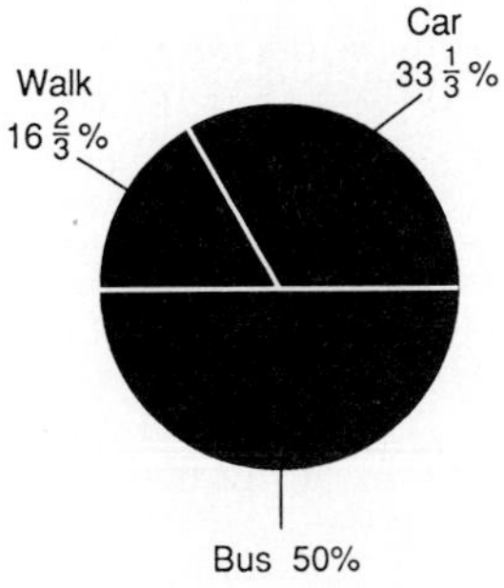

3. Personal income = 90°
Corporate income = 90°
Excise = 54°
Sales = 72°
Highway = 36°
Miscellaneous = 18°

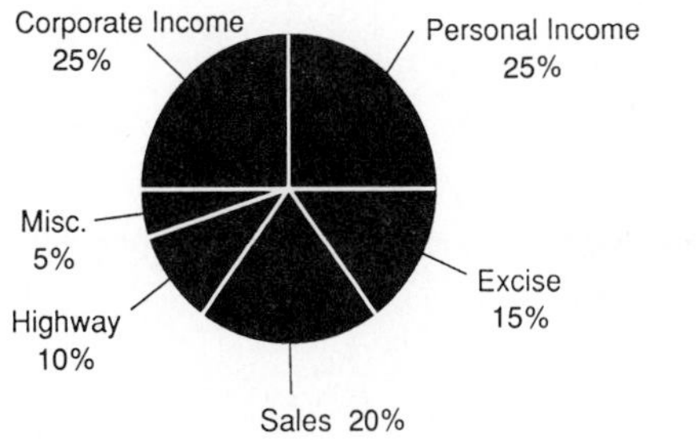

5. Federal aid = 126°
State aid = 54°
Property tax = 108°
Licenses = 18°
Sales tax = 36°
Other = 18°

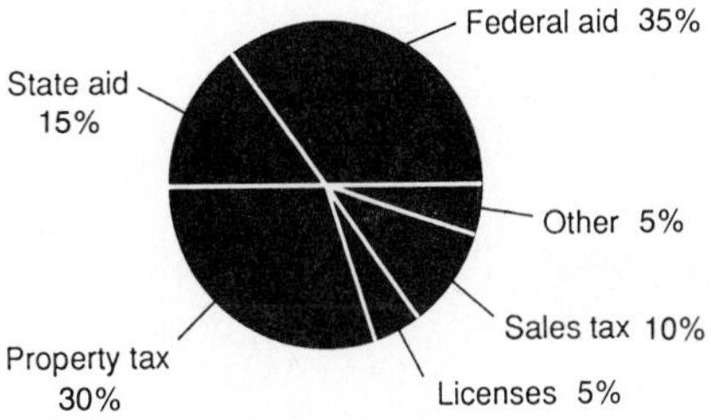

7.

Number	Tally	Frequency
60	//	2
61	/	1
64	//	2
65	//	2
66	//	2
67	////	4
68	//	2
69	//	2
70	~~////~~ /	6
71	//	2
72	//	2
75	/	1

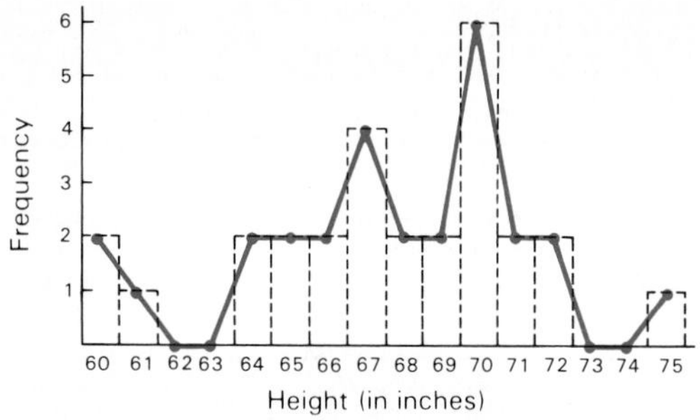

9. a.

Speed	Tally	Frequency
15–19	//	2
20–24	~~////~~ /	6
25–29	~~////~~ //	7
30–34	////	4
35–39	////	4
40–44		0
45–49	/	1
50–54	/	1

b.

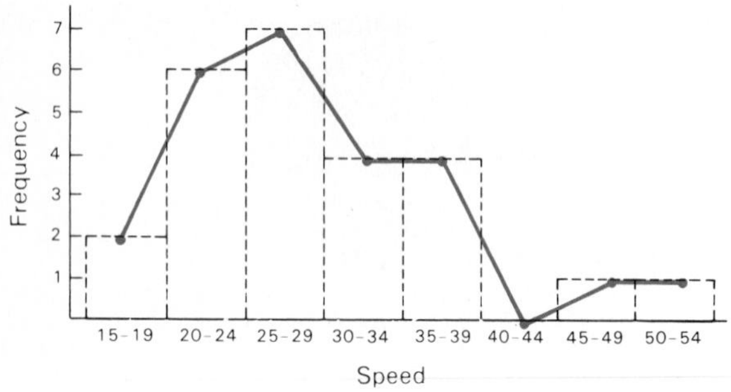

11.

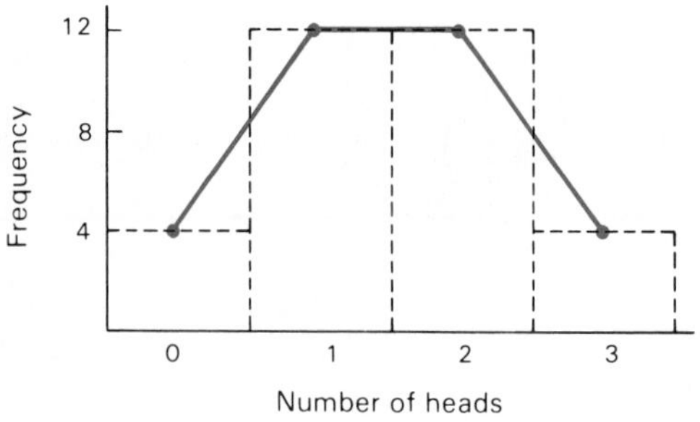

13. a.

Interval	Tally	Frequency
95–99	/	1
90–94	////	4
85–89	////	4
80–84	~~////~~ /	6
75–79	~~////~~ /	6
70–74	~~////~~ /	6
65–69	~~////~~	5
60–64	~~////~~ /	6
55–59	////	4
50–54	///	3
45–49	//	2
40–44	///	3

b.

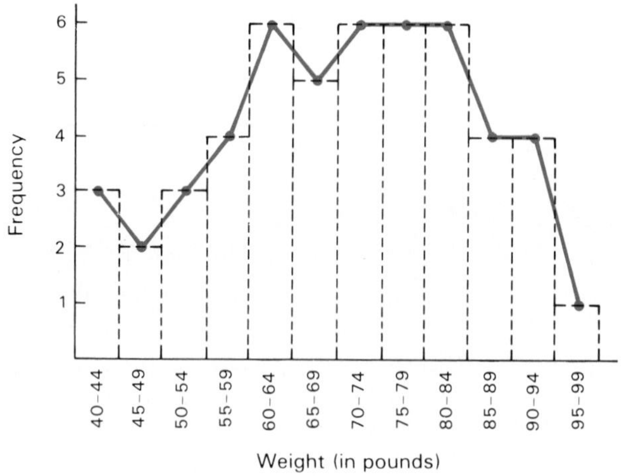

15. No; the graph does not indicate what part or percentage of a dollar is spent.

17. Answers may vary.

Just for Fun

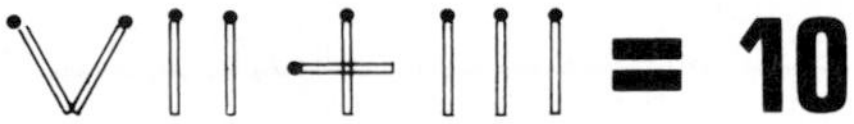

SECTION 4.6

1. a. 68.2% b. 95.4% c. 99.7%
3. a. 50,000 b. 12,000 c. 84.1% d. 318
5. a. 795 b. 115 c. 6.5 d. 4,090

7. a. 2.3% b. 15.9%
   c. 81.8% d. 97.6%

9. a. 4.5% b. 9.1% c. 74.9%
   d. 7.9% e. 3.6%

11. a. 38.4% b. 62.5% c. 2.3%
    d. 69.2% e. 28.5%

13. a. 21.2% b. 3.6% c. 5.5%
    d. 54% e. 17.6%

15. a. 0.6% b. 6.7%

17. a. 38.4% b. 61.6%

19. a. 0.50 or $\frac{50}{100}$
    b. 0.308 c. 0.48

Just for Fun

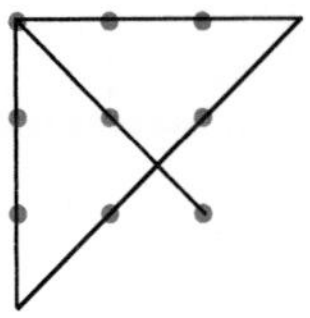

REVIEW EXERCISES FOR CHAPTER 4

1. a. mean, median, mode, midrange
   b. mode c. median d. mean
2. mean = 15.4; mode = 12; median = 16; midrange = 15.5
3. mean = 6; mode = 5; median = 5; midrange = 7.5
4. mean = 9.6; no mode; median = 9; midrange = 10.5
5. mean = 55; no mode; median = 55; midrange = 55
6. 72
7. a. 384 b. 76.8
8. range = 10; midrange = 65; mean = 65
9. range = 42; midrange = 79; mean = 79.7
10. $\sigma = \sqrt{5}$ or 2.2
11. $\sigma = \sqrt{24}$ or 4.9
12. 85 13. Julie
14. a.

Number	Tally	Frequency
6	~~////~~ ~~////~~ /	11
5	~~////~~ ////	9
4	~~////~~ ~~////~~	10
3	////	4
2	~~////~~ ///	8
1	~~////~~ ///	8

b, c.

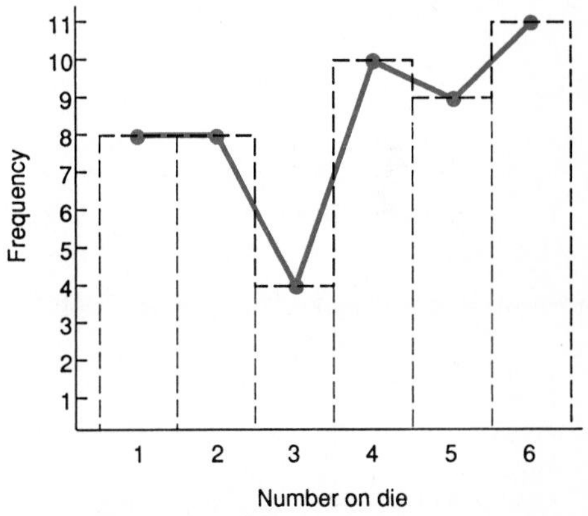

15.

Cigarettes = 144°
Cigars = 108°
Pipe tobacco = 72°
Chewing tobacco = 22°
Snuff = 14°

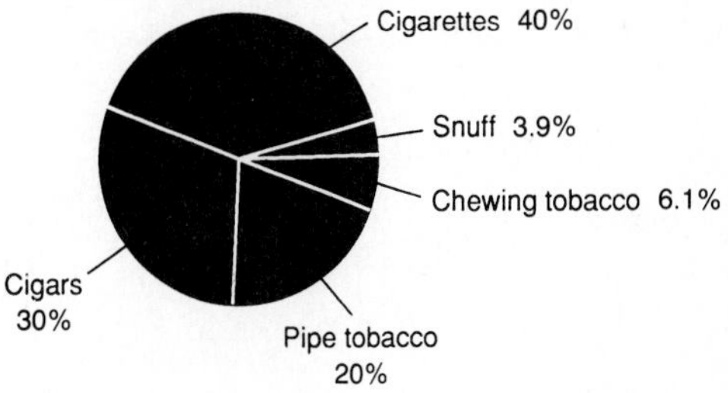

16. a.

Height	Tally	Frequency
75	//	2
74		0
73		0
72	//	2
71	//	2
70	~~////~~///	8
69	////	4
68	//	2
67	///	3
66	//	2
65	//	2
64	///	3

b, c.

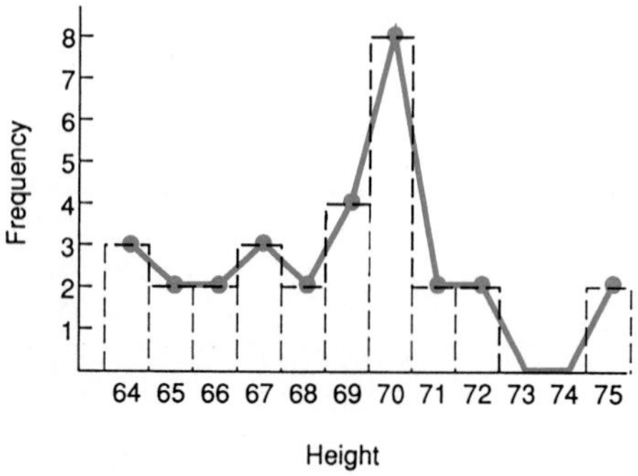

d. No, a normal curve has a normal distribution and the mean, median, and mode all have the same value at the center of distribution.

17. a. 15.9% b. 2.3% c. 15.9% d. 81.8% e. 682

18. a. 5.5% b. 34.5% c. 60% d. 21.2% e. 15.7%

19. a. Everybody but Steve b. Steve c. No one d. Steve

20. a. 32.5 b. 38 c. 46 d. 30 e. 60

21. a. 26 b. 26 c. 26 d. 14 e. 27 f. $\sqrt{20}$ or 4.5

22. a. third b. 75 c. 27 d. 22 e. bimodal, 22 and 20

23. a. 5.8 b. 4 c. 4 d. 6

CHAPTER 4 QUIZ

1. True 2. True 3. True 4. False
5. False 6. True 7. True 8. False
9. False 10. False 11. False 12. False
13. False 14. True 15. False 16. False
17. True 18. True 19. True 20. False
21. False 22. True 23. False 24. True
25. False

CHAPTER 5

SECTION 5.1

Just for Fun

Nautical mile = 6,076 feet; statute mile = 5,280 feet

SECTION 5.2

1. (*i*) To reduce the number of different sizes; (*ii*) to make conversion from one unit to another easier; (*iii*) to create a potential increase in export; (*iv*) because it is the international system of measurement.

3. Cubit, span, palm, digit

5. a. deka b. deci c. kilo d. centi e. hecto f. milli

7. a. 39,370 b. 3,937 c. 393.7 d. 3.937 e. 0.3937 f. 0.03937

9. a. 1,000 b. 10 c. 0.1 d. 0.001 e. 0.01 f. 100

11. a. 100 b. 1,000 c. 10 d. 100 e. 1,000 f. 10

13. a. 2 b. 32,000 c. 3,000 d. 400 e. 300 f. 5,000

15. a. False b. False c. False d. False e. True

Just for Fun

1. meter 2. gram 3. liter 4. dekagram
5. kiloliter 6. decimeter 7. kilogram
8. milliliter 9. centimeter 10. kilometer

SECTION 5.3

1. a. 26 mm b. 60 mm c. 35 mm
 d. 71 mm e. 46 mm f. 34 mm

3. a. 1.8 b. 3,700 c. 320,000
 d. 70,200 e. 0.0423 f. 81.4

5. a. 41.37 b. 4.5 c. 4,700
 d. 0.04378 e. 0.302 f. 98.5

7. 4,200 9. 2,814

11. Yes, 50 mi/hr = 80 km/hr

13. 34,800,000 15. 4 cm^2 17. 2.56

19. a. 1 $in.^2$ b. 1 m^2
 c. 1 m^2 d. 1 mi^2
 e. 1 ha f. 1 are

21. b 23. b 25. c

Just for Fun

151,476

SECTION 5.4

1. a. 1,200 b. 2,500 c. 3.5
 d. 10 e. 6.5 f. 450

3. a. 0.732 b. 2.314 c. 314
 d. 0.01495 e. 490 f. 7,200

5. kiloliter, hectoliter, dekaliter, liter, deciliter, centiliter, milliliter

7. a. 2 kL b. 3 liters c. 3 pt
 d. 2 liters e. 3 tsp f. 5 fl oz

9. 20 liters 11. 1.2 m^3

13. Rice Pudding: 120 mL uncooked rice; 600 mL milk; 60 mL sugar; 80 mL raisins; 7.5 mL cinnamon; 2.5 mL salt

15. b 17. c 19. c

Just for Fun

3,692

SECTION 5.5

1. a. 52,000 b. 3,500 c. 0.25
 d. 0.43 e. 7,200 f. 4,700

3. a. 0.417 b. 342,000 c. 14,300
 d. 0.8859 e. 0.014359 f. 2,710

5. a. 224 b. 220 c. 17.5
 d. 9 e. 90 f. 2.7

7. 0.5 9. 123

11. a. 360 b. 360 t

13. a. 1 kg b. 2 oz c. 5 kg
 d. 3 t e. 2 kg f. 280 g

15. a 17. b 19. b

Just for Fun

13,120

SECTION 5.6

1. a. 40 °C b. 10 °C c. 5 °C
 d. 35 °C e. 30 °C f. −5 °C

3. a. 68 °F b. 122 °F c. 185 °F
 d. 50 °F e. 32 °F f. 149 °F

5. a. −10 °C b. 59 °F c. 77 °F
 d. 30 °C

7. 104 °F 9. 3000 °C 11. −50 °C

13. **58 °C** 15. 57 °C 17. b

19. a 21. b

Just for Fun

a. It's third down and *9 meters* to go.
b. Give them *2.5 cm* and they will take *1.6 km*.

c. The cowboy wore a *38 liter* hat.
d. Did you see the *30-cm* note?
e. A miss is as good as *1.6 km*.
f. He climbed the rope *2.5 cm* by *2.5 cm*.
g. *Twenty-eight grams* of prevention are worth *0.45 kg* of cure.
h. Have you read *"Celsius 233"*?
i. Erskine Caldwell wrote *"God's Little 0.4 Hectare."*
j. A *0.47 liter* is *0.45 kg* the world around.

REVIEW EXERCISES FOR CHAPTER 5

1. The metric system is a system of measurement with basic units of measure for length, area, volume, and weight in decimal relationship to each other. The basic unit is international and relates to units of length, volume, and weight. It is a well-planned, logical system with uniformity, allowing for easier and more precise calculations.
2. a. kilo b. deka c. deci d. hecto e. centi f. milli
3. One ten-millionth of the distance from the north pole to the equator, along the meridian that passes through Dunkirk and Paris.
4. a. 100 b. 2 c. 4 d. 350 e. 32,000 f. 1.5
5. a. 39,370 b. 0.03937 c. 3,937 d. 0.3937 e. 393.7 f. 3.937
6. 2803.2
7. 2532.8
8. a. 60 b. 90 c. 8 d. 11 e. 6 f. 65
9. A liter is defined as the volume of a cube that is 1 decimeter long, 1 decimeter wide, and 1 decimeter high.
10. a. 2,000 b. 30 c. 50 d. 1 e. 200 f. 2
11. 90 liters
12. 240 liters
13. The weight of one gram equals the weight of one milliliter of very cold water.
14. a. 4,000 b. 8,000 c. 3 d. 200 e. 4 f. 1,800
15. a. 1,000 b. 1,000,000 c. 10 d. 100,000 e. 45 f. 50,000
16. −88 °C
17. 136 °F
18. c
19. c
20. b
21. d
22. a
23. c
24. c
25. b
26. c
27. a
28. b
29. c
30. a. 10 b. 0.95 c. 106 d. 56 e. 90 f. 35
31. Sour Cream Cookies

120 mL sour cream	224 g butter
5 mL vanilla	5 mL baking soda
240 mL brown sugar	600 mL flour
2 eggs	Bake at 177 °C

Just for Fun

A	K	E	I	B	C	U	B	I	C	M	E	T	E	R
M	I	L	L	I	G	R	A	M	A	B	C	D	R	E
I	L	O	U	L	A	E	I	O	U	H	F	M	G	T
L	O	C	E	L	S	I	U	S	H	E	I	R	J	E
L	L	E	K	L	M	N	O	P	Q	C	E	T	R	M
I	I	N	J	S	T	U	S	C	O	T	T	R	V	A
M	T	T	O	N	N	E	W	X	E	A	Y	I	Z	K
E	E	I	E	A	G	R	A	M	M	R	M	C	C	E
T	R	M	B	C	D	E	O	A	F	E	G	H	I	D
E	J	E	K	L	M	L	R	D	S	H	T	A	M	E
R	O	T	P	Q	I	G	I	R	N	E	T	E	S	R
T	U	E	V	K	A	W	X	T	Y	A	T	A	R	A
Z	T	R	E	K	L	F	O	U	E	B	L	E	M	U
O	H	T	E	N	O	A	D	D	I	R	E	S	K	Q
T	A	D	C	U	B	I	C	M	E	T	E	R	I	S

CHAPTER 5 QUIZ

1. False
2. True
3. True
4. False
5. True
6. False
7. False
8. False
9. False
10. False
11. True
12. False

13. False 14. True 15. True 16. False
17. True 18. False 19. True 20. True
21. True 22. True 23. True 24. True
25. False

CHAPTER 6

SECTION 6.2

1. a. 6 b. 1 c. 11 d. 10
   e. 9 f. 7
3. a. 9 b. 8 c. 7 d. 6
   e. 1 f. 8
5. a. 10 b. 10 c. 8 d. 11
   e. 9 f. 8
7. a. 12 b. 12 c. 9 d. 4
   e. 7 f. 9
9. a. 2 b. 6 c. 8 d. 9
   e. 6 f. 10
11. a. Closure for addition
    b. Commutative property for multiplication
    c. Commutative property for addition
    d. Associative property for multiplication
    e. Inverse element for multiplication
    f. Identity element for addition
13. a. True b. False c. True
    d. False e. False f. True
15. Yes, it satisfies the required properties.
17. a. 7 b. 1 c. 5 d. 3
    e. 3 f. 2
19. Assuming zero is allowed, when $a = b = c$; or when $a = 9$, $b = 3$, $c = 6$.

Just for Fun

Yes

SECTION 6.3

1. Spring 3. Summer 5. Fall
7. Winter 9. Winter 11. Spring
13. Spring 15. Fall 17. No
19. Monday 21. Wednesday 23. Monday
25. Friday 27. Thursday 29. Thursday
31. Saturday 33. Tuesday 35. Saturday
37. a. Thursday b. Wednesday c. Tuesday
39. Sunday × (Tuesday × Friday)
    $\stackrel{?}{=}$ (Sunday × Tuesday) × Friday
    Sunday × Wednesday
    $\stackrel{?}{=}$ Tuesday × Friday
    Wednesday = Wednesday
41. Yes

Just for Fun

It is practically impossible.

SECTION 6.4

1. a. 4 b. 0 c. 1 d. 2
   e. 4 f. 1
3. a. 3 b. 4 c. 0 d. 2
   e. 0 f. 2
5. a. 3 b. 2 c. 3 d. 4
   e. 0 f. 3
7. a. 2 b. 3 c. 1 d. 4
   e. 3 f. 3
9.

+	0	1	2	3	4	5	6
0	0	1	2	3	4	5	6
1	1	2	3	4	5	6	0
2	2	3	4	5	6	0	1
3	3	4	5	6	0	1	2
4	4	5	6	0	1	2	3
5	5	6	0	1	2	3	4
6	6	0	1	2	3	4	5

a. Yes b. 0 c. Yes
d. For example,
$(1 + 3) + 5 \stackrel{?}{=} 1 + (3 + 5);$
$4 + 5 \stackrel{?}{=} 1 + 1;$
$2 = 2$
e. Element: 0, 1, 2, 3, 4, 5, 6
Inverse: 0, 6, 5, 4, 3, 2, 1
f. Yes

11. a. 4 b. 4 c. 2 d. 6
e. 3 f. 6

13. a. True b. False c. True
d. True e. True f. False

15. a. True b. False c. False
d. False e. True f. True

17. a. 5 b. 4 c. 1 d. 2
e. 6 f. 11

19. 51

Just for Fun

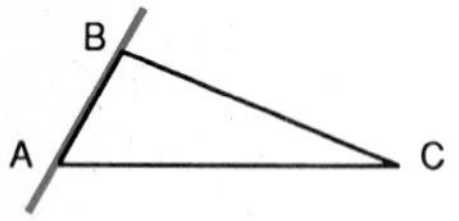

SECTION 6.5

1. Q 3. R 5. S 7. R

9. S

11. Yes; there are no new elements.

13. All of them: the inverse of P is P; the inverse of Q is S; the inverse of R is R; and the inverse of S is Q.

15. ! 17. ! 19. ? 21. !

23. ?

25. No; there is a new element in the table, namely, !.

27. None 29. No 31. a

33. d 35. b 37. e

39. e 41. e 43. a

45. Yes; the symmetric table indicates commutativity.

47. a. Yes b. None c. None d. No

49. a.

*	a	b	c
a	b	c	a
b	c	a	b
c	a	b	c

b. Yes c. c
d. All; the inverse of a is b; the inverse of b is a; and the inverse of c is c.
e. Yes

Just for Fun

6 dozen dozen

SECTION 6.6

1. a, d

3. An axiomatic system consists of four main parts:
   1. Undefined terms
   2. Defined terms (definitions)
   3. Axioms
   4. Theorems

5. 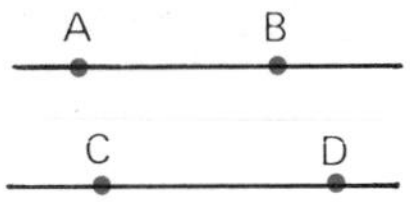

Axiom II tells us there is a line, say line AB. Axiom I tells us there are exactly two points on the line, A and B. Axiom IV tells us there is another line (say line CD) that has no points in common with line AB. Axiom I tells us there are exactly two points on the line, C and D. Axiom III tells us that the lines are distinct because for each pair of points there is one and only one line containing them. We have two pairs of points, or at least four points.

7. a. 1. axiom II; 2. axiom IV
   b. 1. axiom I; 2. axiom IV
   c. 1. axiom I; 2. axiom III
   d. 1. axiom II; 2. axiom IV; 3. axiom V; 4. axiom II
   e. 1. axiom I: 2. axiom IV: 3. axiom V; 4. axiom I: 5. axiom II

Just for Fun

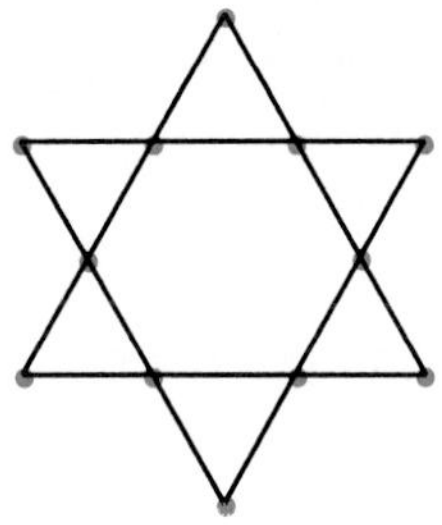

REVIEW EXERCISES FOR CHAPTER 6

1. a. 2 b. 1 c. 3 d. 11 e. 8 f. 5
2. a. 10 b. 8 c. 8 d. 2 e. 8 f. 11
3. a. 12 b. 8 c. 6 d. 3 e. 6 f. 12
4. a. 2 b. 12 c. 6 d. 2 e. 8 f. 4
5. A *mathematical system* is a set of elements together with one or more operations (rules) for combining those elements.
6. a. True b. False c. True d. False
7. a. True b. False c. True d. False
8. a. True b. True
9. a. False b. False
10. a. Winter b. Summer c. Fall d. Fall e. Fall f. Fall
11. a. Spring b. Summer c. Spring d. Fall e. Winter f. Summer
12. a. 2 b. 1 c. 4 d. 3 e. 2 f. 3
13. a. 3 b. 5 c. 3 d. 5 e. 1 f. 1
14. 57
15. The odometer recycles after 100,000 miles, modulo 100,000.
16. a. \$ b. ? c. π d. π e. & f. \$ g. ? h. π
    i. No; there is a new element in the table, namely, π.
    j. ¢ k. \$, ¢ l. no
17. An axiomatic system consists of four main parts:
    1. Undefined terms
    2. Defined terms (definitions)
    3. Axioms
    4. Theorems
18. We must have three squirrels (axiom I). They must be in a tree (axiom II). If they are all in the same tree, then there must be another squirrel in another tree (axioms IV and II). Therefore there are at least four squirrels.

    If only two of the three squirrels are in a given tree (axioms II and III), then the third squirrel is in another tree (axiom II) together with a fourth squirrel (axiom III).
19. $3x + 4 + 2x + 8 + 5x - 4 = 48$

    $10x + 8 = 48$

    $10x = 40$

    $x = 4$

    If $x = 4$, $3x + 4 = 12 + 4 = 16$

    $2x + 8 = 8 + 8 = 16$

    $5x - 4 = 20 - 4 = 16$

    and all three sides are equal.
20. $5x - 5 + 2x + 9 + 4x - 17 + 3x + 11 = 180$

    $14x - 2 = 180$

    $14x = 182$

    $x = 13$

    If $x = 13$, $m \angle DOC = 35$, and $m \angle COB = 35$ and $\overline{OC}$ bisects $\angle BOD$.

Just for Fun

The error occurs when we divide by $(a - b)$. If $a = b$, then $(a - b) = 0$, and we cannot divide by zero.

CHAPTER 6 QUIZ

1.	True	2.	True	3.	True	4.	False
5.	True	6.	False	7.	False	8.	True
9.	False	10.	False	11.	False	12.	True
13.	True	14.	True	15.	False	16.	False
17.	False	18.	False	19.	True	20.	True
21.	False	22.	False	23.	False	24.	True
25.	True						

CHAPTER 7

SECTION 7.2

1. a.  b.  c. 
   d.  e. 
   f. 

3. a.  b. 
   c. 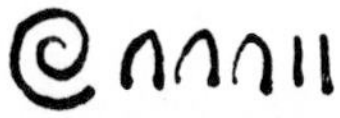
   d. 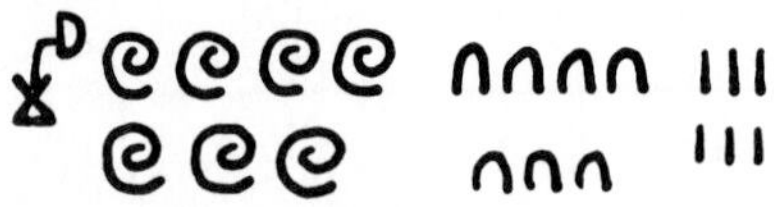
   e. 
   f. 

5. a. 22 b. 1,213 c. 1,101
   d. 12,212 e. 1,222 f. 2,212

7. a.  b. 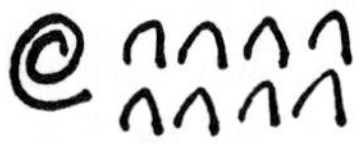
   c.  d. 

9. a.  b. 
   c.  d. 

11. a.  b. 
    c.  d. 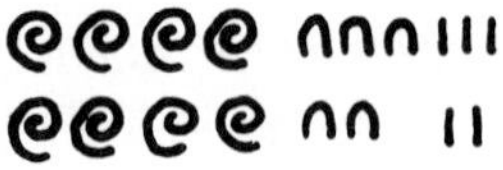

Just for Fun

1.	451	2.	7	3.	40	4.	500
5.	40	6.	4	7.	10	8.	100
9.	20,000	10.	600				

SECTION 7.3

1. a. ΔΓIII b. ΔΔIII c. ΔΔΔIIII
   d. ΔΔΔΔIIII e. HΓΔΔΔΓII
   f. ΓHΓΔΔΔΔΔΓIII

3. a. ΔΔI b. ΓΔΓII
   c. HΔΔΔΓII d. XΓHHHΓΔΔΔΓI
   e. XΓHHHHHΓΔΔΔΔIIII
   f. XXI

5. a. 17 b. 117 c. 11,010 d. 656
   e. 6,601 f. 5,555

7. a. 二十一 b. 五十七 c. 百三十七

d. 千七百七十六

e. 千九百八十四

f. 二千一

9. a. 十六 b. 五十四 c. 百四十七 d. 八百九十七

e. 三千四百七十三 f. 四千百七十六

11. a. 527 b. 315 c. 1,110 d. 8,236

Just for Fun

9,567
+ 1,085
10,652

SECTION 7.4

1. a. 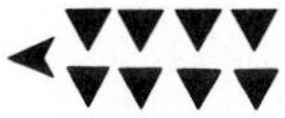 b. 

c.  d. 

e. 

f. 

3. a.  b. 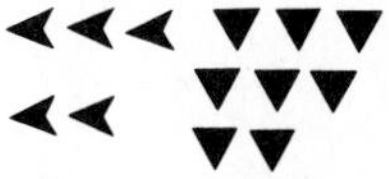

c.  d. 

e.  f. 

5. a. 91 b. 142 c. 71 d. 81

7. a. $(2 \times 10^2) + (4 \times 10^1) + (3 \times 10^0)$
b. $(3 \times 10^2) + (7 \times 10^1) + (8 \times 10^0)$
c. $(1 \times 10^3) + (2 \times 10^2) + (3 \times 10^1) + (4 \times 10^0)$
d. $(2 \times 10^3) + (5 \times 10^1) + (1 \times 10^0)$
e. $(1 \times 10^4) + (4 \times 10^2) + (1 \times 10^0)$

9. a. $(4 \times 10^2) + (2 \times 10^0)$
b. $(1 \times 10^3) + (4 \times 10^2) + (7 \times 10^1) + (6 \times 10^0)$
c. $(2 \times 10^4) + (1 \times 10^2) + (8 \times 10^1) + (2 \times 10^0)$
d. $(2 \times 10^3) + (3 \times 10^2) + (8 \times 10^1) + (1 \times 10^0)$
e. $(1 \times 10^4) + (5 \times 10^1)$

11. a. 240 b. 2,311 c. 1,776
d. 4,213 e. 204 f. 40,301

Just for Fun

Turn the page upside down.

SECTION 7.5

1. a. 7 years 4 months b. 2 years 5 months
c. 25 feet 9 inches d. 3 feet 11 inches

e. 7 gross 3 dozen 2 units
f. 2 gross 9 dozen 8 units

3. a. 10 feet 3 inches
b. 2 feet 5 inches
c. 6 years 3 months
d. 10 months
e. 1 great gross 2 gross 4 dozen 3 units
f. 3 great gross 9 gross 8 dozen 10 units

5. a. 8 b. 24 c. 66 d. 79 e. 25 f. 511

7. a. 11_{five} b. 34_{five} c. 123_{five}
d. 3_{five} e. 441_{five} f. 3442_{five}

9. a. 11_{five} b. 2_{five} c. 21_{five}
d. 143_{five} e. 401_{five} f. 13001_{five}

11. a. 50 b. 64 c. 117 d. 187 e. 339 f. 3,457

13. a. 36_{twelve} b. 45_{twelve} c. 50_{twelve}
d. $E5_{twelve}$ e. 176_{twelve} f. 610_{twelve}

15. a. 37_{twelve} b. 51_{twelve} c. 94_{twelve}
d. 149_{twelve} e. 300_{twelve} f. 2459_{twelve}

17. a. 123 b. 142 c. 568 d. 1,577

19. a. 43 b. 47 c. 22 d. 32
e. 19 f. 240

21. a. 37 b. 66 c. 123
d. 160 e. 280 f. 875

23. 1,915 25. a. 210_5 b. 17_{12}

Just for Fun

1 fathom = 6 feet; 1 hand = 4 inches;
3 barleycorns = 1 inch

SECTION 7.6

1. a. 41_5 b. 41_5 c. 110_5
d. 302_5 e. 1003_5 f. 1010_5

3. a. 113_5 b. 140_5 c. 101_5
d. 311_5 e. 1131_5 f. 3230_5

5. a. 4_5 b. 3_5 c. 3_5
d. 102_5 e. 2424_5 f. 424_5

7. a. 1243_5 b. 1414_5 c. 3333_5
d. 10401_5 e. 14112_5 f. 32343_5

9. a. 1234_5 b. 424_5 c. 4301_5
d. 10111_5 e. 22142_5 f. 34221_5

11. a. 34_5, $R = 1_5$ b. 112_5, $R = 2_5$
c. 44_5, $R = 1_5$ d. 224_5, $R = 2_5$
e. 22_5, $R = 1_5$ f. 22_5

Just for Fun

for, four, fore

SECTION 7.7

1. a. 2 b. 3 c. 5
d. 7 e. 11 f. 27

3. a. 8 b. 9 c. 12
d. 15 e. 16 f. 17

5. a. 110_2 b. 1001_2 c. 1101_2
d. 10101_2 e. 11001_2 f. 100001_2

7. a. 101_2 b. 110_2 c. 100_2
d. 1000_2 e. 1001_2 f. 1100_2

9. a. 1101_2 b. 10000_2 c. 10101_2
d. 11100_2 e. 10100_2 f. 11110_2

11. a. 10_2 b. 101_2 c. 10_2
d. 101_2 e. 110_2 f. 11_2

13. a. 1001_2 b. 110_2 c. 1111_2
d. 10010_2 e. 100011_2 f. 11001_2

15. a. 10101_2 b. 1000_2 c. 1100_2
d. 10100_2 e. 111100_2 f. 101101_2

17. 1313_4 19. 111_8

21. a. False b. False c. True d. True

23. a. 101_6 b. 1060_7 c. 15_7 d. 155_6

25. a. 2036_7 b. 2000_6
c. 1322_8 d. 2146_9

27. a. 20211_3 b. 22452_6
c. 20411_5 d. 22312_7

Just for Fun

Let the Yankee fans be Y_1, Y_2, Y_3, and let the Mets fans be M_1, M_2, M_3.

1. M_1, M_2 go up, leaving Y_1, Y_2, Y_3, M_3.
2. Elevator returns to 1st floor empty.
3. Y_1, Y_2 go up, leaving Y_3, M_3 and joining M_1, M_2.
4. Elevator returns empty.
5. Y_3 and M_3 go up, joining Y_1, Y_2, M_1, M_2.

REVIEW EXERCISES FOR CHAPTER 7

1. a.  b. ΓΔΔΔΓIII

   c.  d. 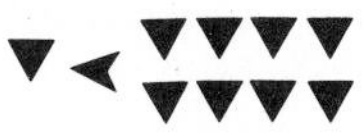

2. a.  b.  c. 

3. a. 1,233 b. 12,022

4. a.  b. 

   c. 

   d. 

5. a. *Simple grouping:* The position of a symbol does not affect the number represented, and a new symbol is used to indicate a certain number or group of things.
   b. *Multiplicative grouping:* Symbols are used for numbers in a basic group, together with a second symbol or notation to represent multiples of the basic group.
   c. *Place-value:* The position of a symbol matters.

6. a. Egyptian
   b. Greek, Chinese-Japanese
   c. Babylonian, Hindu-Arabic

7. a. $(3 \times 10^2) + (4 \times 10^1) + (5 \times 10^0)$
   b. $(3 \times 5^2) + (4 \times 5^1) + (2 \times 5^0)$
   c. $(1 \times 2^4) + (0 \times 2^3) + (1 \times 2^2) + (1 \times 2^1) + (1 \times 2^0)$

8. a. 55 b. 19 c. 71
   d. 30 e. 21 f. 419

9. a. 140_5 b. 201_5 c. 28_{12}
   d. 100001_2 e. 111111_2 f. $2T_{12}$

10. Base 12 is used when buying many items; for example, we use dozen (12^1), gross (12^2), and great gross (12^3). Modern computers work in base 2, 8, or 16.

11. a. 100_5 b. 1040_5 c. 4_5 d. 44_5

12. a. 1243_5 b. 1102_5 c. 20004_5 d. $34_5r = 1_5$

13. a. 1011_2 b. 1010_2 c. 11010_2 d. 11_2

14. a. 100_2 b. 110_2 c. 1110_2 d. 110111_2

15. a. 222_6 b. 211_3 c. 43_5 d. 111111_2

16. a. 12_5, 111_2 b. 23_5, 1101_2 c. 113_5, 100001_2

17. a. 53 b. 25 c. 53
    d. 326 e. 121 f. 243

18. a. 36_9 b. 64_7 c. 401_4
    d. 10102_3 e. 203_6 f. 207_8

19. a. 1200_6 b. 1220_4
    c. 52_7 d. 215_9

20. a. 22112_6 b. 10011_3
    c. 16123_7 d. 25503_9

Just for Fun

4.5 inches

CHAPTER 7 QUIZ

1.	True	2.	False	3.	True	4.	True
5.	False	6.	True	7.	True	8.	True
9.	False	10.	True	11.	True	12.	False
13.	False	14.	False	15.	True	16.	False
17.	False	18.	False	19.	False	20.	False
21.	True	22.	True	23.	False	24.	False
25.	True						

CHAPTER 8

SECTION 8.2

1. a. Cardinal b. Ordinal
   c. Identification d. Identification

3. a. 2,688 is divisible by 2, 3, 4, 8
   b. 73,440 is divisible by 2, 3, 4, 5, 8, 9, 10
   c. 3,290,154 is divisible by 2, 3

5. 2, 3, 5, 7, 11, 13, 17, 19, 23, 29

7. a. Prime b. Prime c. Neither
   d. Composite e. Composite f. Composite

9. a. $2^2 \times 3^2$ b. $2^3 \times 3^2$ c. $2^3 \times 3^3$
   d. $5^2 \times 19$ e. 5^4 f. 3×7^2

11. For example: 11 and 13; 17 and 19; 41 and 43; 59 and 61; 71 and 73

Just for Fun

28; 496; 8,128

SECTION 8.3

1. a. 2 b. 14 c. 3
   d. 26 e. 3 f. 12

3. a. $\frac{5}{6}$ b. $\frac{7}{9}$ c. $\frac{3}{5}$
   d. $\frac{2}{3}$ e. $\frac{147}{152}$ f. $\frac{1}{2}$

5. a. 56 b. 28 c. 120
   d. 156 e. 990 f. 9,879

7. a. $\frac{35}{36}$ b. $\frac{23}{36}$ c. $\frac{11}{12}$
   d. $\frac{19}{24}$ e. $\frac{6}{55}$ f. $\frac{43}{60}$

9. 18

11. a. $\frac{61}{77}$ b. $\frac{3}{4}$ c. $\frac{73}{120}$
    d. $\frac{7}{55}$ e. $\frac{19}{72}$ f. $\frac{13}{45}$

13. a. 24 b. 30 c. 45
    d. 40 e. 42 f. 105

15. a. $\frac{1}{3}$ b. $\frac{1}{4}$ c. $\frac{1}{6}$
    d. $\frac{1}{19}$ e. $\frac{1}{3}$ f. $\frac{1}{10}$

17. a. 4 b. 2 c. 1
    d. 1 e. 9 f. 6

Just for Fun

$4 = 2 + 2$; $6 = 3 + 3$; $8 = 5 + 3$;
$10 = 5 + 5$; $12 = 7 + 5$; $14 = 7 + 7$;
$16 = 13 + 3$; $18 = 13 + 5$; $20 = 13 + 7$;
$22 = 11 + 11$; $24 = 19 + 5$; $26 = 19 + 7$;
$28 = 23 + 5$; $30 = 23 + 7$

SECTION 8.4

1. a. −1 b. −3 c. −3
   d. 2 e. 1 f. 3

3. a. −2 b. −3 c. −14
   d. 11 e. −1 f. −4

5. a. −5 b. 0 c. 25
   d. 0 c. −35 f. 9

7. a. 20 b. −24 c. −24
   d. 16 e. 40 f. 126

9. a. 17 b. 3 c. 6
   d. 20 e. 12 f. 4

11. a. $>$ b. $<$ c. $=$ d. $<$ e. $=$ f. $>$

13. a. Even: $12 = 2 \cdot 6$ b. Odd: $17 = 2 \cdot 8 + 1$
    c. Even: $-6 = 2(-3)$ d. Odd: $101 = 2 \cdot 50 + 1$

15. 833

17. Aristotle

19. $(2k + 1) + 2n = (2k + 2n) + 1 = 2(k + n) + 1$, which is odd.

Just for Fun

It's a fake! No authentic coins would be dated B.C.

SECTION 8.5

1. a. $\frac{3}{8}$ b. $\frac{1}{9}$ c. $\frac{27}{43}$ d. $\frac{27}{224}$

3. a. $\frac{33}{35}$ b. $\frac{11}{12}$ c. $\frac{17}{72}$
   d. $\frac{5}{33}$ e. $\frac{1}{80}$ f. $\frac{5}{9}$

5. a. $\frac{8}{35}$ b. $\frac{12}{55}$ c. $\frac{32}{91}$
d. $\frac{14}{5}$ e. $\frac{27}{44}$ f. $\frac{14}{13}$

7. a. $\frac{27}{44}$ b. $\frac{14}{65}$ c. $\frac{2}{15}$
d. $\frac{9}{10}$ e. $\frac{9}{10}$ f. $\frac{50}{63}$

9. a. $\frac{15}{8}$ b. $\frac{40}{3}$ c. $\frac{4}{15}$
d. $\frac{3}{10}$ e. $\frac{44}{75}$ f. $\frac{6}{11}$

11. a. $\frac{17}{32}$ b. $\frac{17}{12}$ c. $\frac{7}{24}$
d. $\frac{27}{12}$

13. a. False b. True c. False
d. False e. False f. True

15. a. False b. True c. False
d. True e. False

Just for Fun

$\frac{1}{2}$

SECTION 8.6

1. a. 0.375 b. 0.3125 c. $0.\overline{6}$
d. $0.\overline{21}$ e. $0.\overline{09}$ f. $0.\overline{405}$

3. a. 0.875 b. 0.4 c. $0.\overline{81}$
d. 0.0625
e. $0.\overline{714285}$
f. $0.\overline{4705882352941176}$

5. a. 1/8 b. 1/400 c. 7/9 d. 34/99
e. 691/110 f. 1

7. a. $\frac{3}{4}$ b. $\frac{7}{8}$ c. $\frac{8}{9}$
d. $\frac{5}{11}$ e. $\frac{235}{999}$ f. $\frac{1379}{300}$

9. a. $\frac{5}{12}$ b. $\frac{7}{24}$ c. $\frac{9}{40}$
d. $\frac{37}{48}$ e. $\frac{69}{88}$ f. $\frac{277}{352}$

11. a. $\frac{5}{16}$ b. $\frac{9}{18}$ c. $\frac{1}{16}$
d. $\frac{19}{20}$ e. $\frac{61}{112}$ f. 1.9995

13. 0.005

Just for Fun

1,135; that is, 11 + 3 + 5 = 19

SECTION 8.7

1. a. Rational b. Rational
c. Rational d. Irrational
e. Irrational f. Rational

3. a. Rational b. Rational
c. Rational d. Rational
e. Rational f. Irrational

5. a. Terminating decimal
b. Repeating decimal
c. Nonterminating, nonrepeating decimal
d. Terminating decimal
e. Nonterminating, nonrepeating decimal
f. Repeating decimal

7. a. True b. True c. True
d. False e. False

9. a. False b. False c. True
d. False e. False f. True

Just for Fun

We use $\frac{22}{7}$ only for sake of convenience. It is a close approximation of π.

SECTION 8.8

1. 1.6×10^2 3. 2.35×10^5

5. 3.4×10^6 7. 9.374×10^2

9. 1.23×10^{-1} 11. 2.345×10^{-4}

13. 1.2×10^{-3} 15. 9×10^{-8}

17. 4×10^6 19. 1×10^{-13}

21. 0.0041 23. 314,200

25. 48,590,000 27. 0.000048

29. 8.84×10^7 31. 3.24×10^5

33. 20.57×10^1 35. 4.35

37. 3×10^3 39. 0.5×10^2

41. 1.5×10^3 43. 2×10^2

45. 10^{-2} 47. 1×10^7

49. 48

Just for Fun

The Englishman, Mr. Dunn

REVIEW EXERCISES FOR CHAPTER 8

1. a. Composite b. Prime c. Prime
   d. Prime e. Composite f. Composite

2. a. $2 \times 3 \times 13$ b. 3×37 c. $5^2 \times 19$
   d. 3×7^2 e. $3 \times 7 \times 43$ f. 11×101

3. a. 6 b. 1 c. 24
   d. 6 e. 3 f. 18

4. a. 240 b. 2,310 c. 144
   d. 990 e. 8,547 f. 11,628

5. a. -4 b. -1 c. -1
   d. -4 e. 16 f. -13

6. a. 6 b. -16 c. -24
   d. -9 e. -8 f. 40

7. a. 8 b. 4 c. 9 d. 1
   e. 2 f. 36

8. a. $\frac{11}{15}$ b. $\frac{55}{63}$ c. $\frac{37}{143}$
   d. $\frac{5}{21}$ e. $\frac{1}{10}$ f. $\frac{19}{40}$

9. a. $\frac{2}{15}$ b. $\frac{4}{21}$ c. $\frac{6}{55}$
   d. $\frac{7}{6}$ e. $\frac{6}{5}$ f. $\frac{10}{11}$

10. a. $\frac{27}{32}$ b. $\frac{3}{20}$ c. $\frac{20}{3}$
    d. $\frac{25}{28}$ e. $\frac{3}{2}$ f. $\frac{1}{4}$

11. a. 0.875 b. 0.4375 c. 0.18
    d. $0.\overline{351}$ e. $0.\overline{384615}$ f. $0.\overline{428571}$

12. a. $\frac{3}{4}$ b. $\frac{213}{1000}$ c. $\frac{157}{50}$
    d. $\frac{46}{99}$ e. $\frac{247}{99}$ f. $\frac{1373}{333}$

13. a. Rational b. Rational c. Irrational
    d. Irrational e. Irrational f. Rational

14. a. False b. False c. True
    d. True e. True f. True

15.

	a.	b.	c.	d.	e.	f.
I:	yes	no	no	no	no	no
II:	yes	no	yes	no	no	no
III:	yes	yes	yes	no	no	no
IV:	yes	yes	yes	yes	yes	no
V:	no	no	no	no	no	yes
VI:	yes	yes	yes	yes	yes	yes

	g.	h.	i.
I:	yes	no	no
II:	yes	no	no
III:	yes	no	no
IV:	yes	no	yes
V:	no	yes	no
VI:	yes	yes	yes

16. a. 2.84×10^9 b. 5.46×10^{-8}
    c. 1×10^7 d. 4.8×10^1

Just for Fun

The number of letters in the word equals the digit.

CHAPTER 8 QUIZ

1. False 2. False 3. False 4. False
5. False 6. True 7. True 8. False
9. True 10. True 11. True 12. False
13. True 14. True 15. False 16. False
17. False 18. True 19. True 20. False
21. False 22. False 23. False 24. False
25. False

CHAPTER 9

SECTION 9.2

1. $\{3\}$ 3. $\{-1\}$ 5. $\varnothing$
7. $\{40\}$ 9. $\{8\}$
11. $\{1, 2, 3, 4\}$ 13. $\{x \mid x > 1\}$ 15. $\{1, 2\}$
17. $\{0, 1, 2, 3, 4\}$ 19. $\{-1, 0, 1, 2, 3\}$

21. 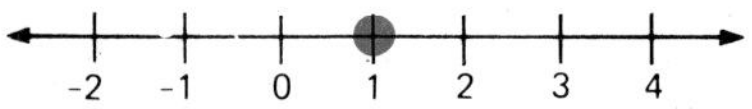

23. 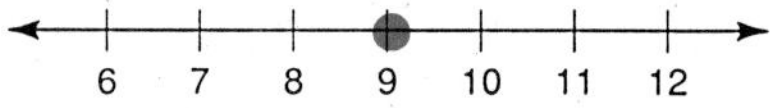

25. 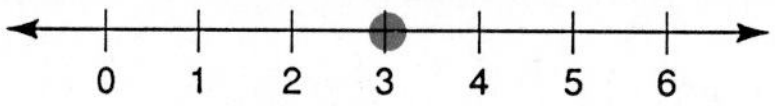

27. 

29. 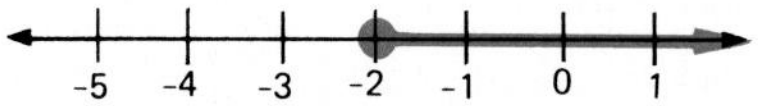

31. 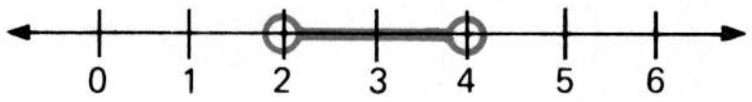

33. 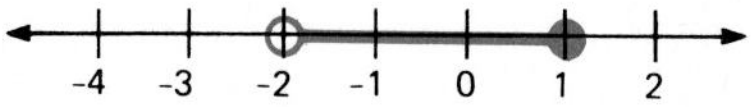

35. 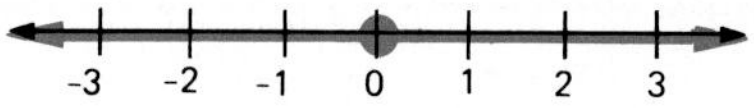

Just for Fun

Write down the names of the numbers, that is, one, two, three, four, and so on.

SECTION 9.3

1. $5x$ 3. $3x$ 5. $2y$
7. $8x$ 9. $14y$ 11. $5x$
13. $2x$ 15. $5x$ 17. $5y + 2$
19. $10y + 4$ 21. $z + 2$ 23. $3y + 1$
25. $12x + 2$ 27. $3x - 6$ 29. $9z + 3$
31. 1 33. $7x + 2$ 35. $15z + 2$
37. x 39. $7z - 2$ 41. $23x$
43. $6x + 1$ 45. $y + 2$ 47. $-7y + 20$

Just for Fun

37

SECTION 9.4

1. $x = 2$ 3. $x = 8$ 5. $y = 6$
7. $y = 6$ 9. $x = 5$ 11. $y = 4$
13. $x = 1$ 15. $x = 3$ 17. $y = 3$
19. $x = 10$ 21. $x = 56$ 23. $x = 2$
25. $z = 4$ 27. $x = 4$ 29. $x = 2$
31. $x = 1$ 33. $x = 4$ 35. $x = 2$
37. $x = 1$ 39. $x = 1$ 41. $x = 6$

SECTION 9.5

1. 7 3. 7 5. 7, 8
7. Lewis is 30, Julia is 23.
9. 11 11. $l = 14$ ft, $w = 11$ ft
13. $l = 34$ m, $w = 16$ m 15. 19, 21
17. 30, 31, 32 19. 5 quarters, 10 dimes
21. 21 dimes, 7 quarters, 32 nickels
23. 10 quarters, 10 dimes, 15 nickels
25. 50°, 50°, 80° 27. 13 dimes, 17 quarters
29. 6 m

Just for Fun

1

SECTION 9.6

Note: Answers may vary. The following are some possible solutions.

1. $(5,0)(0,5)(3,2)$
3. $(3,0)(0,-3)(4,1)$
5. $(0,-6)(-3,0)(2,-10)$
7. $(0,10)(2,16)(-4,-2)$
9. $(0,-4)(4,2)(2,-1)$
11. $(0,-3)(-5,0)(5,-6)$
13. $(0,0)(1,1)(3,3)$
15. $(0,0)(2,-1)(4,-2)$
17. $(2,5)(5,7)(8,9)$

19. $(1, -3)(4, -5)(7, -7)$

21. $(0, 3)(4, -2)(8, -7)$

Just for Fun

Tierce, hogshead, pipe, butt, and tun are names of casks (barrels). They were originally used in measuring amounts of beer and ale.

SECTION 9.7

1.

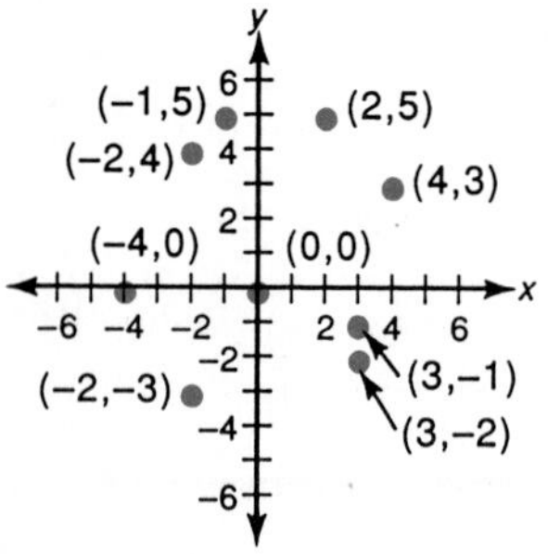

3.

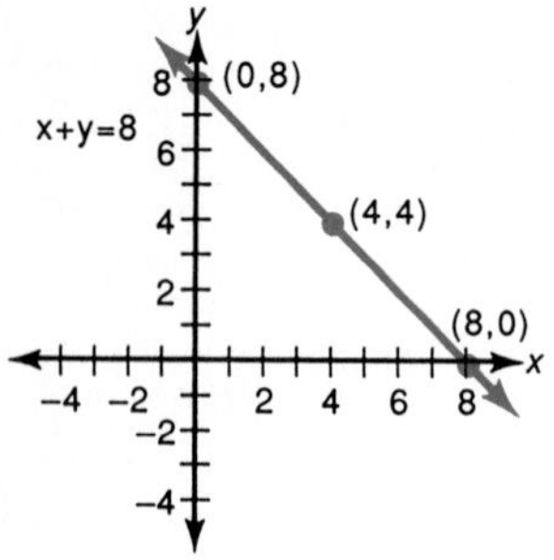

5.

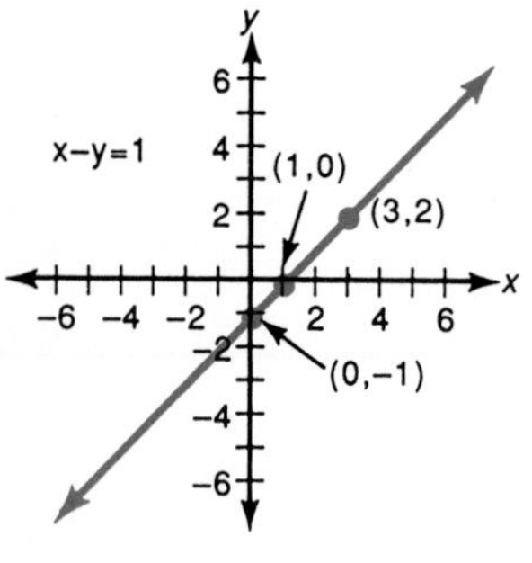

7.

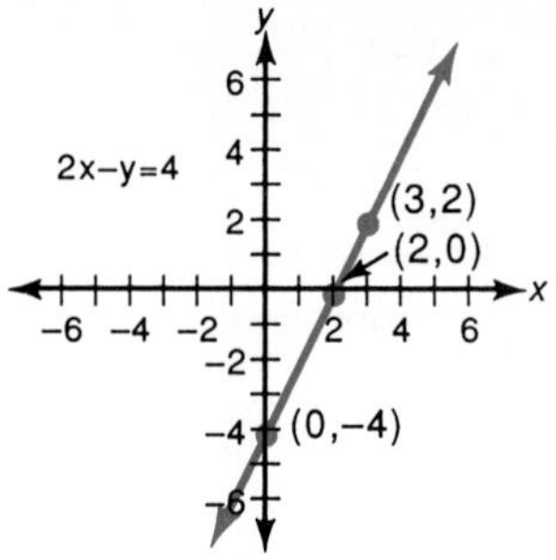

9.

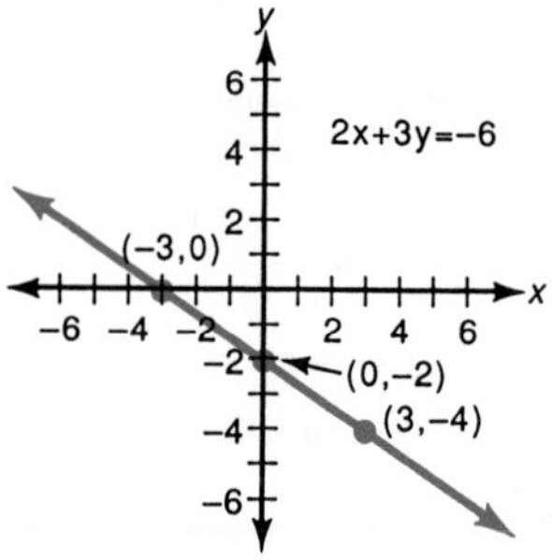

11.

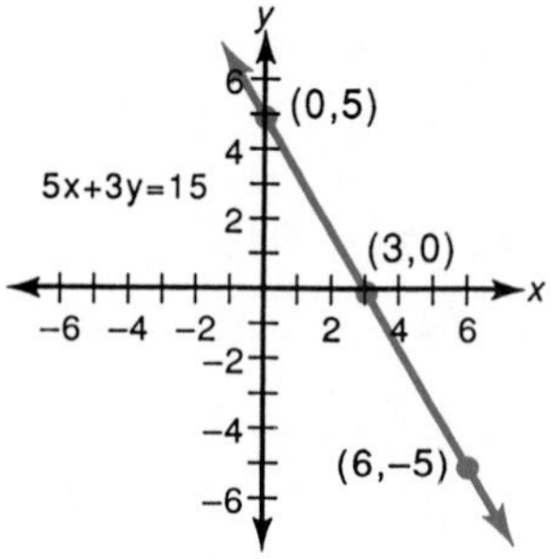

13.

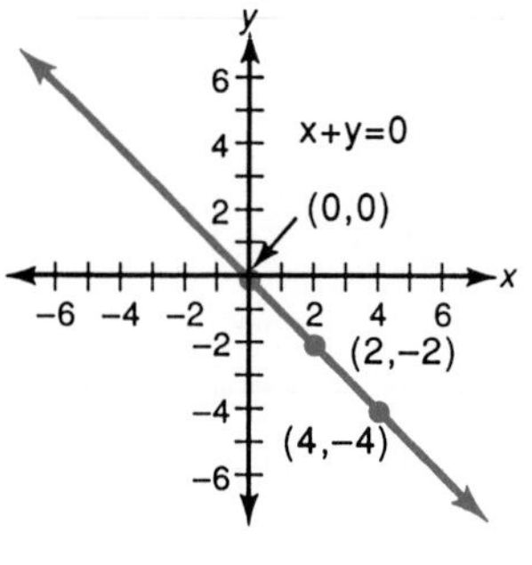

15.

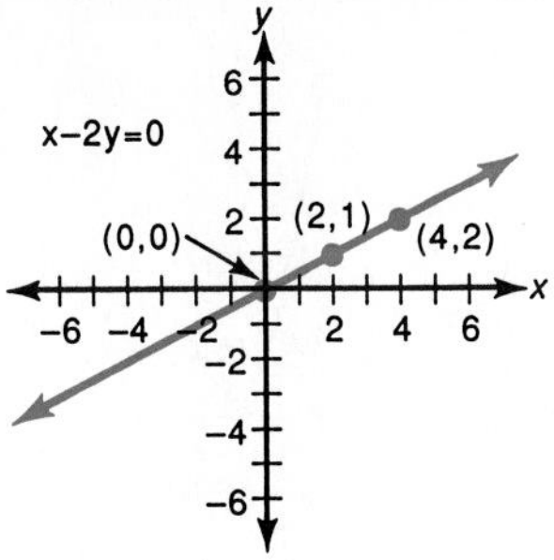

17.

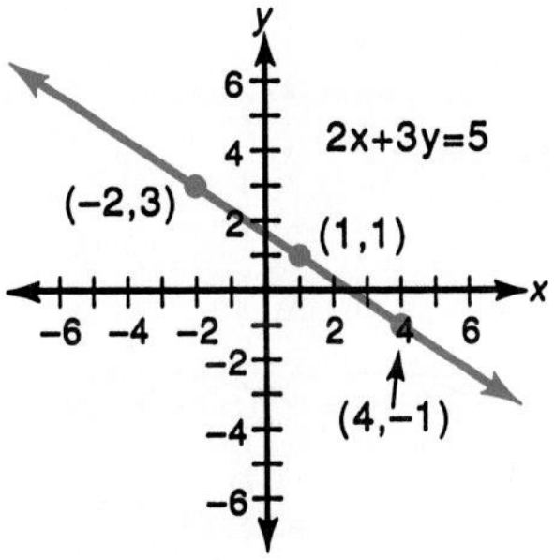

19.

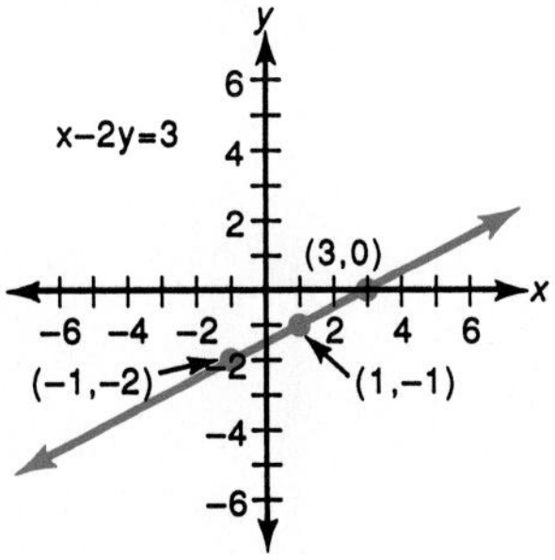

21.

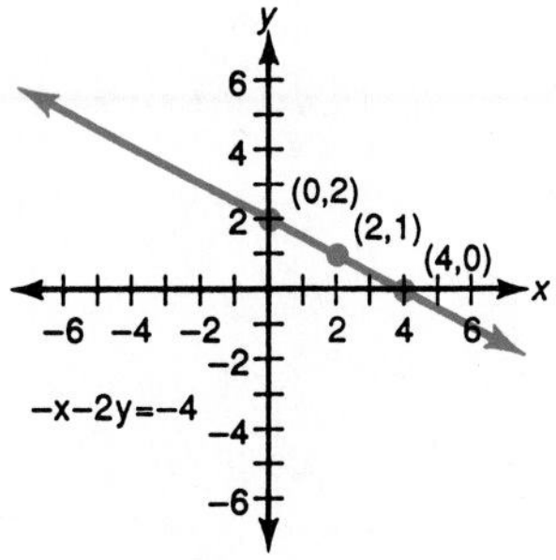

23.

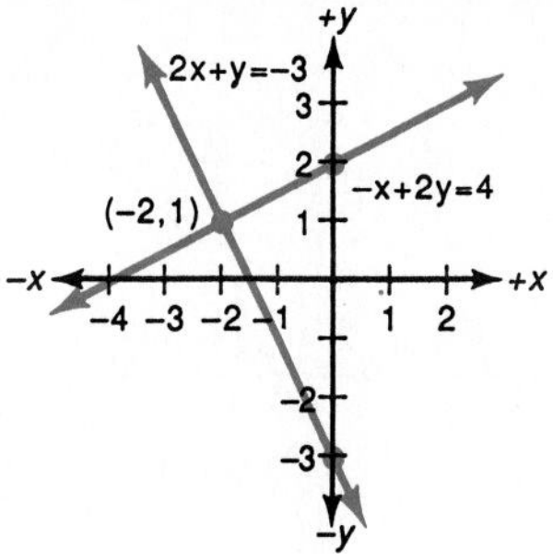

25.

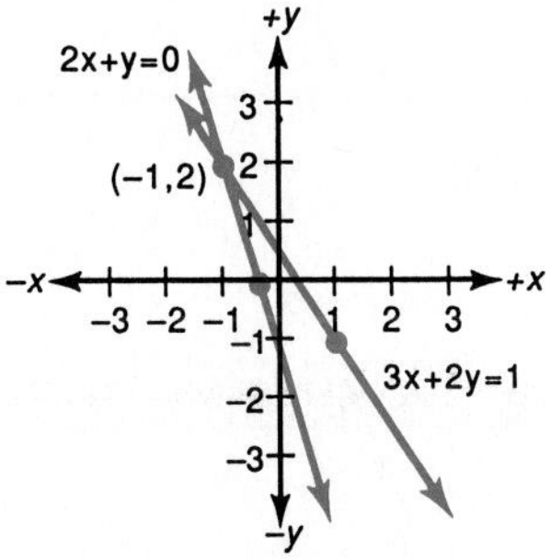

27. ∅

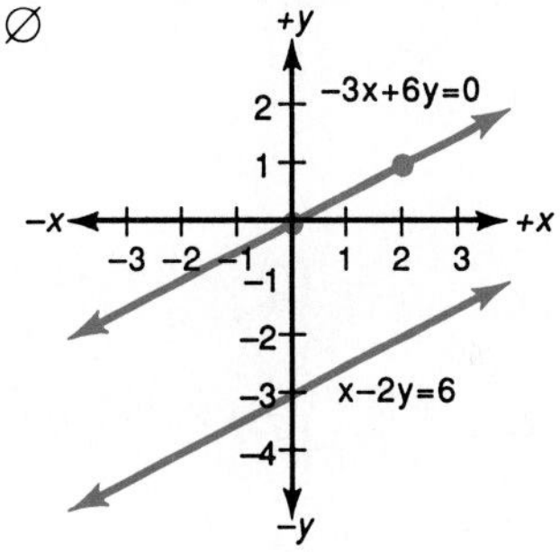

29.

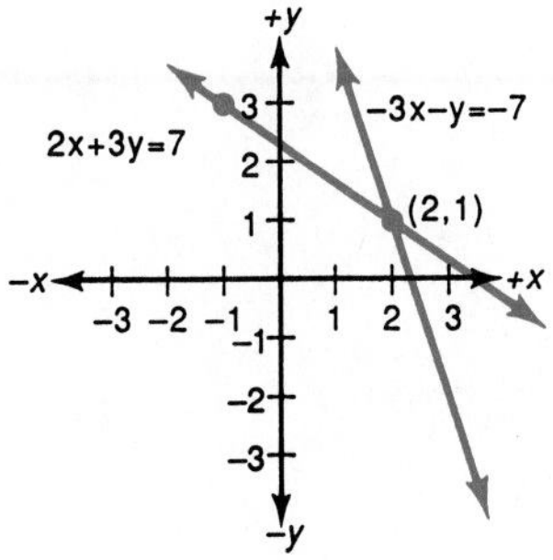

Just for Fun

12

SECTION 9.8

1. $y = x^2 - 1$

x	y
−2	3
−1	0
0	−1
1	0
2	3

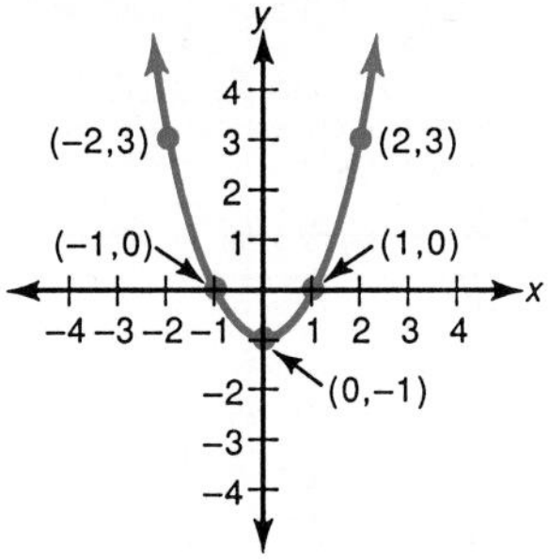

3. $y = -x^2 + 2$

x	y
−2	−2
−1	1
0	2
1	1
2	−2

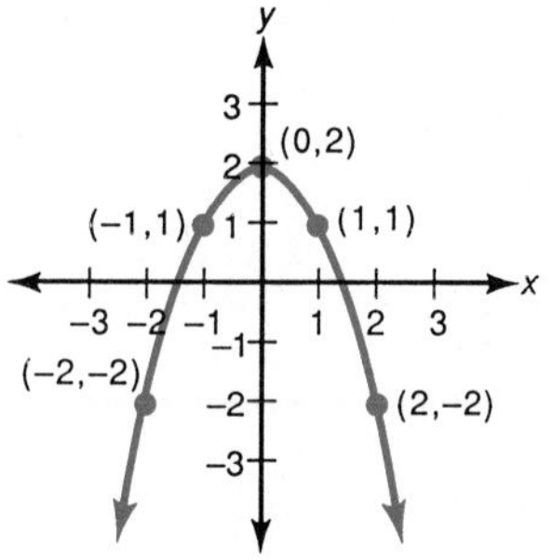

5. $y = -x^2 + 6x$

x	y
0	0
1	5
2	8
3	9
4	8
5	5
6	0

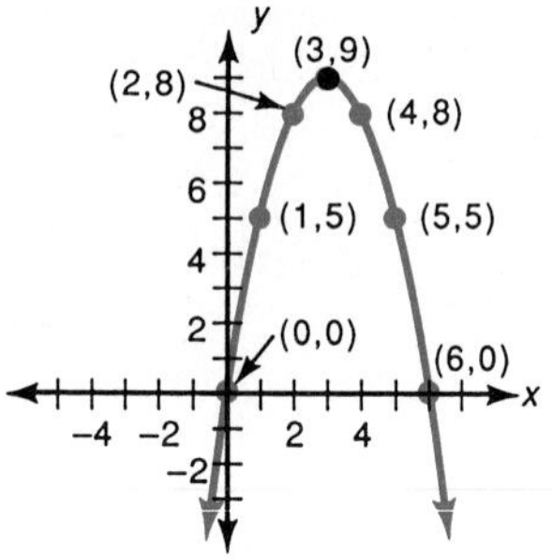

7. $y = x^2 + 2x + 1$

x	y
−4	9
−3	4
−2	1
−1	0
0	1
1	4
2	9

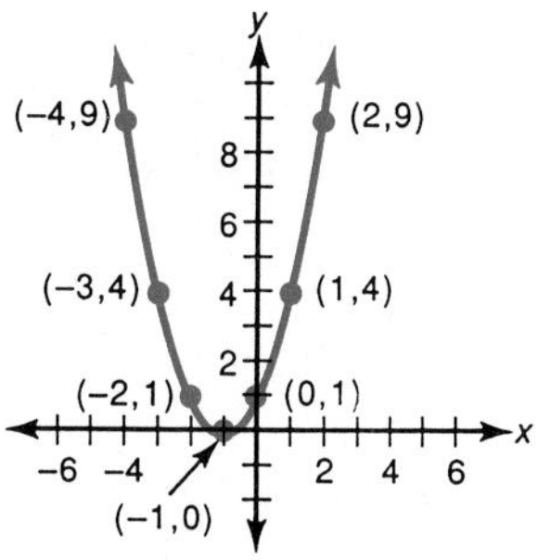

9. $y = x^2 - 4x + 3$

x	y
−1	8
0	3
1	0
2	−1
3	0
4	3
5	8

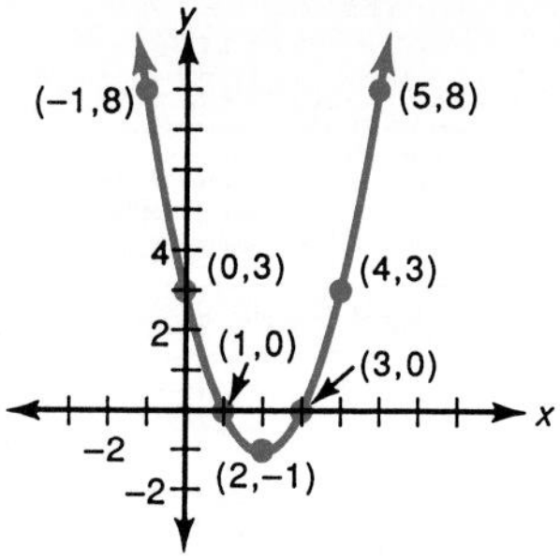

11. $y = -x^2 + 6x - 9$

x	y
1	−4
2	−1
3	0
4	−1
5	−4

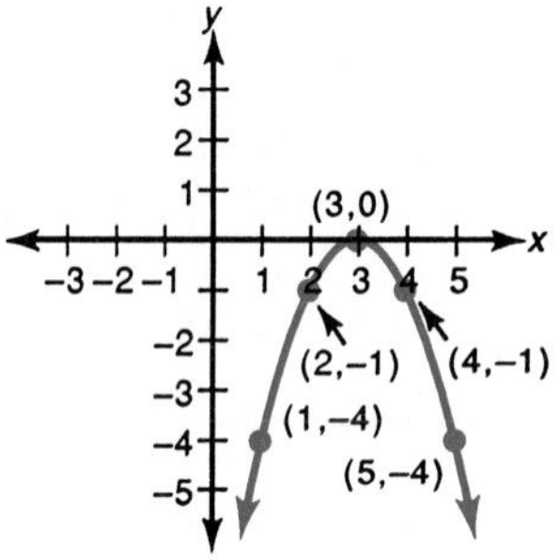

13. $y = x^2 - 6x + 5$

x	y
0	5
1	0
2	−3
3	−4
4	−3
5	0
6	5

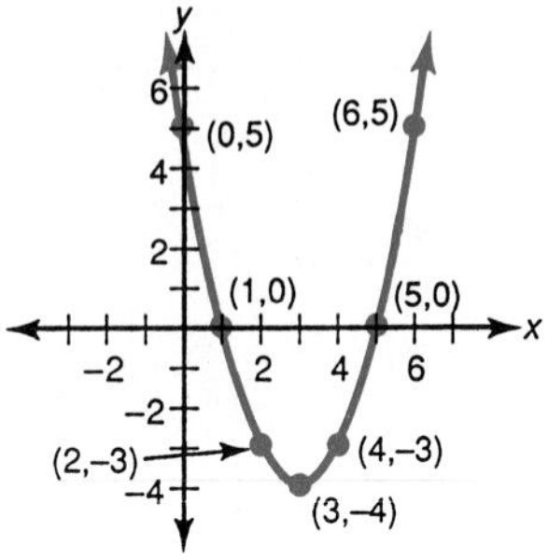

15. $y = -x^2 + 4$

x	y
−1	3
−2	0
0	4
1	3
2	0

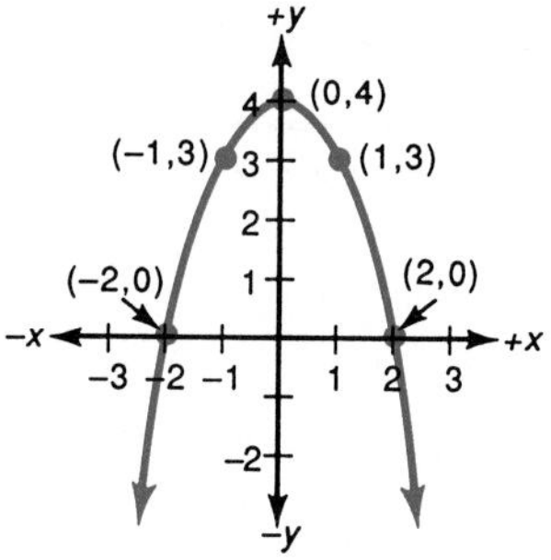

17. $y = x^2 + 2x$

x	y
−3	3
−2	0
−1	−1
0	0
1	3

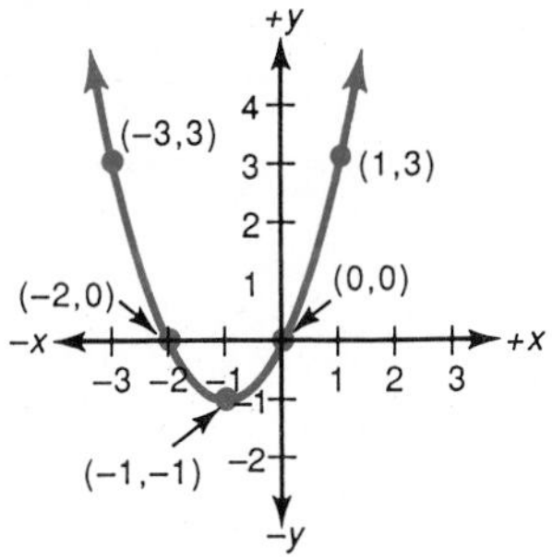

19. $y = x^2 + 3x + 2$

x	y
−3	2
−2	0
−1	0
0	2

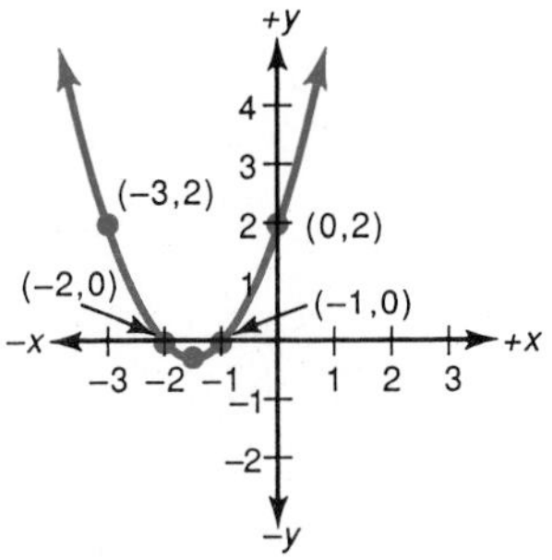

Just for Fun

The box full of $10 gold pieces; the value of gold is determined by weight, not denomination.

SECTION 9.9

1. $x + y > 5$

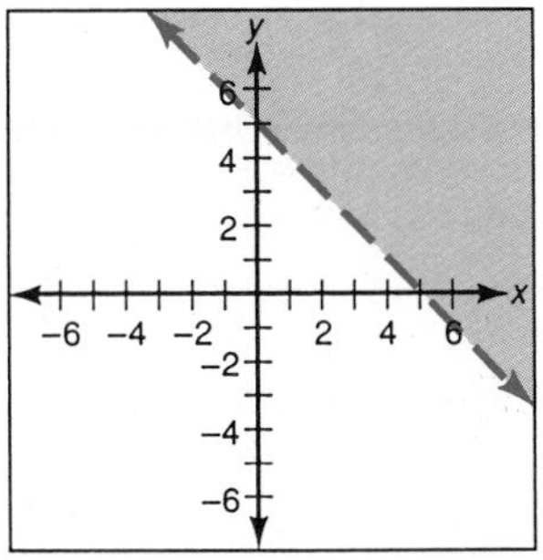

3. $x - y < 3$

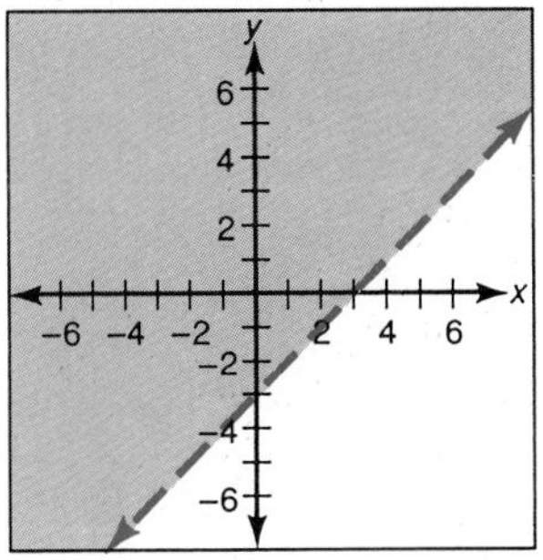

5. $x - y \leq 1$

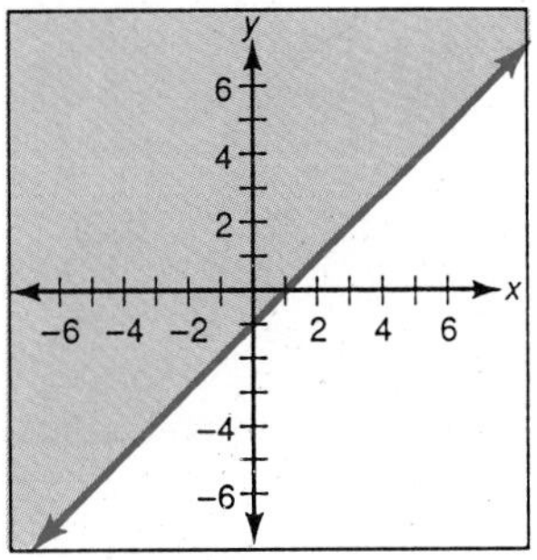

7. $-2x + y \geq 4$

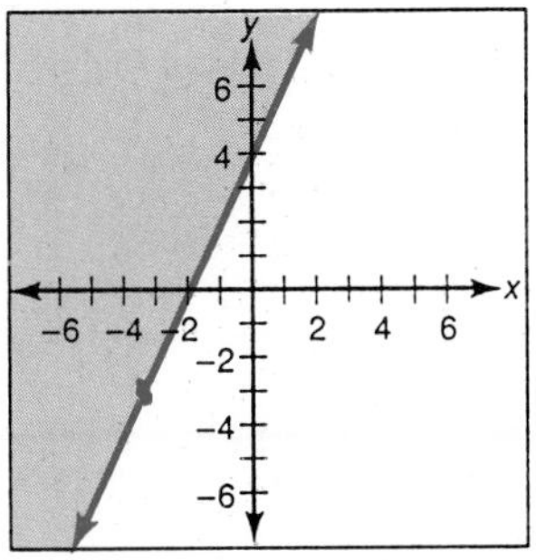

9. $y - x < 0$

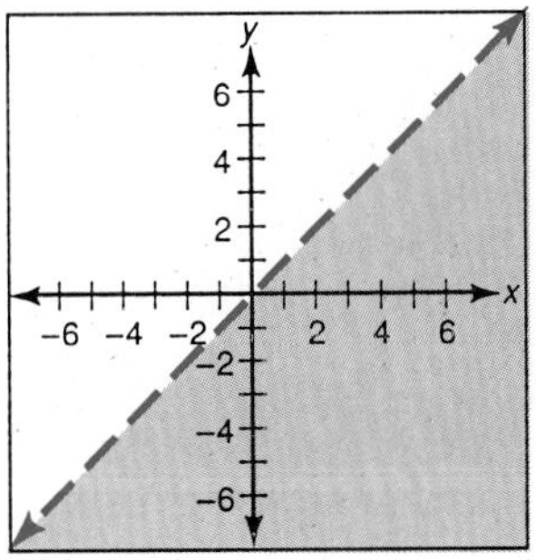

11. $-2x - 2y > 0$

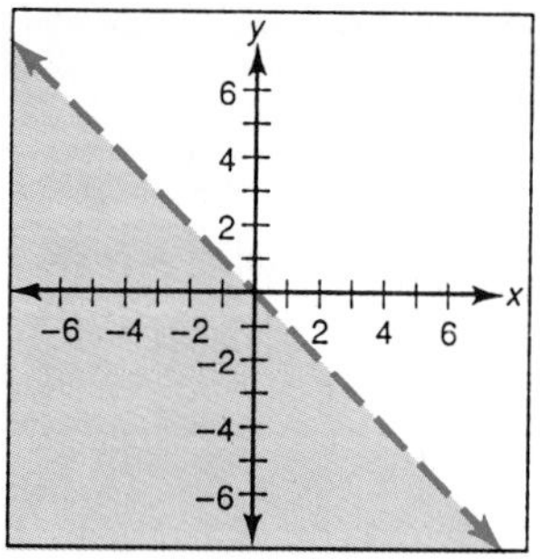

13. $-3x + 5y \leq 15$

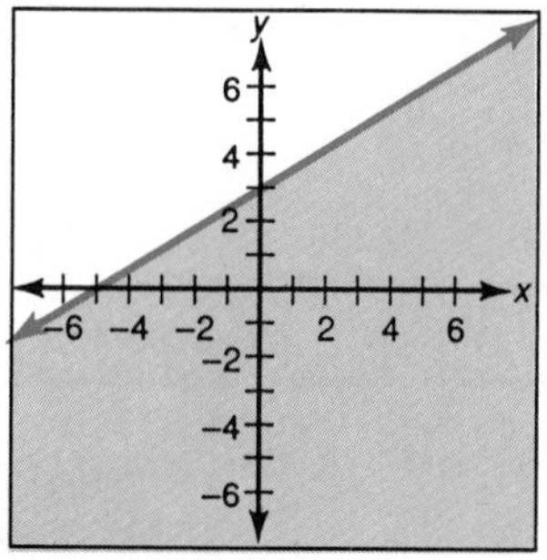

15. $3x + 5y > 15$

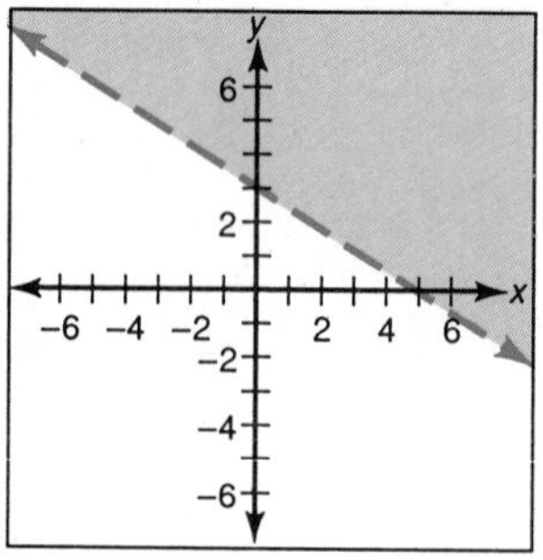

17. $-3x - 2y < -6$

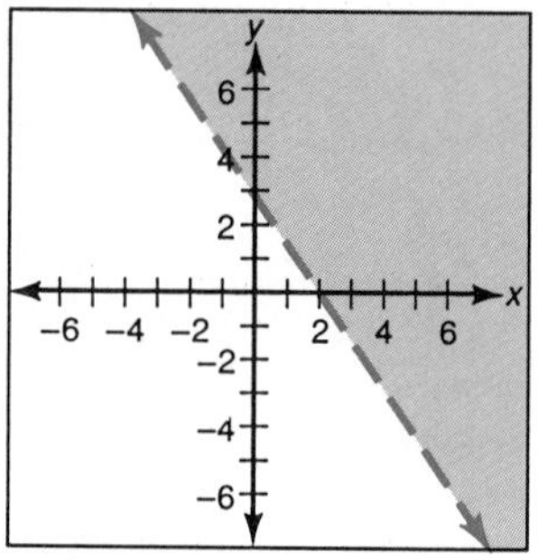

19. $x + y \geq 2$ and $x - y < 2$

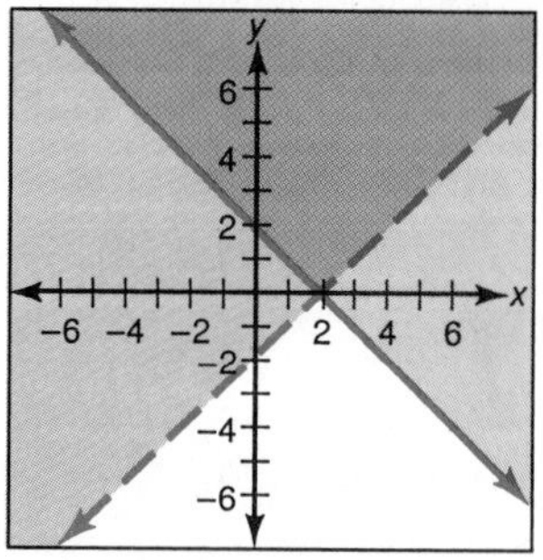

21. $y \geq x$ and $y \geq -x$

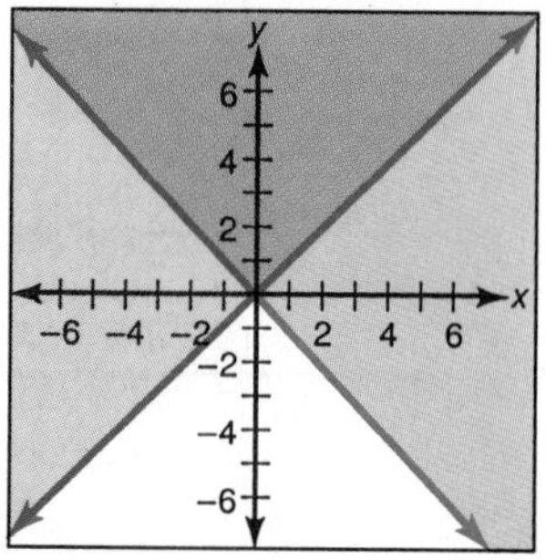

Just for Fun

None; pea*cocks* do not lay eggs.

SECTION 9.10

1. Maximum value of P is 13 at $(3, 2)$.
3. Maximum value of P is 26 at $(5, 2)$.
5. Three bikes and two wagons; maximum profit is \$56.
7. Sixty pheasants and 40 partridges; maximum profit is \$740.
9. 100 bushels of oysters and 500 bushels of clams; maximum profit is \$5,800.

Just for Fun

Two of a set of triplets.

SECTION 9.11

1. $\{3, 2\}$ 3. $\{4, -1\}$ 5. $\{2, -3\}$
7. $\{2, 1\}$ 9. $\{5, -2\}$
11. $\left\{\frac{-1}{2}, \frac{-2}{3}\right\}$ 13. $\left\{\frac{5}{3}, \frac{-1}{2}\right\}$
15. $\left\{\frac{1}{3}, \frac{-1}{2}\right\}$ 17. $\left\{\frac{3}{5}, -2\right\}$
19. $\left\{1, \frac{-1}{2}\right\}$ 21. $\{1 + \sqrt{3}, 1 - \sqrt{3}\}$
23. $\{1 + \sqrt{6}, 1 - \sqrt{6}\}$ 25. $\left\{\frac{\sqrt{3}}{2}, \frac{-\sqrt{3}}{2}\right\}$
27. $\left\{\frac{\sqrt{3}}{3}, \frac{-\sqrt{3}}{3}\right\}$

Just for Fun

a. The tunnel of Corti is a part of the ear.
b. The Island of Reil is a part of the brain.
c. McBurney's Point is the point of incision for an appendectomy.

REVIEW EXERCISES FOR CHAPTER 9

1. a. $\{2\}$ b. $\{-6\}$
c. $\{6\}$ d. $\{1, 2, 3\}$
e. $\{0, 1, 2, 3, 4, 5\}$ f. $\{-2, -1, 0, 1, 2\}$
g. $\{1, 2, 3\}$

2. a. 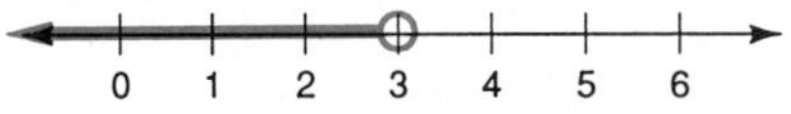

b. 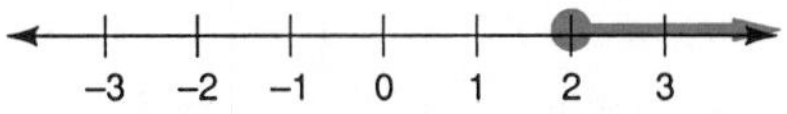

c. 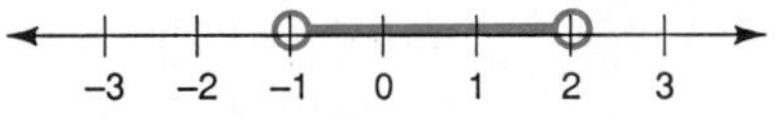

d. 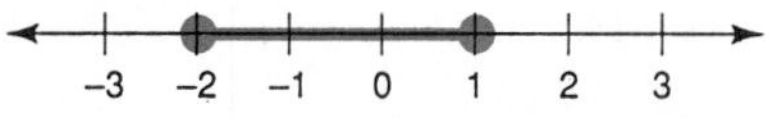

e. 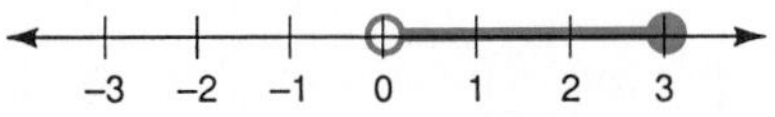

f. 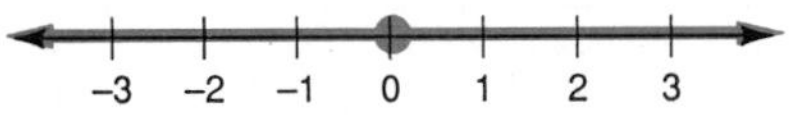

3. a. $x = 6$ b. $y = 2$ c. $z = 2$
d. $y = 3$ e. $x = 10$ f. $z = \frac{6}{4}$
4. 13, 15
5. Bill is 24, Ike is 26.
6. $n = 7$
7. $w = 5$ m; $l = 10$ m
8. 10 quarters; 5 dimes
9. $n = 5$
10. $w = 8$ m, $l = 24$ m
11. 10 nickels, 15 dimes, 25 quarters
12. 350
13. a. $x - y = 2$

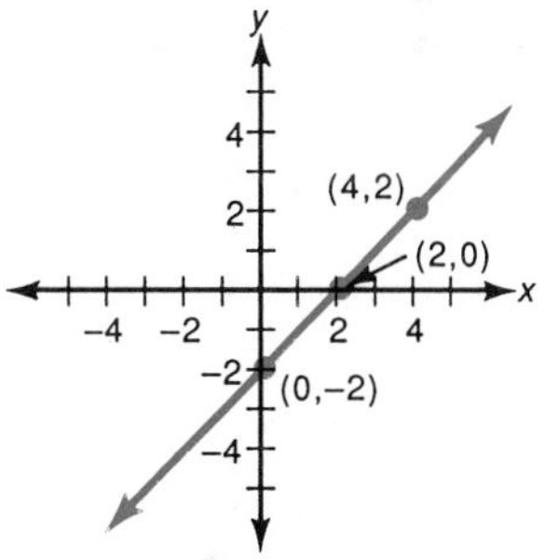

b. $2x + y = 4$

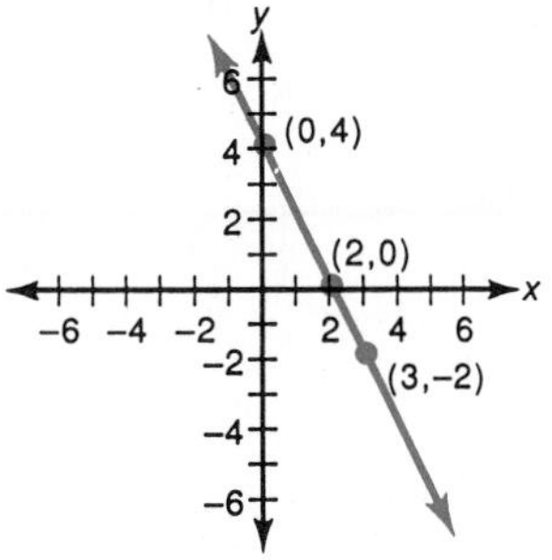

c. $2x - y = 4$

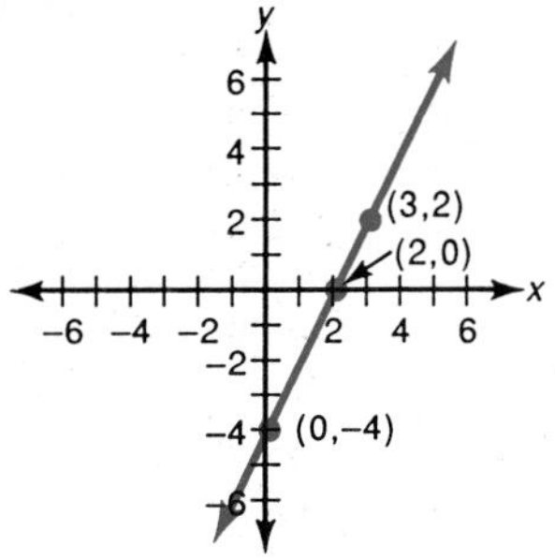

d. $-2x + y = 6$

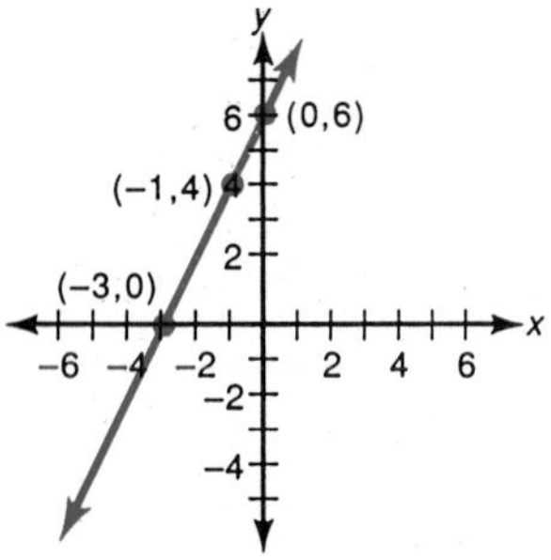

e. $3x - 5y = 15$

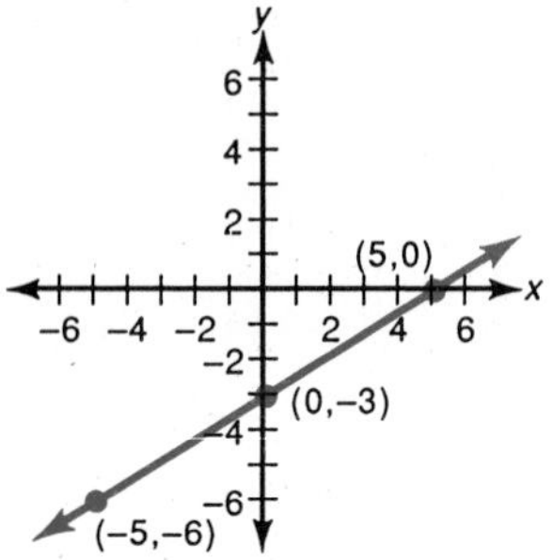

f. $y = x$

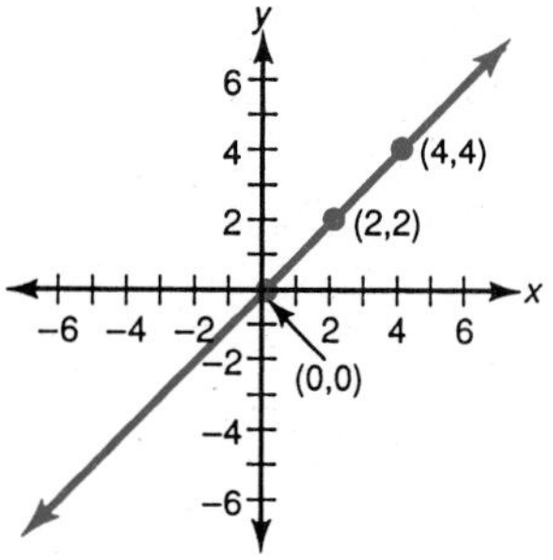

14. a.

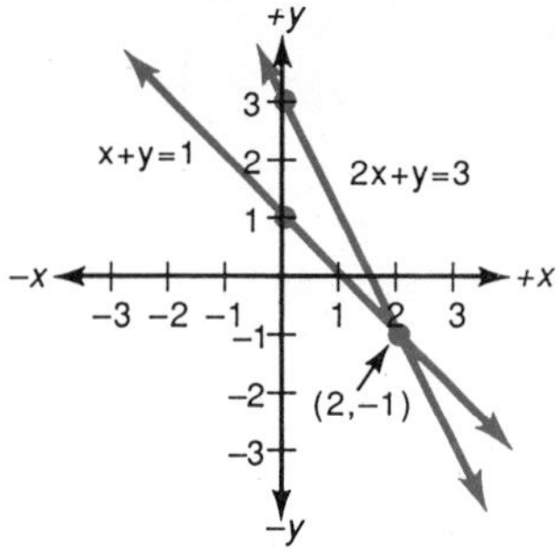

b.

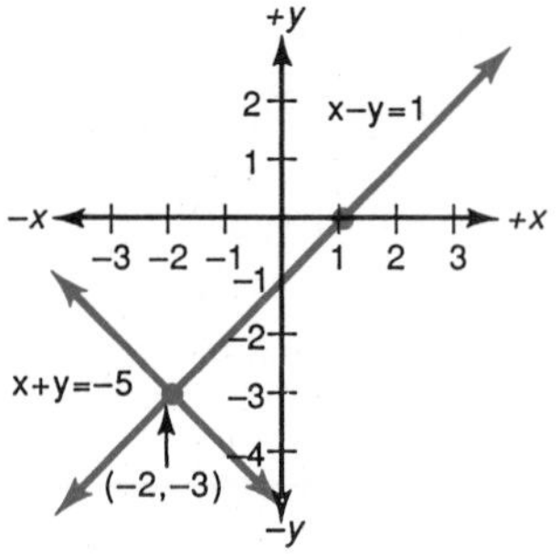

c.

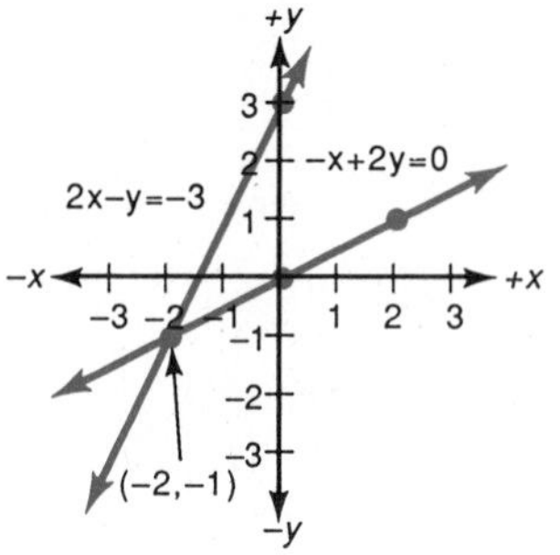

15. a. $y = x^2$

x	y
−2	4
−1	1
0	0
1	1
2	4

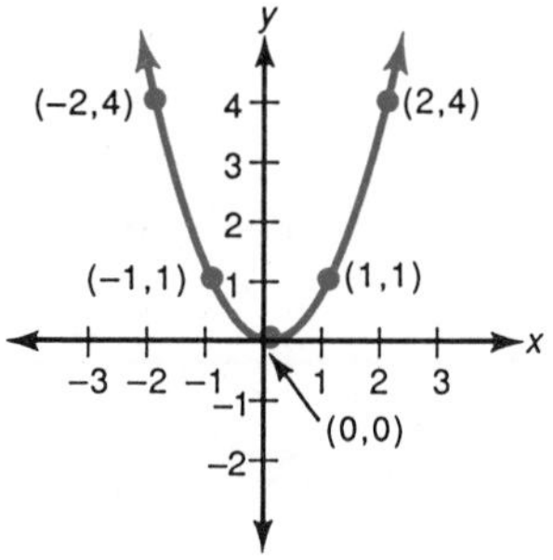

b. $y = x^2 - 4$

x	y
−3	5
−2	0
−1	−3
0	−4
1	−3
2	0
3	5

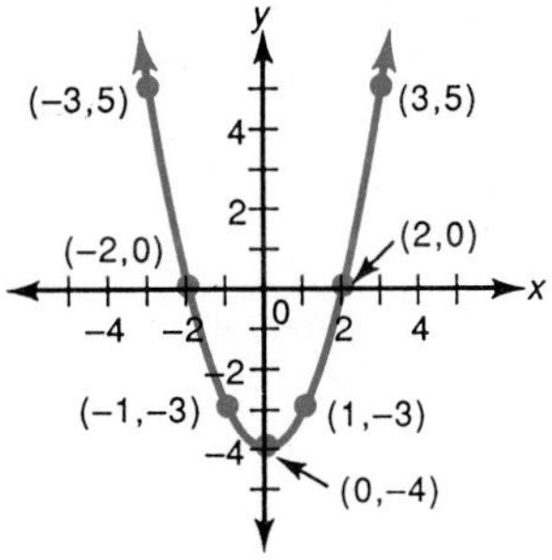

c. $y = -x^2 + 2$

x	y
−2	−2
−1	1
0	2
1	1
2	−2

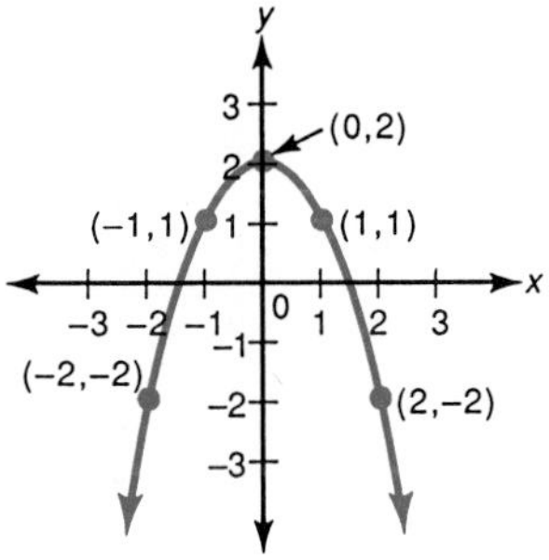

d. $y = x^2 - 6x$

x	y
0	0
1	−5
2	−8
3	−9
4	−8
5	−5
6	0

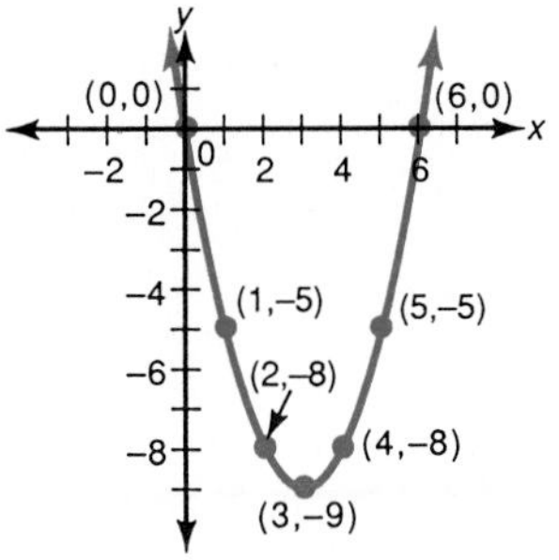

e. $y = x^2 + 2x - 3$

x	y
−4	5
−3	0
−2	−3
−1	−4
0	−3
1	0
2	5

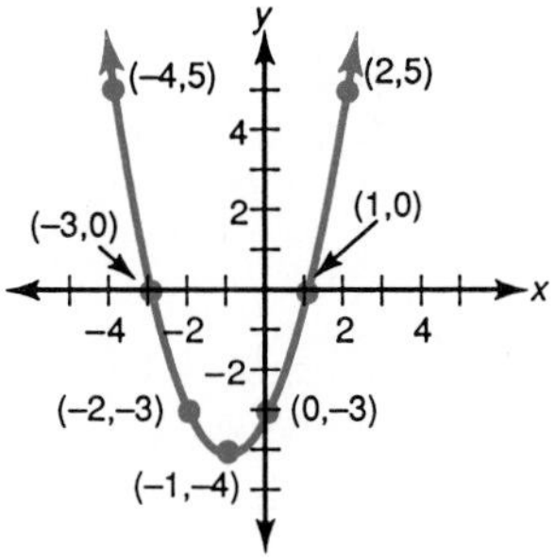

f. $y = -x^2 + 6x - 5$

x	y
0	−5
1	0
2	3
3	4
4	3
5	0
6	−5

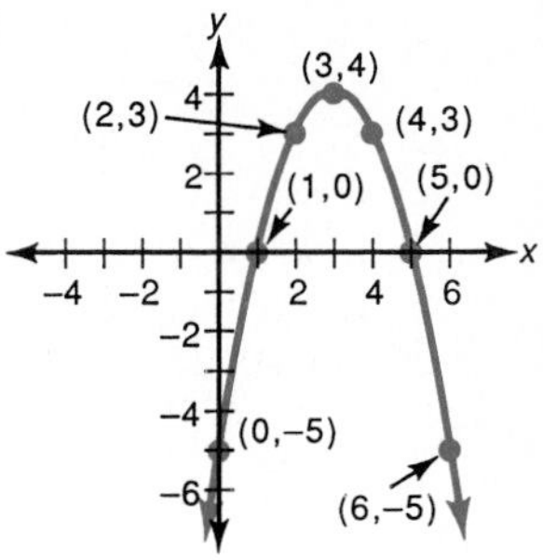

16. a. $x + y > 3$

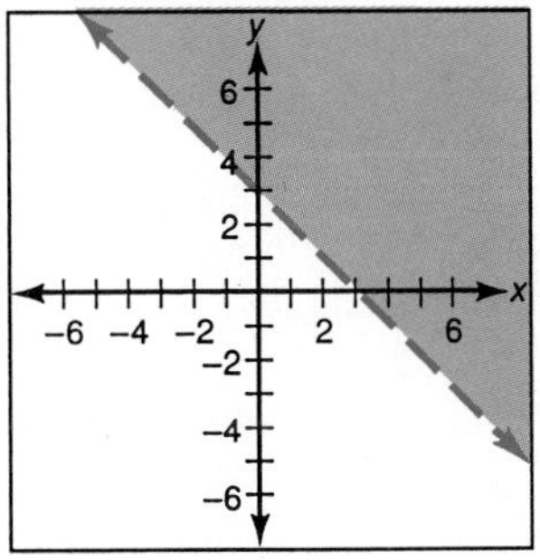

b. $x - y \le 2$

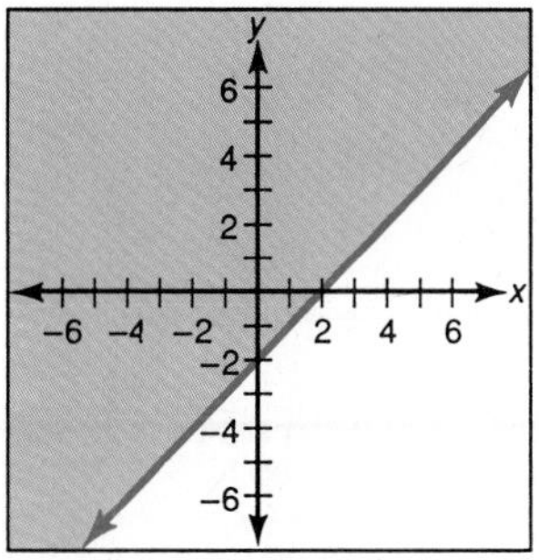

c. $-x + y > 2$

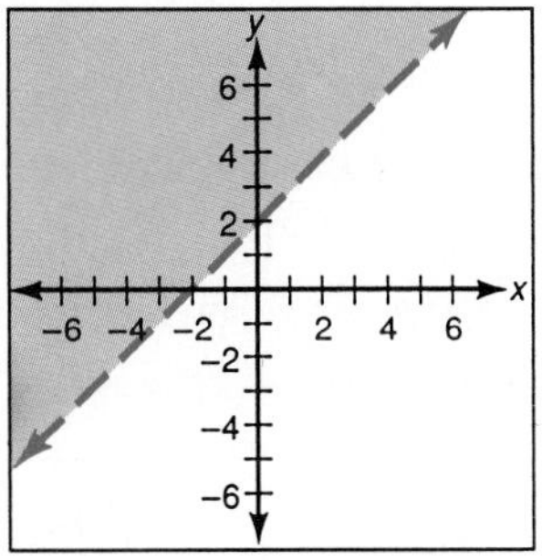

d. $3x + 5y > -15$

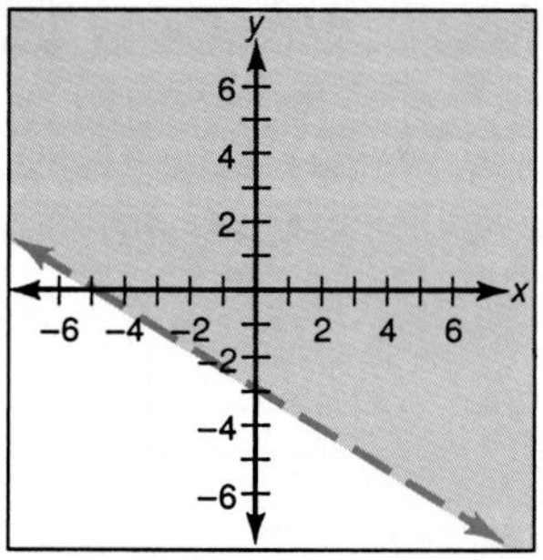

e. $x - 2y \leq 0$

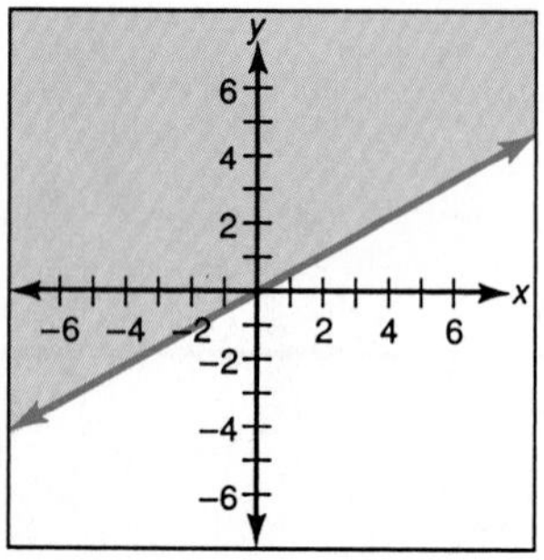

f. $-3x - 2y < -6$

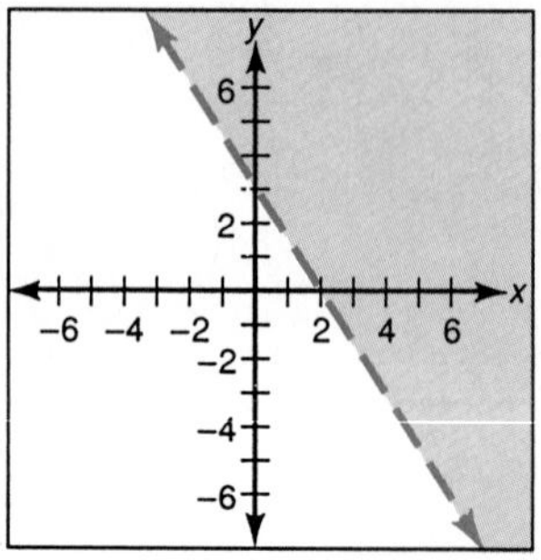

17. Maximum value of P is 18 at $(3, 4)$.
18. Four couches and three recliners; maximum profit is $255.
19. Thirty standard and five deluxe models; maximum profit is $260.
20. $\{2, -\frac{2}{3}\}$
21. $\{\frac{1}{2}, \frac{1}{4}\}$
22. $\{2, -1\}$
23. $\{-1 + \sqrt{3}, -1 - \sqrt{3}\}$
24. $\{1, -1\}$
25. $\{1, -\frac{1}{2}\}$
26. $\{4, 0\}$
27. $\{2, -5\}$

Just for Fun

2 hours

CHAPTER 9 QUIZ

1. False	2. True	3. True	4. False
5. True	6. False	7. True	8. False
9. True	10. False	11. False	12. True
13. False	14. False	15. True	16. False
17. True	18. False	19. True	20. True
21. False	22. False	23. False	24. True
25. True			

CHAPTER 10

SECTION 10.2

1. $\{I\}$
3. $\overline{QK}$
5. QU
7. $\overleftrightarrow{UK}$
9. $\{I\}$
11. $\{I\}$
13. $\overline{QU}$
15. $\{E\}$
17. $\{E\}$
19. $\{E\}$
21. $\angle WET$
23. $\overleftrightarrow{BT}$ or $\angle BET$
25. $\overrightarrow{BE}$
27. $\overrightarrow{EW}$
29. $\{A\}$
31. $\overleftrightarrow{ZP}$ or $\angle ZAP$
33. $\varnothing$
35. $\angle ZAW$
37. $\overrightarrow{AM}$
39. $\{A\}$
41. triangle WAZ

Just for Fun

3

SECTION 10.3

1. R 3. $\overleftrightarrow{AS}$ 5. $\angle DRA$

7. $\overrightarrow{RS}$ 9. $\angle TON$

11. $\overrightarrow{OE}$ 13. Interior $\angle EON$

15. $\varnothing$ 17. Interior $\angle EON$

19. $\angle COT$ 21. $\angle SOJ$

23. Interior $\angle SOC$ 25. $\overrightarrow{OT}$

27. $\{O\}$ 29. Exterior $\angle JOC$

31. True 33. False

35. False 37. True

39.

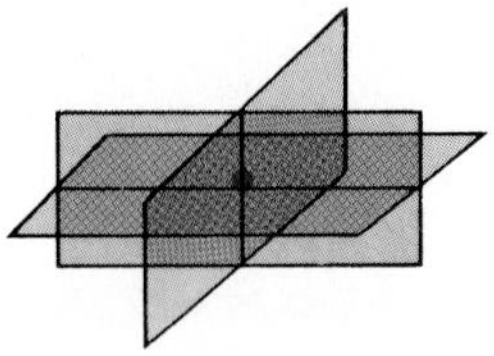

Just for Fun

Irene would receive a larger salary.

SECTION 10.4

1. 6: $\angle ABD$, $\angle ABE$, $\angle ABC$, $\angle DBE$, $\angle DBC$, $\angle EBC$

3. Note: Answers may vary; some possible solutions are
   a. $\angle FOE$, $\angle EOR$ b. $\angle LOE$, $\angle NOR$
   c. $\angle LOF$, $\angle FOR$ d. $\angle WOE$, $\angle LOR$
   e. $\angle FOE$, $\angle EOR$ and $\angle LON$, $\angle NOF$
   f. $\angle LON$, $\angle SOR$ and $\angle NOF$, $\angle BOS$
   g. $\angle FOE$, $\angle EOR$ and $\angle LON$, $\angle NOF$
   h. $\angle NOL$ and $\angle LOS$; $\angle FOE$ and $\angle BOE$

5. a. $90°$ b. $70°$ c. $120°$
   d. $59°40'$ e. $141°30'$ f. $79°9'35''$

7. $60°$ and $120°$ 9. $30°$ and $60°$

11. $x = 28$ 13. $45°$

15. True 17. False 19. False

21. True

23. $m\angle 1 = 130°$, $m\angle 2 = 50°$, $m\angle 4 = 130°$, $m\angle 5 = 130°$, $m\angle 6 = 50°$, $m\angle 7 = 50°$, $m\angle 8 = 130°$

25. $m\angle 1 = 155°$, $m\angle 2 = 25°$, $m\angle 3 = 25°$, $m\angle 4 = 155°$, $m\angle 6 = 25°$, $m\angle 7 = 25°$, $m\angle 8 = 155°$

27. $m\angle 3 = 100°$, $m\angle 5 = 80°$

29. $m\angle 4 = 120°$, $m\angle 8 = 120°$

Just for Fun

Tetrahedron (pyramid)

SECTION 10.5

1. b, d, e

3. a, b, d, e

5. a. True b. False c. True
   d. False e. True

7. 15 9. 26

11. 5 miles 13. 7 feet

15. 3

17. a. True b. False c. False
    d. True e. False

19. Polygon, triangle, right triangle, hypotenuse

21. a. Rectangle, square b. Rhombus, square

23. b 25. $x = 35$

27. $m\angle F = 45°$ 29. a 31. b

Just for Fun

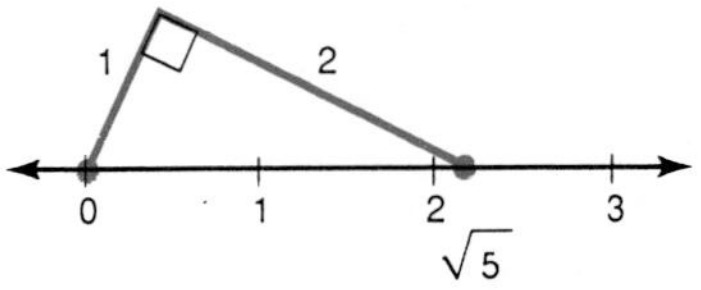

SECTION 10.6

1. a. 24 m b. 24 cm c. 17 dm
3. 30 m 5. 5.6 m 7. 214 cm
9. 34 11. 4 13. 5 in.
15. 60 square units
17. 12.5 in^2 19. 8 in^2
21. 10 m
23. 45.5 square units
25. 7.4 ft
27. 17.5 square units
29. 90 square units
31. 30 square units
33. 25 m^2
35. a. 27.5; $\sqrt{41} + \sqrt{74} + 11$
b. 40; $24 + \sqrt{32}$
c. 25.5; $17 + \sqrt{18} + \sqrt{13}$
d. 148.5; 45.1
e. 102.3; 50.3
37. 94.2 cm 39. 94.2 cm 41. 31.4 m
43. 2826 m^2 45. 78.5 mm^2 47. 7850 m^2
49. Circle, by 336.5

Just for Fun

More than 100

SECTION 10.7

1. 4 3. 4
5. a. 6 b. 8 c. 12
d. $8 - 12 + 6 = 2$
$14 - 12 = 2$
$2 = 2$
7. a. 8 b. 6 c. 12
d. $6 - 12 + 8 = 2$
$14 - 12 = 2$
$2 = 2$
9. 96 cm^2, 42 cm^3
11. 539.7 mm^2, 400 mm^3
13. a. 785 cm^2 b. 1570 cm^3
15. a. 226.1 mm^2 b. 251.2 mm^3
17. a. 282.6 mm^2 b. 314 mm^3
19. No, 1570 cm^3 vs. 600 cm^3
21. a. 1808.6 cm^2 b. 7234.6 cm^3
23. a. 1017.4 cm^2 b. 3052.1 cm^3
25. a. 2.0 m^2 b. 0.3 m^3

Just for Fun

One

SECTION 10.8

1. True 3. False 5. True 7. False
9. $x = 7$ 11. No, $\overline{AC} \neq \overline{DF}$, so $SAS \neq SAS$
13. 4 15. 2 17. 45° 19. 4
21. 3 23. 37° 25. 6 27. 12
29. 36 31. 20 m 33. 27.5 ft
35. 9, 12, 15 37. 60 m

Just for Fun

An angle

SECTION 10.9

1. a. 3 b. 0 c. Yes
d. *A*, *B*, or *C*
3. a. 2 b. 2 c. yes
d. *A* or *C*
5. a. 4 b. 0 c. Yes
d. *A*, *B*, *C*, or *D*
7. a. 2 b. 4 c. No
d. None

9. a. 3 b. 2 c. Yes
d. E or C

11. No, there are four vertices of odd order.

REVIEW EXERCISES FOR CHAPTER 10

1. $\{R\}$ 2. AS 3. $\{R\}$

4. $\overleftrightarrow{AR}$ or $\angle ARS$ 5. $\varnothing$

6. $\varnothing$ 7. $\angle LRA$ 8. $\angle SRJ$

9. $\overrightarrow{RJ}$ 10. $\{R\}$ 11. $\angle SOJ$

12. $\angle TOE$ 13. $\{O\}$ 14. $\overline{OT}$

15. Interior $\angle EOT$

16. Interior $\angle JOS$

17. $\{O\}$ 18. $\overrightarrow{OC}$

19. Interior $\angle JOE \cup$ exterior $\angle SOT$

20. Interior $\angle COE$

21. a. 48°, 138° b. 53°, 143°
c. 11°, 101° d. 7° 30′, 97° 30′
e. 86° 15′, 176° 15′
f. 56° 26′ 27″, 146° 26′ 27″

22. 66°, 24° 23. 50°, 130°

24. d 25. d 26. 17

27. 66° 28. 14.4 cm 29. 50 in.^{2}

30. 91 m^2 31. 8 m 32. 30 cm^2

33. 31.4 cm 34. 254.3 m^2

35. a. 94.2 cm b. 706.5 cm^2

36. a. 148 cm^2 b. 120 cm^3

37. a. 360 cm^2 b. 300 cm^3

38. a. 628 cm^2 b. 1177.5 cm^3

39. a. 282.6 mm^2 b. 314 mm^3

40. a. 1808.6 cm^2 b. 7234.6 cm^3

41. $x = 12$ 42. 40 m

43. a. 3 b. 2 c. Yes d. A or E

44. a. 4 b. 4 c. No d. None

45. a. 0 b. 4 c. No d. None

46. a. 3 b. 2 c. Yes d. A or C

Chapter 10 Quiz

1. False 2. True 3. True 4. False

5. False 6. False 7. False 8. True

9. True 10. True 11. False 12. False

13. True 14. False 15. False 16. False

17. False 18. True 19. False 20. False

21. True 22. False 23. True 24. False

25. True

CHAPTER 11

SECTION 11.2

1. a. $\frac{3}{2}$ b. $\frac{4}{7}$ c. $\frac{8}{5}$ d. $\frac{5}{6}$

3. a. $\frac{3}{1}$ b. $\frac{1}{2}$ c. $\frac{3}{2}$
d. $\frac{4}{1}$ e. $\frac{3}{4}$ f. $\frac{2}{3}$

5. a. $\frac{2}{3}$ b. $\frac{3}{2}$ c. $\frac{2}{5}$

7. $\frac{3}{19}$

9. a. 2 b. 6 c. 30 d. 35

11. a. 32 b. 18
c. $\frac{4}{3}$ d. $\frac{15}{4}$

13. 24.75

15. a. $\frac{1}{3}$ b. $\frac{9}{2}$
c. $\frac{2}{3}$ d. $\frac{1}{10}$

17. 20 kg lead, 40 kg tin

19. 7.5 min. 21. 75 ft. 23. $\frac{7}{12}$

25. a. 16 b. 3.5
c. $2\frac{3}{7}$ or 2.4 d. 9.2

27. 12

Just for Fun

22 minutes

SECTION 11.3

1. a. $\frac{3}{20}$ b. $\frac{1}{4}$ c. $\frac{3}{4}$
3. a. $\frac{9}{200}$ b. $\frac{7}{300}$ c. $\frac{1}{16}$
5. $\frac{13}{200}$ b. $\frac{23}{1000}$ c. $\frac{3}{2}$
7. a. 0.17 b. 0.03 c. 0.045
9. a. 0.065 b. 3.0 c. 0.0025
11. a. 9% b. 214% c. 90%
13. a. 112.5% b. 1% c. 301%
15. a. 75% b. 80% c. 62.5%
17. a. 260% b. 112.5% c. 87.5%
19. a. $66\frac{2}{3}$% b. 275% c. 6.25%
21. 30

Just for Fun

We should subtract the $2 the waiter kept in order to get the cost of the meal ($27 − $2 = $25).

SECTION 11.4

1. a. 50% b. 25% c. 133.3%
   d. 66.7% e. 25% f. 160%
3. a. $12.50 b. $100
   c. $7.50 d. $21.88
5.

	Markup	Percent Markup
a.	$30	150%
b.	$25	33.3%
c.	$3	28.6%
d.	$4	25.1%

7. $89.29 9. $275.38 11. $516.67
13.

	Markup	Cost
a.	$160	$240
b.	$2.20	$8.78
c.	$6.24	$43.71

15. $79; $316
17. 33.3% 19. $895; 33.5%
21. a. $225
    b. $157.50
23. a. $22.12; $338.12 b. $0.07; $1.07
    c. $62.65; $957.65 d. $11.03; $168.53

Just for Fun

Jets

SECTION 11.5

1. $720 3. $120 5. $287.50
7. $945; $3,945
9. $1,275; $106.25
11. $243.60 13. $745 15. $555
17. $425
19. 6 months or $\frac{1}{2}$ year
21. 6%

Just for Fun

Approximately 2,740 years

SECTION 11.6

1. $883.40 3. $5,397.50 5. $13,000
7. $1095.50 9. $1,126 11. $8,955
13. $586 15. $2,191 17. $20,330
19. $5,418; $2,418
21. 12 years
23. $28,242 25. $6.30
27. a. $13,600 b. $13,690
    c. $13,730 d. $13,200
29. $15,179

Just for Fun

$5; $10; $20; $50; $100; $500; $1,000; $5,000; $10,000

SECTION 11.7

1. The *effective annual interest rate* is the annual interest rate which gives the same yield as the nominal interest rate compounded several times a year.
3. 12.36% 5. 10.38%
7. 6.14% 9. 7.98%
11. a. 6% b. 6.09% c. 6.14%
13. 5.5% compounded semiannually (5.58% vs. 5.10%)
15. Pamela (6.71% vs. 6.56%)
17. Julia (7.02% vs. 6.92%)

Just for Fun

Mario should drive Roy's car, and Roy should drive Mario's car.

SECTION 11.8

1. a. $437.50 b. $519.20
   c. $129 d. $2,556.50
   e. $1,021.32
3. a. $151.40 b. $154.42
   c. $157.44 d. $163.56
5. a. $715.20 b. $729.50
   c. $743.80 d. $772.44
7. $4,091.60
9. a. $90.84 b. $61.80
11. a. $596; $17,960
    b. Limited payment; $6,040 more
13. a. $567.40
    b. $9,786.80

Just for Fun

They are the same distance from Miami.

SECTION 11.9

1. a. $37.33 b. 22.2%
3. a. $63.58 b. 30.9%
5. $1.85
7. a. 5 b. $5.47
9. $253.40
11. $275.25
13. $586.85
15. $399.20; $55,808
17. Bank B; $19,548
19. $823.20; $216,352

Just for Fun

Alexander Hamilton; U.S. Treasury Building

REVIEW EXERCISES FOR CHAPTER 11

1. $\frac{2}{5}$ b. $\frac{3}{2}$ c. $\frac{2}{5}$ d. $\frac{2}{3}$
2. a. No b. Yes c. No d. Yes
3. a. 4 b. 3 c. 8 d. 24
4. a. $\frac{19}{50}$ b. $\frac{1}{8}$ c. $\frac{23}{1000}$ d. $\frac{5}{4}$
5. a. 0.48 b. 0.105 c. 0.047 d. 0.0025
6. a. 82% b. 130% c. 78% d. 213.4%
7. a. 60% b. 12.5% c. 220% d. 18.75%
8. 20,000 9. 200 10. $96.15 11. $80
12. Markup = $160; cost = $240
13. $31.25; $93.74 14. $1,650
15. $900 16. $4,000; $333.33
17. $18,644; $14,644 18. $10,165; $5,165
19. a. 5% b. 5.06% c. 5.09%
20. a. $172 b. $262.50 c. $1,077.60
    d. $2,107.60

21. $42 per year

22. Annual percentage rate, which is also known as the true annual interest rate.

23. a. $380.10 b. $411.75 c. $617.40
d. $724.35

24. $127.43

25. a. $132 b. $28.42 c. $732
d. 23.0%

26. $533.50; $142,060

Just for Fun

False; they were invented by people in India.

CHAPTER 11 QUIZ

1. True	2. False	3. False	4. True
5. False	6. True	7. False	8. True
9. True	10. True	11. False	12. True
13. False	14. False	15. False	16. False
17. True	18. False	19. False	20. False
21. False	22. False	23. True	24. True
25. True			

CHAPTER 12

SECTION 12.2

1. The abacus. It required a great deal of skill and the carrying of numbers had to be done by hand.

3. Howard Aiken—Mark I computer
John Von Neumann—EDVAC computer

5. a. First generation: vacuum tubes
b. Second generation: transistors
c. Third generation: integrated circuits
d. Fourth generation: microcomputers, small size

7. U.S. Census

9. 8×97; 776

11. 4×975; 3,900

13. 7×608; 4,256

15. 2,185

17. 982,215

Just for Fun

$$\begin{array}{r} 789 \\ +\ 246 \\ \hline 1035 \end{array}$$

SECTION 12.3

1. A central processing unit (CPU), main memory, auxiliary memory, an input unit, and an output unit.

3. Translates a program written in a high-level language into its equivalent machine language. It also produces a program listing and diagnostics.

5. Keyboard

7. They perform the same function.

9. Beginner's All-purpose Symbolic Instruction Code.

11. Formula translation.

13. 18

15. 14

17.

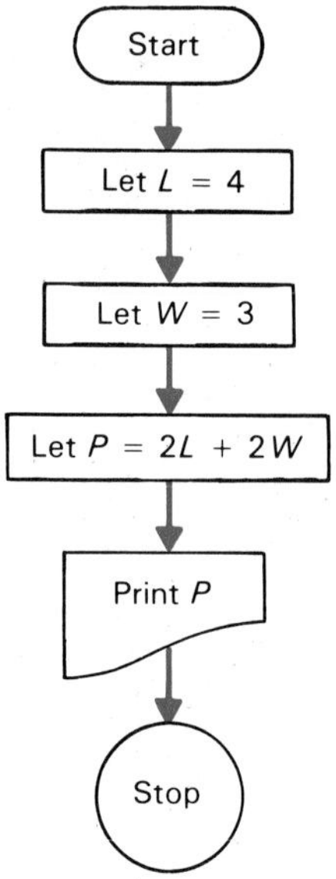

19.

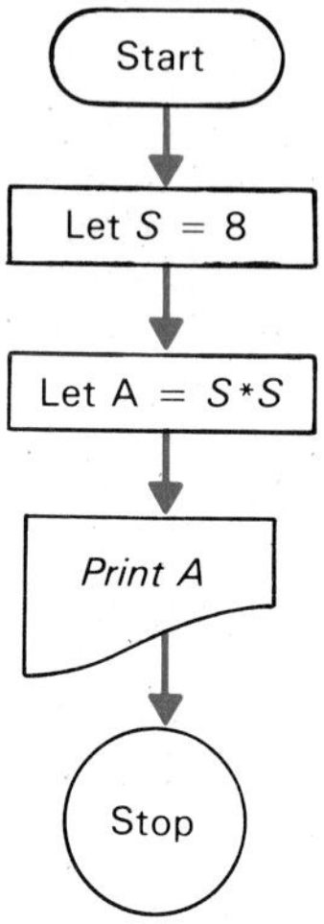

Just for Fun

675

SECTION 12.4

1. 12 3. 11 5. 27 7. 1

9. 88 11. 22 13. 17 15. 7

17. 3 * 5 + 4

19. 2 ↑ 3 − 3

21. 3 * (4 − 2)

23. 3 ↑ 4 − 2 ↑ 2

25. 7 * (3 + 4) ↑ 2

27. 8 * 3/2 + 2 ↑ 3

29. (2 ↑ 2 + 3 ↑ 2)/2 ↑ 3

31. $3x^2 - 2x + 3$

33. $(x + 2)(x + 3)$

35. $\dfrac{(2x - 1)(3x + 1)}{(2x + 1)}$

37. 3 * *x* ↑ 2 + 2 * *x* + 1

39. (*x* ↑ 2 − 8 * *x* + 16)/(*x* + 4)

41. ((*x* + 3) * (2 * *x* + 1)) ↑ 2

43. 14 45. 12 47. 32 49. 13

51.
```
5      REM INTEREST
10  LET P = 1000
15  LET R = 0.085
20  LET T  = 2
25  LET I =P*R*T
30  PRINT I
35  END
```

53.
```
5  REM AREA OF A TRIANGLE
10  LET B = 10
15  LET H = 6
20  LET A = (1/2)*B*H
30  PRINT "THE AREA IS ";A
35  END
```

55.
```
5  REM AVERAGE
10  INPUT A,B,C,D,E
15  LET M  = (A+B+C+D+E)/5
20  PRINT "THE AVERAGE
      IS ";M
25  END
```

57.
```
5  REM EVALUATE
10  LET A = 2
15  LET B = 3
20  LET C = 4
25  LET D = A^3*(B +C)
30  PRINT D
35  END
```

59.
```
5  REM SQUARE ROOT OF A
     NUMBER
10  INPUT X
15  LET Y = X^(1/2)
20  PRINT " THE SQUARE ROOT
      OF X = ";Y
25  END
```

Just for Fun

Zero

SECTION 12.5

1. A sequence of instructions are repeatedly executed until a specified condition has been met.

3. When there is no more data for the READ command.

5. a.
```
HI

MY NAME IS

KETES THE COMPUTER

HI

MY NAME IS

KETES THE COMPUTER
```

(repeated continuously)

b. Yes, the program contains an infinite loop.

7.
```
28.5
39
86
```

9.
```
2     30    58    86
4     32    60    88
6     34    62    90
8     36    64    92
10    38    66    94
12    40    68    96
14    42    70    98
16    44    72    100
18    46    74
20    48    76
22    50    78
24    52    80
26    54    82
28    56    84
```

11.
```
1     -3
2      0
3      5.
4     12
5     21
6     32
7     45
8     60
9     77
10    96
```

13.
```
A = 28
A = 18
A = 10
```

Note: Answers may vary for Exercises 15–25.

15.
```
5  REM - SQUARES
10  FOR X = 1 TO 100
15  LET Y = X ∧ 2
20  PRINT Y
25  NEXT X
30  END
```

17.
```
5  REM SUMS
10  LET N = 100
20  LET S = N * (N + 1) / 2
30  PRINT S
35  END
```

19.
```
5  REM TABLE OF VALUES
10  FOR X = 1 TO 15
15  LET Y = (X∧2 - 1)∧(1/2)
20  PRINT X,Y
25  NEXT X
30  END
```

21.
```
1  REM MEAN
5  PRINT "INPUT X, AND F"
10  INPUT X1,F1
15  LET Y1 = X1 * F1
20  PRINT "INPUT THE NEXT X,
     AND F"
25  INPUT X2,F2
30  LET Y2 = X2 * F2
35  PRINT "INPUT THE NEXT X,
     AND F"
40  INPUT X3,F3
45  LET Y3 = X3 * F3
50  PRINT "INPUT THE NEXT X,
     AND F"
55  INPUT X4,F4
60  LET Y4 = X4 * F4
65  PRINT "INPUT THE NEXT X,
     AND F"
70  INPUT X5,F5
75  LET N = F1 + F2 +
     F3 + F4 + F5
80  LET S = Y1 + Y2 +
     Y3 + Y4 + Y5
85  LET M = S / N
90  PRINT "THE MEAN IS      ";M
95  END
```

Note: Input statements may have to be separate lines.

23.
```
5  REM DISTANCE
10  INPUT X1,Y1
15  INPUT X2,Y2
20  LET D = ((X2 - X1) ∧ 2 +
     (Y2 - Y1) ∧ 2) ∧
     (1 / 2)
25  PRINT "THE DISTANCE BETWEEN
    THESE TWO POINTS IS      ";D
30  END
```

25.
```
5  REM AREA
10  FOR S = 1 TO 10
15  LET A = (S ∧ 2 * (3) ∧
     (1 / 2)) / 4
20  PRINT S,A
25  NEXT S
30  END
```

Just for Fun

This is a satisfactory solution, otherwise the shares are $8\frac{1}{2}$, $5\frac{2}{3}$, and $1\frac{8}{9}$, which means that some cows would have to be slain.

REVIEW EXERCISES FOR CHAPTER 12

1. John Napier: Napier's rods used to perform multiplication; inventor of logarithms.
   Henry Briggs: Introduced common logarithms.
   Edmund Gunther: Invented the slide rule.
   Blaise Pascal: Invented the first mechanical adding machine.
   Gottfried Leibniz: Produced a calculating machine that could add, subtract, multiply, divide, and find square roots.
   Charles Babbage: Devised "difference engine."
   Joseph Jacquard used punched cards to control weaving designs on a loom.
   Herman Hollerith and James Sperry devised ways of entering data on punched cards.

2. Howard Aiken: MARK I
J. Eckert, J. Mauchly: UNIVAC I
John Von Neumann: EDVAC

3. Transistors were replaced by integrated circuits in the 1960s.

4. $8 \times 498 = 3{,}984$ 5. $6 \times 574 = 3{,}444$

6. a. Input information b. Output information
c. Ends flowchart d. Decision, comparison, or question

7. The CPU (central processing unit) performs the fundamental arithmetic and logic functions, and oversees the operation of the entire system.

8. Beginners All-purpose Symbolic Instruction Code

9. a. 375 b. 4 c. 7

10. a. $3 * x \uparrow 2 + 2 * x - 5$
b. $(4 * x \uparrow 2 - 3 * x + 2)/(x \uparrow 2 + 2 * x)$

11. a.
```
1      54
2      57
3      62
4      69
5      78
```
b.
```
4.5
24.5
60.5
112.5
180.5
```

12.
```
5  REM AREA
10  LET B = 8
15  LET H = 5
20  LET A = B * H
25  PRINT "THE AREA IS  ";A;"
      SQUARE CM"
30  END
```

13.
```
5  REM SIMPLE INTEREST
10  LET P = 10000
15  LET R = 0.155
20  LET T = 3
25  LET I = P * R * T
30  PRINT "THE INTEREST IS
      $ ";I
35  END
```

14.
```
5  REM TABLE OF VALUES
10  FOR X = 1 TO 100
20  LET Y = 4 * X ∧ 2 - 5
      * X + 1
25  PRINT X,Y
30  NEXT X
35  END
```

15.
```
5  REM AREA OF A TRIANGLE
6  PRINT "I WILL FIND THE
      AREA OF ANY"
7  PRINT
8  PRINT "TRIANGLE. GIVE ME
      THE BASE AND HEIGHT."
9  PRINT
10  PRINT "WHAT IS THE BASE?"
11  PRINT
12  INPUT B
13  PRINT
14  PRINT " WHAT IS THE
      HEIGHT?"
15  PRINT
16  INPUT H
20  LET A = (1 /2) * B * H
25  PRINT
30  PRINT "THE AREA IS ";A;"
      SQUARE UNITS"
31  PRINT
35  END
```

16.
```
5  REM STANDARD DEVIATION
10  LET A = 82
15  LET B = 80
20  LET C = 81
25  LET D = 87
30  LET E = 86
35  LET F = 86
40  LET G = 88
45  LET H = 84
50  LET I = 82
55  LET J = 84
60  LET M = (A + B + C + D
      + E + F + G + H + I + J)
      / 10
65  LET A1 = (A - M) ∧ 2
70  LET B1 = (B - M) ∧ 2
75  LET C1 = (C - M) ∧ 2
80  LET D1 = (D - M) ∧ 2
85  LET E1 = (E - M) ∧ 2
90  LET F1 = (F - M) ∧ 2
95  LET G1 = (G - M) ∧ 2
100  LET H1 = (H - M) ∧ 2
105  LET I1 = (I - M) ∧ 2
110  LET J1 = (J - M) ∧ 2
115  LET Z = A1 + B1 + C1
      + D1 + E1 + F1 + G1
      + H1 + I1 + J1
120  LET Z1 = Z / 9
```

```
125  LET S = Z1 ∧ (1 / 2)
130  PRINT "THE STANDARD
      DEVIATION IS ";S
135  END
```

Just for Fun

2.5¢

CHAPTER 12 QUIZ

1. True	2. False	3. True	4. True
5. False	6. True	7. False	8. False
9. False	10. True	11. True	12. True
13. False	14. False	15. True	16. False
17. True	18. True	19. True	20. True
21. True	22. False	23. False	24. True
25. False			

INDEX